Emotion and Stress Recognition Related Sensors and Machine Learning Technologies

Emotion and Stress Recognition Related Sensors and Machine Learning Technologies

Editors

Kyandoghere Kyamakya
Fadi Al-Machot
Ahmad Haj Mosa
Hamid Bouchachia
Jean Chamberlain Chedjou
Antoine Bagula

MDPI • Basel • Beijing • Wuhan • Barcelona • Belgrade • Manchester • Tokyo • Cluj • Tianjin

MDPI

Editors
Kyandoghere Kyamakya
University Klagenfurt
Austria

Fadi Al-Machot
Universitaet Klagenfurt
Austria

Ahmad Haj Mosa
Universitaet Klagenfurt
Austria

Hamid Bouchachia
Bournemouth University
UK

Jean Chamberlain Chedjou
Universitaet Klagenfurt
Austria

Antoine Bagula
University of the Western Cape
South Africa

Editorial Office
MDPI
St. Alban-Anlage 66
4052 Basel, Switzerland

This is a reprint of articles from the Special Issue published online in the open access journal *Sensors* (ISSN 1424-8220) (available at: https://www.mdpi.com/journal/sensors/special_issues/emot_stress_sens).

For citation purposes, cite each article independently as indicated on the article page online and as indicated below:

LastName, A.A.; LastName, B.B.; LastName, C.C. Article Title. *Journal Name* **Year**, *Volume Number*, Page Range.

ISBN 978-3-0365-1138-2 (Hbk)
ISBN 978-3-0365-1139-9 (PDF)

Contents

About the Editors

Kyandoghere Kyamakya is currently a full professor of transportation informatics and the deputy director of the Institute for Smart Systems Technologies at Universitaet Klagenfurt in Austria. He is actively conducting research involving modeling, simulation and test-bed evaluations for a series of concepts, amongst others, in the context of intelligent transportation systems. In the research addressing transportation systems, a series of fundamental and theoretical tools from the fields of applied mathematics, electronics and computer science is either extensively exploited or a source of inspiration for innovative solutions and concepts, including nonlinear dynamics, systems science, machine learning/deep learning, nonlinear image processing and neurocomputing. He has co-edited more than 6 books, and has published more than 100 journal papers and some hundreds of conference papers.

Fadi Al-Machot finished his Ph.D. in computer science at Universitaet Klagenfurt in November 2013 and his habilitation in applied computer science at the University of Lübeck in 2020. As a researcher, he developed different algorithms and approaches in the areas of complex event detection in multimodal sensor networks, advanced driver assistance systems, human cognitive reasoning and human activity and emotion recognition. His work is patented and published in different international conferences and journals. He is currently a senior data scientist at Leibniz Lung Center —Research Center Borstel.

Ahmad Haj Mosa is a researcher and AI developer in the team of Digital Services at PwC Austria. He is also a researcher and a lecturer at the Institute for Smart Systems Technologies (IST) at the Universitaet Klagenfurt, Austria. His research area focus lies on augmented intelligence and explainable deep learning, and self-driving cars. His research interests also include machine vision, machine learning, applied mathematics and neurocomputing. He has developed a variety of methods in the scope of human–machine interaction and pattern recognition.

Hamid Bouchachia is Professor of Data Science and Intelligent Systems, leading the Machine Intelligence Research Group, Department of Computing and Informatics, Faculty of Science and Technology, UK. He is holder of an engineering degree, master's, Ph.D. and a habilitation degree. He was also a postdoc at the University of Alberta, Department of Computer and Electrical Engineering, Edmonton, Canada.

His research encompasses various topics of artificial intelligence and data science and their applications (ubiquitous health and medical science, smart environments and industrial monitoring, smart energy, smart agriculture, assistive technologies and pattern recognition and, in particular, scalable machine learning and distributed artificial intelligence, scalable online, active, semi-supervised learning for data streams, scalable pattern recognition including deep learning and hierarchical (graphical) models, reasoning and decision making and big data technologies and high-performance computing for machine learning. He has published more than 140 papers in international journals and conferences and edited a dozen of volumes and special issues as guest editor. He has organized a number of international events (conferences and workshops as well as special sessions). He founded the International Conference on Adaptive and Intelligent Systems (ICAIS). He serves as program committee member for numerous international conferences and is

Associate Editor of Evolving Systems and acts as a member of the Evolving Intelligent Systems (EIS) Technical Committee (TC) of the IEEE Systems, Man and Cybernetics Society, the IEEE Task-Force for Adaptive and Evolving Fuzzy Systems and the IEEE Computational Intelligence Society. He has supervised 7 postdoctoral researchers and 12 Ph.D. students to completion.

Jean Chamberlain Chedjou is currently an Associate Professor with the Institute for Smart Systems Technologies, Universitaet Klagenfurt, Austria. He is conducting research in the field of dynamic systems in traffic engineering. His current research interests include nonlinear dynamics in intelligent transportation systems (ITSs), applications of neural networks and cellular neural networks in ITSs, electronics circuits engineering and graph theory. He has been serving as a reviewer in several journals, including *IEEE Access, The IEEE Transactions on Neural Networks and Learning Systems, The IEEE Transactions on Circuits and Systems, The IEEE Transactions on Communications, Neuro-Computing, Nonlinear Dynamics, Sensors, The International Journal of Bifurcation and Chaos, The Journal of Applied Physics and the AEU—International Journal of Electronics and Communications.*

Antoine Bagula received a Ph.D. degree in Communication Systems from the Royal Institute of Technology (KTH), Stockholm, Sweden, and 2 MSc degrees (Computer Engineering—Université Catholique de Louvain (UCL), Belgium and Computer Science—University of Stellenbosch (SUN), South Africa). He is currently a full professor and head of the Department of Computer Science at the University of the Western Cape (UWC) where he also leads the Intelligent Systems and Advanced Telecommunication (ISAT) laboratory. He is a well-published scientist in his research field. His current research interests include data engineering, including big data technologies, cloud/fog computing and network softwarization (e.g., NFV and SDN); the Internet of Things (IoT), including the Internet of Things in motion and the tactile internet; data science, including artificial intelligence and machine learning with their applications in big data analytics; and next generation networks, including 4G/5G.

Preface to "Emotion and Stress Recognition Related Sensors and Machine Learning Technologies"

Emotions reflect how we feel about something in various circumstances of daily life. For example, facial expressions are one of the ways to identify emotions through some form of nonintrusive sensor/camera. Besides nonintrusive visual or acoustic sensors, various intrusive physiological sensors can be used to collect data, out of which emotion and/or stress can reliably and robustly be extracted through appropriate machine learning schemes. Some examples of such sensors are: pulse rate-measuring devices, electroencephalography (EEG), electrocardiogram (ECG), respiration rate measurements (RESP), etc.

Indeed, automatically recognizing human emotions through the use of electronic machines has always been a challenge although very fascinating. There are numerous applications and use cases in which reliable emotion information will/can be a precious cornerstone for related value-added services. A very positive trend is also that the relevant physiological sensors are becoming cheaper and more accurate over time.

Human emotion recognition systems are of high relevance for various fields of application which include, amongst others, robotics, various areas of medicine, industry, quality control, visual inspection, surveillance, driving assistance systems, etc.

In this book, 25 different contributions involving either visual or physiological sensors for detecting stress or emotions or facial expressions are provided. The book is divided into four core parts:

(1) Stress level recognition
(2) Wearable body sensors
(3) Dermatological sensors
(4) Facial expression recognition

This book will be a valuable read for research experts and professionals in the various above-cited application fields/areas in which human emotion-related information is of precious use. Further, this book will also be very inspiring for graduate students in machine learning and artificial intelligence.

Kyandoghere Kyamakya, Fadi Al-Machot, Ahmad Haj Mosa, Hamid Bouchachia, Jean Chamberlain Chedjou, Antoine Bagula
Editors

sensors

MDPI

Editorial

Emotion and Stress Recognition Related Sensors and Machine Learning Technologies

Kyandoghere Kyamakya [1,*], Fadi Al-Machot [2], Ahmad Haj Mosa [1], Hamid Bouchachia [3], Jean Chamberlain Chedjou [1] and Antoine Bagula [4]

[1] Institute for Smart Systems Technologies, Universitaet Klagenfurt, A9020 Klagenfurt, Austria; ahmad.haj.mosa@pwc.com (A.H.M.); Jean.Chedjou@aau.at (J.C.C.)

[2] Department of Applied Informatics, Universitaet Klagenfurt, 9020 Klagenfurt, Austria; Fadi.AlMachot@aau.at

[3] Machine Intelligence Group, Bournemouth University, Bournemouth BH12 5BB, UK; abouchachia@bournemouth.ac.uk

[4] ISAT Laboratory, University of the Western Cape, 7535 Bellville, South Africa; bbagula@uwc.ac.za

* Correspondence: kyandoghere.kyamakya@aau.at

Citation: Kyamakya, K.; Al-Machot, F.; Haj Mosa, A.; Bouchachia, H.; Chedjou, J.C.; Bagula, A. Emotion and Stress Recognition Related Sensors and Machine Learning Technologies. *Sensors* **2021**, *21*, 2273. https://doi.org/10.3390/s21072273

Received: 11 March 2021
Accepted: 15 March 2021
Published: 24 March 2021

Intelligent sociotechnical systems are gaining momentum in today's information-rich society, where different technologies are used to collect data from such systems and mine this data to make useful insights about our daily activities. These systems range from driver-assistance systems, to medical-patient monitoring systems, to emotion-aware intelligent systems, to complex collaborative robotics systems. They are built around (i) intrusive technologies such as physiological sensors, used for example in EEG, ECG, electrodermal activity and skin conductance and (ii) nonintrusive technologies that use piezo-vibration sensors, facial images, chairborne differential vibration sensors and bed-borne differential vibration sensors. However, despite their undisputable advantages in our daily lives, there are a number of issues relating to the design and development of such systems, as they rely on emotion and stress classification from physiological signals. These issues can be viewed from various perspectives including: (a) quality and reliability of sensor data; (b) classification performance in terms of accuracy, precision, specificity, recall and F1-measure; (c) robustness of subject-independent recognition; (d) portability of the classification systems to different environments and (e) the estimation of the emotional state for dynamic systems.

This book emerging from the Special Issue of the *Sensors* journal on Emotion and Stress Recognition Related Sensors and Machine Learning Technologies emerges as a result of the crucial need for massive deployment of intelligent sociotechnical systems. Such technologies are being applied in assistive systems in different domains and parts of the world to address challenges that could not be addressed without the advances made in these technologies. The Special Issue includes 25 papers submitted in response to the call for papers. The high number of submissions to the Special Issue is an indication of the momentum of the current research in this field. This momentum is driven not only by technological development, but also the need for assistive technologies. The Special Issue includes impactful papers that present scientific concepts, frameworks, architectures and ideas on sensing technologies and machine-learning techniques. These are relevant in tackling the following challenges: (i) the field readiness and use of intrusive sensors systems and devices for capturing biosignals, including EEG sensor systems, ECG sensor systems and Electrodermal activity sensor systems; (ii) the quality assessment and management of sensor data; (iii) data preprocessing, noise filtering and calibration concepts for biosignals; (iv) the field readiness and use of nonintrusive sensor technologies, including Visual sensors, Acoustic sensors, Vibration sensors and Piezo-electric sensors; (v) emotion recognition using mobile phones and smartwatches; (vi) body area sensor networks for emotion and stress studies; (vii) the use of experimental datasets in emotion recognition, including datasets generation principles and concepts, quality insurance and emotion elicitation material and concepts; (viii) machine-learning techniques for robust emotion recognition, including Graphical models, Neural network methods, Deep learning methods, Statistical learning and Multivariate empirical mode decomposition; (ix) subject-independent emotion and stress recognition concepts and systems, including Facial expression-based systems, Speech-based systems, EEG-based systems, ECG-based systems, Electrodermal activity-based systems, Multimodal recognition systems and Sensor fusion concepts and (x) emotion and stress estimation-and-forecasting from a nonlinear dynamical system's perspective.

In general, these papers are grouped into four categories/groups:

1. Stress level recognition
2. Wearable body sensors
3. Dermatological sensors
4. Facial expression recognition

1. Stress Detection

Addressing the issue of stress as a naturally occurring psychological response, identifiable by several body signs, [1] proposed a novel way of discriminating between acute stress and relaxation by using movement and posture characteristics of the foot. The authors used several machine-learning techniques to build models that were used to assess the validity of their method based on data collected from 23 participants performing tasks that induced stress and relaxation. Data collected from an additional sample of 11 participants were used to test their models, with results demonstrating replicability and an overall accuracy of 87%. External validity was also demonstrated by conducting a field study with 10 participants that revealed the robustness of the results.

The research in [2] contributed to bridging the gap between laboratory experimentation and daily life activities. The authors used a laboratory experiment and ecological momentary assessment-based data collection with smartwatches in daily life to propose a stress level detection system. The system pre-processes noisy physiological signals, extracts features and applies machine-learning techniques to classify the levels of stress. The study revealed that the accuracy of the system when tested in daily life improved significantly when machine-learning models were trained in the laboratory instead of with data from daily life.

In [3], regression and classification models were compared for stress detection using both personal and user-independent models' experimentation. The paper used the stress-detection dataset AffectiveROAD, which contained data gathered using Empatica E4 sensor and also continuous target variables—a feature that is missing in the other stress-detection dataset. The two classification models used for stress detection were Random Forest and Bagged tree based ensemble. From conducted experiments and using the AffectiveROAD dataset, the study revealed that regression models outperform classification models when classifying observations as stressed or not-stressed.

The research done in [4] revisited stress by using EEG as an objective measure for cost-effective and personalized stress management in situations where mental health facilities are not available. The study conducted by the paper considered: (i) a scenario in which- long-term stress was classified with machine-learning algorithms using resting state EEG signal recordings and (ii) the labelling for the stress and control groups was performed using two currently accepted clinical practices: the perceived stress scale score and expert evaluation. Support vector machine was found by the authors to be the most suitable classification algorithm for long-term human stress when used with the alpha asymmetry feature.

2. Wearable Body Sensors

The main contribution of [5] was to study electroencephalography (EEG) and galvanic skin response (GSR) together for boredom classification, with the objective of using the potential features of the associated data for emotion classification. The authors investigated the combined effect of these features on boredom classification by: (i) collecting EEG and GSR data from 28 participants using off-the-shelf sensors; (ii) labelling the collected samples using the participants' questionnaire-based testimonies of the various boredom levels experienced; (iii) using the collected data to initially train 30 models with 19 machine-learning algorithms and select the top three candidate classifiers and (iv) tuning the hyperparameters and validating the final models through 1000 iterations of 10-fold cross validation to increase the robustness of the test results. The work revealed the relative efficiency of multilayer perceptron compared to other machine-learning techniques. It also showed the correlation between boredom and the combined features of EEG and GSR.

The research in [6] addressed the issues of features extraction from Electroencephalography (EEG) signals and emotional aspects by considering both intra-subject and inter-subject approaches to EEG-based affect detection. Using three public repositories, the paper analysed both modelling approaches and showed that the subject's influence on the EEG signals is substantially higher than that of the emotion, thus (i) the subject's influence on the EEG signals should be accounted for and (ii) a data transformation that seamlessly integrates individual traits into an inter-subject approach should be performed to improve the classification process.

In [7], the authors suggested a better classification method for detecting stressed states based on raw electrocardiogram (ECG) data and a method for training a deep neural network (DNN) with a smaller data set. The work built an end-to-end architecture to detect stress using raw ECGs, using a multistage architecture that includes convolutional layers. Two kinds of datasets were used to train and validate the model, which were: a driving dataset and a smaller mental arithmetic dataset. A transfer learning method was then used to train the proposed model with a small dataset. It is shown in the paper that: (i) based on receiver operating curves, the proposed model performs better than conventional methods and (ii) compared with other DNN methods using raw ECGs, both the proposed model and the transfer learning method improves accuracy. These findings revealed that the proposed model can significantly contribute to mobile healthcare for stress management in daily life.

The issue of recognizing mental stress with deep ECG-respiration network was addressed in the workplace by proposing a novel stress-detection algorithm that uses multiple physiological signals, such as electrocardiogram (ECG) and respiration (RESP) signals to achieve end-to-end deep learning in [8]. The study mimicked workplace stress by using Stroop and mathematical tasks as stressors, with each stressor being followed by relaxation task(s). It also provided experimental results demonstrating its superiority over conventional machine-learning models.

The authors in [9] focused on the field readiness of low-cost wearable devices, which are increasingly being used in research as well as for personal and private purposes. The goal was to evaluate the accuracy of these devices in comparison to well-calibrated, high-quality devices used in laboratory experiments for physiological and medical research. The study demonstrated an approach for quantification of the accuracy of low-cost wearables in comparison to high-quality laboratory sensors by developing a benchmark framework for physiological sensors. The benchmark covered the entire workflow from sensor data acquisition to computation and interpretation of diverse correlation and similarity metrics. The study showed that the benchmarked wearables provide physiological measurements, such as heart rate and interbeat interval, with an accuracy close to those of the professional/high-end sensors. It was also revealed that accuracy varied more for parameters such as galvanic skin responses.

In [10], the issue of remote patient monitoring was revisited with the perspective of developing a wearable device that was low cost, single channel, dry contact and suitable for in-ear EEG for nonintrusive monitoring. The paper covered all aspects of the designs, engineering and experimenting. By applying machine learning for emotion classification, it was revealed that the proposed device was able to classify basic emotion with results that were comparable to those measured from the more conventional EEG headsets at T7 and T8 scalp positions.

In [11], a deep analysis of features proposed to extract information from the electrocardiogram, thoracic electrical bioimpedance and electrodermal activity signals was carried out with a focus on activities such as neutral, emotional, mental and physical. The study tested a total of 533 features for activity recognition. A comprehensive study was then performed taking into consideration the prediction accuracy, feature calculation, window length and type of classifier. This study enabled the determination of the ideal number of features and the best subset of features among those proposed in literature to obtain good error probability while avoiding over-fitting.

3. Dermatological Sensors

The association between the physiological responses of a driver and driving stress was addressed in [12], where the relationship between driving stress and traffic conditions, and driving stress and road types, respectively, was quantified through research. The study used electrodermal activity (EDA) signals for a male driver collected in real road-driving conditions for 60 min a day and over a 21-day period. Two separate models were used that incorporate the statistical features of the EDA signals, one for traffic conditions and the other for road types to classify the levels of driving stress (low vs. high). The classification results of the two models indicated that the traffic conditions and the road types were important features for driving stress and its related applications.

The work done in [13] addressed the issue of Active and Assisted Living environments for elderly and/or disabled people and the subjectivity of results when training a machine-learning model on a specific group of people while testing on a totally new group of persons. The study relied on electrodermal activity sensors to collect emotions and used a Convolutional Neural Network (CNN) architecture to provide promising robustness-related results for both subject-dependent and subject-independent human emotion recognition. The results revealed that by solely using the nonintrusive EDA sensors, a robust classification of human emotion was possible even without involving additional/other physiological signals.

The research in [14] presented the identification of the level of arousal in older people by monitoring their electrodermal activity (EDA) through a commercial device. The objective was to use the notion of familiarity with a musical genre on emotional induction in order to recognize arousal changes and hence create future therapies that can help older people to improve their mood. This can ultimately contribute to the reduction of depression and anxiety. Using methods based on the process of deconvolution of the EDA signal, two different studies were carried out, the first being a purely statistical study based on the search for statistically significant differences for a series of temporal, morphological, statistical and frequency features of the processed signals. The second study was a machine-learning study using a wide range of classifiers to analyse the possible correlations between the detection of the EDA-based arousal level compared to the participants' responses to the level of arousal subjectively felt. While the first study revealed that Flamenco and Spanish Folklore presented the highest number of statistically significant parameters, the second study showed that the best classifiers are the support vector machines, with 87% accuracy for Flamenco and 83.1% for Spanish Folklore, followed by K-nearest neighbours.

Motivated by the limitations of emotion recognition systems in terms of lack of systematic analysis in literature regarding the selection of classifiers to use, sensor modalities, features and range of expected accuracy, and many other limitations, the work in [15] contributed to the body of work in machine learning by presenting a systematic study across five public datasets commonly used in Emotion Recognition (ER) with the objective of evaluating emotion in terms of low/high arousal and valence classification through Supervised Learning (SL), Decision Fusion (DF) and Feature Fusion (FF) techniques using multimodal physiological data, namely Electrocardiography (ECG), Electrodermal Activity (EDA), Respiration (RESP) or Blood Volume Pulse (BVP). The work considered: (i) Classification performance analysis of ER benchmarking datasets in the arousal/valence space; (ii) Summarising the ranges of the classification accuracy reported across the existing literature;

(iii) Characterising the results for diverse classifiers, sensor modalities and feature set combinations for ER using accuracy and F1-score; (iv) Exploration of an extended feature set for each modality and (v) Systematic analysis of multimodal classification in DF and FF approaches. The study revealed that FF is the most competitive technique in terms of classification accuracy and computational complexity.

Moving away from the affective computing research that has mostly used nonimmersive two-dimensional (2D) images or videos to elicit emotional states, [16] adopted an immersive virtual reality (VR) approach. This allowed the researchers to simulate various environments in controlled laboratory conditions with high levels of sense of presence and interactivity. The paper presented a systematic review of the emotion recognition research undertaken with physiological and behavioural measures using head-mounted displays as elicitation devices. The results highlighted the evolution of the field, gave a clear perspective of the use of aggregated analysis and revealed the current open issues and guidelines for future research works.

Focusing on affecting computing, which is an artificial intelligence area of study that recognizes, interprets, processes and simulates human affect computers, a survey of the pertinent scientific literature on affecting computing from 2015 to 2020 was presented in [17]. The paper presented trends and compared algorithm applications in new implementations from a computer science perspective. The survey provided an overview of datasets, emotion elicitation methods, feature extraction and selection, classification algorithms and performance evaluations.

4. Facial Expression Recognition

Building upon deep transfer learning techniques, facial expression recognition (FER) was addressed in [18]. The authors tackled the challenging issues of: (i) diversity of factors, which are unrelated to facial expressions (ii) the lack of training data for FER and (iii) the intrinsic imbalance in existing facial emotion datasets. The deep transfer contribution to FER was complemented by a novel loss function called weighted-cluster loss used during a fine-tuning phase of the model.

In [19], the authors revisited the analysis of pain-related facial expressions by proposing an end-to-end approach based on attention networks for the analysis and recognition of pain-related facial expressions. The method proposed by the authors combined both spatial and temporal aspects of facial expressions through a weighted aggregation of attention-based neural networks' outputs that use sequences of Motion History Images (MHIs) and Optical Flow Images (OFIs). A combination of Convolutional Neural Network (CNN) and Bidirectional Long Short-Term Memory (BiLSTM) Recurrent Neural Network (RNN) was used to achieve pain recognition.

Building around a human-computer interaction (HCI) setting, [20] addressed the challenging issue of induction of dialog-based HCI relevant emotional and cognitive load states by presenting a multimodal dataset for affective computing research. The dataset used an experimental mobile and interactive scenario design that was implemented based on a gamified generic paradigm. The work consisted of six experimental sequences inducing Interest, Overload, Normal, Easy, Underload and Frustration.

Facial-landmark detection was revisited in [21] in a multistage architecture. At the first stage, the goal was to obtain local pixel-level accuracy for local-context information. The second stage was concerned with integrating obtained information with knowledge of spatial relationships between each key point in a whole image for global-context information. The paper considered a pipeline architecture consisting of two main components: (i) a deep network for local-context subnet used to generate detection heatmaps via fully convolutional DenseNets with additional kernel convolution filters and (ii) a dilated skip convolution subnet consisting of a combination of dilated convolutions and skip-connections networks used to robustly refine the local appearance heatmaps.

Building around the Child–Robot Interaction (CRI), [22] proposed a system for emotion recognition in children by recording facial images using both visual (RGB—red, green and blue) and Infrared Thermal Imaging (IRTI) cameras. Building upon the Viola–Jones algorithm on colour images to detect facial regions of interest (ROIs), the paper proposed as a novel contribution the computation of the error probability for each ROI located over thermal images, using a reference frame manually marked by a trained expert, in order to choose that ROI better-placed according to the expert criteria. The results: (i) show that the proposed approach for ROI locations may track facial landmarks with significant low errors with respect to the traditional Viola–Jones algorithm and (ii) suggest that the proposed system be integrated to a social robot to infer child emotions during a child–robot interaction.

A comparison of machine-learning algorithms applied to the recognition of emotion intensities was proposed in [23] as a solution to the lack of encoding the intensity of observed facial emotion and multifacial behaviour in existing emotion recognition systems. The work compared several algorithms, include (i) Gabor filters, a Histogram of Oriented Gradients (HOG), and Local Binary Pattern (LBP) for feature extraction and (ii) Support Vector Machine (SVM), Random Forest (RF), and Nearest Neighbour Algorithm (KNN) for classification. The experiment suggested that the comparative study could be further used in real-time behavioural facial emotion and intensity of emotion recognition.

A transfer learning approach was adopted for mouth-based emotion recognition in [24]. The study was predicated on the fact that there were only a few datasets available in practice and most of them included emotional expressions simulated by actors, instead of adopting real-world categorisation. By enabling the image of the mouth to be available, even when the whole face was only visible from an unfavourable perspective, the transfer learning approach allowed the authors to use fewer training data. This minimized the effort of training a whole network from scratch and resulted in an improved dynamic emotion recognition when taking into account not only new scenarios but also modified situations to the initial training phase. As presented in the paper, the transfer learning approach and the underlying method proved the relevance of mouth detection in the complex process of emotion recognition.

The authors in [25] proposed a multimodal approach to emotion recognition in the aviation domain with the goal of filling some of the gap between pilots' emotions and their bioreactions during flight procedures such as take-off, climbing, cruising, descent, initial approach, final approach and landing. Building around a sensing architecture and a set of simulated flight experiments, the study showed that it was indeed possible to recognize emotions from different pilots in flight, combining their present and previous emotions.

As we alluded to in our introduction, assistive technology is a research field with a number of open challenges. Some of those are present in this Special Issue, which we think will foster more research. Other fields were not covered, hence leaving room for new ideas to be discovered in this field.

Author Contributions: K.K.: overall coordination. All authors have read and agreed to the published version of the manuscript.

Funding: This research received no external funding.

Conflicts of Interest: The authors declare no conflict of interest.

References

1. Elvitigala, D.S.; Matthies, D.J.C.; Nanayakkara, S. StressFoot: Uncovering the Potential of the Foot for Acute Stress Sensing in Sitting Posture. *Sensors* **2020**, *20*, 2882. [CrossRef]
2. Can, Y.S.; Gokay, D.; Kılıç, D.R.; Ekiz, D.; Chalabianloo, N.; Ersoy, C. How Laboratory Experiments Can Be Exploited for Monitoring Stress in the Wild: A Bridge Between Laboratory and Daily Life. *Sensors* **2020**, *20*, 838. [CrossRef]
3. Siirtola, P.; Röning, J. Comparison of Regression and Classification Models for User-Independent and Personal Stress Detection. *Sensors* **2020**, *20*, 4402. [CrossRef]
4. Saeed, S.M.U.; Anwar, S.M.; Khalid, H.; Majid, M.; Bagci, A.U. EEG Based Classification of Long-Term Stress Using Psychological Labeling. *Sensors* **2020**, *20*, 1886. [CrossRef]
5. Seo, J.; Laine, T.H.; Sohn, K.-A. An Exploration of Machine Learning Methods for Robust Boredom Classification Using EEG and GSR Data. *Sensors* **2019**, *19*, 4561. [CrossRef] [PubMed]

6. Arevalillo-Herráez, M.; Cobos, M.; Roger, S.; García-Pineda, M. Combining Inter-Subject Modeling with a Subject-Based Data Transformation to Improve Affect Recognition from EEG Signals. *Sensors* **2019**, *19*, 2999. [CrossRef] [PubMed]
7. Cho, H.-M.; Park, H.; Dong, S.-Y.; Youn, I. Ambulatory and Laboratory Stress Detection Based on Raw Electrocardiogram Signals Using a Convolutional Neural Network. *Sensors* **2019**, *19*, 4408. [CrossRef] [PubMed]
8. Seo, W.; Kim, N.; Kim, S.; Lee, C.; Park, S.-M. Deep ECG-Respiration Network (DeepER Net) for Recognizing Mental Stress. *Sensors* **2019**, *19*, 3021. [CrossRef] [PubMed]
9. Sagl, G.; Resch, B.; Petutschnig, A.; Kyriakou, K.; Liedlgruber, M.; Wilhelm, F.H. Wearables and the Quantified Self: Systematic Benchmarking of Physiological Sensors. *Sensors* **2019**, *19*, 4448. [CrossRef] [PubMed]
10. Athavipach, C.; Pan-Ngum, S.; Israsena, P. A Wearable In-Ear EEG Device for Emotion Monitoring. *Sensors* **2019**, *19*, 4014. [CrossRef] [PubMed]
11. Mohino-Herranz, I.; Gil-Pita, R.; Rosa-Zurera, M.; Seoane, F. Activity Recognition Using Wearable Physiological Measurements: Selection of Features from a Comprehensive Literature Study. *Sensors* **2019**, *19*, 5524. [CrossRef]
12. Bitkina, O.V.; Kim, J.; Park, J.; Park, J.; Kim, H.K. Identifying Traffic Context Using Driving Stress: A Longitudinal Preliminary Case Study. *Sensors* **2019**, *19*, 2152. [CrossRef]
13. Al Machot, F.; Elmachot, A.; Ali, M.; Al Machot, E.; Kyamakya, K. A Deep-Learning Model for Subject-Independent Human Emotion Recognition Using Electrodermal Activity Sensors. *Sensors* **2019**, *19*, 1659. [CrossRef] [PubMed]
14. Bartolomé-Tomás, A.; Sánchez-Reolid, R.; Latorre, A.F.-S.J.M.; Fernández-Caballero, A. Arousal Detection in Elderly People from Electrodermal Activity Using Musical Stimuli. *Sensors* **2020**, *20*, 4788. [CrossRef] [PubMed]
15. Bota, P.; Wang, C.; Fred, A.; Silva, H. Emotion Assessment Using Feature Fusion and Decision Fusion Classification Based on Physiological Data: Are We There Yet? *Sensors* **2020**, *20*, 4723. [CrossRef] [PubMed]
16. Marín-Morales, J.; Llinares, C.; Guixeres, J.; Alcañiz, M. Emotion Recognition in Immersive Virtual Reality: From Statistics to Affective Computing. *Sensors* **2020**, *20*, 5163. [CrossRef] [PubMed]
17. Torres, E.P.; Torres, E.A.; Hernández-Álvarez, M.; Yoo, S.G. EEG-Based BCI Emotion Recognition: A Survey. *Sensors* **2020**, *20*, 5083. [CrossRef]
18. Ngo, Q.T.; Yoon, S. Facial Expression Recognition Based on Weighted-Cluster Loss and Deep Transfer Learning Using a Highly Imbalanced Dataset. *Sensors* **2020**, *20*, 2639. [CrossRef]
19. Thiam, P.; Kestler, H.A.; Schwenker, F. Two-Stream Attention Network for Pain Recognition from Video Sequences. *Sensors* **2020**, *20*, 839. [CrossRef] [PubMed]
20. Hazer-Rau, D.; Meudt, S.; Daucher, A.; Spohrs, J.; Hoffmann, H.; Schwenker, F.; Traue, H.C. The uulmMAC Database—A Multimodal Affective Corpus for Affective Computing in Human-Computer Interaction. *Sensors* **2020**, *20*, 2308. [CrossRef]
21. Chim, S.; Lee, J.-G.; Park, H.-H. Dilated Skip Convolution for Facial Landmark Detection. *Sensors* **2019**, *19*, 5350. [CrossRef] [PubMed]
22. Goulart, C.; Valadão, C.; Delisle-Rodriguez, D.; Funayama, D.; Favarato, A.; Baldo, G.; Binotte, V.; Caldeira, E.; Bastos-Filho, T. Visual and Thermal Image Processing for Facial Specific Landmark Detection to Infer Emotions in a Child-Robot Interaction. *Sensors* **2019**, *19*, 2844. [CrossRef] [PubMed]
23. Mehta, D.; Siddiqui, M.F.H.; Javaid, A.Y. Recognition of Emotion Intensities Using Machine Learning Algorithms: A Comparative Study. *Sensors* **2019**, *19*, 1897. [CrossRef] [PubMed]
24. Franzoni, V.; Biondi, G.; Perri, D.; Gervasi, O. Enhancing Mouth-Based Emotion Recognition Using Transfer Learning. *Sensors* **2020**, *20*, 5222. [CrossRef]
25. Roza, V.C.C.; Postolache, O.A. Multimodal Approach for Emotion Recognition Based on Simulated Flight Experiments. *Sensors* **2019**, *19*, 5516. [CrossRef] [PubMed]

sensors

MDPI

Article

Two-Stream Attention Network for Pain Recognition from Video Sequences

Patrick Thiam [1,2], Hans A. Kestler [1,†] and Friedhelm Schwenker [2,*,†]

1 Institute of Medical Systems Biology, Ulm University, Albert-Einstein-Allee 11, 89081 Ulm, Germany; patrick.thiam@uni-ulm.de (P.T.); hans.kestler@uni-ulm.de (H.A.K.)

2 Institute of Neural Information Processing, Ulm University, James-Frank-Ring, 89081 Ulm, Germany

* Correspondence: friedhelm.schwenker@uni-ulm.de

† Equally contributing senior authors.

Received: 23 January 2020; Accepted: 2 February 2020; Published: 4 February 2020

Abstract: Several approaches have been proposed for the analysis of pain-related facial expressions. These approaches range from common classification architectures based on a set of carefully designed handcrafted features, to deep neural networks characterised by an autonomous extraction of relevant facial descriptors and simultaneous optimisation of a classification architecture. In the current work, an end-to-end approach based on attention networks for the analysis and recognition of pain-related facial expressions is proposed. The method combines both spatial and temporal aspects of facial expressions through a weighted aggregation of attention-based neural networks' outputs, based on sequences of Motion History Images (MHIs) and Optical Flow Images (OFIs). Each input stream is fed into a specific attention network consisting of a Convolutional Neural Network (CNN) coupled to a Bidirectional Long Short-Term Memory (BiLSTM) Recurrent Neural Network (RNN). An attention mechanism generates a single weighted representation of each input stream (MHI sequence and OFI sequence), which is subsequently used to perform specific classification tasks. Simultaneously, a weighted aggregation of the classification scores specific to each input stream is performed to generate a final classification output. The assessment conducted on both the *BioVid Heat Pain Database (Part A)* and *SenseEmotion Database* points at the relevance of the proposed approach, as its classification performance is on par with state-of-the-art classification approaches proposed in the literature.

Keywords: convolutional neural networks; long short-term memory recurrent neural networks; information fusion; pain recognition

1. Introduction

An individual's affective disposition is often expressed throughout facial expressions. Human beings are therefore able to assess someone's current mood or state of mind by observing his or her facial demeanour. Therefore, an analysis of facial expressions can provide some valuable insight about one's emotional and psychological state. Thus, facial expression recognition (FER) has been attracting a lot of interest from the research community in the recent decades and constitutes a steadily growing area of research, particularly in the domains of machine learning and computer vision. The current work focuses on the analysis of facial expressions for the assessment and recognition of pain in video sequences. More specifically, a two-stream attention network is designed, with the objective of combining both temporal and spatial aspects of facial expressions, based on sequences of motion history images [1] and optical flow images [2], to accurately discriminate between neutral, low, and high levels of nociceptive pain. The current work is organised as follows. An overview of pain recognition approaches based on facial expressions is provided in Section 2, followed by a thorough description of the proposed approach in Section 3. In Section 4, a description of the datasets used for the assessment of the proposed approach as well as the performed experiments is provided, followed by a description of the corresponding results. The current work is subsequently concluded in Section 5 with a short discussion and description of potential future works.

2. Related Work

Recent advances in both domains of computer vision and machine learning, combined with the release of several datasets designed specifically for pain-related research (e.g., *UNBC-McMaster Shouder Pain Expression Archive Database* [3], *BioVid Heat Pain Database* [4], *Multimodal EmoPain Database* [5] and *SenseEmotion Database* [6]), have fostered the development of a multitude of automatic pain assessment and classification approaches. These methods range from unimodal approaches, characterised by the optimisation of an inference model based on one unique and specific input signal (e.g., video sequences [7,8], audio signals [9,10] and bio-physiological signals [11–13]), to multimodal approaches that are characterised by the optimisation of an information fusion architecture based on parameters stemming from a set of distinctive input signals [14–16].

Regarding pain assessment based on facial expressions, several approaches have been proposed, ranging from conventional supervised learning techniques based on specific sets of handcrafted features, to deep learning techniques. These approaches rely on an effective preprocessing of the input signal (which in this case consists of a set of images or video sequences) and involves the localisation, alignment and normalisation of the facial area in each input frame. Moreover, further preprocessing techniques include the localisation and extraction of several fiducial points characterising specific facial landmarks, and in some cases, the continuous extraction of facial Action Units (AUs) [17,18]. The preprocessed input signal, as well as the extracted parameters, are subsequently used to optimise a specific inference model based on different methods. In [19], the authors use an ensemble of linear Support Vector Machines (SVMs) [20] (each trained on a specific AU), in which inference scores are subsequently combined using Logistical Linear Regression (LLR) [21] for the detection of pain at a frame-by-frame level. The authors in [22] apply a k-Nearest Neighbours (k-NN) [23] model on geometric features extracted from a specific set of facial landmarks for the recognition of AUs. Subsequently, the pain intensity in a particular frame is evaluated based on the detected AUs by using a pain evaluation scale provided by Prkachin and Solomon [24]. Most recently, the authors in [25] improve the performance of a pain detection system based on automatically detected AUs by applying a transfer learning approach based on neural networks to map automated AU codings to a subspace of manual AU codings. The encoded AUs are subsequently used to perform pain classification, using an Artificial Neural Network (ANN) [26].

In addition to AU-based pain assessment approaches, several techniques based on either facial texture, shape, appearance and geometry or on a combination of several of such facial descriptors have been proposed. Yang et al. [27] assess several appearance-based facial descriptors by comparing the pain classification performance of each feature with its spatio-temporal counterpart using SVMs. The assessed spatial descriptors consist of Local Binary Patterns (LBP) [28], Local Phase Quantization (LPQ) [29], Binarized Statistical Image Features (BSIF) [30] as well as each descriptor's spatio-temporal counterpart extracted from video sequences on three orthogonal planes (LBP-TOP, LPQ-TOP and BSIF-TOP). In [8,31], the authors propose several sets of spatio-temporal facial action descriptors based on both appearance- and geometry-based features extracted from both the facial area, as well as the head pose. Those descriptors are further used to perform the classification of several levels of pain intensities using a Random Forest (RF) [32] model. Similarly, the authors in [7,14,15,33], propose several spatio-temporal descriptors extracted either from the localised facial area or from the estimated head pose, including, among others, Pyramid Histograms of Oriented Gradients (PHOG) [34] and Local Gabor Binary Patterns from Three Orthogonal Planes (LGBP-TOP) [35], to perform the classification of several levels of nociceptive pain. The classification experiments are also performed with RF models and ANNs.

Alongside handcrafted feature-based approaches, several techniques based on deep neural networks have also been proposed for the assessment of pain induced facial expressions. Such approaches are characterised by the joint extraction of relevant descriptors (from the preprocessed raw input data) and optimisation of an inference model, based on neural networks in an end-to-end manner. In [36–38], the authors propose a hybrid deep neural network pain detection architecture characterised by the combination of a feature embedding network consisting of a Convolutional Neural Network (CNN) [39] with a Long Short-Term Memory (LSTM) [40] Recurrent Neural Network (RNN), to take advantage of both spatial and temporal aspects of facial pain expressions in video sequences. Soar et al. [41] propose a similar approach by combining a CNN with a Bidirectional LSTM (BiLSTM), and using a Variable-State Latent Conditional Random Field (VRS-CRF) [42] instead of a conventional Multi-Layer Perceptron (MLP) to perform the classification. In [43], the authors also use a similar hybrid approach as in [36,37]; however, in this case, the feature embedding CNN is coupled to two distinct LSTM networks. The outputs of the LSTM networks are further concatenated and a MLP is used to perform the classification of the pain intensities in video sequences. Furthermore, Zhou et al. [44] propose a Recurrent Convolutional Neural Network (RCNN) [45] architecture for the continuous estimation of pain intensity in video sequences at the frame-level, whereas Wang et al. [46] propose a transfer learning approach, consisting of the regularisation of a face verification network, which is subsequently applied to a pain intensity regression task.

The current work focuses on the analysis of facial expressions for the discrimination of neutral, low and high levels of nociceptive pain in video sequences. Thereby, an end-to-end hybrid neural network characterised by the integration of spatial and temporal information at both the representational level of the input data (OFI and MHI) and the structural level of the proposed architecture (hybrid CNN-BiLSTM) is proposed. Furthermore, frame attention parameters [47] are integrated into the proposed architecture to generate an aggregated representation of the input data based on an estimation of the representativeness of each single input frame, in relation with the corresponding level of nociceptive pain. An extensive assessment of the proposed architecture is performed on both *BioVid Heat Pain Database (Part A)* [4] and *SenseEmotion Database* [6].

3. Proposed Approach

A video sequence can be characterised by both its spatial and temporal components. The spatial component represents the appearance (i.e., texture, shape and form) of each frame's content, whereas the temporal component represents the perceived motion across consecutive frames due to dynamic changes of the content's appearance through time. Most of the deep neural network approaches designed for the assessment of pain-related facial expressions generate spatio-temporal descriptors of the input data in two distinct and conjoint stages: a specific feature embedding neural network (which is commonly a pre-trained CNN) first extracts appearance based descriptors from the individual input frames (which are greyscale or colour images), and a recurrent neural network is subsequently used for the integration of the input's temporal aspect based on sequences of previously extracted appearance features, thus generating spatio-temporal representations of video sequences that are used for classification or regression tasks. Therefore, both the temporal and spatial aspects of video sequences are integrated uniquely at the structural level (e.g., the architecture of the neural network) of such approaches. The current approach extends this specific technique by additionally integrating motion information at the representational level (e.g., input data) of the architecture throughout sequences of motion history images [1] and optical flow images [2].

3.1. Motion History Image (MHI)

Introduced by Bobick and Davis [48], a MHI consists of a scalar-valued image depicting both the location and direction of motion in a sequence of consecutive images, based on the changes of pixel intensities of each image through time. The intensity of a pixel in a MHI is a function of the temporal motion history at that specific point. A MHI H_τ is computed using an update function $\Psi(x, y, t)$, and is defined as follows,

$$H_\tau(x, y, t) = \begin{cases} \tau & \text{if } \Psi(x, y, t) = 1 \\ max(0, H_\tau(x, y, t-1) - \delta) & \text{otherwise} \end{cases} \tag{1}$$

where (x, y) represents the pixel's location, t the time and τ the temporal extent of the observed motion (e.g., the length of a sequence of images); $\Psi(x, y, t) = 1$ is synonym of motion at the location (x, y) and at the time t; and δ represents a decay parameter. The update function $\Psi(x, y, t)$ is defined as follows,

$$\Psi(x, y, t) = \begin{cases} 1 & \text{if } D(x, y, t) \geq \xi \\ 0 & \text{otherwise} \end{cases} \tag{2}$$

where ξ is a threshold; $D(x, y, t)$ represents the absolute value of the difference of pixel intensity values of consecutive frames and is defined as follows,

$$D(x, y, t) = |I(x, y, t) - I(x, y, t \pm \Delta t)| \tag{3}$$

where $I(x, y, t)$ represents the pixel intensity at the location (x, y) and at the time t; Δt represents the temporal distance between the frames.

Therefore, the computation of a MHI consists in first performing image differencing [49] between a specific, preceding frame and the current tth frame, and detecting the pixel locations where a substantial amount of movement has occurred (depending on the value assigned to the threshold ξ) based on Equation (2). Subsequently, Equation (1) is used to assign pixel values to the MHI. If a motion has been detected at the location (x, y) of the tth frame, a pixel value of τ is assigned at that location. Otherwise, the previous pixel value of that location is reduced by δ, thereby indicating the temporal occurrence of the motion at that specific location, according to the actual time t. This whole process is conducted iteratively until the entire sequence of images has been processed. The temporal history of motion is thereby encoded into the resulting MHI. Therefore, a whole sequence of images can be encoded into a single MHI. However, in the current work, a sequence of MHIs is generated from each single sequence of images by saving each single MHI generated at each single step of the iterative process described earlier. The resulting sequence of images is used as input for the designed deep neural networks. A depiction of such a sequence of MHIs can be seen in Figure 1b, with the corresponding sequence of greyscale images depicted in Figure 1a.

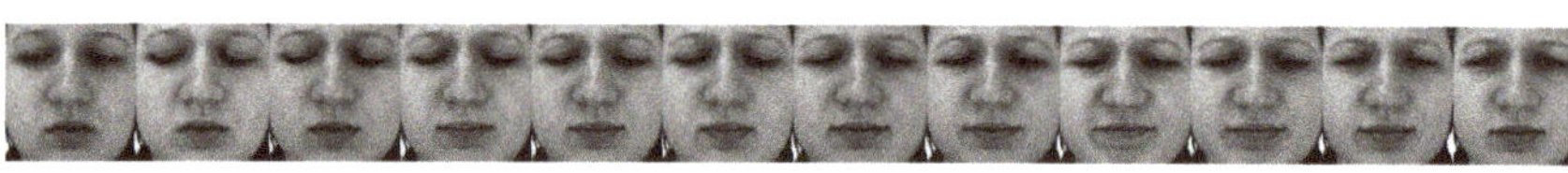

(**a**) Greyscale input sequence.

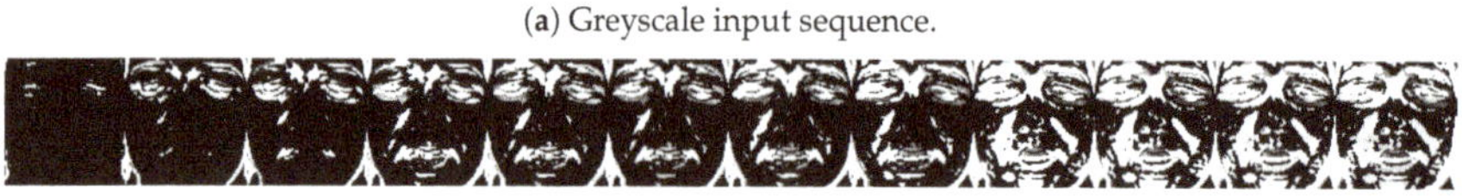

(**b**) MHI input sequence.

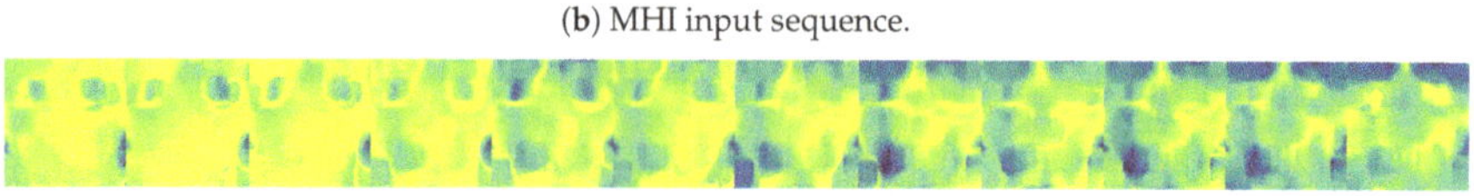

(**c**) OFI input sequence.

Figure 1. Data preprocessing. Following the detection, alignment, normalisation and extraction of the facial area in each frame of a video sequence, the images are converted into greyscale. MHI and OFI sequences are subsequently generated.

3.2. Optical Flow Image (OFI)

Optical flow refers to the apparent motion of visual features (e.g., corners, edges, textures and pixels) in a sequence of consecutive images. It is characterised by a set of vectors (optical flow vectors) defined either at each location (x, y) of an entire image (dense optical flow [50,51]), or at specific locations of a predefined set of visual features (sparse optical flow [2,52]). The orientation of an optical flow vector depicts the direction of the apparent motion, whereas the magnitude of an optical flow vector depicts the velocity of the apparent motion of the corresponding visual feature in consecutive frames. Thus, an OFI provides a compact description of the location, direction and velocity of a specific motion occurring in consecutive frames. The estimation of the optical flow is based on the brightness constancy assumption, which stipulates that pixel intensities are constant between consecutive frames. If $I(x, y, t)$ is the pixel intensity at the location (x, y) and at the time t, the brightness constancy assumption can be formulated as follows,

$$I(x, y, t) = I(x + \delta x, y + \delta y, t + \delta t) \tag{4}$$

where $(\delta x, \delta y, \delta t)$ represents a small motion. By applying a first-order Taylor expansion, $I(x + \delta x, y + \delta y, t + \delta t)$ can be written as follows,

$$I(x + \delta x, y + \delta y, t + \delta t) \approx I(x, y, t) + \frac{\partial I}{\partial x}\delta x + \frac{\partial I}{\partial y}\delta y + \frac{\partial I}{\partial t}\delta t. \tag{5}$$

Thus,

$$\frac{\partial I}{\partial x}\delta x + \frac{\partial I}{\partial y}\delta y + \frac{\partial I}{\partial t}\delta t \approx 0 \tag{6}$$

and by dividing each term by δt, the optical flow constraint equation can be written as follows,

$$\frac{\partial I}{\partial x}\frac{dx}{dt} + \frac{\partial I}{\partial y}\frac{dy}{dt} + \frac{dI}{dt} \approx 0. \tag{7}$$

Resolving the optical flow constraint equation (Equation (7)) consists of the estimation of both parameters $u = \frac{dx}{dt}$ and $v = \frac{dy}{dt}$. Several methods have been proposed to perform this specific task. The authors in [53,54] propose an overview of such approaches. In the current work, dense optical flow is applied, using the method of Farnebäck [50], to compute OFIs from consecutive greyscale images. A depiction of such a sequence of images can be seen in Figure 1c (both motion direction and motion velocity are color encoded).

3.3. Network Architecture

As opposed to still images, the motion component of a video sequence is integrated into both MHIs and OFIs, therefore providing more valuable information for facial expressions analysis. Therefore, the proposed architecture consists of a multi-view learning [55] neural network with both OFIs and MHIs as input channels. An overall illustration of the proposed two-stream neural network can be seen in Figure 2. In a nutshell, an attention network specific to each input channel (OFIs and MHIs) first generates a weighted representation from the jth input sequence (h_j^{ofi} and h_j^{mhi}). The generated representation is subsequently fed into a channel specific classification model (which in this case is a MLP). The resulting class probabilities of each channel ($score_j^{ofi}$ and $score_j^{mhi}$) are further fed into an aggregation layer with a linear output function, where a weighted aggregation of the provided scores is performed as follows,

$$score_j = \alpha_{ofi} \cdot score_j^{ofi} + \alpha_{mhi} \cdot score_j^{mhi} \tag{8}$$

with the constraint

$$\alpha_{ofi} + \alpha_{mhi} = 1. \tag{9}$$

The entire architecture is trained in an end-to-end manner by using the following loss function,

$$\mathcal{L} = \lambda_{ofi} \cdot \mathcal{L}_{ofi} + \lambda_{mhi} \cdot \mathcal{L}_{mhi} + \lambda_{agg} \cdot \mathcal{L}_{agg} \tag{10}$$

where the loss functions of each input channel and of the aggregation layer are respectively depicted with $\mathcal{L}_{ofi}$, $\mathcal{L}_{mhi}$ and $\mathcal{L}_{agg}$. The parameters λ_{ofi}, λ_{mhi} and λ_{agg} correspond to the regularisation parameters of each respective loss function. Once the network has been trained, unseen samples are classified based on the output of the aggregation layer.

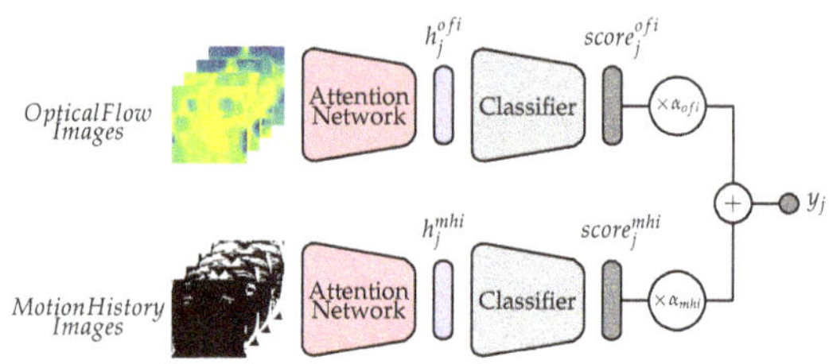

Figure 2. Two-Stream Attention Network with Weighted Score Aggregation.

The attention network (see Figure 3) consists of a CNN coupled to a BiLSTM with a frame attention module [47]. The CNN consists of a time distributed feature embedding network which takes a single facial image $im_{k,j}$ as input and generates a fixed-dimension feature representation $X_{k,j}$ specific to that image. Therefore, the output of the jth input sequence of facial images $\{im_{k,j}\}_{k=1}^{l}$ consists of a set of facial features $\{X_{k,j}\}_{k=1}^{l}$. The temporal component of the sequence of images is further integrated by using a BiLSTM layer. A BiLSTM [56] RNN is an extension of a regular LSTM [40] RNN, to enable the use of context representations in both forward and backward directions.

It consists of two LSTM layers, one processing the input sequence $\left\{X_{1,j}, X_{2,j}, \ldots, X_{l,j}\right\}$ sequentially forward in time (from $X_{1,j}$ to $X_{l,j}$) and the second processing the input sequence sequentially backward in time (from $X_{l,j}$ to $X_{1,j}$). A LSTM RNN is capable of learning long-term dependencies in sequential data, while avoiding the vanishing gradient problem of standard RNNs [57]. This is achieved throughout the use of cell states (see Figure 4), which regulate the amount of information flowing through a LSTM network throughout the use of three principal gates: forget gate (f_t), input gate (i_t) and output gate (o_t). The cell's output h_t (at each time step t) is computed, given a specific input x_t, the previous hidden state h_{t-1}, and the previous cell state C_{t-1}, as follows,

$$f_t = \sigma\left(W_f x_t + U_f h_{t-1} + b_f\right) \tag{11}$$

$$i_t = \sigma\left(W_i x_t + U_i h_{t-1} + b_i\right) \tag{12}$$

$$\tilde{C}_t = tanh\left(W_c s_t + U_c h_{t-1} + b_c\right) \tag{13}$$

$$C_t = f_t \otimes C_{t-1} + i_t \otimes \tilde{C}_t \tag{14}$$

$$o_t = \sigma\left(W_o x_t + U_o h_{t-1} + b_o\right) \tag{15}$$

$$h_t = o_t \otimes tanh(C_t) \tag{16}$$

where σ represents the sigmoid activation function $\sigma(x) = (1 + exp(-x))^{-1}$ and $tanh$ represents the hyperbolic tangent activation function. The element-wise multiplication operator is represented by the symbol $\otimes$. The weight matrices for the input x_t are represented by W_i, W_f, W_o and W_c for the input gate, forget gate, output gate and cell state, respectively. Analogously, the weight matrices for the previous hidden state h_{t-1} for each gate are represented by U_i, U_f, U_o and U_c. The amount of information to be further propagated into the network is controlled by the forget gate (Equation (11)), the input gate (Equation (12)) and the computed cell state candidate $\tilde{C}_t$ (Equation (13)). These parameters are subsequently used to update the cell state C_t based on the previous cell state C_{t-1} (Equation (14)). The output of the cell can subsequently be computed using both Equation (15) and Equation (16).

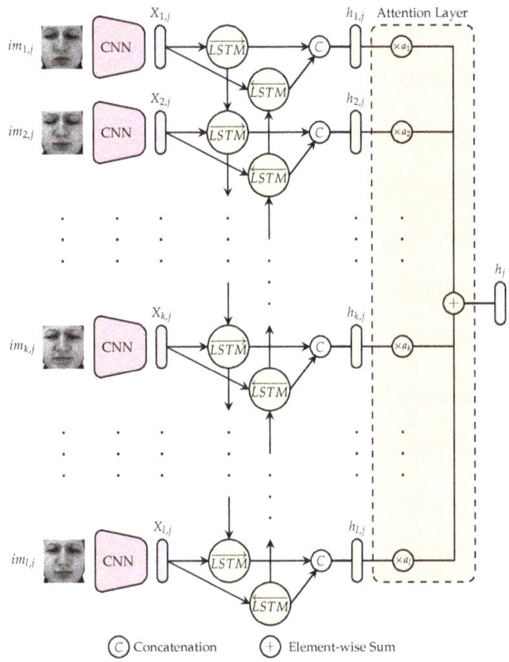

Figure 3. Attention Network.

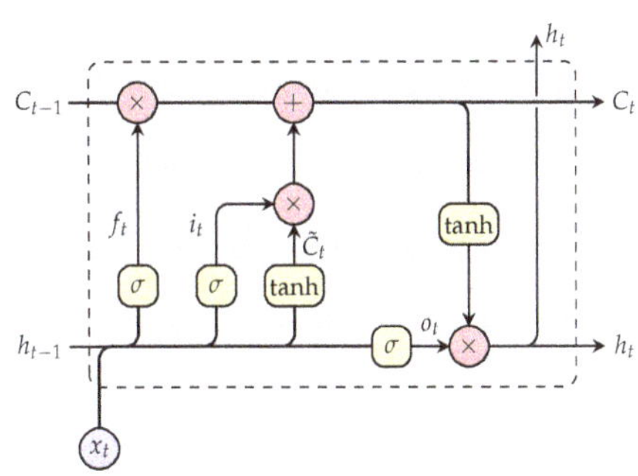

Figure 4. Long Short-Term Memory (LSTM) Recurrent Neural Network (RNN).

In the current work, the hidden representation stemming from the forward pass $\left\{\overrightarrow{h_{1,j}}, \overrightarrow{h_{2,j}}, \ldots, \overrightarrow{h_{l,j}}\right\}$ and the one stemming from the backward pass $\left\{\overleftarrow{h_{1,j}}, \overleftarrow{h_{2,j}}, \ldots, \overleftarrow{h_{l,j}}\right\}$ are subsequently concatenated $\left\{\left[\overrightarrow{h_{1,j}}, \overleftarrow{h_{1,j}}\right], \left[\overrightarrow{h_{2,j}}, \overleftarrow{h_{2,j}}\right], \ldots, \left[\overrightarrow{h_{l,j}}, \overleftarrow{h_{l,j}}\right]\right\}$ and fed into the next layer. For the sake of simplicity, the output of the BiLSTM layer will be depicted as follows, $\left\{h_{1,j}, h_{2,j}, \ldots, h_{l,j}\right\}$ (with $h_{k,j} = \left[\overrightarrow{h_{k,j}}, \overleftarrow{h_{k,j}}\right]$). The next layer consists of an attention layer, where self-attention weights $\{a_k\}_{k=1}^{l}$ are computed and subsequently used to generate a single weighted representation of the input sequence. The self-attention weights are computed as follows,

$$\alpha_k = elu\left(W_k h_{k,j} + b_k\right) \tag{17}$$

$$a_k = \frac{exp(\alpha_k)}{\sum\limits_{i=1}^{l} exp(\alpha_i)} \tag{18}$$

where W_k are the weights specific to the input feature representation $h_{k,j} = \left[\overrightarrow{h_{k,j}}, \overleftarrow{h_{k,j}}\right]$ and elu represents the exponential linear unit activation function [58], which is defined as

$$elu_\alpha(x) = \begin{cases} \alpha \cdot (\exp(x) - 1) & \text{if } x < 0 \\ x & \text{if } x \geq 0 \end{cases} \tag{19}$$

with $\alpha = 1$. Each self-attention weight expresses the relevance of a specific image for the corresponding emotional state expressed within the video sequence. Thereby, relevant images should be assigned significantly higher weight values as irrelevant images. The final representation of the input sequence is subsequently computed by performing a weighted aggregation of the BiLSTM output $\left\{h_{1,j}, h_{2,j}, \ldots, h_{l,j}\right\}$ based on the computed self-attention weights $\{a_k\}_{k=1}^{l}$ as follows,

$$h_j = \sum_{k=1}^{l} a_k \cdot h_{k,j} \tag{20}$$

and is further used to perform the classification task.

4. Experiments

In the following section, a description of the experiments performed for the evaluation of the proposed approach is provided. First, the datasets used for the evaluation are briefly described, followed by a depiction of the conducted data preprocessing steps. The experimental settings as well as the performed experiments are described subsequently. This section is finally concluded with a description and discussion of the experimental results.

4.1. Datasets Description

The presented approach is evaluated on both the *BioVid Heat Pain Database (Part A)* (BVDB) [4] and the *SenseEmotion Database* (SEDB) [6]. Both datasets were recorded with the principal goal of fostering research in the domain of pain recognition. In both cases, several healthy participants were submitted to a series of individually calibrated heat-induced painful stimuli, using the exact same procedure. Whereas the BVDB consists of 87 individuals submitted to four individually calibrated and gradually increasing levels of heat-induced painful stimuli (T_1, T_2, T_3 and T_4), the SEDB consists of 40 individuals submitted to three individually calibrated and gradually increasing levels of heat-induced stimuli (T_1, T_2 and T_3). Each single level of heat-induced pain stimulation was randomly elicited a total of 20 times for the BVDB and 30 times for the SEDB. Each elicitation lasted 4 s, followed by a recovery phase of a random length of 8 to 12 s during which a baseline temperature T_0 (32°C) was applied (see Figure 5). Whereas the elicitations were performed uniquely on one specific hand for the BVDB, the experiments were conducted twice for the SEDB, with the elicitation performed each time on one specific arm (left forearm and right forearm). Therefore, with the inclusion of the baseline temperature T_0, the dataset specific to the BVDB consists of a total of $87 \times 20 \times 5 = 8700$ samples, whereas the SEDB consists of a total of $40 \times 30 \times 4 \times 2 = 9600$ samples. During the experiments, the demeanour of each participant was recorded using several modalities consisting of video and bio-physiological channels for the BVDB, while the SEDB included audio, video and bio-physiological channels. The current work focuses uniquely on the video modality, and the reader should refer to the work in [10,14–16,33,59–64] for more experiments including the other recorded modalities.

4.2. Data Preprocessing

The evaluation performed in the current work is undertaken in both cases (BVDB and SEDB) on video recordings performed by a frontal camera. The recordings were performed at a frame rate of 25 frames per second (fps) for the BVDB and 30 fps for the SEDB. Furthermore, the evaluation is performed uniquely on windows of length 4.5 s with a shift of 4 s from the elicitation's onset, as proposed in [16] (see Figure 5). Once these specific windows are extracted, the facial behaviour analysis toolkit OpenFace [65] is used for the automatic detection, alignment and normalisation of the facial area (with a fixed size of 100×100 pixels) in each video frame. Subsequently, MHI sequences and OFI sequences are extracted using the OpenCV library [66]. Both MHIs and OFIs are generated relatively to a reference frame, which in this case is the very first frame of each video sequence. Concerning MHIs, the temporal extent parameter τ (see Equation (1)) was set to the length of the sequence of images ($25 \times 4.5 \cong 113$ frames for the BVDB and $30 \times 4.5 = 135$ frames for the SEDB). Furthermore, the threshold parameter ξ (see Equation (2)) was set to 1 to capture any single motion from two consecutive frames (in this case, the fluctuation of pixel intensities between the reference frame and the tth frame). Finally, to reduce the computational requirements, the number of samples in each sequence is reduced by sequentially selecting each second frame of an entire sequence for the BVDB (resulting in sequences with a total length of 57 frames), and each third frame of an entire sequence for the SEDB (resulting in sequences of length 45 frames). The dimensionality of the tensor input specific to the BVDB is, respectively, $(bs, 57, 100, 100, 3)$ for OFI sequences and $(bs, 57, 100, 100, 1)$ for MHI sequences (bs representing the batch size). The dimensionality of the tensor input specific to the SEDB is, respectively, $(bs, 45, 100, 100, 3)$ for OFI sequences and $(bs, 45, 100, 100, 1)$ for MHI sequences.

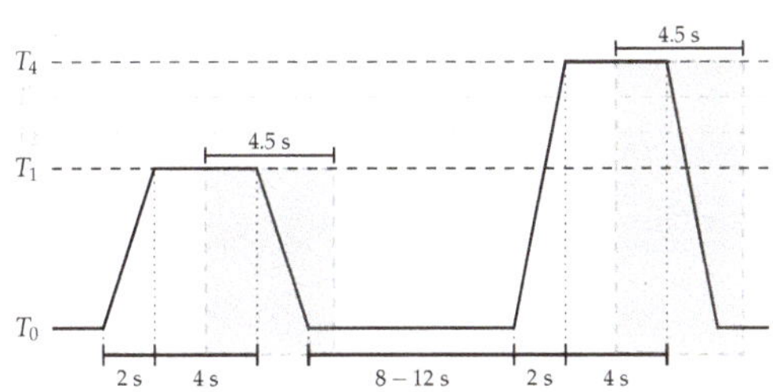

Figure 5. Video Signal Segmentation (BioVid Heat Pain Database (Part A)). Experiments are carried out on windows of length 4.5 s with a temporal shift of 4 s from the elicitations' onsets.

4.3. Experimental Settings

The evaluation performed in the current work consists of the discrimination between high and low stimuli levels. Therefore, two binary classification tasks are performed for each database: $T_0 vs. T_4$ and $T_1 vs. T_4$ for the BVDB, and $T_0 vs. T_3$ and $T_1 vs. T_3$ for the SEDB. Furthermore, the assessment of the proposed approach is conducted by applying a *Leave-One-Subject-Out* (LOSO) cross-validation evaluation, which means that a total of 87 experiments were conducted for the BVDB (40 experiments for the SEDB), during which the data specific to each participant is used once to evaluate the performance of the classification architecture optimised on the data specific to the remaining 86 participants (the data specific to 39 participants is used to optimise the architecture for the SEDB, and the data specific to the remaining participant is used to evaluate the performance of the architecture).

The feature embedding CNN used for the evaluation is adapted from the one proposed by the Visual Geometry Group of the University of Oxford *VGG16* [67]. The depth of the CNN model is substantially reduced to a total of 10 convolutional layers (instead of 13 as in the *VGG16* model), and the number of convolutional filters is gradually increased from one convolutional block to the next starting from 8 filters until a maximum of 64 filters. The activation function in each convolutional layer consists of the *elu* activation function (instead of the rectified linear unit (*relu*) activation function as in the *VGG16* model). Max-pooling and Batch Normalisation [68] are performed after each convolutional block. A detailed description of the feature embedding CNN architecture can be seen in Table 1. The coupled BiLSTM layer consists of two LSTM RNNs with 64 units each. The resulting sequence of spatio-temporal features is further fed into the attention layer in order to generate a single aggregated representation of the input sequence. The classification is further performed based on this representation and the architecture of the classification model is described in Table 2. The exact same architecture is used for the two input sequences (MHIs and OFIs). The outputs of the classifiers are further aggregated based on both Equation (8) and Equation (9). The whole architecture is subsequently trained in an end-to-end manner, using the Adaptive Moment Estimation (Adam) [69] optimisation algorithm with a fixed learning rate set empirically to 10^{-5}. The categorical cross entropy loss function is used for each network ($\mathcal{L}_{mhi} = \mathcal{L}_{ofi} = \mathcal{L}_{agg} = \mathcal{L}$), and is defined as follows,

$$\mathcal{L} = -\sum_{j=1}^{c} y_j log(\hat{y}_j) \tag{21}$$

where $\hat{y}_j$ represents the classifier's output, y_j is the ground-truth label value and $c \in \mathbb{N}$ is the number of classes for a specific classification task.

Table 1. Feature embedding CNN architecture.

Layer	No. Filters
2× Conv2D	8
MaxPooling2D	–
Batch Normalisation	–
2× Conv2D	16
MaxPooling2D	–
Batch Normalisation	–
3× Conv2D	32
MaxPooling2D	–
Batch Normalisation	–
3× Conv2D	64
MaxPooling2D	–
Batch Normalisation	–
Flatten	–

The size of the kernels is identical for all convolutional layers and is set to 3×3, with the convolutional stride set to 1×1. Max-pooling is performed after each block of convolutional layers over a 2×2 window, with a 2×2 stride.

The regularisation parameters of the loss function in Equation (10) are set as follows: $\lambda_{mhi} = \lambda_{ofi} = 0.2$ and $\lambda_{agg} = 0.6$. The value of the regularisation parameter specific to the aggregation layer's loss is set higher than the others in order to enable the architecture to compute robust aggregation weights. The whole architecture is trained for a total of 20 epoches with the batch size set to 40 for the BVDB and 60 for the SEDB. The implementation and evaluation of the whole architecture is done with the Python libraries Keras [70], Tensorflow [71] and Scikit-learn [72].

Table 2. Classifier Architecture.

Layer	No. Units
Dropout	–
Fully Connected	64
Dropout	–
Fully Connected	*c*

The dropout rate is empirically set to 0.25. The first fully connected layer uses the *elu* activation function, while the last fully connected layer consists of a *softmax* layer (whereby *c* depicts the number of classes of the classification task).

4.4. Results

The performance of the classification architectures specific to each input channel (MHIs and OFIs), as well as the performance of the weighted score aggregation approach are depicted in Figure 6. The performance metric in this case is the accuracy, which is defined as

$$Accuracy = \frac{tp + tn}{tp + fp + tn + fn} \tag{22}$$

where tp refers to true positives, tn refers to true negatives, fp refers to false positives and fn refers to false negatives (since we are dealing with a binary classification task with two balanced datasets). For both datasets and both classification tasks, the aggregation approach significantly outperforms the classification architecture based uniquely on MHIs. Furthermore, the classification architecture based uniquely on OFIs outperforms the one based on MHIs for both databases and both classification tasks, with significant performance improvement in the case of the BVDB. The aggregation approach also performs slightly better than the architecture based uniquely on OFIs for both databases, although not significantly in most cases. The only significant performance improvement is achieved for the classification task T_1 vs. T_4 for the SEDB. However, the performance of both channel specific architectures and the performance of the score aggregation approach are significantly higher than chance level (which is 50% in the case of binary classification tasks) pointing at the relevance of the designed approach. Furthermore, the performance of the classification architecture is improved by using both channels and performing a weighted aggregation of the scores of both channel specific deep attention models.

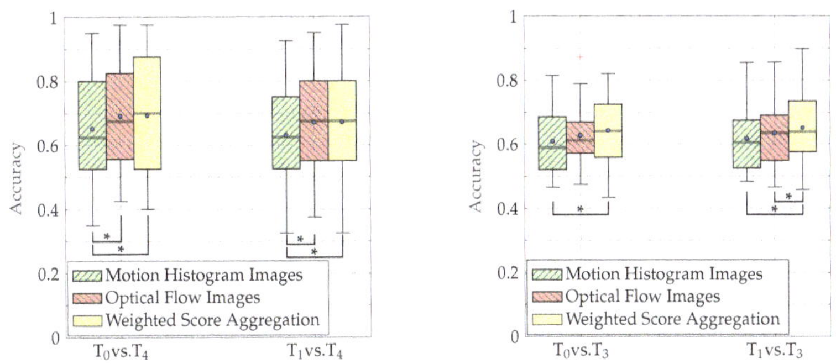

(**a**) BioVid Heat Pain Database (Part A). (**b**) SenseEmotion Database.

Figure 6. Classification performance (Accuracy). An asterisk (*) indicates a significant performance improvement. The test has been conducted using a Wilcoxon signed rank test with a significance level of 5%. Within each boxplot, the mean and the median classification accuracy are depicted respectively with a dot and a horizontal line.

Moreover, to provide more insights into the self attention mechanism, the frame attention weight values computed at each evaluation step during the LOSO cross-validation evaluation process are depicted in Figure 7 for the BVDB and in Figure 8 for the SEDB (uniquely for the classification task T_0 vs. T_4, as the results for the classification task T_1 vs. T_4 are similar). The distribution of the weight values specific to the MHI deep attention models for both databases (Figure 7a,c for the BVDB, Figure 8a,c for the SEDB) is skewed left. It depicts a steady growth of weight values along the temporal axis of each sequence, with the MHIs located at the end of a sequence weighted significantly higher as the others. This is in accordance with the sequential extraction process of MHIs, as each extracted image contains more motion information as the previous one, with the last images accumulating almost the totality of motion information of an entire sequence. Therefore, concerning the actual classification task, the last MHIs are more interesting and relevant than the early images. Thus, such images should be weighted accordingly higher. The designed network is therefore capable of conducting this specific task by using self attention mechanisms.

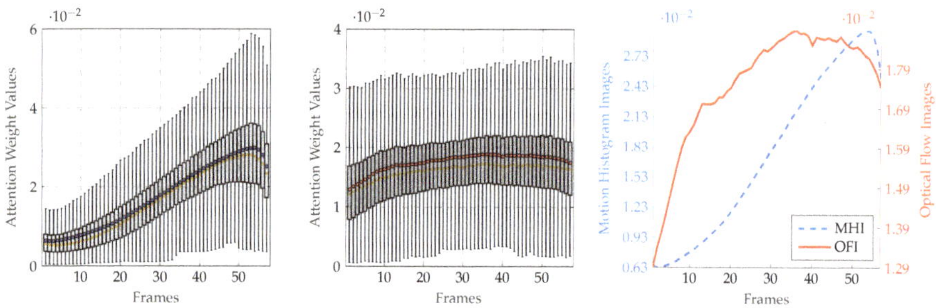

(**a**) Motion History Images. (**b**) Optical Flow Images. (**c**) Mean Attention Weight Values.

Figure 7. BioVid Heat Pain Database (Part A): Attention network weight values for the classification task $T_0 vs. T_4$. Within each boxplot in (**a**,**b**), the mean and the median weight values are depicted, respectively, with a dot and a horizontal line. In (**c**), the average weight values are normalised between the maximum average value and the minimum average value to allow a better visualisation of the values distributions.

A similar observation can be made concerning the distribution of the weight values of OFIs (see Figure 7b,c for the BVDB, Figure 8b,c for the SEDB). Both depicted distributions are also skewed left, with gradually increasing weight values relative to the temporal axis. This shows that the recorded pain-related facial expressions for both BVDB and SEDB consist of gradually evolving facial movements, starting from a neutral facial depiction (not relevant for the actual classification task) to the apex of the facial movement (which is the most relevant frame for the depicted facial emotion) before gradually turning back to the neutral facial depiction. Therefore, the network assigns weight values according to this specific characterisation of pain-related facial movements using attention mechanisms, thus the relevance of such approaches for facial expression analysis.

Furthermore, the performance of the weighted score aggregation approach is further assessed based on the following additional performance metrics,

$$Macro\ Precision = \frac{1}{c}\sum_{i=1}^{c}\frac{tp_i}{tp_i + fp_i} \tag{23}$$

$$Macro\ Recall = \frac{1}{c}\sum_{i=1}^{c}\frac{tp_i}{tp_i + fn_i} \tag{24}$$

$$Macro\ F1\ Score = \frac{2 \times Macro\ Precision \times Macro\ Recall}{Macro\ Precision + Macro\ Recall} \tag{25}$$

where tp_i, fp_i and fn_i refer, respectively, to the true positives, false positives and false negatives of the ith class. The results of the evaluation are depicted in Figure 9, for both the BVDB (see Figure 9a) and the SEDB (see Figure 9b).

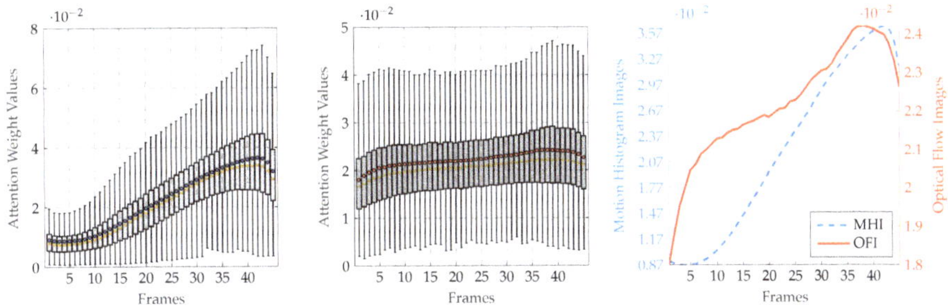

(**a**) Motion History Images. (**b**) Optical Flow Images. (**c**) Mean Attention Weight Values.

Figure 8. SenseEmotion Database: Attention network weight values for the classification task $T_0 vs. T_3$. Within each boxplot in (**a**,**b**), the mean and the median weight values are depicted respectively with a dot and a horizontal line. In (**c**), the average weight values are normalised between the maximum average value and the minimum average value to allow a better visualisation of the values distributions.

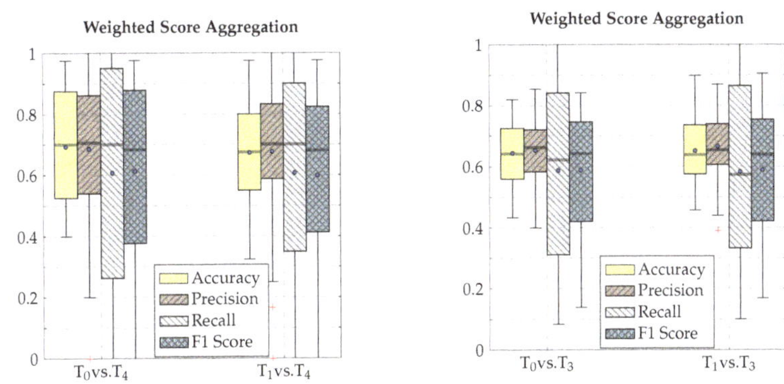

(**a**) BioVid Heat Pain Database (Part A). (**b**) SenseEmotion Database.

Figure 9. Weighted score aggregation classification performance. Within each box plot, the mean and median values of the respective performance evaluation metrics are depicted with a dot and a horizontal line, respectively.

These results depict a huge variance amongst all performance metrics, in particular the *Macro Recall*, which points at the fact that the classification tasks remain difficult. The evaluation on some participants yields a *Macro F1 Score* of null or nearly null, pointing at the fact that the architecture is unable to discriminate between low and high levels of pain elicitation for these specific participants. This is, however, similar and in accordance with previous works on these specific datasets. The authors of the BVDB in [73] were able to identify a set of participants who did not react to the levels of pain elicitation, therefore causing the huge variance in the classification experiments.

Finally, the performance of the weighted score aggregation approach is compared to other pain-related facial expressions classification approaches proposed in the literature. For the sake of fairness, we compare the results of the proposed approach with those results in related works which are based on the exact same dataset and were computed based on the exact same evaluation protocol (LOSO). The results are depicted in Table 3 for the BVDB and in Table 4 for the SEDB.

Table 3. Classification performance comparison to early works on the BioVid Heat Pain Database (Part A) in a LOSO cross-validation setting for the classification task $T_0 vs. T_4$.

Approach	Description	Performance
Yang et al. [27]	BSIF	65.17
Kächele et al. [31,62]	Geometric Features	65.55 ± 14.83
Werner et al. [8]	Standardised Facial Action Descriptors	**72.40**
Our Approach	Motion History Images	65.17 ± 15.49
Our Approach	Optical Flow Images	69.11 ± 14.73
Our Approach	Weighted Score Aggregation	$\underline{69.25 \pm 17.31}$

The performance metric consists of the average accuracy (in %) over the LOSO cross-validation evaluation. The best performing approach is depicted in bold and the second best approach is underlined.

Table 4. Classification performance comparison to early works on the SenseEmotion Database in a LOSO cross-validation setting for the classification task $T_0 vs. T_3$.

Approach	Description	Performance
Kalischek et al. [38]	Transfer Learning	60.10 ± 00.06
Thiam et al. [15]	Standardised Geometric Features	$\mathbf{66.22 \pm 14.48}$
Our Approach	Motion Histogram Images	60.86 ± 09.81
Our Approach	Optical Flow Images	62.70 ± 09.24
Our Approach	Weighted Score Aggregation	$\underline{64.35 \pm 10.40}$

The performance metric consists of the average accuracy (in %) over the LOSO cross-validation evaluation. The best performing approach is depicted in bold and the second best approach is underlined.

In both cases, the performance of the weighted score aggregation approach is on par with the best performing approaches. However, it has to be mentioned that the authors of the best performing approaches for both the BVDB [8] and the SEDB [15] perform a subject-specific normalisation of the extracted feature representations in order to compensate for the differences in expressiveness amongst the participants. Although this specific preprocessing step has proven to significantly improve the classification performance of the architecture [61], it is not realistic as it requires that the whole testing set is already available beforehand. The normalisation parameters should be learned on the available training material and subsequently applied to the testing material during the inference phase. Nevertheless, the proposed approach based on the weighted aggregation of the scores of both MHI- and OFI-specific deep attention models generalises well and is capable of achieving state-of-the-art classification performances.

5. Conclusions

In the current work, an approach based on a weighted aggregation of the scores of two deep attention networks based, respectively, on MHIs and OFIs has been proposed and evaluated for the analysis of pain-related facial expressions. The assessment performed on both BVDB and SEDB shows that the proposed approach is capable of achieving state-of-the-art classification performances and is on par with the best performing approaches proposed in the literature. Moreover, the visualisation of the weight values stemming from the implemented attention mechanism shows that the network is capable of identifying relevant frames in relation with the current level of pain elicitation depicted by a sequence of images, by assigning significantly higher values to the most relevant images in comparison to the weight values of irrelevant images. Furthermore, as the proposed architecture was trained from scratch in an end-to-end manner, it is believed that transfer learning, in particular, for the feature embedding CNN used to generate the feature representation of each frame, could potentially improve the performance of the whole architecture. Such an analysis was not conducted in the current work, as the optimisation of the presented approach was not the goal of the conducted experiments, but rather the assessment of such an architecture for the analysis of pain-related facial expressions. Moreover, a multi-stage training strategy could also potentially improve the overall performance of the architecture, as the end-to-end trained approach is likely to suffer from overfitting, in particular, when considering the coupled aggregation layer. The representation of the input sequences should be further investigated as well. Both MHIs and OFIs have the temporal aspect of the sequences integrated into their properties. The performed evaluation has shown that a model based on OFIs significantly outperforms the one based on MHIs in most cases. However, it has also been shown that most of the interesting frames in MHI sequences are located at the very end of the temporal axis of each sequence. Therefore, single MHIs extracted from entire sequences could also be used as input for deep architectures. Overall, the performed experiments show that the discrimination between lower and higher pain elicitation levels remains a difficult endeavour. This is due to the variety of expressiveness amongst the participants. However, personalisation and transfer learning strategies could potentially help improve the performance of inference models applied in this specific area of research.

Author Contributions: Conceptualisation, P.T. and F.S.; Methodology, P.T.; Software, P.T.; Validation, P.T.; Formal Analysis, P.T.; Investigation, P.T. and F.S.; Writing—Original Draft Preparation, P.T.; Writing—Review and Editing, P.T., H.A.K. and F.S.; Visualisation, P.T.; Supervision, H.A.K. and F.S.; Project Administration, H.A.K. and F.S.; Funding Acquisition, H.A.K. and F.S. All authors have read and agreed to the published version of the manuscript.

Funding: The research leading to these results has received funding from the Federal Ministry of Education and Research (BMBF, SenseEmotion) to F.S., (BMBF, e:Med, CONFIRM, ID 01ZX1708C) to H.A.K., and the Ministry of Science and Education Baden-Württemberg (Project ZIV) to H.A.K.

Acknowledgments: We gratefully acknowledge the support of NVIDIA Corporation with the donation of the Tesla K40 GPU used for this research.

Conflicts of Interest: The authors declare no conflicts of interest.

References

1. Ahad, M.A.R.; Tan, J.K.; Kim, H.; Ishikawa, S. Motion History Image: its variants and applications. *Mach. Vis. Appl.* **2012**, *23*, 255–281. [CrossRef]
2. Horn, B.K.P.; Schunck, B.G. Determining optical flow. *Artif. Intell.* **1981**, *17*, 185–203. [CrossRef]
3. Lucey, P.; Cohn, J.F.; Prkachin, K.M.; Solomon, P.E.; Matthews, I. Painful data: The UNBC-McMaster shoulder pain expression archive database. In Proceedings of the Face and Gesture, Santa Barbara, CA, USA, 21–25 March 2011; pp. 57–64.
4. Walter, S.; Gruss, S.; Ehleiter, H.; Tan, J.; Traue, H.C.; Crawcour, S.; Werner, P.; Al-Hamadi, A.; Andrade, A. The BioVid heat pain database data for the advancement and systematic validation of an automated pain recognition system. In Proceedings of the IEEE International Conference on Cybernetics, Lausanne, Switzerland, 13–15 June 2013; pp. 128–131.
5. Aung, M.S.H.; Kaltwang, S.; Romera-Paredes, B.; Martinez, B.; Singh, A.; Cella, M.; Valstar, M.; Meng, H.; Kemp, A.; Shafizadeh, M.; et al. The automatic detection of chronic pain-related expression: requirements, challenges and multimodal dataset. *IEEE Trans. Affect. Comput.* **2016**, *7*, 435–451. [CrossRef] [PubMed]
6. Velana, M.; Gruss, S.; Layher, G.; Thiam, P.; Zhang, Y.; Schork, D.; Kessler, V.; Gruss, S.; Neumann, H.; Kim, J.; et al. The SenseEmotion Database: A multimodal database for the development and systematic validation of an automatic pain- and emotion-recognition system. In Proceedings of the Multimodal Pattern Recognition of Social Signals in Human-Computer-Interaction, Cancun, Mexico, 4 December 2016; pp. 127–139.
7. Thiam, P.; Kessler, V.; Schwenker, F. Hierarchical combination of video features for personalised pain level recognition. In Proceedings of the 25th European Symposium on Artificial Neural Networks, Computational Intelligence and Machine Learning, Bruges, Belgium, 26–28 April 2017; pp. 465–470.
8. Werner, P.; Al-Hamadi, A.; Limbrecht-Ecklundt, K.; Walter, S.; Gruss, S.; Traue, H.C. Automatic Pain Assessment with Facial Activity Descriptors. *IEEE Trans. Affect. Comput.* **2017**, *8*, 286–299. [CrossRef]
9. Tsai, F.S.; Hsu, Y.L.; Chen, W.C.; Weng, Y.M.; Ng, C.J.; Lee, C.C. Toward Development and Evaluation of Pain Level-Rating Scale For Emergency Triage Based on Vocal Characteristics and Facial Expressions. In Proceedings of the Interspeech 2016, San-Francisco, CA, USA, 8–12 September 2016; pp. 92–96.

10. Thiam, P.; Schwenker, F. Combining deep and hand-crafted features for audio-based pain intensity classification. In Proceedings of the Multimodal Pattern Recognition of Social Signals in Human-Computer-Interaction, Beijing, China, 20 August 2018; pp. 49–58.
11. Walter, S.; Gruss, S.; Limbrecht-Ecklundt, K.; Traue, H.C.; Werner, P.; Al-Hamadi, A.; Diniz, N.; Silva, G.M.; Andrade, A.O. Automatic pain quantification using autonomic parameters. *Psych. Neurosci.* **2014**, *7*, 363–380. [CrossRef]
12. Chu, Y.; Zhao, X.; Han, J.; Su, Y. Physiological signal-based method for measurement of pain intensity. *Front. Neurosci.* **2017**, *11*, 279. [CrossRef]
13. Lopez-Martinez, D.; Picard, R. Continuous pain intensity estimation from autonomic signals with recurrent neural networks. In Proceedings of the 40th Annual International Conference of the IEEE Engineering in Medecine and Biology Society, Honolulu, HI, USA, 18–21 July 2018; pp. 5624–5627.
14. Thiam, P.; Schwenker, F. Multi-modal data fusion for pain intensity assessement and classification. In Proceedings of the 7th International Conference on Image Processing Theory, Tools and Applications, Montreal, QC, Canada, 28 November–1 December 2017; pp. 1–6.
15. Thiam, P.; Kessler, V.; Amirian, M.; Bellmann, P.; Layher, G.; Zhang, Y.; Velana, M.; Gruss, S.; Walter, S.; Traue, H.C.; et al. Multi-modal pain intensity recognition based on the SenseEmotion Database. *IEEE Trans. Affect. Comput.* **2019**. [CrossRef]
16. Thiam, P.; Bellmann, P.; Kestler, H.A.; Schwenker, F. Exploring deep physiological models for nociceptive pain recognition. *Sensors* **2019**, *19*, 4503. [CrossRef]
17. Ekman, P.; Friesen, W.V. *The Facial Action Unit System: A Technique for the Measurement of Facial Movement*; Consulting Psychologist Press: Mountain View, CA, USA, 1978.
18. Senechal, T.; McDuff, D.; Kaliouby, R.E. Facial Action Unit detection using active learning and an efficient non-linear kernel approximation. In Proceedings of the IEEE International Conference on Computer Vision Workshop, Santiago, Chile, 7–13 December 2015; pp. 10–18.
19. Lucey, P.; Cohn, J.; Lucey, S.; Matthews, I.; Sridharan, S.; Prkachin, K.M. Automatically detecting pain using Facial Actions. In Proceedings of the 3rd International Conference on Affective Computing and Intelligent Interaction and Workshops, Amsterdam, The Netherlands, 10–12 September 2009; pp. 1–8.
20. Abe, S. *Support Vector Machines for Pattern Classification*; Springer: Berlin, Germany, 2005.
21. Brümmer, N.; Preez, J.D. Application-independent evaluation of speaker detection. *Comput. Speech Lang.* **2006**, *20*, 230–275. [CrossRef]
22. Zafar, Z.; Khan, N.A. Pain intensity evaluation through Facial Action Units. In Proceedings of the 22nd International Conference on Pattern Recognition, Stockholm, Sweden, 24–28 August 2014; pp. 4696–4701.
23. Cover, T.; Hart, P. Nearest Neighbor Pattern Classification. *IEEE Trans. Inf. Theory* **1967**, *13*, 21–27. [CrossRef]
24. Prkachin, K.M.; Solomom, P.E. The structure, reliability and validity of pain expression: Evidence from patients with shoulder pain. *Pain* **2008**, *139*, 267–274. [CrossRef]
25. Xu, X.; Craig, K.D.; Diaz, D.; Goodwin, M.S.; Akcakaya, M.; Susam, B.T.; Huang, J.S.; de Sa, V.S. Automated pain detection in facial videos of children using human-assisted transfer learning. In Proceedings of the International Workshop on Artificial Intelligence in Health, Stockholm, Sweden, 13–14 July 2018; pp. 162–180.
26. Monwar, M.; Rezaei, S. Pain recognition using artificial neural network. In Proceedings of the IEEE International Symposium on Signal Processing and Information Theory, Vancouver, BC, Canada, 27–30 August 2006; pp. 8–33.
27. Yang, R.; Tong, S.; Bordallo, M.; Boutellaa, E.; Peng, J.; Feng, X.; Hadid, A. On pain assessment from facial videos using spatio-temporal local descriptors. In Proceedings of the 6th International Conference on Image Processing Theory, Tools and Applications, Oulu, Finland, 12–15 December 2016; pp. 1–6.
28. Zhao, G.; Pietikaeinen, M. Dynamic texture recognition using local binary patterns with an application to facial expressions. *IEEE Trans. Pattern Anal. Mach. Intell.* **2007**, *29*, 915–928. [CrossRef]
29. Ojansivu, V.; Heikkilä, J. Blur insensitive texture classification using local phase quantization. In Proceedings of the Image and Signal Processing, Cherbourg-Octeville, France, 1–3 July 2008; pp. 236–243.
30. Kannala, J.; Rahtu, E. BSIF: Binarized Statistical Image Features. In Proceedings of the 21st International Conference on Pattern Recognition, Tsukuba, Japan, 11–15 November 2012; pp. 1363–1366.

31. Kächele, M.; Thiam, P.; Amirian, M.; Werner, P.; Walter, S.; Schwenker, F.; Palm, G. Engineering Applications of Neural Networks. Multimodal data fusion for person-independent, continuous estimation of pain Intensity. In Proceedings of the Engineering Applications of Neural Networks, Rhodes, Greece, 25–28 September 2015; pp. 275–285.
32. Breiman, L. Random Forests. *Mach. Learn.* **2001**, *45*, 5–32. [CrossRef]
33. Thiam, P.; Kessler, V.; Walter, S.; Palm, G.; Scwenker, F. Audio-visual recognition of pain intensity. In Proceedings of the Multimodal Pattern Recognition of Social Signals in Human-Computer-Interaction, Cancun, Mexico, 4 December 2016; pp. 110–126.
34. Bosch, A.; Zisserman, A.; Munoz, X. Representing shape with a spatial pyramid kernel. In Proceedings of the 6th ACM International Conference on Image and Video Retrieval, Amsterdam, The Netherlands, 9–11 July 2007; pp. 401–408.
35. Almaev, T.R.; Valstar, M.F. Local Gabor Binary Patterns from Three Orthogonal Planes for automatic facial expression recognition. In Proceedings of the 2013 Humaine Association Conference on Affective Computing and Intelligent Interaction, Geneva, Switzerland, 2–5 September 2013; pp. 356–361.
36. Bellantonio, M.; Haque, M.A.; Rodriguez, P.; Nasrollahi, K.; Telve, T.; Guerrero, S.E.; Gonzàlez, J.; Moeslund, T.B.; Rasti, P.; Anbarjafari, G. Spatio-temporal pain recognition in CNN-based super-resolved facial images. In Proceedings of the International Conference on Pattern Recognition: Workshop on Face and Facial Expression Recognition, Cancun, Mexico, 4 December 2016; pp. 151–162.
37. Rodriguez, P.; Cucurull, G.; Gonzàlez, J.; Gonfaus, J.M.; Nasrollahi, K.; Moeslund, T.B.; Roca, F.X. Deep Pain: Exploiting Long Short-Term Memory networks for facial expression classification. *IEEE Trans. Cybern.* **2018**. [CrossRef]
38. Kalischek, N.; Thiam, P.; Bellmann, P.; Schwenker, F. Deep domain adaptation for facial expression analysis. In Proceedings of the 8th International Conference on Affective Computing and Intelligent Interaction Workshops and Demos, Cambridge, UK, 3–6 September 2019; pp. 317–323.
39. LeCun, Y.; Kavukcuoglu, K.; Farabet, C. Convolutional networks and application in vision. In Proceedings of the IEEE International Symposium on Circuits and Systems, 2010, Paris, France, 30 May–2 June 2010; pp. 253–256.
40. Hochreiter, S.; Schmidhuber, J. Long Short-Term Memory. *Neural Comput.* **1997**, *9*, 1735–1780. [CrossRef]
41. Soar, J.; Bargshady, G.; Zhou, X.; Whittaker, F. Deep learning model for detection of pain intensity from facial expression. In Proceedings of the International Conference on Smart Homes and Health Telematics, Singapore, 10–12 July 2018; pp. 249–254.
42. Lafferty, J.D.; McCallum, A.; Pereira, F.C.N. Conditional Random Fields: Probabilistic models for segmenting and labeling sequence data. In Proceedings of the 18th International Conference on Machine Learning, Williams College, Williamstown, MA, USA, 28 June–1 July 2001; pp. 282–289.
43. Bargshady, G.; Soar, J.; Zhou, X.; Deo, R.C.; Whittaker, F.; Wang, H. A joint deep neural network model for pain recognition from face. In Proceedings of the IEEE 4th International Conference on Computer and Communication Systems, Singapore, 23–25 February 2019; pp. 52–56.
44. Zhou, J.; Hong, X.; Su, F.; Zhao, G. Recurrent convolutional neural network regression for continuous pain intensity estimation in Video. In Proceedings of the IEEE Conference on Computer Vision and Pattern Recognition Workshops, Las Vegas, NV, USA, 26 June–1 July 2016; pp. 1535–1543.
45. Liang, M.; Hi, X. Recurrent convolutional neural network for object recognition. In Proceedings of the IEEE Conference on Computer Vision and Pattern Recognition, Boston, MA, USA, 7–12 June 2015; pp. 3367–3375.
46. Wang, F.; Xiang, X.; Liu, C.; Tran, T.D.; Reiter, A.; Hager, G.D.; Quaon, H.; Cheng, J.; Yuille, A.L. Regularizing face verification nets for pain intensity regression. In Proceedings of the IEEE International Conference on Image Processing, Beijing, China, 17–20 September 2017; pp.1087–1091.
47. Meng, D.; Peng, X.; Wang, K.; Qiao, Y. Frame attention networks for facial expression recognition in videos. In Proceedings of the IEEE International Conference on Image Processing, Taipei, Taiwan, 22–25 September 2019; pp. 3866–3870.
48. Bobick, A.F.; Davis, J.W. The recognition of human movement using temporal templates. *IEEE Trans. Pattern Anal. Mach. Intell.* **2001**, *23*, 257–267. [CrossRef]
49. Yin, Z.; Collins, R. Moving object localization in thermal imagery by forward-backward MHI. In Proceedings of the Conference on Computer Vision and Pattern Recognition Workshop, New York, NY, USA, 17–22 June 2006; pp. 133–140.

50. Farnebäck, G. Two-frame motion estimation based on polynomial expansion. In Proceedings of the Scandinavian Conference on Image Analysis, Halmstad, Sweden, 29 June–2 July 2003; pp. 363–370.
51. Brox, T.; Bruhn, A.; Papenberg, N.; Weickert, J. High accuracy optical flow estimation based on a theory for warping. In Proceedings of the European Conference on Computer Vision, Prague, Czech Republic, 11–14 May 2004; pp. 25–36.
52. Lucas, B.D.; Kanade, T. An iterative image registration technique with an application to stereo vision. In Proceedings of the 7th International Joint Conference on Artificial Intelligence, University of British Columbia, Vancouver, BC, Canada, 24–28 August 1981; pp. 674–679.
53. Beauchemin, S.S.; Barron, J.L. The computation of optical flow. *ACM Comput. Surv.* **1995**, *27*, 433–466. [CrossRef]
54. Akpinar, S.; Alpaslan, F.N. Chapter 21—Optical flow-based representation for video action detection. In *Emerging Trends in Image Processing, Computer Vision and Pattern Recognition*; Deligiannidis, L., Arabnia, H.R., Eds.; Morgan Kaufmann: Boston, MA, USA, 2015; pp. 331–351.
55. Sun, S. A survey of multi-view machine learning. *Neural Comput. Appl.* **2013**, *23*, 2031–2038. [CrossRef]
56. Schuster, M.; Paliwal, K.K. Bidirectional Recurrent Neural Network. *IEEE Trans. Signal Process.* **1997**, *45*, 2673–2681. [CrossRef]
57. Hochreiter, S.; Bengio, Y.; Frasconi, P. Gradient flow in recurrent nets: The difficulty of learning long-term dependencies. In *Field Guide to Dynamical Recurrent Networks*; IEEE Press: Piscataway, NJ, USA, 2001.
58. Clevert, D.A.; Unterthiner, T.; Hochreiter, S. Fast and accurate deep network learning by exponential linear units (elus). *arXiv* **2016**, arXiv:1511.07289. Available online: https://arxiv.org/abs/1511.07289 (accessed on 3 February 2020) [CrossRef]
59. Werner, P.; Al-Hamadi, A.; Niese, R.; Walter, S.; Gruss, S.; Traue, H.C. Automatic pain recognition from video and biomedical signals. In Proceedings of the International Conference on Pattern Recognition, Stockholm, Sweden, 24–28 August 2014; pp. 4582–4587.
60. Walter, S.; Gruss, S.; Traue, H.; Werner, P.; Al-Hamadi, A.; Kächele, M.; Schwenker, F.; Andrade, A.; Moreira, G. Data fusion for automated pain recognition. In Proceedings of the 9th International Conference on Pervasive Computing Technologies for Healthcare, Istanbul, Turkey, 20–23 May 2015; pp. 261–264.
61. Kächele, M.; Thiam, P.; Amirian, M.; Schwenker, F.; Palm, G. Methods for person-centered continuous pain intensity assessment from bio-physiological channels. *IEEE J. Sel. Top. Sign. Process.* **2016**, *10*, 854–864. [CrossRef]
62. Kächele, M.; Amirian, M.; Thiam, P.; Werner, P.; Walter, S.; Palm, G.; Schwenker, F. Adaptive confidence learning for the personalization of pain intensity estimation systems. *Evol. Syst.* **2016**, *8*, 1–13. [CrossRef]
63. Bellmann, P.; Thiam, P.; Schwenker, F. Computational Intelligence for Pattern Recognition. In *Computational Intelligence for Pattern Recognition*; Pedrycz, W., Chen, S.M., Eds.; Springer International Publishing: Cham, Switzerland, 2018; pp. 83–113.
64. Bellmann, P.; Thiam, P.; Schwenker, F. Using a quartile-based data transtransform for pain intensity classification based on the SenseEmotion Database. In Proceedings of the 8th International Conference on Affective Computing and Intelligent Interaction Workshops and Demos, Cambridge, UK, 3–6 September 2019; pp. 310–316.
65. Baltrusaitis, T.; Robinson, P.; Morency, L.P. OpenFace: An open source facial behavior analysis toolkit. In Proceedings of the IEEE Winter Conference on Applications of Computer Vision, Lake Placid, NY, USA, 7–10 March 2016; pp. 1–10.
66. Bradski, G. The OpenCV library. *Dr Dobb's J. Softw. Tools* **2000**, *25*, 120–125.
67. Simonyan, K.; Zisserman, A. Very deep convolution networks for large-scale image recognition. *arXiv* **2015**, arXiv:1409.1556. Available online: https://arxiv.org/abs/1409.1556 (accessed on 3 February 2020) [CrossRef]
68. Ioffe, S.; Szegedy, C. Batch Normalization: Accelerating deep network training by reducing internal covariate shift. *arXiv* **2015**, arXiv:1502.03167. Available online: https://arxiv.org/abs/1502.03167 (accessed on 3 February 2020) [CrossRef]
69. Kingma, D.P.; Ba, J. Adam: A method for stochastic optimization. *arXiv* **2015**, arXiv:1412.6980. Available online: https://arxiv.org/abs/1412.6980 (accessed on 3 February 2020) [CrossRef]
70. Chollet, F. Keras. 2015. Available online: https://keras.io (accessed on 21 January 2020).

71. Abadi, M.; Agarwal, A.; Barham, P.; Brevdo, E.; Chen, Z.; Citro, C.; Corrado, C.; Davis, A.; Dean, J.; Devin, M.; et al. TensorFlow: Large-Scale Machine Learning on Heterogeneous Systems. 2015. Available online: https://www.tensorflow.org/ (accessed on 21 January 2020).
72. Pedregosa, F.; Varoquaux, G.; Gramfort, A.; Michel, V.; Thirion, B.; Grisel, O.; Blondel, M.; Prettenhofer, P.; Weiss, R.; Dubourg, V.; et al. Scikit-learn: Machine Learning in Python. *J. Mach. Learn. Res.* **2011**, *12*, 2825–2830.
73. Werner, P.; Al-HamadiAl-Hamadi, A.S. Analysis of facial expressiveness during experimentally induced heat pain. In Proceedings of the 7th International Conference on Affective Computing and Intelligent Interaction Workshops and Demos, San Antonio, TX, USA, 23–26 October 2017; pp. 176–180.

Sample Availability: The BioVid Heat Pain Database (Part A) is publicly available for non-commercial research and can be acquired by contacting the authors of the database at the following web-page: http://www.iikt.ovgu.de/BioVid.print.

sensors

MDPI

Article

How Laboratory Experiments Can Be Exploited for Monitoring Stress in the Wild: A Bridge Between Laboratory and Daily Life

Yekta Said Can *, Dilara Gokay, Dilruba Reyyan Kılıç, Deniz Ekiz, Niaz Chalabianloo and Cem Ersoy

Computer Engineering Department, Bogazici University, Bebek, 34342 Istanbul, Turkey; dilara.gokay@boun.edu.tr (D.G.); dilruba.kilic@boun.edu.tr (D.R.K.); deniz.ekiz@boun.edu.tr (D.E.); niaz.chalabianloo@boun.edu.tr (N.C.); ersoy@boun.edu.tr (C.E.)

* Correspondence: yekta.can@boun.edu.tr

Received: 3 January 2020; Accepted: 1 February 2020; Published: 4 February 2020

Abstract: Chronic stress leads to poor well-being, and it has effects on life quality and health. Society may have significant benefits from an automatic daily life stress detection system using unobtrusive wearable devices using physiological signals. However, the performance of these systems is not sufficiently accurate when they are used in unrestricted daily life compared to the systems tested in controlled real-life and laboratory conditions. To test our stress level detection system that preprocesses noisy physiological signals, extracts features, and applies machine learning classification techniques, we used a laboratory experiment and ecological momentary assessment based data collection with smartwatches in daily life. We investigated the effect of different labeling techniques and different training and test environments. In the laboratory environments, we had more controlled situations, and we could validate the perceived stress from self-reports. When machine learning models were trained in the laboratory instead of training them with the data coming from daily life, the accuracy of the system when tested in daily life improved significantly. The subjectivity effect coming from the self-reports in daily life could be eliminated. Our system obtained higher stress level detection accuracy results compared to most of the previous daily life studies.

Keywords: smart band; stress recognition; physiological signal processing; machine learning

---

1. Introduction

Stress, an ever-growing issue in modern societies, has become an inseparable part of people's fast-paced daily lives. Continuously increasing workload, tight deadlines, and the resulting time pressure all contribute to the increasing stress levels. Stress is an organism's reaction mechanism to a stressor. In a stressful state, certain control systems in the human body, such as the autonomic nervous system (ANS), act mostly unconsciously to control the responses to stress by regulating some bodily functions. This mechanism has been constantly improved throughout human evolution, to create prompt canny reactions in life-threatening situations [1]. Stress symptoms can be measured and observed in numerous ways. The sympathetic nervous system (SNS) kicks off the stress reaction, which will appear in the form of psychological, physiological, and behavioral indications [1]. The bidirectional impacts of the mind on the body and vice versa are among the major hallmarks of the autonomic nervous system (ANS), which has evolved in a way to have a direct role in human life and survival [2]. The autonomic nervous system (ANS) is divided into the sympathetic nervous system (SNS) and the parasympathetic nervous system (PNS). Most of the studies investigating the effects of chronic psychological stress on the human body have concluded that the sympathetic and parasympathetic nervous systems become over and under activated respectively, while the individual is under psychological stress. The resulting abnormal activities of the sympathetic and parasympathetic nervous systems cause physical, behavioral, and affective irregularities. In order to regulate the physiological arousal states, a balance of activity is expected between the sympathetic and parasympathetic subclasses of the autonomic nervous system (ANS). It is feasible to measure and evaluate the autonomous nervous system (ANS) function through non-invasive physiological phenomena like the electrodermal activity (EDA) and heart rate variability (HRV). For example, the high frequency (HF) component of the HRV, which is one of the frequency domain characteristics of the heart rate variability (HRV), is an indicator of the vagus nerve and parasympathetic nervous system's (PNS) activity. In contrast, low frequency (LF) reflects the activity of the sympathetic nervous system (SNS) [3–5]. The final goal concerning almost all of the stress detection research is to find ways to notify the user about their stress levels and help them to manage it to avoid the social, economic, and health-related consequences.

Traditional approaches for measuring stress are taken either by a psychologist interviewing the subject or by requesting the study subject to fill in particular questionnaires designed explicitly for self-reporting. Such interviews require the constant presence of a trained psychology specialist. Requesting the subjects to fill out long lists of questionnaires and self-report diaries are the most widely adopted processes to evaluate stress. These techniques are the current gold-standard as well. However, these methods are cumbersome and entirely manual. There are also other concerns and drawbacks regarding the interviews, self-report diaries, and questionnaires due to the way that different subjects commonly behave. Since these reports are based on or influenced by personal feelings, tastes, or opinions, they are highly subjective. As an example, some people tend to respond to questions in a manner that will be viewed favorably by others [6]. Some responders may feel that sharing their private psychological feelings may put them to shame and hide their real feelings by understating them. In contradiction to that, there are numerous cases in which responses are exaggerated [7]. Gender differences also play an essential role in how men and women express and report their stress levels and affect states when confronted with various stressors [8]. However, for everyday life, the self-reports are the closest labels to the ground truth.

Any potential automated stress detection framework for daily life will be developed using a mechanism that tries to surmount the need for any intervention from a psychologist and making the whole process less incommodious. Due to the disadvantages that self-report questionnaires' possess, research has emerged detecting regular psychophysiological signs of stress with machine learning (ML) algorithms utilizing the reliable and proven indicators such as response activities of the sympathetic nervous system, skin temperature (ST), electrodermal activity (EDA), and electrocardiogram (ECG) [9]. The purpose of adopting ML methods for raw psychophysiological objective data is to extract meaningful emotional and affective information. The raw sensor data are collected in this process and transformed into information-containing features. Some of those features would then be used while assigning affective state labels. Subjective data are also recorded by taking records of the known context and/or daily questionnaires and ecological momentary assessments (EMAs). Finally, the primary purpose of the whole process is to train machine learning algorithms to diagnose different types of behavioral and emotional states by using the EMAs, questionnaires, and the known context and test the system using the rest of the features and any feature recorded in the future. Although some of the steps described above are not necessarily required for deep learning models, the whole process for almost all of the traditional machine learning systems is almost the same [10]. In an automated stress detection scheme, these psychophysiological signals are utilized, and analysis of these signals reveals the frequency and intensity of the stress experienced by the subjects.

Preliminary studies of automated stress detection were held in the laboratory environments, and then, the research took a step into daily life since researchers realized that the stress level experienced in the laboratory is different from daily life stress [11]. However, the problems encountered in this new environment were both more complicated and intricate. Achieving precise annotations and identification of the perceived stress in the wild is a difficult task due to the high diversity in human psychological evaluation and the lack of direct observation over subject activities. Furthermore, the unrestricted movements of subjects in the wild may cause misinterpreting of stress detection by causing artifacts in the signal data. Lastly, since the medical-grade devices with cables, electrodes, and boards cannot be used in daily life due to their obtrusive nature, more pervasive and comfortable devices should be used. However, alternatives such as smart bands and smartwatches have lower data quality, and more advanced signal processing techniques should be applied to overcome this problem. Due to the issues mentioned above, the performance of daily life stress detection systems is lower than those proposed for laboratory environments. Smartwatches and smart bands are unobtrusive, easy to use without requiring specific actions, and are more suitable for daily life. They are adopted by the consumer and easily available on the market. Most of them are equipped with photoplethysmography (PPG) sensors [12]. Our solution is easily applicable to consumers due to the availability of these devices.

Given the complexity of data acquisition solely performed in the wild in addition to the advantages of the in-lab data, consequently, the question arises as to whether it is possible to combine these systems somehow and use the high accuracy from the first for the sake of the latter. One of the questions that stands out the most is the feasibility of combinations of these two systems to achieve even better results. In this study, we will represent a framework by which it is possible to design a stress detection mechanism that utilizes both of those methods. The data collected in a laboratory can be used for the long-term performance evaluation of the same system in daily life scenarios.

Our work contributes to state-of-the-art in four different aspects:

- Developing laboratory-based models to improve the performance of daily life stress detection systems;
- Comparing the performance of laboratory, daily life, and hybrid laboratory-daily life models;
- Collection of smartwatch-based EDA and HRV data coming from the laboratory and daily life (14 participants, 1003 h of physiological data with 388 ecological momentary assessments (EMAs)), with self-reports and context information;
- Investigating the effect of using context and self-report labels while training the model on system accuracy in different environments.

These contributions will provide insights for researchers into how to improve the performance of daily life stress detection systems.

The organization of the rest of the paper is as follows. In Section 2, we present the related work in stress detection and alleviation. Our proposed unobtrusive system for stress level monitoring with smart bands is described in Section 3. Data collection procedures are explained in Section 3. Experimental results and discussion are presented in Section 4. We present the conclusions of the study in Section 5.

2. Related Work

Most of the automatic stress detection studies in the literature were conducted either in the laboratory environments or restricted daily life settings. We can examine the studies in the literature as five different classes. The first class develops a laboratory model with known context labels and tests in the same environment. Since the stressor levels (i.e., the context participants are in) are known at any time in laboratory experiments, they could be used as the ground truth labels for machine learning (ML) models, and we called this type laboratory-to-laboratory known context (LLKC) models. The second type uses collected self-report labels in laboratory environments and tests the created model in the same environment. We call this type laboratory-to-laboratory self-report (LLSR) models. The third type is using self-report questionnaires collected in the wild and testing the model in the same environment. We named this model daily-to-daily self-report (DDSR) models. Since we could not monitor the everyday life of participants and get the ground truth all the time, a known context does not exist in daily life environments. The laboratory data could be used to enhance daily life stress detection models. If the known context labels in the laboratory are used for an ML model development, we call this fourth type laboratory-to-daily known context (LDKC) models. On the contrary, if the self-reports in the laboratory are used as labels and the developed models are tested in the wild, we name the fifth type laboratory-to-daily self-report (LDSR). Table 1 illustrates some of the studies conducted either in the laboratory, in daily life, or both. In this section, we will briefly outline some of the research for stress detection that has been conducted in the laboratory and everyday life environments by using the taxonomy developed above.

Table 1. Stress detection experiments in laboratory and daily life settings. EDA, electrodermal activity; IAPS, International Affective Picture System; PPG, photoplethysmography; ACC, Accelerometer; MIST, The Montreal Imaging Stress Task; SCWT, Stroop Color and Word Test; TSST, Trier Social Stress Test; BVP, Blood Volume Pressure; RR, R to R interval.

Article	Stress Signal	Stress Test	Unobtrusive	Model Type					Accuracy	
				LLKC	LLSR	DDSR	LDKC	LDSR	Lab	Daily Life
[13] (2009)	EDA, ECG, ACC , Respiration	MIST	✗	✓	✗	✗	✗	✗	82.8%	-
[14] (2015)	EDA, Bluetooth, ACC	Mixed	✗	✓	✗	✗	✗	✗	91%	-
[15] (2017)	ECG	SCWT	✗	✓	✗	✗	✗	✗	70%	-
[16] (2016)	PPG, EDA, Respiration, Thermal Camera	Lie Detection	✗	✓	✗	✗	✗	✗	73%	-
[17] (2016)	EEG	Arithmetic Task	✗	✓	✗	✗	✗	✗	89%	-
Our Previous Work [18] (2019)	PPG, EDA	Programming Contest	✓	✓	✓	✗	✗	✗	97.92%	-
[19] (2015)	EDA, PPG	TSST	✗	✓	✗	✗	✗	✗	80%	-
[20] (2015)	ECG, Facial recognition	IAPS	✗	✗	✓	✗	✗	✗	83%	-
[21] (2017)	ECG, GSR, Blood Oximeter, Blood Pressure, Respiration	Ice Test, IAPS	✗	✓	✗	✗	✗	✗	95.8%	-
[22] (2016)	Mobile App Usage Pattern, Light, Physical Activity	Daily Life	✓	✗	✓	✓	✗	✗	80%	70%
[23] (2015)	ECG + Respiratory + Accelerometer	Daily Life	✗	✗	✓	✗	✗	✓	90%	72%
[24] (2018)	Usage Data for Different Application Categories	Daily Life	✓	✗	✗	✓	✗	✗	-	68%
[25] (2018)	HR (Heart Rate)-ACC	Daily Life	✓	✗	✗	✓	✗	✗	-	0.76 precision
[26] (2017)	BVP , EDA, Skin Temperature, RR	Daily Life	✓	✓	✗	✗	✓	✗	83%	76%
[27] (2018)	PPG	Daily Life, Arithmetic Tasks	✓	✗	✓	✗	✗	✓	0.7 correlation	0.56 correlation
Our Work	PPG, EDA	TSST, Daily Life	✓	✓	✓	✓	✓	✓	94.40%	73%

LLKC: Laboratory-to-laboratory known context. LLSR: Laboratory-to-laboratory self-report. DDSR: Daily-to-daily self-report. LDKC: Laboratory-to-daily known context. LDSR: Laboratory-to-daily self-report.

Early experimental practices in the laboratory were the first building blocks of this research field. These preliminary studies provided researchers with the idea of choosing the most proper devices, features, and machine learning algorithms that can be used later in everyday life settings. In a trade-off between unobtrusiveness and accurate signals, researchers chose the types of devices with different sensing technologies based on the requirements of their study. Several researchers demonstrated that wearable devices equipped with PPG sensors added more comfort and convenience compared to ECG devices for HRV measurements [28–30]. Nevertheless, single-lead electrocardiogram (ECG) devices are becoming more compact, easy to wear, and commercially available over time. In a recent study by Billeci et al., single-lead ECG was used to study the autonomic nervous system response through monitoring the heart rate and HRV features [31]. In [14], Zubair et al. developed a smart wristband that has EDA, Bluetooth, and accelerometer sensors to detect the stress values of the test subjects. In their designed scheme, EDA and the accelerometer were used together to enhance the detection accuracy. Their experiments resulted in 91% accuracy for two-class stressed-relaxed classification. Castaldo et al. administered a study in a laboratory environment to investigate the effects of ultra-short HRV (heart rate variability) with a two minute duration on stress detection accuracy [15]. By applying the support vector machine (SVM) to HRV features, they could score 70% accuracy on their binary classification problem. Although a 70% accuracy seems low compared to other studies in the same year, they speculated that this could be due to the stress induction methods used and ultra-short-term HRV features. Tivatansakul et al. used HRV features inferred from the ECG signals and facial expressions to recognize negative emotions in the laboratory environment. Subjects were exposed to the International Affective Picture System (IAPS) set of disturbing images for four minutes. After applying the SVM classifier on the extracted features, classification accuracy for the negative emotion detection was 83% [20]. All of the in-lab implemented cases mentioned above used only one biosignal, either EDA or ECG, in their proposed stress detection mechanism. Other studies practiced multimodality by employing multiple biosignals to achieve even higher accuracy levels. In a recent stress detection and alleviation research work, Akmandor et al. [21] recorded EDA, ECG, blood pressure, blood oximeter, and respiration rates of the participants. In order to induce stress, memory games and IAPS were used, and for the alleviation part, they used various stress mitigation techniques such as classical music and micro-meditation. Their classification accuracy using SVM and kNN for their binary class problem was 95.8% [21]. Researchers used a combination of medical-grade devices with the highest possible accuracies in the laboratory settings. For instance, there are cases in which electroencephalogram (EEG) signals were used for stress monitoring [17]. Implementing such a setting in the wild is practically impossible due to the obtrusive nature of the EEG devices and the lack of daily-life suitable wearable devices with EEG capability.

In other cases, researchers decided to mitigate the trade-off between the accuracy and unobtrusiveness and conducted their studies either in the lab or in the wild using unobtrusive wearable devices. The choice of device type (i.e., whether to use medical-grade obtrusive devices with cables, electrodes, boards, or not) is important because it determines the applicability of the system to daily life, which is the main and final goal of all the stress detection studies. In some of the studies, the data were collected in the wild and trained and tested in these environments [22–25] (DDSR model type in the taxonomy). Ciman et al. [22] detected stress by analyzing smartphone usage behaviors. In the controlled laboratory experiment, by developing an Android app with search and typing tasks, participants' tap, swipe, scroll, and text input gestures were recorded. These gestures were expected to induce stress on the participants. They obtained approximately 80% stress detection classification accuracy with SVM, NN, kNN, and decision tree classifiers. The physical activities of the user, the light values of the screen, and the events related to the mobile phone screen were recorded in the wild experiment. The classification accuracy was about 70% with the same classifiers. Vildjounaite et al. [24] used mobile phone usage data and implemented MPM (Maximum Posterior Marginal) and HMM (Hidden Markov Model) for stress recognition. They experimented with 28 subjects for four days and obtained a maximum of 68% accuracy on their semi-personal data. Mishra et al. [25] detected stress by analyzing heart activity signals captured with a Moto360 smartwatch. The data were collected from 23 subjects for three days. They also added the activity type to physiological features to increase the accuracy of the system. Random forest (RF) was used for the classification of stressed and relaxed sessions. They increased their F1 score from 0.50 to 0.76 by adding the activity-related contextual information. A recent daily life study was carried out with 1002 participants using unobtrusive wearables [32]. They extracted heart rate variability features and employed the RF classifier. They could only achieve a 0.43 F1 score (with three classes) and listed the additional difficulties of working in the wild while explaining the reasons for the relatively low performance [32]. The performance of daily life stress monitoring systems is lower than studies conducted in controlled environments due to the mentioned issues such as low-quality physiological signals with artifacts and the difficulty of collecting the ground truth in [33,34] (see Table 1).

The data collected in the experiments conducted in the laboratory could be used for developing models in the wild to improve the reported low accuracies. There is a limited number of studies using laboratory data for creating models with higher performance in daily life. Li et al. [27] built their model in the lab, verified its performance in the lab with different participants, and later verified it in the field test again with new subjects. They used mental arithmetic tasks to induce stress and seven-point scale Likert self-reports as labels in the laboratory (LDSR model type). They evaluated their system performance by using the Elastic-net regression technique. The correlation results of predicted labels and ground truth self-reports were 0.72 and 0.56 for laboratory and daily-life settings, respectively. In another study, Gjoreski et al. proposed a continuous stress detection scheme, which consisted of a base detection mechanism solely using the laboratory data, an activity recognizer for identifying the contextual information, and a real-life stress detection mechanism that utilized the results from the laboratory detector and the context information for real-life stress detection [26]. While training the model, they used only the context label in the laboratory environment (LDKC-type of model). They did not collect self-reports during these tests. The classification accuracy of their stress detection system without using context information was 76% [26] (they increased it to 92% by adding the activity type). Since different methods that were mentioned in our taxonomy were tried in different datasets created for each particular paper in the literature (see Table 1), one could not infer the success of a method over others. Therefore, there is clearly a need for comparing the different proposed techniques for stress detection in the laboratory and daily life settings with unobtrusive devices.

3. Methodology

3.1. Laboratory Data Collection

We collected controlled laboratory data from the participants during a psychological experiment using our implementation of the Trier Social Test (TSST) [35]. The social test has been proven to be an academically effective way of inducing stress after the baseline condition we created at the beginning of the experiment. This experiment was conducted on 14 different participants who were university students aged between 20 and 25. There were nine male and five female participants.

The experiment, which took approximately 1 hour, had the following steps:

1. Setup
2. Pre-stress measurements (baseline)
3. The TSST (Trier Social Stress Test) (inducing stress)
4. Post-stress recovery measurements (recovery)

The communication language between interviewers and the participants was Turkish. Please note that the mother tongue of all participants was Turkish. In addition to that, they knew English as a foreign language. This circumstance affected stress induction.

3.1.1. Setup

The setup was done as follows;

1. Preparation of experiment areas: The camera should be set. Empatica E4 should be ready.
2. Interviewers should keep eye contact with the participant. Their gestures and facial expressions should be neutral.
3. The participant is informed about the procedure and then signs the consent form.
4. The participant wears the smart band (Empatica E4).
5. The participant is asked to turn off his/her phone in order to eliminate distraction.

3.1.2. Pre-Stress Measurements

Before the experiment, the following procedure was applied;

1. The participant filled out the Perceived Stress Scale (PSS) with 14 questions.
2. The participant was told to stay in the waiting area and get rest for 10 min. Reading materials such as magazines with emotionally-neutral contents (home and garden, car magazines) were presented to the participant for this period.
3. The participant filled out the PSS-5 (ambulatory PSS) form. This questionnaire was first created by Cohen et al. [36] and used for measuring the perceived stress in ambulatory settings in [37] (see Figure 1).

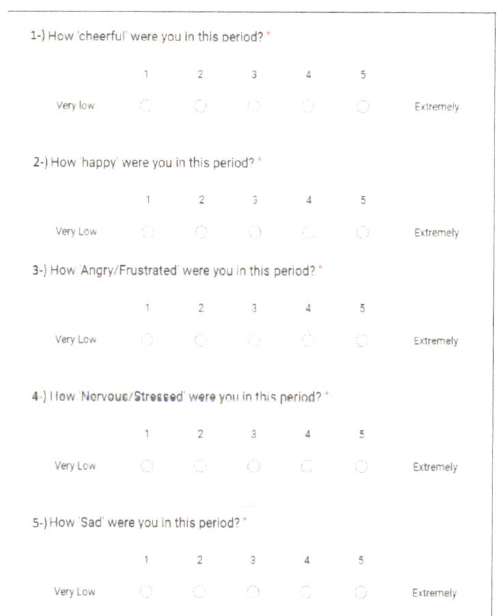
1-) How 'cheerful' were you in this period? *

	1	2	3	4	5	
Very low						Extremely

2-) How 'happy' were you in this period? *

	1	2	3	4	5	
Very Low						Extremely

3-) How 'Angry/Frustrated' were you in this period? *

	1	2	3	4	5	
Very Low						Extremely

4-) How 'Nervous/Stressed' were you in this period? *

	1	2	3	4	5	
Very Low						Extremely

5-) How 'Sad' were you in this period? *

	1	2	3	4	5	
Very Low						Extremely

Figure 1. The Perceived Stress Scale (PSS)-5 questionnaire used in the experiment.

3.1.3. The TSST

Our implementation of the TSST is described as follows:

1. The participant was directed to the interview area.
2. TSST speech preparation period: the following text was read to the participant: "This is the speech preparation portion of the task; you are expected to prepare a five-minute speech describing why you study [name of the degree that the participant studies/studied] and why you would

be a good candidate for your ideal job. Your speech will be videotaped and reviewed by the psychologists that we conduct the research with. You have five minutes to prepare and your time begins now."

3. The participant prepared his/her speech. There should be a digital timer in the room set to five minutes. Interviewers should leave the room.
4. The following text was read to the participant at the end of the speech preparation period: "This is the speech portion of the task. You should speak for the entire five-minute time period. Your time begins now". Interviewers should start the recording of the camera.
5. TSST speech performance period: If the participant stopped during this period, interviewers allowed him/her to stay silent for around 20 s and then prompted: "You still have time remaining."
6. After the first 2 min of the speech period, the participants were interrupted and asked to continue their speech in English by telling them: "Could you continue in English from now on, please?"
7. At the end of 2.5 min, if the participant did not attempt to reply to both questions, interviewers prompted the participant to answer the other question.
8. At the end of the speech performance period, the communication between interviewers and the participant resumed in Turkish. Interviewers reset the timer to 5 min and read the following to the participant: "During the final five-minute math portion of this task, you will be asked to subtract 13 from 1022 sequentially. You will verbally report your answers aloud, and be asked to start over from 1022 if a mistake is made. Your time begins now." If the participant makes any mistake, the interviewer says the following: "That is incorrect, please start over from 1022." (Figure 2)
9. Participant filled out the PSS-5 questionnaire.

Figure 2. An example scene from the TSST phase in our experiment. The participant is presented at this moment in front of the neutral experimenter.

3.1.4. Post-Stress Recovery Measurements

In order to alleviate the stress response, we applied a biofeedback based intervention, which was the built-in breathing application of Apple Watch [38]. The procedure was applied as follows:

1. Participants were directed to the couch as a relaxing place.
2. Participants wore an Apple Watch given to him/her at this stage, followed the breathing exercise built in the Apple Watch for a minute and then followed a mindfulness video, for the remaining four minutes, on a comfortable couch, sitting or lying as the participant preferred.
3. Interviewers should leave the room after giving the Apple Watch.
4. At the end of the five minute long recovery period, interviewers returned to the room, and the participant filled out the PSS-5 questionnaire.

3.2. Daily Life Data Collection and Ecological Momentary Assessment

After the controlled room experiments were finished, we gave the Empatica E4 devices to all participants. They were told to wear the Empatica E4 devices for twelve hours per day, between 9 a.m. and 9 p.m., for seven days. These days were not necessarily consecutive. We applied the EMA to collect information about the subject's stress level [39]. EMA involved the repeated sampling of subjects' current behaviors and experiences in real time, in subjects' natural environments [39]. EMA aims to minimize recall bias, maximize ecological validity, and allow the study of micro processes that influence behavior in real-world contexts. We implemented an online version of the PSS-5 questionnaire. For each three hour session, we asked the participant to fill in the EMA. In order to make sure the collection of self-reports, we sent them e-mails over seven days at the end of each three hour session when they were wearing the wristband. In other words, the participants were reminded to fill in the PSS-5 at 12 p.m., 3 p.m., 6 p.m., and 9 p.m. over seven days. The EMA was delivered to the participants through a survey app. The app was available on both desktop and mobile browsers. The link to the EMA was delivered to the participants through e-mail. Each e-mail that was sent in order to remind participants to fill in the EMA contained the link to the EMA. This questionnaire is strongly correlated with PSS-14 and appropriate for ambulatory settings. In total, we obtained 1003 h of physiological data and 388 EMAs (including 14 h of physiological data and 56 EMAs collected in the lab). There were some sessions with missing EMAs (60 EMAs in total), and we disregarded their physiological data.

The procedure of the methodology used in this study was approved by the Institutional Review Board for Research with Human Subjects of Boğaziçi University with Approval Number 2018/16. Prior to the data acquisition, each participant received a consent form that explained the experimental procedure and its benefits and implications to both society and the subject. The procedure was also explained verbally to the subject. All of the data were stored anonymously.

3.3. Stress Recognition Framework

In order to propose an unobtrusive stress detector suitable for everyday use, we used an Empatica E4 [40] comfortable wristband, which has more than 48 h of battery life and is equipped with a variety of sensors such as the three-dimensional accelerometer (32 Hz), the continuous heart rate monitoring unit based on the photoplethysmography (PPG), the skin temperature (4 Hz), and the EDA sensors (4 Hz). The major difference between the wristbands and contact sensors used in hospital settings is the vulnerability to the motion artifacts due to their design and attachment to the body. A daily life suitable and comfortable stress detection system should consider these artifacts. Thus, it should have a preprocessing unit that detects and removes the artifacts due to contact loss and motion. In daily life, the activity of the individual is important; for example, HRV and EDA can change due to high-intensity activity in short periods. Therefore a single sensor-based system can fail. Hence, a stress recognition system suitable for daily life should be multi-modal in terms of the collected data. We used robust preprocessing and feature extraction modules from our previous work [18] for this purpose (see Figure 3).

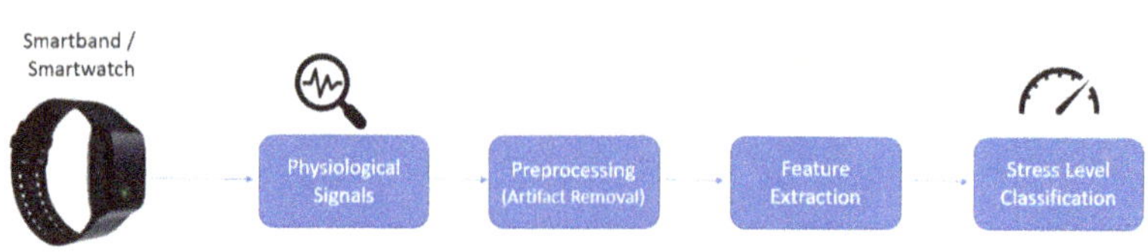

Figure 3. A high level block diagram of the stress level detection system with the Empatica E4 wristband.

3.3.1. Preprocessing

The physiological signals coming from different sensors were segmented into non-overlapping time windows. We selected the window size as two minutes since it was reported that the duration of stress stimulation and recovery processes was approximately a few minutes and these segment sizes could capture such processes [41]. IBI (Interbeat Interval) and EDA signals were sent to the artifact detection units. Therefore, the response time of our system was approximately two minutes.

The artifact detection and removal unit for the heart rate signal applied an artifact detection percentage threshold between the time of the successive R to R interval recordings. The threshold was selected as 20% [42]. After the removal of the artifacts, they were replaced with a cubic spline interpolation function, as applied in Kubios [43]. The interpolation method was selected for applying time and sample constraints on the remaining data since it achieved better results [18]. The windows containing more than 10% of RR interval artifacts were removed. This unit was developed in our previous work [18].

For the EDA artifact detection unit, we used the toolbox developed by Taylor et al. [44]. It uses the SVM classifier and detects the artifacts in the EDA signal with 95% accuracy by analyzing skin temperature, accelerometer, and skin conductance signals. We added a batch processing feature to this toolbox and removed the detected artifacts. Then, the resulting signal was transferred to the EDA feature extraction unit.

3.3.2. Feature Extraction

To create a multi-modal stress recognition framework, we extracted state-of-the-art features from the EDA and RR intervals to create a feature vector. In this section, we describe the feature extraction methodology for each of the physiological signals. We decomposed the EDA signal into phasic and tonic components using the convex optimization-based EDAcvx tool [45]. EDAcvx also cleans the noise in the EDA signal. We extracted seven features from both the phasic and tonic components of the EDA signal. Percentiles are very handy for exploring the distribution of number sets using various EDA graphs. These following EDA features were selected from the literature [13,26].

1. Mean value
2. Standard deviation
3. Number of peaks
4. Number of strong peaks
5. Twentieth percentile
6. Eightieth percentile
7. Quartile deviation

We extracted 13 heart rate variability (HRV) time and frequency domain features from RR intervals. These features were commonly used in the previous works [11,26,43,46,47]. In order to compute frequency domain features, we re-sampled RR intervals at 4 Hz [48] and applied fast Fourier transform (FFT). The computed HRV features are shown below:

1. Mean value of the inter-beat (RR) intervals
2. Standard deviation of the inter-beat interval
3. Root mean square of the successive difference of the RR intervals.
4. Percentage of the number of the successive RR intervals varying more than 50 ms from the previous interval
5. Total number of RR intervals divided by the height of the histogram of all RR intervals measured on a scale with bins of 1/128 s
6. Triangular interpolation of RR interval histogram
7. Power in the low-frequency band (0.04–0.15 Hz)

8. Power in the high-frequency band (0.15–0.4 Hz)
9. Ratio of LF to HF.
10. Prevalent low-frequency oscillation of the heart rate
11. Prevalent high-frequency oscillation of the heart rate
12. Power in the very low-frequency band (0.00–0.04 Hz)
13. Related standard deviation of successive RR interval differences

3.3.3. Feature Selection

We applied correlation-based feature (CBF) [49] selection using the Weka [50] tool. The importance of the features is shown in Figure 4. Since our goal was to develop a stress detection model that works in daily life settings, we conducted experiments with the 1 to 20 best features for the DDSR model. We achieved the best results with ten features for HRV + EDA, five features for HRV, and five features for EDA. We applied the classifiers to the selected features.

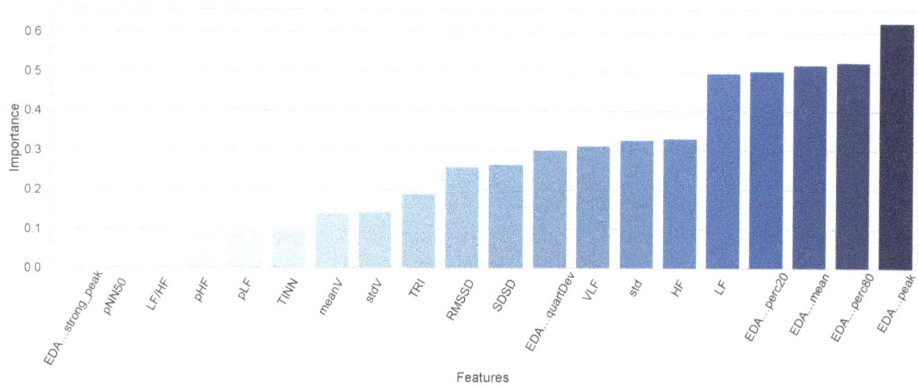

Figure 4. Features listed in order of importance based on correlation-based feature selection for the DDSR model. EDA peaks are the feature that has the highest importance, whereas the EDA strong peak has the lowest.

3.3.4. Preparation of the Data for ML Algorithms

Although we had evenly distributed data in terms of known context labels in the laboratory, we had a class imbalance problem in the self-reported ground truth labels obtained from the daily life and laboratory. In the laboratory, 71.5% of the data was relaxed and 28.5% of data was stressed. In addition, 73% of the daily life data was relaxed, and the remaining 27% was stressed. We overcame this problem by randomly undersampling the extra samples of the majority class, which is the most commonly used procedure for imbalanced datasets [51]. We further applied normalization on features to prevent overfitting. Lastly, we converted the numeric labels to the nominal type as an input to the Weka toolkit classification algorithms.

3.3.5. Forming ML Models

We developed five different models that used different ground truth types and 248 training and test environment combinations (see Figure 5). The data were divided into two minute windows in the preprocessing part. The label of the window extracted from the time was added to the feature vector. The label could be the stressor level of the session or the perceived stress level obtained from the participants. Features extracted from all windows belonging to a session were averaged, and the label was appended to this feature vector. Data coming from all participants were merged and randomly listed. We then formed one general model from our dataset. In other words, separate models for all participants were not formed. In order to evaluate the performance of the classifiers, we applied 10-fold cross-validation by partitioning the original sample into a training set to train the model and a test set to evaluate it, then we changed the test and training until each partition was used for the test set. We further created separate training (80%) and test sets (20%) and evaluated the results to compare with 10-fold cross-validation results. In the 10-fold results, the standard deviations of the folds are provided in parentheses.

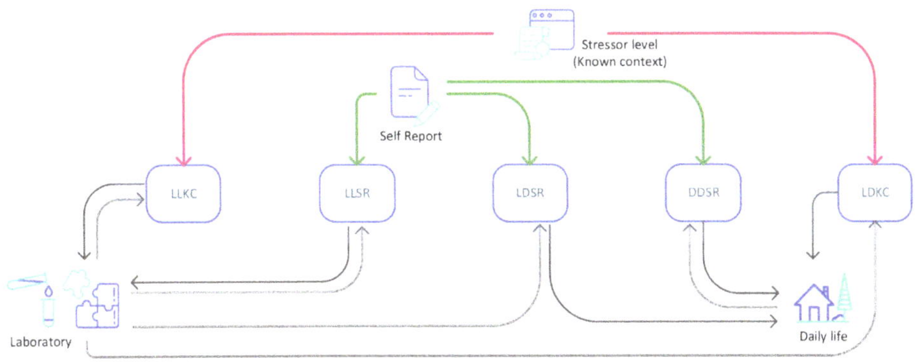

Figure 5. We developed five different stress level classification models with varying ground truth labels and training-test environments. Red and green arrows indicate the ground truth type used. The incoming black arrow shows the training environment, the an outgoing black arrow shows the testing environment.

3.3.6. Classification Algorithms

We used five different machine learning classifiers for recognition of stress events. These classifiers were the ones mostly used and that best performed in the literature [26,42], namely multi-layer perceptron (MLP), random forest (RF), k-nearest neighbor (kNN), support vector machine (SVM), and logistic regression (LR). For the LR, the output probability was divided into two different classes, which were above and below 0.5 [52]. We used the Weka Machine Learning Software [53] for the classification section of our proposed system.

In order to select the best parameter set for the MLP, we experimented with the different numbers of hidden layers (1, 2, and 3) and units (from 1 to 20); one outperformed others, which had two hidden layers, and each layer had five units. We experimented with N for the kNN. After parameter tuning, the best N was selected as 3. We applied the radial basis kernel (RBF) and linear kernel for the SVM. We selected RBF for SVM because it outperformed the linear kernel. For the RF, we selected the number of trees as 100.

4. Experimental Results and Discussion

We divided our tests into two groups: laboratory and in the wild experiments. We examined the results in the following two sections.

4.1. Laboratory Experiments

In this experiment, our aim was to differentiate between the stress and baseline states. In this section, we investigate the performance of our stress detection scheme in two different manners. The first one was using the known context as the ground truth. We further provide these labels as classes to the ML algorithm. The second way was to use the perceived stress levels collected from self-reports as the ground truth. In order to measure the perceived stress levels, we collected PSS-5. For these five questions, positive emotions (happy and cheerful) were evaluated inversely. In other words, if a participant stated six: extremely happy from 1–6, it was evaluated as one because happiness and cheerfulness are inversely proportional to stress levels. On the other hand, anger, sadness, and frustration were evaluated proportionally when calculating the score (see Equation (1)).

$$PercStress = (7 - H_i) + (7 - C_i) + A_i + S_i + F_i \tag{1}$$

where H_i: happiness score, C_i: cheerfulness score, A_i: anger score, S_i: sadness score, and F_i: frustration score. Individual scores on the PSS can range from zero to 30 with higher scores indicating higher perceived stress. Scores ranging from 0–15 would be considered low stress, and scores ranging from 15–30 would be considered high perceived stress. The division was made by adapting the three class division of the PSS-14 class [54]. The performance of our system is presented in Tables 2 and 3.

Table 2. Stress detection accuracies with different ML algorithms: 2 class classification. On the left side, stress recognition results, which only use self-reports as the ground truth labels are presented. On the right side, known context information is used for the ground truth label. LLKC stands for laboratory-to-laboratory known context, whereas LLSR stands for laboratory-to-laboratory self-report. Ten-fold cross-validation is used. Standard deviations are shown in parenthesis. HRV, heart rate variability.

Algorithm	LLSR			LLKC		
	HR	**EDA**	**HRV + EDA**	**HR**	**EDA**	**HRV + EDA**
MLP	83.30 (3.04)	77.30 (8.56)	87.20 (1.19)	62.90 (1.89)	**76.20** (9.96)	**82.90** (1.52)
RF	83.30 (6.10)	**86.70** (6.78)	84.90 (2.98)	57.80 (0.14)	66.70 (9.94)	80.39 (4.22)
kNN	**94.40** (1.79)	86.40 (8.58)	89.70 (6.34)	68.6 (5.45)	73.80 (14.65)	77.10 (5.69)
Logistic Regression	**94.40** (4.76)	72.70 (7.89)	89.70 (6.28)	68.60 (3.12)	**76.20** (13.3)	80.00 (2.01)
SVM	88.90 (6.28)	77.30 (6.10)	**92.30** (5.18)	**74.30** (4.13)	73.80 (9.77)	77.10 (3.49)

Table 3. Stress detection accuracies with different ML algorithms: 2 class classification. On the left side, stress recognition results, which only use self-reports as the ground truth labels are presented. On the right side, known context information is used for the ground truth label. Separate training (80%) and test sets (20%) are used. LLKC stands for laboratory-to-laboratory known context, whereas LLSR stands for laboratory-to-laboratory self-report.

Algorithm	LLSR			LLKC		
	HR	**EDA**	**HRV + EDA**	**HRV**	**EDA**	**HRV + EDA**
MLP	92.59	84.25	94.20	69.90	**78.20**	**86.66**
RF	**93.60**	91.20	91.40	58.60	68.80	82.21
kNN	91.20	**94.00**	**95.60**	67.20	68.80	78.32
Logistic Regression	65.74	66.66	73.14	70.89	**78.20**	79.36
SVM	77.77	84.25	90.74	**75.40**	74.40	81.10

We successfully differentiated stress with baseline states, as seen in Table 2 (with a maximum of 94.4% accuracy). Perceived stress level detection classification performance is always higher than physiological stress level detection in the known context because participants may experience different stress levels than the expected level of the context. Some participants may experience lower stress in the TSST while preparing the presentation, presenting in a foreign language, or counting tasks. This proves that the choice of using the ground truth as known context labels or perceived stress labels has a significant influence on the performance of the system. The combination of the two physiological signals always achieves higher accuracy than the single modality with minimum accuracy. However, it does not give the maximum accuracy in all conditions. We could state that multi-modality has an observable effect on some conditions. Overall, we successfully differentiated between baseline and stress states in the laboratory environment.

4.2. Testing the Models in the Wild

As mentioned, we had two types of class labels in the laboratory environment stress detection experiment: PSS-5 self-reports and the known context (the stressor level). On the other hand, in the wild, we had only self-reports of individuals, which was among the reasons why daily life stress detection performances were low [34], and there is still a room for improvement in daily life stress detection research.

We used both laboratory ground truth labels separately while developing machine learning models for laboratory environments and testing in daily life. The LDSR model was trained with the laboratory self-report labels, whereas the LDKC model was trained with the laboratory known context labels. We expected that laboratory self-report based labeling would have higher performance than laboratory context-based labeling because we had only the perceived stress labels (questionnaires) in the wild, which were more coherent with the laboratory self-reports. Furthermore, participants might experience different perceived stress than the known context implied, which reduced the performance of the LDKC models. We further developed a DDSR model that was solely trained and tested with daily life data. We compare these three models in Tables 4–7. As expected, the LDSR model had the best performance, whereas the DDSR model had the lowest performance. In the laboratory, collected self-reports were more reliable because the environment was controlled. The noise coming from the daily life environment (i.e., forgetting stressful events in a session, unrestricted movements) decreased the performance of DDSR models when compared to LDSR models, which had more clear training data and labels. We could state that collecting laboratory data and training a stress level detection model with that data improved the stress level detection performance in the wild. Furthermore, choosing the self-report label resulted in better performance when compared to that with the known context labels. As far as the performance of modalities is concerned, we achieved the best results with the HRV signal. In most of the daily life cases (15/20 tests), a combination of the signals achieved better results than single modalities alone. In the remaining tests, negative correlations between the selected best features from different modalities could decrease the performance of the system when modalities were combined. In daily life, RF and SVM achieved the best performances, which aligned with the recent literature [26]. Especially with the EDA signal, RF always outperformed other classifiers in daily life tests. Lastly, when the contribution of features were examined (see Figure 4), the best ones included five EDA and five HRV features, which also showed that the combination of these modalities was important.

Table 4. The classification accuracy using the combination of two modalities and a single modality along with the DDSR (Daily-to-Daily-Self-Report) technique was provided. 10-fold cross validation is used. Standard deviations are shown in parenthesis.

Algorithm	Accuracy		
	Combined	**HRV**	**EDA**
MLP	63.50 (8.25)	**68.30 (9.66)**	56.80 (8.89)
RF	61.90 (12.94)	65.10 (14.47)	**63.60 (11.79)**
kNN	65.90 (10.97)	64.30 (15.01)	61.40 (11.26)
Logistic Regression	70.60 (16.33)	**68.30 (8.75)**	59.30 (10.85)
SVM	**71.40** (7.03)	67.50 (8.73)	62.10 (1.53)

Table 5. The classification accuracy using the combination of two modalities and a single modality along with the DDSR (daily-to-daily self-report) technique are provided. Separate training (80%) and test sets (20%) are used.

Algorithm	Accuracy		
	Combined	**HRV**	**EDA**
MLP	68.00	57.30	57.30
RF	52.00	66.30	64.00
kNN	60.00	65.70	56.00
Logistic Regression	64.00	65.40	58.30
SVM	**68.00**	58.20	58.20

Table 6. The classification accuracy using the combination of two modalities and a single modality along with the LDKC (lab-to-daily known context) technique are provided.

Algorithm	Accuracy		
	Combined	**HRV**	**EDA**
MLP	64.73	34.43	35.26
RF	68.87	34.85	68.04
kNN	70.53	57.67	65.14
Logistic Regression	62.65	39.04	52.28
SVM	**71.78**	44.39	42.32

Table 7. The classification accuracy using the combination of two modalities and a single modality along with the LDSR (lab-to-daily self-report) technique are provided.

Algorithm	Accuracy		
	Combined	**HRV**	**EDA**
MLP	72.20	63.41	70.95
RF	**74.61**	**71.78**	72.61
kNN	72.20	71.37	**73.02**
Logistic Regression	73.81	71.78	71.78
SVM	73.44	73.44	72.61

5. Conclusions

Since stress detection systems have lower accuracies in the wild when compared to laboratory environments, there is a need to develop new techniques to improve their performance. In this study, we examined the effect of developing ML models in different environments and with varying ground truth labels. To the best of our knowledge, this was the first work to examine all possible combinations of perceived stress measurements for daily life and laboratory settings along with different ground truth labels. We used EDA, HRV, ST, and ACC signals (ST and ACC were used for artifact detection and removal) in our unobtrusive stress detection system. We first trained and tested our system in the laboratory environment. We obtained a maximum of 94.4% accuracy with HR, 86.70% with EDA, and 92.30% with HRV + EDA, which showed that our system detected the stress levels in the laboratory successfully. These results were aligned with the literature [14]. Choosing the ground truth as self-reports while training the ML model always achieved higher accuracies than using the known context labels, which could be explained by the fact that stressor levels (i.e., known context) might not represent the perceived stress of participants. We further took a step out to daily life environments and tried the DDSR model, which achieved 68.30% accuracy with HRV, 63.60% accuracy with EDA, and 71.40% accuracy with multimodal HRV + EDA. Then, we applied the model trained in the laboratory with self-reports and showed that the performance increased with all types of physiological signals for daily life stress recognition (7% increase for HRV, 14.8% increase for EDA, 2.8% increase for HRV + EDA). We also investigated that the mean accuracy was enhanced with the LDKC model for the HRV + EDA- and EDA-based stress recognition frameworks. The classification performance of the proposed system changed significantly based on the event labeling methodology. Models trained in the laboratory for daily life stress detection outperformed the ones trained with daily life data. We achieved the best results (73.81%) in the LDSR model, where the daily life stress detection system was trained in the laboratory environment with self-report labels. This demonstrated that we could increase the accuracy of the system by training the model in the laboratory with the same kind of ground truth since we also had self-reports as the ground truth in the wild. We also showed that multi-modal sensing provided a more robust framework for all types of session labeling approaches. The performance of the LDKC model was better than the DDSR model and worse than that of the LDSR in the multimodal framework. On the other hand, the accuracies of the LDKC model were lower than both those of the LDSR and DDSR results when only a single modality was used. We could infer that the DDSR method suffered from relatively noisy training labels and data when compared to LDSR, where training data were obtained from the controlled laboratory environment. Using different types of labels in training and testing (known context labels in the laboratory for training and self-reports in the wild for testing) might be responsible for the low performance of LDKC models. RF and SVM classifiers outperformed other classifiers, and these results were aligned with the daily life stress recognition studies mentioned in the Related Work Section. Feature and modality selection is vital for achieving better performances. We selected the best ten features; five of them were from EDA, and five of them were from HRV, which also suggested that a multi-modality approach was crucial for daily life stress detection. In most of the test cases (15/20), the combination of modalities increased the performance of the system. In the remaining tests, anticorrelations between the features of different modalities might be the cause of lower accuracies. There is still room for improvement for daily life stress recognition. Moreover, our study was not without limitations. In order to generalize the conclusions, additional studies based on larger heterogeneous sample groups are needed. As future works, we plan to develop personalized perceived stress models to overcome the subjectivity problem of self-reports. We will try to exploit baseline surveys and daily session-based questionnaires of individuals to prevent the bias caused by subjective self-reports.

Author Contributions: Y.S.C. was the main editor of this work and made major contributions to the data collection, analysis, and manuscript writing. D.G. and D.R.K. made valuable contributions to both data collection and manuscript writing. They designed the experiment and contributed to the related sections regarding data collection. D.E. and N.C. contributed equally to this work in the design, implementation, data analysis, and writing the manuscript. C.E. provided invaluable feedback and technical guidance to interpret the design and the detail of the field study. He also performed comprehensive critical editing to increase the overall quality of the manuscript. All authors have read and agreed to the published version of the manuscript.

Funding: This work is supported by AffecTech: Personal Technologies for Affective Health, Innovative Training Network funded by the H2020 People Programme under Marie Skłodowska-Curie Grant Agreement No. 722022, and by the Turkish Directorate of Strategy and Budget under the TAM Project Number DPT2007K120610.

Conflicts of Interest: The authors declare no conflict of interest.

References

1. Andreassi, J.L. *Psychophysiology: Human Behavior and Physiological Response*; Psychology Press: Hove, UK, 2010.

2. Quintana, D.S.; Guastella, A.J.; Outhred, T.; Hickie, I.B.; Kemp, A.H. Heart rate variability is associated with emotion recognition: Direct evidence for a relationship between the autonomic nervous system and social cognition. *Int. J. Psychophysiol.* **2012**, *86*, 168–172. [CrossRef] [PubMed]
3. Berntson, G.G.; Thomas Bigger, J., Jr.; Eckberg, D.L.; Grossman, P.; Kaufmann, P.G.; Malik, M.; Nagaraja, H.N.; Porges, S.W.; Saul, J.P.; Stone, P.H.; et al. Heart rate variability: Origins, methods, and interpretive caveats. *Psychophysiology* **1997**, *34*, 623–648. [CrossRef] [PubMed]
4. Critchley, H.D. Electrodermal responses: What happens in the brain. *Neuroscientist* **2002**, *8*, 132–142. [CrossRef] [PubMed]
5. Poh, M.Z.; Swenson, N.C.; Picard, R.W. A wearable sensor for unobtrusive, long-term assessment of electrodermal activity. *IEEE Trans. Biomed. Eng.* **2010**, *57*, 1243–1252.
6. Krumpal, I. Determinants of social desirability bias in sensitive surveys: A literature review. *Qual. Quant.* **2013**, *47*, 2025–2047. [CrossRef]
7. Northrup, D.A. *The Problem of the Self-Report in Survey Research*; Institute for Social Research, York University: Toronto, ON, Canada, 1997.
8. Liapis, A.; Katsanos, C.; Sotiropoulos, D.; Xenos, M.; Karousos, N. Stress Recognition in Human-Computer Interaction Using Physiological and Self-Reported Data: A Study of Gender Differences. In Proceedings of the 19th Panhellenic Conference on Informatics, Athens, Greece, 1–3 October 2015; Association for Computing Machinery: New York, NY, USA, 2015; pp. 323–328, doi:10.1145/2801948.2801964. [CrossRef]
9. Sharma, N.; Gedeon, T. Elsarticle-template-1-numObjective measures, sensors and computational techniques for stress recognition and classification: A survey. *Comput. Methods Programs Biomed.* **2012**, *108*, 1287–1301. [CrossRef]
10. Mohr, D.C.; Zhang, M.; Schueller, S.M. Personal sensing: Understanding mental health using ubiquitous sensors and machine learning. *Annu. Rev. Clin. Psychol.* **2017**, *13*, 23–47. [CrossRef]
11. Picard, R.W. Automating the Recognition of Stress and Emotion: From Lab to Real-World Impact. *IEEE MultiMedia* **2016**, *23*, 3–7. [CrossRef]
12. Ghamari, M. A review on wearable photoplethysmography sensors and their potential future applications in health care. *Int. J. Biosens. Bioelectron.* **2018**, *4*, 195. [CrossRef]
13. Setz, C.; Arnrich, B.; Schumm, J.; La Marca, R.; Tröster, G.; Ehlert, U. Discriminating stress from cognitive load using a wearable EDA device. *IEEE Trans. Inf. Technol. Biomed.* **2009**, *14*, 410–417. [CrossRef]
14. Zubair, M.; Yoon, C.; Kim, H.; Kim, J.; Kim, J. Smart wearable band for stress detection. In Proceedings of the 2015 5th International Conference on IT Convergence and Security (ICITCS), Kuala Lumpur, Malaysia, 24–27 August 2015; pp. 1–4.
15. Castaldo, R.; Montesinos, L.; Melillo, P.; Massaro, S.; Pecchia, L. To What Extent Can We Shorten HRV Analysis in Wearable Sensing? A Case Study on Mental Stress Detection. In *EMBEC & NBC 2017*; Springer: Berlin, Germany, 2017; pp. 643–646.
16. Abouelenien, M.; Burzo, M.; Mihalcea, R. Human acute stress detection via integration of physiological signals and thermal imaging. In Proceedings of the 9th ACM International Conference on PErvasive Technologies Related to Assistive Environments, Corfu Island, Greece, 29 June–1 July 2016; p. 32.
17. Vanitha, V.; Krishnan, P. Real time stress detection system based on EEG signals. *Biomed. Res.* **2016**, *1*, S271–S275.
18. Can, Y.S.; Chalabianloo, N.; Ekiz, D.; Ersoy, C. Continuous Stress Detection Using Wearable Sensors in Real Life: Algorithmic Programming Contest Case Study. *Sensors* **2019**, *19*, 1849. [CrossRef] [PubMed]
19. Sandulescu, V.; Andrews, S.; Ellis, D.; Bellotto, N.; Mozos, O.M. Stress detection using wearable physiological sensors. In Proceedings of the International Work-Conference on the Interplay Between Natural and Artificial Computation, Elche, Spain, 1–5 June 2015; pp. 526–532.
20. Tivatansakul, S.; Ohkura, M. Improvement of emotional healthcare system with stress detection from ECG signal. In Proceedings of the 2015 37th Annual International Conference of the IEEE Engineering in Medicine and Biology Society (EMBC), Milan, Italy, 25–29 August 2015; pp. 6792–6795.
21. Akmandor, A.O.; Jha, N.K. Keep the Stress Away with SoDA: Stress Detection and Alleviation System. *IEEE Trans. Multi-Scale Comput. Syst.* **2017**, *3*, 269–282. [CrossRef]
22. Ciman, M.; Wac, K. Individuals' stress assessment using human-smartphone interaction analysis. *IEEE Trans. Affect. Comput.* **2016**, *9*, 51–65. [CrossRef]

23. Hovsepian, K.; Al'Absi, M.; Ertin, E.; Kamarck, T.; Nakajima, M.; Kumar, S. cStress: Towards a Gold Standard for Continuous Stress Assessment in the Mobile Environment. In Proceedings of the 2015 ACM International Joint Conference on Pervasive and Ubiquitous Computing, Osaka, Japan, 7–11 September 2015; ACM: New York, NY, USA, 2015; pp. 493–504.
24. Vildjiounaite, E.; Kallio, J.; Kyllönen, V.; Nieminen, M.; Määttänen, I.; Lindholm, M.; Mäntyjärvi, J.; Gimel'farb, G. Unobtrusive stress detection on the basis of smartphone usage data. *Pers. Ubiquitous Comput.* **2018**, *22*, 671–688. [CrossRef]
25. Mishra, V.; Hao, T.; Sun, S.; Walter, K.N.; Ball, M.J.; Chen, C.H.; Zhu, X. Investigating the Role of Context in Perceived Stress Detection in the Wild. In Proceedings of the 2018 ACM International Joint Conference and 2018 International Symposium on Pervasive and Ubiquitous Computing and Wearable Computers, Singapore, 8–12 October 2018; pp. 1708–1716.
26. Gjoreski, M.; Luštrek, M.; Gams, M.; Gjoreski, H. Monitoring stress with a wrist device using context. *J. Biomed. Inform.* **2017**, *73*, 159–170. [CrossRef] [PubMed]
27. Li, F.; Xu, P.; Zheng, S.; Chen, W.; Yan, Y.; Lu, S.; Liu, Z. Photoplethysmography based psychological stress detection with pulse rate variability feature differences and elastic net. *Int. J. Distrib. Sens. Netw.* **2018**, *14*, 1550147718803298. [CrossRef]
28. Lin, W.H.; Wu, D.; Li, C.; Zhang, H.; Zhang, Y.T. Comparison of heart rate variability from PPG with that from ECG. In Proceedings of the International Conference on Health Informatics, Verona, Italy, 15–17 September 2014; pp. 213–215.
29. Bolanos, M.; Nazeran, H.; Haltiwanger, E. Comparison of heart rate variability signal features derived from electrocardiography and photoplethysmography in healthy individuals. In Proceedings of the 2006 International Conference of the IEEE Engineering in Medicine and Biology Society, New York, NY, USA, 30 August–3 September 2006; pp. 4289–4294.
30. Selvaraj, N.; Jaryal, A.; Santhosh, J.; Deepak, K.K.; Anand, S. Assessment of heart rate variability derived from finger-tip photoplethysmography as compared to electrocardiography. *J. Med. Eng. Technol.* **2008**, *32*, 479–484. [CrossRef]
31. Billeci, L.; Tonacci, A.; Brunori, E.; Raso, R.; Calderoni, S.; Maestro, S.; Morales, M.A. Autonomic nervous system response during light physical activity in adolescents with anorexia nervosa measured by wearable devices. *Sensors* **2019**, *19*, 2820. [CrossRef]
32. Smets, E.; Velazquez, E.R.; Schiavone, G.; Chakroun, I.; D'Hondt, E.; De Raedt, W.; Cornelis, J.; Janssens, O.; Van Hoecke, S.; Claes, S.; et al. Large-scale wearable data reveal digital phenotypes for daily-life stress detection. *NPJ Digit. Med.* **2018**, *1*, 67. [CrossRef]
33. Can, Y.S.; Arnrich, B.; Ersoy, C. Stress detection in daily life scenarios using smart phones and wearable sensors: A survey. *J. Biomed. Inform.* **2019**, *92*, 103139. [CrossRef] [PubMed]
34. Giannakakis, G.; Grigoriadis, D.; Giannakaki, K.; Simantiraki, O.; Roniotis, A.; Tsiknakis, M. Review on psychological stress detection using biosignals. *IEEE Trans. Affect. Comput.* **2019**, doi:10.1109/TAFFC.2019.2927337. [CrossRef]
35. Kirschbaum, C.; Pirke, K.M.; Hellhammer, D.H. The 'Trier Social Stress Test'—A tool for investigating psychobiological stress responses in a laboratory setting. *Neuropsychobiology* **1993**, *28*, 76–81. [CrossRef] [PubMed]
36. Cohen, S.; Kamarck, T.; Mermelstein, R. A global measure of perceived stress. *J. Health Soc. Behav.* **1983**, *24*, 385–396. [CrossRef] [PubMed]
37. Plarre, K.; Raij, A.; Hossain, S.M.; Ali, A.A.; Nakajima, M.; Al'absi, M.; Ertin, E.; Kamarck, T.; Kumar, S.; Scott, M.; et al. Continuous inference of psychological stress from sensory measurements collected in the natural environment. In Proceedings of the 10th ACM/IEEE International Conference on Information Processing in Sensor Networks, Chicago, IL, USA, 12–14 April 2011; pp. 97–108.
38. AppleWatch Series 5—Apple. 2019. Available online: https://www.apple.com/newsroom/2019/09/apple-unveils-apple-watch-series-5 (accessed on 10 December 2019).
39. Shiffman, S.; Stone, A.A.; Hufford, M.R. Ecological Momentary Assessment. *Annu. Rev. Clin. Psychol.* **2008**, *4*, 1–32. [CrossRef] [PubMed]
40. Empatica. 2018. Available online: https://ec.europa.eu/research/participants/documents/downloadPublic/Zjc3emJFS1pUazFRWVZ3RzVDTStGL2R3Z0tJamw2N29nZlZ2RG96c2l2MWNhNXhxVU9tTUZRPT0=/attachment/VFEyQTQ4M3ptUWZXRzJwcjZDUWRSMHlVRmFDUkJoS3A= (accessed on 5 December 2018).

41. Understanding the Stress Response. 2018. Available online: https://www.mentalhealth.org.nz/assets/Working-Well/FS-understanding-stress-reponse.pdf (accessed on 5 December 2018).
42. Cinaz, B.; Arnrich, B.; Marca, R.; Tröster, G. Monitoring of Mental Workload Levels During an Everyday Life Office-work Scenario. *Pers. Ubiquitous Comput.* **2013**, *17*, 229–239. [CrossRef]
43. Tarvainen, M.P.; Niskanen, J.P.; Lipponen, J.A.; Ranta-aho, P.O.; Karjalainen, P.A. Kubios HRV—A Software for Advanced Heart Rate Variability Analysis. In Proceedings of the 4th European Conference of the International Federation for Medical and Biological Engineering, Munich, Germany, 7–12 September 2009; Vander Sloten, J., Verdonck, P., Nyssen, M., Haueisen, J., Eds.; Springer: Berlin/Heidelberg, Germany, 2009; pp. 1022–1025.
44. Taylor, S.; Jaques, N.; Chen, W.; Fedor, S.; Sano, A.; Picard, R. Automatic identification of artifacts in electrodermal activity data. In Proceedings of the 2015 37th Annual International Conference of the IEEE Engineering in Medicine and Biology Society (EMBC), Milano, Italy, 25–29 August 2015; pp. 1934–1937.
45. Greco, A.; Valenza, G.; Lanata, A.; Scilingo, E.P.; Citi, L. cvxEDA: A Convex Optimization Approach to Electrodermal Activity Processing. *IEEE Trans. Biomed. Eng.* **2016**, *63*, 797–804. [CrossRef]
46. Gjoreski, M.; Gjoreski, H.; Luštrek, M.; Gams, M. Continuous Stress Detection Using a Wrist Device: In Laboratory and Real Life. In Proceedings of the 2016 ACM International Joint Conference on Pervasive and Ubiquitous Computing: Adjunct, Heidelberg, Germany, 12–26 September 2016; ACM: New York, NY, USA, 2016; pp. 1185–1193.
47. Alberdi, A.; Aztiria, A.; Basarab, A. Towards an automatic early stress recognition system for office environments based on multimodal measurements: A review. *J. Biomed. Inform.* **2016**, *59*, 49–75. [CrossRef]
48. Singh, D.; Vinod, K.; Saxena, S. Sampling frequency of the RR interval time series for spectral analysis of heart rate variability. *J. Med. Eng. Technol.* **2004**, *28*, 263–272. [CrossRef]
49. Hall, M.A.; Holmes, G. Benchmarking attribute selection techniques for discrete class data mining. *IEEE Trans. Knowl. Data Eng.* **2003**, *15*, 1437–1447. [CrossRef]
50. Holmes, G.; Donkin, A.; Witten, I.H. WEKA: A machine learning workbench. In Proceedings of the ANZIIS'94—Australian New Zealnd Intelligent Information Systems Conference, Adelaide, Australia, 18–20 November 1994; pp. 357–361.
51. Kotsiantis, S.; Kanellopoulos, D.; Pintelas, P. Handling imbalanced datasets: A review. *Gests Int. Trans. Comput. Sci. Eng.* **2006**, *30*, 25–36.
52. Le Cessie, S.; Van Houwelingen, J.C. Ridge estimators in logistic regression. *J. R. Stat. Soc. Ser. Appl. Stat.* **1992**, *41*, 191–201. [CrossRef]
53. Eibe, F.; Hall, M.; Witten, I. The WEKA Workbench. Online Appendix for Data Mining: Practical Machine Learning Tools and Techniques. In *Morgan Kaufmann*; Elsevier: Amsterdam, The Netherlands, 2016.
54. PSS-14. 2019. Available online: http://www.ebserh.gov.br/sites/default/files/concurso/arqconc/2019-12/MINUTA%20EDITAL%20NORMATIVO_02_%20PSS%2014.2019-%20HUAP-UFF-1_ajustado%2025nov2019.pdf (accessed on 5 December 2019).

sensors

MDPI

Article

Activity Recognition Using Wearable Physiological Measurements: Selection of Features from a Comprehensive Literature Study

Inma Mohino-Herranz [1,*], Roberto Gil-Pita [1], Manuel Rosa-Zurera [1] and Fernando Seoane [2,3,4]

1 Department of Signal Theory and Communications, University of Alcala, 28805 Alcala de Henares, Madrid, Spain; roberto.gil@uah.es (R.G.-P.); manuel.rosa@uah.es (M.R.-Z.)
2 Clinical Science, Intervention an Technology, Karolinska Institutet, 17177 Stockholm, Sweden; fernando.seoane@ki.se
3 Department Biomedical Engineering, Karolinska University Hospital, 14186 Stockholm, Sweden
4 Swedish School of Textiles, University of Boras, 50190 Boras, Sweden
* Correspondence: inmaculada.mohino@uah.es

Received: 12 November 2019; Accepted: 11 December 2019; Published: 13 December 2019

Abstract: Activity and emotion recognition based on physiological signal processing in health care applications is a relevant research field, with promising future and relevant applications, such as health at work or preventive care. This paper carries out a deep analysis of features proposed to extract information from the electrocardiogram, thoracic electrical bioimpedance, and electrodermal activity signals. The activities analyzed are: neutral, emotional, mental and physical. A total number of 533 features are tested for activity recognition, performing a comprehensive study taking into consideration the prediction accuracy, feature calculation, window length, and type of classifier. Feature selection to know the most relevant features from the complete set is implemented using a genetic algorithm, with a different number of features. This study has allowed us to determine the best number of features to obtain a good error probability avoiding over-fitting, and the best subset of features among those proposed in the literature. The lowest error probability that is obtained is 22.2%, with 40 features, a least squares error classifier, and 40 s window length.

Keywords: activity recognition; physiological signals; electrocardiogram; thoracic electrical bioimpedance; electrodermal activity

1. Introduction

Activity can be defined as the state or quality of being active, which implies that the activity can be emotional, intellectual, physical, etc. Typical activity recognition systems focus on daily life activities such as walking, running, exercising, scrubbing and cooking [1–5], mental tasks [6,7] or emotion recognition [8]. Activity-state recognition systems can be applied to human error prevention tasks in many professional activities such as first responders, crane operators or train drivers. The present work aims at deeply studying several features found in the literature to characterize the signals of electrocardiogram, thoracic bioimpedance and electrodermal activity, whose objective is to recognize four different activities: emotional, mental, physical and neutral activity (resting).

Currently, there are different methods for detecting activity. For instance, Inertial Measurement Units (IMUs) [9,10] in combination with Global Positioning System (GPS) data for outdoor applications [11] or sensor located indoors for smart homes [3,12] for detecting physical activity. On the other hand, speech and gestures can be useful for assessing emotional activity [13–16]. Another alternative is physiological signals captured through sensors located in the body of the subject. Wearable biomedical sensing through smart clothing [17,18] allows the recording of physiological measurements such as the Electrocardiogram (ECG), the Thoracic Electrical Bioimpedance (TEB) or the Electrodermal Activity (EDA), among others, which contain not only information about specific body functions and physiological states, but also valuable information about the activity and the person's condition regarding emotional state, mental load and physical activity [19].

In the literature, numerous works are found in which these three signals are used to detect stress, emotions, and activity. For instance, ECG is affected by these factors, since the heart rate is directly related to the body and mind condition [20–22]. In this sense, the Heart Rate Variability (HRV) has been widely used to extract information about the status of the autonomous nervous system and emotions [23]. On the other hand, TEB can be used as an indicator of the breathing function, and it has been used in different studies for activity recognition [24] and stress detection [25]. EDA measures the activity of sweating glands on the skin which are directly controlled by the sympathetic nervous system, and thus can also be used for emotion recognition [26–29].

However, few papers provide deep studies including all these three signals with the same purpose, comparing the physiological signals under study and determining which physiological signal provides more relevant information about the individual activity. For instance, the features extracted from TEB signal acquired together with the ECG and the heart sound can be used to study cardiovascular reactivity during emotional activation in men and women [24]. Numerous features have been found for this purpose in the literature, but there is not a clear rule of which ones are more relevant for a given problem. In general, the larger the number of features, the greater the generalization problems, that is, the ability to handle unseen data [30]. Selecting a subset of features results mandatory for many activity recognition application.

Taking all this into account, the present paper aims at assessing the utility of features extracted from ECG, TEB, and EDA in activity recognition systems. These physiological signals have been recorded using sensorized garments combined with wearable instrumentation. We intend to recognize four different activities: emotional activity, mental activity, physical activity, and resting. The paper is structured as follows: Section 1 introduces the problem tackled in this paper; Section 2 is a review of the literature about physiological sensing, window length, features, and possible classifiers; Section 3 summarizes the sensors used to acquire the signals and the mental activity states that are considered; Section 4 presents the experiments carried out; Section 5 includes the obtained results; Section 6 presents the main conclusions. A set of Appendixes A–C are also included with a detailed description of the considered features extracted from the different acquired signals.

2. Background

In this study several parameters have been analyzed: (a) the physiological sensing mode (ECG, TEB and EDA), (b) the window length, (c) the features extracted from each signal, (d) the number of features to obtain the best results, and (e) the type of classifier.

- Selection of Physiological sensing modality: In this part, we compare the physiological signal under study and determine which physiological signal provides more relevant information about the individual activity. The signals used are ECG, TEB, and EDA. It is possible to find numerous works in which these signals are used to detect stress, emotions, and activity in the literature. The ECG signal is used in some papers such as [23], where the obtained results suggest that positive emotions lead to alterations in HRV, which may be beneficial in some illness treatment [19,31,32].

  TEB is also used in some papers, though it is less useful than ECG and EDA signals. The work [25] demonstrated that its use is decisive to detect stress. In addition, most of the studies considered several signals, such as the paper [28] which contains the study on the correlation between heart rate, electrodermal activity and Player Experience in First-Person Shooter Games, concluding that their results indicate correlation between the physiological measures and gameplay experience, even in relatively simple measurement scenarios. Another work, [29] studies the individual differences within the electrodermal activity as subjects' anxiety, which concludes that in normal subjects there are individual electrodermal differences as a function of trait-anxiety scores. However, few papers provide a deep study of features for the three signals, such as the use of these signals with the same purpose.

- In order to obtain the window length, the first limit found in the literature review is imposed by feature calculation. There are some features that require a minimum window length to be calculated, such as, HRV triangular index, which takes at least 20 min to be calculated [33–35], Standard Deviation of NN intervals (SDNN) index, calculated as mean standard deviations of all NN intervals for all 5 min segments of the entire recording [34], and for all derivatives (Standard Deviation of Successive Differences (SDSD), Standard Deviation of sequential 5-min RR interval (SDANN)) found in [34]. In our case, we decided to use window lengths lower than 60 s, as the database could be largely cut down, which would change the study.
- In our study, we have studied a large number of features which have been selected from a deep revision of the literature. The most frequently used with ECG signals were obtained both in the frequency domain and the time domain: frequency bands [23,26,34,36–48], and power ratios [23,43,44,46,47,49], in frequency domain; and Heart Rate Variability (HRV) [23,26,38,39,41,42,45,48,50,51], the SDNN [42,48,49], Number of NNs in 50 ms (NN50), pNN50 [34,42,48] and some statistical parameters, such as mean amplitude rate, mean frequency, standard deviations of the raw signals, [25,37,52–55]. In our study, we have studied all the features found in a literature review of more than 90 papers.

  The features extracted from the TEB signal are used in some works such as, [24] where the approach is to study cardiovascular reactivity during emotional activation in men and women. Here, the TEB has been acquired together with ECG and the heart sound. In [56] the full respiratory signal was derived from the thoracic impedance raw data, like in our case.

  Finally, the EDA signal is studied in several papers, [26,42,49,53,55,57,58]. A complete study about the EDA signal is shown in [41,59].
- Most published papers use the calculated features to feed the classifier. Therefore, the number of features used depends on the particular study. We propose to implement feature selection from all the available features to find the best ones and to avoid generalization problems in classification.
- The classifier is usually determined by the author without comparisons or detailed studies about suitability. In numerous works, the selected classifier is the Support Vector Machine (SVM). We think it is positive to make a comparison of different classifiers with very different characteristics.

3. Materials

A sensor network capable of acquiring the ECG, TEB and EDA signals has been designed, in order to obtain a database of signals to be used in this study. The complete system acquires all the signals described in the literature, that have been mentioned above, which is explained in detail in [25,60]. To acquire the multimodal biosignals a set of sensorized garments were used, which are shown in Figure 1. A glove to acquire EDA measurement in hand, a bracelet to acquire EDA measurement in the arm and a vest to acquire ECG and TEB. These garments are connected to the measurement devices shown in Figure 2. The glove and the bracelet are connected to the device called GSR, which acquires the EDA signal and the vest is connected to the vest through a recorder called ECGZ2, which acquires ECG and TEB signals. The ECGZ2 is capable of sampling each signal with a different sampling frequency. For the TEB and EDA, the sampling frequency is 100 Hz and for the ECG, the sampling frequency is 250 Hz.

Measurements were collected from $k = 40$ subjects, students and climbers aged 20 to 49, including 12 females and 28 males. The total duration of the complete experiment was approximately 90 min per subject. All of the experiments were performed under the conditions of respect for individual rights and ethical principles that govern biomedical research involving human beings, and written informed consent was obtained from all participants. The experiments were approved by the Research Ethics

Committee at the University of Alcala, and the protocol number assigned to this experiment is CEI: 2013/025/201130624.

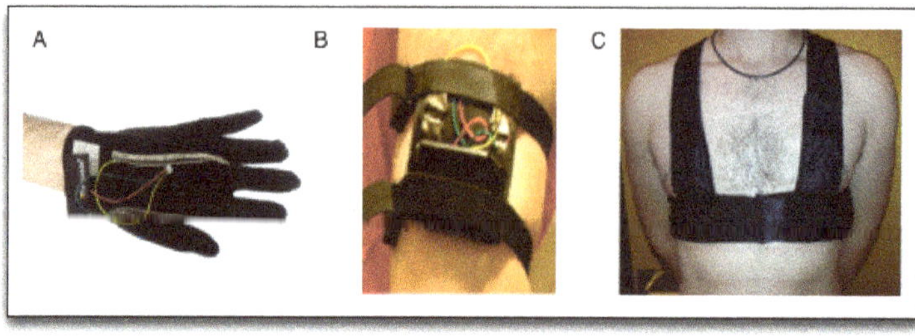

Figure 1. Recording Devices: (**A**) Electrocardiogram (ECG) and Thoracic Electrical Bioimpedance (TEB) device, (**B**) electrodermal activity (EDA) device. (**A**) Glove to acquire EDA signal in hand; (**B**) Arm bracelet to acquire EDA signal. (**C**) Vest to acquire ECG and TEB signals prior published in [25] under license CC by 4.0.

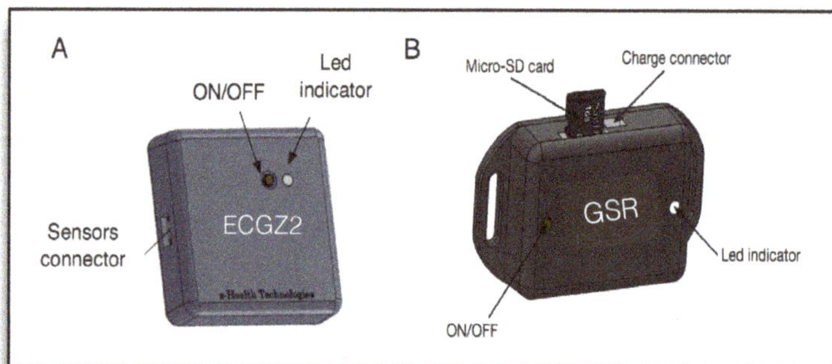

Figure 2. Devices: (**A**) ECGZ2 device, (ECG and TEB recorder prior published in [25] under license CC by 4.0); (**B**) GSR device (EDA recorder published in [60] under license CC by 4.0).

As was stated above, there were four different activities to be recognized: emotional activity, mental activity, physical activity, and neutral activity (resting).

In order to elicit the different activities, we have used a segment documentary called "Earth" to induce Neutral Activity. In order to elicit emotional activity, we used a set of segments extracted from several validated movies [61]. "American History X" (1998) by Savoy Pictures [62], "I am legend" (2007) by Warner Bross [63], "Life is beautiful" (1997) by Miramax [64,65] and "Cannibal Holocaust" (1980) by F.D. Cinematografica [66]. The mental activity was elicited using a set of games based on mental arithmetic and playing the well-known game "Tetris", used several times to elicit mental activity [67].

The designed activity recognition system had to take a decision every 10 s, and each individual generated 28 time slots of each activity (the database is balanced). Thus, the total number of patterns (decisions) for this analysis was 4480, and each class is composed of 1120 different patterns.

In the present analysis, we have used four different activities:

- Neutral activity, registered during the last 140 s of the first movie (the documentary). As each individual watched each movie twice, there are 280 s for each individual in the database
- Emotional activity, registered during the viewing of the last 70 s of the second and third movies (140 s); therefore, we obtained a total of 280 s per individual.

- Mental activity, registered during the last 140 s of both games, producing 280 s in total.
- Physical activity registered during the last 280 s of the physical activity stage. To elicit physical load the participant had to go up and down the stairs for five minutes.

The database particular characteristics can be found in [25]. The full dataset can also be downloaded from the Supplementary Information included in the paper.

4. Methods

The main objective is to extract or calculate all the features found in the literature, applied in different experiments related to activity detection, and after that, to apply a feature selection algorithm to determine the most suitable feature set and the number of features. The acquired signals are processed to identify the activity. The process can be divided into three stages: (a) Feature Extraction, (b) Feature Selection and (c) Classification. An extensive literature review was carried out to find out the typical features used to determine the subject's activity condition identifying a total of 533 features.

Figure 3 shows the main scheme of the activity recognition system.

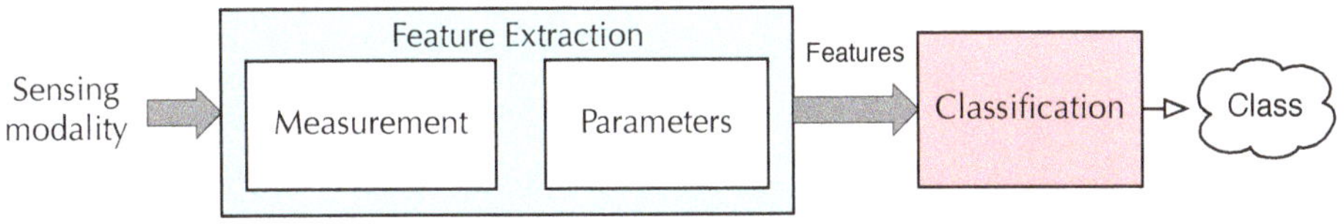

Figure 3. Scheme of the used detection system.

4.1. Feature Extraction

This stage is divided into two sub-stages. The first one carries out time or frequency domains measurements. These measurements can be the signal acquired itself, or preliminary data used to calculate the features. The second one extracts parameters from each measurement with information related to the classification problem.

The measurements are very dependent on the type of signal. For clarity sake, the description of the measurements and parameters strictly related to a given signal is included in the Appendixes A–C. On the other hand, some parameters are common to all the measurements considered in this work, such as the most common statistical parameters. The statistical parameters considered in this work are denoted as the Standard Set of Statistical Parameters (SSSP), and they include: mean, median, standard deviation, 25% trimmed mean, skewness, kurtosis, maximum, minimum, percentile 25%, percentile 75%, geometric mean, harmonic mean and mean absolute deviation.

In addition to these parameters, another parameter has been frequently calculated in almost all the measurements, which tries to model a very important concept in physiological signal analysis: the baseline. To determine the baseline of a measurement under study, we will use an ultra-low pass filter, so that it integrates the average valued of the measurement over a large period of time. The calculation of this baseline is based on the use of an Infinite Impulse Response (IIR) filter, which can achieve a very low cutoff frequency with only a couple of coefficients. Thus, for a given measurement z_i, the baseline y_i is calculated as follows:

$$y_i = z_i \cdot \beta + y_{i-1} \cdot (1 - \beta). \tag{1}$$

The β value controls the speed of variation of the baseline parameter, that is, the cutoff frequency of the equivalent low pass filter. Depending on the sampling frequency, we have chosen a value of β which corresponds to a filter that takes approximately the last 20 min of recording of the measurement to obtain the baseline.

Due to the huge number of features, and so as to avoid distractions about the paper goals, the description of the calculated features has been included in a set of Appendixes A–C at the end of this paper.

4.2. Classification

The literature of activity recognition using physiological signals includes numerous types of classifiers with different characteristics in terms of complexity, intelligence, and generalization. In this work, we compare the performance of four widely used classifiers with different rules aiming at studying the performance of the set of features: the Least Squares Linear Classifier (LSLC); the Least Squares Quadratic Classifier (LSQC); the Support Vector Machines (SVMs), the Multi-layer Perceptrons (MLPs), the k-Nearest Neighbor (kNN), the Centroid Displacement-Based k-Nearest Neighbor (CDNN) and Random Forests (RF) .

4.2.1. Least Squares Linear Classifier (LSLC)

In a linear classifier, given a set of training patterns $\mathbf{x} = [x_1, x_2, \ldots, x_L]^T$, where each pattern has associated a class, denoted as C_i, $i = 1, \ldots, M$, the decision rule is obtained using a set of M linear combinations of the training patterns. In the least squares approach (the LSLC), the values of the weights of the linear combinations are those that minimize the mean squared error (MSE), obtaining the *Wiener-Hopf* equations [68]. These classifiers are fast and simple, and they present a good generalization capability.

4.2.2. Least Squares Quadratic Classifier (LSQC)

Like with the LSLC, the LSQC also renders very good results with a very fast learning process. It slightly increases the intelligence of the LSLC by adding quadratic terms to the linear combinations, thus improving the performance by increasing the complexity, with the consequence of a decrease in generalization.

4.2.3. Support Vector Machines (SVMs)

An SVM projects the observation vector $\mathbf{x}$ to a higher dimension space, using a set of kernel functions, where the patterns can be better linearly separated. The patterns of the design set selected to be the center of these functions are denominated "support vectors" [69]. In the present study, we used linear SVM (LINSVM) and nonlinear SVM using Gaussian Radial Basis Function (RBF) kernels, denoted RBFSVM.

SVMs are essentially binary classifiers, and to implement multi-class classifiers an strategy must be defined. In this paper we used a one-against-all strategy. Furthermore, SVMs present mainly two parameters (the kernel scale and the box constraint) that must be optimized. In this paper a k-fold cross validation strategy over the design set was carried out in order to determine the best values of these hyper-parameters. RBFSVMs are also sensitive to differences in the scaling of the features, thus to avoid scale problems features were normalized by removing the mean value and dividing by the standard deviation, being these values estimated using the design data.

4.2.4. Multi-Layer Perceptrons (MLPs)

MLPs are composed of one or more layers of neurons/perceptrons arranged sequentially so that the outputs of the neurons of a layer are the inputs of the neurons of the next layer. It is a feed-forward network, therefore the outputs of the network can be calculated as explicit functions of inputs and weights. Each neuron implements a linear combination of its inputs applied to a nonlinear function denominated activation function. The complexity of the MLP depends on the number of neurons in the hidden layers, allowing to easily control the intelligence of the classifier.

In this paper we considered MLPs with one hidden layer of 8, 12 and 16 neurons. They were trained with the Levenberg Marquardt algorithm, and 20% of the design data was used to monitor and early-stop the training process, avoiding overfitting.

4.2.5. k-Nearest Neighbor (kNN)

The kNN is a classification method in which no assumptions are made on the underlying data distribution in the learning process [70]. This classifier estimates the value of the posterior probability in $\mathbf{x}$ using the k closest patterns from the design database, being k a hyper-parameter of the classifier. So, a test pattern $\mathbf{x}$ is assigned to the class C_i that maximizes the posterior probability, that is, its class is determined by majority voting over the classes of its k nearest neighbors. To define the proximity a distance must be defined. In this paper we consider the euclidean distance. To determine the best value of k in each case, a k-fold validation process was carried out over the design data, and the value of k that renders the lowest error rate over the k-fold process is selected as the final k value. Data from the individuals included in the design set were used as folds on the process.

Like in the case of the RBFSVM, the distance measurement is sensitive to changes in the scale of the features. Thus, features were normalized using the mean and the standard deviation of the features over the design set. Some advantages of the kNNs are: there are no assumptions about data, and it is an easy to understand algorithm. The disadvantages of this classifier include: high memory requirements, computationally expensive, and sensitive to irrelevant features.

4.2.6. Centroid Displacement-Based k-Nearest Neighborgs (CDNN)

The CDNN is a modified version of the kNN algorithm proposed in [71] that replaces the majority voting scheme of the kNN by a centroid based classification criterion. Considering the k-th nearest patterns in the database, the centroid of the patterns of each class with and without including the test pattern are evaluated, and the class that suffers less change due to the inclusion of the test pattern is selected. Like in the kNN method, the value of k is a hyper-parameter that must be properly determined. Again, k-fold cross validation over the design data is used to estimate the best value of k. Features were also previously normalized.

4.2.7. Random Forests (RFs)

RFs [72] are classifiers consisting of a collection of T tree-structured classifiers $h_T(\mathbf{x}), k = 1, \ldots, T$ where the decision is taken by majority voting over the T independent tree classifiers. Randomization is used in the design of each tree by two factors: first, design data is randomly selected without replacement from the data from the design set. Second, in each node of the tree a subset of F features is randomly selected. In this work we grew the trees using CART methodology without pruning, and the ratio of considered features in each node was $F = \lfloor log_2 M + 1 \rfloor$, as proposed in [72]. A total of $T = 200$ trees were used to generate each RF classifier.

4.3. *Feature Selection*

Feature selection is the process of selecting a subset of the most relevant features. There are mainly two reasons to use feature selection: to reduce the generalization problems by reducing overfitting and to simplify the model. The feature selection process used in the present work follows the wrapper approach [73]. This approach selects the subset of features that minimize the error rate of a predetermined classification algorithm.

In the literature there are numerous algorithms to select the best features of a set, being Genetic algorithms (GAs) widely used. GAs, proposed in [74], combine the principles of survival of the fittest applying evolutionary laws and emulating biological evolution in nature. These algorithms work with a population consisting of several possible solutions to the problem, being each one of them called chromosome. The optimization is carried out applying modifications to the genes of the chromosomes in the population of possible solutions. They constitute a meta-heuristically search algorithm which can be applied to optimization issues in different areas [75], and they can be successfully applied to the problem of feature selection [76,77].

In our problem, we seek the best reduced set of features which is able to obtain the minimum error probability of a classifier. For this purpose, a "population" of possible sets of features is evaluated with the goal of minimizing the classification error probability, with a limited number of features (the number of selected features must be lower than N_{max}). To avoid loss of generalization of the results, the design set is exclusively used to determine the best subset of features by applying a GA, that is, the classification rate optimized by the GA is determined exclusively with the design data.

Since the GA requires the evaluation of many classifiers in the optimization process, the choice of the classifier used in the optimization is crucial. We must consider that for each chromosome in each generation the classifier must be fully trained. Thus, the use of classifiers with a very fast learning process is required. In this work, we rely on the LSLC.

The full process is described as follows:

- A "population" of 100 combinations of features (chromosomes) is randomly generated.
- If there are two combinations with exactly the same set of features, one of them is modified by randomly replacing one of the features.
- For each combination in the population, if the number of features is greater than the maximum N_{max}, then features are randomly removed from the chromosome until the condition is satisfied.
- Each combination is ranked using the mean squared error of a LSLC measured using the design set.
- The best 10 combinations of the population are selected as "parents" that survive and are used to regenerate the remaining 90 chromosomes using a random crossover of the parents.
- Mutations are added to the population by changing a feature with a probability of 1%. It is important to highlight that the best individual of each population remains unaltered. The process iterates in Step 2 until a given number of generations are evaluated.

To achieve less risk of premature stalling of the search, we used a method known as Elimination Tournament of GAs [78], that combines several small GAs in a tournament in which the original population of each GA is generated by a random crossover of the "winner" chromosomes from previous GAs.

For this work, the number of features selected was discretized by group size in 5, 10, 20, 40, 60, 80 and the full set, for instance 174 features in case of using the ECG measurement.

To avoid overfitting in the results (generalization loss) while maximizing accuracy in the estimation of the classification error rate, k-fold cross-validation was used in the experiments, being k the number of subjects available in the design database, 40 subjects. Thus, the data were divided into k folds or subsets containing data from each subject, and each time, the registers from one given subject are used as a test set, with the data from the remaining $k-1$ used for the design task. For each fold, the design process is carried out, including the feature selection process, the choice of the parameters and the training of the classifier. That is, for each fold, features are normalized estimating the mean and standard deviation of the design set (the remaining $k-1$ folds in the dataset), the GA is implemented selecting the best subset of features, the classifier is trained with the corresponding methodology, and the hyper-parameters of the classifiers are estimated (please note here that the hyper-parameters were estimated using exclusively the design set). Once this process is completed, the estimated mean and standard deviation is used to normalize the features selected by the GA, and the classifier is evaluated with the previously determined hyper-parameters. The classification of error is then estimated by analyzing the ratio of patterns wrongly classified in the test fold.

The final classification error rate is estimated by averaging the error rates obtained for all the k folds. Since data from the same subject is not used for designing and testing at the same time, this method guarantees the generalization of the results to subjects different from the ones included in the database.

This whole process is also repeated 20 times to analyze the statistical significancy of the results. So, the error rate measures the average ratio of classification errors over 40 different test folds (40 individuals of the dataset) and 20 full repetitions of the design process (including feature selection and training the classifier). To study the significance of the results we also carry out a hypothesis test, where the null hypothesis is that the method with the lowest error rate (taken as reference) is not really better than the considered method. So, the performance obtained with different methods and parameters is statistically compared using a single-tail paired-sample t-test over the estimated errors. From this t-test we measure the p-value, which can be defined as the level of marginal significance within the statistical hypothesis test [79]. This value represents the probability of obtaining an equal result to or "more extreme" result to what was actually observed when the null hypothesis is true. It is a number between 0 and 1, so that the null hypothesis is rejected if the significance level of the test is less than the significance level (α), which is normally 0.05. The method has been interpreted as follows:

- A small value of p-value (typically ≤ 0.05) implies that the test suggests that the observed data is inconsistent with the null hypothesis, so the null hypothesis must be rejected.
- The hypothesis is not rejected when the p-value is greater than 0.05. This does not imply that the null hypothesis should be accepted, but that it is feasible.

5. Results and Analysis

This section includes the analysis of the results obtained in the experiments described in the previous section, including a detailed study of the window length selection, the classifier, the combination of signals, the number of features and the most selected features.

5.1. Window Length Selection

The first parameter to determine is the window length. In order to analyze the performance of the system with different window sizes, we consider windows of 10 s, 20 s, 40 s, and 60 s. Please note here that the shift between decisions is fixed in 10 s, independently of the window length. It means that the size of the database and the number of decisions are not affected by the variation in the window length. To determine which window length is the most appropriate to extract the features, several experiments were carried out for each feature set. Table 1 shows the results obtained using the simplest classifier (LSLC) for the different signal combinations considered in this work, as function of the window length. The table includes the best error probability and the number of selected features N_{max} that generates this result. To assess the significance of the results obtained with respect to the window length, the p-value [79] has also been included in the table, comparing the best result and the remaining of results for each combination of signals.

The results indicate that the window length for which the obtained error probability is the lowest one is 40 s for all the cases in which the TEB signal is used. We observe that for the ECG signal, the best result was obtained with a window length of 60 s, and for EDA of 10 s. In case of using all the signals, the best result is obtained with a window length of 40 s as well. For this reason, we have fixed the window length to 40 s.

Table 1. Error Probability using a Least Squares Linear Classifier (LSLC) for the best number of features as function of the window length.

Combination of Signals	Par.	Window Length			
		10 s	20 s	40 s	60 s
ECG	Error(%)	43.0%	41.2%	40.1%	39.6%
	N_{max}	174	80	174	80
	p-value	<0.001	<0.001	<0.001	Best
TEB	Error(%)	51.0%	42.2%	34.6%	37.8%
	N_{max}	60	60	40	20
	p-value	<0.001	<0.001	Best	<0.001
ECG+TEB+EDA	Error(%)	26.6%	27.9%	22.2%	24.1%
	N_{max}	20	80	40	20
	p-value	<0.001	<0.001	Best	<0.001
ECG+TEB	Error(%)	41.9%	31.3%	25.7%	27.1%
	N_{max}	80	80	60	40
	p-value	<0.001	<0.001	Best	<0.001
ECG+EDA	Error(%)	26.0%	28.3%	27.9%	29.2%
	N_{max}	40	40	40	10
	p-value	Best	<0.001	<0.001	<0.001
TEB+EDA	Error(%)	29.9%	31.2%	29.7%	30.9%
	N_{max}	20	20	40	20
	p-value	<0.001	<0.001	Best	<0.001
EDA	Error(%)	36.1%	37.3%	36.5%	37.1%
	N_{max}	20	20	20	20
	p-value	Best	0.003	<0.001	<0.001

5.2. Classifier Selection

To select the best classifier, we have trained the different types of classifiers with different combinations of signals, and a different number of maximum features to be selected. Table 2 contains the error probability (%) obtained for each classifier using the different combination of signals. The best combination of signals is the case including all the physiological signals (ECG+TEB+EDA) with $N_{max} = 40$ features, obtaining a 22.2% of error rate, and the second best is the case including ECG and TEB with $N_{max} = 60$ features, that gets a 24.5% of error.

Figure 4 shows the error probability for each feature set and for each activity with the LSLC classifier and $N_{max} = 40$ features, where it is possible to observe the percentages of error, being the lowest value obtained using all signals (ECG+TEB+EDA). Furthermore, we can appreciate that the activity most recognizable for all feature set is the physical activity and the least one the mental activity.

For a more detailed analysis, Figure 5 shows four different figures in which it is possible to observe each activity separately. The first one (top left) refers to the error probability for the neutral activity and for each of the feature set, where we can observe that the best performance of 19.11% is obtained using all feature set (ECG+TEB+EDA), provided by all signals. For the second one (top right) refers to the error probability for the emotional activity, in which the least error probability is 27.14% obtained for TEB+EDA. The third one (bottom left) shows the error probability for the mental activity, in which the minimum error probability is 41.07% using the feature set ECG+TEB+EDA. Finally, the fourth graph (on the bottom right) indicates the error probability for physical activity with errors ranging from 2.95% obtained with only EDA features to 5.45% for ECG. The error obtained for the ECG+TEB+EDA is 4.20%, which is very close to the minimum value.

Table 2. Error probability (%) obtained for each classifier using the different combination of signals with a window length of 40 s.

Classifier		Single Signal				Combination of Signals				
		ECG	TEB	EDA Arm	EDA Hand	ECG TEB EDA	ECG TEB	ECG EDA	TEB EDA	EDA
LSLC	Error	40.1	34.6	45.3	39.0	22.2	25.7	27.9	29.7	36.5
	N_{max}	174	40	10	5	40	60	40	40	20
LSQC	Error	39.3	35.2	71.2	52.8	26.2	25.9	40.6	31.9	51.4
	N_{max}	60	40	5	40	40	80	20	20	20
LINSVM	Error	41.0	34.5	61.7	47.0	22.5	24.5	36.4	28.7	47.2
	N_{max}	174	40	104	104	40	60	382	60	208
RBFSVM	Error	43.3	32.4	61.8	53.3	28.6	27.5	41.9	35.4	55.0
	N_{max}	174	60	40	40	40	325	80	20	40
MLP8	Error	41.3	29.5	61.7	43.9	24.9	26.7	35.8	29.2	46.4
	N_{max}	174	40	60	20	20	20	10	20	10
MLP12	Error	41.4	29.6	61.7	44.4	25.6	26.2	37.7	30.3	46.9
	N_{max}	174	60	60	20	20	325	10	20	10
MLP16	Error	41.6	29.6	61.9	45.1	26.1	25.9	38.2	30.5	47.3
	N_{max}	174	60	20	20	10	325	10	10	10
kNN	Error	45.6	32.4	55.4	49.0	28.7	28.7	40.3	33.1	50.5
	N_{max}	174	10	10	20	10	5	5	10	10
CDNN	Error	44.5	31.4	54.7	47.6	27.0	26.9	38.9	31.3	49.1
	N_{max}	174	80	5	10	5	5	10	20	10
RF	Error	41.0	28.9	54.9	50.9	25.5	26.5	36.7	28.2	46.5
	N_{max}	20	20	10	10	20	20	40	80	20

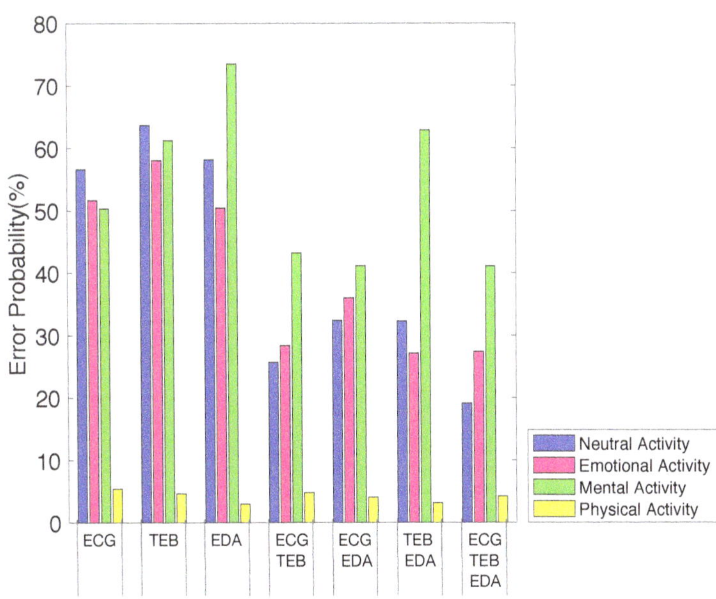

Figure 4. Error probability for each feature set and activity.

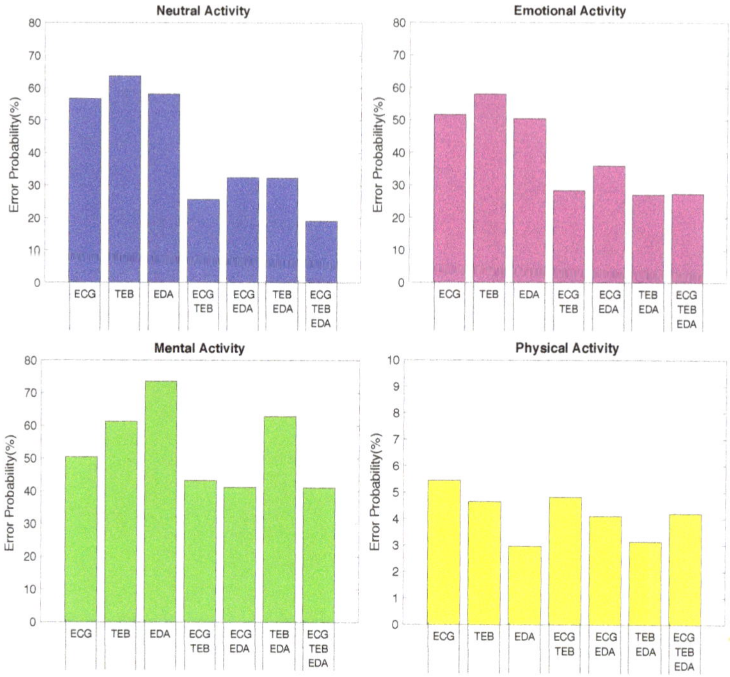

Figure 5. Error probability for each feature set and activity. Neutral (**top left**), Emotional (**top right**), Mental (**bottom left**) and Physical Activities (**bottom right**).

On the other hand, if we analyze the signals separately, we can see that the independent signal which renders the best results is the TEB (29.50% with $N_{max} = 40$ features and an MLP with 8 hidden neurons).

In order to study the main differences in the identification of the activity, the confusion matrix shown in Figure 6 indicates the misclassification between classes obtained using a LSLC and $N_{max} = 40$ features obtained from all 3 signals (ECG+TEB+EDA), where the classes that present more misclassification are emotional and mental activity.

For a more detailed analysis of the performance of the classifiers when the number of features is varied, three figures are presented below. The figures represent the performance of the classifiers in the most significant cases. As with all features, it combines all feature sets. Another case, with the two signals that combined get the best result (ECG+TEB feature set), and the signal that gets the best result independently (TEB feature set).

Figure 7, presents the results obtained with the combination including all signals (ECG+TEB+EDA). We can see that the linear classifiers render the best results, and that the GA-based feature selection process that limits the number of features helps improving the performance of the classifiers. The fact that the complex classifiers (MLPs and RBFSVMs) do not match the results of the linear classifiers might imply the presence of strong generalization problems.

Figure 8 shows the performance of the classifiers when the ECG+TEB feature set is used. In this case again the best results are provide by linear classifiers. However, the classifier that renders the best result for ECG+TEB feature set is the LINSVM with an error probability of 24.5%.

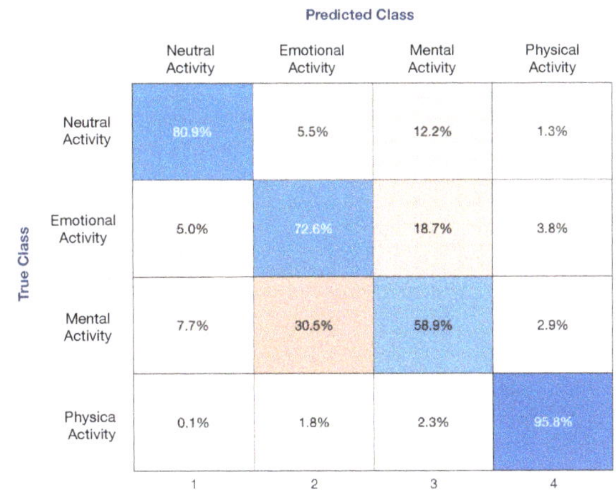

Figure 6. Confusion matrix between classes.

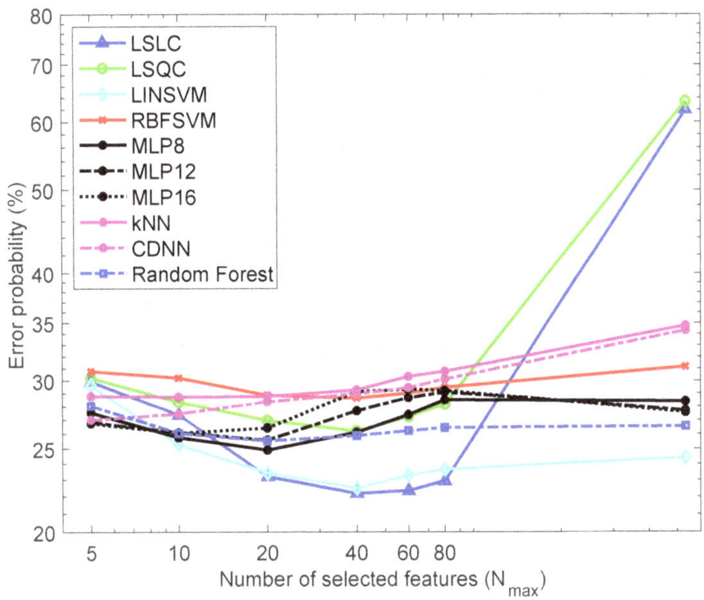

Figure 7. Classifiers comparison using *All feature set* (ECG+TEB+EDA).

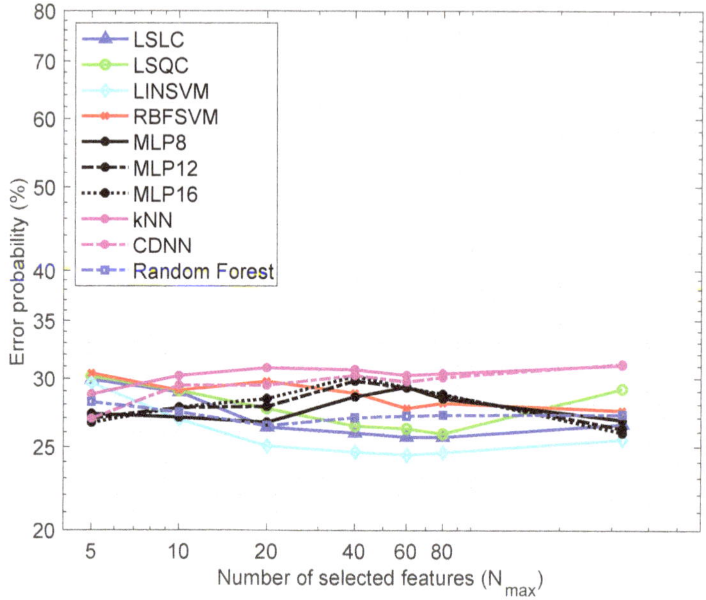

Figure 8. Classifiers comparison using the *ECG+TEB feature set.*

Finally, in case of considering just one signal the best choice is the use of the TEB. Figure 9 shows the performance of the classifiers under study with only features from the TEB signal. In this case the results are somewhat different from the previous ones, since the classifier that gives the best results is the RF, with an error probability of 28.9%.

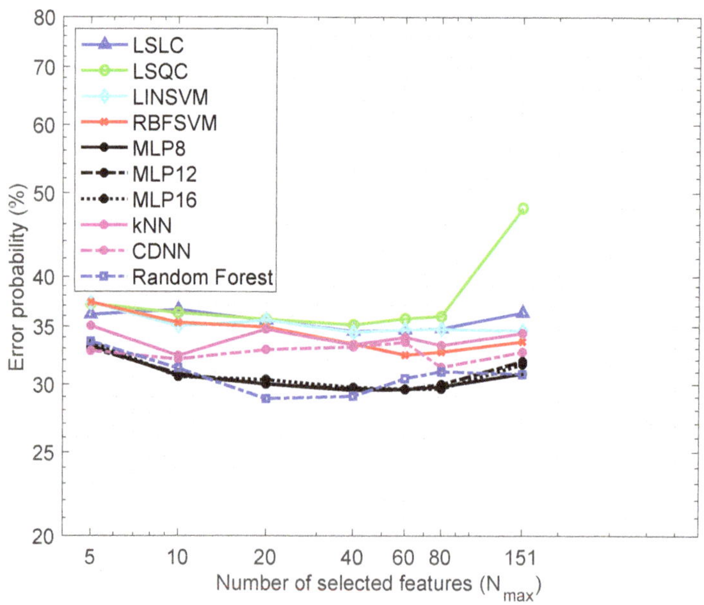

Figure 9. Classifiers comparison using the *TEB feature set.*

5.3. Frequently Selected Features

In order to complete the study, we will show which features and measurements are the most frequently selected and the percentage of selection. Table 3 shows the average number of features selected by the GAs from each measurement and each signal, considering a maximum number of selected features $N_{max} = 40$, for the different combination of signals. As we can see, the most frequently selected measurement from the ECG is the RR. In general, the measurements extracted in the frequency domain for the ECG are not very useful. Concerning the TEB, the RF and the BRV measurements present high ratios in the case of considering all the signals in the combination. And the most selected measurement from the EDA is the processed measurement taken in the hand.

Table 3. Average number of features selected from the measurements of the different signals, with $N_{max} = 40$ features.

	Single Signal				Combination of Signals				
Signal: Measurement	ECG	TEB	EDA Arm	EDA Hand	ECG TEB EDA	ECG TEB	ECG EDA	TEB EDA	EDA
ECG: Original	6.5	-	-	-	1.5	3.5	3.8	-	-
ECG: RR	13.1	-	-	-	4.9	8.3	6.0	-	-
ECG: RA	6.7	-	-	-	2.4	3.5	2.2	-	-
ECG: HR	6.5	-	-	-	1.5	2.9	1.1	-	-
ECG: HRV	2.8	-	-	-	2.4	2.4	2.2	-	-
ECG: PSD	0.6	-	-	-	0.4	0.7	0.3	-	-
ECG: PSD-VLF	0.5	-	-	-	0.3	0.4	0.7	-	-
ECG: PSD-LF	0.6	-	-	-	0.4	0.5	0.8	-	-
ECG: PSD-MF	0.9	-	-	-	0.4	0.6	0.9	-	-
ECG: PSD-HF	1.0	-	-	-	0.4	0.7	0.7	-	-
ECG: PSD-VLLF	0.6	-	-	-	0.3	0.4	0.7	-	-
TEB: Original	-	8.4	-	-	1.4	3.1	-	1.6	-
TEB: LF	-	8.9	-	-	1.4	3.8	-	1.5	-
TEB: RF	-	10.1	-	-	2.0	1.2	-	3.3	-
TEB: BRV	-	5.0	-	-	2.5	3.6	-	3.8	-
TEB: PSD	-	1.9	-	-	0.3	0.6	-	0.2	-
TEB: PSD-VLF	-	0.9	-	-	0.6	0.5	-	0.7	-
TEB: PSD-LF	-	0.8	-	-	1.0	1.0	-	1.1	-
TEB: PSD-MF	-	1.0	-	-	1.0	0.9	-	1.1	-
TEB: PSD-HF	-	2.3	-	-	0.7	0.9	-	0.6	-
TEB: PSD-VLLF	-	0.8	-	-	0.6	0.6	-	0.7	-
EDA-arm: Original	-	-	5.6	-	0.6	-	1.6	1.3	2.6
EDA-arm: Processed	-	-	9.8	-	1.3	-	2.4	2.5	3.5
EDA-arm: LF	-	-	8.9	-	0.6	-	1.5	1.3	4.0
EDA-arm: HF	-	-	7.8	-	0.6	-	0.8	0.8	3.8
EDA-arm: PSD	-	-	3.2	-	0.5	-	0.9	0.7	1.4
EDA-arm: PSD-LF	-	-	1.9	-	0.4	-	0.4	0.7	0.5
EDA-arm: PSD-HF	-	-	2.9	-	0.4	-	0.8	0.6	1.3
EDA-hand: Original	-	-	-	2.8	1.9	-	1.7	2.1	2.3
EDA-hand: Processed	-	-	-	11.6	3.9	-	6.0	6.9	9.2
EDA-hand: LF	-	-	-	6.1	1.5	-	1.5	1.9	2.7
EDA-hand: HF	-	-	-	6.6	1.4	-	1.7	2.0	3.2
EDA-hand: PSD	-	-	-	1.3	0.2	-	0.4	0.7	0.8
EDA-hand: PSD-LF	-	-	-	7.1	0.2	-	0.6	2.4	3.1
EDA-hand: PSD-HF	-	-	-	4.6	0.2	-	0.4	1.2	1.6

To go deeper into the analysis, Table 4 shows the top-40 selected features, again in the case of selecting a maximum of $N_{max} = 40$ features. In this case, we show the percentage of occurrence in the three best combinations of signals: the TEB alone, the TEB and the ECG, and the case of using all the biosignals. We can see that, in general, the mean baseline is one of the most frequent parameters. The most selected features from each signal in the case of considering all possible features in the GA are:

- From the ECG signal: the geometric mean of the HRV, the mean baseline of the RR, the logarithm of the SD of the RR, and the DFA1 of the HR.

- From the TEB signal: the average BR of the RF, the mean baseline of the BRV, and the minimum of the BRV.
- From the EDA measured in the hand: the mean baseline of the original measurement, and the mean baseline of the processed measurement.
- There are no features from the EDA measured in the hand which is used more than 40% of cases in the case of considering all possible biosignals in the GA. The most frequent one from this signal is the skewness of the processed measurement.

Table 4. Top-40 selected features from the different signal, and percentage of occurrence with $N_{max} = 40$ features.

Feature			Combination of Signals		
Signal	Measure	Parameter	TEB	ECG TEB	ECG TEB EDA
TEB	RF	Average BR	100%	0%	100%
TEB	BRV	Mean baseline	100%	0%	100%
EDA-hand	Original	Mean baseline	0%	100%	100%
EDA-hand	Processed	Mean baseline	0%	100%	100%
ECG	HRV	Geom. mean	0%	0%	100%
ECG	RR	Mean baseline	0%	0%	100%
ECG	RR	log(SD())	0%	0%	99%
ECG	RR	DFA1	0%	0%	98%
TEB	BRV	Minimum	100%	0%	94%
ECG	HR	Mean baseline	0%	0%	93%
ECG	HRV	Mean baseline	0%	0%	87%
ECG	RA	Mean baseline	0%	0%	68%
EDA-hand	LF	Mean baseline	0%	43%	56%
TEB	PSD-VLLF	Mean baseline	66%	0%	50%
TEB	PSD-MF	Mean baseline	97%	0%	50%
EDA-hand	Processed	Number SCR	0%	100%	49%
EDA-hand	HF	Mean baseline	0%	57%	48%
TEB	PSD-VLF	Mean baseline	72%	0%	48%
TEB	PSD-LF	Mean baseline	56%	0%	48%
ECG	Original	Skewness	0%	0%	44%
ECG	RA	Mean abs. dev.	0%	0%	40%
EDA-arm	Processed	Skewness	0%	8%	40%
TEB	PSD-HF	HF/LF	78%	0%	39%
TEB	LF	Mean baseline	100%	0%	37%
ECG	RA	SD	0%	0%	36%
TEB	Original	Mean baseline	100%	0%	36%
ECG	RR	25% Trm. mean	0%	0%	36%
TEB	PSD-LF	(LF+MF)/HF	25%	0%	35%
EDA-hand	Processed	PNS	0%	35%	35%
EDA-hand	Processed	NZC	0%	41%	34%
EDA-hand	Processed	PZC	0%	24%	33%
TEB	PSD-MF	MF/HF	5%	0%	33%
TEB	RF	Mean baseline	100%	0%	33%
TEB	LF	Percentile 75%	93%	0%	32%
EDA-hand	Processed	Maximum	0%	82%	31%
ECG	RR	Median	0%	0%	31%
EDA-hand	Processed	Minimum	0%	47%	27%
EDA-hand	Processed	Median	0%	100%	26%
ECG	RR	Geom. mean	0%	0%	25%
TEB	Original	Percentile 75%	16%	0%	23%

6. Discussion and Conclusion

Nowadays, activity recognition based on physiological signals is a relevant research field with a promising future. This paper presents an evaluation of the classification performance of different sensing modes ECG, TEB and EDA for detection of 4 different activities. The evaluation includes typical characterization features for the measured signal within each sensing mode. The characterization features included in the evaluation have been selected from a throughout review of the literature available. The evaluation has been done from several perspectives, the sensing mode perspective, the type of activity targeted and other parameters related to the feature extraction and classifier training. Consequently numerous conclusion can be derived from this work:

- In most of the relevant cases, the best results are obtained with a window length of 40 s. For the used database, the classifier that render the best results is the simplest ones, the LSLCs.
- When evaluating the combination of physiological signals which is better to correctly detect the type of activity, an LSLC trained with the feature set obtained when applying a GA considering all signals (TEB+ECG+EDA) achieves the lowest classification error probability (22.2%). In the case of the system trained with features selected from the ECG+TEB signal, the results are quite similar (24.5%), and there is no need to measure the EDA signal, making this choice very convenient for those cases in which we desire to pay attention to the simplicity of the acquisition system. That is, the comfort of the subject when there is no need to wear any glove or armband is higher, and the performance of the activity detection system is near the same.
- In addition, for each activity separately the feature set that provides the best results depends significantly on the activity under study. While for neutral activity and mental activity, the best result is obtained with ECG+TEB+EDA feature set, for emotional activity, the best result is obtained with TEB+EDA. Finally, the best result for physical activity is provided by the EDA feature set. This may be because the physical activity causes the activation of the sweat glands in a more meaningful way than the rest of the activities studied. In general, the signals working independently obtain worse results that when we make combinations between signals. Although it depends on the activity under study since in the case of physical activity the results are very similar using one or several signals. However, this does not happen in other cases in which the error is reduced in a remarkable way when combinations of signals are used in the training of the classifier. For the other type of activities, combining sensing modes provides similar or better performance than using only one type of sensing mode.
- The GA seems to be very useful in order to select the most relevant features, improving the results in terms of both complexity after training and error rate. From a total of 533 features, only 40 were necessary to achieve the minimum observed error. TEB signal seems to contain more useful information than the other signals.
- The results clearly suggest that the activity most easily identifiable is physical activity. Then the neutral, the emotional and finally the mental activity. This is due to the presence of misclassification between emotional and mental activities, as can be naturally expected.
- As a possible limitation of the study, we should consider that these conclusions might be different with other electronic devices. For instance, improvement on the textile based sensors or the use of gel-based classical sensors might improve the quality of the acquired signals, changing the usefulness of the measured features. Furthermore, the use of a more extensive database might overcome the generalization problems, allowing to obtain better results with more complex classifiers. In this sense, this paper does not try to propose a close solution but a methodology, and the comparison of the features and signals carried out might be conditioned to the actual textile sensor technology.

As a final conclusion, we have demonstrated the suitability of the GAs to select the best features among a wide dataset, containing most of the features identified as useful in the literature. The present study allows to extract significant conclusions concerning the information in each measurement, and determines a set of relevant measurements and features that can lead the research in future studies. On the other hand, the generalization capability of the classifiers has been identified as crucial in order to further improve the results in activity recognition through physiological signals, which opens new opportunities for researching within in the field.

Supplementary Materials: The following are available online at http://www.mdpi.com/1424-8220/19/24/5524/s1, Supplementary Data.

Author Contributions: Conceptualization, I.M.-H. and R.G.-P.; methodology, I.M.-H. and R.G.-P.; software, I.M.-H.; validation, I.M.-H. and R.-G.-P.; formal analysis, I.M.-H. and R.G.-P.; investigation, I.M.-H.; resources, I.M.-H and R.G.-P., data curation, I.M.-H and R.G.-P.; writing–original draft preparation, I.M.-H.; writing–review and editing, I.M.-H., R.G.-P., M.R.-Z. and F.S.; visualization, I.M.-H. and F.S.; supervision, I.M.-H. and F.S., project administration, M.R.-Z. and R.G.-P.; funding acquisition, M.R.-Z. and R.G.-P.

Funding: This research was funded by the Spanish Ministry of Economy and Competitiveness/FEDER under Project RTI2018-098085-B-C42.

Conflicts of Interest: The authors declare no conflict of interest.

Abbreviations

The following abbreviations are used in this manuscript:

ECG	Electrocardiogram
TEB	Thoracic Electrical Bioimpedance
EDA	Electrodermal Activity
HRV	Heart Rate Variability
SDNN	Standard Deviation of NN intervals
SDSD	Standard Deviation of Successive Differences
LSLC	Least Squares Linear Classifier
LSQC	Least Squares Quadratic Classifier
SVM	Support Vector Machine
LINSVM	Linear Support Vector Machine
RBFSVM	Radial Basis Function Support Vector Machine
MLP	Multi-Layer Perceptron
kNN	*k*-Nearest Neighbor
CDNN	Centroid Displacement-based Nearest Classifier
RF	Random Forest
GA	Genetic Algorithm
SSSP	Standard Set of Statistical Parameters

Appendix A. Features from the ECG Signal

The measurements used to characterize the ECG can be divided into two groups, those calculated in the *time domain* and those calculated in the *frequency domain*. Figure A1 shows the features extracted from the ECG signal. As we can appreciate, there are 83, 89 and 2 features for the time domain, frequency domain, and the mixed domain, respectively. That means that the total number of features calculated to characterize the ECG measurement is 174.

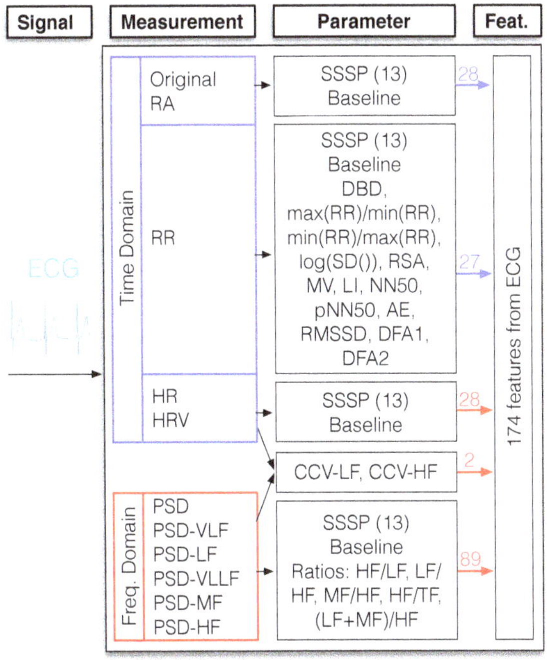

Figure A1. ECG-based feature extraction scheme.

Appendix A.1. Time-Domain

A total of five measurements were considered in the time-domain, directly or indirectly extracted from the QRS analysis:

- The original unprocessed signal [53,55] and the R wave Amplitude (RA) (amplitude of the different R waves in each window). The SSSPs and the baseline parameters were calculated for these measurements."
- The interval between successive Rs (RR) (time lapsed between successive R waves) [35,49,80]. Apart from the SSSP and the baseline, some special features have been extracted from the RR measurement:
  - Deep Breathing Difference (DBD), calculated as the difference between the maximum RR and minimum RR in the window under study [81,82].
  - Ratios maximum RR vs. minimum RR, that is, RRmax/RRmin and RRmin/RRmax [83].
  - Logarithm of the standard deviation of RR in the window under study.
  - Respiratory Sinus Arrhythmia (RSA), calculated as the quotient between the DBD and the mean value of the RR in the window under study. This measurement is related to the function of parasympathetic nervous during spontaneous ventilation [84].
  - Modal Value (MV), defined as the most frequent value in the RR intervals in the window under study [40].
  - Load Index (LI), based on the ratio between the number of occurrences of each Modal Value and DBD [40].
  - NN50, determined as the number of successive RR interval pairs differing by more than 50 ms [40,42,48].
  - pNN50, obtained dividing NN50 by the total number of RR intervals [40,42,48].

- Approximate Entropy (AE), originally proposed in [85], and applied to physiological data in [39,42,86].
- Root Mean Square of Successive Differences (RMSSD), determined by calculating the square root of the mean squared difference between consecutive RR intervals [34,48,49,87]. The RMSSD is the primary time domain used to estimate the high-frequency beat-to-beat variations that represent vagal regulatory activity [48].
- Two parameters of the Detrended Fluctuation Analysis (DFA) [85]. These two parameters (DFA1 and DFA2) have been used to quantify the presence or absence of fractal-like correlation properties of the heart period time series [39].

- Heart Rate (HR). It is measured as the number of pulses per unit of time, usually beats per minute (bpm). It is calculated as the inverse of the RR interval. It is obtained through the inverse of the RR interval. This parameter is highly important, as it is related to physical exercise, anxiety, sleep, illness, food intake, and drugs, among others. The increase or decrease on this speed is the answer of our body or mind condition [34,48]. The SSSPs and the baseline parameters were calculated from this measurement.
- Heart Rate Variability (HRV), which has been widely used to extract information about the status of the autonomic nervous system and emotions [23]. The work [88] provides a review of this measurement. In addition, numerous studies reveal the importance of this parameter [23,26,38,39,41,42,45,48,50,51,89]. We decided to obtain the HRV as proposed in [26], where the HRV is determined from a modified version of the HRV sampled at 256 Hz. Once the HRV is obtained, it is possible to extract different valuables features, using the SSSPs and the baseline parameter.

Many other measurements were found in the literature such as SDNN index, SDANN among others [34]. However, we did not use these measurements because they require at least 5 min to be calculated, since they are often calculated over a 24-h period.

Appendix A.2. Frequency-Domain

The other main group of measurements are evaluated in the frequency domain, through the Discrete Fourier Transform (DFT). The main measurements taken in this part were:

- Power Spectral Density (PSD) of the HRV signal is obtained using Welch's method [23,38,42].
- Power per Bands. From this PSD parameter, several frequency bands were considered: Very-Low-frequency (PSD-VLF), taken from 0.0033–0.05 Hz; Low Frequency (PSD-LF) from 0.05–0.08 Hz; Very-Low and Low-Frequency (PSD-VLLF) from 0.0033–0.08 Hz, Mid Frequency (PSD-MF) from 0.08–0.15 Hz. and High frequency (PSD-HF) from 0.15–0.5 Hz. These values were established taking into account several papers such as [23,26,34,36–38,40–48].

The parameters taken from these spectral measurements were the SSSPs, the baseline parameter, and a set of specific parameters related to ratios between the average power for the different bands: HF/LF, LF/HF, MF/HF, (LF+MF)/HF, and HF/TF, being TF the total power in all frequencies [23,43,44,46,47,49].

Appendix A.3. Mixed Domain

There were also two parameters taken from relationships between time and frequency parameters, denoted as Coefficients of Component Variance (CCV). The CCVs considered were the CCV-LF and the CCV-HF [39], and they were calculated as the square root of LF or HF power divided by the average HR.

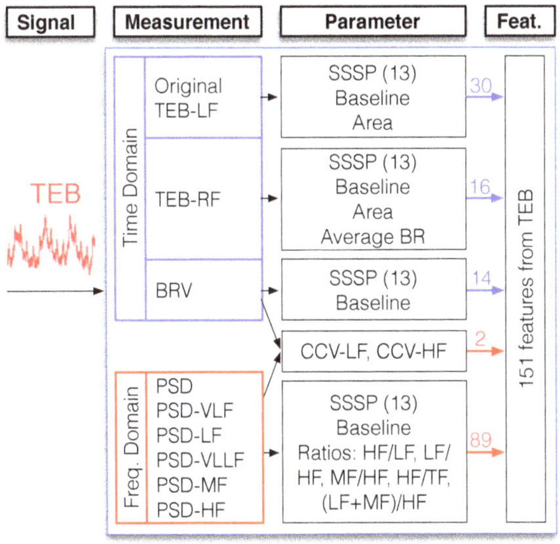

Figure A2. TEB-based feature extraction scheme for the classical set of features.

Appendix B. Features from the TEB Signal

The measurements used in order to extract the most relevant information from TEB signal follow a structure similar to the one described in the case of the ECG signal, being again divided into the time domain, frequency domain, and mixed domain features.

Figure A2 shows the features extracted from the TEB signal. There are 60 time domain features, 89 frequency domain features, and 2 mixed domain features. Therefore, the total number of features calculated to characterize the TEB signal is 151.

Appendix B.1. Time Domain

- TEB-Original Signal: The 13 SSSPs and the baseline parameter aforementioned are calculated from the TEB-Original signal. Apart from these parameters, the area was also calculated, using an approximated segment-based integral of the measurements via a trapezoidal method with unit spacing.
- TEB-LF: the original signal is low-pass filtered (LF block) with a cutoff frequency of 3 Hz, using an FIR filter with order $N_1 = 100$. Again, the 13 SSSPs and the baseline parameter are calculated.
- TEB-RF: Additionally, another new signal is obtained from TEB-LF. The first low pass filter (LF block) acts as an anti-aliasing filter, which allows the use of Interpolated Finite Impulse Response (IFIR) filters [90]. Thus, the output of this anti-aliasing filter is applied to a band-pass filter with cutoff frequencies of 0.1 Hz and 0.5 Hz with a stretch factor of $SF = 10$ and an order $N_2 = 5 \times F_{TEB} - N_1 = 400$, (being $F_{TEB} = 100$ Hz). We denominate TEB Respiration Frequency (TEB-RF) to the measurement obtained. The TEB-RF measurement was used to determine the Breathing Rate (BR). This parameter calculates the number of breaths per minute [91] using a peak detection algorithm. The parameters taken from this measurement, apart from the SSSPs, the baseline, and the area, include the average BR.
- Breath Rate Variability (BRV). Using the BR measured from the TEB-RF, we can calculate the Breath Rate Variability (BRV) in a similar way to HRV. The 13 SSSPs and the baseline parameter are calculated from this measurement.

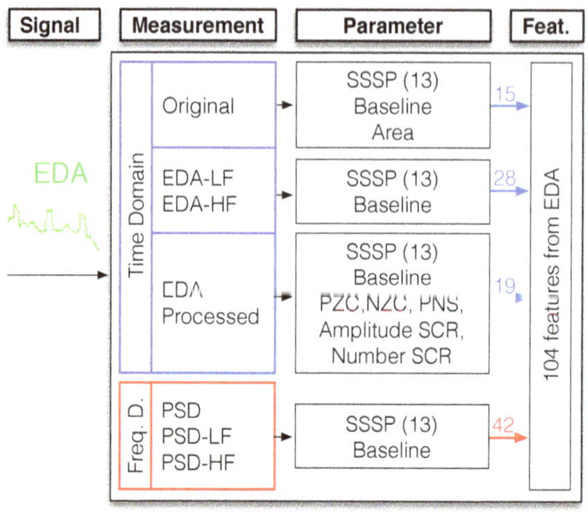

Figure A3. EDA-based feature extraction scheme.

Appendix B.2. Frequency Domain

The frequency features calculated from TEB are similar to those calculated from ECG. So, again the PSD of the original signal was calculated and applied to several filtered versions of the signal. Again, apart from standard parameters, power ratios were also evaluated.

Appendix B.3. Mixed Domain

The features considered are the CCV mixed-domain parameters (in this case, using the average BR to normalize the squared root of the energy).

Appendix C. Features from the EDA Signal

The structure used in order to extract the most relevant information from the EDA signal follows a structure similar to the one described in the case of the ECG signal and the TEB signal.

Figure A3 shows the measurements obtained from the EDA signal and the procedure used to extract the features. In total, we obtained 62 time domain features and 42 frequency domain features. Taking into account that the EDA was registered in both the hand and the arm (as was described above), we obtained 104 features from the arm EDA and 104 features from hand EDA.

Appendix C.1. Time Domain

We obtained four different time domain measurements:

- EDA-Original Signal. The 13 SSSPs and the baseline aforementioned parameter are calculated to the EDA-Original signal. The area is also calculated from this measurement, using an approximated integral of the time segments through a trapezoidal method with unit spacing.
- EDA-LF: The original signal is filtered with a 20-order low-pass FIR filter (LF block) with a cutoff frequency of 0.2 Hz [41]. The 13 SSSPs and the baseline parameter were evaluated.
- EDA-HF: A complementary filter is also applied to obtain the high frequency components (20-order high-pass FIR filter with a cutoff frequency of 0.2Hz), and the same parameters than those from the EDA-LF measurement are evaluated over the obtained EDA-HF measurement.
- EDA-Processed: The work [26] shows the steps to process EDA signal, for Skin Conductance Response (SCR) detection. The process consists in removing the mean value, resampling to 20 Hz, time differentiating, and filtering with a 20-order Bartlett window. From the processed EDA

measurement, typical parameters are extracted using the SSSPs and the baseline parameter, and also some specific parameters:

- Zero Crossing (ZC): Positive ZC (PZC), and Negative ZC (NZC) are interesting for the application at hand [26,41].
- Ratio or proportion of Negative Samples (PNS), evaluated as the quotient between the number of negative samples and the total number of samples [41].
- SCRs were evaluated analyzing the zero crossings in the processed EDA signal. The average amplitude of the SCR occurrences and the number of occurrences in the analysis window were used as parameters [26,42,54,91]. SCRs were determined by finding two consecutive zero-crossings, from negative to positive and from positive to negative.

Appendix C.2. Frequency Domain

The PSDs extracted from the EDA-Original signal, the EDA-LF and the EDA-HF, were taken as spectral measurements, using a Welch's overlapped segment averaging estimator. The SSSPs and the baseline parameter were calculated from these measurements.

References

1. Rueda, F.M.; Lüdtke, S.; Schröder, M.; Yordanova, K.; Kirste, T.; Fink, G.A. Combining Symbolic Reasoning and Deep Learning for Human Activity Recognition. In Proceedings of the 2019 IEEE International Conference on Pervasive Computing and Communications Workshops (PerCom Workshops), Kyoto, Japan, 11–15 March 2019; pp. 22–27.
2. Chen, L.M.; Nugent, C.D. Sensor-Based Activity Recognition Review. In *Human Activity Recognition and Behaviour Analysis*; Springer: Cham, Germany, 2019; pp. 23–47.
3. Tapia, E.M.; Intille, S.S.; Larson, K. Activity recognition in the home using simple and ubiquitous sensors. In Proceedings of the International Conference on Pervasive Computing, Linz and Vienna, Austria, 21–23 April 2004; pp. 158–175.
4. Kim, E.; Helal, S.; Cook, D. Human activity recognition and pattern discovery. *IEEE Pervasive Comput.* **2010**, *9*, 48–53. [CrossRef] [PubMed]
5. Bao, L.; Intille, S.S. Activity recognition from user-annotated acceleration data. In Proceedings of the International Conference on Pervasive Computing, Linz and Vienna, Austria, 21–23 April 2004; pp. 1–17.
6. Handley, T.E.; Lewin, T.J.; Perkins, D.; Kelly, B. Self-recognition of mental health problems in a rural Australian sample. *Aust. J. Rural Health* **2018**, *26*, 173–180. [CrossRef] [PubMed]
7. del R Millan, J.; Mouriño, J.; Franzé, M.; Cincotti, F.; Varsta, M.; Heikkonen, J.; Babiloni, F. A local neural classifier for the recognition of EEG patterns associated to mental tasks. *IEEE Trans. Neural Networks* **2002**, *13*, 678–686. [CrossRef] [PubMed]
8. Horlings, R.; Datcu, D.; Rothkrantz, L.J. Emotion recognition using brain activity. In Proceedings of the 9th International Conference on Computer Systems and Technologies and Workshop for PhD Students in Computing, Gabrovo, Bulgaria, 12–13 June 2008; p. 6.
9. Cloete, T.; Scheffer, C. Benchmarking of a full-body inertial motion capture system for clinical gait analysis. In Proceedings of the 2008 30th Annual International Conference of the IEEE Engineering in Medicine and Biology Society, Vancouver, BC, Canada, 20–25 August 2008; pp. 4579–4582.
10. Fong, D.T.P.; Chan, Y.Y. The use of wearable inertial motion sensors in human lower limb biomechanics studies: A systematic review. *Sensors* **2010**, *10*, 11556–11565. [CrossRef]
11. Kwapisz, J.R.; Weiss, G.M.; Moore, S.A. Activity recognition using cell phone accelerometers. *ACM SigKDD Explor. Newsletter* **2011**, *12*, 74–82. [CrossRef]
12. Mshali, H.; Lemlouma, T.; Moloney, M.; Magoni, D. A survey on health monitoring systems for health smart homes. *Int. J. Ind. Ergon.* **2018**, *66*, 26–56. [CrossRef]
13. Albanie, S.; Nagrani, A.; Vedaldi, A.; Zisserman, A. Emotion Recognition in Speech Using Cross-Modal Transfer in the Wild. Available online: https://arxiv.org/abs/1808.05561 (accessed on 12 December 2019).

14. Wang, S.H.; Phillips, P.; Dong, Z.C.; Zhang, Y.D. Intelligent facial emotion recognition based on stationary wavelet entropy and Jaya algorithm. *Neurocomputing* **2018**, *272*, 668–676. [CrossRef]
15. Mohino, I.; Goni, M.; Alvarez, L.; Llerena, C.; Gil-Pita, R. Detection of emotions and stress through speech analysis. In Proceedings of the Signal Processing, Pattern Recognition and Application-2013, Innsbruck, Austria, 12–14 February 2013; pp. 12–14.
16. Busso, C.; Deng, Z.; Yildirim, S.; Bulut, M.; Lee, C.M.; Kazemzadeh, A.; Lee, S.; Neumann, U.; Narayanan, S. Analysis of emotion recognition using facial expressions, speech and multimodal information. In Proceedings of the 6th International Conference on Multimodal Interfaces, State College, PA, USA, 13–15 October 2004; pp. 205–211.
17. Lymberis, A.; Olsson, S. Intelligent biomedical clothing for personal health and disease management: State of the art and future vision. *Telemed. J. e-Health* **2003**, *9*, 379–386. [CrossRef]
18. Wei, D.; Nagai, Y.; Jing, L.; Xiao, G. Designing comfortable smart clothing: For infants? health monitoring. *Int. J. Des. Creativity Innov.* **2019**, *7*, 116–128. [CrossRef]
19. Jerritta, S.; Murugappan, M.; Nagarajan, R.; Wan, K. Physiological signals based human emotion recognition: A review. In Proceedings of the 2011 IEEE 7th International Colloquium on Signal Processing and its Applications, Penang, Malaysia, 4–6 March 2011; pp. 410–415.
20. Agrafioti, F.; Hatzinakos, D.; Anderson, A.K. ECG pattern analysis for emotion detection. *IEEE Trans. Affective Comput.* **2012**, *3*, 102–115. [CrossRef]
21. Rattanyu, K.; Mizukawa, M. Emotion recognition based on ECG signals for service robots in the intelligent space during daily life. *J. Adv. Comput. Intell. Intell. Inf.* **2011**, *15*, 582–591. [CrossRef]
22. Lara, O.D.; Labrador, M.A. A survey on human activity recognition using wearable sensors. *IEEE Commun. Surv. Tutor.* **2013**, *15*, 1192–1209. [CrossRef]
23. McCraty, R.; Atkinson, M.; Tiller, W.A.; Rein, G.; Watkins, A.D. The effects of emotions on short-term power spectrum analysis of heart rate variability. *Am. J. Cardiol.* **1995**, *76*, 1089–1093. [CrossRef]
24. Neumann, S.A.; Waldstein, S.R. Similar patterns of cardiovascular response during emotional activation as a function of affective valence and arousal and gender. *J. Psychosomatic Res.* **2001**, *50*, 245–253. [CrossRef]
25. Mohino-Herranz, I.; Gil-Pita, R.; Ferreira, J.; Rosa-Zurera, M.; Seoane, F. Assessment of mental, emotional and physical stress through analysis of physiological signals using smartphones. *Sensors* **2015**, *15*, 25607–25627. [CrossRef] [PubMed]
26. Kim, K.H.; Bang, S.; Kim, S. Emotion recognition system using short-term monitoring of physiological signals. *Med. Biol. Eng. Comput.* **2004**, *42*, 419–427. [CrossRef]
27. Prokasy, W. *Electrodermal Activity in Psychological Research*; Elsevier: Amsterdam, The Netherlands, 2012.
28. Drachen, A.; Nacke, L.E.; Yannakakis, G.; Pedersen, A.L. Correlation between heart rate, electrodermal activity and player experience in first-person shooter games. In Proceedings of the 5th ACM SIGGRAPH Symposium on Video Games, Los Angeles, CA, USA, 28–29 July 2010; pp. 49–54.
29. Naveteur, J.; Baque, E.F.I. Individual differences in electrodermal activity as a function of subjects' anxiety. *Person. Ind. Differ.* **1987**, *8*, 615–626. [CrossRef]
30. Bellman, R. *Dynamic Programming*; Princeton University Press: Princeton, NJ, USA, 1957.
31. Myers, K.A.; Bello-Espinosa, L.E.; Symonds, J.D.; Zuberi, S.M.; Clegg, R.; Sadleir, L.G.; Buchhalter, J.; Scheffer, I.E. Heart rate variability in epilepsy: A potential biomarker of sudden unexpected death in epilepsy risk. *Epilepsia* **2018**, *59*, 1372–1380. [CrossRef]
32. Cai, J.; Liu, G.; Hao, M. The research on emotion recognition from ECG signal. In Proceedings of the 2009 International Conference on Information Technology and Computer Science, Kiev, Ukraine, 25–26 July 2009; pp. 497–500.
33. Rumpa, L.D.; Wibawa, A.D.; Attamimi, M.; Sampelawang, P.; Purnomo, M.H.; Palelleng, S. Analysis on Human Heart Signal during Sad Video Stimuli using Heart Rate Variability Triangular Index (HRVi). In Proceedings of the 2018 International Conference on Computer Engineering, Network and Intelligent Multimedia (CENIM), Surabaya, Indonesia, 26–27 November 2018; pp. 25–28.
34. Malik, M.; Bigger, J.T.; Camm, A.J.; Kleiger, R.E.; Malliani, A.; Moss, A.J.; Schwartz, P.J. Heart rate variability: Standards of measurement, physiological interpretation, and clinical use. *Eur. Heart J.* **1996**, *17*, 354–381. [CrossRef]
35. Cripps, T.; Malik, M.; Farrell, T.; Camm, A. Prognostic value of reduced heart rate variability after myocardial infarction: clinical evaluation of a new analysis method. *Br. Heart J.* **1991**, *65*, 14–19. [CrossRef]

36. Healey, J.A.; Picard, R.W. Detecting stress during real-world driving tasks using physiological sensors. *IEEE Trans. Intell. Transp. Syst.* **2005**, *6*, 156–166. [CrossRef]
37. Picard, W.; Healey, J.A. Wearable and Automotive Systems for Affect Recognition from Physiology. Available online: http://citeseerx.ist.psu.edu/viewdoc/summary?doi=10.1.1.30.1519 (accessed on 11 December 2019).
38. Dishman, R.K.; Nakamura, Y.; Garcia, M.E.; Thompson, R.W.; Dunn, A.L.; Blair, S.N. Heart rate variability, trait anxiety, and perceived stress among physically fit men and women. *Int. J. Psychophysiol.* **2000**, *37*, 121–133. [CrossRef]
39. Vuksanović, V.; Gal, V. Heart rate variability in mental stress aloud. *Med. Eng. Phys.* **2007**, *29*, 344–349. [CrossRef] [PubMed]
40. Tkacz, E.; Komorowski, D. An examination of some heart rate variability analysis indicators in the case of children. In Proceedings of the 15th Annual International Conference of the IEEE Engineering in Medicine and Biology Societ, San Diego, CA, USA, 31 October 1993; pp. 794–795.
41. Soleymani, M.; Lichtenauer, J.; Pun, T.; Pantic, M. A multimodal database for affect recognition and implicit tagging. *IEEE Trans. Affect. Comput.* **2012**, *3*, 42–55. [CrossRef]
42. Kim, J.; André, E. Emotion recognition based on physiological changes in music listening. *IEEE Trans. Pattern Anal. Mach. Intell.* **2008**, *30*, 2067–2083. [CrossRef]
43. Billman, G.E. The LF/HF ratio does not accurately measure cardiac sympatho-vagal balance. *Front. Physiol.* **2013**, *4*, 26. [CrossRef]
44. Piccirillo, G.; Vetta, F.; Fimognari, F.; Ronzoni, S.; Lama, J.; Cacciafesta, M.; Marigliano, V. Power spectral analysis of heart rate variability in obese subjects: Evidence of decreased cardiac sympathetic responsiveness. *Int. J. Obes. Relat. Metab. Disord.* **1996**, *20*, 825–829.
45. Malarvili, M.; Rankine, L.; Mesbah, M.; Colditz, P.; Boashash, B. Heart rate variability characterization using a time-frequency based instantaneous frequency estimation technique. In Proceedings of the 3rd Kuala Lumpur International Conference on Biomedical Engineering 2006, Kuala Lumpur, Malaysia, 11–14 December 2006; pp. 455–459.
46. Longin, E.; Schaible, T.; Lenz, T.; König, S. Short term heart rate variability in healthy neonates: normative data and physiological observations. *Early Hum. Dev.* **2005**, *81*, 663–671. [CrossRef]
47. Winchell, R.J.; Hoyt, D.B. Spectral analysis of heart rate variability in the ICU: A measure of autonomic function. *J. Surg. Res.* **1996**, *63*, 11–16. [CrossRef]
48. Von Borell, E.; Langbein, J.; Després, G.; Hansen, S.; Leterrier, C.; Marchant-Forde, J.; Marchant-Forde, R.; Minero, M.; Mohr, E.; Prunier, A.; et al. Heart rate variability as a measure of autonomic regulation of cardiac activity for assessing stress and welfare in farm animals—A review. *Physiol. Behav.* **2007**, *92*, 293–316. [CrossRef]
49. Pan, J.; Tompkins, W.J. A real-time QRS detection algorithm. *IEEE Trans. Biomed. Eng.* **1985**, *32*, 230–236. [CrossRef] [PubMed]
50. Haag, A.; Goronzy, S.; Schaich, P.; Williams, J. Emotion recognition using bio-sensors: First steps towards an automatic system. In Proceedings of the Tutorial and research workshop on affective dialogue systems, Kloster Irsee, Germany, 14–16 June 2004; pp. 36–48.
51. Brennan, M.; Palaniswami, M.; Kamen, P. Do existing measures of Poincare plot geometry reflect nonlinear features of heart rate variability? *IEEE Trans. Biomed. Eng.* **2001**, *48*, 1342–1347. [CrossRef] [PubMed]
52. Picard, R.W.; Vyzas, E.; Healey, J. Toward machine emotional intelligence: Analysis of affective physiological state. *IEEE Trans. Pattern Anal. Mach. Intell.* **2001**, *23*, 1175–1191. [CrossRef]
53. Rigas, G.; Katsis, C.D.; Ganiatsas, G.; Fotiadis, D.I. A user independent, biosignal based, emotion recognition method. In Proceedings of the 11th International Conference on User Modeling, Corfu, Greece, 25–29 July 2007; pp. 314–318.
54. Katsis, C.D.; Katertsidis, N.; Ganiatsas, G.; Fotiadis, D.I. Toward emotion recognition in car-racing drivers: A biosignal processing approach. *IEEE Trans. Syst. Man Cybern. Part A Syst. Humans* **2008**, *38*, 502–512. [CrossRef]
55. Maaoui, C.; Pruski, A.; Abdat, F. Emotion recognition for human-machine communication. In Proceedings of the 2008 IEEE/RSJ International Conference on Intelligent Robots and Systems, Nice, France, 22–26 September 2008; pp. 1210–1215.
56. Lackner, H.K.; Weiss, E.M.; Hinghofer-Szalkay, H.; Papousek, I. Cardiovascular effects of acute positive emotional arousal. *Appl. Psychophysiol. Biofeedback* **2014**, *39*, 9–18. [CrossRef] [PubMed]

57. Wu, G.; Liu, G.; Hao, M. The analysis of emotion recognition from GSR based on PSO. In Proceedings of the 2010 International Symposium on Intelligence Information Processing and Trusted Computing, Huanggang, China, 28–29 October 2010; pp. 360–363.
58. Caruelle, D.; Gustafsson, A.; Shams, P.; Lervik-Olsen, L. The use of electrodermal activity (EDA) measurement to understand consumer emotions–A literature review and a call for action. *J. Bus. Res.* **2019**, *104*, 146–160. [CrossRef]
59. Hernandez, J.; Riobo, I.; Rozga, A.; Abowd, G.D.; Picard, R.W. Using electrodermal activity to recognize ease of engagement in children during social interactions. In Proceedings of the 2014 ACM International Joint Conference on Pervasive and Ubiquitous Computing, Seattle, WC, USA, 13–17 September 2014; pp. 307–317.
60. Seoane, F.; Ferreira, J.; Alvarez, L.; Buendia, R.; Ayllón, D.; Llerena, C.; Gil-Pita, R. Sensorized garments and textrode-enabled measurement instrumentation for ambulatory assessment of the autonomic nervous system response in the atrec project. *Sensors* **2013**, *13*, 8997–9015. [CrossRef]
61. Gross, J.J.; Levenson, R.W. Emotion elicitation using films. *Cognit. Emot.* **1995**, *9*, 87–108. [CrossRef]
62. Merck, M. *America First: Naming the Nation in US Film*; Routledge: Abingdon, UK, 2007.
63. Clasen, M. Vampire apocalypse: A biocultural critique of Richard Matheson's I Am Legend. *Philosophy Lit.* **2010**, *34*, 313–328. [CrossRef]
64. von Jagow, B. Representing the Holocaust, Kertész's Fatelessness and Benigni's La vita è bella. In *Imre Kertész and Holocaust Literature*; Purdue University Press: West Lafayette, IN, USA, 2005; p. 76.
65. Megías, C.F.; Mateos, J.C.P.; Ribaudi, J.S.; Fernández-Abascal, E.G. Validación española de una batería de películas para inducir emociones. *Psicothema* **2011**, *23*, 778–785.
66. Fenton, H.; Grainger, J.; Castoldi, G.L. *Cannibal Holocaust: And the Savage Cinema of Ruggero Deodato*; Fab Press: Surrey, UK, 1999.
67. Denot-Ledunois, S.; Vardon, G.; Perruchet, P.; Gallego, J. The effect of attentional load on the breathing pattern in children. *Int. J. Psychophysiol.* **1998**, *29*, 13–21. [CrossRef]
68. Van Trees, H.L. *Detection, Estimation, and Modulation Theory*; John Wiley & Sons: Hoboken, NJ, USA, 2004.
69. Vapnik, V.N.; Vapnik, V. *Statistical Learning Theory*; Wiley: New York, NY, USA, 1998.
70. Fix, E.; Hodges, J.L. Discriminatory analysis. Nonparametric discrimination: Consistency properties. *Int. Stat. Rev.* **1989**, *57*, 238–247. [CrossRef]
71. Nguyen, B.P.; Tay, W.L.; Chui, C.K. Robust biometric recognition from palm depth images for gloved hands. *IEEE Trans. Hum. Mach. Syst.* **2015**, *45*, 799–804. [CrossRef]
72. Breiman, L. Random Forests. *Mach. Learn.* **2001**, *45*, 5–32. [CrossRef]
73. Kohavi, R.; John, G.H. Wrappers for feature subset selection. *Artif. Intell.* **1997**, *97*, 273–324. [CrossRef]
74. Holland, J.H. *Adaptation in Natural and Artificial Systems: An Introductory Analysis with Applications to Biology, Control, and Artificial Intelligence*; The University of Michigan Press: Ann Arbor, MI, USA, 1975.
75. Golberg, D.E. Genetic algorithms in search, optimization, and machine learning. *Addion Wesley* **1989**, *1989*, 102.
76. Zhuo, L.; Zheng, J.; Li, X.; Wang, F.; Ai, B.; Qian, J. A genetic algorithm based wrapper feature selection method for classification of hyperspectral images using support vector machine. In Proceedings of the Geoinformatics 2008 and Joint Conference on GIS and Built Environment: Classification of Remote Sensing Images. International Society for Optics and Photonics, Guangzhou, China, 28–29 June 2008; p. 71471J.
77. Ferreira, F.L.; Cardoso, S.; Silva, D.; Guerreiro, M.; de Mendonça, A.; Madeira, S.C. Improving Prognostic Prediction from Mild Cognitive Impairment to Alzheimer's Disease Using Genetic Algorithms. In Proceedings of the 11th International Conference on Practical Applications of Computational Biology & Bioinformatics, Orto, Portugal, 21–23 June 2017; pp. 180–188.
78. Bautista-Durán, M.; Garía Gómez, J.; Gil-Pita, R.; Mohíno-Herranz, I.; Rosa-Zurera, M. Energy-efficient acoustic violence detector for smart cities. *Int. J. Comput. Intell. Syst.* **2017**, *10*, 1298–1305.
79. Westfall, P.H.; Young, S.S. *Resampling-based Multiple Testing: Examples and Methods for P-Value Adjustment*; John Wiley & Sons: Hoboken, NI, USA, 1993.
80. Thuraisingham, R. Preprocessing RR interval time series for heart rate variability analysis and estimates of standard deviation of RR intervals. *Comput. Meth. Programs Biomed.* **2006**, *83*, 78–82. [CrossRef]
81. Ekholm, E.M.; Piha, S.J.; Erkkola, R.U.; Antila, K.J. Autonomic cardiovascular reflexes in pregnancy. A longitudinal study. *Clin. Autonomic Res.* **1994**, *4*, 161–165. [CrossRef]

82. Sathyaprabha, T.; Satishchandra, P.; Netravathi, K.; Sinha, S.; Thennarasu, K.; Raju, T. Cardiac autonomic dysfunctions in chronic refractory epilepsy. *Epilepsy Res.* **2006**, *72*, 49–56. [CrossRef]
83. Sundkvist, G.; O Almér, L.; Lilja, B. Respiratory influence on heart rate in diabetes mellitus. *Br. Med. J.* **1979**, *1*, 924. [CrossRef] [PubMed]
84. Loula, P.; Jäntti, V.; Yli-Hankala, A. Respiratory sinus arrhythmia during anaesthesia: Assessment of respiration related beat-to-beat heart rate variability analysis methods. *Int. J. Clin. Monit. Comput.* **1997**, *14*, 241–249. [CrossRef] [PubMed]
85. Richman, J.S.; Moorman, J.R. Physiological time-series analysis using approximate entropy and sample entropy. *Am. J. Physiol. Heart Circul. Physiol.* **2000**, *278*, H2039–H2049. [CrossRef] [PubMed]
86. Melillo, P.; Bracale, M.; Pecchia, L. Nonlinear Heart Rate Variability features for real-life stress detection. Case study: Students under stress due to university examination. *Biomed. Eng. Online* **2011**, *10*, 1. [CrossRef]
87. Yoo, S.K.; Lee, C.K.; Park, Y.J.; Kim, N.H.; Lee, B.C.; Jeong, K.S. Neural network based emotion estimation using heart rate variability and skin resistance. In Proceedings of the International Conference on Natural Computation, Changsha, China, 27–29 August 2005; pp. 818–824.
88. Acharya, U.R.; Joseph, K.P.; Kannathal, N.; Lim, C.M.; Suri, J.S. Heart rate variability: A review. *Med. Biol. Eng. Comput.* **2006**, *44*, 1031–1051. [CrossRef]
89. Soleymani, M.; Chanel, G.; Kierkels, J.J.; Pun, T. Affective ranking of movie scenes using physiological signals and content analysis. In Proceedings of the 2nd ACM workshop on Multimedia semantics, British Columbia, BC, Canada, 31 October 2008; pp. 32–39.
90. Mehrnia, A.; Willson, A.N. On optimal IFIR filter design. In Proceedings of the 2004 IEEE International Symposium on Circuits and Systems (IEEE Cat. No. 04CH37512), Vancouver, BC, Canada, 23–26 May 2004; pp. 3–133.
91. Hahn, G.; Sipinkova, I.; Baisch, F.; Hellige, G. Changes in the thoracic impedance distribution under different ventilatory conditions. *Physiol. Meas* **1995**, *16*, A161. [CrossRef]

sensors

MDPI

Article

Multimodal Approach for Emotion Recognition Based on Simulated Flight Experiments

Válber César Cavalcanti Roza [1,2,*] and Octavian Adrian Postolache [1]

[1] Instituto Universitário de Lisboa (ISCTE-IUL) and Instituto de Telecomunicações (IT-IUL), Av. das Forças Armadas, 1649-026 Lisbon, Portugal; octavian.postolache@gmail.com

[2] Universidade Federal do Rio Grande do Norte (UFRN), Av. Sen. Salgado Filho, 3000, Candelária, Natal, RN 59064-741, Brazil

* Correspondence: valbercesar@gmail.com

Received: 18 October 2019; Accepted: 9 December 2019; Published: 13 December 2019

Abstract: The present work tries to fill part of the gap regarding the pilots' emotions and their bio-reactions during some flight procedures such as, takeoff, climbing, cruising, descent, initial approach, final approach and landing. A sensing architecture and a set of experiments were developed, associating it to several simulated flights ($N_{flights} = 13$) using the Microsoft Flight Simulator Steam Edition (FSX-SE). The approach was carried out with eight beginner users on the flight simulator ($N_{pilots} = 8$). It is shown that it is possible to recognize emotions from different pilots in flight, combining their present and previous emotions. The cardiac system based on Heart Rate (HR), Galvanic Skin Response (GSR) and Electroencephalography (EEG), were used to extract emotions, as well as the intensities of emotions detected from the pilot face. We also considered five main emotions: happy, sad, angry, surprise and scared. The emotion recognition is based on Artificial Neural Networks and Deep Learning techniques. The Root Mean Squared Error (RMSE) and Mean Absolute Error (MAE) were the main methods used to measure the quality of the regression output models. The tests of the produced output models showed that the lowest recognition errors were reached when all data were considered or when the GSR datasets were omitted from the model training. It also showed that the emotion *surprised* was the easiest to recognize, having a mean RMSE of 0.13 and mean MAE of 0.01; while the emotion *sad* was the hardest to recognize, having a mean RMSE of 0.82 and mean MAE of 0.08. When we considered only the higher emotion intensities by time, the most matches accuracies were between 55% and 100%.

Keywords: emotion recognition; physiological sensing; multimodal sensing; deep learning; flight simulation

1. Introduction

With the growth of air safety and accident prevention, especially in the mechanical–structural and avionics aspects, a gap of probable cause of accidents is emerging, which can justify the occurrence of several unwanted situations. This can be referred to as the relationship between emotions and aviation accidents caused by human failure.

The development of research about the relation between emotions and aviation activities is quite new and is mainly based on preliminary and final accident reports. It was important to show the real need of improvements and strategies regarding emotion effects in risky situations of a real flight, mainly on take off, approach and landing.

To know how important are the studies of emotions over the aviation contexts, we first need to understand emotion definitions. Emotion is led by the brain and it can sometimes be the result of chemical processes that join several internal and external factors to produce an output or response that reflects an emotional state [1]. This response can also reflects some physiological changes in our human body [2]. Some emotions e.g., the primary emotion "anger" plays a fundamental role in many cases, such as fear and trust, that are directly related to protection, defense and maintenance of life.

Several methods and techniques can be applied to perform emotion recognition through the use of a couple of hardware devices and software such as: analysis of emotional properties based on two physiological data such as, ECG and EEG [3]; unified system for efficient discrimination of positive and negative emotions based on EEG data [4]; automatic recognizer of the facial expression around the eyes and forehead based on Electrooculography (EOG) data giving support to emotion recognition task [5]; use of GSR and ECG data to develop a study to examine the effectiveness of Matching Pursuit (MP) algorithm in emotion recognition, using mainly PCA to reduce the features dimensionality and Probabilistic Neural Network (PNN) as the recognition technique [6]; emotion recognition system based on physiological data using ECG and respiration (RSP) data, recorded simultaneously by a physiological monitoring device based on wearable sensors [7]; emotions recognition using EEG data and also performed an analyze about the impact of positive and negative emotions using SVM and RBF as the recognition methods [8]; new approach to emotion recognition based on EEG and classification method using Artificial Neural Networks (ANN) with features analysis based on Kernel Density Estimation (KDE) [9]; an application that stores several psychophysiological data based on HR, ECG, SpO2 and GSR, that were acquired while the users watched advertisements about smoking campaigns [10]; experiments based on flight simulator to developed a multimodal sensing architecture to recognize emotions using three different techniques for biosignal acquisitions [11]; multimodal sensing system to identify emotions using different acquisition techniques, based on photo presentation methodology [12]; real-time user interface with emotion recognition that depends on the need for skill development to support a change in the interface paradigm to one that is more human centered [13]; recognize emotions through psychophysiological sensing using a multiple-fusion-layer based on ensemble classifier of stacked auto encoder (MESAE) [14].

In addition, it is also possible to present some research that is more related to emotion analysis e.g., the use of the Friedman test to verify whether the work on exposure and emotional identification influences helps to decrease the levels of anxiety and depression [15]; emotion recognition system based on cross-correlation and the Flowsense database [16]; derived features based on bi-spectral analysis for quantification of emotions using a Valence-Arousal emotion model, to get a way of gaining phase information by detecting phase relationships between frequency components and characterization of the non-Gaussian information from EEG data [17]; a novel real-time subject-dependent algorithm using Stability Intra-class Correlation Coefficient (ICC) with the most stable features that gives a better accuracy than other available algorithms when it is crucial to have only one training session [18]; analysis of emotion recognition techniques used in existing systems to enhance ongoing research on the improvement of tutoring adaptation [19]; and the ensemble deep learning framework by integrating multiple stacked auto-encoder with parsimonious structure to reduce the model complexity and improve the recognition accuracy using physiological feature abstractions [20].

In the present work, we mainly studied the multimodal or multisensing architecture, processing, feature extraction and emotion recognition, regarding the pilots' feelings during a couple of simulated flights. These "pilots" in command, were represented by the beginner users of a flight simulator (not real pilots), following a sequence of steps during the flight experiments. The result of this work can also be applied to several workplaces and contexts e.g., administrative sectors [21], in aviation companies/schools [11] and in urban areas [16], among others.

Main Motivation and Contribution

Among a broad set of possible applications of the developed sensing architecture, the use of emotion recognition applied to an aviation context was the chosen one.

In 2017, Boeing presented a statistical summary [22], about commercial jet airplane accidents confirmed in worldwide operations for 1959 through 2016. It considered airplanes that were heavier than 60,000 pounds maximum gross weight. There was a very clear statistical analysis, in which it was possible to note the impressive evolution of aviation safety along the past years. As well as Boeing, the International Civil Aviation Organization (ICAO) also presented a similar report considering the period between 2008 and 2018, showing the same evolution of aviation safety along this period [23]. Every year, aviation has become safer, reaching lower levels of accidents with fatalities including hull losses or not. Although, there are no reasons to completely relax, because there are other problems to solve: the psychophysiological aspect inside a real flight operation.

According to several reports from the last 5 years, it is easy to observe that the main causes of these accidents were human failures, which some of it were also associated with human emotions. Based on that, we can note that aviation safety is facing a new age of accident factors i.e., the "age" of aviation accidents caused by human failure and it is quite a new and extremely important aspect that might be considered. The lack of a proper attention can provoke many results e.g., serious injuries and fatal accidents. The main causes of these accidents are: stress, drugs, fatigue, high workload and emotional problems [24].

Therefore, this work presents a practical contribution regarding the *on flight* phase, including data acquisition, processing, storage and emotion recognition, analyzing it in offline mode, i.e., non real-time recognition.

2. Proposed Multimodal Sensing System

The multimodal sensing approach it is not a new architecture or new method to aim for a recognition system, but it is a more robust and powerful approach to be applied in situations in which a low amount of inputs (or channels) are not sufficient to reach a good recognition accuracy along the time. This approach is based on several channels (inputs) that come mainly from different sources of data. It is sometimes challenging for researchers due the time and multi sampling rate synchronization.

For some research based on contexts like emotion recognition based on physiological data, it is not recommended to use only one data e.g., heart rate variability, to accurately detect emotions, because it can reflect emotions only in strong or intense emotional situations [25]. According to some studies, when an extended number of physiological data are considered, better results can be reached.

2.1. Flight Experiment

A total of 13 simulated flights ($N_{flights} = 13$) and eight beginner users on flight simulator ($N_{pilots} = 8$) were considered, using the *Microsoft Flight Simulator Steam Edition* (FSX-SE). These flights were labeled as: RC1, RC2, RC3, GC1, GC3, LS1, LS2, VC1, VC2, CR1, CR3, CLX and CL3.

The proposed experiment corresponds to the human behaviour study of the users (pilot in command) along some proposed simulated flight tasks such as: Take off (Task 1), climbing (Task 2), route flight (cruise navigation) (Task 3), descend (Task 4), approach (Task 5), final approach (Task 6) and landing (Task 7). The environment's setup from the main experiment, was the result of two initial Proof of Concepts (PoCs). Several improvements from these PoCs are: a large screen to improve the immersive experience during the simulation; addition of a separated computer to run the flight simulator and record facial emotions; the user must only use the joystick during the experimental flight and must use only one hand to control the flight; the GSR sensors were placed on the free hand, i.e., without movements to avoid motion artefacts; a microcontroller was used to acquire the HR data from HR device (e.g., Arduino board); the supervisor used two softwares, one to receive HR and GSR data from Bluetooth communication, and another to receive the Bluetooth data from EEG device; also a video camera was used to record the users' body gestures.

The users were trained before the experiment regarding the flight tasks and procedures. During the main experiment, they had no contact with the experiment supervisor, who only interfered before and after each simulation. It was also recommended to the users to avoid to talk and to move the hand with GSR electrodes.

All main experiments and training were executed on Visual Meteorological Condition (VMC) and minimum navigation altitude of 1800ft (feet MSL). For each user, a maximum of three flights were executed. The used airplane for this main experiment was the default aircraft model *Cessna 172SP Skyhawk*. Furthermore, the route used in this experiment, was executed with the airplane Cessna 172SP and it have almost 8.4 nm (Nautical Miles) of distance from Lisbon International Airport (ICAO LPPT/374ft/THD ELEV 378ft MSL) to Alverca (ICAO LPAR/11ft/THD ELEV 15ft MSL), as shown in Figure 1.

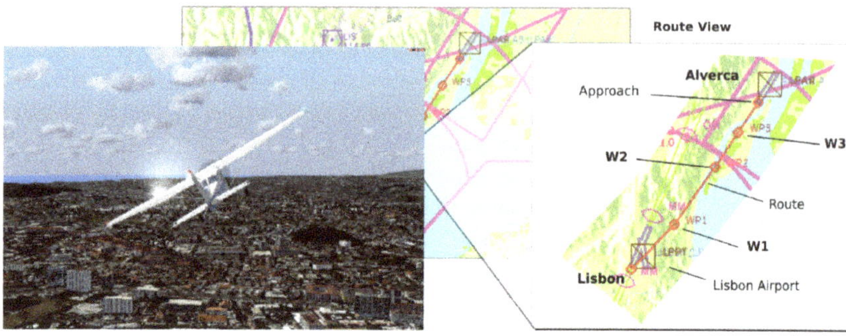

Figure 1. Airplane Cessna 172SP (**left**); flight route (red line) of the experiment (**right**).

2.2. User Profile

The experiment considered users (not real pilots) of both genders between 21 and 40 years old. Considering the 13 valid flights, nine were executed by male users and four were executed by female users.

Regarding the user experience in experimental context, one male user reported to have a more deep experience in flight simulation; the other male users reported to have more experience with electronic games and all female users reported to have low experience in flight simulators and electronic games.

2.3. Acquisition Devices

The multimodal data acquisition was based on Heart Rate (HR), Galvanic Skin Response (GSR) and electroencephalography (EEG). The emotion monitoring system includes a set of smart sensors such as: two shimmer3-GSR+, one Medlab-Pearl100 and one Enobio-N8.

The two Shimmer3-GSR+ units were the devices used to, acquire the GSR data and to act as an auxiliary head shaking indicator, using its embedded accelerometer. It includes: 1 channel GSR (Analog); the measurement range: 10 k and 4.7 MΩ (0.2–100 µS); frequency range: DC-15.9 Hz; input protection RF/EMI filtering, current limiting; auxiliary input: 2 channel analog/I2C; digital input: via 3.5 mm; 24 MHz MSP430 CPU with a precision clock subsystem; 10 DoF inertial sensing via accelerometer integrated, gyroscope, magnetometer and altimeter; low power consumption, light weight and small form factor; also perform the analog to digital conversion and readily connects via Bluetooth or local storage via micro SD card. Furthermore, it is also a highly configurable which can be used in a variety of data capture scenarios [26].

The HR data was acquired by the Medlab-Pearl100 device. It is considered an excellent artefact0suppression device due to PEARL-technology and includes: a compact, portable and attractive design; crisp, easily readable TFT colour display; reliably measures SpO2; pulse rate, and pulse strength; integrated 100 h trend memory; integrated context sensitive help system; intuitive, multi-language user interface; works on mains and from integrated battery; full alarm system with adjustable alarm limits; usable from neonates to adults [27].

To acquire the EEG data, the Enobio Toolkit was used. It is a wearable toolkit with a wireless electrophysiology sensor system for the recording of EEG. Using the Neuroelectrics headcap toolkit (having several dry and wet electrodes), the Enobio-N8 is ideal for out-of-the-lab applications. It comes integrated with an intuitive, powerful user interface for easy configuration, recording and visualization of 24 bit EEG data at 500 sampling rate, including spectrogram and 3D visualization in real time of spectral features. It is ready for research or clinical use. In addition to EEG, triaxial accelerometer data is automatically collected. You can also use a microSD card to save data offline in Holter mode; and as like as Shimmer device, it can use Bluetooth to transmit real time data too [28].

2.4. Facial Emotion Sensing

During the experiment, the users' faces and the flights along the experiments were recorded and outputs were processed after the experiment. To do this, two softwares were used: the *OBS Studio*, to record the flight and face at same time in a synchronized manner; and the software *Face Reader v.7*, a software marketed by Noldus (www.noldus.com) used to recognize the emotions based on the face recording. The last one considers seven emotions: neutral, happy, sad, angry, surprised, scared and disgust.

In offline analysis and processing, the emotions neutral and disgust were omitted. In these experiments, the Face Reader output a neutral emotion as a main emotion along each flight, which seems unrealistic because it almost omitted the amplitudes of another relevant emotions. It was confirmed by the users; they noted to not feel neutral most of the time. For this reason, we decided to omit the neutral emotion in this work. Regarding the disgust emotion, it was also omitted due to not being directly related to the flight context as confirmed by the users who they said that did not feel disgust along the flights.

Some users' faces captured during the main experiment are shown in Figure 2, and it is possible to see some different reactions along the simulated flights.

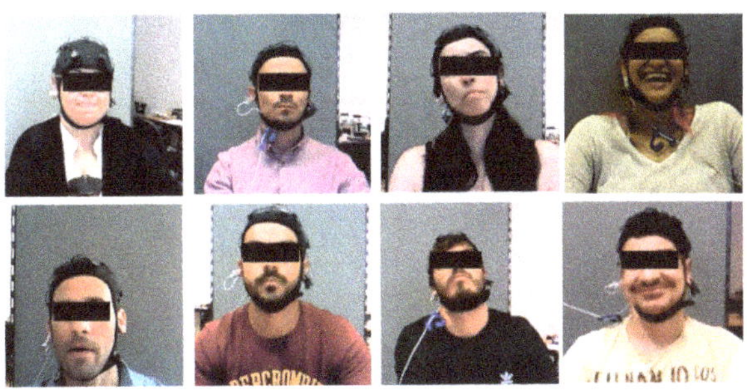

Figure 2. Face recording of some users during experiment.

The efficiency of the Face Reader software is shown in several researches and publications, being used as a reference regarding to emotion detection from facial expressions on several contexts and applications [29–31].

2.5. Physiological Sensing

The proposed multimodal sensing system considered three methods: Heart Rate (HR), Galvanic Skin Response (GSR) and Electroencephalography (EEG). To acquire these, 11 Ag/AgCl dry electrodes and one earclip were used: eight electrodes placed on the scalp (EEG), one placed on the earlobe (EEG reference), one placed on earlobe (HR) and two on the hand of the user (GSR).

The GSR data is based on Electrodermal Activity (EDA) and refers to the electrical resistance between two sensors when a very weak current occurs passed between them. It is typically acquired from the hands or fingers [6]. In this work, it was acquired by the Shimmer3-GSR+ unit, which can measure activity, emotional engagement and psychological arousal in lab scenarios and in remote capture scenarios that are set outside of the lab. It was recommended that these electrodes be kept immobile during the experiment to avoid an additional motion artifacts in GSR data.

Regarding EEG, some studies showed that it is very difficult to find the specific region on the scalp where the brain activity is sufficiently high to detect emotional states [32,33]. According to several studies, the prefrontal cortex or frontal lobe (located near the front of the head) is more involved with cognition and in decision making from emotional responses [34,35]. The 10–20 system or International 10–20 system was the method used to describe and apply the location of scalp electrodes. This way, to better detect emotion artifacts from the scalp, the electrodes were placed on that recommended area: Fp1 (channel 1), F3 (channel 2), C3 (channel 3), T7 (channel 4), Fp2 (channel 5), F4 (channel 6), C4 (channel 7) and T8 (channel 8). The EEG reference electrode (EEGR) was placed on the user's earlobe. It frequency aimed our choice to use the beta rhythms (or band) in this experiment [32,36].

Figure 3, shows the electrodes position used during experiment. Note the use of the frontal cortex to acquire EEG data, which the beta rhythm (β-band) were considered i.e., brain signals between 12 and 30 Hz.

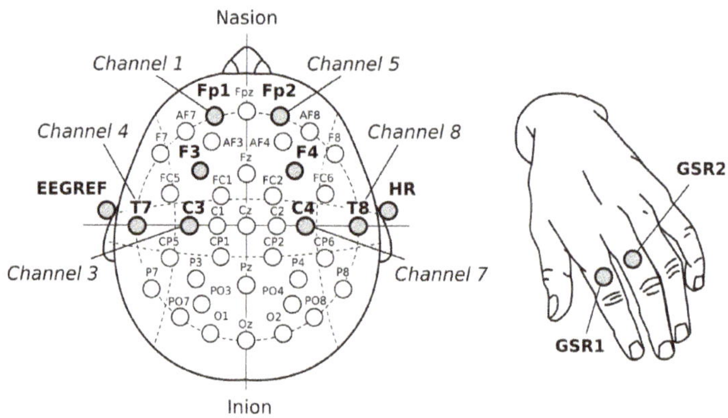

Figure 3. Electrodes placement. EEG and HR, placed on the scalp and earlobe (**left**); and GSR, placed on the indicator and middle fingers (**right**).

Putting all datasets together, it is possible to see the role of each one in this work (Figure 4). One dataset is produced by the Face Reader v7.0 and it outputs in real time the amplitudes of five emotions along the time.

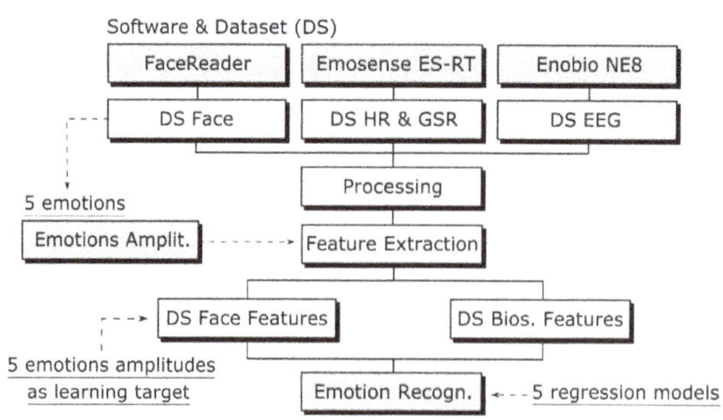

Figure 4. Datasets used in this work.

This research does not process face expression to detect emotions, instead, the Face Reader does it for us and outputs five emotional amplitudes which are used to lead the emotion recognition task during the Deep Learning and ANN training over another dataset based on biosignals, both synchronized in time.

3. Feature Extraction

Feature extraction is the last step before data recognition or classification. It is very important in pattern identification, classification, modeling and general automatic recognition. Its importance is fundamental to minimize the loss of important information embedded in the data [37] and to also optimize a dataset, giving clearer information to recognize any pattern.

This work uses different feature extraction techniques according to the technique used to acquire the physiological data (e.g., HR, GSR and EEG). Its extraction was executed after the processing phase which prepared the data to have more clear features. It was applied over time and frequency contexts are better described in the next section.

3.1. Features Description

In this work, we extracted 15 different features based on time and frequency. Each feature was chosen according to each dataset characteristics as presented in Table 1, which describes all extracted features, as such as the correspondent datasets. If the dataset needed a frequency analysis, it was applied through its features such as EEG datasets which used filtering and other analysis in frequency and time.

Table 1. Features extraction for HR, GSR, EEG and Face datasets.

Ord.	Extracted Features	Feature Description	Applied to Dataset (s)
1	**FEAT_MN**	⋄ Mean of a sample.	HR, GSR, EEG, Face
2	**FEAT_MD**	⋄ Middle value of a sample.	HR, GSR, EEG, Face
3	**FEAT_STD**	⋄ Standard deviation (σ) of a sample.	HR, GSR, EEG
4	**FEAT_VAR**	⋄ Variance (σ^2) of a sample.	HR, GSR, EEG
5	**FEAT_ENT**	⋄ Measure the samples' entropy i.e., irregularities.	HR, GSR, EEG
7	**FEAT_RNG**	⋄ Absolute range ($max - min$) value of a sample.	HR, GSR, EEG
8	**FEAT_RMS**	⋄ Root mean squared of a sample.	HR, GSR, EEG
9	**FEAT_PEK**	⋄ Measure the amount of peaks into a sample.	GSR
10	**FEAT_WAC**	⋄ Mean of the wavelet approximation coefficient.	EEG
11	**FEAT_WDC**	⋄ Mean of the wavelet detailed coefficient.	EEG
12	**FEAT_SD1**	⋄ Short-term HR variability.	HR
13	**FEAT_SD2**	⋄ Long-term HR variability.	HR
14	**FEAT_SCT**	⋄ Vector norm from the Poincaré plot centroid.	HR
15	**FEAT_SAR**	⋄ Ellipse area based on $SD1$ and $SD2$.	HR

Regarding to GSR datasets, it was important to understand its data profile and behaviour to properly relate it to the amount of peaks (peaks frequency) along the time/events; for this reason, one feature that relates peaks by time, was applied. Other peculiarities are also found over the HR datasets as, for instance, the HR variabilities during several emotional events along time. This HR dynamic fluctuation along the time, were mainly represented by three features. Furthermore, several statistical features were also applied over all datasets, along time, considering several sample lengths.

3.2. Wavelets (FEAT_WAC, FEAT_WDC)

The wavelet analysis plays an important role as part of the feature extraction methods. It allows to analyze time and frequency contents of data simultaneously and with high data resolution. When applied over a continuous data, it is called Continuous Wavelet Transform (CWT), and over a discrete data, it is Discrete Wavelet Transform (DWT) [38] (Equation (1)).

$$CWT(a,b) = \int_{-\infty}^{+\infty} x(t)\psi^*_{a,b}(t)dt, \quad (1)$$

where $x(t)$ represents the unprocessed data, a is the dilation, and b is the translation factor.

It lies on the concept of *mother wavelet* (MWT), which is a function used to decompose and describe the analyzed data. The Symlets ('sym7') was the MWT used, due its high similarities and compatibilities with the EEG data on all scalp regions [39].

Furthermore, as shown previously, the CWT method includes a complex conjugate term denoted by $\psi^*_{a,b}$, where $\psi(t)$ means wavelet [37] (Equation (2)).

$$\psi_{a,b}(t) = \frac{1}{\sqrt{|a|}}\psi(\frac{t-b}{a}). \tag{2}$$

3.3. Continuous Entropy (FEAT_ENT)

The continuous entropy or differential entropy is another feature used in this work. It is a concept in data theory to represent the measurement of the average rate of a random variable; it is also understood as a method to measure the quality or class diversity of such datasets. On *continuous probability distributions*, it is based on the expansion from *Shannon entropy* concept, defined by Equation (3),

$$h(X) = -\int_0^{N(S)} f(x) log f(x) dx. \tag{3}$$

where X represents a random variable defined by a probability density function of a subset S.

3.4. Sample Absolute Interval Range (FEAT_RNG)

The range of a sample was also used as a feature. It is defined as the absolute difference between the values compared to the last $f(t)$ and the first position $f(t - \Delta t)$ of a sample in time, as shown in Equation (4), which Δt represents the interval length to displace the interval from the actual position t.

$$R(t) = |f(t) - f(t - \Delta t)| \tag{4}$$

3.5. Poincaré Plots (FEAT_SD1, FEAT_SD2, FEAT_SCT, FEAT_SAR)

The Poincaré plots of RR intervals is one of the methods used in Heart Rate Variability (HRV) analysis. It returns a useful visual map (or cloud), which is capable to summarize the dynamics of an entire RR time series regarding to actual and next one values. It is also a quantitative method to give information over the long- and short-term HRV [40,41].

This method is represented by Poincaré *descriptors*, $SD1$ and $SD2$, which are used to quantify geometrically the produced cloud. It is given in terms of the variance of each RR_i and RR_{i+1} pairs. The i refers to the ith RR value, as shown in Figure 5.

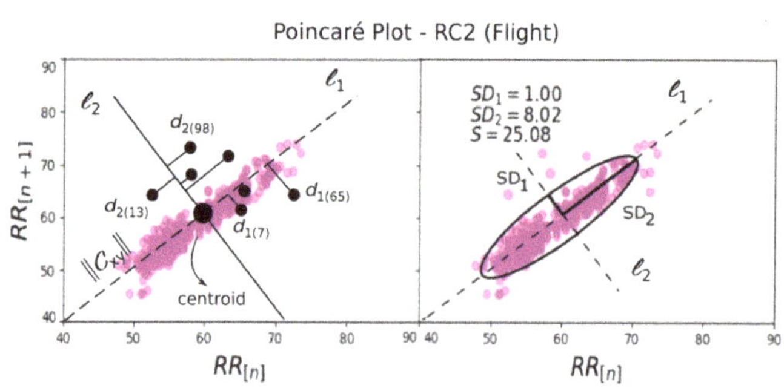

Figure 5. Poincaré plot demonstration over the flight dataset RC2.

Mathematically, let the HRV be defined by the vector $RR = [RR_1, RR_2, \ldots, RR_{n+1}]$ and the position-correlated vectors x and y defined as [41,42],

$$x = [x_1, x_2, \ldots, x_n] \equiv [RR_1, RR_2, \ldots, RR_n], \tag{5}$$

$$y = [y_2, x_3, \ldots, y_{n+1}] \equiv [RR_2, RR_3, \ldots, RR_{n+1}]. \tag{6}$$

For a regular Poincaré plot, the centroid vector $C_{xy} = [x_c, y_c]$ of its cloud representation, is define by,

$$x_c = \frac{1}{n}\sum_{i=1}^{n} x_i, y_c = \frac{1}{n}\sum_{i=1}^{n} y_i. \tag{7}$$

To compute the numerical representation of the centroid, the *vector norm* is applied using the Equation (8).

$$||C_{xy}|| = \sqrt{x_c^2 + y_c^2} \tag{8}$$

To compute the *descriptors* (short-term variability) $SD1$ and $SD2$ of a standard Poincaré plot, the distances d_1 and d_2 of any ith RR from the centroid *interceptors* l_1 and l_2 respectively are defined as,

$$d_{1i} = \frac{|(x_i - x_c) - (y_i - y_c)|}{\sqrt{2}}, d_{2i} = \frac{|(x_i - x_c) + (y_i - y_c)|}{\sqrt{2}} \tag{9}$$

Considering those prior algebraic definitions for a *standard cloud*, it is possible to compute the $SD1$ and $SD2$.

$$SD1_c = \sqrt{\frac{1}{n}\sum_{i=1}^{n} d_{1i}^2}, SD2_c = \sqrt{\frac{1}{n}\sum_{i=1}^{n} d_{2i}^2} \tag{10}$$

The area covered by the resulted ellipse, was also used as a feature for HR dataset, and it can be determined as below.

$$SA = \pi.SD1.SD2 \tag{11}$$

3.6. Singular Value Decomposition: Features Selection

When the features are extracted, some of them can be useless in the recognition process; to select the best set of them (i.e., to do a *dimensionality reduction*), the Singular Value Decomposition (SVD) was used, executing a *matrix decomposition* or *matrix factorization* of the input matrix. It is based on eigenvalues, applied to a bidimensional $m \times n$ matrix A. Mathematically, this method factorizes a matrix into a product of matrices, as shown in Equation (12).

$$A = UDV^*, \tag{12}$$

where D is a non negative diagonal matrix having the singular values of A; U and V are matrices that satisfy the condition $U^*U = I$, $V^*V = I$. The resultant matrix of that decomposition is the new input applied into the recognition process.

4. Emotion Recognition

The emotion recognition uses Artificial Neural Networks (ANN) and Deep Learning techniques (DL). The Multilayer Perceptron (MLP-ANN) architecture, Back-Propagation and Deep Learning algorithms were developed over the *Python3* Toolkits, *PyBrain*, *Keras* and *TensorFlow* having execution support of the Graphics Processing Unit (GPU).

The ANN is a supervised technique, inspired by human brain behaviour; it can process several instructions in short periods of time, taking fast decisions and reactions. Its architecture can be designed according to the problem to be solved. A small number of neurons is recommended for simpler problems. If the problem complexity increases, a new amount of neurons must to be analysed as needed. Each single neuron represents a single function over several parameters of activation and thresholds/biases. The techniques based on neural networks e.g., ANN, CNN, RNN, DNN, are powerful tools due their high capacity to solve complex tasks, being massively used in modern controls, dynamic systems, data mining, automatic bio-patterns identification (e.g., fingerprints or face recognition) and robotics. We can also cite the high capacity of the ANN to produce complex and parallel solutions over the field of features, which each ANN layer can presents different and parallel outputs to converge on final functions or probabilistic outputs. It does not mean that it cannot be applied, combined with other techniques such as, K-Means or SVM, for instance.

The final emotions are recognized from the biosignals and are based on ANN training using the labels produced by the Face Reader. In other words, initially, the system uses the emotions' labels processed by the Face Reader, synchronizes it in time with the biosignals and uses these labels in the training phase to teach the ANN to predict or recognize new emotions using only the biosignals.

4.1. ANN Development and Modeling

The training data (partial set of features) is defined in Equation (13), where τ represents the training-set, $x(n)$ the input-set (or input data features), $d(n)$ the desired output in each iteration n (due the use of a supervised learning method), and N_i that represents the amount of instances from the training-set [43].

$$\tau = \{x(n), d(n)\} \mid_{n=1}^{N_i} \tag{13}$$

The *Induced local field* (for forward computation) was used and can be computed by Equation (14), which x_i goes from input neurons i, w_{ji} and w_b represent the weights connections from the neuron j to i, and b_{ji} is the bias applied for each neuron, by iteration n.

$$v_j(n) = \sum_{i=1}^{N} w_{ji}(n)x_i(n) + b_{ji}w_b, j \geq 1 \tag{14}$$

For each hidden layer, two different activation functions were considered: the sigmoidal and ReLU. The *sigmoidal activation function* $\varphi(\cdot)$ is defined by Equation (15), where a determines the threshold's function. The sigmoid function returns values between 0 and 1.

$$\varphi_j(v_j(n))_{sig} = \frac{1}{1 + e^{-av_j(n)}}, a \geq 1 \tag{15}$$

Another activation function applied in this work, is the *ReLU* or *rectified linear unit*. It is defined by Equation (16), which returns values between 0 and $+\infty$.

$$\varphi_j(v_j(n))_{ReLU} = \begin{cases} 0 & \text{if } x < 0 \\ x & \text{if } x \geq 0 \end{cases} \tag{16}$$

Regarding to the output layer, also two different activation were considered. The prior active functions and the *softmax activation function* $P(y|X)$, which it is applied as defined by Equation (17). It represents the prediction probability for each emotion in all N_o output neurons.

$$P(y = j|X)(n) = \frac{e^{v_j(n)}}{\sum_{k=1}^{N_o} e^{v_k(n)}} \tag{17}$$

the $P(y|X)$ is mainly applied, when we are facing a classification problem i.e., which the outputs return independent probabilities for each considered class in case. Otherwise, when using the $\varphi_j(v_j(n))$, the ANN output can be represented by any amount of neurons, which it must return independent values (not probabilities), being useful when we are working with regression analysis. Since this work lies over the ANN and regression problems, the $\varphi_j(v_j(n))$ was used.

The error data or *instantaneous error* produced by output layer of each neuron j, is defined by Equation (18),

$$\varepsilon_j(n) = d_j(n) - y_k(n) \tag{18}$$

where $d_j(n)$ represents the jth element of $d(n)$ and $y_k(n)$ the kth *instantaneous output*. Furthermore, the $y_k(n)$ and the *instantaneous error energy* (ξ) of each neuron j (Equation (19)) are both considered to reach best network accuracy along epochs (iterations) [43,44].

$$\xi_j(n) = \frac{1}{2}\varepsilon_j^2(n) \tag{19}$$

The local *gradient* applied to each neuron k from the output layer, is described by Equation (20).

$$\delta_k(n) = \varepsilon_k(n) y_k(n)(1 - y_k(n)) \tag{20}$$

The general ANN weights adjustments (for backward computation) applied to each output neuron, is defined by *delta-rule* [43] (Equation (21)),

$$\Delta w_{kj}(n) = \alpha \Delta w_{kj}(n-1) + \eta \delta_k(n) y_k(n) \tag{21}$$

where the *momentum* α ([0; 1]) is used to avoid learning instabilities while increasing the *learning rate* η ([0; 1]), decreasing the mean error; furthermore, both are adjusted during the training phase.

4.2. Cross Validation—Testing Recognition Models

All the emotion recognition test were executed based on the methodology of Leave-One-Out Cross Validation (LOOCV).

The LOOCV was shown to be a good methodology on the proposed multimodal system to support the emotions recognition from each pilot, based on the learned emotions captured from the prior flights. It leaves one flight dataset out, while it trains the ANN using the other flight datasets.

4.3. Realtime Outliers Removal—RTOR

Sometimes, the neurons output abrupt values; wrong values are critical to compute the evaluation metrics (e.g., the absolute mean errors) correctly. To correct these realtime abrupt outputs, the Realtime Outliers Removal (RTOR) method was developed in this work. It is based on the last N output samples (based on a *batch* to store realtime samples acquisition) to eliminate the actual outlier from each output at same time (Figure 6).

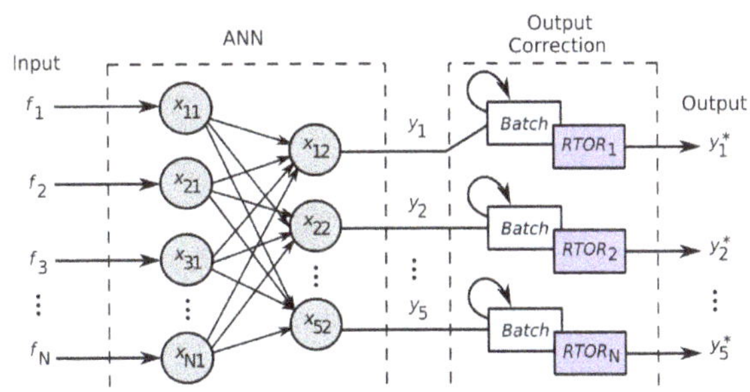

Figure 6. RTOR operation diagram.

Figure 7, shows in practice, the correction of two neuron outputs (top and bottom plot), representing two intensities of emotion outputs. The dotted red line, represents the target (emotion detected from face); the blue line, represents the corrected output (RTOR); and the dotted green line, represents the raw output. The relative output outliers detection and removal, are controlled by a batch length, which represents the amount of samples to be treated in realtime.

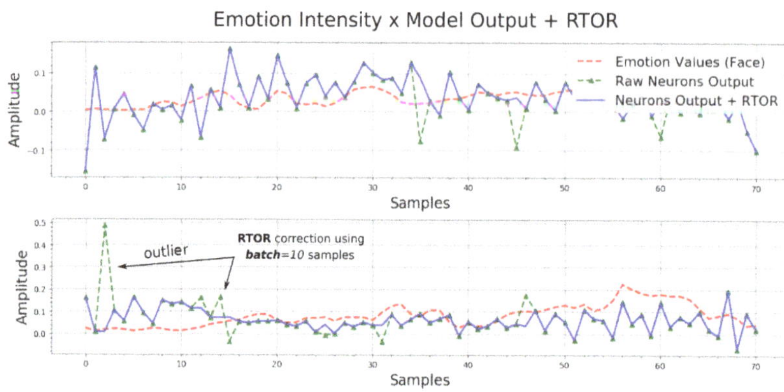

Figure 7. Realtime outlier correction based on RTOR method. Note the corrected output (blue) and the raw output with outliers (green).

4.4. *Evaluation Metrics for Emotion Output: Regression Models*

Before the metrics presentation to evaluate the emotion recognition outputs, it is extremely important to know that this work does not consider one single emotion as the final output, but the intensities of several emotions i.e., five emotions by time, output by each independent output neuron. This is because the most human bodies do not usually feel one single emotion at a time, but several of them, having different intensities and valences. For this reason, the presented evaluation metrics work over regression outputs over all output neurons measured separately.

Each output neuron was designed as a regression function and trained to output emotion intensities using the emotion detected from the face as the target. These outputs are measured to analyze how close the outputs are from the targets.

4.4.1. Root Mean Squared Error (RMSE)

Considering the prior R^2, the Root Mean Squared Error (RMSE) or also called, Root Mean Squared Deviation (RMSD), computes the error distance between the estimated values $\hat{y}(n)$, as defined below.

$$RMSE = \sqrt{\frac{\sum_{n=1}^{N} (\hat{y}(n) - y(n))^2}{N}} \tag{22}$$

4.4.2. Mean Absolute Error (MAE)

The MAE represents the average of the absolute difference between the predicted values and observed value (prediction). In another words, it is a linear representation, which means that all the single differences are weighted equally in the average as shown in Equation (23):

$$MAE = \frac{1}{N} \sum_{n=1}^{N} |y(n) - \hat{y}(n)| \tag{23}$$

5. Result Analysis

This work presented a multimodal solution to recognize emotions from several physiological inputs based on the bio-reactions from beginner users of flight simulator. It is an important contribution regarding aviation and a more general perspective of emotion relationship. It was proposed as one way to contribute to emotion studies and in this work, the context of application was mainly the aviation side of the scope of aviation accidents, which were caused by human failures.

Several tests were executed in this work to try to find recognition results for each pilot, i.e., the best model possible to try to estimate emotions felt by each pilot during the flight experiment. The cross-validation was the method used to aim the emotions recognition process for each pilot dataset obtained during each flight experiment. The recognition tasks were initially based on two different tests: *tests without feature extraction* (i.e., raw data directly applied in ANN inputs, with any treatment or preprocessing) and *processed data with feature extraction*. Were also considered different ANN architectures, amount of training iteration, amount of inputs and hidden neurons, and different flight datasets.

In all emotion recognition tests, the cross validation was applied to recognize the emotions felt by the pilot in a single flight according to the emotions already detected from another flights. In other words, the training was based on 12 flight datasets ($N-1$ flights), to try to recognize the emotions from one single flight. It is important considers, that the dataset having emotions values from the face (5 different emotions), was the reference of the ANN training. For this reason several mistakes from the facial reader software, detecting wrongly several emotions, were not possible to be avoided; the consequence of these wrong matches is several errors under the regression models, outputted from each output neuron.

5.1. Description of the Recognition Tests

The main procedures applied from the processsing and feature extraction, are shown in the test sequence below. It was based on feature selection and data type of treatment. For most of the tests, at least the data normalization and abrupt data correction were used (Table 2). In these tests, we considered all the features from each data, i.e., 11 features from HR, seven features from GSR and 72 features from EEG (9 × 8 Ch), including the best and worst features.

Table 2. Description of each execution test according to preprocessing, processing and feature extraction.

Tests	Preprocessing		Processing, Feature Extraction and Recognition					Data		
	Detrend	Outliers	*FE**	SVD	*CC**	$\varphi_j(v_j(n))$	Optmization	HR	GSR	EEG
Test 1	–	–	–	–	–	sigmoid	'sgd'	×	×	×
Test 2	–	–	–	–	–	sigmoid	'adam'	×	×	×
Test 3	×	×	×	–	×	ReLU	'adam'	×	×	×
Test 4	×	×	×	–	×	sigmoid	'sgd'	×	×	×
Test 5	×	×	×	–	×	sigmoid	'adam'	×	×	×
Test 6	×	×	×	–	×	ReLU	'sgd'	×	×	×
Test 7	×	×	×	–	×	ReLU	'adam'	–	×	×
Test 8	×	×	×	–	×	sigmoid	'sgd'	–	×	×
Test 9	×	×	×	–	×	sigmoid	'adam'	–	×	×
Test 10	×	×	×	–	×	ReLU	'sgd'	–	×	×
Test 11	×	×	×	–	×	ReLU	'adam'	×	–	×
Test 12	×	×	×	–	×	sigmoid	'sgd'	×	–	×
Test 13	×	×	×	–	×	sigmoid	'adam'	×	–	×
Test 14	×	×	×	–	×	ReLU	'sgd'	×	–	×
Test 15	×	×	×	–	×	ReLU	'adam'	×	×	–
Test 16	×	×	×	–	×	sigmoid	'sgd'	×	×	–
Test 17	×	×	×	–	×	sigmoid	'adam'	×	×	–
Test 18	×	×	×	–	×	ReLU	'sgd'	×	×	–

CC: Column Centering—data centering for each data. FE*: Feature Extraction—select all features for each data.*

Between tests 19 and 34, we considered the features selection based on SVD (it means that now, the features are selected in order of importance). There were six features from HR, four features from GSR and 40 (5× 8 Ch) features from EEG, as presented in Table 3.

Table 3. Description of each execution test according to preprocessing, processing and feature selection.

Tests	Preprocessing		Processing, Feature Extraction and Recognition					Data		
	Detrend	Outliers	*FE*	SVD	*CC*	$\varphi_j(v_j(n))$	Optmization	HR	GSR	EEG
Test 19	×	×	×	×	×	ReLU	'adam'	×	×	×
Test 20	×	×	×	×	×	sigmoid	'sgd'	×	×	×
Test 21	×	×	×	×	×	sigmoid	'adam'	×	×	×
Test 22	×	×	×	×	×	ReLU	'sgd'	×	×	×
Test 23	×	×	×	×	×	ReLU	'adam'	–	×	×
Test 24	×	×	×	×	×	sigmoid	'sgd'	–	×	×
Test 25	×	×	×	×	×	sigmoid	'adam'	–	×	×
Test 26	×	×	×	×	×	ReLU	'sgd'	–	×	×
Test 27	×	×	×	×	×	ReLU	'adam'	×	–	×
Test 28	×	×	×	×	×	sigmoid	'sgd'	×	–	×
Test 29	×	×	×	×	×	sigmoid	'adam'	×	–	×
Test 30	×	×	×	×	×	ReLU	'sgd'	×	–	×
Test 31	×	×	×	×	×	ReLU	'adam'	×	×	–
Test 32	×	×	×	×	×	sigmoid	'sgd'	×	×	–
Test 33	×	×	×	×	×	sigmoid	'adam'	×	×	–
Test 34	×	×	×	×	×	ReLU	'sgd'	×	×	–

5.2. Emotion Recognition Tests Based on Raw Data: Test 1 and Test 2

In these tests of emotion recognition, no feature extractions and preprocessing were considered; all raw data were directly applied in the ANN input layer. The ANN activation function was sigmoid and two different optimization algorithms: stochastic gradient descend ('sgd') and 'adam'.

Table 4, presents an emotion recognition result using a raw data approach and no feature extraction. Its results show the importance of feature extraction in a multimodal sensing system in which without it, the recognition will have more undesirable results and high execution time. The RMSE and MAE were used to compare the output models with the emotions from the flight datasets.

Table 4. Emotion recognition results tests 1 and 2. ANN with 6×10^3 train epochs and raw data (no feature extraction).

	Test 1—Emotion Recognition + RTOR $\varphi_j(v_j(n))$ — *Sigmoid*, opt = 'sgd', $N_h = 10 \times 2$, $N_o = 5$										
DS	Happy		Sad		Angry		Surprised		Scared		Match
	RMSE	*MAE*	*RMSE*	*MAE*	*RMSE*	*MAE*	*RMSE*	*MAE*	*RMSE*	*MAE*	Accuracy (%)
RC1	3.64	0.06	4.14	0.06	3.83	0.05	3.43	0.06	5.08	0.08	50.50 (1854/3671)
RC2	4.34	0.06	5.72	0.07	3.59	0.05	3.84	0.06	5.88	0.09	82.13 (3488/4247)
RC3	3.88	0.05	9.58	0.11	3.78	0.06	3.62	0.06	5.57	0.09	58.83 (2342/3981)
GC1	5.68	0.09	8.46	0.13	7.34	0.11	4.58	0.07	5.79	0.09	23.55 (961/4081)
GC3	5.63	0.09	7.45	0.11	7.41	0.11	5.42	0.08	5.84	0.09	99.65 (4240/4255)
LS1	5.70	0.08	6.22	0.08	3.46	0.04	5.18	0.07	6.20	0.08	22.49 (1250/5558)
LS2	5.52	0.09	3.68	0.05	2.93	0.04	5.04	0.08	5.42	0.08	69.43 (2844/4096)
VC1	3.98	0.08	3.38	0.06	4.40	0.08	3.43	0.07	4.79	0.08	15.63 (408/2611)
VC2	3.76	0.08	3.89	0.08	4.27	0.09	2.78	0.06	2.53	0.05	30.12 (615/2042)
CR1	4.46	0.07	17.54	0.24	5.00	0.06	3.58	0.06	1.64	0.02	92.47 (3697/3998)
CR3	1.69	0.08	3.66	0.15	1.16	0.04	1.39	0.07	1.28	0.05	74.18 (339/457)
CLX	4.45	0.16	1.00	0.04	1.73	0.07	1.47	0.06	0.54	0.02	0.00 (0/518)
CL3	3.27	0.04	3.07	0.04	5.58	0.07	5.39	0.08	3.76	0.05	15.57 (735/4722)
	4.31 ± 1.11	0.08 ± 0.02	5.98 ± 4.05	0.09 ± 0.05	4.19 ± 1.77	0.07 ± 0.02	3.78 ± 1.29	0.07 ± 0.00	4.18 ± 1.92	0.07 ± 0.02	48.81 ± 31.67

Table 4. *Cont.*

DS	Test 2—Emotion Recognition + RTOR $\varphi_j(v_j(n)) = Sigmoid$, opt = 'Adam', $N_h = 10 \times 2$, $N_o = 5$										
	Happy		Sad		Angry		Surprised		Scared		Match
	RMSE	*MAE*	*RMSE*	*MAE*	*RMSE*	*MAE*	*RMSE*	*MAE*	*RMSE*	*MAE*	Accuracy (%)
RC1	1.19	0.02	5.44	0.08	3.63	0.05	0.79	0.01	1.76	0.03	50.50 (1854/3671)
RC2	1.26	0.02	5.77	0.07	2.41	0.03	1.02	0.01	1.73	0.03	82.13 (3488/4247)
RC3	4.96	0.06	9.14	0.12	4.81	0.07	0.73	0.01	3.44	0.05	58.83 (2342/3981)
GC1	0.64	0.01	3.97	0.06	2.96	0.05	0.64	0.01	0.74	0.01	23.55 (961/4081)
GC3	0.63	0.01	3.69	0.06	3.34	0.05	0.34	0.01	0.84	0.01	99.65 (4240/4255)
LS1	0.69	0.01	1.71	0.02	3.47	0.04	0.97	0.01	0.35	0.00	22.49 (1250/5558)
LS2	0.49	0.01	3.63	0.04	2.99	0.04	0.44	0.01	0.27	0.00	69.43 (2844/4096)
VC1	0.81	0.01	2.20	0.04	2.39	0.04	0.39	0.01	7.67	0.13	15.63 (408/2611)
VC2	0.28	0.01	1.76	0.03	1.07	0.02	0.96	0.02	4.68	0.09	30.12 (615/2042)
CR1	2.48	0.04	16.56	0.23	5.05	0.07	0.67	0.01	1.93	0.03	92.47 (3697/3998)
CR3	1.03	0.05	2.83	0.12	1.34	0.05	0.48	0.02	1.75	0.06	74.18 (339/457)
CLX	5.66	0.22	0.92	0.04	2.27	0.09	0.32	0.01	1.08	0.05	0.00 (0/518)
CL3	3.49	0.04	4.57	0.05	9.82	0.13	0.20	0.00	2.24	0.03	15.57 (735/4722)
	1.82 ± 1.71	0.04 ± 0.05	4.78 ± 3.98	0.07 ± 0.05	3.50 ± 2.13	0.06 ± 0.02	0.61 ± 0.26	0.01 ± 0.00	2.19 ± 1.97	0.04 ± 0.03	48.81 ± 31.67

5.3. Emotion Recognition Tests Based on Feature Extraction: Test 3 to 34

All tests between 3 and 34 considered feature extraction over the raw input data. In detail, between tests 3 and 18, 90 features were extracted, considering all features. Between tests 19 and 34, the SVD was applied to select the best features to be used. Table 5 presents the tests results, referring to tests 11 and 12, based on feature extraction and, in this case, without features selection.

Table 5. Emotion recognition results tests 11 and 12. ANN with 6×10^3 train epochs and input data with feature extraction.

DS	Test 11—Emotion Recognition + RTOR [HR+EEG] $\varphi_j(v_j(n)) = ReLU$, opt = 'Adam', $N_h = 83 \times 2$, $N_o = 5$										
	Happy		Sad		Angry		Surprised		Scared		Match
	RMSE	*MAE*	*RMSE*	*MAE*	*RMSE*	*MAE*	*RMSE*	*MAE*	*RMSE*	*MAE*	Accuracy (%)
RC1	0.25	0.02	0.46	0.04	0.60	0.06	0.12	0.01	0.82	0.09	22.38 (15/67)
RC2	0.34	0.03	0.90	0.08	0.81	0.08	0.14	0.01	0.30	0.03	34.61 (27/78)
RC3	0.83	0.08	1.67	0.14	0.43	0.04	0.12	0.01	0.51	0.04	38.35 (28/73)
GC1	0.71	0.07	1.28	0.13	0.74	0.08	0.14	0.01	0.22	0.02	21.33 (16/75)
GC3	0.29	0.03	0.64	0.05	0.42	0.04	0.18	0.02	0.04	0.00	65.38 (51/78)
LS1	0.19	0.01	1.26	0.11	0.43	0.03	0.11	0.01	0.16	0.01	25.54 (26/102)
LS2	0.32	0.03	0.43	0.04	0.35	0.03	0.16	0.02	0.23	0.02	42.66 (32/75)
VC1	0.09	0.01	0.35	0.04	0.42	0.05	0.05	0.01	1.09	0.14	16.66 (8/48)
VC2	0.20	0.03	0.61	0.08	0.65	0.10	0.08	0.01	0.57	0.08	21.05 (8/38)
CR1	0.15	0.01	2.53	0.26	0.76	0.07	0.12	0.01	0.51	0.06	47.94 (35/73)
CR3	0.10	0.02	0.65	0.20	0.40	0.12	0.03	0.01	0.27	0.07	44.44 (4/9)
CLX	0.76	0.21	0.31	0.08	0.43	0.12	0.05	0.01	0.18	0.05	0.00 (0/10)
CL3	0.40	0.04	1.10	0.09	0.96	0.09	0.19	0.02	0.26	0.02	19.76 (17/86)
	0.36 ± 0.24	0.05 ± 0.05	0.94 ± 0.60	0.10 ± 0.06	0.57 ± 0.18	0.07 ± 0.03	0.11 ± 0.04	0.01 ± 0.00	0.40 ± 0.28	0.05 ± 0.03	30.78 ± 16.27

Table 5. *Cont.*

DS	**Test 12—Emotion Recognition + RTOR [HR+EEG]** $\varphi_j(v_j(n)) = Sigmoid$, opt = 'sgd', $N_h = 83 \times 2$, $N_o = 5$										
	Happy		Sad		Angry		Surprised		Scared		Match
	RMSE	*MAE*	*RMSE*	*MAE*	*RMSE*	*MAE*	*RMSE*	*MAE*	*RMSE*	*MAE*	Accuracy (%)
RC1	0.17	0.02	0.42	0.04	0.50	0.04	0.11	0.01	0.31	0.04	53.73 (36/67)
RC2	0.23	0.02	0.82	0.07	0.29	0.03	0.13	0.01	0.39	0.04	82.05 (64/78
RC3	0.67	0.06	1.37	0.11	0.29	0.03	0.11	0.01	0.35	0.04	57.53 (42/73)
GC1	0.36	0.04	0.96	0.11	0.69	0.08	0.21	0.02	0.38	0.04	22.66 (17/75)
GC3	0.35	0.04	0.82	0.09	0.70	0.08	0.32	0.04	0.39	0.04	100.00 (78/78)
LS1	0.31	0.03	0.64	0.06	0.30	0.02	0.23	0.02	0.37	0.04	22.54 (23/102)
LS2	0.34	0.04	0.38	0.04	0.22	0.02	0.27	0.03	0.33	0.04	68.00 (51/75)
VC1	0.22	0.03	0.31	0.04	0.35	0.05	0.14	0.02	0.90	0.11	16.66 (8/48)
VC2	0.23	0.04	0.42	0.06	0.37	0.06	0.10	0.02	0.47	0.06	28.94 (11/38)
CR1	0.22	0.02	2.58	0.26	0.90	0.09	0.10	0.01	0.27	0.03	93.15 (68/73)
CR3	0.10	0.03	0.55	0.15	0.17	0.05	0.05	0.02	0.26	0.07	77.77 (7/9)
CLX	0.71	0.20	0.06	0.02	0.36	0.10	0.06	0.02	0.10	0.03	0.00 (0/10)
CL3	0.28	0.02	0.39	0.03	1.07	0.10	0.29	0.03	0.16	0.01	15.11 (13/86)
	0.32 ± 0.17	0.05 ± 0.04	0.75 ± 0.61	0.08 ± 0.06	0.48 ± 0.26	0.06 ± 0.02	0.16 ± 0.08	0.02 ± 0.00	0.36 ± 0.18	0.05 ± 0.02	49.09 ± 32.00

The accuracy of the *match* procedure i.e., the correct match in each sample regarding to the higher emotion amplitude (between five emotions), presented the worst values on recognition from the flight dataset CLX, having any match on most recognition, surely due its high noises and small number of samples analysed before and after the feature extraction, which changed from 518 to 10 samples.

5.4. Emotion Recognition Analysis

Figure 8, presents the barplots correspondent to the errors results from the tests 3–6, with feature extraction but without feature selection and considering all three data.

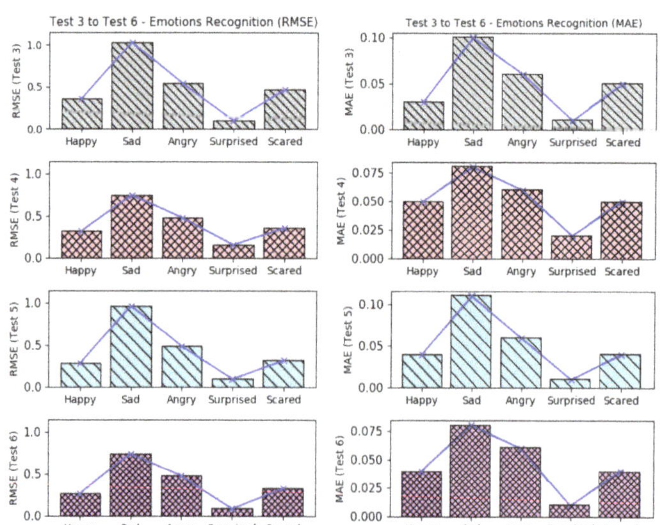

Figure 8. Errors results (RMSE+MAE) from tests 3–6 (with feature extraction).

It is possible to see that in tests 3–6, the emotion *surprised* presented a higher recognition accuracy, having the smallest error level. The *happy* and *scared* were the emotions which also presented low errors. Nevertheless, these error levels can be improved if the training datasets are more coherent. The emotions *sad* and *angry*, presented the worst error levels; it is probably due the misclassifications from the face emotiom detection software, which sometimes confused situations of angry and disappointed rather than sadness. If we compare all tests (from test 3–34), it is possible to note that again, the *surprised* emotion kept with best recognition values (low errors), as shown in Figures 9 and 10, which it present all considered errors along the tests.

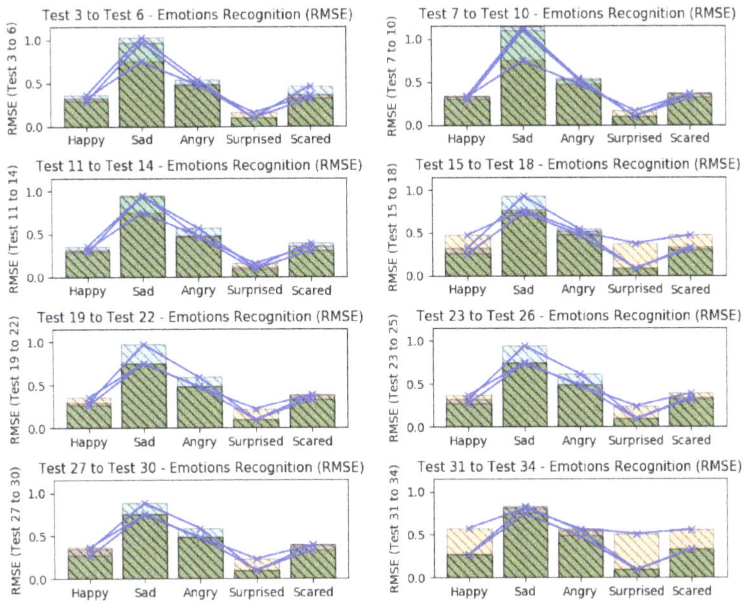

Figure 9. Errors results (RMSE) comparison from tests 3–34 (with feature extraction).

The higher recognition errors were reached when the EEG datasets were omitted in different tests (tests 15–18 and tests 31–34), showing that in these tests, the recognition results were better when all data were considered; when GSR datasets were ommited, the results got good predictions too (tests 11–14 and tests 27–30). The application of *feature selection* based on SVD and the omission of GSR datasets, returned the less recogmition errors (tests 27–30). The *sad* emotion got the worst error levels when HR datasets were omitted (tests 7–10), as like as the *happy* emotion got the worst error levels when the EEG datasets were omitted.

In summary, all tests showed that the smallest error levels can be reached when all datasets were considered or when the GSR datasets were omitted. Also, they showed that the emotion *surprised* was easier to expect, having a mean RMSE of 0.13 and mean MAE of 0.01; while the worst predictions were found to emotion *sad*, having a mean RMSE of 0.82 and mean MAE of 0.08.

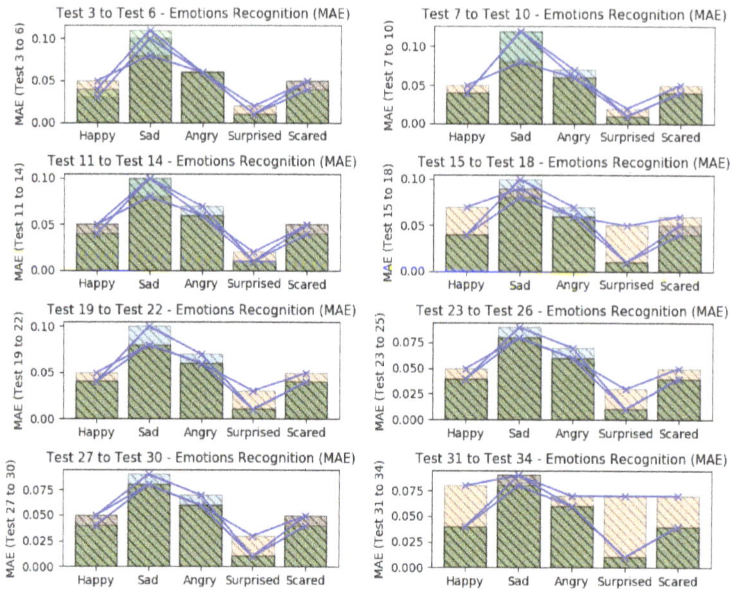

Figure 10. Errors results (MAE) comparison from tests 3–34 (with feature extraction).

5.5. Improvements Coming from the Feature Extraction

In prior discussion, we presented the need to use features extraction in a very dense datasets. One direct benefit of it is the execution time. With the feature extraction, the dataset is sampled to fractions of data which it must continue to represent all raw data with more or equal meaning. For this reason, a featured dataset is smaller if compared to its raw dataset. Another benefit of feature extraction is that it can brings information from a dataset in statistical or frenquency context, e.g., data variances and other tiny patterns of the frenquency domain. Figure 11 shows the errors levels between the use of *raw datasets* (tests 1 and 2) and *featured datasets* (tests 3 to 34).

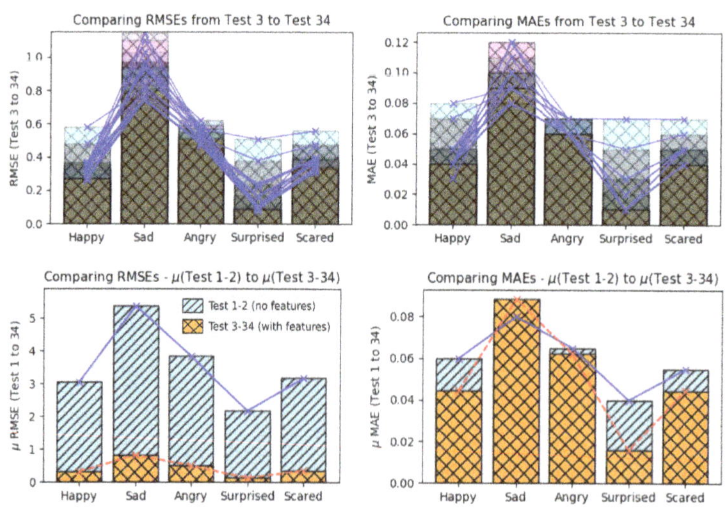

Figure 11. Errors results comparison between RMSE and MAE from tests 1 to 34 (with feature extraction).

Analyzing the RMSE values (left barplot), it is possible to see that the improvements were important over all emotions when feature extraction was used. The emotion *happy* presented an improvement of 89.66% (prior 3.06/actual 0.31); *sad* of 84.58% (5.38/0.82); *angry* of 86.75% (3.84/0.50); *surprised* of 93.89% (2.19/0.13); and *scared* of 88.67% (3.18/0.36).

Analyzing the MAE values (right barplot), it is possible to see that the improvements were good over 4 emotions of 5 (emotion *sad* wasn't improved on MAE values), when feature extraction was used. The emotion *happy* presented an improvement of 26.04% (prior 0.06/actual 0.04); *angry* of 4.32% (0.065/0.062); *surprised* of 60.15% (0.04/0.01); and *scared* of 18.75% (0.05/0.04).

5.6. Considering the Higher Intensities Between Emotions

The higher emotion intensities by time (between five emotion intensities) were also computed and its amount of matches were also analyzed, comparing the correct matches betewen its higher emotion (from the face dataset) with the higher output from the five neurons (output layer).

The benefit to also consider these higher values by time is to understand if the regression models from each output neuron is following the original emotion intensities related to the other emotions. In other words, if a regression model from each neuron, fits to a target, both *errors levels* (RMSA+MAE) and *major/higher values* will improve together.

The corrected amount of matches between these emotions and its relations is shown in Figure 12, presenting the case of tests 3–6.

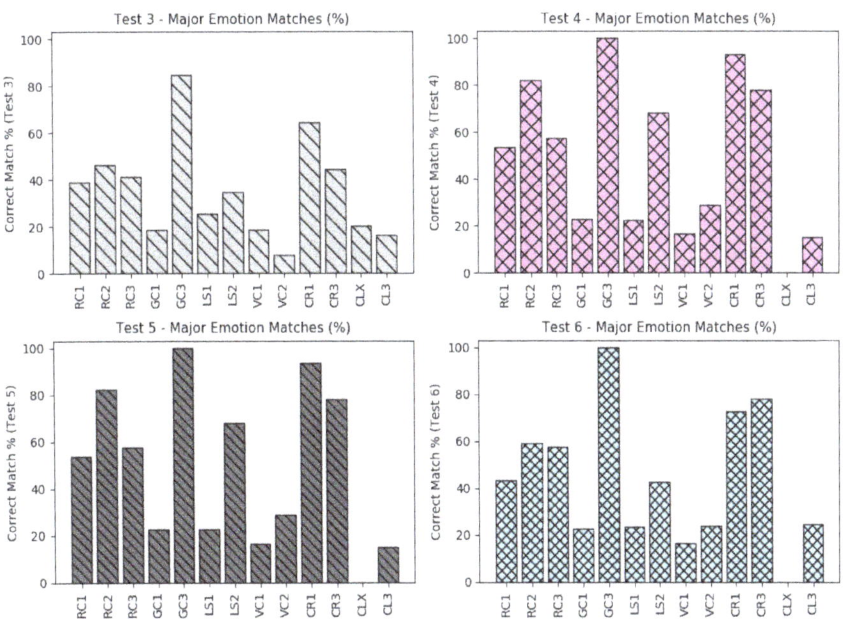

Figure 12. Major emotion accuracies from the tests 3 to 6 (with feature extraction).

Some datasets presented a very low amount of matches during all tests as for instance, *GC1*, *LS1*, *VC1*, *CLX* and *CL3*. These low accuracies are probably due the high misclassification of emotions from the pilots' faces as also presented on prior errors values based o RMSE and MAE. However, if we consider the possibility to improve these results, the next tests can omit these datasets with low accuracies to get better general results.

When comparing all matches (from test 3–34) regarding to the major emotion values, it is possible to see that the accuracy of the dataset *CLX* continues to present the worst accuracies and the dataset *GC3* the best accuracies values.

Figure 13, shows a comparison of all accuracies, regarding to the major emotions from the tests 3 to 34 (top plots) and from tests 1 to 34 (bottom plot). Note that on the top plot, shows that six datasets kept the major emotion accuracies less than 50%.

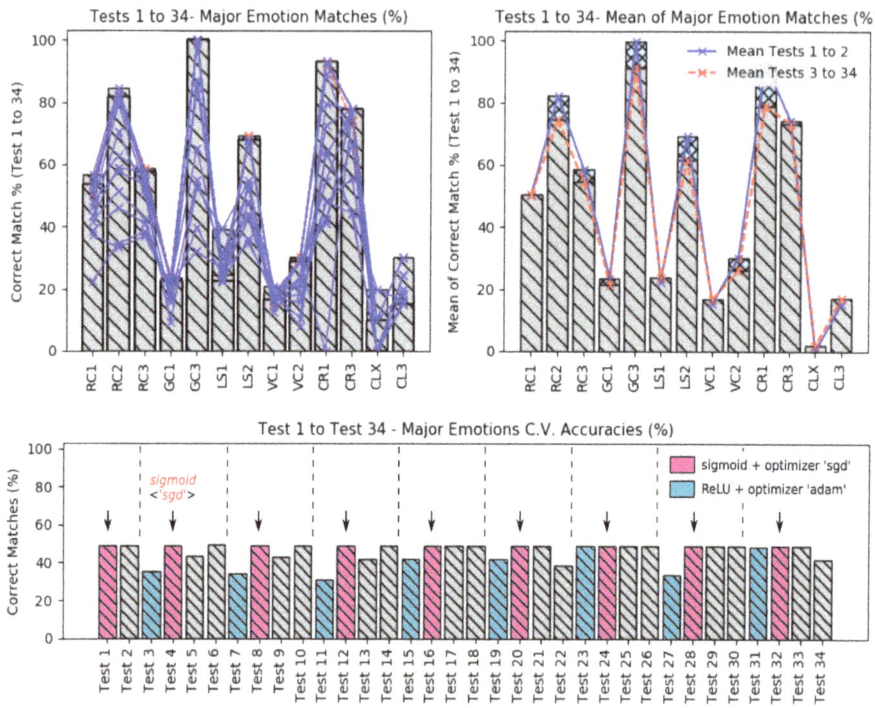

Figure 13. All higher emotion accuracies from tests 1–34. All accuracies (**left**); mean of all accuracies (**right**).

The top left plot, presents the relation between the mean of the raw dataset accuracies (tests 1 and 2) over the featured datasets accuracies (tests 3–34), which the raw data tests seems to have better accuracies over the feaured dataset. It means that the recognition based on raw datasets was the best solution in this proposed work.

The answer for that is *not necessarily*; if we go back a little and observe the error levels during the tests based on raw datasets, we will see that it was extremily bad compared to the others tests based on featured dataset; this way we can easily note that actually, a good regression model must be based on a combination of low error levels and good major emotion accuracies.

Finally, when analizing the bottom plot, it is possible to note that when the activation function was the *sigmoid* together with the *gradient descend* optimization, the general accuracies presented a constant behaviour along the executed tests. The activation function *rectified unit* presented the worst major emotion accuracies in this work.

5.7. Improving These Results

To improve these results, this work shows that is strongly recommended, to first, to optimize the emotions detection from the face. It were undoubtedly, the main reason for several undesirable recognition error levels. Another way to improve it is to omit some datasets which presented bad predictions; it surely will improve the general predictions or emotional recognition.

However, some results were already improved during this work. For instance, when looking to the learning tasks, absolute improvements, were applied, changing the *traditional learning* techniques by the *Deep Learning* techniques. These last improvements optimized the resognition results in *accuracies* of recognition and in *execution time*.

Figure 14, shows the improvement due the use of *Deep Learning* techniques, regarding to the amount of correct matches of the major emotions values, between all emotions considered in this work. It is possible to see, that the dataset CLX kept with worst accuracy also on traditional learning.

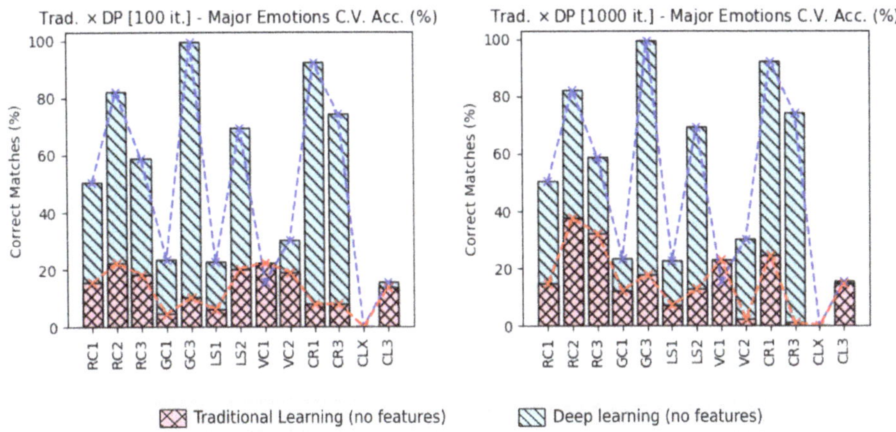

Figure 14. Traditional learning versus Deep Learning (DP). Improvement applied in this work regarding to the major value emotions when applying the traditional learning and Deep Learning (no feature extraction).

Regarding the accuracies of the major value emotions based on 100 training iteration of the traditional learning, the improvement happened in 11 flight datasets from 13: RC1 was improved in 69.52% (prior 15.39/actual 50.50); RC2 72.71% (22.41/82.13); RC3 of 68.97% (18.25/58.83); GC1 of 80.97% (4.48/23.55); GC3 of 89.88% (10.08/99.65); LS1 of 73.63% (5.93/22.49); LS2 of 70.96% (20.16/69.43); VC2 of 37.08% (18.95/30.12); CR1 of 91.40% (7.95/92.47); CR3 of 89.39% (7.87/74.18); and CL3 of 12.13% (13.68/15.57). The higher and lower improvements happened for datasets CR1 and CL3, respectively.

Considering the traditional learning using 1000 training iteration, the improvement happened in 11 flight datasets from 13, as in prior situation: RC1 was improved in 70.77% (14.76/50.50); RC2 of 54.25% (37.57/82.13); RC3 of 45.31% (32.17/58.83); GC1 of 47.77% (12.30/23.55); GC3 of 82.00% (17.93/99.65); LS1 of 68.25% (7.14/22.49); LS2 of 81.17% (12.69/69.43); VC2 of 92.19% (2.35/30.12); CR1 of 73.36% (24.63/92.47); CR3 of 98.53% (1.09/74.18); and CL3 of 5.20% (14.76/15.57). The higher and lower improvements happened for dataset CR3 and CL3 respectively.

The improvements of accuracies over the major emotion values at 100 training iterations were higher, because the execution with 1000 training iterations presented better accuracies (i.e., less difference from Deep Learning); however, due the very high exponential execution time of the tradition learning, it demotivate the execution of it traditional manner, using the same training iteration used with the Deep Learning (6000 training iterations), which can take days or weeks.

If we consider the improvements over the *execution time*, the use of Deep Learning instead the traditional methods, we notice an optimization of 92.17%, having 4406.32 s (mean of the Deep Learning applied on tests 1 and 2) instead of 56,321.40 s (traditional learning), even when the amount of training iteration was 60 times less, i.e., 100 over 6000 the from Deep Learning. When the training interation of the traditional learning was increased to 1000, the improvement with the use of Deep Learning was 99.09%, having 4406.32 s (Deep Learning) instead of 484,586.47 s from traditional learning, even using 6 times less training iterations.

Another way to improve the final results, is to exeute more flight tests, increasing the amount of data in the dataset. Also, applying personal dataset concept, which the emotion recognition should also be based on personal characteristics of each pilot.

5.8. Emotions Instances from Face Expressions

The emotion amplitudes detected by the Face Reader v7.0 were used as the emotion references during the emotion recognition phase using all biosignals based on Deep Learning and ANN. Each emotion instance detected during all flights executed in this work is shown in Figure 15-left, which shows the mean of the percentage of emotion instances along the 13 flights and in Figure 15-right, which shows the total amount of emotion instances detected along all flights.

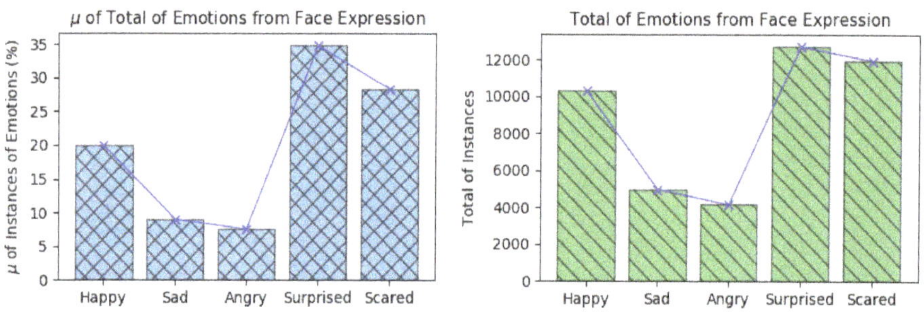

Figure 15. Amount of emotions instances detected by the Face Reader v7.0.

According to the Face Reader outputs, during the flight experiments, the pilots experienced more of: *happy*, *surprised* and *scared*. Emotions *sad* and *angry*, presented less occurrences along the experiments. These outputs are in line with the arguments presented by each pilot during the experiments.

These emotion instances presented a relation with the amount of recognition errors presented during the emotion recognition phase. It is because the emotions *happy*, *surprised* and *scared* presented more instances along the experiments (more instances to train), what it resulted on the lower errors levels along the emotion recognition.

6. Conclusions and Future Work

This work presented a solution to detect emotions from pilots in command (i.e., beginner users on a flight simulator), during simulated flights. These flights were executed by the Microsoft Flight Simulator Steam Edition (FSX-SE), using a Cessna 172SP aircraft. The users from the experiment were beginners in simulated flights and they were trained before. A total of seven flight tasks were defined such as: take off, climbing, navigation, descent, approach, final approach and landing.

We considered three different data from the pilots' bodies: HR, GSR and EEG. They were acquired at the same time during the flight based on several sensors such as Enobio-NE8, Shimmer3-GSR+, MedLab-Pearl100 and Arduino.

After data acquisition, the processing was executed to correct abrupt changes of the data, to detrend, remove outliers, normalize the data and execute filterings and data sampling. The feature extraction was executed over the processed data where several features were extracted to aim for the recognition phase. The ANN was used to recognize emotions using the extracted features such as the ANN inputs, based on traditional and Deep Learning techniques.

The emotion recognition results reached different levels of accuracy. The tests of the produced output models showed that the lowest recognition errors were reached when all data were considered or when the GSR datasets were omitted from the model training. It also showed that the emotion *surprised* was the easiest to analyse, having a mean RMSE of 0.13 and mean MAE of 0.01; while the emotion *sad* was the hardest to recognize, having a mean RMSE of 0.82 and mean MAE of 0.08. When were considered only the higher emotion intensities by time, the most matches accuracies were between 55% and 100%. It can be partially explained by the amount of emotion instances detected by the Face Reader, which the emotions *happy*, *surprised* and *scared* presented more instances along the experiments.

As part of future work, we intend to execute more emotion recognition tests, omitting the datasets that presented the lowest accuracies (considering the matches with the higher emotions by time), to optimize the total mean accuracies. Also, we aim to optimize the quality of the face emotion dataset, processed by the Face Reader software, and then to obtain higher accuracies and lower error levels. Increasing the number of flight experiments is another improvement that can be applied in future work; it would generate more data for training during the recognition phase.

Author Contributions: V.C.C.R. conceived and developed the experiment architecture, methodologies and techniques used along the acquired datasets; also developed the software used to acquire the data from accelerometer, GSR and HR. O.A.P. revised the experiment architecture and methodologies, provided the laboratory and all devices used along the experiments; also provided the software used to detect emotions from the users' faces and signals from the brain.

Funding: This research was partially supported by Fundação para a Ciência e a Tecnologia (FCT) (Project UID/EEA/50008/2019), Instituto Universitário de Lisboa (ISCTE-IUL) and Instituto de Telecomunicações (IT-IUL), from Lisbon, Portugal.

Conflicts of Interest: The authors declare no conflict of interest.

References

1. Misky, M. *The Emotion Machine: Commonsense Thinking, Artificial Intelligence and the Future of the Human Mind*; Simon and Schuster: New York, NY, USA, 2006; Volume 1.
2. Roberson, P.N.; Shorter, R.L.; Woods, S.; Priest, J. How health behaviors link romantic relationship dysfunction and physical health across 20 years for middle-aged and older adults. *Soc. Sci. Med.* **2018**, *201*, 18–26. [CrossRef] [PubMed]
3. Alhouseini, A.M.A.; Al-Shaikhli, I.F.; bin Abdul Rahman, A.W.; Dzulkifli, M.A. Emotion Detection Using Physiological Signals EEG & ECG. *Int. J. Adv. Comput. Technol. (IJACT)* **2016**, *8*, 103–112.
4. Bozhkov, L.; Georgieva, P.; Santos, I.; Pereira, A.; Silva, C. EEG-based Subject Independent Affective Computing Models. *Procedia Comput. Sci.* **2015**, *53*, 375–382. [CrossRef]
5. Cruz, A.; Garcia, D.; Pires, G.; Nunes, U. Facial Expression Recognition Based on EOG Toward Emotion Detection for Human-Robot Interaction. *Comput. Sci.* **2015**, 31–37. [CrossRef]
6. Goshvarpour, A.; Abbasi, A.; Goshvarpour, A. An accurate emotion recognition system using ECG and GSR signals and matching pursuit method. *Biomed. J.* **2017**, *40*, 355–368. [CrossRef]
7. He, C.; Yao, Y.J.; Ye, X.S. An Emotion Recognition System Based on Physiological Signals Obtained by Wearable Sensors. In *Wearable Sensors and Robots*; Springer; Singapore, 2017; Volume 399, pp. 15–25. [CrossRef]
8. Kaur, B.; Singh, D.; Roy, P.P. EEG Based Emotion Classification Mechanism in BCI. *Procedia Comput. Sci.* **2018**, *132*, 752–758. [CrossRef]
9. Lahane, P.; Sangaiah, A.K. An Approach to EEG Based Emotion Recognition and Classification Using Kernel Density Estimation. *Procedia Comput. Sci.* **2015**, *48*, 574–581. [CrossRef]
10. Reis, E.; Arriaga, P.; Postolache, O.A. Emotional flow monitoring for health using FLOWSENSE: An experimental study to test the impact of antismoking campaigns. In Proceedings of the 2015 E-Health and Bioengineering Conference (EHB), Iasi, Romania, 19–21 November 2015; pp. 1–4. [CrossRef]

11. Roza, V.C.C.; Postolache, O.A. Emotion Analysis Architecture Based on Face and Physiological Sensing Applied with Flight Simulator. In Proceedings of the 2018 International Conference and Exposition on Electrical And Power Engineering (EPE), Iasi, Romania, 18–19 October 2018; pp. 1036–1040. [CrossRef]
12. Roza, V.C.C.; Postolache, O.A. Design of a Multimodal Interface based on Psychophysiological Sensing to Identify Emotion. In Proceedings of the 22nd IMEKO TC4 International Symposium & 20th International Workshop on ADC Modelling and Testing, Iaşi, Romania, 14–15 September 2017; Volume 1, pp. 1–6.
13. Shin, D.; Shin, D.; Shin, D. Development of emotion recognition interface using complex EEG/ECG bio-signal for interactive contents. *Multimedia Tools Appl.* **2017**, *76*, 11449–11470. [CrossRef]
14. Yin, Z.; Zhao, M.; Wang, Y.; Yang, J.; Zhang, J. Recognition of emotions using multimodal physiological signals and an ensemble deep learning model. *Comput. Methods Programs Biomed.* **2017**, *140*, 93–110. [CrossRef]
15. Capuano, A.S.; Karar, A.; Georgin, A.; Allek, R.; Dupuy, C.; Bouyakoub, S. Interoceptive exposure at the heart of emotional identification work in psychotherapy. *Eur. Psychiatry* **2017**, *41*, S783. [CrossRef]
16. Roza, V.C.C.; Postolache, O.A. Citizen emotion analysis in Smart City. In Proceedings of the 2016 7th International Conference on Information, Intelligence, Systems & Applications (IISA), Chalkidiki, Greece, 13–15 July 2016; Volume 1, pp. 1–6. [CrossRef]
17. Kumar, N.; Khaund, K.; Hazarika, S.M. Bispectral Analysis of EEG for Emotion Recognition. *Procedia Comput. Sci.* **2016**, *84*, 31–35. [CrossRef]
18. Lan, Z.; Sourina, O.; Wang, L.; Liu, Y. Real-time EEG-based emotion monitoring using stable features. *Vis. Comput.* **2016**, *32*, 347–358. [CrossRef]
19. Petrovica, S.; Anohina-Naumeca, A.; Ekenel, H.K. Emotion Recognition in Affective Tutoring Systems: Collection of Ground-truth Data. *Procedia Comput. Sci.* **2017**, *104*, 437–444. [CrossRef]
20. Yin, Z.; Wang, Y.; Zhang, W.; Liu, L.; Zhang, J.; Han, F.; Jin, W. Physiological Feature Based Emotion Recognition via an Ensemble Deep Autoencoder with Parsimonious Structure. *IFAC-PapersOnLine* **2017**, *50*, 6940–6945. [CrossRef]
21. Mishra, B.; Mehta, S.; Sinha, N.; Shukla, S.; Ahmed, N.; Kawatra, A. Evaluation of work place stress in health university workers: A study from rural India. *Indian J. Community Med.* **2011**, *36*, 39–44. [CrossRef]
22. Boeing. *Statistical Summary of Commercial Jet Airplane Accidents*; Worldwide Operations | 1959–2017; Boeing Aerospace Company: Seattle, WA, USA, 2017; pp. 1–26.
23. ICAO. Accident Statistics. In *Aviation Safety*; International Civil Aviation Organization: Montreal, QC, Canada, 2017.
24. McKay, M.P.; Groff, L. 23 years of toxicology testing fatally injured pilots: Implications for aviation and other modes of transportation. *Accid. Anal. Prev.* **2016**, *90*, 108–117. [CrossRef]
25. Choi, K.H.; Kim, J.; Kwon, O.S.; Kim, M.J.; Ryu, Y.H.; Park, J.E. Is heart rate variability (HRV) an adequate tool for evaluating human emotions?—A focus on the use of the International Affective Picture System (IAPS). *Psychiatry Res.* **2017**, *251*, 192–196. [CrossRef]
26. Shimmer3. Shimmer GSR+ Unit. Available online: https://www.shimmersensing.com/products/shimmer3-wireless-gsr-sensor (accessed on 13 December 2019).
27. Medlab, P. PEARL100L Medlab—Pulse Digital Desktop Pulse Oximeter. Medlab medizinische Diagnosegerte GmbH. Available online: https://www.medical-world.co.uk/p/pulse-oximeters/medlab-nanox/pulse-oximeter-medlab-pearl-100l-desktop/3791 (accessed on 13 December 2019)
28. Quesada Tabares, R.; Cantero, A.; Gomez Gonzalez, I.M.; Merino Monge, M.; Castro, J.; Cabrera-Cabrera, R. Emotions Detection based on a Single-electrode EEG Device. In *PhyCS 2017: 4th International Conference on Physiological Computing Systems (2017)*; SciTePress: Madrid, Spain, 2017. [CrossRef]
29. Stockli, S.; Schulte-Mecklenbeck, M.; Borer, S.; Samson, A. Facial expression analysis with AFFDEX and FACET: A validation study. *Behav. Res. Methods* **2017**, *50*. [CrossRef]
30. Danner, L.; Sidorkina, L.; Joechl, M.; Dürrschmid, K. Make a face! Implicit and explicit measurement of facial expressions elicited by orange juices using face reading technology. *Food Qual. Prefer.* **2014**, *32*, 167–172. [CrossRef]
31. Den Uyl, M.; van Kuilenburg, H. The FaceReader: Online Facial Expression Recognition. In Proceedings of the Measuring Behavior 2005, Wageningen, The Netherlands, 30 August–2 September 2005.

32. Murugappan, M.; Nagarajan, R.; Yaacob, S. Discrete Wavelet Transform Based Selection of Salient EEG Frequency Band for Assessing Human Emotions. In *Discrete Wavelet Transforms—Biomedical Applications*; IntechOpen: Seriab, Malaysia; Kangar, Malaysia, 2011; pp. 33–52. [CrossRef]
33. Min, Y.K.; Chung, S.C.; Min, B.C. Physiological Evaluation on Emotional Change Induced by Imagination. *Appl. Psychophysiol. Biofeedback* **2005**, *30*, 137–150. [CrossRef]
34. Umeda, S.; Emotion, Personality, and the Frontal Lobe. In *Emotions of Animals and Humans: Comparative Perspectives*; Watanabe, S., Kuczaj, S., Eds.; Springer: Tokyo, Japan, 2013; pp. 223–241. [CrossRef]
35. Rosso, I.; Young, A.D.; Femia, L.A.; Yurgelun-Todd, D.A. Cognitive and emotional components of frontal lobe functioning in childhood and adolescence. *Ann. N. Y. Acad. Sci.* **2004**, *1021*, 355–362. [CrossRef] [PubMed]
36. Othman, M.; Wahab, A.; Karim, I.; Dzulkifli, M.A.; Alshaikli, I.F.T. EEG Emotion Recognition Based on the Dimensional Models of Emotions. *Procedia Soc. Behav. Sci.* **2013**, *97*, 30–37. [CrossRef]
37. Al-Fahoum, A.; A Al-Fraihat, A. Methods of EEG Signal Features Extraction Using Linear Analysis in Frequency and Time-Frequency Domains. *ISRN Neurosci.* **2014**, *2014*. [CrossRef] [PubMed]
38. Mallat, S. *A Wavelet Tour of Signal Processing*; Elsevier: Amsterdam, The Netherlands, 2009; pp. 1–805.
39. Al-Qazzaz, N.; Bin Mohd Ali, S.H.; Ahmad, S.A.; Islam, M.S.; Escudero, J. Selection of Mother Wavelet Functions for Multi-Channel EEG Signal Analysis during a Working Memory Task. *Sensors* **2015**, *15*, 29015–29035. [CrossRef]
40. Golinska, A.K. Poincaré Plots in Analysis of Selected Biomedical Signals. *Stud. Logic Gramm. Rhetor.* **2013**, *35*, 117–126. [CrossRef]
41. Piskorski, J.; Guzik, P. Geometry of the Poincaré plot of RR intervals and its asymmetry in healthy adult. *Physiol. Meas.* **2007**, *28*, 287–300. [CrossRef]
42. Tayel, M.B.; AlSaba, E.I. Poincaré Plot for Heart Rate Variability. *World Acad. Sci. Eng. Technol. Int. J. Med. Health Biomed. Bioeng. Pharm. Eng.* **2015**, *9*, 708–711.
43. Haykin, S.O. *Neural Networks and Learning Machines*; Pearson Education: Newmarket, ON, Canada, 2011; pp. 1–936.
44. Marsland, S. *Machine Learning: An Algorithmic Perspective*; Chapman and Hall/CRC: Boca Raton, FL, USA, 2015; pp. 1–457.

sensors

MDPI

Article

Dilated Skip Convolution for Facial Landmark Detection

Seyha Chim, Jin-Gu Lee and Ho-Hyun Park *

School of Electrical and Electronics Engineering, Chung-Ang University, 84 Heukseok-ro, Dongjak-gu, Seoul 06974, Korea; seyhachim@cau.ac.kr (S.C.); dlwlsrn21@cau.ac.kr (J.-G.L.)

* Correspondence: hohyun@cau.ac.kr; Tel.: +82-10-3354-9180

Received: 26 September 2019; Accepted: 2 December 2019; Published: 4 December 2019

Abstract: Facial landmark detection has gained enormous interest for face-related applications due to its success in facial analysis tasks such as facial recognition, cartoon generation, face tracking and facial expression analysis. Many studies have been proposed and implemented to deal with the challenging problems of localizing facial landmarks from given images, including large appearance variations and partial occlusion. Studies have differed in the way they use the facial appearances and shape information of input images. In our work, we consider facial information within both global and local contexts. We aim to obtain local pixel-level accuracy for local-context information in the first stage and integrate this with knowledge of spatial relationships between each key point in a whole image for global-context information in the second stage. Thus, the pipeline of our architecture consists of two main components: (1) a deep network for local-context subnet that generates detection heatmaps via fully convolutional DenseNets with additional kernel convolution filters and (2) a dilated skip convolution subnet—a combination of dilated convolutions and skip-connections networks—that are in charge of robustly refining the local appearance heatmaps. Through this proposed architecture, we demonstrate that our approach achieves state-of-the-art performance on challenging datasets—including LFPW, HELEN, 300W and AFLW2000-3D—by leveraging fully convolutional DenseNets, skip-connections and dilated convolution architecture without further post-processing.

Keywords: face landmark detection; fully convolutional DenseNets; skip-connections; dilated convolutions

1. Introduction

In computer vision, facial landmark detection is known as face alignment and is a crucial part of face recognition operations. Its algorithms attempt to predict the locations of the fiducial facial landmark coordinates that vary owing to head movements and facial expressions. These landmarks are located at major parts of the face, such as the contours, tip of the nose, chin, eyes, corners of the mouth (see [1] in review). Facial landmark detection has sparked much interest recently as it is a prerequisite in many computer vision applications, including facial recognition [2], facial emotion recognition [3,4], face morphing [2,5], 3D face modelling [6] and human-computer interactions [7]. In recent years, considerable research works [8–10] have developed remarkable networks to predict facial landmark location more accurately even under challenging conditions, such as large appearance variations, facial occlusion and difficult illumination. Facial landmark detection is classified into three types of methods: holistic, constrained local model (CLM), and regression-based. Among these, regression-based approaches [5,11] have demonstrated superiority in both efficiency and accuracy, even in challenging scenarios. Regression-based methods contain two stages: early and updated. The inceptive key points are located on the predicted face shape in the early stage and gradually refined in the updated stage. However, [1] points out two main issues of this approach. The first issue is the sensitivity of the face detector. Commonly, the face is initially determined by the face bounding box. In the case it fails to detect the face in the first place, the accuracy also declines. Another issue is that the algorithms apply a fixed number of predictions, so it is impossible to judge the quality of the landmark prediction and adapt the necessary stages for different image tests.

Before the success of deep learning [9,12] for computer vision problems, [13] used a scale-invariant feature transform (SIFT) algorithm to learn appearance models from current landmarks. The algorithm iteratively regresses the models until the convergence criteria are reached. Recently, discriminative models such as convolutional neural networks (CNNs) and recurrent neural networks (RNNs) have dominated the field of facial landmark detection. Deep learning based models have been shown to outperform SIFT based models, which use hand-crafted features, for many vision tasks [14]. Hierarchical deep learning structures, in particular CNNs, can generate feature descriptors that capture more complex image characteristics and learn task specific features. In contrast, SIFT is not robust to non-linear transformations, particularly where SIFT cannot match sufficient feature points. It is unsuitable for data with large intra-class shape variations. Consequently, deep learning has attracted more attention than SIFT for computer vision applications. In early research [15], a probabilistic deep model for facial landmark detection that captured facial shape variations caused by poses and expressions was used. Also, [16] proposed to extract shape-indexed deep features from fully convolutional networks (FCNs) and refine the landmark locations recurrently via recurrent attentive-refinement (RAR) networks. In the early stage of [16]'s study, the network employed direct methods to regress key points directly on given images that are highly non-linear and difficult to estimate key point positions.

The research in [17] argues that learning indirectly to extract discriminative features from images yields more advantages over direct mapping. Accordingly, [17] applies an indirect prediction framework based on heatmap regression at individual body key points over the raw image. Furthermore, [17] mentions that adding several large convolutions (e.g., 13×13 kernel convolution) would improve estimation performance, although this increases the number of parameters and makes optimizations more difficult.

To address this problem, [18] pursued dilated convolutions that increase the effective receptive fields without introducing additional parameters. Intuitively, applying heatmap regression methods in a network of large convolutional kernels and deeper models enhances the performance of overall networks. Thus, we propose a deep end-to-end model which leverages fully convolutional DenseNets (FC-DenseNets) [19] that use heatmap regression to learn deep feature maps from the given image. Moreover, inspired by [18], we carefully designed a network that can extract more complex data dependencies by building extra skip-connections in the stacked dilated convolutions network. In doing so, we expect that our network will obtain different sizes of receptive fields and informative feature maps, which will boost prediction accuracy.

The main contributions of this work are as follows:

- To the best of our knowledge, this is the first work to exploit FC-DenseNets as a local detector with a heatmap regression to predict dense heatmaps from the given image.
- We designed a thorough dilated skip convolution (DSC) network that can refine the estimated heatmaps of the facial key points by combining a stack of dilated convolutions and a skip-connections method.
- We developed a robust method to estimate the initial facial shape to work in challenging conditions.
- We evaluated our framework's performance with other state-of-the-art networks on LFPW [20], HELEN [21], 300 W [22] and AFLW2000-3D [23] datasets.

The rest of this paper is organized as follows. First, a summary of our paper's relevant works is given in Section 2. Next, we present in detail our proposed methodology in Section 2. Then, the results of our experiments are presented in Section 4. Finally, the conclusions are drawn in Section 5.

2. Related Works

Facial landmark detection is divided into three types of methods: holistic, constrained local model (CLM) and regression-based. Holistic methods build a global model to learn the facial appearance and obtain shape information during training to estimate the best fits of any given test face image during testing via the model parameters. CLM methods use independent local appearance information around each landmark combined with a global face shape model for facial landmark detection, outperforming holistic methods for capturing illumination and occlusion. Unlike the first two methods, which build a global shape model, regression-based methods directly map the local facial appearance and regress the landmark locations between individual inputs and outputs.

2.1. Regression-Based Methods

Regression based methods have recently demonstrated outstanding performance compared with holistic and CLM methods. Regression based methods effectively build a parametric face shape or appearance model to extract feature maps from an image and infer a facial shape. Regression functions initially focus on holistic picture details, subsequently updating those features using finer image details to provide more accurate predictions. Using typical approaches, [5,11] proposed a regression function to predict landmark coordinates from shape indexed feature maps from the input image. Subsequently, [24] proposed a combined regression network to initially detect facial landmarks and then refine landmark locations using their scoremaps at progressively finer detail; and [25] proposed a cascade stacked auto-encoder network to produce finer images from low resolution input images; and [26] proposed multiple cascaded regressors to learn discriminative features around each facial landmark. Extending this early work, [27] proposed a two-step facial segmentation network to estimate head pose, gender and expression. The system first segmented face images into semantically small regions, for example hair, skin, nose, eyes, background, mouth and so forth.; and then classified thee regions using support vector machines (SVMs). The [27] process is effectively an extended version of the FASSEG dataset [28]. Rather than directly manipulating images in the spatial domain, [3,4] represented images as signals in the frequency domain with high time-frequency resolution. They then extracted useful feature maps from the decomposed image and employed supervised learning algorithms to classify facial expressions in the images. Ref. [3] applied stationary wavelet entropy to extract features in the frequency domain followed by a single hidden layer feedforward neural network, using the Jaya algorithm, a gradient-free optimizer. Similarly, [4] proposed biorthogonal wavelet entropy to extract multi-scale information and employed fuzzy multiclass SVM classifiers. Heatmap regression has also been used to estimate human pose, [29,30] and detect facial landmarks, [8–10]. Ref. [29] employed multiple regressors to predict human poses. The first regressor crops the input image to focus only on the human torso, reducing required computational resources for background analysis. [29] used subsequent regressors to roughly estimate joint locations and then crop joint centers and repeatedly regress the image. This not only considerably reduces the number of network parameters but also increases prediction accuracy since there is no information loss compared with using pooling layers to reduce data size. Ref. [30] proposed a stacked hourglass network to capture information from local to global scale and hence enable the network to learn spatial relationships between joints. Similarly, [8] cascaded four stacked hourglass networks in heatmaps regression to extract discriminative features from images, which were subsequently used to detect facial landmarks. Ref. [9] proposed a three step regression network based on convolutional response maps and component based models to robustly detect facial landmarks. Ref. [10] proposed combining heatmap and coordination contextual information into a feature representation that was subsequently refined by an arbitrary convolutional neural network (CNN) model.

2.2. Fully Convolutional Heatmap Regression Methods

Early methods used heatmap regression as an approach for 2D pose estimation [5,8,17,31]. Unlike the holistic regression methods, heatmap regression methods have the benefit of providing higher output resolutions that assist in accurately localizing the key points in the image via per-pixel predictions. To leverage this advantage, [17,31] regress a heatmap over the image for each key point and then obtain the key point position as a mode in this heatmap. Ref. [31] presents a convolutional network architecture incorporating motion features as a cue for body part localization and [17] proposes a CNN model to predict 2D human body poses in an image. The model regresses a heatmap representation for each body key point, learning and representing both partial appearances and the context of those partial configurations. In contrast, [5,8] exploit FCNs to estimate dense heatmaps for facial landmark detection. Ref. [5] proposes a two-step detection followed by a regression network to create the detection score map for each landmark, whereas [8] uses a stacked hourglass network for 2D and 3D face alignment.

Fully Convolutional DenseNets

Densely connected convolutional networks (DenseNets) [32] introduce a connectivity pattern that proves the gradient-vanishing problem can be solved even though the depth of CNN is increased. At the same time, the number of parameters can be reduced by connecting each layer with additional inputs from all preceding layers and reusing its feature maps in all subsequent layers. Recently, FC-DenseNets [19] extend DenseNets to be a fully convolutional network that achieves state-of-the-art results by tackling problem semantics with image segmentation. The resulting network is a deep network between 56 and 103 layers that has very few parameters. The goal of FC-DenseNets is to further exploit feature reuse by extending the more sophisticated DenseNets architecture while avoiding feature explosion at the upsampling path of the network. To recover the input spatial resolution, FC-DenseNets implicitly inherit the advantages of DenseNets that use pooling operations and dense blocks (DBs) to perform iterative concatenation of feature maps. The feature maps have a sufficiently large amount of detailed spatial information. To some extent, heatmap regression through FC-DenseNets is especially useful for multiple outputs per input (e.g., multiple faces).

FC-DenseNets are constructed from two symmetric parts where the downsampling part is an exact mirror of the upsampling part as shown in Figure 1. FC-DenseNets consist of 11 DBs: 5 DBs in the downsampling part followed by its own transitions down (TD), 5 DBs in the upsampling part followed by its own transitions up (TU) and one DB in the middle and so-called "bottleneck". Each DB layer is composed of dense layers followed by batch normalization [33] and ReLU [34]. The solid line in Figure 1 represents the connection between each dense block of the fully convolutional DenseNet (FC-DenseNet), which passes output feature maps forward from one dense block to the next, whereas the dashed line indicates skip connections between FC-DenseNet downsampling and upsampling paths. The overall FC-DenseNet goal is to capture spatially detailed information from the downsampling path and recover it in the upsampling path by reusing the features maps. The last layer in the network is a 1×1 convolution followed by a softmax nonlinearity function to predict the class label.

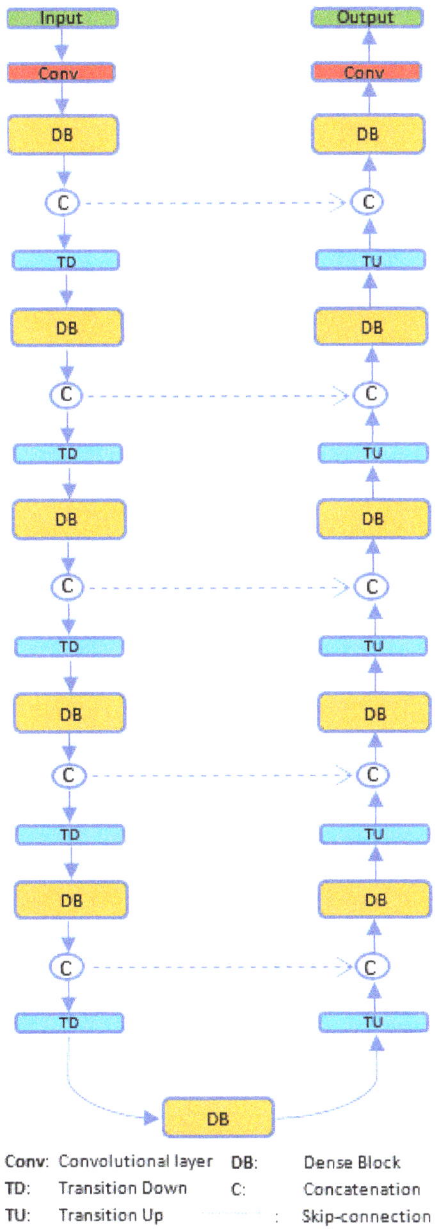

Figure 1. FC-DenseNet architecture.

2.3. *Dilated Convolutions*

Dilated (or atrous) convolutions have been widely utilized for various dense prediction and generation applications. As indicated in Reference [35], dilated convolutions enlarge exponentially receptive fields without loss of resolution or convergence while the number of parameters grows linearly. Larger kernel receptive fields can increase network capability to capture spatial context, which is beneficial to reconstruct large and complex edge structures. However, ordinary convolutions require a large number of parameters to expand their receptive fields. In contrast to ordinary convolutions, dilated convolution has zero-padding inside its kernels, injecting zeros into defined gaps to expand receptive field size, as shown in Figure 2. Thus, dilated convolutions can view larger input image portions without requiring a pooling layer, resulting in no spatial dimension loss and reduced computational time.

For semantic segmentation tasks, Reference [35] presents a new convolutional architecture that fully exploits dilated convolutions for multi-scale context aggregation. Reference [36] proposes two simple, yet effective, gridding methods by studying the decomposition of dilated convolutions. In these studies, dilated convolutions replace the need to upsample parts to keep the output resolutions the same as the input size. For other tasks such as audio generation [37], video modeling [38] and machine translation [39], dilated convolutions are used to capture global views of inputs with fewer parameters. WaveNet [37] was proposed by Google DeepMind and employs dilated convolutions to generate and recognize speech from raw audio waveforms. The dilation factor in Reference [37] is doubled, starting from 1 to a fixed factor number for every forward layer; then, the pattern is repeated.

Figure 2 illustrates how dilated convolutions enlarge the receptive fields by altering dilation factors (d). When dilation factors are increased exponentially, the gap pixels between the original kernel elements get progressively wider; this causes the receptive field to expand. In Figure 2a, a dilation factor of 1 (1-Dilated convolution) is performed in a dense 3×3 field on a feature map. We observed that the 1-Dilated convolution is the same as the 3×3 standard convolution filter. When the dilation factor is set to 2 as shown in Figure 2b, the region of the receptive field is increased dramatically to 7×7 pixels. The same occurs in Figure 2c when the dilation factor is changed to 4 and the receptive field is 15×15 pixels. In Figure 2, the group of red boxes is a 3×3 input filter that captures the receptive field (represented by the gray area) and the blue number indicates the meaning of the dilation factors that are applied to the kernels. The most important factor is the number of space pixels between the original kernel elements. In our work, we stack 7 dilated convolution layers with different dilation factors together to perceive a wider range for capturing global contexts of input feature maps.

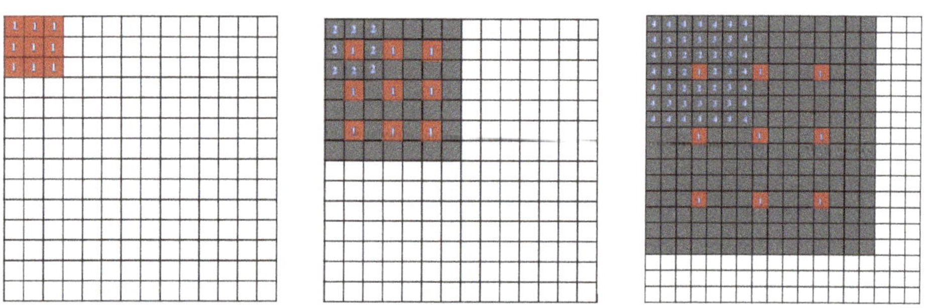

(**a**) Conventional convolution $d = 1$ (**b**) Dilated convolution with $d = 2$ (**c**) Dilated convolution with $d = 4$

Figure 2. Conventional convolution and dilated convolution.

3. Methods

The proposed facial landmark detection architecture is illustrated in Figure 3. We divide our approach into two connected sub-parts: the local appearance initialization (LAI) subnet and the dilated skip convolution (DSC) subnet for shape refinement. LAI pursues a heatmap regression approach convolved with kernel convolution to serve as a local detector of facial landmarks and the DSC subnet is designed to refine the local prediction of the first subnet.

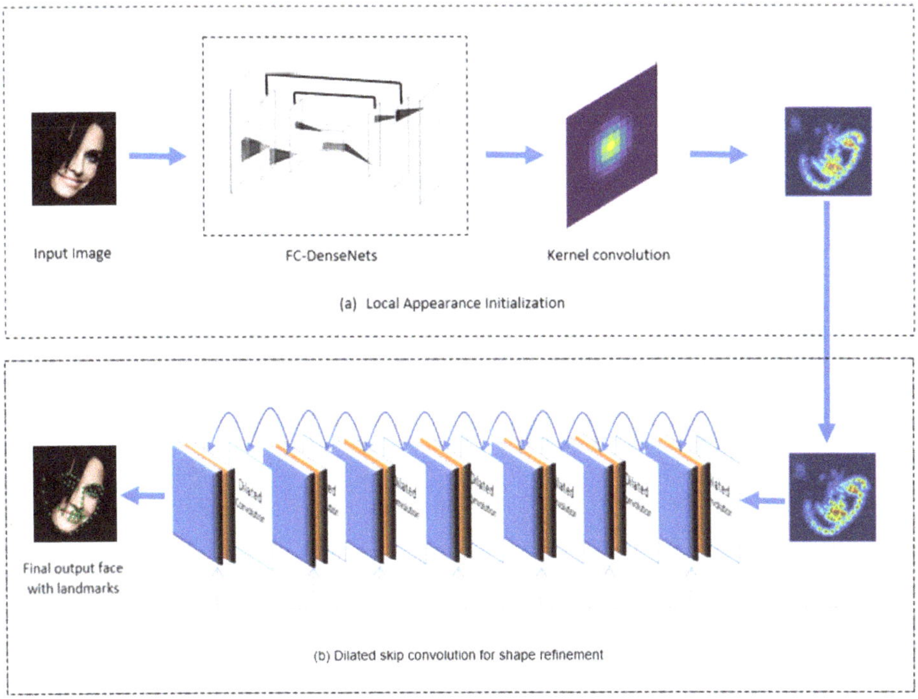

Figure 3. Overview of the proposed approach for facial landmark detection.

3.1. Local Appearance Initialization Networks

It is well known that facial landmark detection uses single specific pixel location data $p(x, y)$ as a training label where x and y are pixel coordinates in 2D images. However, using the training label data as a single-pixel point $p(x, y)$ is inefficient for learning features from the input data. Even though the model returns a result close to the ground-truth pixel, a result that does not comply with the exact pixel location data $p(x, y)$ may be considered wrong; as a result, the model may search for another pattern despite being close to the answer.

Recently, Gaussian distribution has come into play for manipulating the training label into a Gaussian heatmap label. It modifies the training label, not as a single specific point $p(x, y)$ but rather as probabilities near the given training label pixel point. References [24,40] present several successful heatmap implementations in facial alignment. As presented by both papers, using heatmaps as a training label allows the network to learn faster. Furthermore, heatmaps demonstrate how the network is thinking during training since heatmaps are more visible to the naked eye. The correct point will have the highest probability in the distribution, whereas the neighboring pixels close to the correct pixel will also have high probabilities but not as high as that of the correct pixel. In Equation (1), the value of $\hat{p}_i$ will let the network know whether or not it is making a guess close to the ground-truth rather than penalizing a guess that deviates by a small number of pixels. During the training, network weight $\boldsymbol{w}$ and bias $\boldsymbol{b}$ are learned in predicted heatmaps $h_i(p; w, b)$.

$$\hat{p}_i = \arg\max_p h_i(p; w, b). \tag{1}$$

The output of a network will now be a continuous probability distribution on an input image plane, making it easier to see where the network's guess is confident; in contrast, having a single position as an output does not show how the network is guessing.

Our goal in the first part of the network is to obtain the output feature maps that contain sufficient pixel-level details, high-resolution outputs that remain the same size as the input image (no resolution loss) and less extensive computation. A FCNs-based heatmap regression, followed by a kernel convolution, is used to meet our goal. To do so, we initially transform the facial landmarks' ground-truth location $p_i^{gt}(x,y)$ of ith key point into target heatmap $h_i^{gt}(p)$ of ith key point (Figure 4a) via 2D Gaussian kernel (Equation (2)). Then, the target heatmap $h_i^{gt}(p)$ are fed into FC-DenseNets and finally convolved with a kernel convolution as illustrated in Figure 4b. In fully convolutional heatmap regression fashion, the task becomes one of predicting per-pixel likelihood of each key point's heatmap from the image. It regresses the target heatmap of each landmark $h_i^{gt}(p)$ directly to obtain the response map $M(p)$ stated in Equation (3), which has the same resolution as the input image.

We transform ground-truth location $p_i^{gt}(x,y)$ to target heatmap $h_i^{gt}(p)$ as

$$h_i^{gt}(p) = \frac{1}{\sigma\sqrt{2\pi}} \exp\left(-\frac{\|p - p_i^{gt}\|}{2\sigma^2}\right), p \in \Omega, \tag{2}$$

where σ is the standard deviation for the heatmaps used to control the response scope and Ω is the set of all pixel locations in image I.

We set the FC-DenseNet architecture to include 56 layers following Reference [19], which had FC-DenseNet56 with 4 layers per dense block and growth rate = 12. We adopted the smallest FC-DenseNet to reduce network computational complexity, as shown in Table 1, while still achieving notable outcomes compared with current popular architectures. We also applied fully convolutional ResNets with 50 layers (FC-ResNets50 [41]), available in the PyTorch framework [42] (Torchvision) and then compared the outcomes with fully convolutional DenseNets with 56 layers (FC-DenseNet56). As expected, FC-DenseNets56 outperformed FC-ResNets50 due to more depth and hence more parameters.

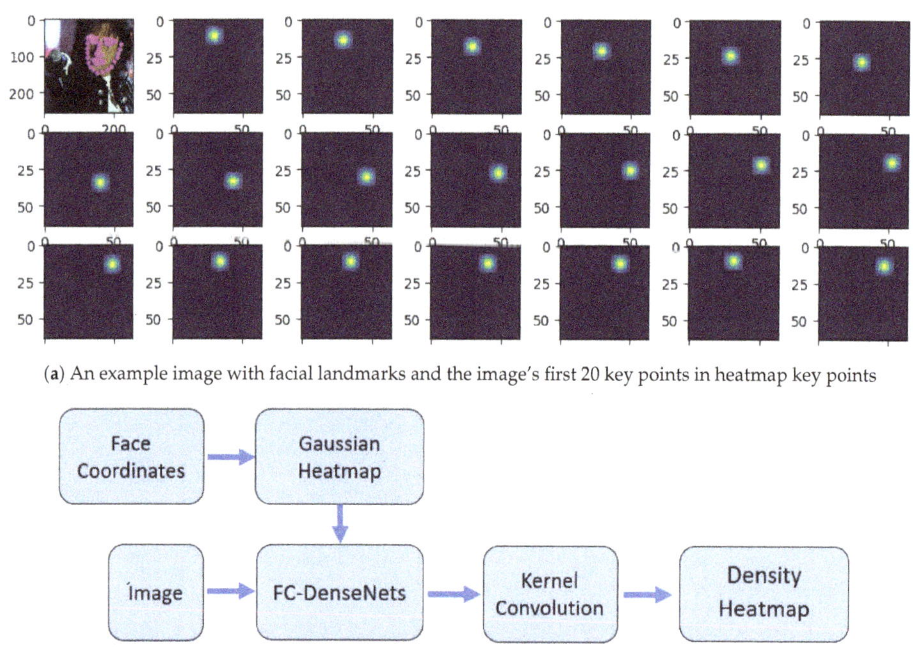

(**a**) An example image with facial landmarks and the image's first 20 key points in heatmap key points

(**b**) Local Appearance Initialization Diagram

Figure 4. Local appearance initialization network.

Table 1. Architecture of FC-DenseNet56 used in the LAI network.

Layer	Number of Feature Maps
Input	3
3 × 3 convolution	36
DB (4 layers) + TD	84
DB (4 layers) + TD	144
DB (4 layers) + TD	228
DB (4 layers) + TD	348
DB (4 layers) + TD	492
DB (4 layers)	672
DB (4 layers) + TU	816
DB (4 layers) + TU	612
DB (4 layers) + TU	434
DB (4 layers) + TU	288
DB (4 layers) + TU	192
1 × 1	68 (keypoints)

Kernel Convolution

The output of FC-DenseNets is in a channel-wise fashion that has the same resolution as the input image. After reaching the output resolution of the network, an implicit 45 × 45 pixel kernel convolution K_σ is applied to produce a clear shape output of the feature maps. For computational efficiency, the kernel convolution K_σ was generated by the Gaussian function in Equation (2). Here, the kernel convolution filter acts as a point-spread function to blur the input feature maps as shown in Figure 5. The kernel convolution filter K_σ removes the detail and noise and provides gentler smoothing by preserving the edges of the feature maps. Without the kernel convolution, landmarks' sub-pixel positions are neglected [43].

The kernel convolution filter convolves with the entire image using grouped convolution [44], which allows for more efficient learning and improved representation. In grouped convolutions, each input channel is convolved with its own filter. The final output of the network is a set of heatmaps that contain the probability of each key point's presence at each pixel. With the convolved response maps $M(p) = [h_i^{gt}(p)|i = 1...N]$ and a kernel convolution filter K_σ, we can obtain the density heatmap H^0 as follows:

$$H^0 = M(p) * K_\sigma \tag{3}$$

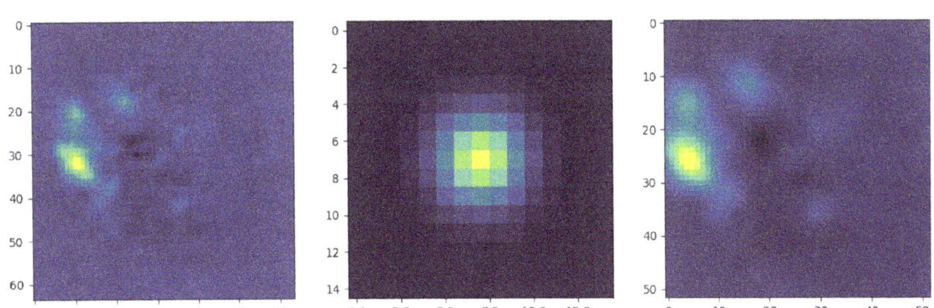

Figure 5. Best viewed in color. **Left**: Output of FC-DenseNets. **Middle**: Visualization of kernel convolution filter (K_σ). **Right**: Feature map after applying the filter (K_σ).

3.2. Dilated Skip Convolution Network for Shape Refinement

To enable networks to learn the spatial relationships between each key point and make better guesses, it must be able to view large portions of the input images. The portion of the input image viewed by the network is called the receptive field. Using the vanilla convolution filter [45] is a challenge when using a large receptive field: it is computationally expensive and can be easily overfitted due to the vast number of parameters. This problem is usually tackled by using pooling layers in conventional CNNs. Pooling layers choose one pixel from its field and discard other information, thereby reducing information and resolution of the input image. This degrades the performance of the network because some important information is lost when the resolution is decreased. Fortunately, dilated convolutions [37] solve this problem by using sparse kernels to alternate the pooling and convolutional layer, which dilates the kernels with zeros as a result of not only affecting the number of parameters but also increasing the size of the receptive field. In practice, kernels with different dilation factors are convoluted to the input and the outputs of those kernels are concatenated for subsequent layers [9]. Subsequent layers have no missing information from the input image and fewer parameters with different receptive fields. To apply this concept, References [18,46] introduced a stack of dilated convolutions in their network that can enlarge the receptive field exponentially while keeping the number of parameters low. Inspired by this design, we constructed a dilated skip convolution network that combined seven consecutive zero-padded dilated convolutions and skip-connections to overcome the issue of scale variations. In the network, our dilation factors ranged from d = 1 to d = 32 as stated in Table 2.

This module was carefully designed to increase the performance of our dense prediction architecture and ensure accurate spatial information by aggregating multi-scale contextual information. Our objective was to combine intermediate feature representations to learn global-context information and improve the final heatmap predictions. We exploited dilated convolutions to extract the global-context from input feature maps and then progressively updated the initial heatmap (H^0). Due to the capacity to capture texture information at the pixel level, concatenating dilated convolutions of sub-layers together aids the network-extracting features from different scales concurrently. We also built extra skip-connections and embedded them in our dilated convolutions network to add global information from the entire image to common knowledge of the network from the previous feature map $(H_{\tau-1}[.])$. During the training, skip-connections concatenated output feature maps from previous and current layers together. Thus, our dilated skip convolution's feature map $H^{DSC}[.]$, which has current feature map H_τ, previous feature map $H_{\tau-1}[.]$, kernel filter $k[.]$ and dilation factor d, is defined as:

$$H^{DSC}[x,y] = \sum_i \sum_j k[i,j] \cdot H_\tau[x-di, y-dj] + H_{\tau-1}[x,y]. \tag{4}$$

Intuitively, Equation (4) shows that the model learns from each dilated convolution layer and the input initial heatmap, H^0, providing robustness against appearance changes. This is achieved through skip connections, which are extra connections between H^0 and dilated layers with different dilation factors, d. Consider the output feature map for the nth layer, H^n and a non-linear transformation of the nth layer, $T_n(.)$. At each stage, the kernel, $k[.]$, convolves with H^n and then concatenates it with H^0. Thus, from Table 2, the network from the initial to the final output feature map for a DSC subnet with 7 dilation factors can be formulated as

$$\begin{cases} H^1 = T_1(H^0) \\ H^2 = T_2([H^0, H^1]) \\ \vdots \\ H^7 = T_7([H^0, \cdots, H^6]), \end{cases} \tag{5}$$

where $[H^0, H^n]$ donates feature map concatenation.

Rather than having the dilated skip convolution network predicting the landmark locations from scratch in Equation (6), it is beneficial to refine the LAI subnet predictions. This was achieved by summing H^0 and H^{DSC} to obtain the final feature map of the architecture,

$$H^f = H^0 + H^{DSC}. \tag{6}$$

To better understand how the heatmap is regressed in a real image, we transferred back H^0, H^{DSC}, H^f to S^0, S^{DSC}, S^f. Thus, Equation (6) was replaced as follows:

$$S^f = S^0 + S^{DSC}. \tag{7}$$

Figure 6 compares visualizations of landmark coordinates (green dots) in the real face image for both stages. Landmark coordinates from Figure 6a are improved in the second stage, for example the green dots with red circles in Figure 6b locate more correctly on the face contour and there is no missed landmark on the left eyebrow compared to Figure 6a.

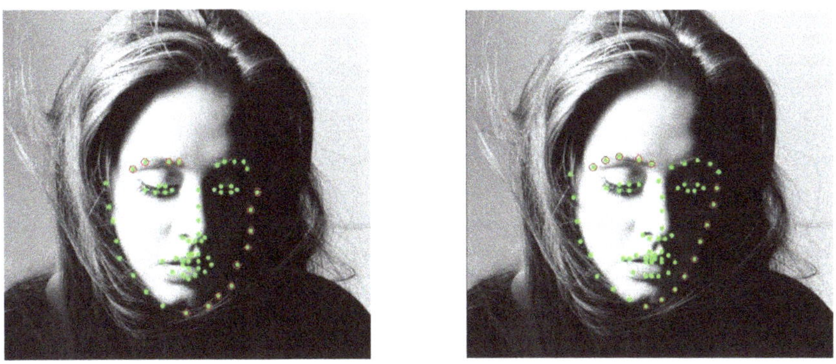

(**a**) First stage: Initial shape (S^0) from LAI subnet (**b**) Second stage: Final shape refinement (S^f)

Figure 6. Dilated skip convolution network for shape refinement.

Thus, dilated convolutions offer a method to increase global view exponentially on input image, hence the dilation factors should be set as exponential values following [35],

$$d_{(i+1)} = 2^i, \quad \text{for} \quad i = (0, 1, 2, \cdots, n-2), \tag{8}$$

where $d_{(i+1)}$ is the dilation factor for the $(i+1)^{th}$ layer and n is the number of layers. In this case, the dilated convolution has 7 layers, hence optimal dilation factors $d_{(i+1)} \leq 32$, for $i = (0, 1, \cdots, 5)$. Table 2 shows dilation factors = 1, 1, 2, 4, 8, 16, 32, where the first two layers serve as conventional convolution layers.

Table 2. Structure of dilated convolutions.

Filter Size	Dilation Factor	Activation Function
3 × 3	d = 1	ReLU
3 × 3	d = 1	ReLU
3 × 3	d = 2	ReLU
3 × 3	d = 4	ReLU
3 × 3	d = 8	ReLU
3 × 3	d = 16	ReLU
3 × 3	d = 32	ReLU

Table 2 compares the proposed method's using the mean error rate of the datasets, which should ideally be as small as possible. Thus, we need to find the optimal number of dilated layers most suitable for our entire network. Table 2 shows the optimal number of dilated layers = 7. Increasing the number of layers beyond that does not significantly improve the mean error rate, while introducing more parameters for the network and aggressively widening the receptive field via dilation factors would be detrimental to local features of small objects.

Algorithm 1 Dilated skip convolution for facial landmark detection

```
for t ← 1 to N_step do
    for all training images (I, H) do
        Feed I into FC-DenseNets and get the response maps M
        Obtain the density map H^0 by using Equation (3)
        Using Equation (4) to calculate H^DSC
        Regress H^0 to get H^f by using Equation (6)
        Optimize parameter Θ in Equation (9) with RMSprop, using loss L and target correction H
    end
end
```

4. Experiments

4.1. Datasets and Data Augmentation

4.1.1. Datasets

To evaluate the proposed algorithms, various datasets were created to investigate the robustness of the algorithms for imitating landmark detection in real-life situations. The datasets contained independent variations in pose, expression, illumination, background, occlusion and image quality. For instance, the 300W dataset [22] consisted of a wide range of head pose images and AFLW2000-3D [43] contained large-scale images in 3D. For training and validation, we used 300W-LP [23], a synthetically expanded version of 300W, as a basis to train our model. The model was fine-tuned with LFPW, HELEN and 300W datasets. To observe how the network was flexible with unseen datasets, we analyzed the AFLW2000-3D dataset without training it in advance, as presented in Table 3. In our evaluation experiments, we implemented our proposed algorithm (Algorithm 1) in "in-the-wild" datasets as follows:

- 300W-LP [23]: 300W Large Pose (300W-LP) dataset consists of 61,225 images with 68 key points for each facial image in both 2D landmarks and the 2D projections of 3D landmarks. It is a synthetically-enlarged version of the 300W for obtaining face appearance in larger poses.
- LFPW [20]: The Labeled Face Parts in-the-Wild (LFPW) dataset has 1035 images divided into two parts: 811 images for training and 224 images for testing.
- HELEN [21]: HELEN consists of 2000 training and 330 test images with highly accurate, detailed and consistent annotations of the primary facial components. It uses annotated Flickr images.
- 300W [22]: The 300 faces in-the-Wild (300W) dataset consists of 3148 images with 68 annotated points on each face for training sets collected from three wild datasets such as LFPW [20], AFW [47] and HELEN [21]. There are three subsets for testing: challenging, common and full set. For the challenging subset, we collected the images from iBUG [48] dataset which contains 135 images; for the common subset, we collected 554 images from the testing sets of HELEN and LFPW datasets; for the full set subset, we merged the challenging and common subsets (689 images).
- AFLW 2000-3D [23]: Annotated Facial Landmarks in the Wild with 2000 three-dimensional images (AFLW 2000-3D) is a 3D face dataset constructed with 2D landmarks from the first 2000 images with yaw angles between ±90° of AFLW [49] samples. It varies expression and illumination conditions. However, some annotations, especially larger poses or occluded faces, are not very accurate.

Table 3. The list of face datasets used for training and testing.

Dataset	Landmark	Pose	Image
	Training		
HELEN	68	±45°	2000
LFPW	68	±45°	811
300W	68	±45°	3148
300W-LP	68	±90°	61,225
	Testing		
HELEN	68	±45°	330
LFPW	68	±45°	224
300W	68	±45°	689
AFLW2000-3D	68	±90°	2000

4.1.2. Data Augmentation

For data augmentation (e.g., randomly flipping, resizing and cropping images, etc.), PyTorch framework [42] leaves the original input images untouched, returning only a changed copy at every batch generation.

To reduce overfitting in our model, we artificially expanded the amount of training data using random augmentation including cropping, rotation, flipping, color jittering, scale noise and random occlusion. We rotated the input image with a random angle of $\pm 50^\circ$ and scale noise from 0.8 to 1.2. We also scaled the longest side to 256 resulting in a $256 \times H$ or $H \times 256$ image, where $H \leq 256$.

4.2. Experimental Setting

4.2.1. Implementation Detail

We implemented our model based on the open source PyTorch framework [42], which is a dynamic program that runs on a GPU. First, we cropped an input image to 256×256 resolution and generated an output set of response maps with the same resolution. Then, we transferred the image's facial key points to heatmap key points using the 2D Gaussian kernel. In our method, the variance (sigma) of the 2D Gaussian kernel in the ideal response map was set to 0.25. For training, we optimized the network parameters by RMSprop [50] with a momentum of 0.9 and a weight decay of 10^{-4}. We trained our model for 100 epochs with an initial learning rate of 10^{-4}. We reduced it subsequently to 10^{-5} after 50 epochs and to 10^{-6} after another 80 epochs.

For loss function in our network, we chose the Euclidean distance loss function for our network,

$$L(\Theta) = \frac{1}{N} \sum_{i=1}^{N} \|Z(X_i; \Theta) - Z_i^{gt}\|_2^2 \tag{9}$$

where N is the size of the training batch and $Z(X_i, \Theta)$ is the output generated by the DSC network with parameters shown as Θ. X_i represents the input images and Z_i^{gt} is the ground-truth result of input image X_i.

During training, $L(\Theta)$ calculates the difference between the estimated and corresponding ground-truth feature map to update weight parameter Θ, to ultimately identify a set of parameters that make $L(\Theta)$ as small as possible.

4.2.2. Evaluation

We evaluated for accuracy with three popular metrics: the normalized mean error (NME), the cumulative error distribution (CED) curve and the area under the curve (AUC). The NME was evaluated by measuring the distance between the detected landmark coordinates and the ground-truth facial landmark coordinates. It calculates the mean of the inter-pupil distance of multiple images which can be represented by

$$NME = \frac{1}{n}\sum_{i=1}^{n}\frac{\|x_i - x_i^{gt}\|^2}{d}, \tag{10}$$

where x^i is the predicted coordinates and x_i^{gt} is the ground-truth coordinates for i^{th} image, d donates inter-ocular distance (Euclidean distance between two eye centres) and n is the total number of facial landmarks.

The CED is the cumulative distribution function of the normalized error which is larger than l and is reported as a failure. Thus, CED at the error is defined as

$$CED = \frac{N_{NME \leq l}}{n}, \tag{11}$$

where N_{NME} is the number of images in which the error NME_i is no higher than l.

AUC calculates the percentages of images that lie under certain thresholds. It is defined as:

$$AUC_\alpha = \int_0^\alpha f(e)de, \tag{12}$$

where e is the normalized error, $f(e)$ is the CED function and α is the upper bound used to calculate the definite integration.

In this study, we present our evaluations using mean error rate and CED curves. We calculated additional statistics from the CED curves such as the AUC which is up to an error of 0.07. CED curves for our experiments on the 300W and AFLW2000-3D testing sets are illustrated in Figure 7. Furthermore, as clearly stated in the figure, the AUC of 300W dataset is 72.49% and 65.99% for AFLW200-3D dataset.

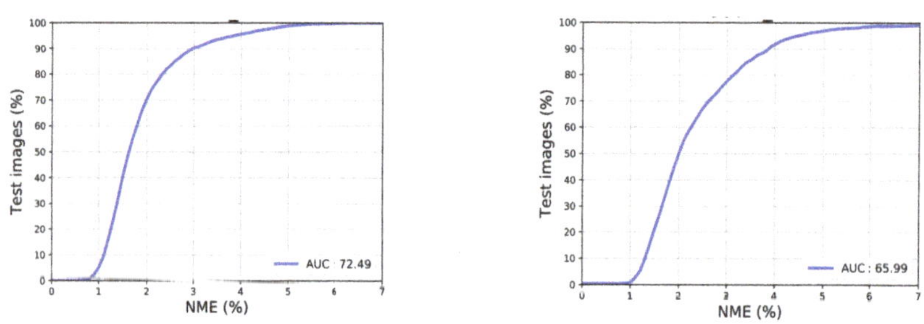

(**a**) 300W (**b**) AFLW2000-3D

Figure 7. Cumulative error distribution (CED) curve and area under the curve (AUC).

4.3. Comparison with State-of-the-Art Algorithms

4.3.1. Comparison with LFPW Dataset

The goal of the LFPW dataset was to study the problem of unconstrained face conditions that were trained on 811 images and tested on 224 images. Images were collected from Google, Flickr and Yahoo using text queries.

Comparisons of different methods versus the proposed method are listed in Table 4. Our proposed method substantially reduced the mean error rate. The second-best mean error rate in the table is the CFSS [51] method, which has a mean error of 4.87%. Our method is considerably superior with an error rate of only 3.52%. Furthermore, compared to the SDM [5] method, which uses cascaded regressions and has an error rate of 5.67%, our method also prevails by 2.15%.

Table 4. Mean error in LFPW dataset.

Method	68 pts
Zhu et al. [52]	8.29
DRMF [53]	6.57
RCPR [43]	5.67
SDM [5]	5.67
GN-DPM [54]	5.92
CFAN [25]	5.44
CFSS [51]	4.87
CFSS Practical [51]	4.90
Ours	3.52

4.3.2. Comparison with HELEN Dataset

Similar to the LFPW dataset, images were taken under unconstrained conditions with high resolutions and collected from Flickr using text queries. The dataset contained 2000 images for training and 330 images for testing.

Mean error comparisons of different methods on the HELEN dataset are presented in Table 5. Our method successfully achieved the lowest mean error percentage among all mentioned methods, with a mean error rate of 3.11% compared to the second-best, TCDCN [55], which achieved only a 4.60% error rate.

Table 5. Mean error on HELEN dataset.

Method	68 pts
Zhu et al. [52]	8.16
DRMF [53]	6.70
ESR [11]	5.70
RCPR [43]	5.93
SDM [5]	5.50
GN-DPM [54]	5.69
CFAN [25]	5.53
CFSS [51]	4.63
CFSS Practical [51]	4.72
TCDCN [55]	4.60
Ours	3.11

4.3.3. Comparison with 300W Dataset

The 300W is an extremely challenging dataset that is widely used to compare the performance of different algorithms for facial landmark detection under the same evaluation protocol. Table 6 presents the comparison results of the mean error rate of the 300W dataset. Our method reduced the mean error rate by 3.60%, 8.69% and 3.90% for the common subset, challenging subset and full set subset. Moreover, our proposed method performed significantly better than the previous methods in full set subsets with an error reduction of 0.46% when compared to the second-best method, CPM [56]. Our method for common, challenging and full set subsets also demonstrated significant improvement compared to the current state-of-the-art method DeFA [57]. Its error rate was 5.37% for the common subset, 9.38% for the challenging subset and 6.10% for the full set subset, which are higher than in our proposed method. The example landmark detection results of our method are illustrated in Figure 8, which is a collection of example results from the common, challenging and full set subsets.

Table 6. Mean error on 300W dataset.

Method	Common	Challenging	Fullset
RCPR [43]	6.18	17.26	7.58
SDM [5]	5.57	15.40	7.50
LBF [58]	4.95	11.98	6.32
CFSS [51]	4.73	9.98	5.76
CFSS Practical [51]	4.79	10.92	5.99
RAR [59]	4.12	8.35	4.94
3DDFA [23]	6.15	10.59	7.01
DeFA [57]	5.37	9.38	6.10
CPM [56]	3.39	8.14	4.36
Ours	3.60	8.69	3.90

Figure 8. Landmark detection examples from the 300W dataset.

4.3.4. Comparison of the AFLW2000-3D Dataset

The goal of the AFLW2000-3D dataset is to evaluate the algorithms on a large-pose dataset. In this dataset, we compared our proposed method with several state-of-the-art methods as presented in Table 7. The results show that our method had a mean error of 4.04%.

In comparison to 3DSTN [60], our method successfully reduced the mean error by 0.45% for the AFLW2000-3D dataset. The third best result in the dataset was DeFA [57], with an error rate of 4.50%. Our method has significantly and effectively improved errors in the dataset. The example landmark detection results of our method are illustrated in Figure 9.

Figure 9. Landmark detection examples from AFLW2000-3D dataset.

Table 7. Mean error on AFLW2000 dataset.

Method	68 pts
ESR [11]	7.99
RCPR [43]	7.80
MDM [61]	6.41
SDM [5]	6.12
3DDFA [23]	5.42
3DSTN [60]	4.49
DeFA [57]	4.50
Ours	4.04

5. Conclusions

In this paper, we presented a deep heatmap regression approach for facial landmark detection. We employed FC-DenseNets to extract dense feature maps along with an explicit kernel convolution for early-stage facial shape prediction. Starting with a suitable shape in the first stage, the detected shapes were refined to match the ground-truth shape during the last stage of the architecture. Our local appearance initialization subnet pursued a heatmap regression approach convolved with kernel convolution to serve as a local detector of facial landmarks in the first stage and the dilated skip convolution subnet was carefully designed to increase the performance of our dense prediction architecture and accurate spatial information by aggregating multi-scale contextual information for the sake of refining the local prediction of the first subnet. The proposed method achieved superior, or at least comparable, performance in comparison to state-of-the-art methods for challenging datasets, including LFPW, HELEN, 300W and ALFW2000-3D.

Author Contributions: The work presented here was completed with collaboration among all authors. Conceptualization, S.C.; Methodology, S.C., J.-G.L.; software, S.C.; Validation, S.C., J.-G.L. and H.-H.P.; Formal analysis, J.-G.L.; Writing—original draft preparation, S.C.; Writing—review and editing, S.C., H.-H.P.; Visualization, S.C.; Supervision, H.H.P.; Funding acquisition, H.-H.P.

Funding: This research was supported by the Chung-Ang University Young Scientist Scholarship (CAYSS), the Ministry of Education (Project number: NRF-2016R1D1A1B03933895, Project name: Face recognition and searching robust in pose, illumination and expression utilizing video big data) and the Ministry of Trade, Industry and Energy (Project number: P0002397, Project name: Advanced Expert Training Program for Industrial Convergence of Wearable Smart Devices.

Conflicts of Interest: The authors declare no conflicts of interest.

Abbreviations

The following abbreviations are used in this manuscript:

2D	Two-dimensional
300W	300 faces in-the-Wild
3D	Three-dimensional
AFLW 2000-3D	Annotated Facial Landmarks in the Wild with 2000 three-dimension
AUC	Area Under the Curve
CED	Cumulative Error Distribution
CLM	Constrained Local Model
CNN	Convolutional Neural Network
DCNN	Deep Convolutional Neural Network
DenseNets	Densely Connected Convolutional Networks
DSC	Dilated Skip Convolution
FCNs	Fully Convolutional Networks
GPU	Graphics Processing Unit
LAI	Local Appearance Initialization
LFPW	The Labeled Face Parts in-the-Wild
NME	Normalized Mean Error
ReLU	Rectified Linear Unit

References

1. Wu, Y.; Ji, Q. Facial Landmark Detection: A Literature Survey. *Int. J. Comput. Vis.* **2017**, 1–28. [CrossRef]
2. Corneanu, C.A.; Simón, M.O.; Cohn, J.F.; Guerrero, S.E. Survey on RGB, 3D, Thermal, and Multimodal Approaches for Facial Expression Recognition: History, Trends, and Affect-Related Applications. *IEEE Trans. Pattern Anal. Mach. Intell.* **2016**, *38*, 1548–1568. [CrossRef] [PubMed]
3. Wang, S.H.; Phillips, P.; Dong, Z.C.; Zhang, Y.D. Intelligent facial emotion recognition based on stationary wavelet entropy and Jaya algorithm. *Neurocomputing* **2018**, *272*, 668–676. [CrossRef]
4. Zhang, Y.; Yang, Z.; Lu, H.; Zhou, X.; Phillips, P.; Liu, Q.; Wang, S. Facial Emotion Recognition Based on Biorthogonal Wavelet Entropy, Fuzzy Support Vector Machine, and Stratified Cross Validation. *IEEE Access* **2016**, *4*, 8375–8385. [CrossRef]
5. Xiong, X.; De la Torre, F. Supervised Descent Method and Its Applications to Face Alignment. In Proceedings of the IEEE Conference on Computer Vision and Pattern Recognition, Portland, OR, USA, 23–28 June 2013; pp. 532–539.
6. Koppen, P.; Feng, Z.H.; Kittler, J.; Awais, M.; Christmas, W.; Wu, X.J.; Yin, H.F. Gaussian mixture 3D morphable face model. *Pattern Recognit.* **2018**, *74*, 617–628. [CrossRef]
7. Sinha, G.; Shahi, R.; Shankar, M. Human Computer Interaction. In Proceedings of the 2010 3rd International Conference on Emerging Trends in Engineering and Technology, Goa, India, 19–21 November 2010; pp. 1–4. [CrossRef]
8. Bulat, A.; Tzimiropoulos, G. How far are we from solving the 2D & 3D Face Alignment problem? (and a dataset of 230, 000 3D facial landmarks). *CoRR* **2017**. [CrossRef]
9. Zhang, H.; Li, Q.; Sun, Z.; Liu, Y. Combining data-driven and model-driven methods for robust facial landmark detection. *IEEE Trans. Inf. Forensics Secur.* **2018**, *13*, 2409–2422. [CrossRef]
10. Shi, H.; Wang, Z. Improved Stacked Hourglass Network with Offset Learning for Robust Facial Landmark Detection. In Proceedings of the 2019 9th International Conference on Information Science and Technology (ICIST), Hulunbuir, China, 2–5 August 2019; pp. 58–64. [CrossRef]
11. Cao, X.; Wei, Y.; Wen, F.; Sun, J. Face alignment by explicit shape regression. *Int. J. Comput. Vis.* **2014**, *107*, 177–190. [CrossRef]
12. Luo, P.; Wang, X.; Tang, X. Hierarchical face parsing via deep learning. In Proceedings of the 2012 IEEE Conference on Computer Vision and Pattern Recognition, Providence, RI, USA, 16–21 June 2012; pp. 2480–2487.

13. Wu, W.; Yang, S. Leveraging Intra and Inter-Dataset Variations for Robust Face Alignment. In Proceedings of the The IEEE Conference on Computer Vision and Pattern Recognition (CVPR) Workshops, Honolulu, HI, USA, 21–26 July 2017.
14. Fischer, P.; Dosovitskiy, A.; Brox, T. Descriptor Matching with Convolutional Neural Networks: A Comparison to SIFT. *arXiv* **2014**, arXiv:1405.5769.
15. Wu, Y.; Wang, Z.; Ji, Q. Facial feature tracking under varying facial expressions and face poses based on restricted boltzmann machines. In Proceedings of the IEEE Conference on Computer Vision and Pattern Recognition, Portland, OR, USA, 23–28 June 2013; pp. 3452–3459.
16. Lai, H.; Xiao, S.; Pan, Y.; Cui, Z.; Feng, J.; Xu, C.; Yin, J.; Yan, S. Deep recurrent regression for facial landmark detection. *IEEE Trans. Circuits Syst. Video Technol.* **2018**, *28*, 1144–1157. [CrossRef]
17. Belagiannis, V.; Zisserman, A. Recurrent human pose estimation. In Proceedings of the 2017 12th IEEE International Conference on Automatic Face & Gesture Recognition (FG 2017), Washington, DC, USA, 30 May–3 June 2017; pp. 468–475.
18. Merget, D.; Rock, M.; Rigoll, G. Robust Facial Landmark Detection via a Fully-Convolutional Local-Global Context Network. In Proceedings of the IEEE Conference on Computer Vision and Pattern Recognition, Salt Lake City, UT, USA, 18–23 June 2018; pp. 781–790.
19. Jégou, S.; Drozdzal, M.; Vazquez, D.; Romero, A.; Bengio, Y. The one hundred layers tiramisu: Fully convolutional densenets for semantic segmentation. In Proceedings of the 2017 IEEE Conference on Computer Vision and Pattern Recognition Workshops (CVPRW), Honolulu, HI, USA, 21–26 July 2017; pp. 1175–1183.
20. Belhumeur, P.N.; Jacobs, D.W.; Kriegman, D.J.; Kumar, N. Localizing Parts of Faces Using a Consensus of Exemplars. *IEEE Trans. Pattern Anal. Mach. Intell.* **2013**, *35*, 2930–2940. [CrossRef] [PubMed]
21. Le, V.; Brandt, J.; Lin, Z.; Bourdev, L.; Huang, T.S. Interactive Facial Feature Localization. In *Computer Vision—ECCV 2012*; Fitzgibbon, A., Lazebnik, S., Perona, P., Sato, Y., Schmid, C., Eds.; Springer: Berlin/Heidelberg, Germany, 2012; pp. 679–692.
22. Sagonas, C.; Tzimiropoulos, G.; Zafeiriou, S.; Pantic, M. 300 Faces in-the-Wild Challenge: The First Facial Landmark Localization Challenge. In Proceedings of the 2013 IEEE International Conference on Computer Vision Workshops, Sydney, Australia, 2–8 December 2013; pp. 397–403. [CrossRef]
23. Zhu, X.; Lei, Z.; Liu, X.; Shi, H.; Li, S.Z. Face Alignment Across Large Poses: A 3D Solution. *CoRR* **2015**. [CrossRef]
24. Bulat, A.; Tzimiropoulos, Y. Convolutional aggregation of local evidence for large pose face alignment. In *Proceedings of the British Machine Vision Conference (BMVC), 19–22 September 2016*; Richard, C., Wilson, E.R.H., Smith, W.A.P., Eds.; BMVA Press: York, UK, 2016; pp. 86.1–86.12. [CrossRef]
25. Zhang, J.; Shan, S.; Kan, M.; Chen, X. Coarse-to-Fine Auto-Encoder Networks (CFAN) for Real-Time Face Alignment. In *Computer Vision–ECCV 2014*; Fleet, D., Pajdla, T., Schiele, B., Tuytelaars, T., Eds.; Springer International Publishing: Cham, Switzerland, 2014; pp. 1–16.
26. Xu, X.; Kakadiaris, I.A. Joint Head Pose Estimation and Face Alignment Framework Using Global and Local CNN Features. In Proceedings of the 2017 12th IEEE International Conference on Automatic Face Gesture Recognition (FG 2017), Washington, DC, USA, 30 May–3 June 2017; pp. 642–649. [CrossRef]
27. Benini, S.; Khan, K.; Leonardi, R.; Mauro, M.; Migliorati, P. Face analysis through semantic face segmentation. *Signal Process. Image Commun.* **2019**, *74*, 21–31. doi:10.1016/j.image.2019.01.005. [CrossRef]
28. Benini, S.; Khan, K.; Leonardi, R.; Mauro, M.; Migliorati, P. FASSEG: A FAce semantic SEGmentation repository for face image analysis. *Data Brief* **2019**, *24*, 103881. doi:10.1016/j.dib.2019.103881. [CrossRef] [PubMed]
29. Toshev, A.; Szegedy, C. DeepPose: Human Pose Estimation via Deep Neural Networks. *CoRR* **2013**. [CrossRef]
30. Newell, A.; Yang, K.; Deng, J. Stacked Hourglass Networks for Human Pose Estimation. *arXiv* **2016**, arXiv:1603.06937.
31. Jain, A.; Tompson, J.; LeCun, Y.; Bregler, C. MoDeep: A Deep Learning Framework Using Motion Features for Human Pose Estimation. *arXiv* **2014**, arXiv:1409.7963.
32. Huang, G.; Liu, Z.; Weinberger, K.Q. Densely Connected Convolutional Networks. *arXiv* **2016**, arXiv:1608.06993.

33. Ioffe, S.; Szegedy, C. Batch Normalization: Accelerating Deep Network Training by Reducing Internal Covariate Shift. *arXiv* **2015**, arXiv:1502.03167.
34. Nair, V.; Hinton, G.E. Rectified Linear Units Improve Restricted Boltzmann Machines. In Proceedings of the 27th International Conference on International Conference on Machine Learning, Haifa, Israel, 21–24 June 2010; Omnipress: Madison, WI, USA, 2010; pp. 807–814.
35. Yu, F.; Koltun, V. Multi-Scale Context Aggregation by Dilated Convolutions. *arXiv* **2015**, arXiv:1511.07122.
36. Wang, Z.; Ji, S. Smoothed Dilated Convolutions for Improved Dense Prediction. *CoRR* **2018**. [CrossRef]
37. van den Oord, A.; Dieleman, S.; Zen, H.; Simonyan, K.; Vinyals, O.; Graves, A.; Kalchbrenner, N.; Senior, A.W.; Kavukcuoglu, K. WaveNet: A Generative Model for Raw Audio. *arXiv* **2016**, arXiv:1609.03499.
38. Kalchbrenner, N.; van den Oord, A.; Simonyan, K.; Danihelka, I.; Vinyals, O.; Graves, A.; Kavukcuoglu, K. Video Pixel Networks. *arXiv* **2016**, arXiv:1610.00527.
39. Kalchbrenner, N.; Espeholt, L.; Simonyan, K.; van den Oord, A.; Graves, A.; Kavukcuoglu, K. Neural Machine Translation in Linear Time. *arXiv* **2016**, arXiv:1610.10099.
40. Pfister, T.; Charles, J.; Zisserman, A. Flowing ConvNets for Human Pose Estimation in Videos. *arXiv* **2015**, arXiv:1506.02897.
41. He, K.; Zhang, X.; Ren, S.; Sun, J. Deep Residual Learning for Image Recognition. *arXiv* **2015**, arXiv:1512.03385.
42. Paszke, A.; Gross, S.; Chintala, S.; Chanan, G.; Yang, E.; DeVito, Z.; Lin, Z.; Desmaison, A.; Antiga, L.; Lerer, A. Automatic differentiation in PyTorch. In Proceedings of the Neural Information Processing Systems (NIPS 2017) Workshop on Autodiff, Long Beach, CA, USA, 8 December 2017.
43. Burgos-Artizzu, X.P.; Perona, P.; Dollár, P. Robust face landmark estimation under occlusion. In Proceedings of the IEEE International Conference on Computer Vision, Sydney, NSW, Australia, 1–8 December 2013; pp. 1513–1520.
44. Krizhevsky, A.; Sutskever, I.; Hinton, G.E. ImageNet Classification with Deep Convolutional Neural Networks. *Commun. ACM* **2017**, *60*, 84–90. [CrossRef]
45. Mairal, J.; Koniusz, P.; Harchaoui, Z.; Schmid, C. Convolutional Kernel Networks. In *Advances in Neural Information Processing Systems 27, 8–13 December 2014*; Ghahramani, Z., Welling, M., Cortes, C., Lawrence, N.D., Weinberger, K.Q., Eds.; MIT Press: Montreal, QC, Canada, 2014; Volume 2, pp. 2627–2635.
46. Wang, L.; Yin, B.; Guo, A.; Ma, H.; Cao, J. Skip-connection convolutional neural network for still image crowd counting. *Appl. Intell.* **2018**, *48*, 3360–3371. [CrossRef]
47. Zhu, X.; Ramanan, D. Face detection, pose estimation, and landmark localization in the wild. In Proceedings of the 2012 IEEE Conference on Computer Vision and Pattern Recognition, Providence, RI, USA, 16–21 June 2012; pp. 2879–2886. [CrossRef]
48. Sagonas, C.; Tzimiropoulos, G.; Zafeiriou, S.; Pantic, M. A Semi-automatic Methodology for Facial Landmark Annotation. In Proceedings of the 2013 IEEE Conference on Computer Vision and Pattern Recognition Workshops, Portland, OR, USA, 23–28 June 2013; IEEE Computer Society: Washington, DC, USA, 2013; pp. 896–903. [CrossRef]
49. Köstinger, M.; Wohlhart, P.; Roth, P.M.; Bischof, H. Annotated Facial Landmarks in the Wild: A large-scale, real-world database for facial landmark localization. In Proceedings of the 2011 IEEE International Conference on Computer Vision Workshops (ICCV Workshops), Barcelona, Spain, 6–13 November 2011; pp. 2144–2151. [CrossRef]
50. Tieleman, T.; Hinton, G. Lecture 6.5-rmsprop: Divide the gradient by a running average of its recent magnitude. *COURSERA Neural Netw. Mach. Learn.* **2012**, *4*, 26–31.
51. Zhu, S.; Li, C.; Loy, C.C.; Tang, X. Face alignment by coarse-to-fine shape searching. In Proceedings of the 2015 IEEE Conference on Computer Vision and Pattern Recognition (CVPR), Boston, MA, USA, 7–12 June 2015; pp. 4998–5006. [CrossRef]
52. Chen, X.; Zhou, E.; Mo, Y.; Liu, J.; Cao, Z. Delving Deep Into Coarse-To-Fine Framework for Facial Landmark Localization. In Proceedings of the The IEEE Conference on Computer Vision and Pattern Recognition (CVPR) Workshops, Honolulu, HI, USA, 21–26 July 2017.
53. Kowalski, M.; Naruniec, J.; Trzcinski, T. Deep Alignment Network: A convolutional neural network for robust face alignment. *arXiv* **2017**, arXiv:1706.01789.

54. Tzimiropoulos, G.; Pantic, M. Gauss-Newton Deformable Part Models for Face Alignment In-the-Wild. In Proceedings of the 2014 IEEE Conference on Computer Vision and Pattern Recognition, Columbus, OH, USA, 23–28 June 2014; pp. 1851–1858. [CrossRef]
55. Yang, J.; Liu, Q.; Zhang, K. Stacked Hourglass Network for Robust Facial Landmark Localisation. In Proceedings of the IEEE Conference on Computer Vision and Pattern Recognition (CVPR) Workshops, Honolulu, HI, USA, 21–26 July 2017.
56. Wei, S.; Ramakrishna, V.; Kanade, T.; Sheikh, Y. Convolutional Pose Machines. *arXiv* **2016**, arXiv:1602.00134.
57. Liu, Y.; Jourabloo, A.; Ren, W.; Liu, X. Dense Face Alignment. *arXiv* **2017**, arXiv:1709.01442.
58. Ren, S.; Cao, X.; Wei, Y.; Sun, J. Face Alignment at 3000 FPS via Regressing Local Binary Features. In Proceedings of the 2014 IEEE Conference on Computer Vision and Pattern Recognition, Honolulu, HI, USA, 21–26 July 2017; pp. 1685–1692. [CrossRef]
59. Xiao, S.; Feng, J.; Xing, J.; Lai, H.; Yan, S.; Kassim, A. Robust Facial Landmark Detection via Recurrent Attentive-Refinement Networks. In *Computer Vision—ECCV 2016*; Leibe, B., Matas, J., Sebe, N., Welling, M., Eds.; Springer International Publishing: Cham, Switzerland, 2016; pp. 57–72.
60. Bhagavatula, C.; Zhu, C.; Luu, K.; Savvides, M. Faster Than Real-time Facial Alignment: A 3D Spatial Transformer Network Approach in Unconstrained Poses. *arXiv* **2017**, arXiv:1707.05653.
61. Yan, J.; Lei, Z.; Yi, D.; Li, S.Z. Learn to Combine Multiple Hypotheses for Accurate Face Alignment. In Proceedings of the 2013 IEEE International Conference on Computer Vision Workshops, Sydney, Australia, 2–8 December 2013; pp. 392–396. [CrossRef]

sensors

Article

An Exploration of Machine Learning Methods for Robust Boredom Classification Using EEG and GSR Data

Jungryul Seo [1], Teemu H. Laine [2] and Kyung-Ah Sohn [1,*]

[1] Department of Computer Engineering, Ajou University, Suwon 16499, Korea; jrseojr@naver.com
[2] Department of Computer Science, Electrical and Space Engineering, The Luleå University of Technology, Skellefteå 93187, Sweden; teemu@ubilife.net
* Correspondence: kasohn@ajou.ac.kr; Tel.: +82-31-219-2434

Received: 12 September 2019; Accepted: 17 October 2019; Published: 20 October 2019

Abstract: In recent years, affective computing has been actively researched to provide a higher level of emotion-awareness. Numerous studies have been conducted to detect the user's emotions from physiological data. Among a myriad of target emotions, boredom, in particular, has been suggested to cause not only medical issues but also challenges in various facets of daily life. However, to the best of our knowledge, no previous studies have used electroencephalography (EEG) and galvanic skin response (GSR) together for boredom classification, although these data have potential features for emotion classification. To investigate the combined effect of these features on boredom classification, we collected EEG and GSR data from 28 participants using off-the-shelf sensors. During data acquisition, we used a set of stimuli comprising a video clip designed to elicit boredom and two other video clips of entertaining content. The collected samples were labeled based on the participants' questionnaire-based testimonies on experienced boredom levels. Using the collected data, we initially trained 30 models with 19 machine learning algorithms and selected the top three candidate classifiers. After tuning the hyperparameters, we validated the final models through 1000 iterations of 10-fold cross validation to increase the robustness of the test results. Our results indicated that a Multilayer Perceptron model performed the best with a mean accuracy of 79.98% (AUC: 0.781). It also revealed the correlation between boredom and the combined features of EEG and GSR. These results can be useful for building accurate affective computing systems and understanding the physiological properties of boredom.

Keywords: boredom; machine learning; emotion; EEG; GSR; classification; sensor

1. Introduction

Sensors, machine learning, artificial intelligence, and other kinds of information technologies have recently been advancing rapidly. Based on these trends, several studies have been conducted on the acquisition and processing of information, aiming at a higher-level understanding of the collected information. In particular, detection of the user's context, such as their emotional or physical state, is of particular importance because it enables the creation of context-aware systems that adapt their behavior to match the context in which they are used. This branch of computer science is known as context-aware computing. A sub-branch of it, which is the focus of this study, concentrates on classifying emotions using physiological data.

The study of emotion classification belongs to the area of affective computing (AC) that aims to build computer systems capable of detecting and reacting to the user's emotions. The area of AC in computer science is considered to have been established when a seminal paper by Picard [1] was published, and it has since become a vibrant field of study, with some example studies being [2–4]. To classify emotions in these systems, researchers have three data collection (i.e., measurement) strategies at their disposal [5]: (i) neurological/physiological measurement, which uses sensors to detect changes in the user's body; (ii) subjective self-reporting by questionnaires, diaries or interviews; and (iii) behavioral measurement that is based on expert observations of the participant's behavior. While all these approaches have their specific advantages and disadvantages, as Kim and Fesenmaier [5] suggest, physiological measurement is considered to be particularly objective. In our literature review, we found that a large body of AC studies exists on classifying emotions from physiological data such as electroencephalography (EEG) [2,6,7], galvanic skin response (GSR) [2,8–11], heart rate [2,10,11], and others [12,13]. However, a higher validity can be achieved by combining more than one measurement strategy. For example, a viable approach, which is employed in this study, is to combine a physiological approach with self-reporting, where the latter is used to verify the existence of the target emotion.

Accurate classification of boredom can be considered of particular importance because boredom affects multiple facets of our lives. In a technical report published by the United States Air Force, unmanned aerial vehicle pilots' reaction times were longer when they felt bored [14]. Furthermore, boredom can contribute to serious medical issues such as cardiovascular disease [15]. Additionally, it can have negative effects on learning [16–19]. If computing devices could accurately classify the occurrence of the user's boredom and administer a suitable intervention to compensate for it, they could be used to tackle the aforementioned boredom-related issues.

Several previous studies have built boredom classification models using different physiological data as summarized in Table 1. However, to the best of our knowledge, no previous studies have used both EEG and GSR data for boredom classification. In this study, we performed a joint analysis of both data by collecting EEG and GSR data from 28 participants who also answered a questionnaire surveying their perceived level of boredom. The participants watched two types of video stimuli that were prepared to elicit boredom and to entertain, respectively. Based on the collected data and questionnaire results, we ran an initial test of 19 machine learning algorithms and selected the best three candidate classification models. After hyperparameter tuning, we measured the final performance of the selected models with 1000 iterations of 10-fold cross validation. The best performance with a mean accuracy of 79.98% (min: 71.43%, max: 93.93%) was obtained using a Multilayer perceptron (MLP) model. Furthermore, we analyzed the used features to investigate the correlation between EEG data, GSR data, and boredom. This study, therefore, has three major contributions: (i) revealing a correlation between EEG, GSR, and boredom; (ii) conducting a reliable performance comparison among 19 machine learning algorithms through repeated cross validations; and (iii) proposing a robust boredom classification model based on MLP.

Table 1. Studies on boredom classification using physiological data.

Study	Data Source	Number of Participants
Shen et al. [2]	HR, GSR, BP, EEG	1
Mandryk and Atkins [11]	HR, GSR, Facial	12
Kim et al. [6]	Eye-tracking, EEG	16
Giakoumis et al. [9]	ECG, GSR	19
Giakoumis et al. [8]	ECG, GSR	21
Seo et al. [7]	EEG	28
D'Mello et al. [12]	Facial, Gesture	30

Table 1. *Cont.*

Study	Data Source	Number of Participants
Jaques et al. [13]	Eye-tracking	67
Jang et al. [10]	HR, GSR, Temperature, PPG	217

HR - Heart Rate, BP - Blood pressure, ECG - Electrocardiogram, PPG - Photo Plethysmo Graphy.

2. Background

Emotion detection based on physiological data are a vibrant research field that has produced a large body of studies focusing on different emotions and analysis approaches. This section provides an overview of the previous physiology-based emotion detection research relevant to our study. In our literature review, we searched previous studies in Google Scholar with searching keywords of "Emotion", "Boredom", "Classification", "Physiology", and "Sensor". We did not set limitations on the publication year or forum; however, we excluded studies that classified emotions solely by interviews or surveys.

Several AC studies focusing on emotion classification from physiological data [2,6,20–25] referred to Russell's Circumplex model (Figure 1) to explain the target emotion [26]. The model categorizes emotions into four groups by the dimensions of valance and arousal. According to the model, boredom is categorized into the low-valence low-arousal group (red box in Figure 1).

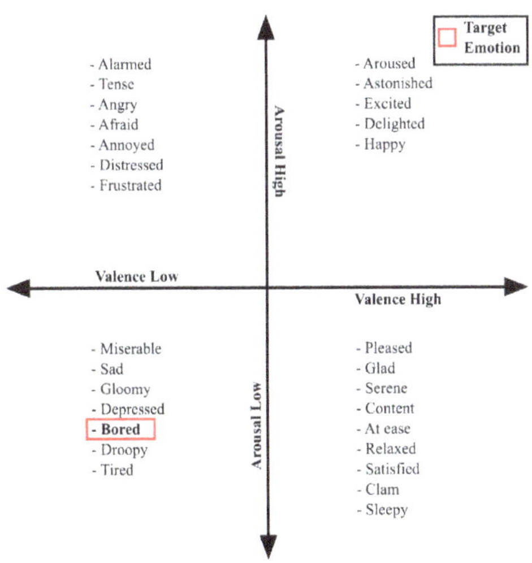

Figure 1. Circumplex model [26].

Despite the simplicity of the way in which the Circumplex model assigns boredom to the third quadrant, boredom is considered a complex emotion as various studies have defined it differently. Vogel-Walcutt et al.'s literature review resulted in 37 definitions of boredom, and concluded that "boredom occurs when an individual experiences both the neurological state of low arousal and the psychological state of dissatisfaction, frustration, or disinterest in response to the low arousal." [27]. The conclusions of Russell's model [26] and Vogel-Walcutt et al.'s definition [27] are thus similar. In contrast, the range of boredom in Eastwood et al.'s definition [28] is wider than that of Russell's model. According to their study, people who are in a low-valence state can feel boredom regardless of the level of arousal. Considering these studies, we conclude that a universally accepted definition of boredom does not exist.

Some previous studies regarding boredom categorized it as a trait, while others handled it as a state. The meaning of trait in boredom-related studies is the proneness of an individual to become bored, thus there is a difference between easily becoming bored, and being able to resist boredom. Conversely, the meaning of a state is the current state of boredom that the person is experiencing. Fahlman et al. [29] handled boredom as a trait, while Eastwood et al. [28] and Kim et al. [6] treated it as a state. In this study, we approach boredom as a state. The reason for this is that we hypothesize that, when a person feels bored, changes in their physiological signals can be identified.

Our literature review identified nine studies on boredom classification from physiological data sources as listed in Table 1. It reveals that seven studies used more than one data source; this approach of sensor fusion is a common technique to increase the detection accuracy. The median number of participants in these studies was 21, which is relatively small compared to other cases where machine learning methods are typically applied. In the individual source perspective, EEG was used by three studies, and GSR was utilized by four studies. However, to the best of our knowledge, no previous study has used both EEG and GSR data for classifying boredom.

Sanei and Chambers [30] and Ashwal and Rust [31] showed that EEG data correlates with emotion states of humans. Furthermore, GSR is related to the autonomic nervous system [32], which is also known to correlate with emotion states, thus GSR can be utilized as a potential source for emotion classification [33]. In the physiological perspective, EEG data are captured from the activity of the brain, which belongs to the nervous system together with GSR. Moreover, the analysis on the characteristics of boredom conducted by Bench and Lench [34] suggested that boredom should be associated with the increased autonomic nervous system activity. This linkage implies that a correlation may exist between boredom, EEG and GSR, but so far it has not been investigated in previous studies.

Table 2 presents the methods and the accuracy results of the previous studies that classified boredom using physiological data. Mandryk and Atkins [11], D'Mello et al. [12], Giakoumis et al. [8], and Kim et al. [6] focused on finding correlations between boredom and physiological data using statistical approaches, thus they did not generate classification models. The other reviewed boredom classification studies built classification models using machine learning algorithms and measured the performance of the models. However, these studies did not address the issue of overfitting carefully, making it difficult to guarantee the robustness of the results. Moreover, many previous studies lacked the discussion on the choice of the classification algorithms and only considered a few limited algorithms. Therefore, it is necessary to consider the potential of a wider range of classification algorithms to classify boredom. Considering these facts and shortcomings of previous studies, our study aims to produce reliable performance results based on more diverse machine learning methods.

Table 2. Methods and accuracies of previous boredom classification studies.

Study	Accuracy	Method
Jaques et al. [13]	73.0%	Random Forest
Jang et al. [10]	84.7%	Discriminant Function Analysis
Shen et al. [2]	86.3%	Support Vector Machine (SVM)
Seo et al. [7]	86.7%	k-Nearest Neighbors (kNN)
Giakoumis et al. [9]	94.2%	Linear Discriminant Analysis
Mandryk and Atkins [11], D'Mello et al. [12], Giakoumis et al. [8], and Kim et al. [6]	Not available	Statistical approaches

3. Data Collection Methodology

This section describes the methods that we used to collect and analyze physiological data for the classification of boredom. We collected data according to the guidelines of the Declaration of Helsinki [35].

Specifically, we obtained written informed consents from the participants before the data collection, advertised data collection for inviting voluntary participants, explained to the participants that they could quit the experiment anytime they want, and provided snacks as a reward for their participation. The details of the data collection procedure are explained in the following sections.

3.1. Participants

We collected the EEG and GSR data from 28 Korean participants (13 males and 15 females) who were either students or staff at a university in the Republic of Korea. The participants' ages ranged from 20 to 34, with an average age of 23.62. We collected the data in two sessions: the first session was carried out with 18 participants (6 males and 12 females), and the second session was carried out with 10 participants (7 males and 3 females). All data collection procedures were designed and executed with careful consideration of legal and ethical issues. All participants elected to join the experiment voluntarily, and they were instructed to quit the experiment at any time if they felt the urge to do so. To secure the safety of the participants, an emergency kit was prepared as a countermeasure for accidents. All collected data were anonymized and stored securely in a password-protected data storage. Finally, a data collection protocol (see Section 3.3) was designed with consideration of the participants' legal rights.

3.2. Sensors

An EEG sensor produced by Muse [36] and a Grove GSR sensor produced by Seeed [37] were used in this study (Figure 2). The upper section of Figure 2 shows the EEG sensor, which has four electrodes. According to the instructions from the EEG sensor manufacturer, the electrodes are located at the positions FP1, FP2, TP9, and TP10 of the head [36]. The mapping of head locations was defined by Jasper [38], and it is used in neuroscience. The EEG data were captured with four electrodes; however, only the data from FP1 and FP2 were utilized. This is because TP9 and TP10 were not attached well on the participants' heads during the data collection, which caused instability in the output data from these electrodes. The EEG sensor captured raw EEG at a 220 Hz sampling rate and provided power spectral density (PSD) values for each electrode. According to the sensor manufacturer, the types of PSD data were absolute band power (ABP), relative band power, and others [36]. The sampling rate of these data was 10 Hz. To calculate ABP, we applied the fast Fourier transform algorithm. We calculated PSD by the following EEG frequency bands [36]:

- Delta: (1–4) Hz,
- Theta: (4–8) Hz,
- Alpha: (7.5–13) Hz,
- Beta: (13–30) Hz,
- Gamma: (30–44) Hz.

The lower section of Figure 2 illustrates the Grove GSR sensor that was used in this study, along with the finger band electrodes. These electrodes were attached to the index and middle fingers of the participants. The GSR sensor captures micro voltages (MV) between the fingers using the attached electrodes. Furthermore, the sensor calculates skin resistance (SR) utilizing the MV input in ohms. The formula for calculating SR using MV is provided by the sensor manufacturer [37], and is replicated in Equation (1). The sensor was connected to an Arduino Uno device, and the captured data were transmitted to a computer at a 192 Hz sampling rate:

$$SR = ((1024 + 2 * MV) * 10{,}000)/(512 - MV). \tag{1}$$

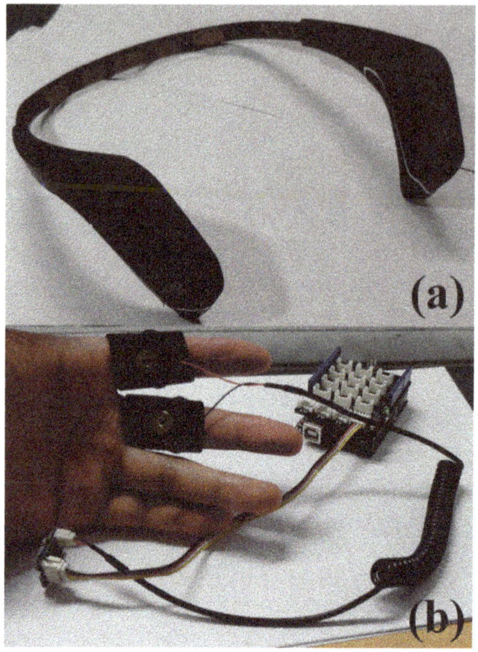

Figure 2. (**a**) EEG sensor, and (**b**) GSR sensor.

3.3. Protocol

Figure 3 presents the protocol of data collection. In the introduction stage, the participants were presented a page showing a consent form of the experiment. The information on the consent form stated that we would use the collected data only anonymously for academic purposes. Furthermore, the participants were instructed that they could stop the experiment anytime when they feel uncomfortable.

In the stage for showing non-boredom videos, a cinematic trailer of Blizzard's Starcraft 2: Heart of the Swarm, and a Korean comedy video clip were used to evoke non-boredom. These video clips were chosen to entertain the participants so that they would not become bored. To evoke boredom, a looping video was played in which a small circle moved slowly tracing the boundary of a bigger circle. Furthermore, to neutralize the emotion state of the participants, a cloud image from the International Affective Picture System [39] was shown for 30 seconds before showing each video stimulus. When the participants watched the videos, they were instructed to stop watching (by pressing a button) at any time they chose. Thus, each participant's data length was different, with the shortest watching time being 7.13 s.

The data collection consisted of two sessions. The sessions were otherwise identical except for the non-boredom video stimulus (i.e., the game trailer and the comedy show clip). The reason for carrying out data collection in two sessions was to get a content-independent classification result. In other words, we wanted to see whether the change of non-boredom video has an effect on the classification. After watching the video stimuli, the participants answered a questionnaire to measure the strength of boredom that they felt. The questionnaire had two questions: one for the boredom video, and another for the non-boredom video. The questionnaires were designed to be answered on a 5-point scale, and the range of the scale was from "None" to "Very much."

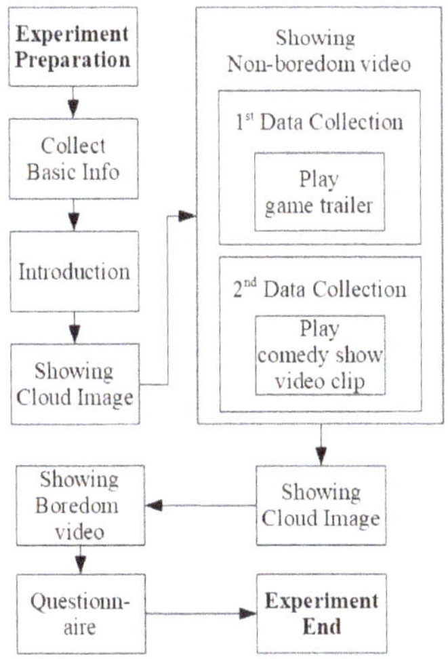

Figure 3. Protocol of data collection.

4. Machine Learning Methods

In this section, we describe the procedure for feature extraction and machine learning techniques for analyzing the collected data to classify boredom.

4.1. Window Size

As explained in Section 3.3, each participant's data length was different because they were instructed to stop the video stimuli playback at the time they chose. The shortest data length was 7.13 s, thus only the last 7 s of each participant's data were extracted to build the models. With the window size of 7 s, the number of samples was 56. Other window sizes, such as 1 s and 0.5 s, were also tested; however, these potentially caused overfitting because two or more samples would be generated from the same data with the same label.

4.2. Features

This section explains the feature extraction methods that we used for the EEG and GSR datasets. MATLAB (R2017a) was used for data analysis pertaining to feature extraction.

4.2.1. EEG

Similar to our previous study [7], we extracted five EEG features: (1) ABP, (2) Normalized ABP (NABP), (3) differential entropy (DE), (4) rational asymmetry (RASM), and (5) differential asymmetry (DASM). As explained in Section 3.2, ABP is a PSD value that is produced at 10 Hz from each electrode and frequency band. NABP is a normalized value of ABP using the following equation:

$$x' = \frac{x - min(x)}{max(x) - min(x)}, \tag{2}$$

where x is the original value, x' is the normalized value, and $max(x)$ and $min(x)$ are the maximum value and the minimum value of the dataset, respectively.

To calculate the DE, RASM, and DASM features, we used the formulae proposed by Zheng et al. [4], which are replicated in Equations (3)–(5):

$$DE = \frac{1}{2}\log 2\pi e\sigma^2, \tag{3}$$

$$DASM = DE_{(left)} - DE_{(right)}, \tag{4}$$

$$RASM = DE_{(left)} / DE_{(right)}. \tag{5}$$

In Equation (3), σ^2 is the variance of the bandpass-filtered EEG data of each frequency band, and π and e are constants. Furthermore, as the equations show, DASM and RASM are based on DE and utilize the electrical current asymmetry between the electrodes of the EEG sensors. Therefore, in this study, we use the asymmetry between the electrodes and the values from each electrode at the same time to train classification models.

As explained in Section 4.1, we set the window size to 7 s for feature extraction. Therefore, 70 units of ABP data and 70 units of NABP data of each electrode and frequency band, and 1540 units of EEG data of each electrode were extracted from each participant's data. We calculated the average and the standard deviation for each 70 units of ABP and NABP data and used the results as features. In the calculation of DE, we used all 1540 units of EEG data. In more detail, we applied the bandpass filter on EEG data. Regarding the frequency range of filtering, we followed the frequency band ranges of our EEG sensor (see Section 3.2). Finally, RASM and DASM were calculated using the extracted DE values.

As a result of EEG feature extraction, we secured 40 features from ABP and NABP (five frequency bands times two electrodes times two summary statistics). Moreover, 10 features were extracted from DE (five frequency bands times two electrodes). Finally, as indicated by Equations (4) and (5), RASM and DASM utilize locational symmetry for each electrode and DE value. Thus, five features were secured from RASM and DASM (five frequency bands).

4.2.2. GSR

As mentioned in Section 3.2, we used MV and SR data features from the GSR sensor. Additionally, normalization of MV with feature scaling was also performed (see Equation (2)). As a result, MV, normalized MV (NMV), and SR were secured from the GSR data. Similar to the feature extraction of the EEG data, we calculated the average and the standard deviation for each feature of the GSR data and used them as the final feature values. Consequently, six features in total were secured from the GSR data.

4.3. Machine Learning Model Selection

Weka, which is an open-source software for data mining that provides several machine learning algorithms, was used for building and testing the machine learning models [40]. In this study, we considered a wide range of machine learning algorithms supported by Weka as candidate algorithms. These candidates were used for initial testing to select the best algorithms for hyperparameter tuning.

Table 3 presents the evaluated algorithms and their options for training. Most of the algorithms were set to default parameters, i.e., the parameters that were preconfigured for the respective algorithms in Weka. Some algorithms (IBk, MLP, SVM) were used several times with different configurations, but these were considered as the same algorithm with different parameters. Therefore, although the number of algorithms was 19, we had 30 models in total to be trained for each dataset (EEG, GSR, and EEG-GSR combined).

Table 3. List of algorithms used for training models.

Algorithm	Option	Algorithm	Option	Algorithm	Option
IBk	Default 1/distance 1-distance	Multilayer Perceptron (MLP)	t i a	SVM	Linear Polynomial Radial
Decision Stump	Default		o		Sigmoid
Decision Table	Default		t,a	LMT	Default
Hoeffding Tree	Default		t,a,o	PART	Default
J48	Default		t,i,a,o	Logistic	Default
Random Tree (RT)	Default	Random Forest (RF)	Default	Simple Logistic	Default
JRip	Default	REP Tree	Default	Zero R	Default
Naïve Bayes (NB)	Default	KStar	Default	One R	Default

Network design parameter of MLP (Number of node per layer). a = (number of features + number of labels)/2, i = number of features, o = number of labels, t = number of features + number of labels, Ex) if 10 features and 2 labels are used, a, i, o, and t are 6, 10, 2, and 12, respectively (single hidden layer).

IBk, which is a k-Nearest Neighbor classifier, has a parameter for getting a weight from the distance between samples. By default, no weight is assigned. In MLP, the number of layers and the number of nodes in each layer can be defined. For example, "t,i" means that the MLP has two layers, with the first layer consisting of "t" nodes and the second layer consisting of "i" nodes. Table 3 notes explain "a", "i", "o" and "t". Finally, SVM provides options for selecting the SVM kernel type to be used. The MLP, IBk and SVM algorithms also have additional parameters than the ones listed in Table 3; however, we did not adjust them in this study.

4.4. Feature Refinement

To increase the models' performance, we applied a feature selection algorithm provided by Weka called Wrapper Subset Evaluator (WSE). WSE was proposed by Kohavi and John [41] to find a classifier-optimized feature subset from a dataset to increase model performance. To use WSE, a searching method (forward or backward searching) is required. The target classifier information is also provided as an input to WSE. We used forward searching and the optimized feature set for each algorithm during initial testing and hyperparameter tuning. In the next section, we present the selected features that produced the highest accuracies.

5. Results

5.1. Questionnaire

We first analyzed the questionnaire results, summarized in Figure 4, to be used for labeling the datasets. We regrouped the questionnaire answers into two groups as follows: "None" and "Little" were merged into the weak boredom group, and the remaining answers were merged into the strong boredom group. The total number of questionnaire answers was 56, which comprised 28 participants' answers for two video stimuli. The number of answers in the weak boredom group was 30, while 26 were assigned to the strong boredom group. These regrouped questionnaire results were then used for labeling the collected samples. Based on Figure 4, the game trailer and the comedy clip did not induce boredom among most participants; however, the circle video was successful in evoking boredom among the participants.

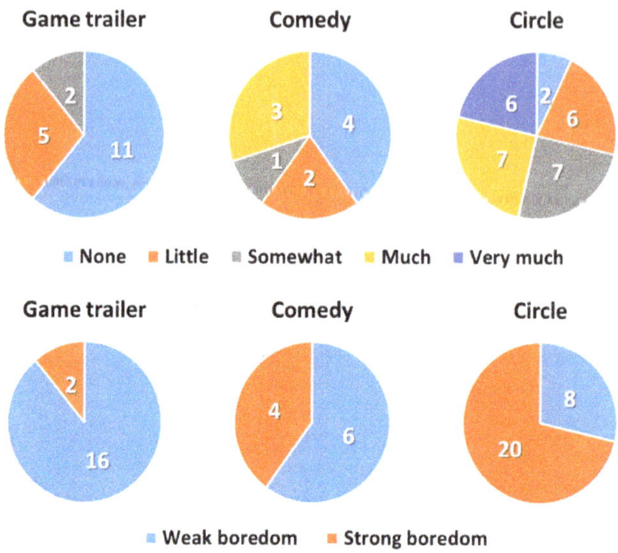

Figure 4. Questionnaire results (Question: How much boredom did you feel from "stimulus name"?).

5.2. Initial Test for Model Selection

We performed initial testing to compare the 19 candidate algorithms and to select the best models for further analysis. Table 4 presents the top ten models for each dataset. Based on the results, we selected RF, MLP, and NB for further investigation. In particular, RF was ranked as the best algorithm for the EEG-GSR combined and EEG datasets. MLP was ranked in the top ten more than other algorithms in all datasets. We also chose NB because it has a relatively low time complexity of $O\left(np\right)$, where n is the number of training samples and p is the number of features, whereas other algorithms with similar performance, such as IBk or J48, have time complexity of $O\left(n^2\right)$.

Table 4. Initial testing results for model selection.

EEG-GSR		EEG		GSR	
Algorithm	Accuracy (%)	Algorithm	Accuracy (%)	Algorithm	Accuracy (%)
RF	83.93	RF	80.36	MLP (t)	75.00
PART	80.36	MLP (a)	78.57	Simple Logistic	73.21
IBk	80.36	MLP (i)	78.57	MLP (a)	71.43
J48	80.36	KStar	78.57	MLP (i)	71.43
NB	80.36	MLP (o)	73.21	SVM (Radial Kernel)	71.43
RT	78.57	NB	71.43	MLP (o)	69.64
Hoeffding Tree	78.57	Hoeffding Tree	71.43	KStar	69.64
MLP (o)	76.79	MLP (t)	71.43	PART	69.64
MLP (a)	76.79	IBk	71.43	Decision Stump	69.64
MLP (t)	76.79	Logistic	69.64	J48	69.64

5.3. Hyperparameter Tuning

5.3.1. Random Forest

Weka's RF API has three major hyperparameters that are related to performance: the number of features to randomly investigate, the number of trees, and the maximum depth of trees. In order to find the best hyperparameters, we trained models with all possible combinations of the parameters within predefined ranges. As a result, we trained 107,100 models and measured these performances with 10-fold cross validation. The predefined ranges of the tuning parameters were as follows:

- Number of features to randomly investigate: 1–6, default = $\text{int}(\log_2(p) + 1)$,
- Number of trees: 1–100,
- Maximum depth of trees: 1–50, no limit.

Table 5 presents the top three hyperparameter combinations for the Random Forest algorithm in each dataset. To select the best hyperparameters, we established the following prioritization: (1) Accuracy, (2) Area Under the receiver operating characteristics Curve (AUC), and (3) expected classification cost. Consequently, we selected 7, 14, and 7 as the number of features, trees, and the maximum depth value, respectively, for the EEG-GSR combined dataset. In the case of EEG and GSR datasets, the default values for the number of features (6, and 3, respectively) were optimal, and 18 and 11 were selected as the number of trees and depth, respectively. The "no limit" in depth means unlimited search on each tree of random forest. Considering the expected classification cost, 11 appears to be a suitable value.

Table 5. RF hyperparameter tuning results.

	Features	Trees	Depth	Accuracy (%)	AUC
EEG-GSR	default (7)	14	7	87.50	0.842
	2	14	7	87.50	0.842
	3	14	7	87.50	0.842
EEG	default (6)	18	no limit	76.79	0.780
	default (6)	18	11	76.79	0.780
	default (6)	18	12	76.79	0.780
GSR	default (3)	18	no limit	76.79	0.780
	default (3)	18	11	76.79	0.780
	default (3)	18	12	76.79	0.780

5.3.2. Multilayer Perceptron

We tuned the neural network design, learning rate, and epoch parameters for MLP in the Weka API. We started by evaluating the neural network design parameters by testing all possible cases, whilst keeping the other parameters at default values. For flexible neural network design, we followed the Weka API's MLP neural network design parameter rule (see Table 3). The general design concept of a neural network was to decrease the number of each layer's nodes gradually, from the first layer to the last layer.

To decide the optimal learning rate and epoch values for MLP, we tested all possible combinations of these within predefined ranges. As a result, 3,600,000 models were trained and their performances were measured using 10-fold cross validation. The predefined ranges of each tuning parameter were as follows (the default value of MLP's momentum parameter is 0.2; however, we fixed it as 0.1):

- Learning rate: 0.01–1.00 (0.01 unit),
- Epoch: 1–2000 (1 unit).

Table 6 presents the hyperparameter tuning results. To select the best hyperparameters, we established the following prioritization: (1) Accuracy, (2) AUC, and (3) low epoch. Considering the characteristics of MLP and to avoid overfitting, the learning needs to stop when the accuracy does not increase. Regarding the network design, models with three hidden layers could not be tested because WSE produced a model that contained only the label data. According to the results of Table 6, we selected "t", 0.76, and 73 as the values of network design, learning rate, and epoch, respectively, for the EEG-GSR combined dataset. For the EEG dataset, "i", 0.19, and 489 were found to the best values of network design, learning rate, and epoch, respectively. Finally, "t", 0.95, and 321 were designated as the values of the three hyperparameters for the GSR dataset.

Table 6. MLP hyperparameter tuning results.

	Layer and Node	Learning Rate	Epoch	Accuracy (%)	AUC
EEG-GSR	a	0.47	444	76.79	0.771
	i	0.90	215	82.14	0.794
	o	0.47	444	76.79	0.771
	t	0.76	73	83.93	0.765
	i, a	0.59	572	82.14	0.795
	i, o	0.49	1351	82.14	0.764
	t, i	0.91	192	76.79	0.737
EEG	a	0.70	452	76.79	0.733
	i	0.19	489	83.93	0.822
	o	0.21	654	80.36	0.751
	t	0.48	163	82.14	0.791
	i, a	0.43	405	78.57	0.706
	i, o	0.71	265	75.00	0.710
	t, i	0.99	320	78.57	0.692
GSR	a	0.44	312	75.00	0.767
	i	0.44	312	75.00	0.767
	o	0.44	312	75.00	0.767
	t	0.95	321	76.79	0.759
	i, a	0.64	1120	73.21	0.663
	i, o	0.90	231	71.43	0.642
	t, i	0.91	255	71.43	0.641

5.3.3. Naïve Bayes

Table 7 presents the tuning results of the NB hyperparameters for each dataset. Weka's NB API has two major options, which are mutually exclusive: whether to use a kernel density estimator rather than the normal distribution for numeric attributes ("Kernel" in Table 7), and whether to use a supervised discretization to process numeric attributes ("Discretization" in Table 7).

As in the parameter tuning processes for RF and MLP, we also tested all possible combinations of the NB hyperparameter options. Thus, we trained and tested nine models in total, and found that the parameters should be disabled for the EEG-GSR combined, EEG, and GSR datasets.

Table 7. NB hyperparameters' tuning results.

	Kernel	Discretization	Accuracy (%)	AUC
EEG-GSR	FALSE	FALSE	82.14	0.785
	TRUE	FALSE	76.79	0.819
	FALSE	TRUE	60.71	0.569
EEG	FALSE	FALSE	67.86	0.653
	TRUE	FALSE	67.86	0.603
	FALSE	TRUE	53.57	0.454
GSR	FALSE	FALSE	69.64	0.681
	TRUE	FALSE	64.29	0.626
	FALSE	TRUE	60.71	0.569

5.4. Final Performance Analysis

5.4.1. Performance Measurement

A common method of evaluating a classification model's performance is to use k-fold cross validation. However, k-fold cross validation is based on a random split of data; thus, when the validation is executed, the produced performance results can be different each time. Therefore, to obtain more reliable performance results, we measured the final models' performance by repeating 10-fold cross validation 1000 times with different seed values. Table 8 presents the mean, maximum and minimum accuracies and AUCs produced by 1000 iterations of the 10-fold cross validation runs for each parameter combination and dataset. Considering the mean accuracy, the MLP algorithm produced the best performance in all datasets. However, when considering the mean AUC values, the RF model outperformed the MLP model on the EEG-GSR combined dataset. The last column of Table 8 reports the average computation time for each run. We measured the average time per cross validation using the last 100 iterations of 10-fold cross validation. MLP took the longest time overall, for example, it took 71 ms on the EEG-GSR combined dataset, which was about 2.7 times longer than RF and 7.5 times longer than NB. The performance on the EEG-GSR combined dataset was better than each individual data performance in all cases.

Table 8. Final performance comparison—1000 runs of 10-fold cross validation.

		Accuracy (%)			AUC			Time (ms)
		Mean	Max	Min	Mean	Max	Min	Mean
EEG-GSR	RF	77.53	89.29	66.07	0.815	0.900	0.722	26.52
	NB	79.39	85.71	67.86	0.783	0.819	0.733	9.41
	MLP	79.98	83.93	71.43	0.781	0.815	0.723	70.98
EEG	RF	64.77	78.57	51.79	0.695	0.833	0.569	44.38
	NB	70.85	78.57	64.29	0.710	0.760	0.640	9.89
	MLP	77.04	83.93	66.07	0.775	0.833	0.641	251.22
GSR	RF	68.33	78.57	53.57	0.731	0.831	0.591	30.84
	NB	66.86	71.43	64.29	0.709	0.751	0.655	11.9
	MLP	70.03	76.79	60.71	0.744	0.796	0.683	163.49

Figure 5 shows the models' final performances using box plots. We observe that the MLP and NB models' performances were more stable than that of RF in 1000 runs of 10-fold cross validation. The mean accuracies of MLP were higher than those of the other models in general. The RF model on the EEG-GSR combined dataset showed a high mean AUC but had a large variance. Muller et al. [42] defined a model's discriminatory ability with AUC as follows: (1) excellent discrimination (AUC $>=$ 0.90), (2) good discrimination (0.80 $<=$

AUC < 0.90), (3) fair discrimination (0.70 <= AUC < 0.80), and (4) poor discrimination (0.60 <= AUC < 0.70). For many of our final models, the AUC was over 0.7, and thus these models can be classified as fair discrimination models.

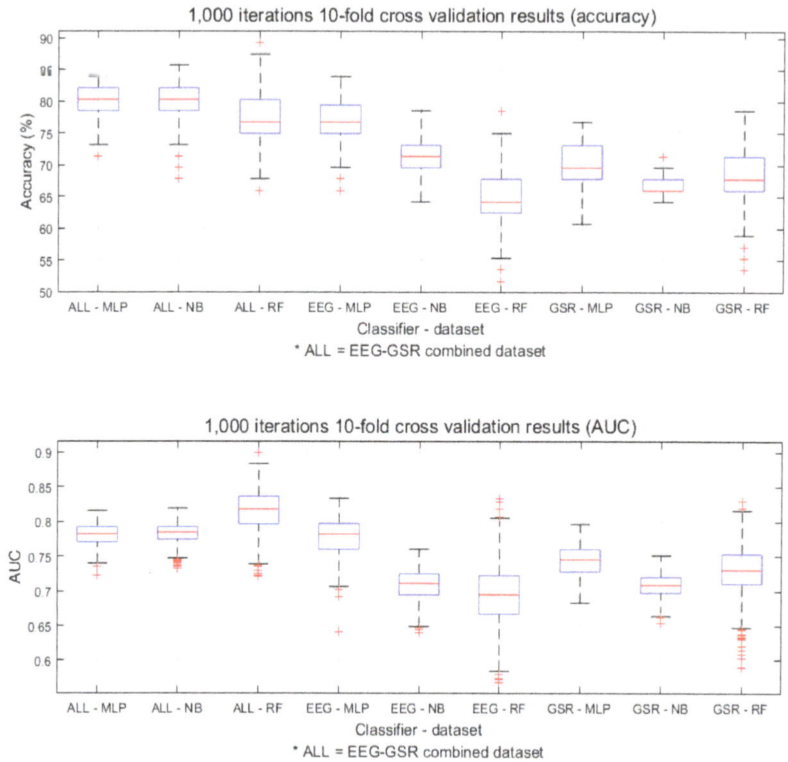

Figure 5. Box plots for the final performance comparison.

Considering all aspects of our model performance validation, the MLP model classified boredom most reliably on all the datasets. Furthermore, the model using the EEG-GSR combined dataset showed the highest performance, while MLP on the EEG dataset and MLP on the GSR dataset ranked second and third, respectively.

5.4.2. Analysis of the Selected Features

Table 9 presents the features that the WSE algorithm selected for each model. These results indicate that the standard deviation of MV was selected for all classifiers that used the GSR datasets. For a more detailed analysis of the selected features, we illustrate the distribution of the selected EEG features by frequency bands and electrodes in Figure 6. As the left pie chart indicates, the features related to the Alpha and Beta bands were selected more frequently than those of the Delta and Gamma bands, and the Theta band features were not selected by WSE at all. The right pie chart in Figure 6 illustrates the distribution of features by electrodes, where the distribution among FP1 and FP2 is nearly balanced; however, FP2 features were picked slightly more frequently.

Table 9. Selected features by WSE in each model.

	EEG-GSR	EEG	GSR
MLP	ABP Delta FP1 std ABP Gamma FP2 mean NABP Delta FP2 mean MV std	ABP Beta FP2 mean ABP Gamma FP2 mean NABP Alpha FP1 std DASM Alpha	MV std NMV mean
NB	ABP Beta FP2 mean ABP Gamma FP2 mean NABP Alpha FP1 mean NABP Beta FP1 mean MV std	NABP Alpha FP1 std DE Beta FP1	MV std NMV std
RF	NABP Beta FP2 mean MV std	ABP Gamma FP2 mean NABP Beta FP2 mean Alpha RASM	MV std SR mean

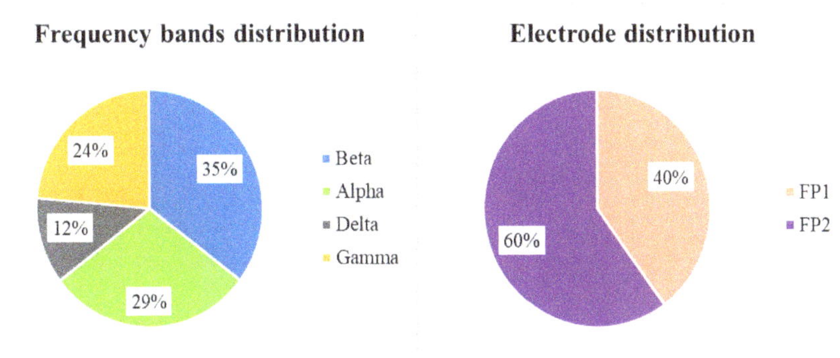

Figure 6. Distributions of EEG features by frequency bands and electrodes.

Figure 7 illustrates the distribution of EEG features by frequency bands for each electrode separately. We find that all selected Alpha band features are concentrated on FP1. In contrast, all selected Gamma band features occur on FP2 as the right pie chart shows.

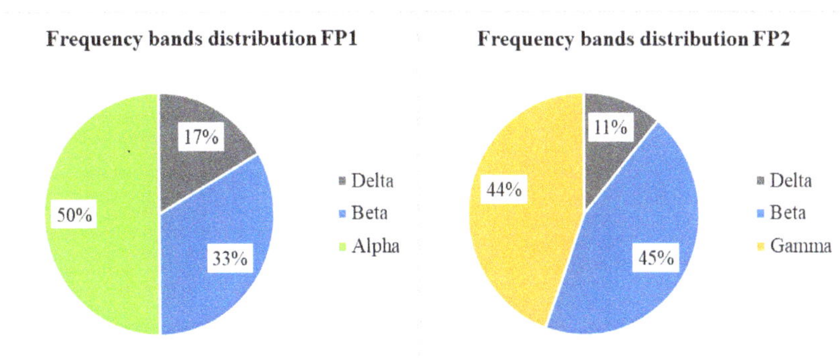

Figure 7. Distributions of EEG features by frequency bands for each electrode.

Based on our analysis of Table 9, Figures 6 and 7, we conclude that EEG and GSR show some indicators for the classification of boredom. First, the standard deviation of MV strongly correlated with boredom because this feature was selected from both GSR datasets. Second, the Gamma, Alpha and Beta bands have a strong correlation with boredom because these were selected more frequently than the Delta and Theta bands. Third, each frequency band has some correlation with a specific electrode location in boredom classification; for example, all selected Gamma band features belonged to FP2 and all selected Alpha band features belonged to FP1.

6. Discussion

In our experiment, the MLP models' performances were more stable than those of the other models. Considering Table 8, the RF model for the EEG-GSR combined dataset produced a maximum accuracy of 89.29%, which is the highest of all accuracies; however, its minimum accuracy was 66.07%, which was the lowest accuracy among all the models on the same dataset. An important property of a good classification model is the robustness of performance over multiple executions. To evaluate this for the selected models, we executed 1000 iterations of 10-fold cross validation with different random seeds. Thus, training data and testing data splits were changed randomly. A good model should be able to produce good classification results for different training and testing datasets. From this aspect, the MLP models classified boredom more robustly than the other models. Furthermore, the EEG-GSR dataset's MLP model (mean accuracy of 79.98%) outperformed the other MLP models in the aspect of mean accuracy. Therefore, our analysis suggests that MLP is generally recommended for classifying boredom from EEG and GSR.

Comparing our main results with the previous research presented in Table 2, our model's performance is better than those of Jaques et al. [13] and lower than Jang et al. [10]. However, as Table 1 shows, previous studies did not utilize EEG and GSR for classifying boredom. Thus, a direct comparison between our model and previous studies' performance is not reasonable. Furthermore, we collected EEG and GSR data from 28 participants whereas many previous studies, with the exception of Jang et al. [10], Jaques et al. [13], and Seo et al. [7], collected data from less than 28 participants (see Table 1); thus, our model is based on data acquired from a sufficient number of participants.

Moreover, this study executed 1000 iterations of 10-fold cross-validation on each model to reduce the effect of randomness on the results and to produce more reliable performance scores. Among previous studies, Jaques et al. [13], Jang et al. [10], and Seo et al. [7] also validated their models with 10-fold cross validation; however, they did not mention the number of repetitions of validation so we assume that cross validation was only executed once. Shen et al. [2] separated their data into training and testing sets but did not consider the random effect. Giakoumis et al. [9] validated their model with the use of leave-one-out cross validation that validates a model without random effect; however, as we mentioned above, a direct comparison of this study to our study is not reasonable because we used EEG and GSR datasets, while Giakoumis et al. [9] used ECG and GSR datasets.

We note that the proposed MLP model's performance (79.98%) is lower than that of the model proposed in our previous study (86.73%) [7]. One of the reasons contributing to this difference is that, as we explained in the paragraph above as well as in Sections 4.1 and 4.2.1, the experimental setting and the way we evaluated the models' performance was modified from the previous study. These changes were made to improve the robustness and generalizability of the results. For example, in our previous study [7], we acquired multiple samples from one participant's data by splitting them into one-second windows and used these for training; in the current study, we acquired only one sample from each participants' dataset by increasing the window size, thus aiming to increase the independence between the samples. This can help to reduce overfitting and achieve more generalizable results. Moreover, the previous study conducted only one iteration of cross validation, whereas, in the current study, mean accuracies were recorded after

1000 iterations of 10-fold cross-validation to increase the robustness of the results. This approach provides more reliable performance scores especially in the applications where the number of available samples is relatively small as in this study. Although the aforementioned steps taken decreased the accuracy of the final model, the generalizability and reliability of the result were increased.

Another novelty of this study is the identification of correlation between EEG, GSR, and boredom through the interpretation of features. As we explained in Section 2, this correlation was not revealed by previous studies. Moreover, our findings are aligned with Bench and Lench [34]'s suggestion that boredom should increase the autonomic nervous system activity, which directly relates to EEG and GSR as data sources. In our feature refinement results, the WSE algorithm recommended the EEG and GSR features on the best performing model for increasing performance. This suggests that the combination of these data correlates with boredom. In particular, Gamma band features were selected for the combined EEG-GSR and the EEG datasets. This indicates that the Gamma band may correlate with boredom, whereas the Alpha, Beta and Delta bands have weaker correlations, and the Theta band has no correlation at all with boredom. Furthermore, the WSE algorithm selected the standard deviation of MV among the GSR features from the combined EEG-GSR and GSR datasets. Consequently, the standard deviation of MV can also be an indicator of boredom.

7. Conclusions

In this study, we classified boredom using features from EEG and GSR datasets that were trained and tested by 30 models based on 19 different machine learning algorithms as an initial test for finding a suitable classification algorithm. We picked MLP, RF, and NB as the most suitable candidate algorithms. After tuning the selected algorithms' hyperparameters, we executed 1000 iterations of 10-fold cross validation with different random seed values to identify the most robust model among these. As a result, we recommended the MLP model which had a mean accuracy of 79.98% on the EEG-GSR combined dataset. Another major finding is that EEG and GSR appear to correlate with boredom, thus supporting the conclusion of Bench and Lench [34] that boredom and autonomic nervous system are linked.

Although this study produced novel contributions, there are noteworthy limitations. First, we collected physiological data from young and healthy participants. Thus, the recommended models may not be applicable to other age groups and to people with health issues. In addition, we hypothesize that emotion elicitation, and possibly also the manifestation of experienced emotions, is related to culture. We collected the data from Korean participants using non-boredom stimuli that were purposefully picked for this cultural context; therefore, the model may not be applicable to participants coming from other cultures. Regarding the protocol, we did not consider the effect of the order of showing the stimuli because the number of participants was deemed to be insufficient for dividing them into comparison groups. Finally, we used only one type of content to elicit boredom. Other types of contents may give different results about the intensity of the experienced emotion. In our future work, we aim to solve these limitations by collecting more data from a diverse group of users who are exposed to different boredom-evoking stimuli.

These results can be of use to developers building accurate affective computing systems as well as to researchers who seek to understand the physiological properties of boredom. As noted above, the current results still have limited applicability due to the experiment design that used only one type of boredom stimulus and a fairly homogeneous participant population. We plan to use diverse stimuli and extend the data collection to children, elderly people, patients suffering from different medical conditions, and participants representing other cultures to overcome these limitations.

Author Contributions: J.S. contributed on data collection, performed the experiments, and wrote the original draft; J.S., T.H.L., and K.S. conceived and designed the methodology; T.H.L. and K.S. supervised this study, and contributed on analysis and writing; K.S. administrated the overall project.

Funding: This research was supported by the MSIT (Ministry of Science and ICT), Korea, under the ITRC (Information Technology Research Center) support program (IITP-2019-2018-0-01431) supervised by the IITP (Institute for Information & Communications Technology Promotion), and also by the National Research Foundation of Korea grant funded by the Korea government (MSIT) (No. NRF-2019R1A2C1006608).

Acknowledgments: We would like to thank the participants who joined the experiment. Furthermore, we extend a special thanks to Byungkon Kang and Jaesik Kim at Ajou University.

Conflicts of Interest: The authors declare no conflict of interest.

References

1. Picard, R.W. *Affective Computing*; MIT Press: Cambridge, MA, USA, 1995; pp. 1–16.
2. Shen, L.; Wang, M.; Shen, R. Affective e-Learning: Using "Emotional" Data to Improve Learning in Pervasive Learning Environment Related Work and the Pervasive e-Learning Platform. *Educ. Technol. Soc.* **2009**, *12*, 176–189.
3. Zheng, W.L.; Lu, B.L. Investigating Critical Frequency Bands and Channels for EEG-Based Emotion Recognition with Deep Neural Networks. *IEEE Trans. Auton. Mental Dev.* **2015**, *7*, 162–175. [CrossRef]
4. Zheng, W.L.; Zhu, J.Y.; Lu, B.L. Identifying Stable Patterns over Time for Emotion Recognition from EEG. *IEEE Trans. Affect. Comput.* **2017**, *417–429*. [CrossRef]
5. Kim, J.J.; Fesenmaier, D.R. Measuring emotions in real time: Implications for tourism experience design. *J. Travel Res.* **2015**, *54*, 419–429. [CrossRef]
6. Kim, J.; Seo, J.; Laine, T.H. Detecting Boredom from Eye Gaze and EEG. *Biomed. Signal Process. Control* **2018**, *46*, 302–313. [CrossRef]
7. Seo, J.; Laine, T.H.; Sohn, K.A. Machine learning approaches for boredom classification using eeg. *J. Ambient Intell. Human. Comput.* **2019**, *10*, 3831–3846. [CrossRef]
8. Giakoumis, D.; Vogiannou, A.; Kosunen, I.; Moustakas, K.; Tzovaras, D.; Hassapis, G. Identifying Psychophysiological Correlates of Boredom and Negative Mood Induced During HCI. In Proceedings of the 1st International Workshop on Bio-Inspired Human-Machine Interfaces and Healthcare Applications, Valencia, Spain, 21 January 2010; pp. 3–12, ISBN 9789896740207.
9. Giakoumis, D., Tzovaras, D., Moustakas, K.; Hassapis, G. Automatic recognition of boredom in video games using novel biosignal moment-based features. *IEEE Trans. Affect. Comput.* **2011**, *2*, 119–133. [CrossRef]
10. Jang, E.H.; Park, B.J.; Park, M.S.; Kim, S.H.; Sohn, J.H. Analysis of physiological signals for recognition of boredom, pain, and surprise emotions. *J. Physiol. Anthropol.* **2015**, *34*, 1–12. [CrossRef]
11. Mandryk, R.L.; Atkins, M.S. A fuzzy physiological approach for continuously modeling emotion during interaction with play technologies. *Int. J. Hum. Comput. Stud.* **2007**, *65*, 329–347. [CrossRef]
12. Sidney, K.D.; Craig, S.D.; Gholson, B.; Franklin, S.; Picard, R.; Graesser, A.C. Integrating Affect Sensors in an Intelligent Tutoring System. In Proceedings of the Affective Interactions: The Computer in the Affective Loop Workshop at 2005 International Conference on Intelligent User Interfaces, San Diego, CA, USA, 10–13 January 2005; pp. 7–13.
13. Jaques, N.; Conati, C.; Harley, J.M.; Azevedo, R. Predicting affect from gaze data during interaction with an intelligent tutoring system. In *Lecture Notes in Computer Science*; 8474 LNCS; Springer: Berlin, Germany, 2014; pp. 29–38.
14. Thompson, W.T.; Lopez, N.; Hickey, P.; DaLuz, C.; Caldwell, J.L.; Tvaryanas, A.P. *Effects of Shift Work and Sustained Operations: Operator Performance in Remotely Piloted Aircraft (Op-Repair)*; Technical report; 311th Human Systems Wing Brooks Air Force Base: San Antonio, TX, USA, 2006.
15. Britton, A.; Shipley, M.J. Bored to death? *Int. J. Epidemiol.* **2010**, *39*, 370–371. [CrossRef]
16. Kanevsky, L.S. A comparative study of children's learning in the zone of proximal development. *Eur. J. High Ab.* **1994**, *5*, 163–175. [CrossRef]

17. Oroujlou, N.; Vahedi, M. Motivation, attitude, and language learning. *Proc. Soc. Behav. Sci.* **2011**, *29*, 994–1000. [CrossRef]
18. Sottilare, R.; Goldberg, B. Designing adaptive computer-based tutoring systems to accelerate learning and facilitate retention. *J. Cogn. Technol* **2012**, *17*, 19–33.
19. Yeager, D.S.; Henderson, M.D.; Paunesku, D.; Walton, G.M.; D'Mello, S.; Spitzer, B.J.; Duckworth, A.L. Boring but important: A self-transcendent purpose for learning fosters academic self-regulation. *J. Personal. Soc. Psychol.* **2014**, *107*, 5592014. [CrossRef] [PubMed]
20. Baker, R.; D'Mello, S.; Rodrigo, M.; Graesser, A. Better to be frustrated than bored: The incidence and persistence of affect during interactions with three different computer-based learning environments. *Int. J. Hum. Comput. Stud.* **2010**, *68*, 223–241. [CrossRef]
21. Fagerberg, P.; Ståhl, A.; Höök, K. EMoto: Emotionally engaging interaction. *Pers. Ubiquitous Comput.* **2004**, *8*, 377–381. [CrossRef]
22. Feldman, L. Variations in the circumplex structure of mood. *Personal. Soc. Psychol. Bull.* **1995**, *21*, 806–817. [CrossRef]
23. Li, M.; Lu, B.-L. Emotion classification based on gamma-band EEG. In proceedings of the 2009 Annual International Conference of the IEEE Engineering in Medicine and Biology Society, Minneapolis, MN, USA, 3–6 September 2009; pp. 1223–1226.
24. Lin, Y.P.; Wang, C.H.; Jung, T.P.; Wu, T.L.; Jeng, S.K.; Duann, J.R.; Chen, J.H. EEG-based emotion recognition in music listening. *IEEE Trans. Biomed. Eng.* **2010**, *57*, 1798–1806.
25. Shen, L.; Leon, E.; Callaghan, V.; Shen, R. Exploratory research on an Affective e-Learning Model. In Proceedings of the Workshop on Blended Learning, Edinburgh, UK, 15–17 August 2007; pp. 267–278, ISBN 978-3-540-78138-7.
26. Russell, J.A. A circumplex model of affect. *J. Personal. Soc. Psychol.* **1980**, *39*, 1161–1178. [CrossRef]
27. Vogel-Walcutt, J.J.; Fiorella, L.; Carper, T.; Schatz, S. The definition, assessment, and mitigation of state boredom within educational settings: A comprehensive review. *Educ. Psychol. Rev.* **2012**, *24*, 89–111. [CrossRef]
28. Eastwood, J.D.; Frischen, A.; Fenske, M.J.; Smilek, D. The Unengaged Mind: Defining Boredom in Terms of Attention. *Perspect. Psycholog. Sci.* **2012**, *7*, 482–495. [CrossRef] [PubMed]
29. Fahlman, S.A.; Mercer-Lynn, K.B.; Flora, D.B.; Eastwood, J.D. Development and Validation of the Multidimensional State Boredom Scale. *Assessment* **2013**, *20*, 68–85. [CrossRef] [PubMed]
30. Sanei, S.; Chambers, J.A. *EEG Signal Processing*; John Wiley & Sons: Hoboken, NJ, USA, 2013, ISBN 978-0-470-02581-9.
31. Ashwal, S.; Rust, R. Child neurology in the 20th century. *Pediatr. Res.* **2003**, *53*, 345. [CrossRef] [PubMed]
32. Martini, F.H.; Bartholomew, E.F. *Essentials of Anatomy and Physiology*; Benjamin Cummings: San Francisco, CA, USA, 2002; ISBN 978-0-13-061567-1.
33. Carlson, N.R. *Physiology of Behavior*; Allyn & Bacon: Boston, MA, USA, 2012; ISBN 978-0-205-23939-9.
34. Bench, S.W.; Lench, H.C. On the function of boredom. *Behav. Sci.* **2013**, *3*, 459–472. [CrossRef]
35. World Medical Association. World Medical Association Declaration of Helsinki. Ethical principles for medical research involving human subjects. *Bull. World Health Organ.* **2001**, *79*, 373–374.
36. MUSE. MUSE TM Headband. Available online: http://www.choosemuse.com/ (accessed on 19 October 2019).
37. Seeed. Grove—GSR Sensor. Available online: http://wiki.seeedstudio.com/Grove-GSR_Sensor/ (accessed on 19 October 2019).
38. Jasper, H. Report of the committee on methods of clinical examination in electroencephalography: 1957. *Electroencephalogr. Clin. Neurophysiol.* **1958**, *10*, 370–375.
39. Lang, P.J.; Bradley, M.M.; Cuthbert, B.N. *International Affective Picture System (IAPS): Affective Ratings of Pictures and Instruction Manual*; Technical Report A-8; University of Florida: Gainesville, FL, USA, 2008.
40. Hall, M.; Frank, E.; Holmes, G.; Pfahringer, B.; Reutemann, P.; Witten, I.H. The WEKA data mining software: An update. *ACM SIGKDD Explor. Newsl.* **2009**, *11*, 10–18. [CrossRef]

41. Kohavi, R.; John, G.H. Wrappers for feature subset selection. *Artif. Intell.* **1997**, *97*, 273–324. [CrossRef]
42. Muller, M.P.; Tomlinson, G.; Marrie, T.J.; Tang, P.; McGeer, A.; Low, D.E.; Detsky, A.S.; Gold, W.L. Can Routine Laboratory Tests Discriminate between Severe Acute Respiratory Syndrome and Other Causes of Community-Acquired Pneumonia? *Clin. Infect. Dis.* **2005**, *40*, 1079–1086. [CrossRef]

sensors

MDPI

Article

Wearables and the Quantified Self: Systematic Benchmarking of Physiological Sensors

Günther Sagl [1,*], Bernd Resch [1,2,*], Andreas Petutschnig [1], Kalliopi Kyriakou [1], Michael Liedlgruber [3] and Frank H. Wilhelm [3]

1 Department of Geoinformatics—Z_GIS, University of Salzburg, 5020 Salzburg, Austria; Andreas.Petutschnig@sbg.ac.at (A.P.); Kalliopi.Kyriakou@sbg.ac.at (K.K.)
2 Center for Geographic Analysis, Harvard University, Cambridge, MA 02138, USA
3 Department of Psychology, University of Salzburg, 5020 Salzburg, Austria; Michael.Liedlgruber@sbg.ac.at (M.L.); Frank.Wilhelm@sbg.ac.at (F.H.W.)
* Correspondence: guenther.sagl@gmail.com (G.S.); bernd.resch@sbg.ac.at (B.R.)

Received: 22 August 2019; Accepted: 10 October 2019; Published: 14 October 2019

Abstract: Wearable sensors are increasingly used in research, as well as for personal and private purposes. A variety of scientific studies are based on physiological measurements from such rather low-cost wearables. That said, how accurate are such measurements compared to measurements from well-calibrated, high-quality laboratory equipment used in psychological and medical research? The answer to this question, undoubtedly impacts the reliability of a study's results. In this paper, we demonstrate an approach to quantify the accuracy of low-cost wearables in comparison to high-quality laboratory sensors. We therefore developed a benchmark framework for physiological sensors that covers the entire workflow from sensor data acquisition to the computation and interpretation of diverse correlation and similarity metrics. We evaluated this framework based on a study with 18 participants. Each participant was equipped with one high-quality laboratory sensor and two wearables. These three sensors simultaneously measured the physiological parameters such as heart rate and galvanic skin response, while the participant was cycling on an ergometer following a predefined routine. The results of our benchmarking show that cardiovascular parameters (heart rate, inter-beat interval, heart rate variability) yield very high correlations and similarities. Measurement of galvanic skin response, which is a more delicate undertaking, resulted in lower, but still reasonable correlations and similarities. We conclude that the benchmarked wearables provide physiological measurements such as heart rate and inter-beat interval with an accuracy close to that of the professional high-end sensor, but the accuracy varies more for other parameters, such as galvanic skin response.

Keywords: wearable sensors; psychophysiology; sensor data analysis; time series analysis; signal analysis; similarity measures; correlation statistics; quantitative analysis; benchmarking

1. Introduction

In the last decade, the body of literature about physiological sensing and deriving emotions from physiological parameters has grown significantly. One reason for this is the rapid increase in variety of affordable wearable sensors that measure a broad range of physiological parameters such as heart rate, galvanic skin response, skin temperature, and others. With this increase, the "Quantified Self" community that promotes the idea of 24/7 tracking and monitoring has been growing significantly [1–3].

These new low-cost wearables are increasingly used in scientific studies in a variety of areas like health research, well-being assessment, disaster management, emotion information extraction and spatial emotion analysis, and stress detection [4–13]. However, some research efforts have used wearable physiological sensors without prior investigation of the sensor's exact quality parameters, i.e., how accurately a sensor actually measures a given parameter or how reliable a sensor is in producing continuously high-quality measurement results.

Understanding a sensor's quality and accuracy is critical because the research results may otherwise be unreliable: while traditional professional wired sensor devices, which have been used for some time in laboratory and ambulatory studies in the fields of psychological and medical research, are proven to be highly accurate, most wearable sensors used in previous studies are not. In fact, most of them are not medically and/or electronically certified, which compromises the reliability of the measurement results. However, recently, some wearable sensors have been released that are certified and comply with a number of international standards (sensor technology, wireless communication, data transmission, etc.), which makes them a viable alternative to traditional wired equipment.

In the context of this research, we aim to investigate the measurement quality of two wearable sensor devices, namely the Zephyr BioHarness 3 and the Empatica E4, by comparing their measurements to those of calibrated laboratory sensors. Concretely, we are interested in the similarity and correlation of univariate time series from two different sensors that measure the same physiological parameters at the same time on the same participant. To evaluate the accuracy of the low-cost sensors, we perform benchmark testing between low-cost sensors against high-quality and well-calibrated sensors that act as the trusted gold standard. The second aim of this research is to detect and quantify relationships and dependencies between pairs of the same and different physiological parameters measured by different sensors. Our study assesses the parameters heart rate (HR), inter-beat interval (IBI), and galvanic skin response (GSR).

The remaining part of the paper is structured as follows. In Section 2, we provide a concise summary of related work regarding sensor benchmarking, followed by an overview of the physiological parameters of interest and the sensors used for this research (Section 3). The benchmarking methodology is presented in Section 4, where we also explain the entire workflow from sensor data acquisition to the analysis results. Section 5 descriptively illustrates the results, including a variety of statistical visualisations of similarity and correlation patterns. Finally, we discuss the results obtained and close the paper with our core conclusions.

2. Sensor Benchmark Methods—Related Work

The analysis of physiological signals from wearable sensors in order to better understand the human emotional response to the immediate surroundings has been investigated for several years. In recent years, a variety of affordable wearable sensors that measure well-established physiological parameters, such as heart rate and galvanic skin response, has reached the market. As a logical consequence—and as already mentioned in the introduction—the "Quantified Self" community is growing faster than ever, and inspiring scientific research, especially related to emotion and stress detection [5,7–9,14–17]. In any case, the basis for any further advanced analyses is adequate data quality in terms of accuracy, reliability, and validity [7,9,18]. However, scientific literature about the similarity and correlation of the measurements from such affordable wearables compared to those from well-calibrated and high-quality sensors from scientific laboratories is rare.

2.1. Similarity Measures

Generally speaking, the term 'similarity' is not rigorously mathematically defined. A variety of similarity measure families exist, for instance, distance-based (e.g., Euclidean distance), feature-based (e.g., Fourier coefficients), model-based (e.g., autoregressive), and elastic measures such as Dynamic Time Warping (DTW) and Edit Distance on Real sequence EDR [19–22]. A comprehensive review, however, is out of the scope of this paper—the interested reader may refer to [19,20,23,24], among other work.

In this research, we go beyond global measures and linear models to assess similarity. To uncover local similarity characteristics of time series, we thus follow a moving window approach combined with more informative distance metrics. Elastic measures, such as DTW and the Fréchet distance, allow for a one-to-many comparison of time series elements, while so-called "Lock-Step" measures, such as Euclidian and Manhattan distance, only allow comparison of fixed pairs, making them very sensitive to local time-shifts and noise [23].

DTW temporally aligns two time series using the shortest path in a distance matrix, i.e., the path with the minimal global warping distance [25,26], thereby finding the most representative distance of the overall difference [20]. However, a comprehensive experimental comparison of representation methods and distance measures of time series reveals inconsistencies and even contradictions in the observations reported in individual studies [23]. An important consequence of this is that experimental results cannot be generalised without critically reviewing the assumptions made for a particular research context and study design. As concluded in [23], "there is no clear evidence that one similarity measure exists that is superior to others in the literature in terms of accuracy. While some similarity measures are more effective on certain data sets, they are usually inferior on some other data sets" (p. 297). The DTW distance outperforms Euclidian distance in a variety of studies [27]. Other types of measures are "Edit measures" and "Threshold measures". The former type includes, for instance, Longest Common Sub-Sequence LCSS, Edit Distance on Real sequence EDR and Edit Distance with Real Penalty ERP. The latter type includes Tightness of Lower Bounds TLB. The accuracy of the aforementioned other types is close to the accuracy of DTW, but DTW is much simpler [23,28]. We thus concluded to use DTW to assess the temporal similarity of the physiological time series.

To assess the geometric shape of a curve or curve segment, other distance measures, such as the Fréchet distance [29], can be used [30–32]. "The Fréchet distance is typically explained as the relationship between a person and a dog connected by a leash walking along the two curves and trying to keep the leash as short as possible. The maximum length the leash reaches is the value of the Fréchet distance" [33] (p. 7). We thus use the Fréchet distance to assess the geometric similarity of time series of sensor measurements of the same physiological parameter (e.g., GSR) on the same participant at the same time but with different sensors.

2.2. Correlation Statistics

The correlation of time series has been investigated for decades, in diverse fields. Herein, our focus on time series correlation is twofold: first, the correlation between equal-type physiological parameters measured by different sensors at the same time on the same participant in order to quantify differences between low-cost and un-calibrated sensors versus high-end and calibrated laboratory sensor equipment; second, the correlation between physiological parameters of different types, for instance, IBI and GSR, to explore potentially hidden relationships.

According to [34], the Pearson's correlation coefficient is the most robust metric when measuring the similarity in physiological time series—where robustness is understood as insensitivity to small variations. However, Pearson's r is highly sensitive to outliers and only considers linear relationships. Spearman's rank correlation coefficient (rho) is—as the name says—based on the rank of the values rather than on the values themselves; thus, it measures monotonicity rather than linearity. Therefore, using Spearman's rho to measure the strength of the associations between two variables leaves room for interpretation [35].

The human cardiovascular system and the autonomic nervous system are highly non-linear systems. In order to explore possible underlying non-linear interactions in the relationship between different physiological parameter, we herein, use the Maximum Information Coefficient (MIC) [36,37]. Several studies show the possibility of gaining new insights into such non-linear interactions when applying the MIC, for instance, in the interactions between neural and respiratory dynamics [38].

Further, one method to assess the temporal lag (or lead) between pairs of time series is the cross-correlation function in the time domain [39]. To get meaningful cross-correlation results, the time series need to be stationary, i.e., have a constant mean and variance. Time series stationarity can be tested using, for instance, the Augmented Dickey-Fuller test [40]. Unless the time series is stationary, it needs to be differenced and tested for stationarity.

3. Physiological Parameter of Interest and Sensors used for Benchmarking

Herein, we describe the physiological parameters we investigated, and the sensors used to measure them. We investigated three sensors and four physiological parameters (Table 1):

- HR: heart rate, i.e., heartbeat frequency, unit: beats per minute
- IBI: inter beat interval, also known as the RR interval, i.e., the time between two R-peaks in the ECG's QRS complex, unit: milliseconds
- ECG: electrocardiogram, i.e., electrical activity of the heart, unit: millivolt
- GSR: galvanic skin response, i.e., the level of electric conductance of the skin, unit: microSiemens [μS]

Table 1. Benchmarked sensors and physiological parameters of interest.

	VarioPort	Zephyr BioHarness 3	Empatica e4
		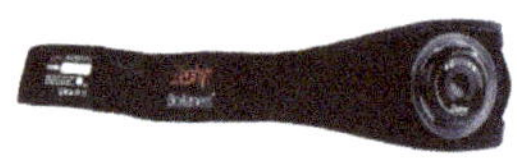	
HR	X	X	–
IBI	X	X	X
ECG	X	X	–
GSR	X	–	X

3.1. *VarioPort*

The VarioPort (http://www.bisigma.de) is a small, lightweight, and highly flexible recording system that is used for multi-channel physiology recordings in laboratory and ambulatory setups. The standard version of the device can record up to 16 signals from connected pre-amplifiers (e.g., electromyography, electrocardiography, electrodermal activity, or respiration). The device has two built-in marker buttons that can be used to signal certain events occurring over time, resulting in an additional channel of data. We used these buttons to mark changes in the activity phases (for details refer to Section 4.1). Available sensors are either wire-connected to the device or are directly integrated into the device. The recorded data are stored on an SD card inside the device. The VarioPort allows setting different sampling rates for different channels, thus effectively reducing the required storage, especially in case of slowly changing signals (e.g., such as the skin conductance). For rapidly changing signals, such as ECG, sampling rates of up to 1024 Hz can be set. Since the VarioPort is the platform used for scientific studies at our Psychology Department, we used it as the gold standard in our benchmarking. In the remaining part of the paper, the VarioPort sensor is called VP.

3.2. *Zephyr BioHarness 3*

The Zephyr BioHarness 3 (https://www.zephyranywhere.com/) is a multivariable physiological monitoring device with a chest belt sensor that measures a wide variety of physiological parameters. The BH is a certified medical product (FDA Class II). Due to its design as a chest belt, the BioHarness 3 can measure ECG, RR intervals, respiration frequency, and other parameters such as 3D acceleration on a single sensor platform. Furthermore, parameters such as heart rate can be derived from the directly measured parameter, for instance, from ECG (HR within 0–240 BPM and an accuracy of ±1 BPM). The sampling rate for ECG is 250 Hz. The BH and the smartphone are connected wirelessly via Bluetooth. The raw data are accessible via a free SDK in binary format, and the device has been extensively tested and validated in practical applications [41,42]. In the remaining part of the paper, the Zephyr BioHarness 3 sensor is called BH.

3.3. Empatica E4

The Empatica E4 (https://www.empatica.com/research/e4/) is a wrist band sensor that measures HR and GSR, as well as other parameters. The E4 is medically certified according to CE Medical 93/42/EEC Directive, class 2a, FCC. The sampling rate for GSR is 4Hz and for IBI 64Hz. According to Gradl et al. [18], the E4 is a wearable sensor that has the potential to measure mental stress. The E4 allows access to the raw data via smartphones through a comprehensive SDK and a Bluetooth connection. In the remaining part of the paper, the Empatica E4 sensor is called E4.

4. Benchmark Method

4.1. Study Setup and Participants

Our study included 18 participants who were recruited via e-mail. The test group comprised nine females and nine males in an age range of 24 to 40 years. All test persons were physically in decent shape and did not suffer from any illness at the time of the study.

The study was carried out at the University of Salzburg's Department of Psychology. After the study leaders attached the sensors, the participants were instructed to sit on an ergometer and follow the following routine:

- Resting phase: 5 min of rested seating on the ergometer, not performing any physical activity; used as a calibration phase for the measurements
- Cycling phase: 10 min of cycling at a constant 50 rpm; stepwise increase of physical load (5 steps of 2 min each, during which the resistance/power of the ergometer was increased from 35–65–100–133–165 W)
- Cool down phase: 5 min cool down, rested seating like in the resting phase

Participants were told not to interact with other people in the room, to focus on their task, and not to perform any physical activity other than as instructed. This was observed by the study leaders. For each 'run', two test persons were doing the lab study in parallel next to each other. Before commencing the actual exercise, we checked that all sensors were well positioned according to the participants' individual body shape. Additionally, we used surgical tape as necessary to hold the devices in place to make sure that we receive plausible measurements. We conducted these checks for each participant individually. All participants were aware of the aim of this research, and we obtained informed consent from all participants prior to commencement of the study.

4.2. Data Acquisition

The basic data acquisition workflow is illustrated in Figure 1. Each participant was equipped with diverse sensors to measure the physiological parameter using different platforms, namely VarioPort (VP), Zephyr BioHarness3 (BH), and Empatica E4 (E4). For each run, which refers to a participant exercising while their physiological parameters are measured, the raw sensor data are either stored in an SQLite database directly on the smartphone, or as files in a proprietary format on an SD card. The measurements from VP were extracted to flat files using the Software ANSLAB [43]. In contrast, the measurements from BH and E4 were sent to and fused by the e-Diary App into an SQLite database. The "raw sensor data" serves as input for the pre-processing, which is necessary to prepare the data for further analyses. The e-Diary App is herein purely used for sensor data collection and data management. During real-world field studies, however, the e-Diary App collects additional data such as GPS positions and contextual user feedback used for ground-truthing, thereby enabling the investigation of moments of stress in a spatio-temporal and contextual manner [4,44].

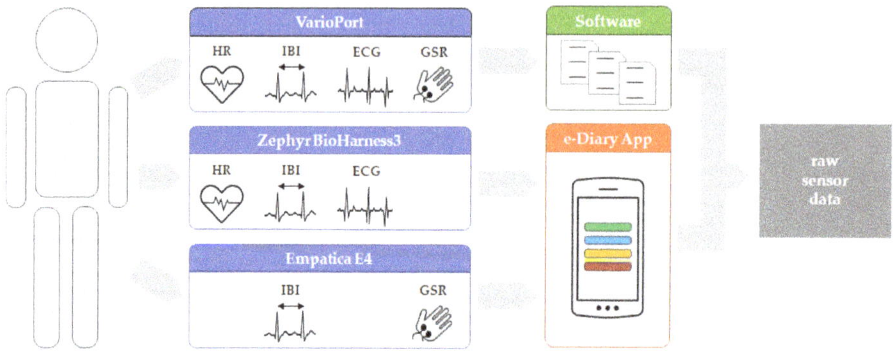

Figure 1. Data acquisition workflow—from the human participant (**left**) to raw sensor data (**right**).

4.3. Data Pre-Processing

The "raw sensor data" from the previous step serves as input for the data pre-processing phase, which is illustrated in Figure 2.

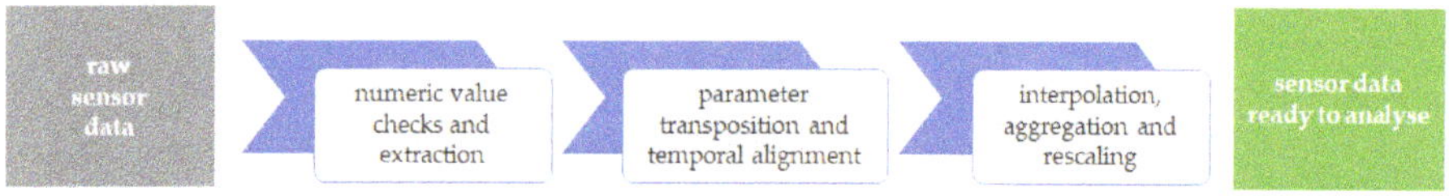

Figure 2. Data pre-processing phase—from raw sensor data (**left**) to sensor data ready to analyse (**right**).

This pre-processing phase consists of three main steps:

1. Extracting numeric values from differently encoded strings of values:

Data in the SQLite DB are stored in 1 s intervals in different formats due to various sampling rates of each of the sensors measuring different physiological parameters. For instance, the Empatica E4 measures GSR at a sampling rate of 4 Hz, while the VarioPort measures GSR at a sampling rate of 25 Hz.

The result is a table with sensor values where each single measurement has a correct timestamp.

2. Transposition of parameters and temporal alignment of measurements:

First, the vertical parameter structure (one row consists of a timestamp and a single sensor measurement, the next row consists of the same timestamp and with another single sensor measurement) needs to be transposed to a horizontal structure (a common timestamp and individual values as columns: one row consists of a timestamp and all sensor measurements that occurred at that timestamp).

Second, the irregular timestamps of all measurements are aligned to the millisecond in order to ensure the best possible time matching to the sensors' synchronized time. Since a 1 millisecond resolution is below the original sampling period, the measurements are aggregated depending on the parameters.

The result is a regular multivariate time series with 10 or 100 millisecond resolutions where some parameters at some timestamps may be missing values while other parameters are averaged within the given millisecond interval.

3. Interpolation, moving average and rescaling:

To fill missing values introduced by the temporal alignment in step 2, we applied spline interpolation, because it tends to greatly reduce oscillation by taking into account data points before and after the gap to be interpolated for a continuous representation [45]. In addition to the raw data, we calculated a moving averaged version with a window of ±5 s to eliminate high local variations.

For the correlation analysis of same type parameters and exploratory plots, we keep the original scaling of individual time series to identify potential offsets of measurements. For the similarity analysis, however, we rescale the measurements of individual time series from minimum and maximum to 0 and 1 in order to compare similarity distance metrics.

The data pre-processing was mainly carried out directly in the database using SQL and Java.

4.4. Statistical Signal Analysis—Time Series Correlation and Similarity Analysis

This sub-section illustrates how we assessed the correlations and similarities between time series of the same physiological parameters measured by different sensor platforms. Additionally, for some selected statistics, such as the MIC, we also run the analysis between different physiological parameters in order to explore potentially unknown relationships. The basic analytical workflow is shown in Figure 3. Note that we use the original signal scaling for exploratory plots, linear regression, and cross-correlation, while we use the rescaled signal (minimum $\rightarrow$ 0 ... maximum $\rightarrow$ 1) to get similarity measures such as Fréchet and DTW distance.

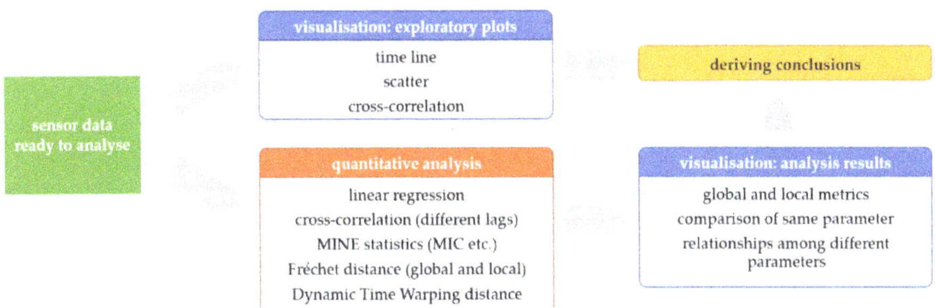

Figure 3. Data analysis workflow—from sensor data ready to analyse (**left**) to visualizations of exploratory plots and quantitative analysis results to deriving conclusions (**right**).

Of the complete time series derived from the pre-processing workflow (see Section 4.2), we focused on the following physiological parameters: HR, GSR, IBI, ECG. Additionally, from the ECG signal, we again derived the IBI and the Complex Demodulation amplitudes for the following frequency bands to estimate the heart rate variability [46]:

- Very low frequency VLF (0.025–0.07 Hz)
- Low frequency LF (0.07–0.14 Hz)
- High frequency HF (0.14–0.5 Hz)

In order to quantify pairwise correlations and similarities, we focused on:

- The linear regression coefficient of determination R^2, to assess the fit of pairwise timer series to a linear model [23]
- Cross-correlation, to assess temporal shifts [47]
- Maximal Information-based Nonparametric Exploration MINE statistics, in particular, the MIC, to assess functional associations, and MIC-R^2 to assess non-linear associations [36,48]
- Fréchet distance, to explore geometric similarity [30,49]
- DTW distance, to explore temporal similarity [26,28]

Signal analysis and most plots were done using the statistical computing software R, while some other plots were produced with the data visualisation software Tableau Desktop.

5. Results

For each of the 18 participants, we captured 15 parameters either measured directly or derived from ECG. For all these parameters, we also computed a moving average for an outlier-smoothed version of the same signal in order to get a better understanding of the signal's overall robustness and reliability. Additionally, we computed low/high-pass filtered versions of the GSR signals, as well as the complex demodulation amplitudes from the ECG signals. For each participant, we analysed 22 pairs of physiological parameters of the same type regarding similarity (e.g., heart rate from BioHarness sensor and heart rate from VarioPort sensor), and another 136 pairs of parameters of different type regarding correlations (e.g., heart rate from BioHarness sensor and galvanic skin response from VarioPort sensor).

Since there are many different parameters, we defined a naming convention that includes the physiological parameter of interest, the platform used to measure it, plus an indication of whether a time series is a moving averaged and/or a filtered version. For the naming of these parameters, we use the following notation:

For direct measurements:

<parameter> <platform> [filt.] [(mv. avg)]

where *parameter* can be GSR, HR, or IBI and *platform* can be BH, E4, or VP; *(mv. avg.)* indicates that this is the moving averaged version; *filt.* indicates that a first order high-pass (0.05 Hz) and first order low-pass (0.5 Hz) Butterworth filter has been applied to the original signal (this filter setting is used for further analysis to identify moments of stress [4,11]; however, this is not within the scope of this paper).

example: *GSR: VP (mv. avg.)* refers to the moving averaged version of galvanic skin response measured by VarioPort

For derived measurements:

<derived parameter> from <original parameter> <platform> [(mv. avg.)]

where *derived parameter* can be HF, IBI, LF, VLF, *original parameter* can be ECG, and *platform* can be BH, or VP; *(mv. avg.)* indicates that this is the moving averaged version example: *IBI: from ECG BH* refers to the inter beat interval derived from the electrocardiogram measured by BioHarness

The following subsections are structured according to Figure 3. We use two representative time series, one HR and one GSR, as examples to guide the reader through the high number of physiological parameters investigated herein. These two examples are cross-referenced between several figures and thus provide views on the same data from different perspectives, thereby fostering the consolidation of a more holistic picture.

5.1. Visualisation: Exploratory Plots

The aim of the exploratory plots is to obtain a basic understanding of the temporal behaviour and the relationship of equal-type physiological parameters. For this first insight, we investigate three complementary plots that provide different views on the same data: a time series plot, a scatter plot, and a cross-correlation plot. These plots show two versions of the same parameter, namely a data-as-is version and moving averaged version. Note that for the cross-correlation plots the second parameter is used as the independent one. To illustrate the methodology and exemplary results, we only show representative sample plots, which highlight the characteristic patterns of about 80% of all plots. Overall, we produced more than 1000 plots based on unscaled and rescaled data.

Figures 4 and 5 each show a time plot (a), a scatter plot (b), and a cross-correlation plot (c,d). The physiological parameter of interest is HR, measured by HB and VP. The HR time plot (Figure 4a) shows two highly similar, almost identical curves. The blue curve has an offset, which is maximal in the low range and converges to zero in the high range. The corresponding error term seems to include a reciprocal component: the higher the actual measurement, the lower the error. The HR scatter plot (Figure 4b) shows a high positive linear relationship with an R^2 of 0.971 for raw data and 0.997 for the moving averaged data. This means that 97.1% and 99.7%, respectively, of the data's total variance can be explained by a linear model. This plot also confirms what is seen in the time plot, namely that in the low range the residuals are higher than in the high range. Note that the higher residuals in red in the upper right quarter of the plot refer to the time plot at ~750 s, where the blue curve drops below the red curve (indicated by a black arrow in Figure 4a,b). The HR cross-correlation plots show the highest cross-correlation for the as-is version (Figure 4c) at a lag of 1 s, and the highest cross-correlation for the moving averaged version (Figure 4d) at a lag of 2 s. In other words, the local trend of the BH is lagging 1 and 2 s, respectively, "behind" the local trend of the VP on average.

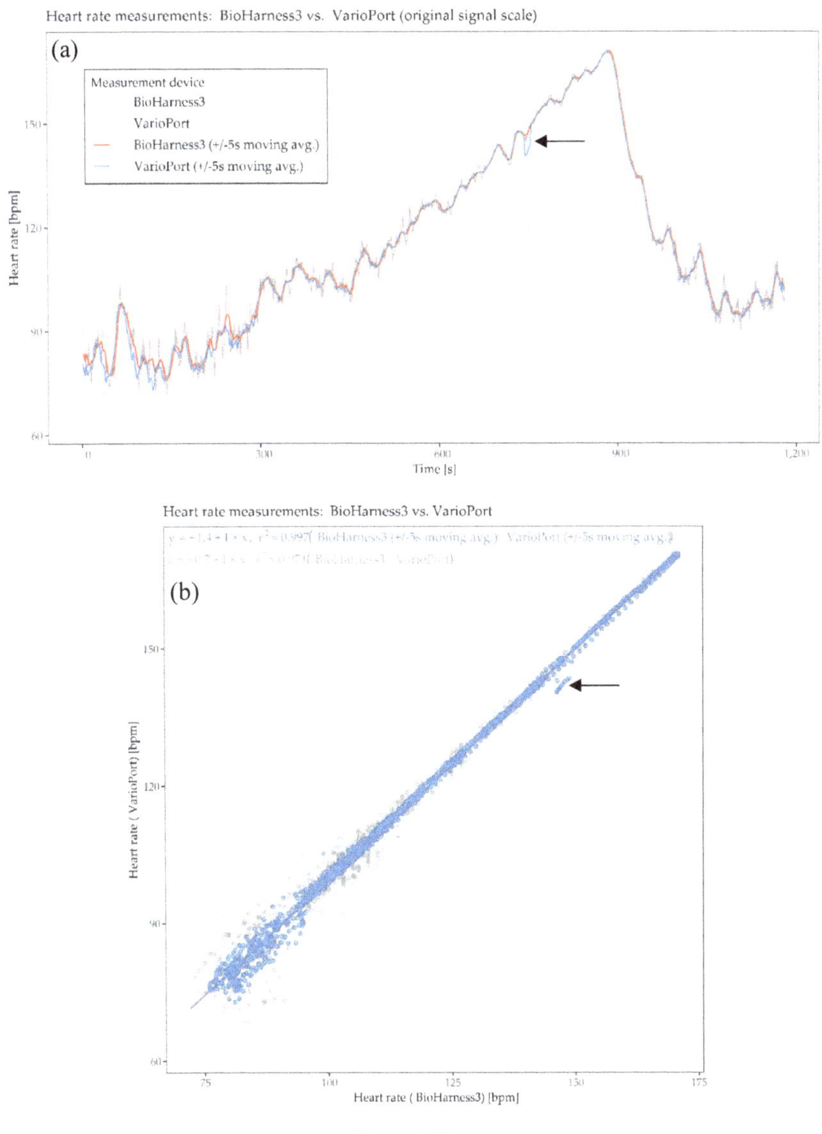

Figure 4. *Cont.*

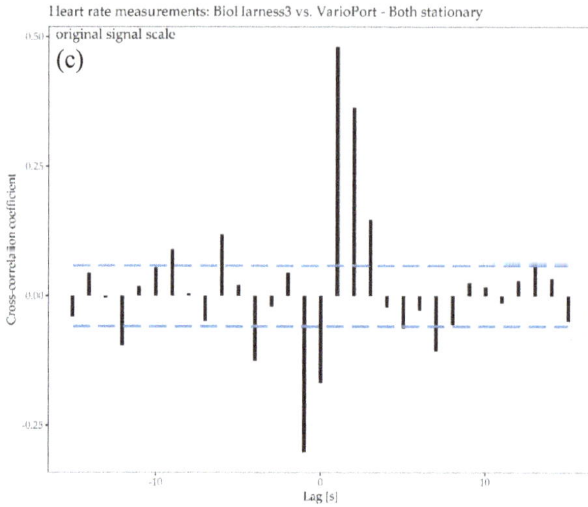

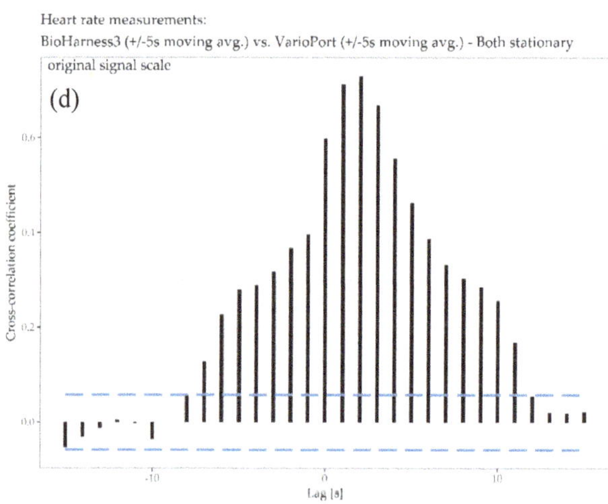

Figure 4. Participant RP 5–14: time plot (**a**), scatter plot (**b**), and cross-correlation plot (**c**), and cross-correlation plot of moving averages (**d**) of heart rate HR [beats per minute] measured by Bioharness3 BH sensor and VarioPort VP.

In Figure 5, the physiological parameter of interest is GSR, measured by the E4 and the VP. Generally speaking, measuring GSR is, in comparison to HR, a delicate undertaking due to the measurement principle, which is solely based on the electrical conductivity of the skin. This conductivity is highly dependent on (1) the participant's skin characteristics and (2) the contact between the sensor electrodes and the skin, especially during physical activity. Thus, these two factors can have a significant impact on the reliability, and thus on the comparability, of the sensor measurements. Additionally, the mounting of the sensors' electrodes can differ as well. For instance, the VP electrodes need an isotonic electrolyte gel to ensure reliable measurements, while E4 does not require anything. The GSR time plot (Figure 5a) shows that the blue curve (VP) increases gradually. The red curve (E4) increases faster than the blue one and shows a local maximum after the 5-min warm-up phase (at around 300 s). This increase is followed by a decrease for another 5 min (until around 600 s). Aside from a little drop at around 900 s, the red curve increases until the end of the cool down phase. In general, the E4 seems to be more responsive to sweating associated with physical effort than the VP, which may be due to its lack of stabilizing isotonic electrolyte gel. The GSR scatter plot (Figure 5b) shows a positive correlation with an R^2 of 0.882 for raw data and 0.896 for the moving averaged data. This means that 88.2% and 89.6%, respectively, of the data's total variance can be explained by a linear model. The cross-correlation plots (Figure 5c) show the highest cross-correlation for the as-is version at lag of 2 s, and the highest cross-correlation for the moving averaged version at a lag of 1 s. In other words, the local trend of the E4 sensor is lagging 2 and 1 s, respectively, "behind" the local trend of the VP sensor on average.

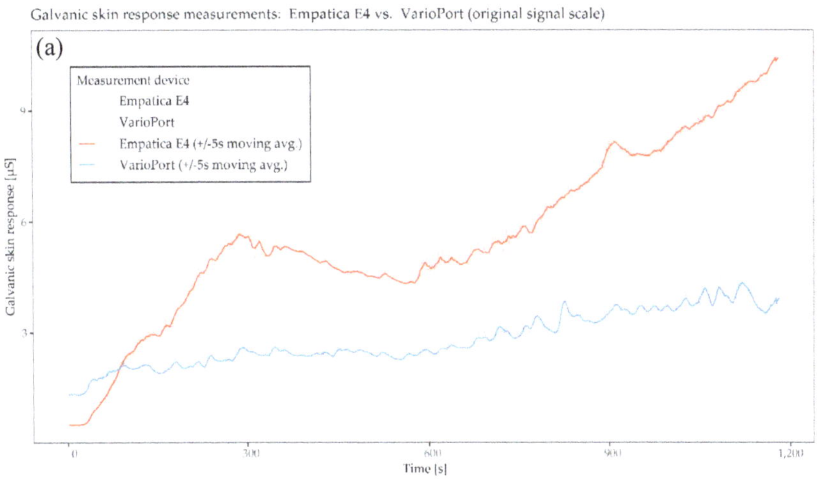

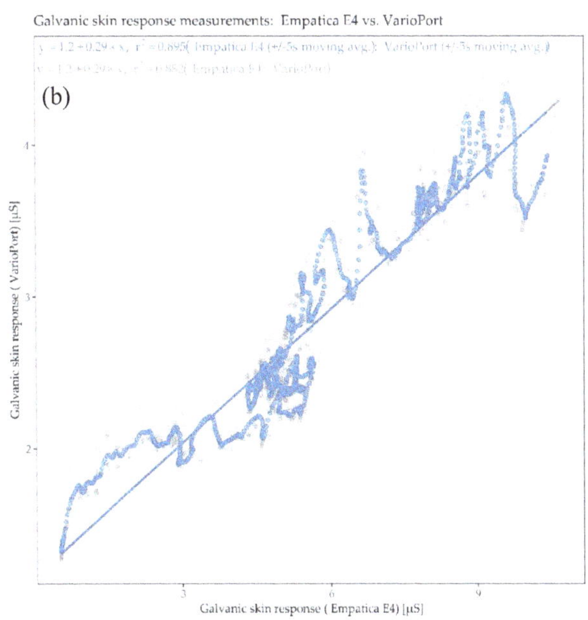

Figure 5. *Cont.*

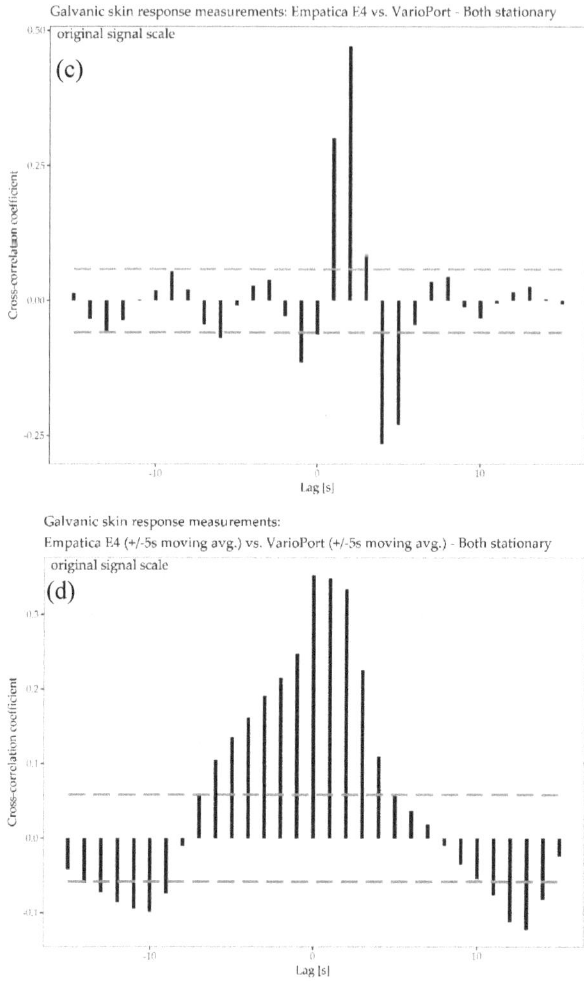

Figure 5. Participant RP 3–8: time plots (**a**), scatter plot (**b**), cross-correlation plot (**c**), and cross-correlation plot of moving averages (**d**) of galvanic skin response GSR measured by Empatica E4 and VarioPort.

5.2. Quantitative Analysis

The aim of the quantitative analysis is twofold: first, we assess the correlation and the similarity of equal-type parameters. Second, we assess potential associations in pairs of parameters of both equal-type and different types. In addition to global statistics, we also apply local measures to derive new information about the relationship between and among the different parameter pairs. This combination of global and local similarity and correlation measures on the individual level further enables a roll-up view on relationship patterns of physiological parameters among participants. Note that for similarity distance metrics, we rescaled measurements from min ... max of the original scale to 0 ... 1. For the correlation analyses we used the original values of the given parameter at the given range and the given unit in order to identify potential offsets.

5.2.1. Linear Regression and Coefficient of Determination R^2

The first statistic of interest is the coefficient of determination R^2, which quantifies the percentage of the variance of the two given parameters that can be explained by a linear regression model. In addition to the R^2 of individual pairs of parameters, as shown in the scatter-plots (refer to Section 5.1, Figures 4b and 5b), we now investigate all pairs among all participants and explore the corresponding R^2 pattern. This pattern can be derived from the R^2 matrix shown in Figure 6. Furthermore, for each group of parameters, e.g., all IBI related parameter, we calculate the total average per participant in order to get an impression of the impact of each participant's individual overall measured activity (GSR base level, skin contact of electrodes, etc.).

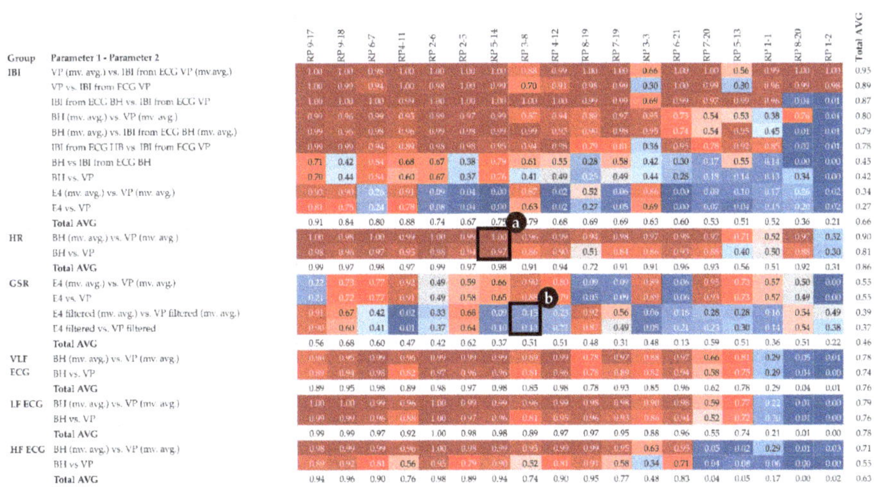

Figure 6. R^2 matrix of pairs of parameter and participants; detail (**a**) complements Figure 4, detail (**b**) complements Figure 5; total average of individual pairs among participants is shown in the last column; total average of participants among individual parameter pairs is shown in the last row of each parameter group; colour: red indicates high correlation, blue indicates low correlation.

In the upper half of the R^2 matrix, the pairs of equal-type parameters, measured by different sensors (or derived from another signal of the same sensor) show a high linear relationship across the majority of the participants. This relationship also indicates that these parameters are rather robust from a measuring point of view. However, the matrix also shows some cases with no relationship at all, see for instance IBI derived from ECG BH (moving averaged version), and IBI derived from ECG VP (moving averaged version) at row three for participant RP 8-20 and RP 1-2. This may indicate that one of the sensors did not have proper contact between the electrodes and the skin and thus failed to collect valid data.

The matrix shows that GSR in general, and IBI measured by Empatica E4, tend to have rather low or even no correlation, while some participants demonstrate the exact opposite (compare instance RP 4-11 and RP 2-5).

Note that the R^2 matrix in Figure 6 is organized as follows: for each group of parameters, the top row shows the highest correlation among all participants, while the bottom row shows the lowest correlation. Further, the left column shows the participant with the highest correlations among all parameters, while the right column shows the participant with the lowest correlations among all parameters.

Figure 6 detail (a) refers to the HR example shown in Figure 4 and detail (b) refer to the GSR example shown in Figure 5.

In summary, the overall pattern shown in Figure 6 confirms that both the type of the parameter measured and the individual parameters, such as skin characteristics of the participant, significantly influence the reliability of the measurements and thus the quality of further analysis.

5.2.2. Cross-Correlation

In order to assess potential temporal shifts between the measurements of the same physiological parameter, we investigate the cross-correlation at different lags. The corresponding cross-correlation pattern in Figure 7 shows that some pairs of parameters (especially the top five rows of the IBI group) have a rather low variance among the lags and tend to correlate positively (i.e., the lags within a cell show rather homogeneous coefficients). Other pairs of parameters (especially the lower half of the IBI group) have a rather high variance among the lags, indicating positive correlations around lag 0 and negative correlations at lag +15 and −15, respectively.

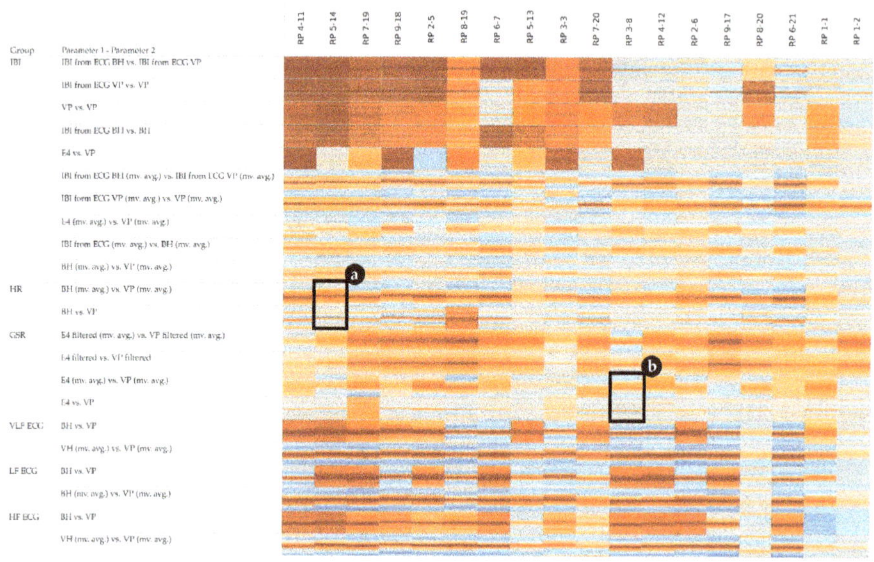

Figure 7. Cross-correlation matrix of pairs of parameters and participants; detail (**a**) complements Figure 4, detail (**b**) complements Figure 5; colour: orange indicates positive cross-correlation, blue indicates negative cross-correlation; a cell detail shows lags as small horizontal bars: lag −15 at top, and lag +15 at bottom.

Note that the cross-correlation matrix in Figure 7 is organized as follows: for each group of parameters, the top row shows the highest cross-correlations (i.e., lowest variance) among all participants, while the bottom row shows the lowest cross-correlation (i.e., highest variance). Further, the left column shows the participant with the highest cross-correlations among all parameters, while the right column shows the participant with the lowest cross-correlations among all parameters.

Figure 7 detail (a) refers to the HR example shown in Figure 4 and detail (b) refers to the GSR example shown in Figure 5.

5.2.3. MINE Statistics

By using MINE statistics, we investigate relationships between and among parameter pairs of same as well as different type, for instance, GSR versus HR. In addition to the linear correlation coefficient of determination R^2 (see Section 5.2.1), we use the Maximal Information Coefficient MIC to identify all functional relationships, also including linear ones as issued by R^2. Although a functional relationship between certain combinations of parameters might be obvious (e.g., IBI [ms] = 60,000/HR [beats per hour]), we nonetheless include such combinations herein for reasons of confirmation.

Figure 8 shows 3 k-means clusters of pairs of different parameter types. We tested with k [1,5] and chose 3 because the result is most intuitive—the clusters show low, moderate, and high correlations. Figure 8a–f highlights some particularly interesting parts of the clusters:

- a, b: LF and VLF derived from ECG measured by BioHarness show a rather low correlation with almost all other parameters
- c, d: IBI measurements, either directly measured or derived from ECG, show a rather high correlation with HR and with IBI from other sensors
- e, f: GSR measurements from VP (high- and low-pass filtered) show a rather low correlation with all other parameters; however, GSR measurements from E4 (high- and low-pass filtered) show a low to moderate correlation with all other parameters.

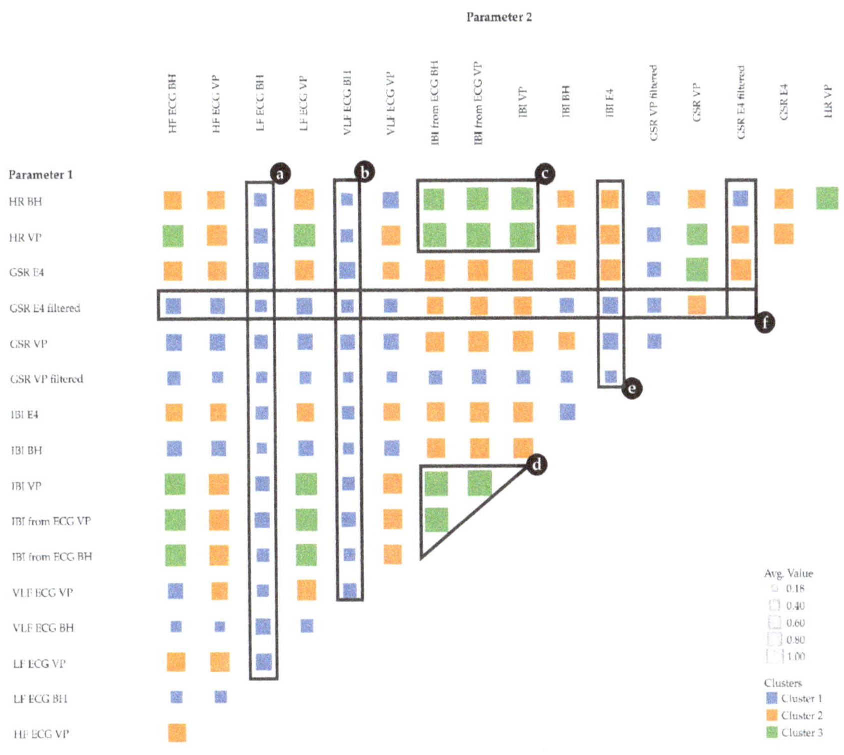

Figure 8. Maximum Information Coefficient (MIC) cluster matrix of pairs of parameters (size show averages among all participants, moving averaged versions only); colors: blue (cluster 1): low correlations; orange (cluster 2): moderate correlations; green (cluster 3): high correlations; symbol size in a matrix cell: average MIC among participants; (**a**–**f**) highlight special characteristics described in the text.

When focusing on the level of individual participants, Figure 9 shows MIC correlations among pairs of different parameters. Within a single cell, the small vertical bars represent participants (one bar per participant). Figure 9 complements Figure 8 by adding participant information to the corresponding clusters.

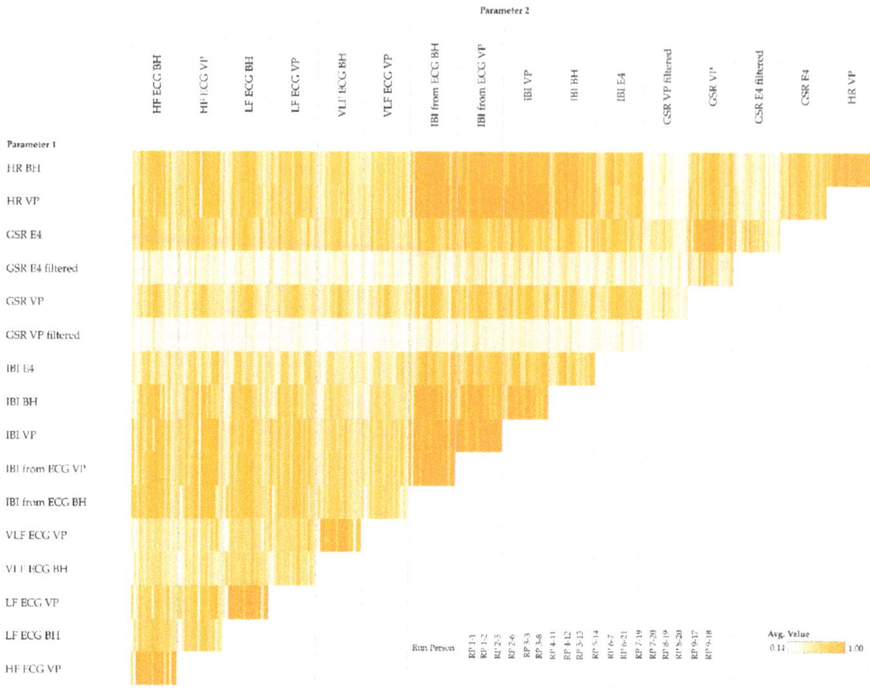

Figure 9. MIC participant-level matrix of pairs of parameters (moving averages only); details of a matrix cell show participants as small vertical bars (order of participants is shown in the legend).

The MIC is used to quantify the strength of any functional relationships, i.e., including linear ones, while the R^2 coefficient can only quantify linear relationships. By subtracting R^2 from the MIC, we compute a measure of nonlinearity [36], which we use to identify the following three classes of relationships as shown in Figure 10:

- "false" linear relationships (R^2 is not confirmed by MIC): MIC–$R^2 < 0$
- "true" linear relationships (R^2 is confirmed by MIC): MIC–$R^2 \sim 0$
- functional but not linear relationships: MIC–$R^2 > 0$

On the individual level, the MIC–R^2 matrix shown in Figure 11 provides additional detail to the clustering view. Interestingly, GSR measurements from E4 and VP show rather strong functional but not linear relationships with almost all other parameters (see third and fifth row in Figures 10 and 11). Particularly interesting is the relationship between GSR VP (filtered and moving averaged version, fourth row) and GSR E4 (filtered and moving averaged version, fifth-last column), which shows some highly negative values (see black arrow). These cases indicate a "false" linear relationship. For instance, participant RP 9–17: MIC 0.2 minus R^2 0.91 results in −0.71. In other words, the MIC does not confirm the highly linear relationship indicated by R^2; in fact, the MIC indicates that there is almost no relationship. From a physiological point of view, this relationship might be obvious; however, the quantification of this relationship from a data-driven perspective is, to our best knowledge, novel.

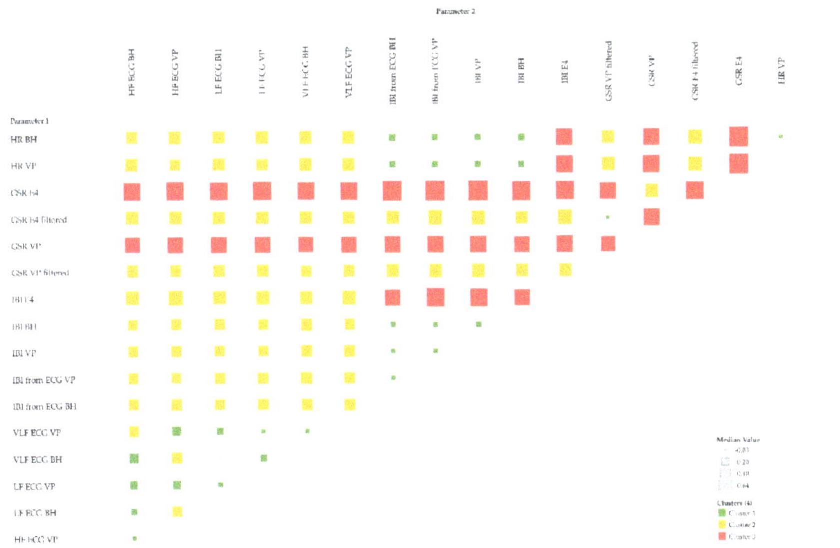

Figure 10. MIC–R^2 cluster matrix of pairs of parameters (moving averages only); colors: green (cluster 1): "false" linear relationships; yellow (cluster 2): "true" linear relationships; red (cluster 3): functional but not linear relationships.

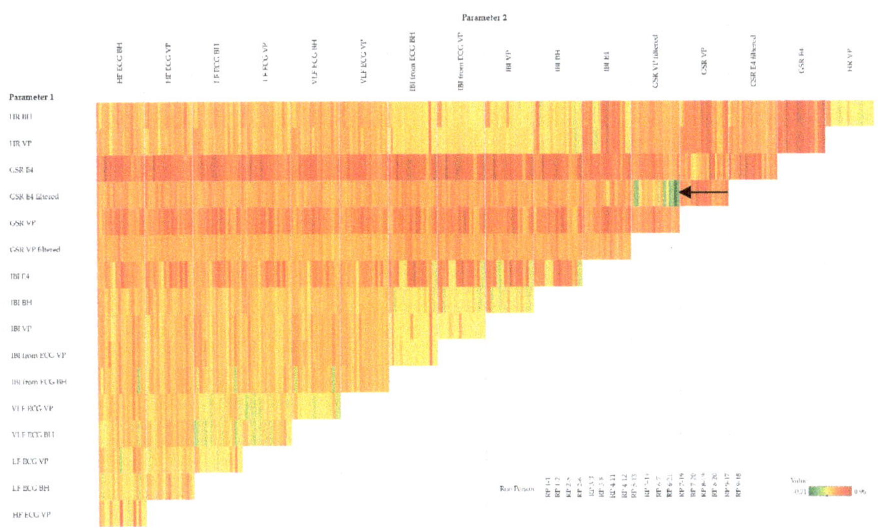

Figure 11. MIC–R^2 individual matrix of pairs of parameters (moving averages only); details of a matrix cell show participants as small vertical bars (order of participants is shown in the legend); back arrow points to pairs of parameters with very weak association.

5.2.4. Fréchet Distance (Global and Local)

The Fréchet distance is a measure of how different two curves are from each other in terms of geometric structure [32]. Herein we use the Fréchet distance to measure the geometric similarity of two time series of a physiological parameter measured by different sensor platforms, one being professional and well-calibrated while the other is low-cost and wearable. In addition to the standard global Fréchet distance, we also compute local versions using a moving window approach.

The global Fréchet distance matrix (Figure 12) shows two expectable general aspects. First, cardiac parameters such as IBI, HF, LF and VLF derived from ECG seem to be more similar than GSR. This is likely because, from a measuring point of view, ECG-related measurements are simply more robust than, for instance, GSR-related ones. Second, the moving averaged versions of the time series also tend to be more similar than their non-averaged counterparts, which include more local fluctuations. Figure 12 detail (a) refers to the HR example shown in Figure 4 and detail (b) refers to the GSR example shown in Figure 5. In both detail (a) and detail (b), the moving averaged time series causes a smoothing effect, thus indicating a higher similarity as compared to the original (non-smoothed) time series.

Group	Parameter 1 - Parameter 2	RP 9-18	RP 9-17	RP 4-11	RP 6-7	RP 2-5	RP 5-14	RP 7-19	RP 4-12	RP 3-8	RP 2-6	RP 6-21	RP 8-19	RP 3-3	RP 7-20	RP 8-20	RP 5-13	RP 1-1	RP 1-2	Total AVG
IBI	IBI from ECG VP (mv. avg.) vs. VP (mv. avg.)	0.05	0.04	0.02	0.49	0.01	0.05	0.07	0.32	0.59	0.34	0.02	0.01	0.28	0.03	0.04	0.47	0.14	0.10	0.17
	IBI from ECG BH (mv. avg.) vs. IBI from ECG VP (mv. avg.)	0.03	0.04	0.09	0.05	0.05	0.04	0.12	0.05	0.08	0.08	0.08	0.51	0.36	0.36	0.45	0.19	0.26	0.86	0.21
	BH (mv. avg.) vs. VP (mv. avg.)	0.40	0.30	0.31	0.23	0.47	0.22	0.30	0.23	0.27	0.37	0.35	0.54	0.44	0.39	0.42	0.29	0.49	0.73	0.38
	IBI from ECG BH (mv. avg.) vs. BH (mv. avg.)	0.40	0.27	0.28	0.38	0.47	0.20	0.30	0.40	0.42	0.40	0.33	0.25	0.37	0.40	0.44	0.32	0.35	0.90	0.38
	IBI from ECG VP vs. VP	0.33	0.28	0.08	0.49	0.12	0.32	0.24	0.50	0.63	0.55	0.04	0.08	0.58	0.19	0.19	0.77	0.83	0.71	0.39
	ibi_from_ecg_bh - ibi_from_ecg_vp	0.12	0.13	0.24	0.10	0.15	0.16	0.23	0.11	0.20	0.45	0.13	0.35	0.55	0.50	0.99	0.65	0.88	0.89	0.39
	E4 (mv. avg.) vs. VP (mv. avg.)	0.30	0.31	0.34	0.89	0.61	0.90	0.75	0.80	0.59	0.73	0.77	0.60	0.60	0.68	0.73	0.81	0.85	0.70	0.67
	BH vs. VP	0.73	0.73	0.53	0.34	0.84	0.49	0.55	0.67	0.68	0.70	0.99	0.72	0.65	0.93	0.81	0.80	0.85	0.76	0.71
	IBI from ECG BH vs. BH	0.73	0.70	0.53	0.48	0.87	0.49	0.57	0.66	0.73	0.72	0.99	0.82	0.55	0.84	0.91	0.86	0.67	0.97	0.73
	E4 vs. VP	0.52	0.44	0.52	0.92	0.72	0.94	0.98	0.82	0.77	0.79	0.81	0.64	0.74	0.77	0.82	0.86	0.87	0.77	0.76
	Total AVG	0.36	0.32	0.30	0.44	0.43	0.38	[illegible]41	0.46	0.50	0.51	0.43	0.47	0.51	0.51	0.58	0.60	0.62	0.74	0.48
HR	BH (mv. avg.) vs. VP (mv. avg.)	0.32	0.19	0.15	0.11	0.18	0.12	0.22	0.30	0.30	0.07	0.16	0.57	0.34	0.20	0.24	0.52	0.69	0.82	0.31
	BH vs. VP	0.50	0.69	0.27	0.30	0.43	0.30	0.42	0.67	0.42	0.60	0.31	0.67	0.49	0.52	0.35	0.90	0.71	0.98	0.53
	Total AVG	0.41	0.44	0.21	0.21	0.32	0.21	0.32	0.48	0.36	0.34	0.23	0.62	0.42	0.36	0.30	0.71	0.70	0.90	0.42
GSR	E4 filtered (mv. avg.) vs. VP filtered (mv. avg.)	0.14	0.14	0.74	0.25	0.25	0.69	0.39	0.49	0.53	0.40	0.38	0.13	0.65	0.59	0.27	0.41	0.65	0.42	0.42
	E4 (mv. avg.) vs. VP (mv. avg.)	0.42	0.71	0.30	0.57	0.38	0.64	0.57	0.41	0.25	0.55	0.93	0.65	0.38	0.43	0.50	0.47	0.50	0.83	0.53
	E4 filtered vs. VP filtered	0.24	0.16	0.80	0.42	0.40	0.80	0.47	0.64	0.7[illegible]	[illegible]48	0.44	0.28	0.78	0.66	0.29	0.50	0.84	0.73	0.34
	E4 vs. VP	0.51	0.74	0.32	0.58	0.40	0.66	0.64	0.48	0.27	0.57	0.96	0.77	0.41	0.46	0.64	0.69	0.52	0.91	0.38
	Total AVG	0.33	0.44	0.54	0.46	0.36	0.70	0.52	0.50	0.45	0.50	0.68	0.46	0.35	0.54	0.42	0.51	0.63	0.72	0.52
VLF ECG	BH (mv. avg.) vs. VP (mv. avg.)	0.11	0.17	0.16	0.09	0.09	0.10	0.12	0.08	0.14	0.11	0.11	0.45	0.29	0.49	0.74	0.54	0.72	0.73	0.29
	BH vs. VP	0.11	0.17	0.31	0.10	0.29	0.16	0.16	0.10	0.21	0.35	0.13	0.59	0.43	0.77	0.95	0.73	0.96	0.91	0.41
	Total AVG	0.11	0.17	0.23	0.10	0.19	0.13	0.14	0.09	0.17	0.23	0.12	0.52	0.36	0.63	0.85	0.64	0.84	0.82	0.35
LF ECG	BH (mv. avg.) vs. VP (mv. avg.)	0.05	0.06	0.17	0.09	0.07	0.09	0.12	0.10	0.14	0.04	0.10	0.21	0.25	0.51	0.74	0.36	0.75	0.79	0.26
	BH vs. VP	0.10	0.10	0.30	0.10	0.28	0.11	0.16	0.12	0.21	0.06	0.13	0.30	0.32	0.72	1.00	0.54	0.97	0.93	0.36
	Total AVG	0.08	0.08	0.24	0.10	0.17	0.10	0.14	0.11	0.17	0.05	0.11	0.26	0.28	0.61	0.87	0.45	0.86	0.86	0.31
HF ECG	BH (mv. avg.) vs. VP (mv. avg.)	0.05	0.08	0.15	0.10	0.09	0.08	0.13	0.07	0.14	0.05	0.14	0.05	0.35	0.28	0.58	0.45	0.45	0.48	0.21
	BH vs. VP	0.20	0.13	0.44	0.12	0.16	0.11	0.22	0.25	0.23	0.17	0.50	0.17	0.76	0.84	0.91	1.00	0.99	0.88	0.45
	Total AVG	0.13	0.10	0.30	0.11	0.13	0.10	0.17	0.16	0.18	0.11	0.32	0.11	0.36	0.56	0.74	0.72	0.72	0.68	0.33

Figure 12. Global Fréchet distance matrix of pairs of parameter and participants; detail (**a**) complements Figure 4, detail (**b**) complements Figure 5; color: green indicates low distance thus high similarity, blue indicates high distance thus low similarity.

In addition to the global geometric similarity, Figure 13 shows local similarity characteristics of the time series using a moving windows approach. The figure shows that the local Fréchet distance of a 1-min moving window indeed reveals differences in similarity at different intensities of physical activity (0–300 s: no activity; 301–900 s: cycling with increasing intensity; 901–1200 s: no activity–cool down; for details refer to Section 4.1). For instance, IBI derived from ECG tends to have a rather constant similarity over the entire measurement period (Figure 13, fourth row), and it tends to be more similar than IBI measured "directly" (Figure 13, first and second row).

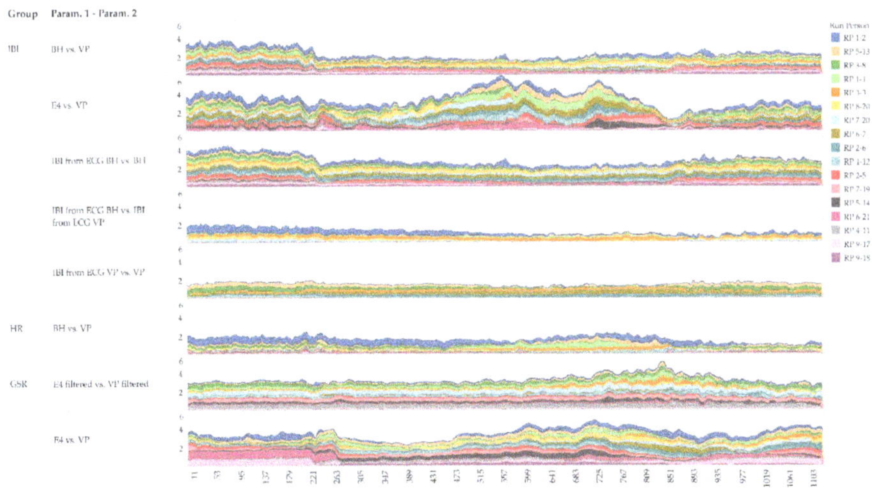

Figure 13. Local Fréchet distance of a moving time window (1 min) of selected pairs of parameters (inter beat interval IBI, heart rate HR, galvanic skin response GSR; moving averaged only).

5.2.5. DTW Distance

The Dynamic Time Warping (DTW) distance is a measure typically used to assess the similarity of time series [50,51]. Simply speaking, DTW tries to optimize the alignment of one time series (test) with another (reference) by stretching or shrinking it in a non-linear fashion along its time axis. The overall distance is the sum of all distances between pairs of points. Identical time series have a distance of zero. Figure 14 shows the DTW distance between pairs of parameters and participants, detail (a) refers to the HR example shown in Figure 4 and detail (b) refer to the GSR example shown in Figure 5.

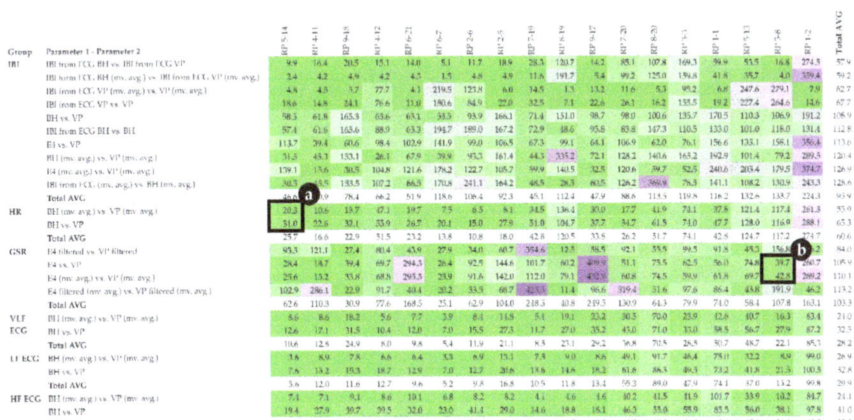

Figure 14. Dynamic Time Warping (DTW) distance matrix of pairs of parameter and participants; detail (**a**) complements Figure 4, detail (**b**) complements Figure 5; colour: green indicates low distance thus high similarity, purple indicates high distance thus low similarity.

The global DTW distances from Figure 14 can be illustrated as an individual pairwise comparison of time series. For instance, Figure 15 shows an example of the DTW distance between two time series of one participant's HR measurements, which are highly similar (low DTW distance). The corresponding exploratory plots of the HR example are shown in Figure 4. Figure 16 shows an example of two time series of GSR measurements with rather low similarity (high DTW distance); however, the overall trend is highly similar. The corresponding exploratory plots of the GSR example are shown in Figure 5.

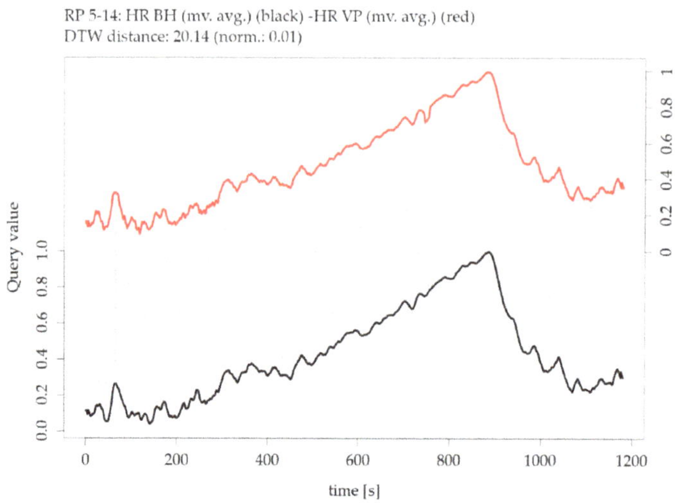

Figure 15. Illustration of the Dynamic Time Warping (DTW) distance between two parameters of participant RP 5-14: moving averaged version of heart rate HR from BioHarness BH versus moving averaged version of heart rate HR from VarioPort VP (note the offset of the two y-axes).

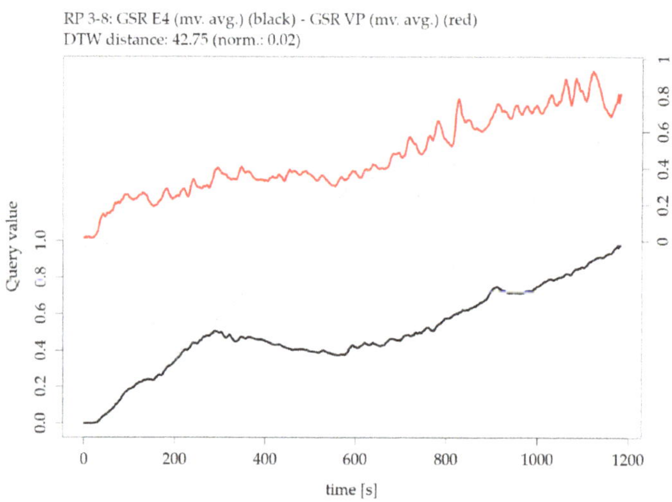

Figure 16. Illustration of the Dynamic Time Warping (DTW) distance between two parameters of participant RP 3-8: moving averaged version of galvanic skin response GSR from E4 versus moving averaged version of galvanic skin response GSR from VarioPort VP (note the offset of the two y-axes).

6. Discussion and Limitations

Overall, the sensor benchmarking worked well, both from a standardized laboratory study and data acquisition viewpoint, as well as from the data analysis methodology perspective. The high correlations between the cardiovascular parameters HR and IBI were as expected because these parameters are comparably simple to measure through a range of methodologies (electrical, optical).

The high correlations between the other ECG-derived measurements were a little more surprising because (1) ECG is measured through a multi-channel electric current-based system, which is a complex procedure; (2) the use of contact electrodes of the BioHarness sensor (in contrast to the sticky electrodes of high-quality sensors) may cause contact (and thus measurement-) problems; 3) ECG is measured at a very high frequency (at least 200 Hz), which is technologically challenging for low-cost wearables.

For GSR, our experiment resulted in lower, but still reasonable similarities, which may be caused by a number of factors like different measurement methods (sticky electrodes vs. plate electrode), and different placement of the sensors (hand palm vs. wrist), etc.

From a more general point of view, it is a known issue that low-cost wearable sensors tend to be prone to producing datasets that suffer from reduced data quality—even though we checked the appropriate positioning of the sensors before we started the exercise.

A particular issue arose with participant RP 1–2. As the results show, the measurements for this participant indicate low correlations for almost all physiological parameters. This may be due to problems with the contact between the electrodes and the skin, which may have been compromised by the person's physical characteristics.

A vital part of the analysis is the visualisation of results on two complementary levels: First, on the individual level, the data from different sensors measuring the same physiological parameter at the same time on the same participant provide a direct comparison between the two-time series of interest. This enables reaching conclusions on the sensors' measuring behaviour. Second, on the collective level, the consolidation of global metrics allows for comparing signals between participants. This further provides useful insights into the influence of the participants' individual components (physical constitution, individual baseline level of skin conductance, etc.).

These complementary visualizations enable a flexible method of interpretation. For instance, it allows starting the interpretation on the individual level on a particular pair of physiological parameters of interest (e.g., HR of participant RP 5–14) using the corresponding exploratory plot as shown in Figure 4, then rolling-up using the R^2 matrix (Figure 6a) together with the cross-correlation matrix (Figure 7a) and comparing the individual result between participants. Further, it allows checking whether that particular pair of parameters has a functional relationship and whether that relationship is stable among other participants by using the MIC–R^2 individual matrix (Figure 11). The focus on a particular pair of parameters and participant can be continued to the similarity measures, namely the Fréchet distance and the DTW distance (Figures 12 and 14). Another method of interpretation is to begin at the collective level using the MIC–R^2 cluster matrix of pairs of parameters (Figure 10), then drilling-down on a specific parameter combination of interest using the MIC–R^2 individual matrix (Figure 11) and contextualizing this matrix with the corresponding exploratory plots as shown in Figure 5.

This kind of visualisation provides the central advantage regarding the sensor benchmarking from a "big picture" view, i.e., to serve as a basis for visual analysis of the correlations between the measurements of one parameter as measured by two different sensors (each row in the matrix) and the correlations between the different parameters for a single participant (each column in the matrix). Furthermore, the matrix allows the simple assessment of each single cell to trace back particularities of each measurement to a test person, which makes it easier to single out anomalies that may be caused by usage errors, a user's characteristics, single sensor failures, or violations of the benchmark protocol.

The cross-correlation analysis shows that groups of physiological parameters can be associated with different patterns of temporal shifts. As illustrated in Figure 7, the cross-correlation also varies between participants: from overall positive for IBI derived from ECG for more than 60% of the participants to highly positive at small lags and highly negative at larger lags at the heart rate variability parameters VLF, LF, and HF. Although the clocks of the sensors were synchronized right before the study, we observed a lag of 1–2 s between HR measurements (BH versus VP) and GSR measurements (E4 versus VP), as exemplarily shown in Figure 4a,c and Figure 5c,d, respectively. In all cases, the VP time series were "leading", which may indicate that the response characteristics of the VP sensors are more sensitive compared to the other sensors.

Measuring the strength of association between pairs of physiological parameters was of particular interest. Herein, we contrasted the coefficient of determination R^2 with the MIC (Figures 10 and 11). In other words, we confronted a statistic that measures linear relationships against a statistic that measures all types of functional relationships, including linear ones, and thereby classified the relationship as 'false linear', 'true linear', or 'functional but not linear'. The results are outstanding: on the one hand, some already expected linear relationships have been confirmed by a purely data-driven approach (for instance, relationships between IBI and VLF, LF, HF); on the other hand, some relationships that were expected to be linear are in fact not linear or functional. For instance, the relationships between GSR measured by E4 and GSR measured by VP (both filtered and moving averaged versions).

7. Conclusions and Future Work

In this paper, we performed a benchmark of two wearable physiological sensors (Zephyr BioHarness 3 and Empatica E4) by comparing their measurements (heart rate, inter-beat interval, and galvanic skin response, and derived heart rate variability parameters) to highly-calibrated high-end professional equipment. In our study, we used the measurements from 18 participants to compare the correlations (Pearson's r), cross-correlations at different temporal lags from −15 sec to +15 sec, the (sub-)linearity of functional dependencies (MIC), the difference of two measurement time series with respect to their geometric structure (Fréchet distance), local time series similarities (moving window), and time series similarity with respect to their temporal alignment (DTW).

The results of our study show that the measured cardiovascular parameters yield very high similarities between the low-cost wearable and the calibrated professional sensors. Although cardiovascular parameters are simple to measure (technically and phenomenon-wise), the obtained similarities are remarkable. For GSR, our experiment resulted in lower similarities, which may be caused by a number of factors like different measurement methods, different placement of the sensors (hand palm vs. wrist), conduction characteristics between skin and sensor surface (use of electrolyte gel or not), and others. It should be noted that the use of isotonic electrolyte gel is a scientific standard for measurement of electrodermal activity [52] and was used with the Varioport GSR measure but not with the other devices.

We demonstrated that our methodological approach to quantify correlations and similarities on both the individual and the aggregated level can provide interesting insights into the relationships between and among physiological parameters. The many figures generated (only the most essential ones are presented in this paper) enable different points of view on the same data and thus a more holistic interpretation for the benchmark of physiological sensors. Our research contributes to such a holistic interpretation in two ways: 1) the confrontation of the coefficient of determination R^2 against the Maximal Information Coefficient MIC, in particular, the classification of non-linear correlations, and 2) the quantification of the signals' temporal and geometric similarity based on well-established distance metrics (DTW distance and Fréchet distance).

Our future work will focus on two main research challenges. First, to continue fine-tuning the methodology and integrate additional similarity measures, for instance, the Time Warp Edit Distance (TWED) [53]. Second, to evaluate the transferability of the methodology to other time series benchmarking challenges, not necessarily physiological measurements. In the long run we want to expand the methodology to the geospatial domain, i.e., integrating the location in addition to the timestamp and the measurement of mobile sensors. This approach will likely warrant an additional field study that addresses the suitability of measurement devices and measurement quality on moving subjects, e.g., persons riding a bicycle or walking, and relating sensor data to subjective experience self-report data.

Author Contributions: Conceptualization, G.S., B.R., A.P., K.K., M.L. and F.H.W.; Data curation, G.S., B.R., A.P., K.K. and M.L.; Formal analysis, G.S. and B.R.; Investigation, G.S., B.R., A.P. and M.L.; Methodology, G.S., B.R.

and K.K.; Project administration, A.P.; Software, G.S., A.P., K.K. and M.L.; Validation, G.S., B.R., M.L. and F.H.W.; Visualisation, G.S.; Writing—Original draft, G.S., B.R. and M.L.; Writing—Review & editing, G.S., B.R., A.P., K.K., M.L. and F.H.W.

Funding: This research was supported by the Austrian Science Fund (FWF) through the project "Urban Emotions" (FWF I-3022) and by the Austria Research Promotion Agency (FFG) through the project "Walk&Feel" (FFG 865208). This research was partly funded by the Austrian Science Fund (FWF) through the Doctoral College GIScience (DK W 1237-N23).

Acknowledgments: Open Access Funding by the Austrian Science Fund (FWF). We would like to thank all participants of the case study who made this work possible by donating several hours of their free time.

Conflicts of Interest: The authors declare no conflict of interest.

References

1. Swan, M. Sensor mania! The internet of things, wearable computing, objective metrics, and the quantified self 2.0. *J. Sens. Actuator Netw.* **2012**, *1*, 217–253. [CrossRef]
2. Crawford, K.; Lingel, J.; Karppi, T. Our metrics, ourselves: A hundred years of self-tracking from the weight scale to the wrist wearable device. *Eur. J. Cult. Stud.* **2015**, *18*, 479–496. [CrossRef]
3. Piwek, L.; Ellis, D.A.; Andrews, S.; Joinson, A. The rise of consumer health wearables: Promises and barriers. *PLoS Med.* **2016**, *13*, e1001953. [CrossRef]
4. Werner, C.; Resch, B.; Loidl, M. Evaluating urban bicycle infrastructures through intersubjectivity of stress sensations derived from physiological measurements. *ISPRS Int. J. Geo-Inf.* **2019**, *8*, 265. [CrossRef]
5. Basu, S.; Jana, N.; Bag, A.; Mahadevappa, M.; Mukherjee, J.; Kumar, S.; Guha, R. Emotion recognition based on physiological signals using valence-arousal model. In Proceedings of the 2015 Third International Conference on Image Information Processing (ICIIP), Waknaghat, India, 21–24 December 2015; pp. 50–55.
6. Resch, B.; Summa, A.; Sagl, G.; Zeile, P.; Exner, J.-P. Urban emotions—Geo-semantic emotion extraction from technical sensors, human sensors and crowdsourced data. In *Progress in Location-Based Services 2014*; Gartner, G., Huang, H., Eds.; Springer International Publishing: Cham, Switzerland, 2015; pp. 199–212.
7. Taj-Eldin, M.; Ryan, C.; O'Flynn, B.; Galvin, P. A review of wearable solutions for physiological and emotional monitoring for use by people with autism spectrum disorder and their caregivers. *Sensors* **2018**, *18*, 4271. [CrossRef]
8. Healey, J. Physiological sensing of emotion. In *The Oxford Handbook of Affective Computing*; Oxford University Press: Oxford, UK, 2014; p. 204.
9. Peake, J.M.; Kerr, G.; Sullivan, J.P. A critical review of consumer wearables, mobile applications, and equipment for providing biofeedback, monitoring stress, and sleep in physically active populations. *Front. Physiol.* **2018**, *9*, 743. [CrossRef] [PubMed]
10. Birenboim, A.; Dijst, M.; Scheepers, F.E.; Poelman, M.P.; Helbich, M. Wearables and location tracking technologies for mental-state sensing in outdoor environments. *Prof. Geogr.* **2019**, *71*, 449–461. [CrossRef]
11. Kyriakou, K.; Resch, B.; Sagl, G.; Petutschnig, A.; Werner, C.; Niederseer, D.; Liedlgruber, M.; Wilhelm, F.H.; Osborne, T.; Pykett, J. Detecting moments of stress from measurements of wearable physiological sensors. *Sensors* **2019**, *19*, 3805. [CrossRef] [PubMed]
12. Zeile, P.; Resch, B. Combining biosensing technology and virtual environments for improved urban planning. *GI_Forum* **2018**, *1*, 344–357. [CrossRef]
13. Dörrzapf, L.; Kovács-Győri, A.; Resch, B.; Zeile, P. Defining and assessing walkability: A concept for an integrated approach using surveys, biosensors and geospatial analysis. *Urban Dev. Issues* **2019**, *62*, 5–15. [CrossRef]
14. Guo, R.; Li, S.; He, L.; Gao, W.; Qi, H.; Owens, G. Pervasive and unobtrusive emotion sensing for human mental health. In Proceedings of the 2013 7th International Conference on Pervasive Computing Technologies for Healthcare and Workshops, Venice, Italy, 5–8 May 2013; pp. 436–439.
15. Majumder, S.; Mondal, T.; Deen, M. Wearable sensors for remote health monitoring. *Sensors* **2017**, *17*, 130. [CrossRef] [PubMed]
16. Kenry, Y.J.C.; Lim, C.T. Emerging flexible and wearable physical sensing platforms for healthcare and biomedical applications. *Microsyst. Nanoeng.* **2016**, *2*, 16043. [CrossRef] [PubMed]
17. Giakoumis, D.; Tzovaras, D.; Hassapis, G. Subject-dependent biosignal features for increased accuracy in psychological stress detection. *Int. J. Hum. Comput. Stud.* **2013**, *71*, 425–439. [CrossRef]

18. Gradl, S.; Wirth, M.; Richer, R.; Rohleder, N.; Eskofier, B.M. An overview of the feasibility of permanent, real-time, unobtrusive stress measurement with current wearables. In Proceedings of the 13th EAI International Conference on Pervasive Computing Technologies for Healthcare, ACM, Trento, Italy, 20–23 May 2019; pp. 360–365.
19. Aghabozorgi, S.; Shirkhorshidi, A.S.; Wah, T.Y. Time-series clustering—A decade review. *Inf. Syst.* **2015**, *53*, 16–38. [CrossRef]
20. Serrà, J.; Arcos, J.L. An empirical evaluation of similarity measures for time series classification. *Knowl. Based Syst.* **2014**, *67*, 305–314. [CrossRef]
21. Shin, K. An alternative approach to measure similarity between two deterministic transient signals. *J. Sound Vib.* **2016**, *371*, 434–445. [CrossRef]
22. Toohey, K.; Duckham, M. Trajectory similarity measures. *Sigspatial Spec.* **2015**, *7*, 43–50. [CrossRef]
23. Wang, X.; Mueen, A.; Ding, H.; Trajcevski, G.; Scheuermann, P.; Keogh, E. Experimental comparison of representation methods and distance measures for time series data. *Data Min. Knowl. Discov.* **2013**, *26*, 275–309. [CrossRef]
24. Chen, L.; Özsu, M.T.; Oria, V. Robust and fast similarity search for moving object trajectories. In Proceedings of the 2005 ACM SIGMOD International Conference on Management of Data, Baltimore, MD, USA, 14–16 June 2005; pp. 491–502.
25. Keogh, E.; Ratanamahatana, C.A. Exact indexing of dynamic time warping. *Knowl. Inf. Syst.* **2005**, *7*, 358–386. [CrossRef]
26. Folgado, D.; Barandas, M.; Matias, R.; Martins, R.; Carvalho, M.; Gamboa, H. Time alignment measurement for time series. *Pattern Recognit.* **2018**, *81*, 268–279. [CrossRef]
27. Jiang, G.; Wang, W.; Zhang, W. A novel distance measure for time series: Maximum shifting correlation distance. *Pattern Recognit. Lett.* **2019**, *117*, 58–65. [CrossRef]
28. Kate, R.J. Using dynamic time warping distances as features for improved time series classification. *Data Min. Knowl. Discov.* **2016**, *30*, 283–312. [CrossRef]
29. Fréchet, M.M. Sur quelques points du calcul fonctionnel. *Rendiconti del Circolo Matematico di Palermo (1884–1940)* **1906**, *22*, 1–72. [CrossRef]
30. Shahbaz, K. Applied similarity problems using fréchet distance. *arXiv* **2013**, arXiv:preprint/1307.6628.
31. De Carufel, J.-L.; Gheibi, A.; Maheshwari, A.; Sack, J.-R.; Scheffer, C. Similarity of polygonal curves in the presence of outliers. *Comput. Geom.* **2014**, *47*, 625–641. [CrossRef]
32. Aronov, B.; Har-Peled, S.; Knauer, C.; Wang, Y.; Wenk, C. Fréchet distance for curves, revisited. In *European Symposium on Algorithms*; Springer: Berlin/Heidelberg, Germany, 2006; pp. 52–63.
33. Wylie, T.R. The Discrete Fréchet Distance with Applications. Ph.D. Thesis, Montana State University-Bozeman, College of Engineering, Bozeman, MT, USA, 2013.
34. Kianimajd, A.; Ruano, M.G.; Carvalho, P.; Henriques, J.; Rocha, T.; Paredes, S.; Ruano, A.E. Comparison of different methods of measuring similarity in physiologic time series. *IFAC-PapersOnLine* **2017**, *50*, 11005–11010. [CrossRef]
35. Hauke, J.; Kossowski, T. Comparison of values of pearson's and spearman's correlation coefficients on the same sets of data. *Quaest. Geogr.* **2011**, *30*, 87–93. [CrossRef]
36. Reshef, D.N.; Reshef, Y.A.; Finucane, H.K.; Grossman, S.R.; McVean, G.; Turnbaugh, P.J.; Lander, E.S.; Mitzenmacher, M.; Sabeti, P.C. Detecting novel associations in large data sets. *Science* **2011**, *334*, 1518–1524. [CrossRef]
37. Speed, T. A correlation for the 21st century. *Science* **2011**, *334*, 1502. [CrossRef] [PubMed]
38. Morelli, M.S.; Greco, A.; Valenza, G.; Giannoni, A.; Emdin, M.; Scilingo, E.P.; Vanello, N. Analysis of generic coupling between EEG activity and $P_{ET}CO_2$ in free breathing and breath-hold tasks using maximal information coefficient (MIC). *Sci. Rep.* **2018**, *8*, 4492. [CrossRef]
39. Brillinger, D.R. *Time Series: Data Analysis and Theory*; Siam: San Francisco, CA, USA, 2001; Volume 36.
40. Dickey, D.A.; Fuller, W.A. Distribution of the estimators for autoregressive time series with a unit root. *J. Am. Stat. Assoc.* **1979**, *74*, 427–431.
41. Johnstone, J.A.; Ford, P.A.; Hughes, G.; Watson, T.; Garrett, A.T. Bioharness (™) multivariable monitoring device: Part. II: Reliability. *J. Sports Sci. Med.* **2012**, *11*, 409–417. [PubMed]
42. Johnstone, J.A.; Ford, P.A.; Hughes, G.; Watson, T.; Garrett, A.T. Bioharness (™) multivariable monitoring device: Part. I: Validity. *J. Sports Sci. Med.* **2012**, *11*, 400–408. [PubMed]

43. Blechert, J.; Peyk, P.; Liedlgruber, M.; Wilhelm, F.H. Anslab: Integrated multichannel peripheral biosignal processing in psychophysiological science. *Behav. Res. Methods* **2016**, *48*, 1528–1545. [CrossRef] [PubMed]
44. Bluemke, M.; Resch, B.; Lechner, C.; Westerholt, R.; Kolb, J.-P. Integrating geographic information into survey research: Current applications, challenges and future avenues. *Surv. Res. Methods* **2017**, *11*, 307–327.
45. Bar-Joseph, Z.; Gerber, G.K.; Gifford, D.K.; Jaakkola, T.S.; Simon, I. Continuous representations of time-series gene expression data. *J. Comput. Biol.* **2003**, *10*, 341–356. [CrossRef]
46. Wilhelm, F.H.; Grossman, P.; Roth, W.T. Assessment of heart rate variability during alterations in stress: Complex demodulation vs. Spectral analysis. *Biomed. Sci. Instrum.* **2005**, *41*, 346–351.
47. Li, L.; Caldwell, G.E. Coefficient of cross correlation and the time domain correspondence. *J. Electromyogr. Kinesiol.* **1999**, *9*, 385–389. [CrossRef]
48. Reshef, D.N.; Reshef, Y.A.; Sabeti, P.C.; Mitzenmacher, M. An empirical study of the maximal and total information coefficients and leading measures of dependence. *Ann. Appl. Stat.* **2018**, *12*, 123–155. [CrossRef]
49. Alt, H. The computational geometry of comparing shapes. In *Efficient Algorithms: Essays Dedicated to Kurt Mèhlhorn on the Occasion of His 60th Birthday*; Albers, S., Alt, H., Näher, S., Eds.; Springer: Berlin/Heidelberg, Germany, 2009; pp. 235–248.
50. Zhu, Q.; Batista, G.; Rakthanmanon, T.; Keogh, E. A novel approximation to dynamic time warping allows anytime clustering of massive time series datasets. In Proceedings of the 2012 SIAM International Conference on Data Mining, Davis, FL, USA, 26–28 April 2012; pp. 999–1010.
51. Tormene, P.; Giorgino, T.; Quaglini, S.; Stefanelli, M. Matching incomplete time series with dynamic time warping: An algorithm and an application to post-stroke rehabilitation. *Artif. Intell. Med.* **2009**, *45*, 11–34. [CrossRef]
52. Fowles, D.C.; Christie, M.J.; Edelberg, R.; Grings, W.W.; Lykken, D.T.; Venables, P.H. Publication recommendations for electrodermal measurements. *Psychophysiology* **1981**, *18*, 232–239. [CrossRef] [PubMed]
53. Marteau, P.-F. Time warp edit distance with stiffness adjustment for time series matching. *IEEE Trans. Pattern Anal. Mach. Intell.* **2009**, *31*, 306–318. [CrossRef] [PubMed]

sensors

Article

Ambulatory and Laboratory Stress Detection Based on Raw Electrocardiogram Signals Using a Convolutional Neural Network

Hyun-Myung Cho [1,2], Heesu Park [1,3], Suh-Yeon Dong [4,*] and Inchan Youn [1,*]

1 Center for Bionics, Biomedical Research Institute, Korea Institute of Science and Technology, Seoul 02792, Korea; wisjmeng@gmail.com (H.-M.C.); pheesoo417@gmail.com (H.P.)
2 Division of Bio-Medical Science & Technology, KIST School, Korea University of Science and Technology, Daejeon 02792, Korea
3 Department of Biomedical Science, College of Medicine, Korea University, Seoul 02841, Korea
4 Department of Information Technology Engineering, Sookmyung Women's University, Seoul 04310, Korea
* Correspondence: suhyeon.dong@gmail.com (S.-Y.D.); iyoun@kist.re.kr (I.Y.)

Received: 24 September 2019; Accepted: 10 October 2019; Published: 11 October 2019

Abstract: The goals of this study are the suggestion of a better classification method for detecting stressed states based on raw electrocardiogram (ECG) data and a method for training a deep neural network (DNN) with a smaller data set. We suggest an end-to-end architecture to detect stress using raw ECGs. The architecture consists of successive stages that contain convolutional layers. In this study, two kinds of data sets are used to train and validate the model: A driving data set and a mental arithmetic data set, which smaller than the driving data set. We apply a transfer learning method to train a model with a small data set. The proposed model shows better performance, based on receiver operating curves, than conventional methods. Compared with other DNN methods using raw ECGs, the proposed model improves the accuracy from 87.39% to 90.19%. The transfer learning method improves accuracy by 12.01% and 10.06% when 10 s and 60 s of ECG signals, respectively, are used in the model. In conclusion, our model outperforms previous models using raw ECGs from a small data set and, so, we believe that our model can significantly contribute to mobile healthcare for stress management in daily life.

Keywords: stress detection; electrocardiogram; deep neural network; convolutional neural network

1. Introduction

As interest in health care increases, the importance of stress management has grown. As many people are exposed to the stressful environments, they are more likely to suffer from physical and mental disorders. Indeed, stress has been shown to cause diseases such as depression, asthma, and autoimmune diseases [1]. To observe the changes in our body caused by stress, many researchers have focused on physiological signals, such as electrocardiography (ECG) signals and galvanic skin response [2].

When a person receives stress stimulation, his/her autonomic nervous system reacts to the stress, which results in physiological changes [3]. Among the physiological signals, the ECG enables us to observe how our bodies react to stress. An ECG is an electrical signal which is generated by heart activity. An ECG signal has three main components: The P-wave, QRS-complex, and T-wave. Among them, the time-series intervals between successive R peaks are used to calculate the heart rate variability (HRV) [4]. The HRV can be represented by various parameters that are calculated along the time, frequency, and non-linear domains. These HRV parameters have often been used for stress recognition [5–7].

With the recent development of mobile sensors for ECG recording, HRV analysis and stress studies have been actively carried out. However, due to the inherent limitations of HRV analysis, which requires sufficient data to observe the variability, the longer the time window of the ECG record for an HRV analysis is, the more accurate the statistical characteristics can be. In other words, it is difficult to perform an HRV analysis with short-term ECG measurements. Some previous research has demonstrated the minimum time window required for an HRV analysis. Camm et al. [7] recommended at least a 5 min ECG measurement to analyze the HRV. Moreover, the R peaks along the ECG time series must be detected. To do so, computational algorithms based on the Pan-Tompkins algorithm [8] can be used. These algorithms contain preprocessing steps, such as filtering and differential operations, to find the QRS-complex adequately. Namely, the classical HRV approach requires additional preprocessing steps with a limited window length.

To recognize stress states, several previous studies [9–11] have reported classical HRV analysis using machine learning algorithms, such as the support vector machine (SVM), k-nearest neighbors (kNN), adaptive boosting (AB), and logistic regression (LR) methods, along with the ECG and the other physiological signals. Other physiological signals include skin conductance (SC) [9,11], respiration [9,11], and skin temperature (ST) [11].

Table 1 shows a comparative summary of previous studies which used conventional machine learning algorithms, as well as DNN models. The column "Window length" refers to the duration of the ECG measurement used to extract the HRV parameters. The column "Performance" indicates the reported accuracy of the classifier mentioned in the column "Classifier". Most studies used more than one classifier, but only those that gave the best results are shown in Table 1. The study [11] used different stressors to the other studies [9,10], including the stroop color word test (SCWT), mental arithmetic (MA), and counting numbers. These stimuli are considered to be laboratory-environmental stress stimulation. Castaldo et al. [10] designed an experiment where an ECG was acquired on two different days. One was a day when the participants were undergoing a university verbal examination, and the other day was after a vacation. Although these studies (which were performed in a laboratory-controlled environment) showed an accuracy of about 85%, outside of the laboratory the accuracy reached only 80% [10]. These results might indicate the limitations of the stressors used in the laboratory environment and of the conventional machine learning methods. A conventional machine learning method based on HRV features involves not only the preprocessing of the ECG signals, but also feature selection among the HRV parameters.

Table 1. Previous studies on stress detection. ECG, Electrocardiogram; SC, Skin Conductance; ST, Skin Temperature; HR, Heart Rate; BN, Bayesian Network; AB, Adaptive Boosting; CNN, Convolutional Neural Network; RNN, Recurrent Neural Network; MT-CNN, Multitask CNN; AUC, Area Under the Curve; SCWT, Stroop Color Word Test; MA, Mental Arithmetic.

Ref	# of Subjects	Signal	Window Length (s)	Classifier	Performance (%)	Stressor
[9]	13	ECG, SC, Respiration	10	BN	84	Driving
[10]	42	ECG	180	AB	80	Verbal examination
[11]	20	ECG, SC, Respiration, ST	30	BN	84.6	SCWT, math, counting
[12]	20, 30	Raw ECG	10	CNN+RNN	87.39, 73.96	MA, interview, SCWT, visual stimuli, cold pressor
[13]	10	HR, SC	-	MT-CNN	0.918 (AUC)	Driving [14]

As the deep neural network (DNN) approach has recently demonstrated outperforming pattern recognition accuracy [15], some studies [12,13,16–21] have utilized DNNs for biomedical engineering applications, including heart arrhythmia classification, medical image classification and enhancement, stress detection, and for other medical diagnoses. U-net [21], which consists of a convolutional neural network (CNN), forms an autoencoder architecture using skip-connection between the encoder and decoder. This architecture could be used in medical imaging applications, including augmentation, classification, and detection. Hannun et al. [16] achieved a performance in arrhythmia detection using a deep convolutional architecture that was similar to or exceeding that of cardiologists.

Hwang et al. [12] proposed the DeepECGNet method, which detects stress using an ultra-short-term (10 s) ECG. They used raw ECG signals for the input of the DNN without extracting the HRV parameters. A model was configured with one CNN and two long short-term memory (LSTM) [22] models. They suggested an optimal convolution filter width and pooling window width, respectively, of 0.6 s and 0.8 s. They recommended selecting the proper hyperparameter values for the convolution filter width and pooling window width, both of which are capable of covering the QRS-complex of an ECG. They designed an experiment for inducing mental and physical stress in participants through MA, SCWT, visual stimuli, and cold pressor. There were 20 and 30 participants, separated into two cases. In the two cases, the model [12] reached 87.39% and 73.96% accuracy. Saeed et al. [13] suggested using a multitask convolutional neural network (MT-CNN). Raw physiological signals configured with the heart rate (HR) derived from an ECG and SC are fed to the MT-CNN to detect stress. There was no mention of how long the duration of the ECG measurement was, but only 300 samples of ECG signals were reported. They achieved 0.918 for the area under the receiver operating characteristic curve (AUROC).

Though these studies suggested models that detect stress automatically using conventional machine learning algorithms, some issues remain unsolved. One of these is the complexity of the classification steps. Four steps are involved in conventional methods: Preprocessing, feature selection, classifier training, and classification. Preprocessing, including R-peak extraction, is required to extract the proper R-peak and calculate the HRV. Most of these studies used other physiological signals (i.e., respiration, SC, and ST) as well as an ECG, and needed to select proper features among the both physiological signals and the HRV parameters. These processes make it difficult to apply these models in a practical environment, from the perspective of real-time classification. Additionally, a person must measure the ECG for at least 1 min to detect a stress state, even though it is a short-term HRV analysis.

In this paper, we suggest a DNN model to detect stress with a raw ECG signal, which possibly overcomes the limitations mentioned above. An end-to-end method using a raw ECG signal does not require preprocessing (i.e., filtering and R-peak extraction). Additionally, this method does not require additional feature selection. The other contribution of our research is a method for training the DNN with a pretrained model. The DNN requires a large amount of data to train the model, but it is difficult to acquire a large data set of physiological signals. We trained the proposed model based on a pretrained model, which learned a large amount of data [14]. We evaluated the performance of the proposed model by calculating evaluation metrics (e.g., accuracy and area under the curve) which are widely used in the evaluation of DNN models. We also assessed the proposed model by comparing it with conventional machine learning methods [9–11] which used only the ECG to detect stress, and with other DNN methods [12,13].

2. Material and Methods

2.1. Subjects and Data Acquisition

We used two kinds of data sets, which were obtained from two different experiments, to train and evaluate the proposed model. The two different data sets can be considered as ambulatory and laboratory stress, respectively. One of the data sets consisted of ECG measurements collected from drivers who drove through a city and on a highway [14]. The other data set was recorded in an experimental environment, where mental stress was induced by arithmetic tasks in the participants. Figure 1 shows detailed information about both protocols, including duration and task.

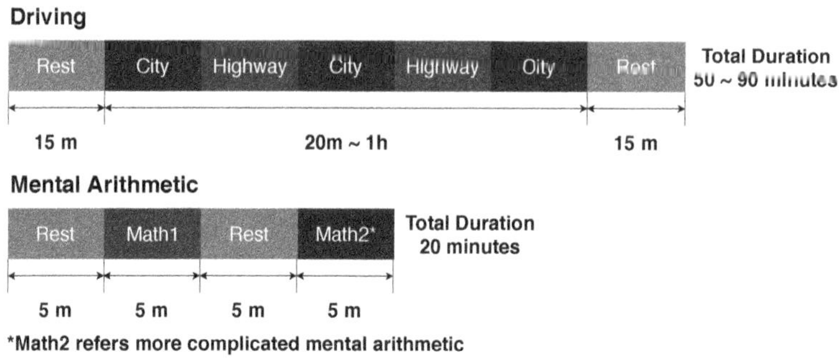

Figure 1. The experimental procedures and their durations.

2.1.1. Driving Data Set

PhysioNet [23] offers free access to a large number of physiological signals recorded under various conditions. We selected the driver stress data set [14] from among the free accessible databases. It consists of various physiological signals, including ECG, electromyography (EMG), SC, and respiration signals, which were recorded under the conditions of driving and resting. Healey et al. [14] tried to monitor real-world stress during driving situations. Among the recorded physiological signals, we chose the ECG, which was sampled at 496 Hz. Modified lead II configuration electrodes were used to measure the ECG. Sixteen participant's records were uploaded to PhysioNet. We excluded 2 subject's data, as they did not contain a record of the marker for when the participants changed the driving region to highway or city. Finally, we selected 14 subject's records for use in our study. All the selected records included approximately 50–85 min of ECG measurements during driving and resting. Due to differences in traffic conditions, the total duration of experiments differed by subject. The participants were made to take a rest for 15 min before and after driving.

2.1.2. Mental Arithmetic Data Set

Seventeen people (6 female and 11 male, 27.9 $\pm$ 3.3 years old) participated in our experiment. We designed an experiment to induce mental fatigue in the participants using a mental arithmetic task. The mathematical tests were developed based on the Montreal Imaging Stress Task (MIST) to elicit two different levels of mental stress in participants. To do so, we simplified the MIST paradigm by two levels of difficulty. The participants had to try to solve the arithmetic problems and push the keypad to answer the questions. The problem consists of two levels: Moderate and high. The moderate level included three integers with plus and minus operations. For the high level, four integers and all of the arithmetic operations were used. All participants encountered the same level of complexity for the arithmetic problems. The ECG data were sampled at 256 Hz and measured by a T-REX TR100A sensor; the electrodes were placed in a modified lead II configuration, which was the same as for the driving data set [14]. First, we measured the baseline ECG for 5 min while the participants took a rest. After the baseline measurement, the participants took the mental arithmetic test two times (5 min each) at a moderate and a high level. They encountered more complicated mathematical problems during the high level test. We provided a 5 min rest between the mental arithmetic tests. To measure whether the mental arithmetic induced mental stress in the participants, we used two questionnaires, including self-assessment manikin (SAM) [24] and distress thermometer (DT) [25]. The self-reports were written after the first rest period and after two repetitions of the tasks. The mental arithmetic experiments were approved by the institutional review board at the Korea Institute of Science and Technology (2017-030).

2.2. Data Preprocessing and Annotation Procedures

We performed some preprocessing procedures before using the data sets for training the neural network. As the ECGs of the two data sets were recorded by different sensors, we scaled them to the same range (0–1) using z-score normalization,

$$x_i^s \leftarrow \frac{x_i^s - \mu^s}{\sigma^s}, \quad \text{for } 0 \leq i < n$$
$$s \in \{\text{Driving, MA}\}$$

where n denotes the number of data sets for each stressor and window. We calculated the mean (μ) and standard deviation (σ), along with all the data for each stressor (i.e., the driving and the mental arithmetic). We normalized the ECG using both the mean and standard deviation. After normalization, the driving data set was downsampled to match the sampling rate of the mental arithmetic data set, which was sampled at 256 Hz. We needed a fixed input dimension, due to using the same neural network for training the model and detecting the stress based on the two different data sets. The ECG signals were sliced into 10 s, 30 s, and 60 s windows to detect stress in short-term windows. The reason for setting a short-term window was to try to recognize stress in nearly real-time.

We needed to annotate the data with specific labels to train a neural network by a supervised method. We segmented the driving data set, based on the boundaries between driving on the highway, driving on the city, and resting. The ECG measurements recorded during driving on the highways and in the cities were labeled as stress, and the other measurements were labeled as rest. In the case of the mental arithmetic data set, we labeled the ECG measurements recorded during the mathematical task as stress. The other ECG measurements, recorded during the rest period, were annotated as rest.

Table 2 shows the numbers of the data sets and their label distribution for each window. The number of data labeled as stress in the driving data set was much larger than those labeled as rest, while the mental arithmetic data set shows a balanced label distribution. The drivers took a rest for approximately 30 min, including an initial and final rest, but they drove for over 45 min, resulting in an imbalanced distribution. The participants who took the mental arithmetic test were exposed to the same amount of time for the stress task and resting; 10 min each. The driving data set and the mental arithmetic data set contained over 72,000 and 16,000 ECG cycles, respectively.

Table 2. Number of samples in the data sets.

Stressor	Window Length (s)	Number of Samples		Total
		Rest	Stress	
Driving	10	2161	3731	5892
	30	712	1227	1939
	60	349	598	947
Mental arithmetic	10	1020	1020	2040
	30	340	340	680
	60	170	170	340

2.3. The Deep Neural Network

We propose a deep convolutional neural network to detect stress events. Figure 2 shows the architecture of the proposed model. It obtains input of only raw ECG signals, not HRV parameters or other physiological signals. The input dimension can be defined as $x \in \mathbb{R}^{m \times w \times c}$, where m and c denote the size of the mini-batch and the number of channels, respectively, and w refers to the width of the ECG, which is defined as the multiplication of the sampling rate and window. Our model contains successive stages ($N = 8$) to extract features from the ECG. Table 3 shows the list of the operations and the detailed parameters used in the stages. The number of filters in the convolutional layer is defined as 8×2^k, where k begins at 0 and is increased by one every second stage. Each stage consists of two convolutional layers and one pooling layer. A convolutional layer performs a convolution operation with its filter and a specific stride. A stride is defined as how much the filter moves within a layer (i.e., the convolutional and pooling layer). An output of the first convolutional layer is fed to both a strided convolutional layer and a pooling layer. The stride values of the strided convolutional layer and pooling layer are set to 2. The inputs of these two layers are subsampled by a factor of 2. Max-pooling, which chooses the maximum value among the filter widths, is used in the pooling layers. Both the strided convolutional layers and the pooling layers subsample their inputs, followed by each input being concatenated along its channels. If the output of a previous stage has a $\sigma^{(N-1)} \in \mathbb{R}^{m \times w \times c}$ dimension, both the strided convolutional layer and pooling layer produce outputs as $\{C^{(N)}, P^{(N)}\} \in \mathbb{R}^{m \times (w/2) \times (c/2)}$, where $C^{(N)}$ and $P^{(N)}$ denote the output of the strided convolutional layer and pooling layer in the N stage. Concatenating along its channels gives it a dimension of $\sigma^{(N)} \in \mathbb{R}^{m \times (w/2) \times c}$. When passing through the stages ($N = 1, 2, \ldots, 8$), dimension reduction along its width is performed. For example, if the input (raw ECG) has dimensions of $x \in \mathbb{R}^{m \times w \times 1}$,

$$\begin{aligned}
\sigma^{(N)} &\in \mathbb{R}^{m \times (w/2^N) \times c}, \quad N \in \{1, 2, \ldots, 8\} \\
w &= 256 \times \text{window}, \quad \text{window} \in \{10, 30, 60\} \\
c &= 8 \times 2^k, \quad k = \begin{cases} \frac{N}{2} - 0.5, & \text{when } N \text{ is odd} \\ \frac{N}{2} - 1, & \text{when } N \text{ is even} \end{cases}
\end{aligned}$$

The extracted features ($\sigma^{(8)}$) generated by the last stage are fed to the softmax classifier, which performs a binary classification between stress and rest:

$$h_j = \frac{\exp(\sigma_j^{(8)})}{\sum_{k=1}^{2} \exp(\sigma_k^{(8)})}, \quad j = 1, 2 \tag{1}$$

where $j = 1, 2$ for the binary classification. The output of the softmax classifier, h_j, represents the probabilistic distributions of each class—stress and rest—the sum of which is 1. Table A1 shows more detailed information of the proposed network, including the shape of the output for each operation with the input of 10 s of ECG.

We used a rectified linear unit (ReLU) as an activation function to generate a non-linear decision boundary from the successive linear combinations of the weighted inputs. The activation function produces a maximum value between zero and its input. We applied dropout [26] with a drop rate of 0.3 and batch normalization [27] to prevent overfitting. A neural network can be easily overfitted to the training data when a model learns within a small number of data sets. Many studies have made use of dropout and batch normalization to overcome overfitting. Dropout requires a drop rate, which represents how many neurons are dropped in each layer. Batch normalization makes the input data follow a specific distribution, based on a normalized input distribution. The distribution can be changed during training through the trainable variables [27] γ and β, where γ scales the normalized input and β shifts it.

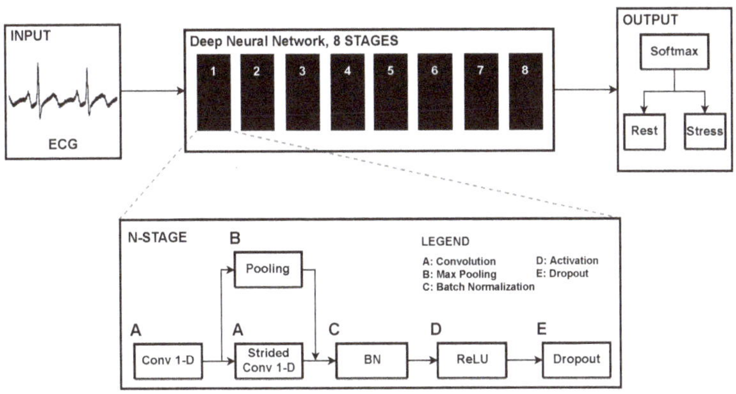

Figure 2. Deep neural network architecture and the components of each stage. Raw ECG signals are provided into the input layer. The successive stages extract features from an output of a previous stage. After the last stage, a softmax classifier performs a binary classification between the rest and stress.

Table 3. List of operations and hyperparameters used in each stage.

Order	Operation	Filter Width	Number of Filters	Stride
1	Conv 1-D	16	8×2^k	1
2	Conv 1-D Pooling	16 16	8×2^k -	2 2
3	Concat	Concatenating		
4	BN	Batch normalization		
5	Activation	ReLU		
6	Dropout	Drop rate: 0.3		

2.4. Training the Neural Network

There are three types of training method for the proposed model: Type I generates a pretrained model that trains using the driver data set; Type II trains a model with the mental arithmetic data set; and Type III trains a pretrained model (i.e., Type I) with the mental arithmetic data set.

All three types of training use the same end-to-end architecture, using the raw ECG signals from each data set. A loss function needs to be set to train the DNN model. We utilized the cross-entropy loss function:

$$L = -\frac{1}{m}\sum_{i=1}^{m}(y_i \log(h_i) + (1 - y_i)\log(1 - h_i)), \tag{2}$$

where h_i and y_i denote the prediction results from the proposed model and the true labels from the data set, respectively, and m represents the size of the mini-batch, which is set to 64. When the model predicts a state (i.e., stress or rest) properly, the loss function becomes nearly zero. However, it diverges from zero in the opposite situation; that is, when the model produces an output different from the data set label. A proper optimizer must be selected to train the DNN stably, because the optimizer ensures that the loss function converges to zero. We used the Adam optimizer ($\beta_1 = 0.9, \beta_2 = 0.999, \epsilon = 10^{-8}$) [28] to train the proposed model. This optimizer calculates the gradients of the loss function by back-propagation, which adjusts the weights of the neurons in an end-to-end model. All the weights of the neurons are initialized by the He initializer [29].

$$W \sim N(0, Var(W)). \tag{3}$$

The weight distribution is initialized to the normal distribution, which has a mean of zero and standard deviation of the weight variance. The weight variance is defined as follows:

$$Var(W) = \sqrt{\frac{2}{n_{in}}}, \tag{4}$$

where n_{in} denotes the number of input weights.

There are many hyperparameters (e.g., the number of layers, number of neurons, size of the mini-batch, filter width, number of channels, among others) to be decided, in order to train a model properly. We first considered how many stages are adequate for the proposed model. The number of the stages began with 1, and the performance for accuracy showed improvement as the number of stages increased by 1, up until 8. Within the proper number of stages, we tuned the filter width and the number of filters to find the best fitting parameters. We searched the hyperparameters of the DNN through a grid search method and a manual search. Finally, we chose the model that achieved the highest accuracy for the test data set along all the maximum 10 epochs. Section 2.5.1 shows how we split the data set into a training set and testing set, based on cross-validation.

2.4.1. Type I Training

We generated a pretrained model with the driving data set. As the DNN requires a large amount of data for training, we used the driving data set, which was larger than the mental arithmetic data set. The learning rate was set to 1×10^{-3} and was reduced by a factor of 10 every 5 epochs.

2.4.2. Type II Training

Using the mental arithmetic data set, the same method of end-to-end training was applied as in Type I training. Additionally, we used all the same hyperparameters to observe how the size of the data set affected training the neural network.

2.4.3. Type III Training

We applied a transfer learning method, using the pretrained model generated by Type I training. As mentioned above, it is difficult to train a neural network with a small data set. If a model is trained based on a pre-trained model, the model can then easily be fine-tuned. We hypothesized that the ECG measurements obtained from the participants who had taken the mental arithmetic task were similar to those obtained from drivers. It is effective to apply transfer learning when the distribution of data set to be learned is similar to that of the pretrained model. The softmax classifier required a re-training process, because there is little difference between the data used in pretraining and the data to be trained, although their distributions were similar. However, re-training only the softmax classifier did not show an acceptable performance. The number of layers to retrain was, thus, considered as a hyperparameter. We applied the grid search method to find the proper stages (Stage 1, Stage 2, ..., Stage 8) to be retrained. The start stage to be retrained was changed until a satisfactory performance was achieved. We kept the pretrained model, except for the last stage ($N = 8$) and the softmax layer ($N = 9$). The trainable variables in the softmax classifier were initialized before training.

$$W^{(N)} \sim N(0, Var(W^{(N)})). \quad N = 9 \text{ (softmax)} \tag{5}$$

We utilized the Adam optimizer [28] to update the trainable variables of the last stages and softmax classifier through backpropagation, but the variables in the other stages were kept constant:

$$W^{(N)} \leftarrow W^{(N)} - \alpha_t \cdot m_t / (\sqrt{v_t} + \epsilon), \quad N = 8, 9 \tag{6}$$

where m and v denote the first moment and second moment of the Adam optimizer, respectively. We used a different learning rate (α), which started at 1×10^{-4} with the same decay rate (decreasing by a factor of 10 every 5 epochs), to train the model. As the pretrained model was already fine-tuned, it was better to use a lower learning rate. The results of three training types and comparisons between them are shown in Section 3.

2.5. Model Evaluation

In this section, we describe how to evaluate the proposed model. All three training types, as mentioned above, performed training with a training set using cross-validation. We tested each type of end-to-end model with its test set and calculated the evaluation metrics (i.e., the receiver operating characteristic curves). Additionally, we observed the features not only at the end of the neural network, but also in the middle stages. The T-distributed stochastic neighbor embedding (t-SNE) [30] makes high-dimensional features visible in a two- or three-dimensional domain.

2.5.1. Cross-Validation

We used k-fold cross-validation ($k = 10$) to evaluate the proposed model. Both the driving and the mental arithmetic data sets were split by subject by cross-validation. The DNN should not have seen data in the test set presented during training. It is obvious that a neural network achieves a high performance with the data used in training. In other words, we needed to divide the data set into both a training set and a test set, and perform training and testing based on each set individually. In the case of the physiological signals, it is difficult to acquire a satisfactory amount of data to train a neural network. There are several limitations in a laboratory environment, such as the portability of the sensors and the inconvenience of the person to be measured. However, cross-validation makes it possible to generate both the training and testing sets with only a small amount of data. We randomly split the data set into individual subjects that make k folds using cross-validation. Each fold consists either of one subject or more than one subject. We trained the models with $k - 1$ folds and assessed them with the one fold left. Thus, the training and test sets had a 9:1 ratio. All these processes were iterated k times with the individual end-to-end models, which produced k models. Therefore, each model was trained by an individual training set and also validated by a test set which had never been seen during training. All of the three training types were evaluated with the data set which was included in the same data set used during the training session, but the model never had seen it. For example, the Type I model was trained and tested with the driving data set. In the case of Type III, although the pretrained model was made based on the driving data set, it was retrained using the mental arithmetic data set. Therefore, the mental arithmetic data set was used for the evaluation. All the evaluation metrics were cross-validation results, which are the mean values of all the folds.

All training and validation was performed on a personal computer (CPU; AMD Ryzen 7 2700X, GPU; NVIDIA GeForce GTX 1080 Ti 11 Gb, Memory; 32 Gb). With the use of GPU, it took less than two and one seconds per epoch for training the model with the driving and mental arithmetic data sets, respectively.

However, without GPU, the driving data set required 22 s per epoch and the mental arithmetic data set needed 5 s per epoch to train the model.

2.5.2. Statistical Analysis

A softmax classifier placed at the end of the DNN produces probabilistic outputs, which indicate how likely it is that the inputs are related to the true labels. Among the outputs, a classifier selects the highest probability for its predictions. It is an important way to compare these predictions with the true labels to evaluate the model performance. Many metrics are used to assess such models, such as receiver operating characteristic (ROC) [31] curves and precision–recall (PR) curves [32]. An ROC curve plots sensitivity against 1-specificity with a changing threshold value. Similarly, the precision against the sensitivity (recall) is plotted in the PR curves, which gives an additional analysis to the ROC curves for an imbalanced data set [32]. We calculated the area under the curve (AUC) for the ROC curves, which was nearly 1.0 when the model had successfully operated. We also computed the F_1 score, which represents the mean of sensitivity and precision. The sensitivity, also called recall, refers to how well a model detects stress among the true stress events. The specificity shows the correct detection rate of the rest state. The precision represents the ratio of the number of true-positives to the number of cases in which a model predicted stress. We compared the proposed model to other models [9–13], and to itself, for each type of training (i.e., Types I, II, and III) using the evaluation metrics. We utilized one-way analysis of variance (ANOVA) and Tukey's test to assess the model itself within each training type.

3. Results

We collected two self-reports (e.g., SAM and DT) from the participants after two levels of the mental arithmetic task and after the initial rest. Lower SAM and higher DT scores refer to stronger negative emotions and higher perceived stress, respectively. Table 4 shows the results of the self-reports. We calculated the difference of score based on thebaseline measurement (i.e., after initial rest) after the mental arithmetic task. SAM decreased after the tasks and the difference for the high level task was larger than that for the moderate level. The DT score increased, compared to the baseline measurement. Similar to the SAM score, a large difference in the DT score occurred after the high level arithmetic task.

Table 4. Difference in self-reported scores, compared to baseline measurement.

Task	SAM	DT
Math1	−0.37	0.37
Math2	−0.58	0.89

In this section, we show the results of the proposed model. It consists of the extracted feature maps and evaluation metrics, including a comparison with the other models and within the proposed model itself. As mentioned in Section 2.4, Type I training indicates the pretrained model using the driving data set. For Type II, the model was trained using the mental arithmetic data set without the pretrained model. In the case of Type III training, we used the same data set as in Type II to train the model, but based it on the pretrained model.

Firstly, we tested the conventional machine learning methods before evaluating the proposed model. We used conventional algorithms, including decision tree (DT), k-nearest neighbors (kNN), logistic regression (LR), random forest (RF), and support vector machine (SVM). All the algorithms were trained and validated with the same data set as the proposed model. Table 5 shows the accuracy of the conventional methods. All the machine learning algorithms could not reach a satisfying performance, in terms of accuracy, which means that trainable algorithms cannot learn the proper features using a raw ECG input.

Table 5. Accuracy of the conventional methods. (DT; Decision Tree, kNN; k-Nearest Neighbors, LF; Logistic Regression, RF; Random Forest, SVM; Support Vector Machine)

Stressor	Classifier	Window Length (s)		
		10	30	60
MA	DT	0.539 (0.050)	0.517 (0.062)	0.490 (0.066)
	kNN	0.497 (0.030)	0.511 (0.040)	0.535 (0.058)
	LR	0.493 (0.029)	0.537 (0.076)	0.508 (0.055)
	RF	0.512 (0.075)	0.505 (0.062)	0.515 (0.041)
	SVM	0.483 (0.025)	0.516 (0.071)	0.520 (0.082)
Driving	DT	0.487 (0.210)	0.457 (0.234)	0.512 (0.208)
	kNN	0.361 (0.051)	0.423 (0.150)	0.451 (0.208)
	LR	0.447 (0.188)	0.443 (0.235)	0.434 (0.225)
	RF	0.528 (0.187)	0.486 (0.215)	0.523 (0.193)
	SVM	0.514 (0.155)	0.533 (0.177)	0.498 (0.205)

3.1. Feature Representation

We observed all the extracted features from each stage using the t-SNE method, which converts high-dimensional features (the number of components, width, and channel) to 2-dimensional features, which we can analyze using a scatter plot. Figure 3 shows the t-SNE scatter plots for the input (raw ECG) and the extracted features from each stage. Each point represents states of the label (i.e., rest and stress). The input is from a subject who participated in the mental arithmetic task and is sliced using a 10 s window. The proposed model, trained by Type III training, generated features in each stage. As shown in Figure 3, there was almost no difference between the stress- and rest-labeled ECGs. By considering the features passed through the stages, a distinction could be observed between the labels. The t-SNE plots imply that it is possible to distinguish the two labels clearly through the softmax classifier after the last stage.

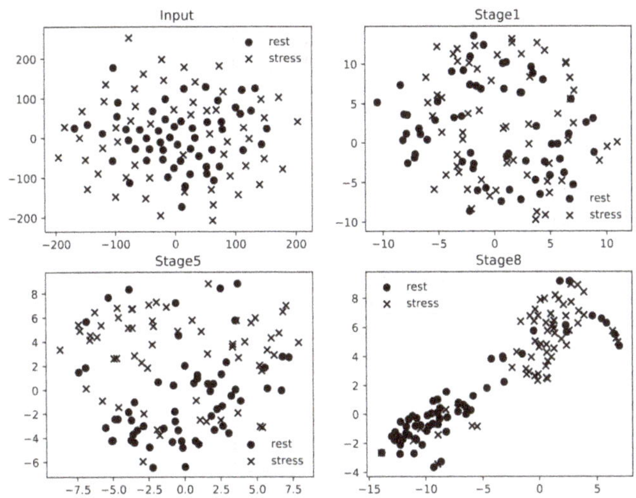

Figure 3. The t-SNE plots of raw ECG and extracted features from the stages. Round points denote features of ECG labeled as rest, and crosses represent stress-labeled features. This figure shows only the extracted features from stage 1, stage 5, and the last stage.

3.2. Performance of the End-to-End Model

Figure 4 shows the accuracy of the proposed model for the binary classification of rest and stress. We compared the results of the three training types, based on the input windows. Overall, Type I, which was trained using the driving data set, showed the best performance for all the windows. It reached the highest mean accuracies at 89.38%, 87.16%, and 79.12% for the 10 s, 30 s, and 60 s windows, respectively. The accuracy of Type I training, 89.38%, was significantly different from both the Type II accuracy, 61.33%, and Type III accuracy, 69.71%, for the 10 s window ($p < 0.001$). Additionally, there was a significant difference between the accuracy of Type II and Type III training ($p < 0.05$). In the case of the 30 s window, Type I training achieved an accuracy of 87.16%, whereas Type II and Type III training achieved 68.38% and 72.13%, respectively ($p < 0.001$). For the 60 s window, the accuracies of Type I, 79.12%, and Type III, 79.50%, training were slightly different, but the accuracy of Type II training, 71.50%, was significantly different from that of Type III training ($p < 0.05$). Considering Type II and Type III training, which were both trained with the same data set (mental arithmetic), there were improvements of 12.01% and 10.06% in accuracy for the 10 s and 60 s windows ($p < 0.05$), respectively; while there was no significant improvement in the 30 s window at the 0.05 level.

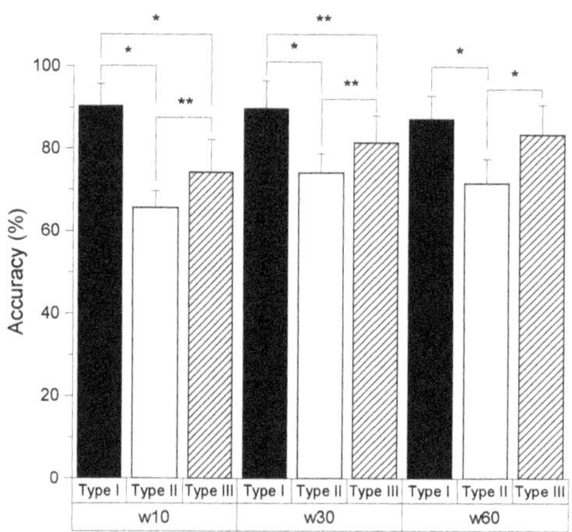

Figure 4. Accuracy of the end-to-end model in binary classification. Types are grouped by each raw ECG window (i.e., 10 s, 30 s, and 60 s) fed to the model. * and ** indicates that difference of the means is significant at the 0.001 and 0.05 level, respectively.

We plotted the ROC and PR curves by each type and window in Figure 5. The ROC curves need to be located above the baseline ($y = x$) to satisfy the model performance. We can observe that Type I training showed the best performance in the ROC and PR curves. Type III training demonstrated little improvement over Type II training, based on the curves. However, both the ROC and PR curves of Type III training are generally positioned higher than the curves of Type II. It is difficult to evaluate the performance of the model with the ROC and PR curves only. Therefore, we calculated the AUC of the ROC curves. It shows the performance on the numerical results, which makes it possible to compare the models.

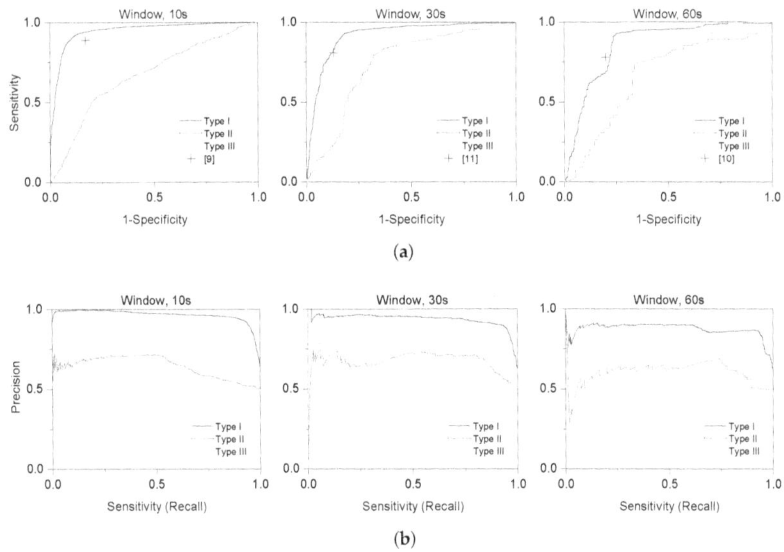

Figure 5. ROC and PR curves. Each line represents a curve from Type I, Type II, and Type III training, respectively. A cross refers to the performances of the conventional model. (**a**) ROC curves and (**b**) PR curves.

Table 6 shows the evaluation metrics, including the AUC, F_1 score, sensitivity, and specificity. Both the mean and standard deviation were calculated based on cross-validation. We compared the performance between Type I and Type II training, which were both trained without any pretrained model, but trained with the different sizes of data sets (i.e., the driving and the mental arithmetic data sets). Type I training for the 10 s window shows the best performance for the AUC, F_1 score, sensitivity, and specificity. It had a value of 0.938 for the AUC ($p < 0.001$), 0.922 for the F_1 score ($p < 0.001$), and 0.930 for sensitivity ($p < 0.001$). Although the specificity of Type I training for the 10 s window, 0.854, showed the highest value, it did not show a significant difference from Type II training. Based on the mean values, Type III training showed an improvement over Type II training, except for specificity with the 10 s window. For the 10 s window, the improvements were 8.00% for the AUC, 19.90% for the F_1 score ($p < 0.001$), and 29.77% for sensitivity ($p < 0.05$). For the 30 s window, the improvements were 5.07% for the AUC, 7.42% for the F_1 score, 1.81% for sensitivity, and 16.61% for specificity. The 60 s window showed improvements of 18.66% for the AUC ($p < 0.05$), 13.23% for the F_1 score ($p < 0.05$), 7.32% for sensitivity, and 20.71% for specificity. In summary, the transfer learning method improved performances by 11.57%, 10.57%, 13.52%, 12.96%, and 9.41% on average for accuracy, the AUC, the F_1 score, sensitivity, and specificity, respectively, along all window lengths.

Table 6. Evaluation metrics.

Type	Window Length (s)	Evaluation Metrics			
		AUC	F_1 Score	Sensitivity	Specificity
I	10	0.938 (0.053)	0.922 (0.044)	0.930 (0.035)	0.854 (0.094)
II		0.701 (0.069)	0.602 (0.094)	0.552 (0.186)	0.759 (0.173)
III		0.761 (0.088)	0.752 (0.079)	0.787 (0.117)	0.696 (0.144)
I	30	0.924 (0.072)	0.922 (0.050)	0.949 (0.039)	0.788 (0.161)
II		0.766 (0.049)	0.755 (0.050)	0.815 (0.143)	0.665 (0.165)
III		0.807 (0.131)	0.815 (0.063)	0.830 (0.130)	0.797 (0.170)
I	60	0.857 (0.141)	0.901 (0.036)	0.923 (0.044)	0.755 (0.214)
II		0.679 (0.113)	0.717 (0.078)	0.760 (0.227)	0.670 (0.258)
III		0.835 (0.095)	0.826 (0.089)	0.820 (0.162)	0.845 (0.161)

3.3. Comparison with Different Models

We compared the proposed end-to-end model with conventional methods [9–11]. Rigas et al. [9] used physiological signals including HRV, SC, and respiration while using 10 s length of window. Smets et al. [11] additionally utilized skin temperature. Castaldo et al. [10] used non-linear HRV parameters, including the sample entropy (SampEn), recurrence plot mean line length (RPlmean), and shannon entropy (ShanEn). Figure 5a shows the comparison results to the proposed model using the ROC curves. Each blue cross is positioned at the best performance in [9–11]. To assess the model exactly, we compared it with [9,11], which used 10 s and 30 s windows, to the proposed model with the same window lengths. To best match Castaldo et al. [10], which used a 3 m window to extract the HRVs from the ECG, we compared the proposed model with the 1 m window. Based on Figure 5a, all blue crosses are positioned lower than Type I, or are similar to it. From the perspective of sensitivity and specificity, the proposed model shows better performance than the conventional methods for a certain range of thresholds.

Both Hwang et al. [12] and Saeed et al. [13] utilized DNNs to classify stress. The comparison results are shown in Table 7. Hwang et al. [12] used a CNN and LSTM with a raw ECG signal and achieved an 87.39% and 73.96% accuracy for each case. Their architecture consisted of one convolutional layer and two LSTM layers. Our proposed model shows improvements in accuracy of 3.10% and 18.00% for each case, with the same window (10 s). Saeed et al. [13] used raw HR signals derived from the ECG and raw SC signals. They used the same driving data set from Healey et al. [14] to train and evaluate their model. The model [13] showed the best performance, with a value of 0.918 for the area under the ROC curve, while the proposed model reached 0.938.

Table 7. Comparison with models featuring a DNN algorithm.

	[12]	[13]	Proposed
Window	10 s	-	10 s
Input	Raw ECG	Raw HR and SC	Raw ECG
Accuracy	87.39%, 73.96%	-	90.19%
AUC	-	0.918	0.938

4. Discussion and Conclusions

We have proposed a novel end-to-end architecture that uses raw ECG signals for stress detection and validated its performance with two different data sets. We believe that our model could replace the conventional machine learning-based methods in several ways. First, in terms of model simplicity, our model has an advantage over conventional methods, which require a few additional steps, such as preprocessing, feature selection, and feature extraction, before classification. As our model was built with an end-to-end architecture, it does not necessarily require such additional steps. The end-to-end architecture enables the detection of stress by automatically extracting features without feature selection. We observed that the successive deep convolutional layers extract distinguishable features, as shown in Figure 3. Second, in the same vein, our model may not depend on the performance of these steps. The methods that use HRV parameters depend highly on the performance of the R-peak detection algorithm. Considering stress management in daily life, R-peak detection in ECG signals recorded in real-world environments may require additional steps, as proposed in [33]. In addition to the independence of the model, our results showed that the detection performance of the proposed model was superior to that of the conventional methods [9–11], as shown in Figure Figure 5a and Table 5. With raw ECG signals, conventional machine learning methods did not show acceptable performance in detecting stress, and rarely can be trained with non-linear inputs (i.e., raw signals). Finally, whereas the HRV parameters require at least a short-term (5 m) or long-term window (24 h) to properly reflect the stress response, our model used much shorter windows (10 s, 30 s, and 60 s). Our approach demonstrates a practically applicable system for daily stress management. As it takes an average of 2.490 ms to estimate stress state by inputs of raw ECGs, it is possible to apply the proposed model in the real-world to detect stress in real-time.

Despite the advantages described above, the performance of the DNN depends highly on the size of the data set used to train a neural network. To investigate the effect of the data set size on stress detection, we compared three different types of models with different training strategies. As expected, the model Type I, which was trained using a larger data set, showed better performance than the model Type II, trained using a smaller data set. There was a size difference of more than four times between the driving [14] and the mental arithmetic data sets. For the last type of model (Type III), we utilized the pretrained model, trained using the driving data set, to train the model with a smaller data set. From the comparison between Type II and Type III training, Type III, which used the pretrained model, showed an improvement over Type II, which did not. Although the size of the mental arithmetic data set might not be large enough to train the neural network, it is possible to achieve a fine-tuned model, based on pretraining with a larger data set. However, it could not reach the performance of the Type I model, trained using a larger data set, which presented the best performance. Unlike other domains of data, such as speech or image, a sufficient amount of physiological data may not be easily accessed or obtained. Thus, our approach can be utilized to train a DNN with a smaller data set, based on the pretrained model.

In this study, we used two different data sets (i.e., driving and mental arithmetic) under the ambulatory and laboratory environments for model development and validation. Mental arithmetic is one of the representative test paradigms used to assess mental stress. It was proved, by two questionnaires (self-assessment manikin and distress thermometer), that mental arithmetic induced a mental load in the participants. However, to develop a stress management method for daily life, there is a need to validate the method out-of-laboratory, as well. Thus, we also chose the driving data set to assess stress out-of-laboratory. Although these data sets cannot represent all of the stress situations that can occur in everyday life, such as workload stress, physical stress, anxiety, and so on, we demonstrated an end-to-end architecture to detect mental stress for both in- and out-of-laboratory environments. However, there were still limitations in this study. Although the two sensors used in the two data sets were individual, in view of generalization, the model needs to be validated by using ECG from diverse sensors, including other electrode configurations. We have fed other data sets, which were different from those using during training, into the model. This showed high-biased results about a specific type of stress and recording sensor dependency. Even though bias or dependency remains, transfer learning from one data set to other may provide a solution to break the limited applicability in real-world settings. As mentioned above, all of the data sets used in this study were acquired during specific stressful tasks. However, ECGs during daily activities are necessary for considering daily monitoring of stress. In future studies, we will apply this model to detect other stressful events, such as workload stress or anxiety, and will apply it to multi-class problems or continuous level recognition. Additionally, we will investigate how to augment physiological signals to train a neural network to overcome the limitations of the data set.

Author Contributions: Conceptualization, H.-M.C., S.-Y.D., and I.Y.; data curation, S.-Y.D., H.P., and H.-M.C.; funding acquisition, I.Y.; investigation, H.P. and H.-M.C.; methodology, H.-M.C. and S.-Y.D.; software, H.-M.C.; supervision, S.-Y.D. and I.Y.; writing–original draft, H.-M.C.; writing–review and editing, S.-Y.D. and I.Y.

Funding: This research was supported in part by the Bio Medical Technology Development Program of the National Research Foundation (NRF) funded by the Korean government, MSIP (2014M3A9D7070128); a grant of the Korea Health Technology R&D Project through the Korea Health Industry Development Institute (KHIDI), funded by the Ministry of Health & Welfare, Republic of Korea (grant number : HI14C3477); and the National Research Council of Science & Technology (NST) grant by the Korea government (MSIT) (No. CAP-18-01-KIST).

Conflicts of Interest: The authors declare no conflict of interest.

Appendix A

Table A1. More detailed information about the proposed architecture. "conv" denotes "conv(filter width)-(filter channel)". Similar to "conv", "maxpool16" refers to max pooling with 16 lengths of the filter.

Order	Operation	Output	Stride	# of Parameters
0	input	(?, 2560, 1)	-	-
1-1	conv16-8	(?, 2560, 8)	1	128
1-2	conv16-8	(?, 1280, 8)	2	1024
1-2	maxpool16	(?, 1280, 8)	2	-
1-3	concatenating	(?, 1280, 16)	-	-
1-4	batch normalization	(?, 1280, 16)	-	32
1-5	activation & dropout	(?, 1280, 16)	-	-

Table A1. *Cont.*

Order	Operation	Output	Stride	# of Parameters
2-1	conv16-8	(?, 1280, 8)	1	2048
2-2	conv16-8	(?, 640, 8)	2	1024
2-2	maxpool16	(?, 640, 8)	2	-
2-3	concatenating	(?, 640, 16)	-	-
2-4	batch normalization	(?, 640, 16)	-	32
2-5	activation & dropout	(?, 640, 16)	-	-
3-1	conv16-16	(?, 640, 16)	1	4096
3-2	conv16-16	(?, 320, 16)	2	4096
3-2	maxpool16	(?, 320, 16)	2	-
3-3	concatenating	(?, 320, 32)	-	-
3-4	batch normalization	(?, 320, 32)	-	64
3-5	activation & dropout	(?, 320, 32)	-	-
4-1	conv16-16	(?, 320, 16)	1	8192
4-2	conv16-16	(?, 160, 16)	2	4096
4-2	maxpool16	(?, 160, 16)	2	-
4-3	concatenating	(?, 160, 32)	-	-
4-4	batch normalization	(?, 160, 32)	-	64
4-5	activation & dropout	(?, 160, 32)	-	-
5-1	conv16-32	(?, 160, 32)	1	16,384
5-2	conv16-32	(?, 80, 32)	2	16,384
5-2	maxpool16	(?, 80, 32)	2	-
5-3	concatenating	(?, 80, 64)	-	-
5-4	batch normalization	(?, 80, 64)	-	128
5-5	activation & dropout	(?, 80, 64)	-	-
6-1	conv16-32	(?, 80, 32)	1	32,768
6-2	conv16-32	(?, 40, 32)	2	16,384
6-2	maxpool16	(?, 40, 32)	2	-
6-3	concatenating	(?, 40, 64)	-	-
6-4	batch normalization	(?, 40, 64)	-	128
6-5	activation & dropout	(?, 40, 64)	-	-
7-1	conv16-64	(?, 40, 64)	1	65,536
7-2	conv16-64	(?, 20, 64)	2	65,536
7-2	maxpool16	(?, 20, 64)	2	-
7-3	concatenating	(?, 20, 128)	-	-
7-4	batch normalization	(?, 20, 128)	-	256
7-5	activation & dropout	(?, 20, 128)	-	-
8-1	conv16-64	(?, 20, 64)	1	131,072
8-2	conv16-64	(?, 10, 64)	2	65,536
8-2	maxpool16	(?, 10, 64)	2	-
8-3	concatenating	(?, 10, 128)	-	-
8-4	batch normalization	(?, 10, 128)	-	256
8-5	activation & dropout	(?, 10, 128)	-	-
	Total			435K

References

1. Cohen, S.; Janicki-Deverts, D.; Miller, G.E. Psychological stress and disease. *JAMA* **2007**, *298*, 1685–1687. [CrossRef] [PubMed]
2. Smets, E.; De Raedt, W.; Van Hoof, C. Into the Wild: The Challenges of Physiological Stress Detection in Laboratory and Ambulatory Settings. *IEEE J. Biomed. Health Inform.* **2018**, *23*, 463–473. [CrossRef] [PubMed]

3. Sztajzel, J. Heart rate variability: A noninvasive electrocardiographic method to measure the autonomic nervous system. *Swiss Med. Wkly.* **2004**, *134*, 514–522. [PubMed]
4. Shaffer, F.; Ginsberg, J. An overview of heart rate variability metrics and norms. *Front. Public Health* **2017**, *5*, 258. [CrossRef] [PubMed]
5. McCraty, R.; Atkinson, M.; Tiller, W.A.; Rein, G.; Watkins, A.D. The effects of emotions on short-term power spectrum analysis of heart rate variability. *Am. J. Cardiol.* **1995**, *76*, 1089–1093. [CrossRef]
6. Appelhans, B.M.; Luecken, L.J. Heart rate variability as an index of regulated emotional responding. *Rev. Gen. Psychol.* **2006**, *10*, 229–240. [CrossRef]
7. Camm, A.; Malik, M.; Bigger, J.; Breithardt, G.; Cerutti, S.; Cohen, R.; Coumel, P.; Fallen, E.; Kennedy, H.; Kleiger, R.; et al. Heart rate variability: Standards of measurement, physiological interpretation and clinical use. Task Force of the European Society of Cardiology and the North American Society of Pacing and Electrophysiology. *Circulation* **1996**, *93*, 1043–1065.
8. Pan, J.; Tompkins, W.J. A real-time QRS detection algorithm. *IEEE Trans. Biomed. Eng* **1985**, *32*, 230–236. [CrossRef]
9. Rigas, G.; Goletsis, Y.; Fotiadis, D.I. Real-time driver's stress event detection. *IEEE Trans. Intell. Transp. Syst.* **2012**, *13*, 221–234. [CrossRef]
10. Castaldo, R.; Xu, W.; Melillo, P.; Pecchia, L.; Santamaria, L.; James, C. Detection of mental stress due to oral academic examination via ultra-short-term HRV analysis. In Proceedings of the 2016 38th Annual International Conference of the IEEE Engineering in Medicine and Biology Society (EMBC), Orlando, FL, USA, 16–20 August 2016; pp. 3805–3808.
11. Smets, E.; Casale, P.; Großekathöfer, U.; Lamichhane, B.; De Raedt, W.; Bogaerts, K.; Van Diest, I.; Van Hoof, C. Comparison of machine learning techniques for psychophysiological stress detection. In Proceedings of the International Symposium on Pervasive Computing Paradigms for Mental Health, Milan, Italy, 24–25 September 2015; Springer: Cham, Switzerland, 2015; pp. 13–22.
12. Hwang, B.; You, J.; Vaessen, T.; Myin-Germeys, I.; Park, C.; Zhang, B.T. Deep ECGNet: An Optimal Deep Learning Framework for Monitoring Mental Stress Using Ultra Short-Term ECG Signals. *Telemed. e-Health* **2018**, *24*, 753–772. [CrossRef] [PubMed]
13. Saeed, A.; Ozcelebi, T.; Lukkien, J.; van Erp, J.; Trajanovski, S. Model Adaptation and Personalization for Physiological Stress Detection. In Proceedings of the 2018 IEEE 5th International Conference on Data Science and Advanced Analytics (DSAA), Turin, Italy, 1–4 October 2018; pp. 209–216.
14. Healey, J.A.; Picard, R.W. Detecting stress during real-world driving tasks using physiological sensors. *IEEE Trans. Intell. Transp. Syst.* **2005**, *6*, 156–166. [CrossRef]
15. Krizhevsky, A.; Sutskever, I.; Hinton, G.E. Imagenet classification with deep convolutional neural networks. In *Advances in Neural Information Processing Systems*; Curran Associates Inc.: Red Hook, NY, USA, 2012; pp. 1097–1105.
16. Hannun, A.Y.; Rajpurkar, P.; Haghpanahi, M.; Tison, G.H.; Bourn, C.; Turakhia, M.P.; Ng, A.Y. Cardiologist-level arrhythmia detection and classification in ambulatory electrocardiograms using a deep neural network. *Nat. Med.* **2019**, *25*, 65. [CrossRef] [PubMed]
17. Manawadu, U.E.; Kawano, T.; Murata, S.; Kamezaki, M.; Muramatsu, J.; Sugano, S. Multiclass Classification of Driver Perceived Workload Using Long Short-Term Memory based Recurrent Neural Network. In Proceedings of the 2018 IEEE Intelligent Vehicles Symposium (IV), Changshu, China, 26–30 September 2018; pp. 1–6.
18. Xu, S.S.; Mak, M.W.; Cheung, C.C. Towards end-to-end ECG classification with raw signal extraction and deep neural networks. *IEEE J. Biomed. Health Inform.* **2018**, *23*, 1574–1584. [CrossRef] [PubMed]
19. Acharya, U.R.; Fujita, H.; Lih, O.S.; Hagiwara, Y.; Tan, J.H.; Adam, M. Automated detection of arrhythmias using different intervals of tachycardia ECG segments with convolutional neural network. *Inf. Sci.* **2017**, *405*, 81–90. [CrossRef]
20. Kiranyaz, S.; Ince, T.; Gabbouj, M. Real-time patient-specific ECG classification by 1-D convolutional neural networks. *IEEE Trans. Biomed. Eng.* **2016**, *63*, 664–675. [CrossRef]

21. Ronneberger, O.; Fischer, P.; Brox, T. U-net: Convolutional networks for biomedical image segmentation. In Proceedings of the International Conference on Medical Image Computing And Computer-Assisted Intervention, Munich, Germany, 5–9 October 2015; Springer: Cham, Switzerland, 2015, pp. 234–241.
22. Hochreiter, S.; Schmidhuber, J. Long short-term memory. *Neural Comput.* **1997**, *9*, 1735–1780. [CrossRef]
23. Goldberger, A.L.; Amaral, L.A.; Glass, L.; Hausdorff, J.M.; Ivanov, P.C.; Mark, R.G.; Mietus, J.E.; Moody, G.B.; Peng, C.K.; Stanley, H.E. PhysioBank, PhysioToolkit, and PhysioNet: Components of a new research resource for complex physiologic signals. *Circulation* **2000**, *101*, e215–e220. [CrossRef]
24. Bradley, M.M.; Lang, P.J. Measuring emotion: The self-assessment manikin and the semantic differential. *J. Behav. Therapy Exp. Psychiatry* **1994**, *25*, 49–59. [CrossRef]
25. Jacobsen, P.B.; Donovan, K.A.; Trask, P.C.; Fleishman, S.B.; Zabora, J.; Baker, F.; Holland, J.C. Screening for psychologic distress in ambulatory cancer patients: A multicenter evaluation of the distress thermometer. *Cancer* **2005**, *103*, 1494–1502. [CrossRef]
26. Srivastava, N.; Hinton, G.; Krizhevsky, A.; Sutskever, I.; Salakhutdinov, R. Dropout: A simple way to prevent neural networks from overfitting. *J. Mach. Learn. Res.* **2014**, *15*, 1929–1958.
27. Ioffe, S.; Szegedy, C. Batch normalization: Accelerating deep network training by reducing internal covariate shift. *arXiv* **2015**, arXiv:1502.03167.
28. Kingma, D.P.; Ba, J. Adam: A method for stochastic optimization. *arXiv* **2014**, arXiv:1412.6980
29. He, K.; Zhang, X.; Ren, S.; Sun, J. Delving deep into rectifiers: Surpassing human-level performance on imagenet classification. In Proceedings of the IEEE International Conference on Computer Vision, Santiago, Chile, 7–13 December 2015; pp. 1026–1034.
30. van der Maaten, L.; Hinton, G. Visualizing data using t-SNE. *J. Mach. Learn. Res.* **2008**, *9*, 2579–2605.
31. Fawcett, T. An introduction to ROC analysis. *Pattern Recognit. Lett.* **2006**, *27*, 861–874. [CrossRef]
32. Saito, T.; Rehmsmeier, M. The precision-recall plot is more informative than the ROC plot when evaluating binary classifiers on imbalanced datasets. *PLoS ONE* **2015**, *10*, e0118432. [CrossRef]
33. Lee, M.; Park, D.; Dong, S.Y.; Youn, I. A Novel R Peak Detection Method for Mobile Environments. *IEEE Access* **2018**, *6*, 51227–51237. [CrossRef]

sensors

MDPI

Article

A Wearable In-Ear EEG Device for Emotion Monitoring

Chanavit Athavipach [1], Setha Pan-ngum [1,*] and Pasin Israsena [2]

[1] Department of Computer Engineering, Faculty of Engineering, Chulalongkorn University, Phayathai Road, Wang Mai, Pathumwan, Bangkok 10330, Thailand; chantaywa0@gmail.com
[2] National Electronics and Computer Technology Center, 112 Thailand Science Park, Phahonyothin Road, Khlong Nueng, Khlong Luang, Pathumthani 12120, Thailand; pasin.israsena@nectec.or.th
* Correspondence: setha.p@chula.ac.th

Received: 30 July 2019; Accepted: 10 September 2019; Published: 17 September 2019

Abstract: For future healthcare applications, which are increasingly moving towards out-of-hospital or home-based caring models, the ability to remotely and continuously monitor patients' conditions effectively are imperative. Among others, emotional state is one of the conditions that could be of interest to doctors or caregivers. This paper discusses a preliminary study to develop a wearable device that is a low cost, single channel, dry contact, in-ear EEG suitable for non-intrusive monitoring. All aspects of the designs, engineering, and experimenting by applying machine learning for emotion classification, are covered. Based on the valence and arousal emotion model, the device is able to classify basic emotion with 71.07% accuracy (valence), 72.89% accuracy (arousal), and 53.72% (all four emotions). The results are comparable to those measured from the more conventional EEG headsets at T7 and T8 scalp positions. These results, together with its earphone-like wearability, suggest its potential usage especially for future healthcare applications, such as home-based or tele-monitoring systems as intended.

Keywords: EEG; in-ear EEG; emotion classification; emotion monitoring; elderly caring; outpatient caring; machine learning

---

1. Background

As societies around the world increasingly face with the issue of aging population, how to take care of these elderly people effectively becomes an important challenge. This is true especially for the less-fortunate ones who live alone. In order to ensure their physical and mental well-being and provide emergency assistance, monitoring technology could potentially be part of the solution. Particularly, wearable devices or smart sensors could be employed for effective and practical monitoring. Apart of conventional physiological signals, such as heart rate or EKG, that can be monitored to analyze the wearer's health conditions, emotional state is one of the factors which reflects mental states and can greatly impact decision-making [1]. Emotion monitoring could therefore also be used as another piece of information for elderly and remote-patient caring supporting systems.

Emotion itself is very complex [2]. There are different interpretations for the many kinds of emotions, making emotion recognition far from straight forward. For research purposes, several simplified models have been proposed that can be categorized into two approaches; defining basic emotions and using a dimensional model. The most widely used basic emotions are the six basic emotions (i.e., anger, disgust, fear, joy, sadness, and surprise) generally used in facial expression recognition [3]. For the second approach, the common dimensional model is characterized by two main dimensions (i.e., valence and arousal). The valence emotion ranges from negative to positive, whereas the arousal emotion ranges from calm to excited [4]. This model has been used in a number of studies, because it is easier to express an emotion in terms of valence and arousal rather than basic emotions that can be confused by emotion names [5].

For a long time, most emotion recognition studies have focused on using facial expressions and speech. For continuous monitoring purposes, these approaches may not be the most suitable, as they may suffer from practical issues, such as ambient light and noises. Especially for camera-based facial recognition, the privacy issue is also a concern. Alternatively, physiological signals, such as galvanic skin response (GSR), electrocardiogram (ECG), skin temperature (ST), and electroencephalogram (EEG), which occur continuously and are harder to conceal, have been considered. As emotions are thought to be related with activity in brain areas that direct our attention, motivate our behavior, and determine the significance of what is going on around us, EEG, which is the signal from voltage fluctuations in the brain that are generated continuously at the level of cellular membranes [6], has been especially of interest.

Emotion classification by EEG has been shown to achieve high accuracy [1,7–16]. However, most of those works employed multiple channel EEG headsets. In reality, these conventional multiple channel EEG standard headsets are not suitable for continuous monitoring due to their size and setup difficultly. Ideally, the EEG recording device used for emotion monitoring should be small, take little time to setup, and be comfortable to wear.

For such requirements, an in-ear EEG which is an EEG recording device introduced by Looney et al. in 2012 [17] could be of interest. Generally, the potential benefits of using an EEG of the in-ear type include the fact that it does not obstruct the visual field. It is also positionally robust, as it is generally fixed inside the ear canals. It is unobtrusive, as it is similar to devices people commonly use, such as earphones, earbuds, and earplugs. It is unlikely to encounter sweat, and also user-friendly for setup and maintenance. Unlike scalp EEG devices, which may require some experienced assistants to help, in-ear EEG devices could be simply put into users' ears. However, an in-ear EEG also has some drawbacks. An in-ear EEG has much fewer electrodes and covers a much smaller area than what the scalp EEG can. So, its application accuracy is expected to be less than that of the scalp EEG.

Our work was aimed at building an in-ear EEG device and evaluating it in terms of signal quality compared to those measured via scalp EEG at comparable positions (i.e., T7 and T8 based on the international 10–20 system [18]). The international 10–20 system is an internationally recognized system for labelling scalp locations for EEG measurement. The T7 position is located above the left ear, while T8 is positioned above the right ear. The prospect of an in-ear EEG usage for emotion classification was also investigated by experiments.

The paper is organized into six sections. Related works are discussed in Section 2. Section 3 describes material selections and system design. Detailed experimental protocols are included in Section 4. Experimental results and analysis are presented in Section 4. Significant findings from the results are discussed in Section 5. Finally, the conclusions are presented in Section 6.

2. Related Work

2.1. Scalp-Based EEG Emotion Classification

Scalp-based emotion classification by multi-channel EEG has been an active field of research [1,7–16]. A review of some of those works can be found in [7]. The majority of the works have focused on signal processing techniques to improve accuracy. For example, Koelstra et al. [19] presented methods for single trial classification using both EEG and peripheral physiological signals. The power spectrum density (PSD) of EEG signals was used as the primary feature. A support vector machine (SVM) classifier was used to classify two levels of valence states and two levels of arousal states. For EEG analysis results, average and maximum classification rates of 55.7% and 67.0% were obtained for arousal and 58.8% and 76.0% for valence. Huang et al. [20] developed an asymmetry spatial pattern (ASP) technique to extract features for an EEG-based emotion recognition algorithm. The system employed k-nearest neighbor (K-NN), naive Bayes (NB), and support vector machine (SVM) methods for emotion classification. The average accuracy rates for valence and arousal were 66.05% and 82.46%, respectively. We note here that several studies [7,21–23] have targeted the PSD of EEG data as the input features and performed emotion classification by using SVM. Other machine learning techniques, such as naive Bayes, K-NN, LDA, and ANN, have been applied in other studies [9,24–26].

Other areas of focus for scalp-based EEG emotion classification include those in [15,27], which look to develop wearable headband solutions. However, for monitoring purposes, these designs may suffer in conditions such as a warm climate; it might be uncomfortable to wear headband for a long duration due to sweating. Moreover, the sweat could affect the electrode impedance, resulting in noisy signal and inaccurate monitoring

2.2. In-Ear EEG Development

Originally, an in-ear EEG, which is an EEG recording device introduced by Looney et al. in 2012 [17], was demonstrated to have wearable characteristics that could potentially fulfill monitoring requirements [28]. It is small and could be worn around the ears, and is similar to earplugs or hand-free devices. Since then, research works have focused on areas such as materials; system design, especially in terms of practicality; and the verification of signal quality [17,27,29–31]. For example, Goverdovsky et al. [30] suggested a new prototype called Ear-EEG that consists of a viscoelastic substrate memory foam earplug and conductive cloth electrodes to insure conformance with the ear canal surface for motion artifacts' reduction. Kullkami et al. [27] designed a soft and foldable electrode that can capture the EEG from different outer complex surfaces of the ear and the mastoid using the epidermal electronics with fractal mesh layouts. Recent work by Kappel et al. [31] developed an in-ear EEG with a soft earpiece, which required customized molding to fit individual ears. The prototype showed good signal quality and the potential for long term EEG monitoring.

2.3. In-Ear EEG for Control

In the field of brain–computer interface, artifacts in EEG signals created through muscle activity, such as eye blinks or other facial expressions, have been studied as a means for controlling external devices. For in-ear implementations, major works in the area include; Matthies et al. [32] which reported an in-ear headset based on a hacked NeuroSky EEG sensor. The prototype utilizes eye winking and ear wiggling for explicit control of the function of a smartphone. Additionally, in 2017, Matthies et al. [33] placed multi electrodes onto a foam earplug to detect 25 facial expressions and head gestures with four different sensing technologies. Five gestures could be detected with accuracy above 90%, and 14 gestures with accuracy above 50%. The prototype was also shown to be robust under practical situations, such as walking.

2.4. In-Ear EEG for Medical and Healthcare Applications

Medical and healthcare applications have also been a major theme for in-ear EEG research, especially for monitoring purposes [34]. Sleep has been particularly of interest [35,36]. For example, Nguyen et al. [35] proposed a dual channel EEG in the form of an earplug that showed a stable sleep stage classification with an average of 95%+ accuracy. In terms of emotion monitoring, which is closely related to this work, previous work [17,37] showed that an in-ear EEG signal measured was similar to T7 and T8 channels on the 10–20 system [18]. Moreover, one of the previous works also showed that T7 and T8 provided some informative data for emotion classification [7]. These results suggest that an in-ear EEG has the potential to classify emotions, which our work was to investigate.

3. Materials and Methods

In this work, to achieve the goal of realizing an in-ear EEG, we looked to find answers to these questions:

(1) What type of in-ear EEG should be studied (physically, design-wise, and engineering-wise)?

(2) What kind of EEG signal quality would we be getting?
(3) How good it is specifically for emotion classification?

For (1) we reviewed previous works and built some prototypes to evaluate their suitability. Once we decided upon the solution, we then moved on to verify the quality of measured signals compared to standard measurements to answer (2). It is important to do this before the main experiment as the result should be relatively comparable before we could move on to emotion measurement. To achieve that, we used the mismatch negativity (MMN) to compare auditory ERP measured via our ear EEG with those measured with a conventional headband EEG at T7 and T8 positions. Finally, for emotion classification, we needed reference to benchmark our measured results, so the DEAP dataset was used to calculate the accuracy of emotion classification at T7 and T8. It results were then used as reference for comparing with our own in-ear EEG measurements. All of this is explained in more detail in the following sections.

3.1. In-Ear EEG Development

3.1.1. Earpieces Selection

Recent research on in-ear EEG devices were studied [17,30,31,37]. There are currently 2 types of in-ear EEG devices; one is a personally customized earpiece, as illustrated in Figure 1, and the other is generic or non-customized. The first type is based on earmolds created from wax impressions, 3D scanning, CAD editing, 3D-printing, and a wiring process, respectively. This type of an in-ear device EEG is robust as it fits completely to the owners' ear canal. However, it is relatively costly. Hence, this type of an in-ear EEG device was not considered in this study, as we would like a generic and low-cost device.

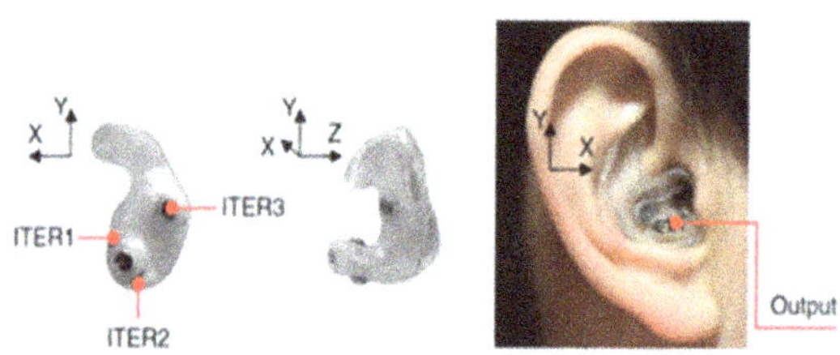

Figure 1. The first in-ear EEG prototype introduced by David Looney et al. in 2012 [17].

The generic prototype is usually based on a cylinder-shaped material. The first generic in-ear EEG device was based on a cylinder of silicone, as illustrated in Figure 2 [37]. However, it has a flexibility disadvantage, as it is not guaranteed to fit into all ear canals [30]. The improved prototype used a cylinder-shaped memory foam instead of silicone.

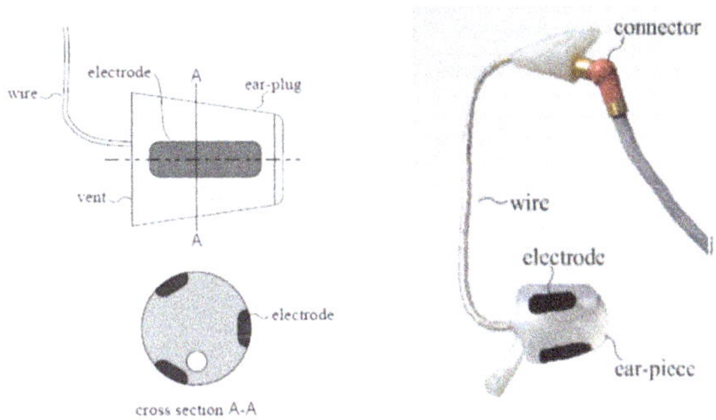

Figure 2. Generic in-ear EEG prototype [37]. The left side illustrates a drawing whereas the right side illustrates a model prototype.

Nevertheless, from our test, the in-ear EEG device built from memory foam ear-plugs could not fit into small ear canals. Furthermore, once fit in, it could also gradually slip out of the ear canal. Thus, in this study, the main body of the in-ear EEG device was changed to earphone rubbers, which were tested and found to have high flexibility. Additionally, they come in different sizes which could be properly selected to fit different ear canals, as shown in Figure 3.

Figure 3. Different sizes of earphone rubbers.

3.1.2. Electrode Selection

Three different materials were considered and tested for the in-ear EEG device electrodes, a half-sphere shaped silver, aluminum foil, and silver-adhesive fabric. The half-sphere shaped silver is probably one of the most widely-used materials for EEG electrodes. However, according to [30] the electrodes should be as similarly flexible as possible to the earpieces to achieve robust contact. Half-sphere silver is solid and not as flexible as the earphone rubbers. Therefore, the half-sphere silver was not selected. For aluminum foil, although it has low impedance and good flexibility, it could not be easily attached to electrical wires. This is because the aluminum foil is not adhesive to soldering.

The silver-adhesive fabric, which was used with memory foam as in-ear EEG prototype [30], has flexibility similarly to memory foam and earphone rubber. It could also be glued and sewed to the wires without soldering. Therefore, the silver-adhesive fabric was considered suitable material for the electrodes for our in-ear EEG device.

In this study, the size of the fabric was made slightly larger than in the previous study [30] for better contact. The fabric was glued to the ear rubbers, and the shield wires were then sewed to the fabrics. The number of the electrodes was also reduced to one channel per ear as the EEG signals among channels in the same in-ear from the previous studies were very similar [17]. The shield wire was slightly larger and heavier than a normal wire. However, it significantly reduced signal noise. Therefore, it was preferable to standard wire.

Our final prototype of in-ear EEG device is shown in Figure 4. The total material cost per piece is approximately 10 US Dollars. Our in-ear EEG device's impedance was measured to be between 0.05 and 5.5 ohms which was comparable to that of OpenBCI electrodes: one of the commercial EEG electrodes [38].

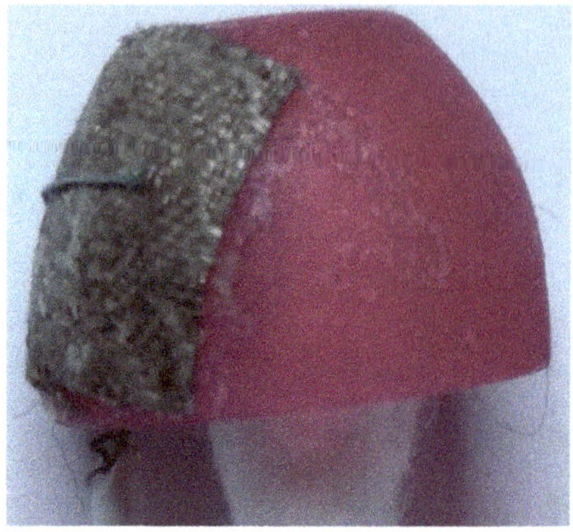

Figure 4. Single channel electrode used in the experiment using earphone rubber and silver-adhesive fabric electrode.

3.2. *In-Ear EEG Signal Verification*

After the in-ear EEG devices were assembled, signal verification was performed. Mismatch negativity (MMN) is one of the widely-used methods for EEG verification [39,40]. It was used to verify in-ear EEG signals in the previous study [41]. Hence, it was also applied in our work. MMN is an experiment which observes the auditory event-related potential (ERP). ERP is a subject's EEG signal response to an unexpected change of sensory stimulation.

Our MMN experiment started by playing a short beep tone repeatedly until the subject was familiar to the tone. Unexpected mismatch tones were then inserted among the familiar tone. Unexpected mismatch tones could have a change of frequency (lower or higher), duration (unusually longer beep duration), intensity (unusually louder or lighter), or phase. The mismatch tone, if acknowledged, will provide an ERP response as a negative peak. The mismatch responses usually give a negative peak between 90 and 250 milliseconds after the beep [40]. The ERP latency may be varied according to personal musical experience [42].

The MMN experiment parameters in this study were set according to the previous study [40]. A combination of three pure tonal frequencies: 500, 1000, and 1500 Hz lasting for 75 milliseconds, were used as a standard tone, whereas two types of mismatch tones were applied. The first type was frequency mismatch containing 10% lower or higher pitch randomly applied to each frequency. The other type was a duration mismatch tone which lasted for 100 milliseconds, 25 milliseconds longer than the standard tone. The standard tone was beeped 15 times in order to make the subject familiar with the tone, before the mismatch tones were inserted. Mismatched tones arrived at the probability of 0.5, but no consecutive mismatch tones were allowed.

The tones were played through an earphone. The in-ear EEG device was inserted to the right ear while the earphone was inserted to the left ear. The ground electrode was placed on the forehead and the reference electrode was placed on the right cheek, as suggested by [43]. An OpenBCI's electrode was also placed at T8 as a comparison electrode. A Butterworth filter was used to notch 50 Hz powerline noise. It was also applied as a bandpass to filter the EEG signal between 2 and 30 Hz. The signal correlation between T8 and in-ear EEG was also calculated.

3.3. Emotion Model Emotion Stimuli

The valence and arousal emotion model [4], as in Figure 5, was used in this research, as it is a widely used simplified emotion model. Four emotions (happiness, calmness, sadness, and fear) will be classified according to the quadrants, respectively.

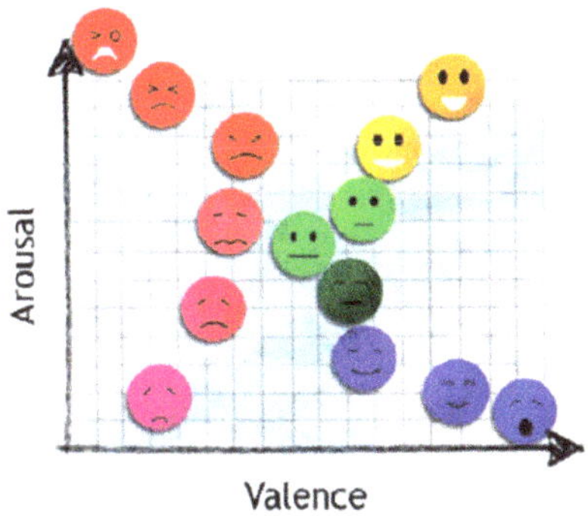

Figure 5. Valence and arousal model. Anger and fear have high valence and arousal. Happiness and excitement have high arousal and valence. Sadness and depression have low arousal and valence. Relaxation and pleasure have low arousal but high valence [4].

The International Affective Picture System (IAPS) [44], and the Geneva Affective Picture Database (GAPED) [45] were used as visual emotional stimuli. IAPS was the most widely used among previous research [1]. IAPS was developed at the Center for the Study of Emotion and Attention, University of Florida, by Lang, et al. [44]. IAPS pictures were standardized, and publicly available for use in emotional stimulation. The emotions elicited were based on two primary dimensions, which were valence and arousal. Valence ranged from unpleasant to pleasant, while arousal ranged from calm to excited. Every picture has valence and arousal rating from the scale 1 (lowest) to 9 (highest). However, IAPS contains fewer numbers of pictures stimulating low valence and low arousal than needed, so additional pictures from GAPED were used.

The GAPED database was developed by Dan-Glauser, et al. at the University of Geneva [45]. It was intended to provide additional pictures to a limited number of IAPS for experimental researchers. GAPED provided a 730 picture database for emotion stimulation, which was also rated based on valence–arousal parameters as used in IAPS [44]. Moreover, four classical music pieces from auditory emotional research [46] were also applied as stimuli. The four musical pieces were also chosen based on the valence–arousal model, which corresponded to the IAPS and GAPED pictures.

3.4. Feasibility

Most previous studies on emotion classification used multiple EEG channels. The feasibility of emotion classification using a single-channel in-ear EEG should be evaluated first. The feasibility evaluation was conducted by performing an emotion classification experiment using secondary data from the Dataset for Emotion classification using Physiological and Audiovisual Signals (DEAP) [47]. DEAP data set is a publicly available dataset for Brain Computer Interface (BCI) based emotion study provided by Koelstra S., et al. [47]. 32 channel EEG data from 32 subjects was collected, while they watched music video clips that were chosen to elicit emotions. The emotions elicited were based on the valence–arousal model. Valence was associated with emotion positivity which ranged from unpleasant to happy/pleasant. Arousal was associated with excitement which ranged from calm to excited. The subjects rated the music video clips on valence–arousal scales. The DEAP dataset was hence labelled, and the classification accuracy on the data could be evaluated by the subjects' rating. Out of 32 channels, only T7 and T8, which were stated to be close and correlate to the in-ear EEG were used for our emotion classification. Our emotion classification using DEAP dataset will be used for evaluating and comparing to the in-ear EEG emotion classification accuracy.

Support vector machine (SVM) which was widely used for emotion classification [1,7,10,16] was used as a classifier. SVM has good generalization and overfitting prevention properties. Therefore, it is considered suitable for this work. Six statistical parameters by Picard et al. [48] were used for signal feature extraction on a 3 s time-lapsed window. The Butterworth filter was used to notch 50 Hz noise, and filter EEG signals into five frequency bands; namely, delta, theta, alpha, beta, and gamma bands [6]. Ten-folded cross validation was applied to suppress biases [49].

3.5. Experiment Setup

This experiment was designed to collect EEG data using our in-ear EEG electrodes when subjects' emotions were stimulated by pictures and sounds, described in Section 3.3. The results would be analyzed to assess the performance of in-ear EEG on emotion classification.

Twelve male and one female subjects aged between 20 to 30 years with an average age of 24, were recruited for emotion classification experiments. Before the experiment started, the impedances of the in-ear EEG were re-measured as quality assurance. An in-ear EEG device was then inserted into either the right ear or left ear according to each subject's preference, whereas earphones were inserted into the other ears. Earwax was cleaned by alcohol before the in-ear EEG insertion.

Unless the subjects preferred to put the in-ear EEG on the left, it was put on the right ear as the left ear is shown to be better for listening to music [50]. The ground electrode was placed at forehead and the reference electrode was placed at either cheek inferior to the ear. A small amount of saline was used as electrolyte gel. Forty trials were recorded per subject. IAPS and GAPED pictures were randomly displayed to the subjects. The total number of pictures used for each emotion was as suggested by IAPS and GAPED datasheets.

Each picture was displayed for 30 s. Subjects were recommended not to move during each picture viewing. Fifteen seconds of black screen was displayed after each picture in order to neutralize subjects' emotions before the next picture was displayed. During the black screen subjects were free to mobilize. After eight pictures, subjects could have a small break and were free to move around before they were ready to continue.

After the experiments were finished, the subjects were asked to evaluate their emotional response on each picture for emotion classification. This is because the emotional response to each picture may be different among subjects or different from the IAPS and GAPED datasets.

Statistical analyzes for any group comparison were performed using either t-tests or ANOVA, depending on the number of groups. A p-value of less than 0.05 was considered statistically significant. All statistics were performed using SPSS (IBM Corp., New York, USA)

4. Results

4.1. MMN Results

Examples of frequency and duration mismatch responses compared to a standard tone are illustrated in Figure 6. In Figure 6a,b negative peaks between 200-400 ms which indicated mismatched ERP responses were found in both T8 and in-ear EEG signals. Different types of mismatched ERP signals, such as frequency and duration mismatched may vary in amplitudes, but general shapes of signals contain significant negative peaks around 200–400 ms [40]. These negative peaks of mismatch duration (Figure 6d) and frequency of mismatch (Figure 6e) from traditional MMN experiments, shown in the dotted line, from the previous study [40] are also shown in Figure 6 for comparison. The dotted lines in Figure 6d,e also show negative peaks between 200 and 400 ms. The examples of ERP responses to standard beeps are shown in Figure 6c. In contrast to the mismatch responses, the negative peaks are not present between 200 and 300ms. This conforms to the theory in [39].

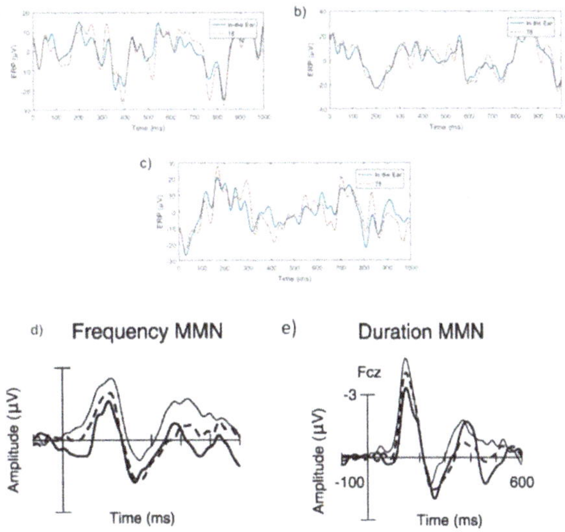

Figure 6. Examples of EEG after mismatch trials. (**a**) Example of frequency mismatch EEG event-related potential (ERP) response. (**b**) An example of duration mismatched EEG ERP response. (**c**) An example of an EEG ERP response after a standard beep. The blue and red lines in (**a–c**) show the in-ear and T8 EEG signal, respectively. (**d**,**e**) Duration and frequency mismatch responses from [40] for comparison. The dotted line in (**d**,**e**) show the ERP responses from similar traditional mismatch negativity (MMN) experiments to our work. The thin and thick lines in (**d**,**e**) show the MMN responses for specially designed experiments from [40].

Furthermore, the similarity between red and blue lines in all the plots in Figure 6a–c shows a high correlation between in-ear and T8 EEG signals. The correlation between T8 and in-ear EEG was approximately 0.8530 across all trials. These MMN results indicated that the signal measured by in-ear device was EEG, as its ERP response characteristics conformed to those of scalp EEGs. Additionally, in-ear EEG signal quality was similar to EEG measured at the nearby T8 scalp location.

The average frequency mismatch response compared to the standard tone is displayed in Figure 7. The red and blue lines showed similar patterns (signs of slopes) between T8, and in-ear EEG. This result supports the findings of [16,36], which reports a high correlation between in-ear, and T7 and T8 EEG signals. It was noted that different amplitudes exist for the red and blue lines, because the signals shown were averaged across all trials, rather than raw data comparison (as shown in Figure 7a–c).

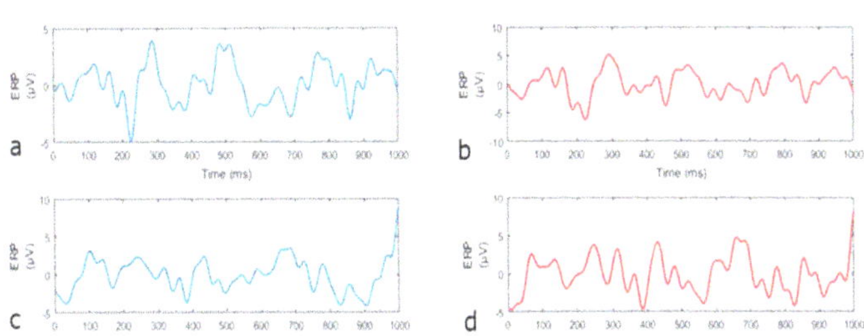

Figure 7. Average EEG of mismatch and standard trials. (**a**) Average in-ear EEG ERP responses from all mismatch trials. (**b**) Average T8 EEG from all mismatch trials. (**c**) Average in-ear EEG after standard beeps. (**d**) Average T8 EEG after standard beeps.

The MMN results show that in-ear EEG highly correlates with T7 and T8 EEG signals. Furthermore, similar signal response to the theory in [39] shows that in-ear EEG signal could be accurately used in a standard ERP test. Hence the validity of in-ear EEG signal was substantiated.

4.2. DEAP Data Analysis

The emotion classification using T7 and T8 signals from DEAP dataset by SVM, as described in Section 2.4, was performed. Data from 32 subjects consisting of 40 trials per each subject were used for the classification. Ten-folded cross-validation was applied to suppress biases. In each classification, 36 trials were used as the training set and the other four were used for the test set. Ten different sets were trained and tested for each subject.

The accuracy achieved was approximately 69.85 percent for valence classification and 78.7 percent for arousal classification. The overall accuracy for classifying four emotions was approximately 58.12 percent.

Furthermore, the analysis of emotion classification using the T7 or T8 channel was conducted and compared. The accuracies of emotion classification using T7 were approximately 71.30% for valence, 76.67% for arousal, and 57.56% for 4 emotions (valence and arousal combined); and the accuracy from emotion classification using T8 were approximately 70.93% for valence, 77.20% for arousal, and 57.34% for 4 emotions (valence and arousal combined) accordingly.

The t-test result from SPSS (IBM Corp., New York, USA) indicated that there was no statistically significant difference in classifying emotions between T7 and T8. The accuracy of T7 was approximately 57.56 ± 15.19 and T8 was 57.34 ± 16.40. The p-value was 0.955 on both tails, which was less than 0.955, indicating that there was no significance difference between classifying emotion using T7 and T8.

The results show that T7 and T8 data could be used as a single channel for valence, arousal, and the simple emotion classification, as the classification accuracy is comparable to the multichannel classification model in [7].

4.3. In-Ear EEG Emotion Classification

Only two out of thirteen subjects, subjects four and 10, decided to put an in-ear EEG on the left. The measurement of raw EEG data showed no statistically significant difference between EEG collected from left and right ear (p-value = 0.95).

In-ear EEG signals were recorded while subjects were watching stimulating pictures during experiment, described in Section 3.5. The EEG signal was filtered using a 4th order Butterworth filter to notch out power line noise at 50 Hz. The signal was then separated into four frequency bands that were theta (4–8 Hz), alpha (8–12 Hz), beta (12–32 Hz), and gamma (30–48 Hz) by Butterworth bandpass filters. Six statistical parameters by Picard et al. [48] were used for signal feature extraction on a 3 s time-lapsed window. The SVM model described in Section 3.4 was used for classification. Ten-fold cross-validation was applied for classifying each subject's data. All the signal processing and classification was performed offline using Matlab (The MathWorks, Inc., Natick, MA, USA)

Binary classification was done by SVM on valence (positive or negative) and arousal (high or low). The four emotion classification was performed using the valence and arousal classification results, mapped onto the simplified valence–arousal emotional model in Figure 5. For example, positive valence and high arousal was classified as happy. Hence the simplified emotions could be classified into four groups: positive valence/high arousal, positive valence/low arousal, negative valence/high arousal, or negative valence/low arousal. Classification accuracy was calculated by comparing SVM classifications with subjects' own evaluations. The classification accuracy of in-ear EEG is shown in Table 1.

Table 1. Emotion classification result from each subject.

Subject	Valence	Arousal	4 Emotions
1	75.00%	69.64%	55.36%
2	89.58%	58.33%	47.92%
3	56.82%	77.27%	43.18%
4	75.00%	85.71%	71.43%
5	75.00%	59.37%	46.87%
6	86.54%	88.46%	76.92%
7	61.76%	70.59%	45.59%
8	86.11%	86.11%	72.22%
9	69.44%	91.67%	66.67%
10	38.64%	43.18%	22.73%
11	57.50%	62.50%	37.50%
12	75.00%	77.27%	54.54%
13	77.50%	77.50%	57.50%
Average	71.07%	72.89%	54.89%

The emotion classification accuracy based on the valence–arousal emotion model was approximately 73.01% for valence, 75.70% arousal, and 59.23% for all four emotions. Subjects four and 10 inserted the in-ear EEG on the left while the rest inserted it on the right. Subject 12 was female.

The accuracy of emotion classification using the in-ear EEG from our experiment, and the T7 and T8 EEG signals from the DEAP dataset were comparable. According to multiple comparison using Bonferroni test, there was no statistical significance difference between emotion classification using T7, T8, or in-ear EEG. The two-tailed p-values were 0.449 and 0.456, which was over the 0.05 threshold, indicating no significant classifying emotion using in-ear and T7/T8. The box-plot of the classification results are shown in Figures 8–10.

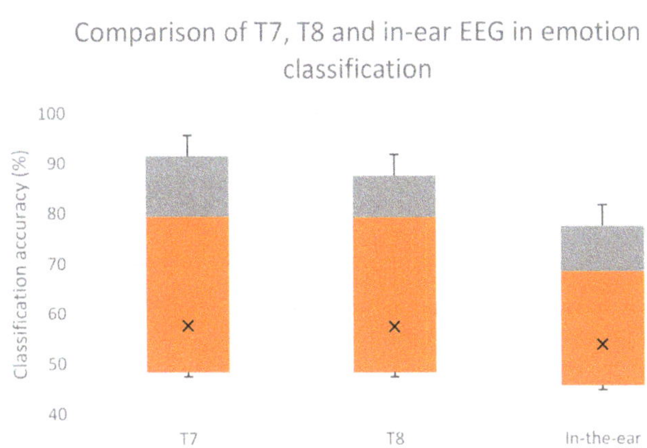

Figure 8. Box plot comparison among emotion classification using single channel T7, T8, and in-ear EEGs. Grey areas indicate proportions of classification accuracy above the median. Orange areas indicate proportions of classification accuracy below the median. X indicates the mean accuracy.

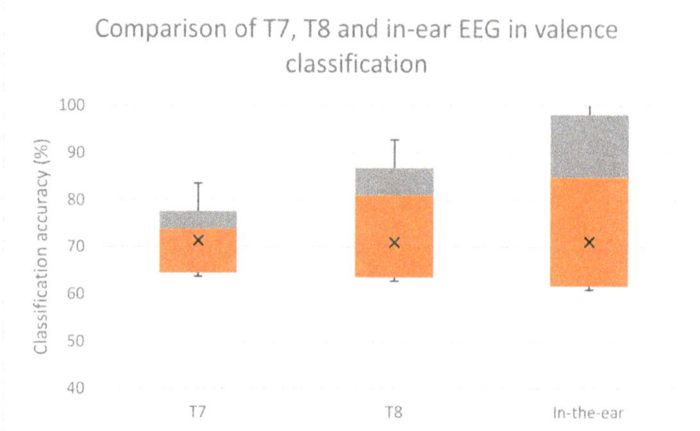

Figure 9. Box plot comparison among valence classifications using single channel T7, T8, and in-ear EEGs. Grey areas indicate proportions of classification accuracy above the median. Orange areas indicate proportion of classification accuracy below the median. X indicates the mean accuracy.

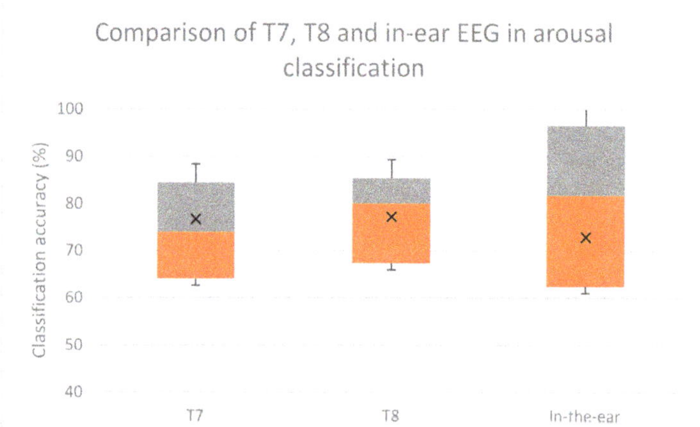

Figure 10. Box plot comparison among arousal classification using single channel T7, T8, and in-ear EEG. Grey areas indicate proportions of classification accuracy above the median. Orange areas indicate proportions of classification accuracy below the medians. X indicates the mean accuracy.

Overall four emotion classification accuracies were approximately 53.72% for in-ear EEG and 58.12% for T7 T8 EEG. Valence classification accuracies were 71.07% and 69.85% for in-ear and T7 T8 EEG, respectively. Arousal classification accuracies were 72.89% and 78.7% for in-ear and T7 T8 EEG, respectively. These comparable accuracies indicate that in-ear EEG has potential for emotion classification as T7 and T8 electrodes do.

5. Discussion

From the MMN results, in-ear EEG signal was verified to be highly correlated to the nearby T7 and T8 scalp EEG signals (correlation between T8 and in-ear EEG was approximately 0.853). This was expected as the 10–20 system scalp positions of T7 and T8 are just above left and right ears, respectively. They are in close proximity to ear canals. The results also correspond to the finding in previous work [17,37].

DEAP data analysis results show that using single electrode at T7 or T8 could achieve valence and arousal classification accuracies above 70 percent. This is comparable to classification accuracies obtained from using multiple EEG electrodes [7]. The results suggest that T7 and T8 could achieve a satisfactory emotion classification level.

The results from 4.3 show that emotion classification accuracy from in-ear EEG was comparable to that of T7 and T8 (71.07% and 69.85% for valence, and 72.89% and 78.7% for arousal). The four emotion classification and arousal accuracies of in-ear EEG were slightly lower than those of T7 and T8 (53.72% and 58.12%). The valence classification accuracy was almost equal.

Furthermore, the differences in accuracies in emotion, valence and arousal classifications between the in-ear EEG, and T7 and T8, are not statistically significant (p-values = 0.74, 0.99, and 0.65, respectively). Hence, an in-ear EEG is considered comparable to T7 and T8 in emotion classification.

From the above findings, in-ear EEG was found to be highly correlated to T7 and T8. Their emotion classification results are also compatible. Hence, in-ear EEG could be considered as an alternative to scalp EEG in positions close to the ears.

In terms of wearability, in-ear EEG could be set up within five minutes and could be put on by the users themselves. During experiments most subjects did not complain of being uncomfortable or being disturbed during usage. It is also unaffected by sweat, which makes it suitable for long term monitoring in a warm climate.

The additional benefits of the in-ear EEG are also in its compatibility and familiarity to users. Earplugs, earphones, and wireless handsfree earpieces have been around for many years and people are used to them. Wearing an earpiece is considered normal, so an in-ear EEG could allow the user's acceptance much easier than conventional scalp EEG headsets. Another benefit of using in-ear EEG is the signal obtained has less artifacts from electrode movement compared to conventional scalp EEG. Scalp EEG headsets are susceptible to artifacts from the user's movement, because contacts between the scalp and electrodes could easily become loose. With an in-ear EEG that fits tightly in the ear canal, body movement causes significantly less artifacts caused by loose contact between electrode and skin [30].

Compared to conventional scalp versions, the in-ear EEG is only a single channel device, with a similar signal to T7 and T8 scalp position near the ears. That limits in-ear EEG usage. Some EEG applications are not viable, such as for attention monitoring to measure the EEG from the frontal lobe [51]. Though this has never been investigated, in-ear EEG is not expected to achieve good accuracy in attention monitoring.

A higher number of EEG channels could achieve higher accuracy in emotion classification [7], so it is a valid point to consider adding channels to the in-ear EEG. This could be done by adding more electrodes to the same earbud or wearing two in-ear EEGs on both ears. The former approach was developed in [17] with the use of a custom made earmold which is similar to the one used in a hearing aid. However, earmolds are much more costly than the generic earbuds used in this work, so additional signals would be gained at much higher costs. Furthermore, due to limited space in an ear canal, two electrodes placed there would be close together, hence similar signals are expected to be measured. The latter approach of wearing two in-ear EEGs on both ears is an alternative. It is probable that emotion classification accuracy would improve. The trade-off here is practicality for long term usage. A user who wears in-ear EEG on both ears will not be able to hear well, since both ear canals are blocked. Earbud redesign is needed to provide a gap in the middle to let sound through the ear canal.

Despite its potential, the in-ear EEG monitoring device would need to be further developed to be more practical. An additional feature required is wireless connectivity, possibly via Bluetooth. This would make it more convenient to use without cumbersome wires. However, the challenge is in the integrated circuit design, which needs to be able to fit into an ear canal. This point was also raised in [31].

6. Conclusions

An in-ear EEG device was developed. Earphone rubber was used as the in-ear EEG device main body. Silver-adhesive fabric was used as an in-ear EEG electrode. The in-ear EEG signals were verified to be close to T7 and T8 on MMN ERP responses, with a correlation of approximately 0.8530. The emotion classification results were approximately 71.07% for valence, 72.89% for arousal, and 53.72% for four emotions, compared to those of the DEAP emotion classification results using T7 and T8, which were about 69.85 % for valence, and 78.7 % for arousal, while the accuracy for classifying four simplified emotions was about 58.12%. Classification accuracies between in-ear EEG, and T7 and T8 electrodes, are not statistically significant. These results together with its earphone-like wearability, suggest its potential for novel healthcare applications, such as home-based or tele-monitoring systems.

Author Contributions: Conceptualization, P.I. and S.P.; methodology, S.P. and C.A.; validation, S.P. and C.A.; resources, P.I.; data curation, C.A.; writing—original draft preparation, S.P. and C.A.; writing—reviewing and editing, P.I. and S.P.; supervision, P.I. and S.P.

Funding: This study was funded by Department of Computer Engineering, Chulalongkorn University, via a Graduate Scholarship for Alumni.

Conflicts of Interest: The Authors declare that there is no conflict of interest.

Abbreviations

EEG	Electroencephalogram
MMN	Mismatch negativity
IAPS	International Affective Picture System
GAPED	Geneva Affective Picture Database
DEAP	Dataset for Emotion Analysis using Physiological Signals
SVM	Support Vector Machine

References

1. Zheng, W.L.; Zhu, J.Y.; Lu, B.L. Identifying Stable Patterns Over Time for Emotion Recognition from EEG. *IEEE Trans. Affect. Comput.* **2017**, *10*, 417–429. [CrossRef]
2. Picard, R.W. *Affective Computing*; MIT Press: Cambridge, MA, USA, 2000.
3. Ekman, P.; Friesen, W. Measuring facial movement with the facial action coding system. In *Emotion in the Human Face*, 2nd ed.; Cambridge University Press: New York, NY, USA, 1982.
4. Russell, J.A. A circumplex model of affect. *J. Personal. Soc. Psychol.* **1980**, *39*, 1161–1178. [CrossRef]
5. Horlings, R. Emotion Recognition Using Brain Activity. Ph.D. Thesis, Department of Mediamatics, Delft University of Technology, Delft, The Netherlands, 2008.
6. Blinowska, K.; Durka, P. Electroencephalography (EEG). *Wiley Encycl. Biomed. Eng.* **2006**.
7. Jatupaiboon, N.; Pan-Ngum, S.; Israsena, P. Real-Time EEG-Based Happiness Detection System. *Sci. World J.* **2013**, *2013*. [CrossRef] [PubMed]
8. Li, M.; Lu, B.L. Emotion classification based on gammaband EEG. In Proceedings of the Annual International Conference of the IEEE Engineering in Medicine and Biology Society (EMBC'09), Minneapolis, MN, USA, 3–6 September 2009; pp. 1223–1226.
9. Chanel, G.; Kronegg, J.; Grandjean, D.; Pun, T. Emotion assessment: Arousal evaluation using EEG's and peripheral physiological signals. In *Multimedia Content Representation, Classification and Security*; Gunsel, B., Jain, A., Tekalp, A.M., Sankur, B., Eds.; Springer: Berlin, Germany, 2006; Volume 4105, pp. 530–537.
10. Takahashi, K. Remarks on Emotion Recognition from Bio-Potential. In Proceedings of the 2nd International Conference on Autonomous Robots and Agents, Palmerston North, New Zealand, 13–15 December 2004; pp. 186–191.
11. Oude Bos, D. EEG-Based Emotion Recognition—The influence of visual and auditory Stimuli. Available online: https://www.semanticscholar.org/paper/EEG-based-Emotion-Recognition-The-Influence-of-and-Bos/5097b37a30b8d7a8d2bb03b307be5bf5deab73c4 (accessed on 17 September 2019).

12. Petrantonakis, P.C.; Hadjileontiadis, L.J. A Novel Emotion Elicitation Index Using Frontal Brain Asymmetry for Enhanced EEG-Based Emotion Recognition. *IEEE Trans. Inf. Technol. Biomed.* **2011**, *15*, 737–746. [CrossRef] [PubMed]
13. Chung, S.C.; Yang, H.K. A Real-Time Emotionality Assessment (RTEA) System Based on Psycho-Physiological Evaluation. *Int. J. Neurosci.* **2008**, *118*, 967–980. [CrossRef] [PubMed]
14. Heraz, A.; Frasson, C. Predicting the Three Major Dimensions of the Learner's Emotions from Brainwaves. *Int. J. Comput. Sci.* **2007**, *2*, 187–193.
15. Wei, Y.; Wu, Y.; Tudor, J. A Real-Time Wearable Emotion Detection Headband Based on EEG measurement. *Sens. Actuators A Phys.* **2017**, *263*, 614–621. [CrossRef]
16. Zhang, Q.; Wang, P.; Liu, Y.; Peng, B.; Zhou, Y.; Zhou, Z.; Tong, B.; Qiu, B.; Zheng, Y.; Dai, Y. A Real-Time Wireless Wearable Electroencephalography System Based on Support Vector Machine for Encephalopathy Daily Monitoring. *Int. J. Distrib. Sens. Netw.* **2018**, *14*, 1550147718779562. [CrossRef]
17. Looney, D.; Kidmose, P.; Park, C.; Ungstrup, M.; Rank, M.L.; Rosenkranz, K.; Mandic, D.P. The In-ear Recording Concept: User-Centered and Wearable Brain Monitoring. *IEEE Pulse* **2012**, *3*, 32–42. [CrossRef]
18. Sharbrough, F.; Chatrian, G.E.; Lesser, R.P.; Luders, H.; Nuwer, M.; Picton, T.W. American electroencephalographic society guidelines for standard electrode position nomenclature. *J. Clin. Neurophysiol.* **1991**, *8*, 200–202.
19. Koelstra, S.; Yazdani, A.; Soleymani, M.; Mühl, C.; Lee, J.S.; Nijholt, A.; Pun, T.; Ebrahimi, T.; Patras, I. Single trial classification of EEG and peripheral physiological signals for recognition of emotions induced by music videos. In Proceedings of the International Conference on Brain Informatics, Toronto, ON, Canada, 28–30 August 2010.
20. Huang, D.; Guan, C.; Ang, K.K.; Zhang, H.; Pan, Y. Asymmetric spatial pattern for EEG-based emotion detection. In Proceedings of the International Joint Conference on Neural Networks (IJCNN '12), Brisbane, Australia, 10–15 June 2012; pp. 1–7.
21. Nie, D.; Wang, X.W.; Shi, L.C.; Lu, B.L. EEG-based emotion recognition during watching movies. In Proceedings of the 5th International IEEE/EMBS Conference on Neural Engineering (NER '11), Cancun, Mexico, 27 April–1 May 2011; pp. 667–670.
22. Chanel, G.; Kierkels, J.J.M.; Soleymani, M.; Pun, T. Short-term emotion assessment in a recall paradigm. *Int. J. Hum. Comput. Stud.* **2009**, *67*, 607–627. [CrossRef]
23. Wang, X.W.; Nie, D.; Lu, B.L. EEG-based emotion recognition using frequency domain features and support vector machines. In *Neural Information Processing*; Lu, B.L., Zhang, L., Kwok, J., Eds.; Springer: Berlin, Germany, 2011; Volume 7062, pp. 734–743.
24. Chanel, G.; Rebetez, C.; Bétrancourt, M.; Pun, T. Emotion assessment from physiological signals for adaptation of game difficulty. *IEEE Trans. Syst. Man Cybern. A Syst. Hum.* **2011**, *41*, 1052–1063. [CrossRef]
25. AlZoubi, O.; Calvo, R.A.; Stevens, R.H. Classification of EEG for affect recognition: An adaptive approach. In *AI 2009: Advances in Artificial Intelligence*; Nicholson, A., Li, X., Eds.; Springer: Berlin, Germany, 2009; Volume 5866, pp. 52–61.
26. Wijeratne, U.; Perera, U. Intelligent emotion recognition system using electroencephalography and active shape models. In Proceedings of the 2nd IEEE-EMBS Conference on Biomedical Engineering and Sciences (IECBES '12), Langkawi, Malaysia, 17–19 December 2012; pp. 636–641.
27. Kulkarni, A.; Rao, P.; Natarajan, S.; Goldman, A.; Sabbisetti, V.S.; Khater, Y.; Korimerla, N.; Chandrasekar, V.; Mashelkar, R.A.; Sengupta, S. Soft, curved electrode systems capable of integration on the auricle as a persistent brain-computer interface. *Proc. Natl. Acad. Sci. USA* **2015**, *112*, 3920–3925.
28. Casson, A.J.; Yates, D.C.; Smith, S.J.M.; Duncan, J.S.; Rodriguez-Villegas, E. Wearable Electroencephalography. *IEEE Eng. Med. Biol. Mag.* **2010**, *29*, 44–56. [CrossRef] [PubMed]
29. Looney, D.; Kidmose, P.; Mandic, D. *Ear-EEG: User-Centered and Wearable BCI*; Springer: Berlin/Heidelberg, Germany, 2014; Volume 6, pp. 41–50.
30. Goverdovsky, V.; Looney, D.; Kidmose, P.; Mandic, D.P. In-Ear EEG From Viscoelastic Generic Earpieces: Robust and Unobtrusive 24/7 Monitoring. *IEEE Sens. J.* **2016**, *16*, 271–277. [CrossRef]
31. Kappel, S.; Rank, M.; Toft, H.; Andersen, M.; Kidmose, P. Dry-Contact Electrode Ear-EEG. *IEEE Trans. Biomed. Eng.* **2018**, *66*, 150–158. [CrossRef] [PubMed]

32. Matthies, D.J.C. InEar BioFeedController: A headset for hands-free and eyes-free interaction with mobile devices. In *CHI'13 Extended Abstracts on Human Factors in Computing Systems*; ACM: New York, NY, USA, 2013; pp. 1293–1298.
33. Matthies, D.J.C.; Strecker, B.A.; Urban, B. Earfieldsensing: A novel in-ear electric field sensing to enrich wearable gesture input through facial expressions. In Proceedings of the 2017 CHI Conference on Human Factors in Computing Systems, Denver, CO, USA, 6—11 May 2017; pp. 1911–1922.
34. Poh, M.; Kim, K.; Goessling, A.; Swenson, N.; Picard, R. Heartphones: Sensor Earphones and Mobile Application for Non-obtrusive Health Monitoring. In Proceedings of the International Symposium on Wearable Computers (ISWC), Linz, Austria, 4–7 September 2009; pp. 153–154.
35. Nguyen, A.; Alqurashi, R.; Raghebi, Z.; Banaei-Kashani, F.; Halbower, A.C.; Vu, T. A lightweight and inexpensive in-ear sensing system for automatic whole-night sleep stage monitoring. In Proceedings of the 14th ACM Conference on Embedded Network Sensor Systems CD-ROM, Stanford, CA, USA, 14—16 November 2016; pp. 230–244.
36. Mikkelsen, K.B.; Villadsen, D.B.; Otto, M.; Kidmose, P. Automatic sleep staging using ear-EEG. *Biomed. Eng. Online* **2017**, *16*, 111. [CrossRef]
37. Kidmose, P.; Looney, D.; Jochumsen, L.; Mandic, D.P. Ear-EEG from generic earpieces: A feasibility study. In Proceedings of the 35th Annual International Conference of the IEEE EMBS, Osaka, Japan, 3—7 July 2013.
38. OpenBCI. Available online: http://openbci.com (accessed on 15 June 2019).
39. Näätänen, R.; Gaillard, A.W.K.; Mäntysalo, S. Early selective-attention effect on evoked potential reinterpreted. *Acta Psychol.* **1978**, *42*, 313–329. [CrossRef]
40. Näätänen, R.; Pakarinen, S.; Rinne, T.; Takegata, R. The mismatch Negativity (MMN): Towards the Optimal Paradigm. *Clin. Neurophysiol.* **2004**, *115*, 140–144. [CrossRef]
41. Mikkelsen, K.B.; Kappel, S.L.; Mandic, D.P.; Kidmose, P. EEG Recorded from the Ear: Characterizing the Ear-EEG Method. *Front. Neurosci.* **2015**, *9*, 438. [CrossRef] [PubMed]
42. Vuust, P.; Brattico, E.; Glerean, E.; Seppänen, M.; Pakarinen, S.; Tervaniemi, M.; Näätänen, R. New fast mismatch negativity paradigm for determining the neural prerequisites for musical ability. *Cortex* **2011**, *47*, 1091–1098. [CrossRef] [PubMed]
43. BIOPAC Systems Inc. Ground vs. Reference for EEG Recording. Available online: https://www.biopac.com/knowledge-base/ground-vs-reference-for-eeg-recording (accessed on 27 March 2019).
44. Lang, P.J.; Bradley, M.M.; Cuthbert, B.N. *International Affective Picture System (IAPS): Affective Ratings of Pictures and Instruction Manual*; Technical Report A-6; University of Florida: Gainesville, FL, USA, 2005.
45. Dan-Glauser, E.S.; Scherer, K.R. The Geneva Affective Picture Database (GAPED): A new 730-Picture Database Focusing on Valence and Normative Significance. *Behav Res Methods* **2011**, *43*, 468–477. [CrossRef] [PubMed]
46. Vempala, N.N.; Russo, F.A. Predicting Emotion from Music Audio Features Using Neural Networks. In Proceedings of the 9th International Symposium on Computer Music Modelling and Retrieval (CMMR 2012), Queen Mary University of London, London, UK, 19–22 June 2012.
47. Koelstra, S.; Muhl, C.; Soleymani, M.; Lee, J.S.; Yazdani, A.; Ebrahimi, T.; Pun, T.; Nijholt, A.; Patras, I. DEAP: A Database for Emotion Analysis Using Physiological Signals. *IEEE Trans. Affect. Comput.* **2012**, *3*, 18–31. [CrossRef]
48. Picard, R.W.; Vyzas, E.; Healey, J. Toward Machine Emotional Intelligence: Analysis of Affective Physiological State. *IEEE Trans. Pattern Anal. Mach. Intell.* **2001**, *23*, 1175–1191. [CrossRef]
49. Stone, M. Cross-Validatory Choice and Assessment of Statistical Predictions. *J. Royal Stat. Soc.* **1974**, *36*, 111–147. [CrossRef]
50. University of California—Los Angeles. Left and Right Ears Not Created Equal as Newborns Process Sound, Finds UCLA/UA Research. Available online: www.sciencedaily.com/releases/2004/09/040910082553.htm (accessed on 20 January 2019).
51. Sun, J.; Yeh, K. The effects of attention monitoring with EEG biofeedback on university students' attention and self-efficacy: The case of anti-phishing instructional materials. *Comput. Educ.* **2017**, *106*, 73–82. [CrossRef]

sensors

Article

Deep ECG-Respiration Network (DeepER Net) for Recognizing Mental Stress

Wonju Seo [1,†], Namho Kim [1,†], Sehyeon Kim [1], Chanhee Lee [2] and Sung-Min Park [1,*]

1 Department of Creative IT Engineering, Pohang University of Science and Technology (POSTECH), Pohang 37673, Korea
2 Research Center of ONESOFTDIGM, Pohang 37673, Korea
* Correspondence: sungminpark@postech.ac.kr
† These authors contributed equally to this work.

Received: 16 June 2019; Accepted: 7 July 2019; Published: 9 July 2019

Abstract: Unmanaged long-term mental stress in the workplace can lead to serious health problems and reduced productivity. To prevent this, it is important to recognize and relieve mental stress in a timely manner. Here, we propose a novel stress detection algorithm based on end-to-end deep learning using multiple physiological signals, such as electrocardiogram (ECG) and respiration (RESP) signal. To mimic workplace stress in our experiments, we used Stroop and math tasks as stressors, with each stressor being followed by a relaxation task. Herein, we recruited 18 subjects and measured both ECG and RESP signals using Zephyr BioHarness 3.0. After five-fold cross validation, the proposed network performed well, with an average accuracy of 83.9%, an average F1 score of 0.81, and an average area under the receiver operating characteristic (ROC) curve (AUC) of 0.92, demonstrating its superiority over conventional machine learning models. Furthermore, by visualizing the activation of the trained network's neurons, we found that they were activated by specific ECG and RESP patterns. In conclusion, we successfully validated the feasibility of end-to-end deep learning using multiple physiological signals for recognition of mental stress in the workplace. We believe that this is a promising approach that will help to improve the quality of life of people suffering from long-term work-related mental stress.

Keywords: mental stress detection; electrocardiogram; respiration; machine learning; deep learning

1. Introduction

Mental health is being recognized as an important issue in the workplace [1]. If mental stress is not treated in a timely manner (i.e., left unmanaged), employees can experience serious physical problems, such as heart disorders, diabetes, cancer, and stomachaches [2,3]. Stress also causes mental disorders such as depression and anger, and can even lead to suicide [2,4]. Such problems can seriously reduce productivity owing to absences and work disability [1], with the medical and socioeconomic costs in the United States adding up to $300 billion annually [5]. Detecting and relieving stress in a timely manner could thus improve overall healthcare substantially.

Stress is typically evaluated using a stress indicator questionnaire, where individuals answer questions such as the perceived stress scale (PSS) [6] and sleep quality assessment (PSQI) [7], and healthcare professionals evaluate the stress score based on those answers. Because these methods rely on expert evaluations, they are not suitable for continuously monitoring stress in the workplace. This limitation makes it difficult or impossible to recognize stress rapidly and intervene appropriately to help people suffering from it. Consequently, there is a growing need for ways to continuously and objectively monitor stress.

The autonomic nervous system comprises the sympathetic nervous system (SNS) and the parasympathetic nervous system (PNS). When an individual is mentally stressed, the PNS activity decreases and the SNS starts to dominate. These neurological changes lead to physiological changes in heart rate (HR), skin conductance, respiration (RESP), and pupil diameter [2] that can be accurately measured by conventional biomedical instruments. Unfortunately, conventional instruments for measuring physiological signals are not optimal for continuous use in the workplace owing to their bulky size and associated cables. However, the recent advancement of wearable technology has made it practical to continuously measure various physiological signals with minimal disturbances, leading to increased research interest in continuous stress detection based on physiological signals.

Similar to the importance of the developments in monitoring devices, developing algorithms to analyze the collected data and accurately recognize the occurrences of stress is also crucial. Several machine learning models have been proposed to recognize stress based on multiple physiological signals [8–12]. Although these models have demonstrated the feasibility of recognizing stress, they have one serious limitation, namely that machine learning approaches require us to extract well-defined, handcrafted features and find the best way to combine them, both very challenging tasks [13]. Furthermore, because the dependence of such approaches on handcrafted features means they cannot find new stress-related features, it can limit their maximum generalization performance. Overcoming this limitation will require a breakthrough.

Recently, deep learning approaches have made great strides in image processing and natural language processing [14]. This is because they not only automatically extract features from data, but also learn new high-level features based on low-level ones owing to their hierarchical structure, something that simple machine learning models cannot do. In particular, convolutional neural networks (CNNs) and long short-term memories (LSTMs) have led to great successes in numerous fields. Owing to these advantages, attempts have also been made to use this approach to recognize stress [5,12,15]. However, these have only considered one type of physiological signal. Because a single signal cannot capture all possible responses to stress, this may degrade their generalization performance. Conversely, the performance degradation can be solved as well as the diversity of individual physiological characteristics be considered using multiple physiological signals. It is thus essential to study the validity and feasibility of deep learning approaches based on multiple physiological signals.

Our goal in this study is to propose an end-to-end deep neural network based on combining two types of physiological signals, namely electrocardiogram (ECG) and RESP data, which have been proposed as meaningful stress-related signals [10]. In addition, we compare the proposed network with conventional machine learning models and visualize the results to see the activation patterns produced by the ECG and RESP signals.

The remainder of this paper is organized as follows. First, in Section 2, we review the literature on both machine learning approaches using multiple physiological signals and deep learning approaches using one type of physiological signal. Then, our experiment's protocol, a machine learning approach, and a procedure developing the networks will be covered in Section 3. In Section 4, we provide statistical results, evaluate our proposed network, and compare it with the benchmark machine learning models. Finally, in Section 5, we visualize the activation patterns in our network and compare our study with previous ones that have proposed deep learning approaches. Then, we discuss the use of multiple datasets and conclude the paper by discussing potential limitations and future work.

2. Related Works

2.1. Machine Learning Approaches

Numerous studies have proposed machine learning approaches for recognizing mental stress based on various types of physiological signals [8,9,11,16]. Of these signals, ECGs and photoplethysmograms (PPGs) have been used to extract handcrafted features related to heart activity, such as the HR and HR variability (HRV). In addition to these, other physiological signals have been investigated, such as RESP, electrodermal activity (EDA), galvanic skin response (GSR), pupil diameter, acceleration, electroencephalograms, electromyograms (EMGs), and electrooculograms. Then, with collected physiological signals, developing such machine learning models requires the following main steps: (1) preprocess and de-noise the data with a digital noise filter; (2) extract well-defined features from the multiple physiological signals and find the best feature set; (3) use these features to train a machine learning model; and (4) evaluate the model on a test dataset.

Sirampraka et al. [8] proposed a stress evaluation model using multiple physiological signals such as ECGs and GSR. In this study, a simulated workplace's stress was considered to replicate workplace stress and signal data were collected. Then, a support vector machine (SVM) was trained and evaluated with either well-defined features or combinations of features. The model was able to recognize stress with greater than 90% accuracy, leading the authors to conclude that HR, HRV, and GSR features in the time and frequency domains were sufficient to accurately detect stress.

In addition to workplace stress, recognizing stress during driving has also been studied. Here, stress is considered to be a risk factor as it can cause aggressive driving behavior and reduced concentration [16]. In [16], the authors developed two main machine learning models, namely an SVM and a K-nearest neighbors (KNN) approach, to identify three distinct stress levels (low, medium, and high). Using Stress Recognition in Automobile Drivers dataset (DRIVERDB) [10] in PHYSIONET, they collected multiple physiological signals including foot GSR, hand GSR, EMGs, ECGs, and RESP, then extracted well-defined features. By finding the feature set that minimized the error rates, the SVM achieved 99% accuracy with a 5-min time window. Their analysis found that selecting the right model, preprocessing steps, and feature set all helped to maximize its generalization performance.

Betti et al. [11] proposed a wearable physiological sensor system for monitoring stress. They conducted Maastricht Acute Stress Tests and collected multiple physiological signals, including ECGs, EDA, and EEGs. After training, the proposed SVM achieved 86.0% accuracy and found correlations between the handcrafted features and the measured cortisol level, which is regarded as a biomarker of stress. By finding these correlations, the study validated the feasibility of monitoring stress with the proposed wearable sensor system.

2.2. Deep Learning Approaches

Although deep learning approaches are heavily used in the image processing and natural language processing fields, and a few studies have used them to detect or recognize stress [5,12,15], no study has yet applied this approach to analyzing multiple signals. Researchers have developed deep neural networks using the following main steps: (1) process the physiological signals with a digital noise filter; (2) design a unique deep neural network based on domain knowledge; (3) train the network; and (4) evaluate it on a test dataset.

Cho et al. [15] proposed a promising approach to recognizing stress with a cheap thermal imaging camera. The collected thermal images of people breathing were preprocessed to create spectrum sequences, and then a CNN was used to extract features from these. To increase the number of data points, a sliding window method was used to augment the data. The proposed CNN achieved greater than 80% accuracy on average for classifying the images as stressed or unstressed. Their main contribution is that they were the first to use spectrum sequences taken from thermal images as input.

Hwang et al. [12] presented the Deep ECG Net for recognizing stress based on short-term ECGs (10 s). The authors proposed a 1D CNN with optimized filter size and pooling length that used domain knowledge of ECG PQRST waveforms. The proposed model achieved better performance than conventional machine learning models. Visualizing the process showed that it detected spiky patterns around ECG P waves, meaning that it was able to automatically extract ECG waveform characteristics. Their network achieved about 80% accuracy on average in classifying the data as stressed or unstressed for their two experiments. Their main contribution is showing that optimized networks based on domain knowledge can provide better performance than conventional machine learning approaches and deep neural networks that are designed without domain knowledge.

He et al. [5] developed a ten-layer CNN to detect acute cognitive stress based on short-term ECGs. Here, spectrum information was used to extract consecutive ECG R-peaks for use as input instead of raw ECG data. This study also used a sliding window method to increase the number of data points. Their results showed that the proposed CNN achieved a lower error rate than conventional machine learning models. In addition, the authors found that the meaningful information related to stress lay in the 0.4–20 Hz range by visualizing the activation maps of multiple layers. This demonstrates that deep learning approaches can benefit from having data-driven features that are not used by conventional machine learning approaches.

3. Methods

3.1. Subjects

We recruited 18 subjects (8 females and 10 males, aged 24.6 $\pm$ 4.6) from Pohang University of Science and Technology (POSTECH), South Korea, via open recruiting. The study subjects were selected based on the following criteria: (1) they had no cardiovascular disease or mental problems and (2) they had not undertaken intense exercise before the day of the test and (3) they had not had caffeinated beverages on the day of the test. This study was approved by the POSTECH Ethics Committee (PIRB-2019-E001).

The experiments were conducted in parallel with the recruiting process. Although most experiments proceeded normally, unexpected technical problems (namely a subject's carelessness and an unexpected Windows OS update) occurred during two experiments, meaning that these two subjects' data were not captured correctly. Thus, we only considered the datasets collected from the remaining 16 subjects.

3.2. Experiments

3.2.1. Protocol

Each experiment set took about 1 h, and comprised two stages: (1) an initial relaxation stage before the experiment began (about 15–20 min) and (2) the main experimental stage (about 45 min). These are described in more detail in Figure 1.

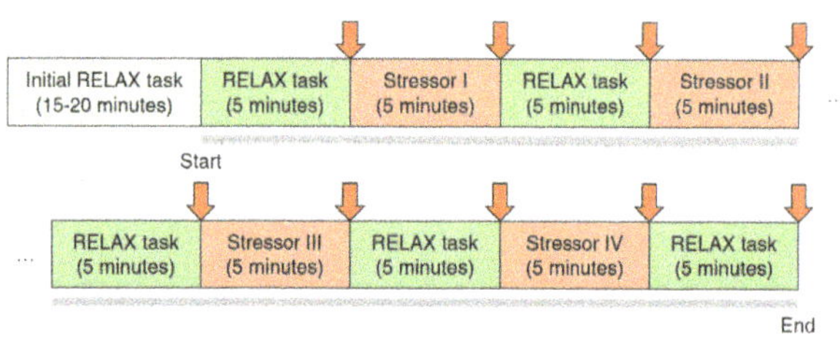

Figure 1. An experiment process. There are two stages for the whole experiment: an initial relaxation stage (colored with light green), and a regular experiment stage (indicated by gray bold lines). The regular experiment stage can be segmented into 5 min of relax tasks (colored with dark green) and stress tasks (colored with dark orange). From the start to end of the regular experiment stage, subjects' physiological signals were captured by a wearable device. At the end of each short a relax or a stress task, a mental stress level assessment was carried (indicated with an orange arrow).

During the initial relaxation stage, each subject was asked to wear a wearable device that collected ECG and RESP data and to completely relax, eliminating any excitement or nervousness regarding the experiment. In addition, we explained our protocol in detail to prevent any mistakes by the subject. During the main experimental stage, the subjects alternately experienced simulated relaxing states (called RELAX tasks in Figure 1) and stressful states. During the relaxation tasks, the subject was asked to sit on a chair in a comfortable position without any mental activity. The first relaxation task aims to build a psychological baseline and remove unwanted excitement before the regular experiment stage. Similarly, the other relaxation tasks aim to remove stress after a stressful task and prepare the next task by setting the psychological baseline. This design improves the reliability of the experiment's results [5]. During the stressful tasks, the subject was provided with one of two types of stressors: (1) a math task, namely a quiz requiring the subject to solve a series of subtraction problems via mental arithmetic, or (2) a Stroop task, namely a quiz where the subject was asked to respond with the color of a given word and ignore its meaning. Because all the subjects were Korean, the words were presented in the Korean language. These are typical tasks that have been used to induce stress in previous studies [12,15,17].

The tasks also varied in difficulty, based on the results of a previous study by Cho et al. [15]. For example, an easy math task might be to repeatedly subtract 1 from a four-digit number, responding within 7.5 s, while a hard math task might involve repeatedly subtracting a two-digit number, rather than 1, from a four-digit number with the same time limit. Likewise, easy Stroop tasks involved words with the same color and meaning, while these were mismatched for hard Stroop tasks. In either case, the time limit for each Stroop problem was 1.5 s. Appropriate sound feedback was also provided to indicate whether or not the entered answer was correct, encouraging the subject to pay attention to the task and inducing additional stress.

The subjects were presented with these four stress-inducing tasks (two types with two difficulties) in random order. Although a previous study [15] used a fixed order, we chose not to do this, for two reasons: (1) a fixed order could bias the stress level, and (2) a random order would better replicate real stress-inducing situations. All of the relaxation and stressor tasks lasted for 5 min. At the end of each task, the subject was asked to evaluate how mentally stressed he or she felt, based on a visual analogue scale (VAS) [15], which is used as a evaluation method of an individual's subjective stress score with value from 0 (not at all) to 10 (extreme stress). For example, if a subject is relaxed or stressed, the VAS score will be close to zero or 10, respectively. In this study, the purpose of the use of VAS is to confirm the average effects (e.g., induce stress or relieve stress) of each task on the 16 subjects. During the experimental stage, the subjects were asked not to use their cellphones and to minimize unexpected mental stimuli. When the experimental stage was complete, the subject took off the wearable device to complete the experiment.

3.2.2. Experimental Setup

A BioHarness module 3.0 (Zephyr Technology, Annapolis, MD, USA) was used to collect the subjects' ECG and RESP data. This wearable device is compact and can be tightened with a strap, making it a good choice to minimize movement disturbance during the experiment. Because making the strap too tight could induce unnecessary stress or pain, we asked the subjects not to tighten it so much that it was uncomfortable to wear.

The experiments were conducted on a laptop computer (with a 2.8 GHz Intel Core i7 processor (Santa Clara, CA, USA) and 16 GB of RAM) in a closed room. During each experiment, the subject was alone in the room, as shown in Figure 2. On the laptop computer, a graphical user interface application was installed, which we developed with MATLAB R2016a (MathWorks, Natick, MA, USA). This was designed to be as simple as possible so as not to confuse the subjects.

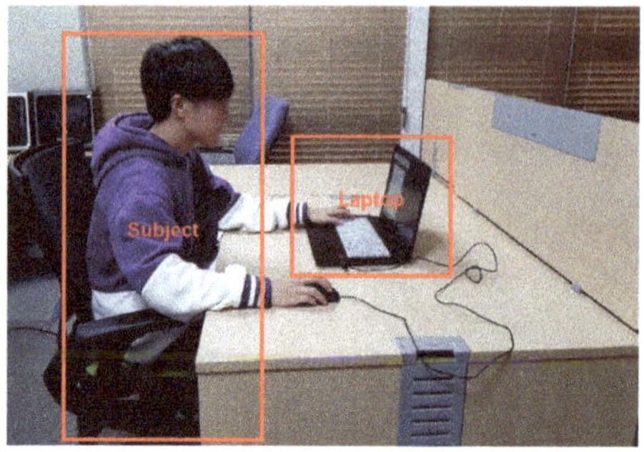

Figure 2. Setup of the experiment in a closed room. A subject proceeds with the experiment with a laptop computer. There was not only no one else except the subject, but also no camera not to make the subject nervous or embarrassed.

3.2.3. Data Preprocessing

After running the experiment a total of 16 times, we collected 16 datasets, consisting of ECG and RESP data, and stress level indices (VAS scores).

During the preprocessing step, the captured ECG signal was first filtered by a 2000th-order finite impulse response notch filter with 58–62 Hz bandwidth, and a second by a 3000th-order finite impulse response bandpass filter with 1.5–150 Hz bandwidth [12]. This de-noising process makes it easy to find the R-peaks of ECG. In contrast, during the preprocessing of the RESP signal, we did not filter this because it was captured from torso expansion and contraction, and thus any motion noise might not be independent of the subject's breathing.

We divided the segment for each task into six clips, consisting of 50-s windows with no overlap. We chose 50-s windows because at least 50 s of physiological data are required to extract several important features [17]. The ECG and RESP segments' start and end times were all clearly synchronized. Here, there was only one data point of overlap between one segment and the next. Then, we excluded the first clip from each segment owing to the initialization time needed for each task. After preprocessing, we obtained a total of 720 segments (16 subjects, each recording nine segments, with five clips per segment). Finally, we labeled each segment with its binary class (stressed or unstressed) according to the task type (relaxation or stressor).

3.3. *Machine Learning Approaches*

To compare our deep learning approach with conventional machine learning approaches, we also developed several machine learning models for use as benchmarks. Here, we selected ECG and RESP features that have been used in many previous studies [11,12,17–19].

We extracted 11 handcrafted features from the ECG data, including four time-domain features and seven frequency-domain features (Table 1). As time-domain features, we extracted the mean HR (HR mean), standard deviation of the Normal-to-Normal (NN) interval (sdNN), root mean square of successive difference of R peak-to-R peak (RR) intervals (rmssd), and percentage of the differences between adjacent RR intervals that were greater than 50 ms (pNN50). As frequency-domain features, we extracted the NN interval powers in the following ranges: 0.00–0.04 Hz (VLF), 0.04–0.15 Hz (LF), 0.15–0.40 Hz (HF), and 0.14–0.40 Hz (TF). In addition, we included the ratios of LF to LF+HF (nLF), HF to LF+HF (nHF), and LF to HF (LF2HF) as frequency-domain features.

We also extracted a total of eight handcrafted RESP features: three time-domain features and five frequency-domain features (Table 1). As time-domain features, we used the square root of the mean squared RESP (RMS), interquartile range (IQR), and mean difference between adjacent elements of each RESP segment (MDA). As frequency-domain features, we used the powers in the 0.00–1.00 Hz (LF1), 1.00–2.00 Hz (LF2), 2.00–3.00 Hz (HF1), and 3.00–4.00 Hz (HF2) ranges, as well as the LF1 + LF2 to HF1 + HF2 ratio (L2H). As with the ECG frequency-domain features, the RESP features were also computed using Welch's method of estimating the data's power spectral density.

Then, we developed several machine learning models that have previously been proposed to classify stress states [20]. While the models were being trained and evaluated, the features were normalized by using a MinMax scaler to bring them into the 0–1 range. To prevent data leakage during training, the scaler parameters were fitted using only the training set features, but used to normalize both the training and test set features. We tuned the models' hyper-parameters via grid search and calculated their average performance using five-fold cross validation.

Table 1. A list of features extracted from ECG and RESP. We computed the power spectral density of ECG's NN interval and RESP, using Welch's method, to extract frequency domain features. Abbreviations: ECG, electrocardiogram; RESP, respiration; NN, normal-to-normal; RR, R peak-to-R peak.

Signal	Domain	Features	Description
ECG	Time	HR mean	Mean of heartrate
		sdNN	Standard deviation of NN intervals
		rmssd	Root mean square of successive difference of RR intervals
		pNN50	Percentage of differences between adjacent RR intervals that are greater than 50 ms
ECG	Frequency	VLF	Power of NN interval (0.00–0.04 Hz)
		LF	Power of NN interval (0.04–0.15 Hz)
		HF	Power of NN interval (0.15–0.40 Hz)
		TF	Power of NN interval (0.14–0.40 Hz)
		nLF	LF to (LF + HF) ratio
		nHF	HF to (LF + HF) ratio
		LF2HF	LF to HF ratio
RESP	Time	RMS	Square root of mean of squared RESP
		IQR	Interquartile range of RESP
		MDA	Square root of mean of squared differences between adjacent elements
RESP	Frequency	LF1	Power of RESP (0.00–1.00 Hz)
		LF2	Power of RESP (1.00–2.00 Hz)
		HF1	Power of RESP (2.00–3.00 Hz)
		HF2	Power of RESP (3.00–4.00 Hz)
		L2H	(LF1+LF2) to (HF1 + HF2) ratio

3.4. Deep Learning Approaches

Unlike machine learning approaches, deep learning approaches are based on deep neural networks that can directly extract features from the data, and are not reliant on well-defined handcrafted features. As the name implies, deep neural networks are artificial neural networks with two or more hidden layers. Having many hidden layers enables such networks to learn more complex nonlinear patterns and hierarchical information than would be possible with shallow networks. Despite these advantages, however, deep neural networks also usually have a large number of parameters, which can lead to over-fitting, and they can experience issues with the gradient vanishing when they have a large number of layers. These problems can result in a failure to learn and an increase in generalization errors. To overcome these limitations, recent algorithmic advances (e.g., rectified linear units, batch normalization, dropout, stochastic gradient descent, and data augmentation), more powerful computational hardware (e.g., general-purpose graphical processor units), and innovative network architectures, such as CNNs and LSTMs, have partially resolved these over-fitting and gradient vanishing problems, enabling high performance to be achieved.

These developments have encouraged the use of deep learning approaches in numerous fields, including physiological signal analysis [21] and stress recognition [5,12,15].

We designed our proposed network based on Deep ECG Net's structure [12]. First, a batch-normalization layer is used to normalize each physiological signal, so that the network can learn to normalize the signals based on the data itself. Then, there is a 1D convolutional layer and a 1D max-pooling layer for each signal, which extract stress-related waveform patterns from the ECG and RESP data. Here, a rectified linear unit (ReLU) is used as the activation function. Next, comes another 1D convolutional layer. There is no additional max-pooling layer this time because the previous max-pooling process has greatly reduced the dimensionality. After that, there are multiple LSTM layers, in order to obtain sequential information about the features extracted from the previous convolutional layer. Next, we concatenate the extracted ECG and RESP features and add a dense layer. Finally, there is a fully-connected layer with a sigmoid activation function, which classifies the data as stressed or unstressed. To prevent over-fitting, we also add dropout and batch-normalization layers. Figure 3 shows the structure of the proposed DeepER (ECG–RESP) Net.

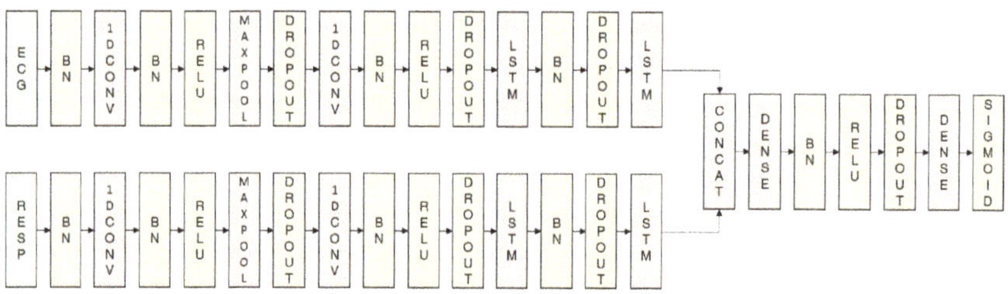

Figure 3. The structure of the proposed DeepER Net. The different signals were processed in each network branch and then concatenated for recognizing the stress. The basic structure is based on the structure of Deep ECG Net [12].

As noted by the developer of Deep ECG Net [12], both the first 1D convolutional layer's kernel length and 1D max-pooling layer's pooling length are important factors. They determined that a kernel length of 0.6 s (i.e., 600 points at a sampling frequency of 1 kHz) and a pooling length of 0.8 s (i.e., 800 points) were optimal. These choices are very plausible. First, the length of the PQRST of the ECG is the sum of its PR and QT intervals that is between 0.57 and 0.67 [12]. Thus, selecting a value between them is reasonable as a kernel length. Furthermore, to apply a max-pooling operation of an interval including at least one R peak that is related to HR and HRV, an average heart rate period (about 0.8 s) can be a considerable candidate. Based on these heuristic choices, we designed our first 1D convolutional layer to have the same kernel and max-pooling lengths (0.6 s and 0.8 s, respectively) for processing the ECG data. The kernel and max-pooling lengths of the network designed to process RESP data were designed similarly: a single respiration period was used for the kernel and max-pooling lengths. Because the RESP pattern is simple and split into by an expiration (e.g., nadir) and an inspiration (e.g., peak), the size is sufficient to extract RESP's features. Because adults normally respire 12–20 times per min [22], we set both lengths to 5 s (i.e., 125 points at a sampling frequency of 25 Hz).

Our proposed network has 50 filters in each of the initial 1D convolutional layers, which has a stride of 1. For the ECG network, there are 50 filters in the second 1D convolutional layer, which has a kernel length of 25 and a stride of 1. For the RESP network, there are 50 filters in the second 1D CNN layer, which has a kernel length of 4 and a stride of 1. Zero-padding was used in all the convolutional layers to maintain the input size. There are 32 and 16 units in the first and second LSTM layers, respectively, and 512 units in the dense layer. The second 1D convolutional layers in the ECG and RESP networks have kernel lengths of 25 and 4, respectively, so as to focus on the same time interval (20 s). All dropout layers have a dropout rate of 0.5 and the weight decay's regularization strength is 10^{-4}.

For training, we used the Adam [23] optimizer with a learning rate of 10^{-3} and a step decay scheduler (i.e., the learning rate is halved every 50 epochs). The binary cross-entropy loss was used to calculate the losses between the labels and predictions, as follows:

$$L = -\frac{1}{M}\sum_{i=1}^{M} y_i \log(\hat{y}_i) + (1 - y_i)\log(1 - \hat{y}_i). \quad (1)$$

We used a total of 250 epochs, a batch size of 32, and a 0.3 validation split (i.e., 30% of the training set). Finally, the model with the lowest loss on the validation set after 250 epochs was used for evaluation. As with the machine learning models, we used five-fold cross validation to evaluate the performance of the network.

All training processes were conducted using the well-known Keras deep learning library, with Python 2.7 running under Ubuntu 16.01.5, on a PC with a 3.6 GHz Intel Core i7 processor, 128 GB of RAM, and 4 NVIDIA GeForce GTX1080 Ti (Santa Clara, CA, USA).

3.5. Metrics

Because this is a binary classification problem (i.e., the subject is stressed or unstressed), we used the following metrics to evaluate both the deep learning network and the machine learning models:

$$Accuracy = \frac{TP + TN}{TP + TN + FP + FN} \times 100\ (\%), \quad (2)$$

$$\text{F1 score} = 2 \times \frac{TP}{2 \times TP + FP + FN}. \quad (3)$$

Here, TP (true positive) is the number of cases correctly classified as "stressed," while TN (true negative) is the number of cases correctly classified as "unstressed." Likewise, FP (false positive) is the number of cases that were classified as "stressed" but were actually "unstressed," while FN (false negative) is the number of cases that were classified as "unstressed" but were actually "stressed." The first metric (accuracy) is the percentage of cases that were correctly predicted, while the second (F1 score) is the harmonic mean of the precision and recall, which indicates the trade-off between these two metrics.

In addition to the accuracy and F1 score, we also used the area under the receiver operating characteristic (ROC) curve to evaluate the models. The area under the ROC curve (AUC) is a well-known model accuracy metric [24]. By calculating each sensitivity and specificity according to probability thresholds, which is within 0 to 1, the ROC is independent of the different thresholds and thus the metric is reliable and reflects the average performance with the thresholds. Models with AUCs above 0.9 are considered to be accurate [24].

4. Results

In our experiments, we collected a total of 144 VAS scores for individual tasks from 16 subjects. These were evaluated after each relaxation and stressor task (Figure 1). To eliminate any inter-subject variability, the scores were normalized for each subject. Table 2 shows the average normalized scores.

As Table 2 shows, the scores are significantly lower for the relaxation tasks than for the stressor ones. Among the stressor tasks, the hard and easy math tasks yielded the highest and lowest average scores, respectively. Contrary to our expectations, the average score was lower for the hard Stroop task than for the easy one, possibly because easy but tedious tasks may be more stressful than difficult tasks. However, if the task is too difficult, as with the hard math task, it appears to be more stressful than a tedious task.

Table 2. Average normalized visual analogue scale (VAS) scores for all tasks. These have been normalized to a range of 0–1 with a MinMax scaler.

Task	Average Value
Relax	0.24
Easy math	0.51
Easy stroop	0.61
Hard math	0.80
Hard stroop	0.52

Because our experiments involved alternating relaxation and stressor tasks, we also calculated the average difference between the normalized VAS score recorded immediately before a stressor task (i.e., after a relaxation task) and that recorded immediately after the stressor (Table 3). Here, it is clear that all the stressor tasks induced stress, and that the most and least stressful tasks were the hard and easy math tasks, respectively, as in Table 2. Again, the easy Stroop task was a stronger stressor than the hard one.

Table 3. Average differences between the normalized VAS scores before and after each task. Here, the relaxation tasks were used as a baseline before stressor tasks.

Task	Average Value
Easy math	0.12
Easy Stroop	0.42
Hard math	0.55
Hard Stroop	0.32

4.1. Performance

Among the 720 segments, one of the ECG segments was significantly distorted by a motion noise; we excluded this segment and its label for further analysis. To evaluate the performance of a model, five-fold cross validation was used on both machine learning models and DeepER Net. This method commonly evaluates the predictability of a model [5]. In particular, the 719 segments were randomly shuffled and split into five folds. Furthermore, five-fold cross validation was applied for evaluation. The use of this cross validation scheme is independent of subjects, indicating that the segments extracted from a subject can be in both test set and training set. Because of the similarity between the test set and training set, this might lead to higher accuracy on the testing set.

We calculated the average performance of the machine learning models, as shown in Table 4, and then selected the best model for comparison purposes. Of these models, the random forest (RF) yielded the highest average accuracy (71.8 ± 2.3%), F1 score (0.67 ± 0.04), and AUC (0.80 ± 0.02). This was followed by the decision tree (DT), then SVM, the KNN, and finally the logistic regression (LR) showed the lowest performance. This highlights the fact that different models can give different performance, even when trained on the same handcrafted feature set, and that we need to find the most suitable model for each problem. In addition, the fact that the RF and LR demonstrated the highest and lowest performance, respectively, suggests that an ensemble model can be suitable for recognizing stress. However, the RF's AUC was less than 0.9, so it cannot be considered to be highly accurate [24].

Turning now to the performance of the proposed DeepER Net, we find that it showed the highest average accuracy (83.9 ± 2.3%), F1 score (0.81 ± 0.05), and AUC (0.92 ± 0.01). Compared with the RF, its average accuracy was 12.1% higher (p-value < 0.05 with paired t-test), its average F1 score was 0.14 higher (p-value < 0.05 with paired t-test), and its average AUC was 0.12 higher (p-value < 0.05 with paired t-test), clearly indicating that our deep learning approach was a substantial improvement. In addition, DeepER Net's AUC was greater than 0.9, so we can conclude that it is highly accurate for recognizing stress [24]. These results thus suggest that our deep learning approach is a promising option for accurately recognizing stress. Loss and accuracy information of the proposed DeepER Net during training is shown in Figure S1.

Table 4. Average metrics after five-fold cross validation. We used Equations (2) and (3) to calculate the average accuracy, F1 score, and AUC, as well as their standard deviations, and show these results as average ± standard deviation. Abbreviations : SVM, support vector machine; RF, random forest; KNN, k-nearest neighbors; LR, logistic regression; DT, decision tree; AUC, area under the ROC curve; ROC, receiver operating characteristic.

Model	Accuracy (%)	F1 Score	AUC
DeepER Net	83.9 ± 2.3	0.81 ± 0.05	0.92 ± 0.01
SVM	61.7 ± 3.4	0.62 ± 0.04	0.68 ± 0.05
RF	71.8 ± 2.3	0.67 ± 0.04	0.80 ± 0.02
KNN	64.0 ± 3.2	0.60 ± 0.02	0.67 ± 0.04
LR	59.1 ± 2.5	0.55 ± 0.05	0.63 ± 0.04
DT	68.8 ± 1.6	0.66 ± 0.02	0.70 ± 0.02

4.2. Visualization

Although numerous studies have considered machine learning models based on handcrafted features, Table 4 shows that our deep learning approach provided superior performance. This suggests that data-driven features can capture more general information than handcrafted ones. Visualizing the neurons' activation is a potentially useful way to further analyze these results, as it can help researchers to understand how the network is making its decisions and find new stress-related features. Here, we selected the network trained during the first fold of cross validation and a sample of the ECG and RESP data. Then, after calculating the activation in both parts of the network, we compared the first batch-normalization layer's output with the activation after the first ReLU for each signal. Because we used zero-padding in the convolutional layer to maintain the input length, we also applied zero-padding to the first batch-normalization layer's output. The activations of the ECG and RESP networks are shown in Figures 4 and 5, respectively.

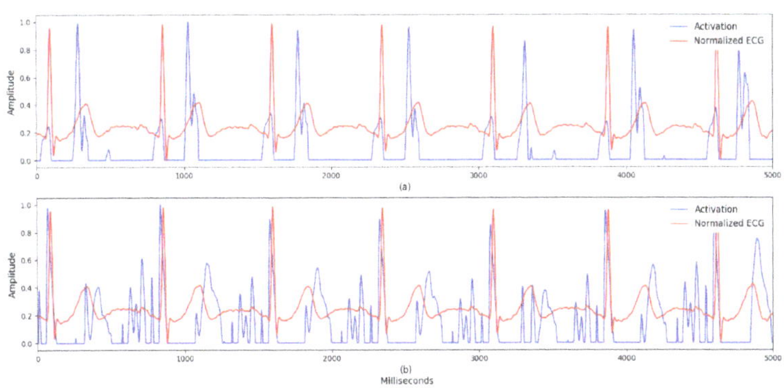

Figure 4. The activations on the first ReLU of electrocardiogram (ECG) signal. To easily see which signal patterns were activated, the activations and the first batch-normalization layer's output were normalized with MinMax Scaler having a range from 0 to 1. The blue line indicates the activations and the red line indicates the output. Activations around (**a**) ECG Q and T's ascending waveform and (**b**) ECG QRS and T's descending waveform.

Figure 4 shows how the neurons in the proposed Deep ER Net were activated by periodic and comprehensive ECG waveform patterns, for (a) Q and T's ascending waveform, and (b) QRS and T's descending waveform. These results indicate that the filters were able to extract these unique ECG waveforms, unlike the machine learning approaches considering only ECG's R-peaks. In Figure 5, we find that neurons were activated around the RESP peaks and troughs. This is clear because their periodic patterns are closely related to stressed or relaxed states. These results show specific patterns, including peaks, troughs, and waveforms, from which we can conclude that the proposed DeepER Net was making decisions based on information about the periods of specific ECG and RESP patterns.

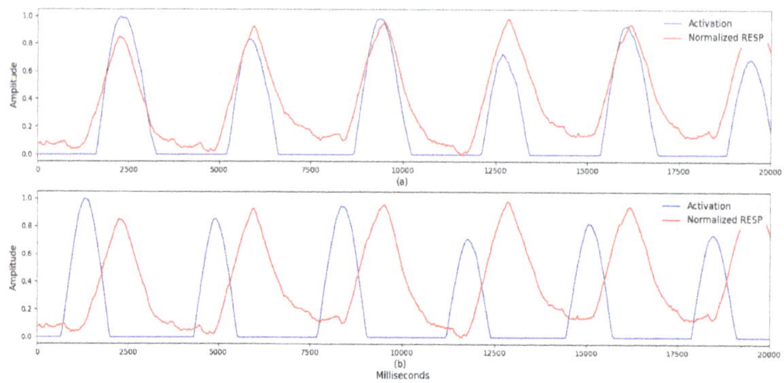

Figure 5. The activations on the first ReLU of respiration (RESP) signal. To easily see which signal patterns were activated, the activations and the first batch-normalization layer's outputs were normalized with MinMax Scaler having a range from 0 to 1. The blue line indicates the activations and the red line indicates the output. Activations around (**a**) RESP peak (e.g., inspiration) and (**b**) RESP nadir (e.g., expiration).

5. Discussion

5.1. Visualization

Visualization is a promising approach to finding evidence for how networks make decisions. Of the various visualization tools, we elected to look at the activation of the network's neurons to identify which ECG and RESP patterns they focused on. As Figure 4 shows, the neurons were activated around ECG QRS, and T waveforms. These are unique patterns, specific to ECG data, and the network's convolutional layers were able to consider changes in their shape and amplitude. Likewise, Figure 5 shows that the network was able to process patterns extracted around the RESP peaks and troughs.

These findings indicate that our network can extract a more comprehensive range of features than simple handcrafted ones that consider specific waveform (e.g., R-peaks), frequency-domain, or time-domain features. This is possible because the network learned meaningful stress-related features from the data. From this point of view, we can understand why the network performed better than the machine learning models (Table 4). We can therefore conclude that this deep learning approach is more promising than the previously proposed machine learning approaches.

5.2. Comparison with Previous Studies

Three studies [5,12,15] have proposed deep learning approaches to stress recognition. Deep ECG Net [12]'s structure was optimized using domain knowledge about the ECG PQRST waveforms, enabling it to achieve a high average accuracy of 80.7% on two different datasets and perform better than conventional machine learning models. Consequently, we used this optimized network structure as the basis for our proposed DeepER Net. Next, because good experimental protocol design is important for obtaining reliable datasets and results, we adapted Cho et al.'s [15] well-designed protocol for use in our study. They proposed a cheap thermal imaging-based stress detection method, which extracts multiple spectrum images from the thermal respiration images and then augments the data using a sliding window method. The resulting CNN achieved 84.6% accuracy for classifying two stress levels (binary classification). Finally, He et al. [5] proposed a deep CNN for detecting acute cognitive stress from 10-s ECGs. They used spectrum images extracted around ECG R-peaks as input and applied data augmentation. Their CNN achieved an average error rate of 17.3%, equivalent to an average accuracy of 82.7%.

In this study, we have proposed the first end-to-end deep neural network (DeepER Net) to recognize stress using multiple signals (ECG and RESP). Because we needed to consider two different signals, we developed a unique network structure that could extract features from both signals. The network achieved an average accuracy of 83.9%, which is comparable to the results achieved by the other proposed models [5,12,15] as summarized in Table 5. For a fair comparison, evaluating the models on a public dataset via the same training conditions and evaluation method can be useful. We proceeded with an experiment validating the models using the DRIVERDB [10] including ECG, RESP, and stress label information. The dataset [10] was collected with the different driving sections (e.g., rest, city, and highway) and each section indicates different stress level. For example, the rest section, city section, and highway section indicate low, high, and medium stress levels, respectively. Among a 17 drivers dataset in [10], we considered only 11 drivers having an existence of the clear marker [25]. The preprocessing including noise filtering and clipping was the same presented in the Methods section. After preprocessing, 801 labeled segments including ECG, RESP, and Lomb Periodogram spectrum [5] were obtained. Finally, the last layer of networks was replaced with a softmax layer for classifying three classes (e.g., low, medium, and high) and then we trained and evaluated the three networks with five-fold cross validation on the segments. As a result, the proposed DeepER Net showed the highest average accuracy of 83.0%; the Deep ECG Net [12] showed the average accuracy of 75.0% and the network [5] showed the average accuracy of 38.5% which may be owing to under-fitting caused by the small capacity of the network. This result means that the use of the multi physiological signals improves the performance of recognizing stress. However, we guess that there may be performance degradation in the open dataset because several important hyper-parameters of networks have been optimized in their dataset, not the open dataset. Thus, more open and reliable data needs to be disclosed. The hyper-parameters, learning rule, and structure of networks [5,12] are shown in Tables S1 and S2.

Table 5. Comparison with the-state-of-the-art deep learning approaches using physiological signals for recognizing stress. Abbreviations: CNN, convolutional neural network; LSTM, long short-term memory.

Models	Physiological Signal	Model	Accuracy
Hwang et al. [12]	ECG	CNN and LSTM	80.7%
Cho et al. [15]	Thermal respiration images	CNN	84.6%
He et al. [5]	Lomb Periodogram spectrum extracted from zero-one transformed NN intervals	CNN	82.7%
Proposed DeepER Net	ECG and RESP	CNN and LSTM	83.9%

By visualization, we also identified the activation patterns produced by the ECG and RESP data and analyzed their meanings. Although previous studies have analyzed ECG activation patterns [5,12], ours is the first to analyze the various ECG and RESP activation patterns related to stress recognition, which we believe makes it distinctly different from previous work.

5.3. Possibility of Personalized Models

Although this study did not focus on personalized models that can adapt to individual stress responses, such models could be developed based on the proposed network. Because DeepER Net's last layer is a sigmoid function, the probability of stress is calculated within a 0–1 range and the model then makes a decision using the default threshold (0.5). Increasing the threshold would make the model stricter when determining stress states, while lowering the threshold would make it more generous. This suggests that we could change the threshold based on individual stress responses, and hence develop personalized models. Alternatively, personalized models could be developed by fine-tuning the network based on data from a single individual. Unlike with conventional machine learning approaches, there is no need to retrain the network from scratch, so it can be trained rapidly and avoid over-fitting issues.

5.4. Multiple Physiological Datasets

The main reasons for using multiple physiological datasets are as follows. First, a small number of subjects can cause over-fitting problems that reduce generalization performance. Such over-fitting issues can be overcome by increasing the amount of data (e.g., by involving more subjects or augmenting the data) or using features based on other independent types of data. Because increasing the number of subjects is difficult, extracting independent features can help to deal with over-fitting problems. In addition, each person's stress responses may vary slightly, leading to the problem of inter-variability, which has the effect of lowering generalization performance for new subjects. Therefore, considering multiple data related to stress could help to reduce the problem.

However, using too many different types of data could reduce the stress recognition system's usability by requiring a variety of monitoring devices to be worn to collect all the different physiological signals, which would be burdensome in practice. Researchers should thus consider the trade-offs involved between usability and performance.

5.5. Limitations and Future Work

Our study has two main limitations: the experimental setting and the use of a respiration monitoring device. Although the setting was intended to simulate a real workplace, the actual experiments were conducted in a more controlled manner because recruiting working subjects is difficult and an uncontrolled experimental setting would have reduced the quality of the data. Once we have established our model's validity, we plan to perform experiments in a real workplace setting. In this study, we used a chest strap-based wearable device to measure the physiological signals, but we are aware that such devices can be hard to wear in the workplace and thus plan to use a patch-type ECG device and a wearable device to measure RESP in a later study.

6. Conclusions

In this study, we have proposed the first end-to-end deep learning approach to stress recognition based on ECG and RESP data. Our protocol involved collecting ECG and RESP data and recording subjective stress scores while the subjects conducted alternating stressor and relaxation tasks. Using this multiple dataset, our proposed DeepER Net performed better than conventional machine learning models that require the extraction of handcrafted features. By visualizing the network's activation, we found that its neurons were being activated by unique and specific patterns. In conclusion, we believe that our proposed DeepER Net will be of benefit to people who suffer from stress in the workplace.

Supplementary Materials: The following are available online at http://www.mdpi.com/1424-8220/19/13/3021/s1. Figure S1: Loss and accuracy information of the proposed DeepER Net during training, Table S1: The structure of the network [12] and its training condition, Table S2: The structure of the network [5] and its training condition.

Author Contributions: W.S. conceived and design this study. W.S. and N.K. performed these experiments; S.K. analyzed the collected data; W.S. and N.K. wrote the paper; C.L. and S.-M.P. revised this paper.

Funding: This research was supported by the Ministry of Science and ICT (MSIT), Korea, under the ICT Consilience Creative program (IITP-2019-2011-1-00783) supervised by the Institute for Information and communications Technology Promotion (IITP), the Basic Science Research Program through the National Research Foundation of Korea (NRF) funded by the Ministry of Science and ICT (NRF-2017R1A5A1015596), and the Technology Innovation Program (or Industrial Strategic Technology Development Program, 20001841, Development of System for Intelligent ContextAware Wearable Service based on Machine Learning) funded By the Ministry of Trade, Industry and Energy (MOTIE, Korea).

Conflicts of Interest: The authors declare no conflict of interest.

Ethical Statements: This study was approved by the Ethics Committee of POSTECH (PIRB-2019-E001).

References

1. Joyce, S.; Modini, M.; Christensen, H.; Mykletun, A.; Bryant, R.; Mitchell, P.B.; Harvey, S.B. Workplace interventions for common mental disorders: A systematic meta-review. *Psychol. Med.* **2016**, *46*, 683–697. [CrossRef] [PubMed]
2. Hajera, S.; Ali, M.M. A Comparative analysis of psychological stress detection methods. *IJCEM* **2018**, *21*, 1–8.
3. Elzeiny, S.; Qaraqe, M. Machine learning approaches to automatic stress detection: A review. In Proceedings of the 2018 IEEE/ACS 15th International Conference on Computer Systems and Applications (AICCSA), Aqaba, Jordan, 28 October–1 November 2018; pp. 1–6.
4. Mozos, O.M.; Andrews, S.; Ferrandez, J.M.; Ellis, D.; Bellotto, N.; Sandulescu, V.; Dobrescu, R. Stress detection using wearable physiological and sociometric sensors. *Int. J. Neural Syst.* **2017**, *27*, 1650041. [CrossRef] [PubMed]
5. He, J.; Li, K.; Liao, X.; Zhang, P.; Jiang, N. Real-time detection of acute cognitive stress using a convolutional neural network from electrocardiographic signal. *IEEE Access* **2019**, *7*, 42710–42717. [CrossRef]
6. Reis, R.S.; Hino, A.A.; Añez, C.R. Perceived Stress Scale. *J. Health Psychol.* **2010**, *15*, 107–114. [CrossRef] [PubMed]
7. Buysse, D.J.; Reynolds, C.F.; Monk, T.H.; Berman, S.R.; Kupfer, D.J. The Pittsburgh sleep quality index: A new instrument for psychiatric practice and research. *Psychiatry Res.* **1989**, *28*, 193–213. [CrossRef]
8. Sriramprakash, S.; Prasanna, V.D.; Murthy, O.V.R. Stress detection in working people. *Proc. Procedia Comput. Sci.* **2017**, *115*, 359–366. [CrossRef]
9. Cheon, D.; Choi, I.; Lee, J.; Moon, J.; Kye, S.; Lee, K. Multimodal data collection framework for mental stress monitoring. In Proceedings of the 2017 ACM International Joint Conference on Pervasive and Ubiquitous Computing and Wearable Computers, Maui, HI, USA, 11–15 September 2017.
10. Healey, J.A.; Picard, R.W. Detecting stress during real-world driving tasks using physiological sensors. *IEEE Trans. Intell. Transp. Syst.* **2005**, *6*, 156–166. [CrossRef]
11. Betti, S.; Lova, R.M.; Rovini, E.; Acerbi, G.; Santarelli, L.; Cabiati, M.; Del Ry, S.; Cavallo, F. Evaluation of an integrated system of wearable physiological sensors for stress monitoring in working environments by using biological markers. *IEEE Trans. Biomed. Eng.* **2017**, *65*, 1748–1758.
12. Hwang, B.; You, J.; Vaessen, T.; Myin-Germeys, I.; Park, C.; Zhang, B.T. Deep ECGNet: An optimal deep learning framework for monitoring mental stress using ultra short-term ECG signals. *Telemed. e-Health* **2018**, *24*, 753–772. [CrossRef] [PubMed]
13. Acharya, U.R.; Fujita, H.; Oh, S.L.; Hagiwara, Y.; Tan, J.H.; Adam, M. Application of deep convolutional neural network for automated detection of myocardial infarction using ECG signals. *Inf. Sci.* **2017**, *415–416*, 190–198. [CrossRef]
14. Lecun, Y.; Bengio, Y.; Hinton, G. Deep learning. *Nature* **2015**, *521*, 436–444. [CrossRef] [PubMed]
15. Cho, Y.; Bianchi-Berthouze, N.; Julier, S.J. DeepBreath: Deep learning of breathing patterns for automatic stress recognition using low-cost thermal imaging in unconstrained settings. In Proceedings of the 2017 7th International Conference on Affective Computing and Intelligent Interaction, ACII 2017, San Antonio, TX, USA, 23–26 October 2017.
16. Ghaderi, A.; Frounchi, J.; Farnam, A. Machine learning-based signal processing using physiological signals for stress detection. In Proceedings of the 2015 22nd Iranian Conference on Biomedical Engineering (ICBME), Tehran, Iran, 25–27 November 2015; pp. 93–98.

17. Salahuddin, L.; Cho, J.; Jeong, M.G.; Kim, D. Ultra short term analysis of heart rate variability for monitoring mental stress in mobile settings. In Proceedings of the Annual International Conference of the IEEE Engineering in Medicine and Biology, Lyon, France, 22–26 August 2007.
18. Ciabattoni, L.; Ferracuti, F.; Longhi, S.; Pepa, L.; Romeo, L.; Verdini, F. Real-time mental stress detection based on smartwatch. In Proceedings of the 2017 IEEE International Conference on Consumer Electronics, ICCE 2017, Las Vegas, NV, USA, 8–11 January 2017.
19. Wijsman, J.; Grundlehner, B.; Liu, H.; Penders, J.; Hermens, H. Wearable physiological sensors reflect mental stress state in office-like situations. In Proceedings of the 2013 Humaine Association Conference on Affective Computing and Intelligent Interaction, ACII 2013, Washington, DC, USA, 2–5 September 2013.
20. Munla, N.; Khalil, M.; Shahin, A.; Mourad, A. Driver stress level detection using HRV analysis. In Proceedings of the 2015 International Conference on Advances in Biomedical Engineering, ICABME 2015, Beirut, Lebanon, 16–18 September 2015.
21. Faust, O.; Hagiwara, Y.; Hong, T.J.; Lih, O.S.; Acharya, U.R. Deep learning for healthcare applications based on physiological signals: A review. *Comput. Methods Programs Biomed.* **2018**, *161*, 1–13. [CrossRef] [PubMed]
22. Sherwood, L. *Fundamentals of Physiology: A Human Perspective*; Thomson Brooks/Cole Belmont: New York, NY, USA, 2006; Volume 380.
23. Kingma, D.P.; Ba, J. Adam: A method for stochastic optimization. *arXiv* **2014**, arXiv1412.6980.
24. Akobeng, A.K. Understanding diagnostic tests 3: Receiver operating characteristic curves. *Acta Paediatr.* **2007**, *69*, 644–647. [CrossRef] [PubMed]
25. Liu, Y.; Du, S. Psychological stress level detection based on electrodermal activity. *Behav. Brain Res.* **2018**, *341*, 50–53. [CrossRef] [PubMed]

sensors

MDPI

Article

Combining Inter-Subject Modeling with a Subject-Based Data Transformation to Improve Affect Recognition from EEG Signals

Miguel Arevalillo-Herráez *, Maximo Cobos, Sandra Roger and Miguel García-Pineda

Departament d'Informàtica, Universitat de València, Avda. de la Universidad, s/n, 46100-Burjasot, Spain
* Correspondence: miguel.arevalillo@uv.es; Tel.: +34-963-543-962

Received: 4 June 2019; Accepted: 5 July 2019; Published: 8 July 2019

Abstract: Existing correlations between features extracted from Electroencephalography (EEG) signals and emotional aspects have motivated the development of a diversity of EEG-based affect detection methods. Both intra-subject and inter-subject approaches have been used in this context. Intra-subject approaches generally suffer from the small sample problem, and require the collection of exhaustive data for each new user before the detection system is usable. On the contrary, inter-subject models do not account for the personality and physiological influence of how the individual is feeling and expressing emotions. In this paper, we analyze both modeling approaches, using three public repositories. The results show that the subject's influence on the EEG signals is substantially higher than that of the emotion and hence it is necessary to account for the subject's influence on the EEG signals. To do this, we propose a data transformation that seamlessly integrates individual traits into an inter-subject approach, improving classification results.

Keywords: EEG; arousal detection; valence detection; data transformation; normalization

1. Introduction

Affect recognition has been an active topic of research for the last two decades, and attempts have been made to detect emotions from many different sources of information, including text [1], facial expressions [2], speech [3], physiological signals [4–6] or interaction data [7], among others.

Electroencephalography (EEG) signals were initially used in medicine to diagnose a diversity of disorders and pathological conditions, such as epilepsy [8,9], alcoholism [10,11], detection of suicidal ideation [12] or monitoring the depth of anesthesia [13]. However, the large quantity of information that EEG signals encode about the subject has motivated their use in other application areas, such as biometric recognition [14,15], gender identification [16] and emotion detection [17,18].

Previous neuropsychological studies [19] have shown a relation between emotions and the electrical activity of the brain, and reported on EEG correlates of emotions [19]. This relation has motivated a large number of attempts to detect emotions by processing EEG signals, sometimes in combination with other sources of information (e.g., [20]). However, EEG signals are relatively complex, and affected by physiologic and extraphysiologic artifacts such as eye movement, pulse, respiration or measurement equipment. Therefore, there is an intrinsic difficulty associated with making this relation explicit. This includes the use of appropriate signal processing methods to cancel undesired artifacts [21,22]; the extraction and selection of the most informative features and channels [23,24]; and the development of techniques that are able to detect patterns that can be linked to specific emotional states (e.g., [25]).

Previous works in the field of psychology suggest that there are significant differences in the way individuals feel and express emotions [26]. Many typical setups use a set of training samples to build a general, subject-independent (inter-subject) model, which is shared by all users (e.g., [27–30]). In this case, a single model is built by considering all data as if it were coming from the same subject, without taking the user's particularities into consideration. Despite the high prediction rates obtained in some cases, these can be significantly improved by using individual models adapted to each user [30–32]. However, subject-dependent (intra-subject) approaches suffer from two severe drawbacks. First, they require the collection of a large amount of data to adequately model the relation between the EEG signal and the emotion for each person. Second, they cannot be used for unseen subjects as they only use data related to the particular individual. These two drawbacks make the approach impractical in many cases.

In this paper, we first study the suitability of intra-subject and inter-subject modeling approaches in an EEG-based affect recognition context, by analyzing the available data in three public databases, namely the Database for Emotion Analysis using Physiological Signals (DEAP) [33], MAHNOB-HCI [34] and DREAMER [35]. The analysis performed clearly indicates that the contribution of the subject to the EEG signal is far larger than the effect of the emotion, hence limiting the applicability of inter-subject models and suggesting a better behavior of subject-dependent models that only use training data associated with the same subject. An in-depth analysis using the DEAP dataset also reveals that many positive results for subject-independent models reported in some previous works may in part be due to the use of imbalanced datasets. As a second and more important contribution, we propose an approach that combines an inter-subject model with a subject-based normalization of the EEG signals, making it possible to effectively generate a single model, which is valid across the entire population. This approach integrates data related to personality traits into the model, encoding a person's individuality in feeling emotions without affecting data capturing needs. The gains achieved open the door for using a single model for unseen subjects, which can be progressively adapted as more personalized data are gathered.

This paper is structured as follows. First, related previous work is described in Section 2, covering both modeling approaches and existing public databases. Then, in Section 3, the three repositories considered are analyzed by computing an embedding that reveals key issues related to the topological structure of the data. After, we present our proposal to partially cancel the subject-related component from the signal to achieve an inter-subject model with comparable performance to typical intra-subject models. This method is evaluated in Section 5. Finally, the main conclusions from this work are presented in Section 6.

2. Related Previous Work

2.1. Modeling Approaches

Computational methods for affect detection attempt to relate features extracted from certain signals measured on a subject to emotional processes. These features may be captured from, e.g., facial expressions, voice, body language and posture, physiological states, functional Magnetic Resonance Imaging (fMRI), Magnetoencephalogram (MEG) brain signals and/or EEG. In general, machine learning algorithms are used to identify signal patterns that are associated with the expression of different emotions, and to build models that enable the automatic detection of a concrete set of states (see, e.g., [36–38] for extensive reviews of the field). These machine learning approaches can be classified as inter-subject or intra-subject. Methods in the first category aim at constructing a model that is valid for all users. Techniques in the second one consider that the appraisal of one's emotional state is strongly related to personal factors, such as one's circumstances [39]. Hence, they aim to construct an individual model for each user, generally increasing performance at the cost of increasing data collection needs [40,41].

The prediction of a subject's emotion/mental state from brain signals has been widely studied, including both EEG and MEG signals [42,43]. In the particular case of EEG, the signals from selected channels are usually pre-processed with noise reduction algorithms and filtering methods to enhance the signal-to-noise power ratio. Feature extraction is then used to determine variables which correlate well with the target emotional states, according to the specific emotional model that is used [19]. Typical feature extraction methods include wavelet transform [44], spectral power features [45], higher order crossings [46], short-time Fourier transform [47], asymmetry index [48] and/or statistical features [49], e.g., mean, standard deviation, variance, quadratic mean, skewness, power or entropy. Finally, a classification method is used to discriminate a particular emotional state from the features. Support Vector Machines (SVM) [47,48,50,51], nearest neighbour classification [45,50], Naive Bayes [50] or Linear Discriminant Analysis (LDA) [52] are examples of an extensive list of methods that are applied in this context.

Most EEG-based emotion recognition studies use Rusell's two dimensional bipolar emotional model to label and represent emotional states, which is based on valence and activation/arousal [53]. This representation relies on the fact that these two variables account for the major proportion of variance in affect scales. In such models, each emotion is found as a combination of values for valence and arousal, falling meaningfully around the perimeter of the space. The valence dimension represents whether the emotion corresponds to a positive or a negative feeling; and the arousal refers to the level of excitement. The valence/arousal representation was extended to a 3D space in [54], by also considering whether the subject feels controlled or in control of the situation (dominance).

2.2. Public Databases

The intensive work in emotion recognition using EEG data has been supported by the existence of a number of public datasets. A first large database is DEAP, which is presented in [33]. DEAP contains EEG and peripheral physiological signals of 32 people who were recorded as each watched 40 one-minute long excerpts of music videos. These were stored along with the levels of arousal, valence, like/dislike, dominance, and familiarity reported by the subjects. The dataset also contains frontal face video for 22 of the participants. In addition, methods and results are presented for single-trial classification of arousal, valence, and like/dislike ratings using the modalities of EEG, peripheral physiological signals, and multimedia content analysis. EEG signals were recorded by using a Biosemi ActiveTwo system. Despite its relatively recent publication, DEAP [33] has been extensively used in the affect recognition field, to evaluate a number of proposals (e.g., [55–57]).

Another large database is presented in [34]. In this case, the repository contains data for 27 people, recorded while watching 20 movie fragments and pictures in a very similar setting as in DEAP. In this case, video data are provided for all participants, from six different cameras. The database also contains eye gaze information, as well as other physiological signals (including EEG). Data are stored along with the emotional state reported by the subject, both using emotional keywords and on a scale of valence, arousal and dominance. EEG signals were recorded by using active AgCl electrodes placed according to the international 10–20 system (32 channels).

More recently, a third dataset of similar characteristics as the previous with regard to the EEG data provided has been published [35]. Under the name of DREAMER, this dataset contains EEG data from 23 participants as they watched 18 music videos. The main difference with respect to the previous two databases refers to the type of equipment that was employed. In DREAMER, 128 Hz EEG signals were recorded using an Emotive EPOC system, a device that offers a considerably lower precision than the Biosemi Active Two. Table 1 summarizes the characteristics of these three databases.

Major problematic issues that have hindered the development of practical applications that use EEG signals are related to the cost, time resolution, and complexity of setting up experimental protocols that resemble real-world activities [36]. This has motivated a track of work that focuses on mobile/low cost devices (e.g., [55,58,59]). Although these devices may be less accurate at the signal acquisition phase, they may offer a comparative performance at detecting emotional changes in the subject. This has led other authors to develop their own datasets to validate their results in specific contexts that use low cost devices [55,59]. However, these datasets have not been made public and are hence not usable in other research works.

Table 1. Summary of characteristics for the databases in the study.

Database	Subjects	Videos	Stimuli	Duration	Device	Channels	Sampling Frequency	Features
DEAP	32	40	Music videos	60 s	Biosemi Active II	32	512 Hz *	230
MAHNOB	27	20	Excerpts from movies	34.9–117 s (M = 81 s)	Biosemi Active II	32	512 Hz *	230
DREAMER	23	18	Music videos	65–393 s (M = 199 s)	Emotive EPOC	14	128 Hz	105

* downsampled to 256 Hz

3. Data Analysis

3.1. Problem Formulation

Let us assume a set of subjects $\mathcal{S} = \{s_i\}$, $i = 1, 2 \ldots m$. Let us also assume that there exist a set of n_i labeled training samples for each subject $\forall s_i : \mathcal{T}_{s_i} = \{(\mathbf{t_{s_i,j}}, l_{s_{i,j}})\}$, $j = 1, 2 \ldots n_i$, where $\mathbf{t_{s_i,j}}$ is conveniently represented in a particular feature space $\mathbb{F}$ and corresponds to the feature vector for the jth sample of subject s_i, and $l_{s_{i,j}}$ refers to the corresponding emotional label.

Current emotion recognition approaches can be classified into inter-subject and intra-subject. In practice, both types of models are typically built by using classification approaches on training data. In general, this training data (the sets $\mathcal{T}_{s_i}$, $i = 1, 2 \ldots m$) consist of a number of labeled entries that relate features to emotions. The fundamental difference between the two approaches is whether the labeled training data refer to a single individual (intra-subject) or to a group of people who are collectively treated as if there were no particularities that make individuals different from each other (inter-subject). In inter-subject methods, a global model which is valid for all users is built, by using the training data $\mathcal{T}_{s_1} \bigcup \mathcal{T}_{s_2} \bigcup \ldots \bigcup \mathcal{T}_{s_m}$. This is, in fact, equivalent to treating all training data for different individuals as if they belong to the same subject [28–30,59]. In intra-subject approaches, an independent model is built for each subject s_i [27,31,32,40,41], by considering only training data that belong to that particular subject ($\mathcal{T}_{s_i}$). The high accuracy achieved by some subject-independent models (e.g., [28,29,59]) suggests that some relations between features and emotions hold for most individuals. At the same time, the usually better prediction performance achieved by intra-subject models [40,41] suggests that the relation between the EEG features and the emotions are, in reality, subject-dependent. Hence, relations between features and emotions can be better established when the user's particularities are taken into consideration. However, intra-subject models require exhaustive data collection from the same subject to build the model. Furthermore, they cannot be used on previously unseen individuals, unlike with inter-subject models.

3.2. Topological Structure of the Data

For the purpose of this work, we replicated feature extraction as described in the original publications describing each database. First, we calculated the Power Spectral Density (PSD) using Welch's method with a Hamming window of 128 samples and 50% overlapping. The spectral power was averaged over the θ (4–8 Hz), slow α (8–10 Hz), α (8–12 Hz), β (12–30 Hz), and γ (>30 Hz) bands from all electrodes. In addition, we computed the difference between the spectral power of all the symmetrical pairs of electrodes on the right and left hemisphere in the same bands, to measure the possible asymmetry in the brain activities due to emotional stimuli. This yielded 230 features for DEAP and MAHNOB-HCI (32 electrodes $\times$ 5 bands + 14 pairs $\times$ 5 bands), and 105 features in DREAMER (14 electrodes $\times$ 5 bands + 7 pairs $\times$ 5 bands), as reported in Table 1.

The resulting features were used to plot a 2-D (two dimensional) map after a space transformation using t-Distributed Stochastic Neighbor Embedding (t-SNE) [60]. t-SNE is an unsupervised dimensionality reduction method that is particularly well suited for the visualization of high-dimensional datasets. t-SNE is capable of capturing and preserving much of the topological structure of the high-dimensional data, while also revealing global structure such as the presence of clusters at several scales [60]. We reduced the data to two dimensions, so that we could easily display and analyze it using a scatterplot.

Figure 1 shows the result produced by t-SNE method on the three databases used in this work. We plotted samples from each subject using a different colored marker, to easily observe that EEG data samples from the same subject are topologically located close to each other in the 2-D space. These plots reveal that the contribution of the subject to the EEG signal is clearly higher than the effect of the emotion, a fact which has been extensively exploited in biometrics (e.g., [61–63]).

Although it seems clear that the topological structure of the maps presented in Figure 1 is not the best for the construction of inter-subject models, other previous works have obtained positive results when applying subject-independent models using a typical classification setting. For example, the affect recognition results reported in [35] refer to accuracies of 0.62 in valence and arousal, using a SVM with a Radial Basis Function (RBF) kernel. However, they used an imbalanced dataset, with a proportion of 56–44% in arousal and 61–39% in valence. Considering Figure 1c, it is possible that the positive accuracy reported is in part due to this fact, rather than to the existence of emotion-evoked specific EEG patterns that are shared by multiple subjects.

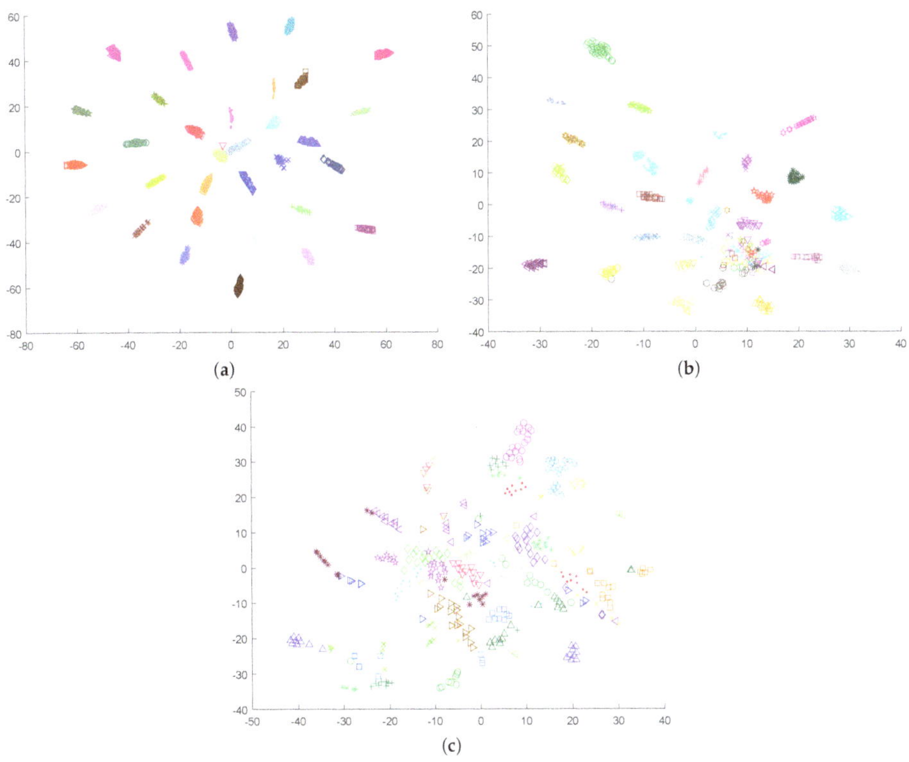

Figure 1. Dimensionality reduction by t-SNE on original data: (**a**) DEAP; (**b**) MAHNOB-HCI; and (**c**) DREAMER. Each subject has been represented with a different colored marker.

4. Proposed Approach

4.1. Typical Data Transformations

The construction of inter-subject models is a harder problem due to the high EEG variability between individuals [64]. The three plots in Figure 1 clearly indicate that classification approaches that use these data would benefit from the removal of the subject's contribution to the EEG signal. Instead of producing an intra-subject model with personalized data coming from a single individual, the subject's particularities can be incorporated into an inter-subject global model by normalizing the data from each subject according to a subject-dependent baseline that summarizes the contribution of the individual to the EEG signal. Other previous works have implicitly attempted this by applying a subject-based normalization of the data. For example, in [33,65], the features were normalized for each participant by scaling them between 0 and 1 to reduce inter-participant variability. The effect of this normalization is shown in Figure 2, for the three databases considered in this work. The effect of such a linear normalization on the subject related component is somehow limited and the latent clustered structure of the original data remains, but the lower distance between the clusters suggests that the subject component in the EEG signals has at least been reduced. This fact outlines the potential of subject-dependent normalizations, and suggests that other more elaborated data transformations may be applied to further reduce or eliminate the subject-related component from the EEG signals.

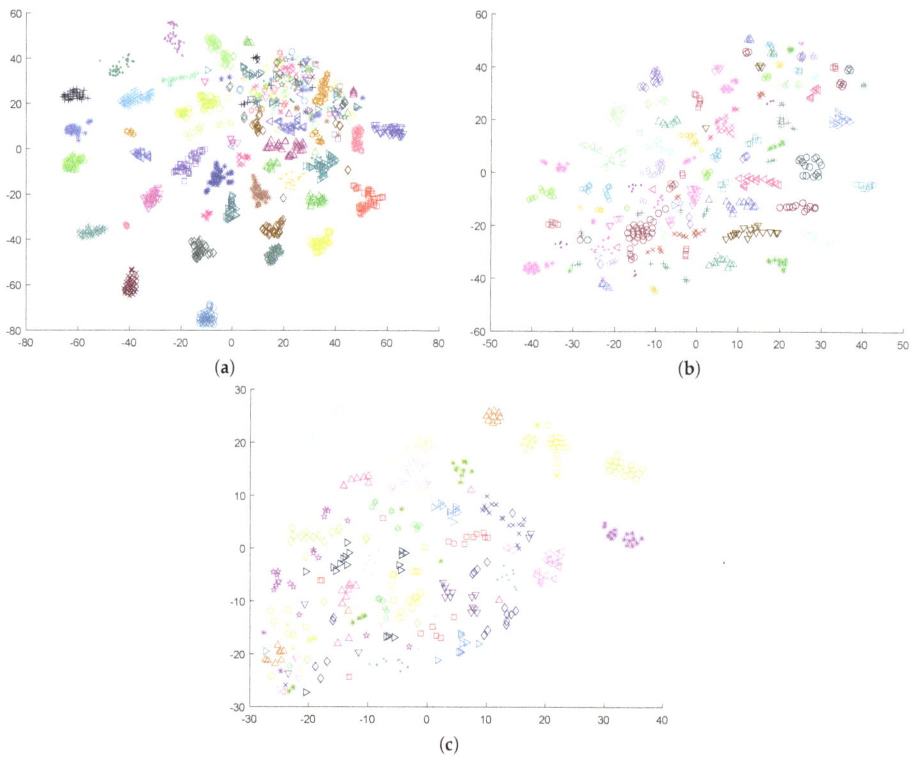

Figure 2. Dimensionality reduction by t-SNE, after normalizing the data by scaling each feature between the maximum and minimum values for the particular subject: (**a**) DEAP; (**b**) MAHNOB-HCI; and (**c**) DREAMER.

4.2. Nonlinear Data Transformation

In particular, and to explore the potential of subject-dependent methods other than a linear scaling, we tested a simple nonlinear transformation of the original data. First, we independently considered each subject, and computed the median value for each feature. Then, the original feature vector was codified as a binary vector of the same size, where components take values 0 or 1 depending on whether the feature value is lower or higher than the median, respectively. More specifically, for any subject s_i, we considered all feature vectors $\mathbf{t_{s_i,j}}$, $j = 1, 2 \ldots n_i$ in the set of training samples $\mathcal{T}_{s_i}$ and computed the median vector $\tilde{\mathbf{t}}_{\mathbf{s}_i}$ across each feature. Then, all feature vectors $\mathbf{u}$ for the same user s_i were transformed according to Equation (1)

$$\widehat{u}[k] = \begin{cases} 1 & u[k] > \tilde{t}_{s_i}[k], \\ 0 & u[k] \leq \tilde{t}_{s_i}[k], \end{cases} \quad (1)$$

where $[k]$ denotes the kth element (feature) of the corresponding vector.

Figure 3 contains the t-SNE representation for the data when this transformation is applied to the entire set. As can be observed, and despite the information loss that is inherent to this operation, the data samples from a same group now appear more sparse, and these plots suggest a more effective reduction of the subject-related component of the signals. A further analysis of the data topology with regard to the labels also revealed a certain level of grouping, more suitable for classification purposes. As an example, Figure 4 shows the positive and negative samples in the MAHNOB dataset in the t-SNE space, according to self-reported arousal levels. An inspection of this plot in relation to the one in Figure 3b suggests that the samples for certain groups of subjects may have been split according to their label.

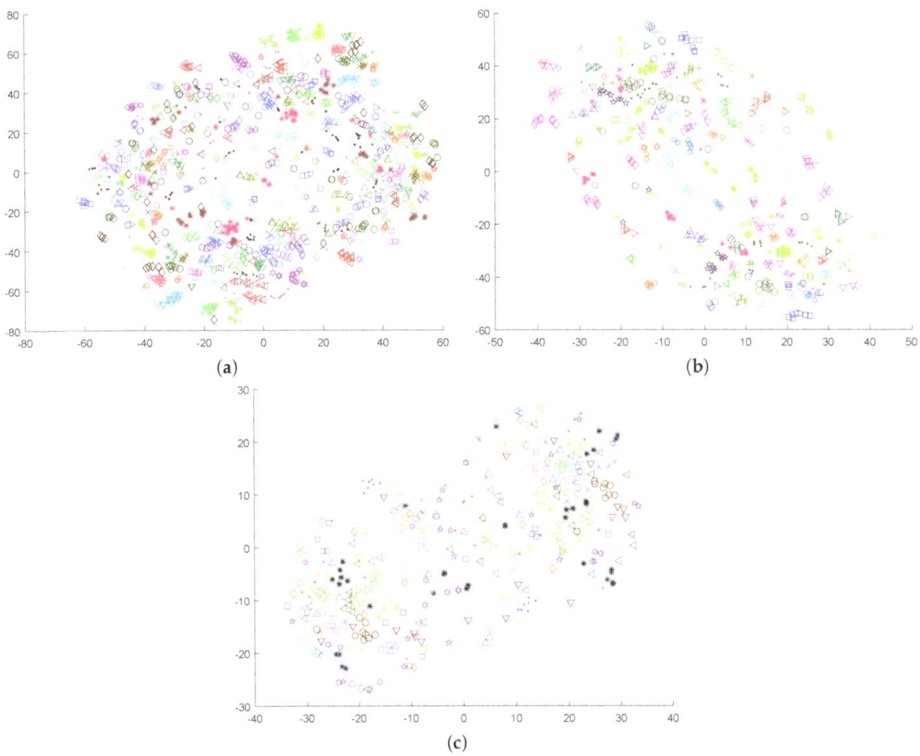

Figure 3. Dimensionality reduction by t-SNE, after transforming the data by binarizing values according to whether they are lower or greater than the median: (**a**) DEAP; (**b**) MAHNOB-HCI; and (**c**) DREAMER.

The proposed transformation allowed us to train the classifier using data from all available subjects, avoiding the small sample case and the need for the personalized training that is typically required when using intra-subject approaches. The only data required by the proposed transformation are the median for each feature, and these can easily be computed and progressively refined from unlabeled data as soon as the EEG capturing device is connected.

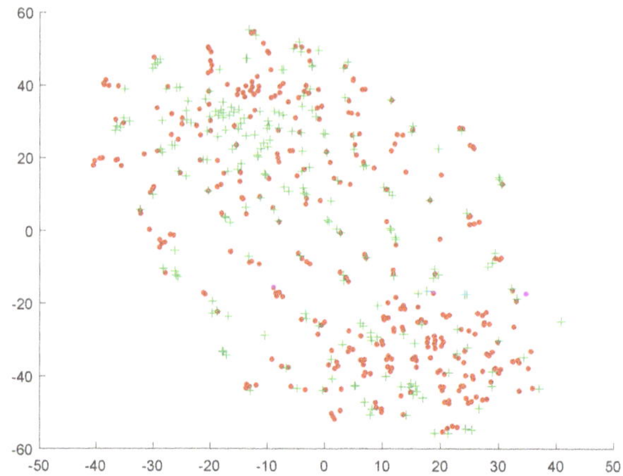

Figure 4. Positive (green plus markers) and negative (red dots) arousal samples in the MAHNOB database, on the representation space produced by t-SNE.

5. Experimental Results

5.1. Improvement on Classification Accuracy

To exhaustively assess the effect of the proposed data transformation, we ran a number of experiments aimed at testing the prediction performance on previously unseen subjects. Results obtained with the proposed data transformation were compared using z-score standardization, a typical data normalization commonly used in machine learning contexts. To this end, we computed the mean and standard deviation vectors μ and σ from the samples in the training set, and normalized each feature vector x according to Equation (2).

$$\hat{x}[k] = \frac{x[k] - \mu[k]}{\sigma[k]} \tag{2}$$

For a comprehensive evaluation, we applied several classification methods, namely SVM with polynomial and Gaussian kernels and Naive Bayes, to be consistent with the previous literature in the field [18,33–35]. All experiments were run in a Matlab R2017a environment, using Matlab's own implementation of the classification algorithms.

All datasets were pre-processed as in [66] to appropriately compare the methods and avoid misleading results caused by different degrees of imbalance in the intra-subject and inter-subject cases. In each database and for each of the labels analyzed (arousal and valence), we randomly selected the same number of samples per class for each user. The number of samples was decided to simultaneously achieve sufficiently populated training sets and minimize the number of subjects that had to be discarded because they did not have sufficient samples in the minority class. Table 2 summarizes the resulting number of users and the samples per user after processing the datasets in this way.

Table 2. Number of subjects and samples per subject in each dataset, after pre-processing.

Database	Valence		Arousal	
	Number of Subjects	Samples per Subject	Number of Subjects	Samples per Subject
DEAP	24	32	16	32
MAHNOB	5	18	10	18
DREAMER	9	16	7	14

In each dataset, and for every combination of normalization and classification method, we ran 20 experiments per subject. In each experiment, all data for one subject $\mathcal{T}_k$ were used as the test set, and 90% of the data from the rest of the individuals, i.e., $(\mathcal{T}_1 \cup \mathcal{T}_2 \cup \ldots \cup \mathcal{T}_m) - \mathcal{T}_k$, were employed for training. As the classes in the three datasets were balanced and had equal importance, the performance was assessed using classification accuracy. This was computed as the proportion of instances that were correctly categorized according to the self-reported binary labels for arousal and valence provided as a ground-truth in each database.

Table 3 compares the classification accuracy when using a typical *z*-score normalization and when the proposed subject-based normalization was applied. To effectively rank the two algorithms according to their general performance, and measure the statistical significance of the results, their classification accuracy was evaluated separately for each test and training pair. With these measurements, a multiple comparison Friedman test [67] was conducted, considering the null hypothesis that the two methods obtained similar results with non-significant differences. This non-parametric test requires computing the average ranks of all methods, which are shown in Table 4, along with the *p*-values and the number of pairwise comparisons that allowed their computation. The *p*-values were calculated using software available from http://sci2s.ugr.es/sicidm [67].

Table 3. Results obtained with a typical *z*-score normalization and with the proposed data transformation.

		Valence			Arousal		
		SVM Cubic	**SVM Radial**	**Naive Bayes**	**SVM Cubic**	**SVM Radial**	**Naive Bayes**
DEAP	*z*-score	0.51	0.50	0.51	0.52	0.50	0.50
	proposed	0.54	0.58	0.57	0.54	0.56	0.55
MAHNOB	*z*-score	0.50	0.56	0.56	0.55	0.52	0.57
	proposed	0.51	0.65	0.65	0.59	0.61	0.62
DREAMER	*z*-score	0.50	0.52	0.51	0.55	0.53	0.50
	proposed	0.54	0.59	0.59	0.58	0.57	0.57

Table 4. Results of Friedman test on data reported in Table 3.

		Valence			Arousal		
		SVM Cubic	**SVM Radial**	**Naive Bayes**	**SVM Cubic**	**SVM Radial**	**Naive Bayes**
DEAP	pairwise comparisons	480	480	480	320	320	320
	average rank *z*-score	1.65	1.73	1.71	1.57	1.78	1.78
	average rank proposed	1.35	1.27	1.29	1.43	1.22	1.22
	p-value	$<10^{-10}$	$<10^{-22}$	$<10^{-18}$	0.01	$<10^{-23}$	$<10^{-23}$
MAHNOB	pairwise comparisons	100	100	100	200	200	200
	average rank *z*-score	1.61	1.82	1.84	1.58	1.84	1.70
	average rank proposed	1.39	1.18	1.16	1.42	1.16	1.30
	p-value	0.02	$<10^{-10}$	$<10^{-11}$	0.02	$<10^{-21}$	$<10^{-8}$
DREAMER	pairwise comparisons	180	180	180	140	140	140
	average rank *z*-score	1.66	1.81	1.82	1.54	1.66	1.73
	average rank proposed	1.34	1.19	1.18	1.46	1.34	1.27
	p-value	$<10^{-4}$	$<10^{-15}$	$<10^{-17}$	0.31	$<10^{-3}$	$<10^{-7}$

When using a radial SVM or the Naive Bayes classifier, the improvement achieved by the proposed subject-based normalization was always statistically significant with *p*-values below 10^{-3} in all cases, which allowed us to reject the null hypothesis. When using a cubic SVM, *p*-values were generally higher, and above 0.05 in one case. Nevertheless, all entries in the table support the performance increase achieved by the proposed data transformation.

As a reference, we also provide in Table 5 the classification accuracy achieved when using an intra-subject model, which was obtained using a different setting. To compute these values, we averaged the results of 100 experiments for each user. In each of these experiments, we selected one positive and one negative sample from the concrete user as the test set, and used the remaining samples for training. This yielded a total of $2 \times m \times 100$ judgments, with m the number of subjects in the pre-processed dataset.

Table 5. Results when using an intra-subject model, in the three databases.

	Valence			Arousal		
	SVM Cubic	**SVM Radial**	**Naive Bayes**	**SVM Cubic**	**SVM Radial**	**Naive Bayes**
DEAP	0.62	0.64	0.61	0.55	0.54	0.59
MAHNOB	0.59	0.59	0.58	0.56	0.66	0.62
DREAMER	0.50	0.52	0.46	0.49	0.51	0.51

When using a standard z-score normalization, it can be observed that the accuracy for intra-subject models was generally better, except in the DREAMER database, which showed very poor results in all cases. This was despite using considerably fewer training data. In general, the accuracy of inter-subject models that use z-score standardization remained close to 50% in most cases, a result that is consistent with the data topology shown in Figure 1, in which samples are grouped by subject rather than their emotional label. On the contrary, the intra-subject models showed reasonable accuracies that are consistent with results reported in previous works [18,33,34], ranging from 0.54 to 0.66 in the DEAP and MAHNOB databases.

When using the proposed data transformation, a significant performance improvement was achieved with regard to the z-score normalization. The results are clearly outperformed in all cases. On many occasions, the inter-subject model on the normalized data performed better than the corresponding intra-subject model. Even in DREAMER, the data transformation led to a reasonable classification accuracy, close to that obtained in other repositories. Rather than a clear performance advantage, the results reported in Table 3 show a comparable performance between using an intra-subject model and the suggested data transformation. However, the proposed approach can be used for previously unseen subjects despite not having additional data available for that specific individual, and offers a performance which is significantly better than that obtained by using a typical z-score normalization.

The behavior reported is consistent with the plots in Figures 1 and 3. When the subject-based data transformation was not applied, the intrinsic subject-dependent component in the signal dominated the data topology, leading to a highly inefficient model for previously unseen subjects. However, the intra-subject model performed reasonably well, as this component equally affected all samples and inherently canceled out. The proposed subject-dependent data normalization removed a significant part of the subject-related component, but it did not cancel it completely. The remnant component can easily be observed in Figure 3, in the form of small clusters of samples that belong to the same individual.

5.2. Scalability

To further test the scalability and generalization capacity of the proposed method, we designed a second experiment that aimed to test how predictions improve as more subjects are incorporated into the training. As an example, Figure 5 shows the results obtained in the three repositories for different approaches, when using a Naive Bayes classifier to predict valence. The methods compared were the z-score normalization, the proposed data transformation and the subject-based scaling proposed in [33,65], in which features were scaled to the range $[0, 1]$. The latter method is labeled as max-min in the figure.

In this plot, the classification accuracy reported for a number of training subjects p is the average of as many trials as subjects there are in the dataset. In each trial, a different subject was considered, and all his/her samples were included in the test set. The training set was composed of all samples from p subjects other than the test subject, chosen at random but maintained across the different algorithms to allow for a fair comparison.

Both the proposed normalization and the subject-based max-min scaling used in [33,65] showed better results when more subjects were used for learning. On the contrary, the z-score normalization did not seem to benefit from learning when the number of training subjects increased. The higher performance of the proposed data transformation can easily be observed in all databases. This shows up as a positive trend that implies a reliability increase as more users are incorporated into the model, and further supports the validity of inter-subject models when they are combined with a suitable transformation function that takes individual traits into consideration.

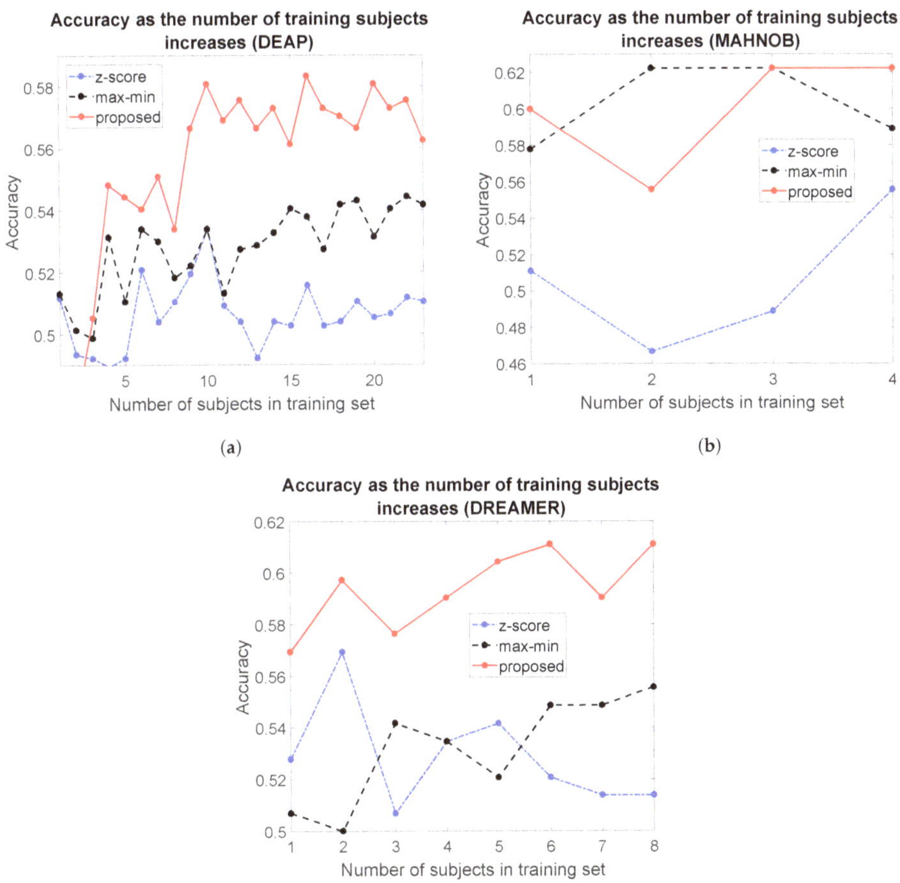

(a) (b) (c)

Figure 5. Classification accuracy as the number of training users is increased: (**a**) DEAP; (**b**) MAHNOB-HCI; and (**c**) DREAMER.

6. Conclusions

Subject-independent models fail to consider that the appraisal of one's emotional state is strongly related to personal factors, such as one's circumstances [39]. Subject-dependent models aim to tackle this weakness, but they do so at the cost of significantly increasing data collection needs [40,41]. This implies that they have to be individually trained for each user and hence cannot be used with previously unseen subjects.

In this paper, a mixed framework to support automatic emotion recognition is proposed. Unlike most typical subject-dependent modeling approaches, the method uses data from all users to build the model, and can be used to make predictions for previously unseen users in an adaptive way, increasing performance as more training data become available. We first show that the existence of an inherent subject-related component in the EEG signals is a major obstacle when attempting to build a user independent model that is simultaneously valid for all subjects. Then, we propose a subject-based normalization procedure that is able to reduce the magnitude of this component when using PSD features. This straightforward normalization procedure is not intended to be a solution to remove this component, but rather a demonstration of the potential benefits of reducing its magnitude. The removal of the subject-dependent component in the signal is indeed feature and problem dependent, and an optimum approach cannot be generalized at this stage. This implies that there is still room for improvement by designing other normalization methods that are more efficient at this task.

The impact of the proposed method goes beyond the construction of inter-subject models for emotion detection from EEG signals. First, the same principles can be exported to other sources of information other than EEG, e.g., physiological, audio, and video. Second, these principles are not limited to the particular problem of emotion recognition. On the contrary, the subject-related component is intrinsic to the signal, and it is present regardless of the problem context.

Author Contributions: All authors were involved at all stages of the research. The original idea was proposed in a group meeting and progressively developed by all authors, who have all equally contributed from the conceptualization to the paper writing.

Funding: This research was partly supported by the Spanish Ministry of Economy and Competitiveness through projects TIN2014-59641-C2-1-P, PGC2018-096463-B-I00 and RTI2018-097045-B-C21; and the Ramón y Cajal grant RYC-2017-22101.

Conflicts of Interest: The authors declare no conflict of interest.

Abbreviations

The following abbreviations are used in this manuscript:

DEAP	Database for Emotion Analysis using Physiological Signals
EEG	Electroencephalography
fMRI	Functional Magnetic Resonance Imaging
LDA	Linear Discriminant Analysis
PSD	Power Spectral Density
RBF	Radial Basis Function
SVM	Support Vector Machine
t-SNE	t-Distributed Stochastic Neighbor Embedding

References

1. Zhang, Y.; Ren, W.; Zhu, T.; Faith, E. MoSa: A Modeling and Sentiment Analysis System for Mobile Application Big Data. *Symmetry* **2019**, *11*, 115. [CrossRef]
2. Samadiani, N.; Huang, G.; Cai, B.; Luo, W.; Chi, C.H.; Xiang, Y.; He, J. A Review on Automatic Facial Expression Recognition Systems Assisted by Multimodal Sensor Data. *Sensors* **2019**, *19*, 1863. [CrossRef] [PubMed]
3. Hajarolasvadi, N.; Demirel, H. 3D CNN-Based Speech Emotion Recognition Using K-Means Clustering and Spectrograms. *Entropy* **2019**, *21*, 479. [CrossRef]
4. Shu, L.; Xie, J.; Yang, M.; Li, Z.; Li, Z.; Liao, D.; Xu, X.; Yang, X. A Review of Emotion Recognition Using Physiological Signals. *Sensors* **2018**, *18*, 2074. [CrossRef] [PubMed]

5. Zhang, X.; Xu, C.; Xue, W.; Hu, J.; He, Y.; Gao, M. Emotion Recognition Based on Multichannel Physiological Signals with Comprehensive Nonlinear Processing. *Sensors* **2018**, *18*, 3886. [CrossRef] [PubMed]
6. Abadi, M.K.; Kia, M.; Subramanian, R.; Avesani, P.; Sebe, N. Decoding affect in videos employing the MEG brain signal. In Proceedings of the 2013 10th IEEE International Conference and Workshops on Automatic Face and Gesture Recognition (FG), Shanghai, China, 22–26 April 2013; pp. 1–6.
7. Arevalillo-HerrÁez, M.; Marco-Giménez, L.; Arnau, D.; González-Calero, J.A. Adding sensor-free intention-based affective support to an Intelligent Tutoring System. *Knowl.-Based Syst.* **2017**, *132*, 85–93. [CrossRef]
8. Wang, X.; Gong, G.; Li, N. Automated Recognition of Epileptic EEG States Using a Combination of Symlet Wavelet Processing, Gradient Boosting Machine, and Grid Search Optimizer. *Sensors* **2019**, *19*, 219. [CrossRef] [PubMed]
9. Zhang, Y.; Yang, S.; Liu, Y.; Zhang, Y.; Han, B.; Zhou, F. Integration of 24 Feature Types to Accurately Detect and Predict Seizures Using Scalp EEG Signals. *Sensors* **2018**, *18*, 1372. [CrossRef] [PubMed]
10. Patidar, S.; Pachori, R.B.; Upadhyay, A.; Acharya, U.R. An integrated alcoholic index using tunable-Q wavelet transform based features extracted from EEG signals for diagnosis of alcoholism. *Appl. Soft Comput.* **2017**, *50*, 71–78. [CrossRef]
11. Mumtaz, W.; Vuong, P. L.; Xia, L.; Malik, A.S.; Rashid, R.B.A. Automatic diagnosis of alcohol use disorder using EEG features. *Knowl.-Based Syst.* **2016**, *105*, 48–59. [CrossRef]
12. Prasad, D.K.; Liu, S.; Chen, S.H.A.; Quek, C. Sentiment analysis using EEG activities for suicidology. *Expert Syst. Appl.* **2018**, *103*, 206–217. [CrossRef]
13. Gu, Y.; Liang, Z.; Hagihira, S. Use of Multiple EEG Features and Artificial Neural Network to Monitor the Depth of Anesthesia. *Sensors* **2019**, *19*, 2499. [CrossRef] [PubMed]
14. Yang, S.; Deravi, F. On the Usability of Electroencephalographic Signals for Biometric Recognition: A Survey. *IEEE Trans. Hum. Mach. Syst.* **2017**, *47*, 958–969. [CrossRef]
15. Zeng, Y.; Wu, Q.; Yang, K.; Tong, L.; Yan, B.; Shu, J.; Yao, D. EEG-Based Identity Authentication Framework Using Face Rapid Serial Visual Presentation with Optimized Channels. *Sensors* **2018**, *19*, 6. [CrossRef] [PubMed]
16. Hu, J. An approach to EEG-based gender recognition using entropy measurement methods. *Knowl.-Based Syst.* **2018**, *140*, 134–141. [CrossRef]
17. Chao, H.; Dong, L.; Liu, Y.; Lu, B. Emotion Recognition from Multiband EEG Signals Using CapsNet. *Sensors* **2019**, *19*, 2212. [CrossRef]
18. Arnau-GonzÁlez, P.; Arevalillo-Herráez, M.; Ramzan, N. Fusing highly dimensional energy and connectivity features to identify affective states from EEG signals. *Neurocomputing* **2017**, *244*, 81–89. [CrossRef]
19. Kim, M.K.; Kim, M.; Oh, E.; Kim, S.P. A Review on the Computational Methods for Emotional State Estimation from the Human EEG. *Comput. Math. Methods Med.* **2013**, *2013*, 573734. [CrossRef]
20. Lu, Y.; Zheng, W.L.; Li, B.; Lu, B.L. Combining Eye Movements and EEG to Enhance Emotion Recognition. In *International Joint Conference on Artificial Intelligence (IJCAI)*; Yang, Q., Wooldridge, M., Eds.; AAAI Press: Palo Alto, CA, USA, 2015; pp. 1170–1176.
21. Jiang, X.; Bian, G.B.; Tian, Z. Removal of Artifacts from EEG Signals: A Review. *Sensors* **2019**, *19*, 987. [CrossRef]
22. Mur, A.; Dormido, R.; Duro, N. An Unsupervised Method for Artefact Removal in EEG Signals. *Sensors* **2019**, *19*, 2302. [CrossRef]
23. Chen, D.W.; Miao, R.; Yang, W.Q.; Liang, Y.; Chen, H.H.; Huang, L.; Deng, C.J.; Han, N. A Feature Extraction Method Based on Differential Entropy and Linear Discriminant Analysis for Emotion Recognition. *Sensors* **2019**, *19*, 1631. [CrossRef] [PubMed]
24. Jenke, R.; Peer, A.; Buss, M. Feature Extraction and Selection for Emotion Recognition from EEG. *IEEE Trans. Affect. Comput.* **2014**, *5*, 327–339. [CrossRef]
25. Campos, J.A.D. Improving BCI-based emotion recognition by combining EEG feature selection and kernel classifiers. *Expert Syst. Appl.* **2016**, *47*, 35–41.
26. Gross, J.J.; John, O.P. Revealing Feelings: Facets of Emotional Expressivity in Self-Reports, Peer Ratings, and Behavior. *J. Pers. Soc. Psychol.* **1997**, *72*, 435–448. [CrossRef] [PubMed]

27. Chen, J.; Hu, B.; Wang, Y.; Moore, P.; Dai, Y.; Feng, L.; Ding, Z. Subject-independent emotion recognition based on physiological signals: A three-stage decision method. *BMC Med. Inform. Decis. Mak.* **2017**, *17*, 167. [CrossRef] [PubMed]
28. Wang, X.W.; Nie, D.; Lu, B.L. Emotional state classification from EEG data using machine learning approach. *Neurocomputing* **2014**, *129*, 94–106. [CrossRef]
29. Hadjidimitriou, S.; Charisis, V.; Hadjileontiadis, L. Towards a Practical Subject-Independent Affective State Recognition Based On Time-Domain EEG Feature Extraction. *Int. J. Herit. Digit. Era* **2015**, *4*, 165–178. [CrossRef]
30. Li, X.; Song, D.; Zhang, P.; Zhang, Y.; Hou, Y.; Hu, B. Exploring EEG Features in Cross-Subject Emotion Recognition. *Front. Neurosci.* **2018**, *12*, 162. [CrossRef]
31. Salmeron-Majadas, S.; Arevalillo-Herráez, M.; Santos, O.C.; Saneiro, M.; Cabestrero, R.; Quirós, P.; Arnau, D.; Boticario, J.G. Filtering of Spontaneous and Low Intensity Emotions in Educational Contexts. In *Artificial Intelligence in Education*; Conati, C., Heffernan, N., Mitrovic, A., Verdejo, M.F., Eds.; Springer International Publishing: Cham, Switzerland, 2015; pp. 429–438.
32. Ayesh, A.; Arevalillo-Herráez, M.; Ferri, F. Cognitive reasoning and inferences through psychologically based personalised modelling of emotions using associative classifiers. In Proceedings of the IEEE 13th International Conference on Cognitive Informatics Cognitive Computing (ICCI*CC), London, UK, 18–20 August 2014; pp. 67–72. [CrossRef]
33. Koelstra, S.; Muhl, C.; Soleymani, M.; Lee, J.S.; Yazdani, A.; Ebrahimi, T.; Pun, T.; Nijholt, A.; Patras, I. DEAP: A Database for Emotion Analysis using Physiological Signals. *IEEE Trans. Affect. Comput.* **2012**, *3*, 18–31. [CrossRef]
34. Soleymani, M.; Lichtenauer, J.; Pun, T.; Pantic, M. A Multimodal Database for Affect Recognition and Implicit Tagging. *IEEE Trans. Affect. Comput.* **2012**, *3*, 42–55. [CrossRef]
35. Katsigiannis, S.; Ramzan, N. DREAMER: A Database for Emotion Recognition Through EEG and ECG Signals from Wireless Low-cost Off-the-Shelf Devices. *IEEE J. Biomed. Health Inform.* **2017**, *22*, 98–107. [CrossRef] [PubMed]
36. Calvo, R.A.; D'Mello, S. Affect detection: An interdisciplinary review of models, methods, and their applications. *IEEE Trans. Affect. Comput.* **2010**, *1*, 18–37. [CrossRef]
37. Zeng, Z.; Pantic, M.; Roisman, G.I.; Huang, T.S. A survey of affect recognition methods: Audio, visual, and spontaneous expressions. *IEEE Trans. Pattern Anal. Mach. Intell.* **2009**, *31*, 39–58. [CrossRef] [PubMed]
38. Schuller, H.G.B. Categorical and dimensional affect analysis in continuous input: Current trends and future directions. *Image Vis. Comput.* **2013**, *31*, 120–136.
39. Smith, C.A.; Ellsworth, P.C. Patterns of cognitive appraisal in emotion. *J. Pers. Soc. Psychol.* **1985**, *48*, 813. [CrossRef] [PubMed]
40. Sohaib, A.T.; Qureshi, S.; Hagelbäck, J.; Hilborn, O.; Jerčić, P. Evaluating Classifiers for Emotion Recognition Using EEG. In *Foundations of Augmented Cognition*; Schmorrow, D.D., Fidopiastis, C.M., Eds.; Springer: Berlin/Heidelberg, Germany, 2013; pp. 492–501.
41. Petrantonakis, P.; Hadjileontiadis, L. Adaptive Emotional Information Retrieval From EEG Signals in the Time-Frequency Domain. *IEEE Trans. Signal Process.* **2012**, *60*, 2604–2616. [CrossRef]
42. Olivetti, E.; Kia, S.M.; Avesani, P. MEG decoding across subjects. In Proceedings of the 2014 International Workshop on Pattern Recognition in Neuroimaging, Tubingen, Germany, 4–6 June 2014; pp. 1–4.
43. Kia, S.M.; Pedregosa, F.; Blumenthal, A.; Passerini, A. Group-level spatio-temporal pattern recovery in MEG decoding using multi-task joint feature learning. *J. Neurosci. Methods* **2017**, *285*, 97–108. [CrossRef]
44. Murugappan, M.; Nagarajan, R.; Yaacob, S. Combining spatial filtering and wavelet transform for classifying human emotions using EEG Signals. *J. Med. Biol. Eng.* **2011**, *31*, 45–51. [CrossRef]
45. Brown, L.; Grundlehner, B.; Penders, J. Towards wireless emotional valence detection from EEG. In Proceedings of the Annual International Conference of the IEEE Engineering in Medicine and Biology Society, EMBC, Boston, MA, USA, 30 August–3 September 2011; pp. 2188–2191. [CrossRef]
46. Petrantonakis, P.; Hadjileontiadis, L. Emotion Recognition From EEG Using Higher Order Crossings. *IEEE Trans. Inf. Technol. Biomed.* **2010**, *14*, 186–197. [CrossRef]
47. Lin, Y.P.; Wang, C.H.; Jung, T.P.; Wu, T.L.; Jeng, S.K.; Duann, J.R.; Chen, J.H. EEG-Based Emotion Recognition in Music Listening. *IEEE Trans. Biomed. Eng.* **2010**, *57*, 1798–1806. [CrossRef]

48. Petrantonakis, P.C.; Hadjileontiadis, L.J. A novel emotion elicitation index using frontal brain asymmetry for enhanced EEG-based emotion recognition. *IEEE Trans. Inf. Technol. Biomed.* **2011**, *15*, 737–746. [CrossRef]
49. Kaundanya, V.; Patil, A.; Panat, A. Performance of k-NN classifier for emotion detection using EEG signals. In Proceedings of the International Conference on Communications and Signal Processing (ICCSP), Melmaruvathur, India, 2–4 April 2015; pp. 1160–1164. [CrossRef]
50. AlZoubi, O.; Calvo, R.A.; Stevens, R.H. Classification of EEG for affect recognition: An adaptive approach. In *AI 2009: Advances in Artificial Intelligence*; Springer: Heidelberg, Germany, 2009; pp. 52–61.
51. Wang, Q.; Sourina, O. Real-time mental arithmetic task recognition from EEG signals. *IEEE Trans. Neural Syst. Rehabil. Eng.* **2013**, *21*, 225–232. [CrossRef] [PubMed]
52. Murugappan, M.; Ramachandran, N.; Sazali, Y. Classification of human emotion from EEG using discrete wavelet transform. *J. Biomed. Sci. Eng.* **2010**, *3*, 390–396. [CrossRef]
53. Russell, J.A. Affective Space is Bipolar. *J. Personal. Soc. Psychol.* **1979**, *37*, 345. [CrossRef]
54. Mehrabian, A. Pleasure-arousal-dominance: A general framework for describing and measuring individual differences in Temperament. *Curr. Psychol.* **1996**, *14*, 261–292. [CrossRef]
55. Liu, Y.; Sourina, O. EEG-based subject-dependent emotion recognition algorithm using fractal dimension. In Proceedings of the IEEE International Conference on Systems, Man and Cybernetics (SMC), San Diego, CA, USA, 5–8 October 2014; pp. 3166–3171. [CrossRef]
56. Jirayucharoensak, S.; Pan-Ngum, S.; Israsena, P. EEG-based emotion recognition using deep learning network with principal component based covariate shift adaptation. *Sci. World J.* **2014**, *2014*, 627892. [CrossRef] [PubMed]
57. Rozgic, V.; Vitaladevuni, S.; Prasad, R. Robust EEG emotion classification using segment level decision fusion. In Proceedings of the IEEE International Conference on Acoustics, Speech and Signal Processing (ICASSP), Vancouver, BC, Canada, 26–31 May 2013; pp. 1286–1290. [CrossRef]
58. Aspinall, P.; Mavros, P.; Coyne, R.; Roe, J. The urban brain: Analysing outdoor physical activity with mobile EEG. *Br. J. Sports Med.* **2015**, *49*, 272–276. [CrossRef]
59. Liu, Y.; Sourina, O.; Nguyen, M.K. Real-Time EEG-Based Human Emotion Recognition and Visualization. In Proceedings of the International Conference on Cyberworlds (CW), Singapore, 20–22 October 2010; pp. 262–269. [CrossRef]
60. van der Maaten, L.; Hinton, G. Visualizing data using t-SNE. *J. Mach. Learn. Res.* **2008**, *9*, 2579–2605.
61. Ruiz-Blondet, M.V.; Jin, Z.; Laszlo, S. CEREBRE: A novel method for very high accuracy event-related potential biometric identification. *IEEE Trans. Inf. Forensics Secur.* **2016**, *11*, 1618–1629. [CrossRef]
62. Armstrong, B.C.; Ruiz-Blondet, M.V.; Khalifian, N.; Kurtz, K.J.; Jin, Z.; Laszlo, S. Brainprint: Assessing the uniqueness, collectability, and permanence of a novel method for ERP biometrics. *Neurocomputing* **2015**, *166*, 59–67. [CrossRef]
63. Thomas, K.P.; Vinod, A.P.; Robinson, N. Online Biometric Authentication Using Subject-Specific Band Power features of EEG. In Proceedings of the 2017 International Conference on Cryptography, Security and Privacy, Wuhan, China, 17–19 March 2017; pp. 136–141.
64. Bozhkov, L.; Georgieva, P.; Santos, I.; Pereira, A.; Silva, C. EEG-based subject independent affective computing models. *Procedia Comput. Sci.* **2015**, *53*, 375–382. [CrossRef]
65. Jatupaiboon, N.; Pan-ngum, S.; Israsena, P. Real-time EEG-based happiness detection system. *Sci. World J.* **2013**, *2013*, 618649. [CrossRef] [PubMed]
66. Arnau-González, P.; Arevalillo-Herráez, M.; Katsigiannis, S.; Ramzan, N. On the influence of affect in EEG-based subject identification. *IEEE Trans. Affect. Comput.* **2018**. [CrossRef]
67. Garcia, S.; Herrera, F. An Extension on Statistical Comparisons of Classifiers over Multiple Data Sets for all Pairwise Comparisons. *J. Mach. Learn. Res.* **2008**, *9*, 2677–2694.

sensors

MDPI

Article

EEG Based Classification of Long-Term Stress Using Psychological Labeling

Sanay Muhammad Umar Saeed [1], Syed Muhammad Anwar [2,3,*], Humaira Khalid [4], Muhammad Majid [1] and Ulas Bagci [3]

1 Department of Computer Engineering, University of Engineering and Technology, Taxila 47050, Pakistan; sanay.muhammad@uettaxila.edu.pk (S.M.U.S.); m.majid@uettaxila.edu.pk (M.M.)
2 Department of Software Engineering, University of Engineering and Technology, Taxila 47050, Pakistan
3 Department of Computer Science, University of Central Florida, Orlando, FL 32816, USA; bagci@ucf.edu
4 Department of Psychology, Benazir Bhutto Hospital, Rawalpindi 46000, Pakistan; merimalik18@gmail.com
* Correspondence: s.anwar@knights.ucf.edu

Received: 18 February 2020; Accepted: 25 March 2020; Published: 29 March 2020

Abstract: Stress research is a rapidly emerging area in the field of electroencephalography (EEG) signal processing. The use of EEG as an objective measure for cost effective and personalized stress management becomes important in situations like the nonavailability of mental health facilities. In this study, long-term stress was classified with machine learning algorithms using resting state EEG signal recordings. The labeling for the stress and control groups was performed using two currently accepted clinical practices: (i) the perceived stress scale score and (ii) expert evaluation. The frequency domain features were extracted from five-channel EEG recordings in addition to the frontal and temporal alpha and beta asymmetries. The alpha asymmetry was computed from four channels and used as a feature. Feature selection was also performed to identify statistically significant features for both stress and control groups (via *t*-test). We found that support vector machine was best suited to classify long-term human stress when used with alpha asymmetry as a feature. It was observed that the expert evaluation-based labeling method had improved the classification accuracy by up to 85.20%. Based on these results, it is concluded that alpha asymmetry may be used as a potential bio-marker for stress classification, when labels are assigned using expert evaluation.

Keywords: long-term stress; electroencephalography; machine learning; perceived stress scale; expert evaluation

---

1. Introduction

The response of the human body to a demand for change is considered as stress [1]. A balance exists between the sympathetic and parasympathetic arms of the autonomic nervous system in healthy people. A fight-or-flight response is invoked when there is an exposure to a threatening situation. Daily routine stress does not pose any danger to life, however, the fight-or-flight response may still be invoked. A persistence of this short-term stress for a longer duration can cause long lasting effects on the neurology of an individual and may give rise to depression [2]. Long-term stress is a better predictor of depressive symptoms as compared to short-term stress [3]. Long term stress is considered a risk factor for many health conditions such as cardiovascular diseases [4,5].

The prevention of the onset of depression requires a timely detection of long-term stress symptoms. Conventional psychological methods and analysis of hormones such as cortisol and alpha-amylase are widely used in long-term stress studies [6]. These methods are practical but they are affected by various factors, such as language and objectivity. For instance, the Perceived Stress Scale (PSS) is a widely used questionnaire to measure the level of chronic stress, validated extensively across diverse samples [7]. Though in general, a self-administered checklist cannot equal the precision of an interviewer trained to elicit aspects of events critical to examine stress. Such interviews have shown to provide substantially better information in comparison to relatively unassisted self-reporting mechanisms [8]. Respondents have been found to report minor or positive events in response to questions designed to elicit negative and undesirable events [9].

Psychological methods alone are not enough to assess stress-related conditions [10]. Stress can be quantified objectively from bio-markers like electroencephalography (EEG), galvanic skin response, and electrocardiography [11]. Recently, wearable systems were developed that can record electro-physiological signals (such as EEG and heart rate variability) to detect acute stress [12]. EEG is one of the most common source of information for studying brain function [13–17]. The oscillations generated by the variation of electric potential in the brain are recorded using low resistance electrodes placed on the human scalp [18]. It is a widely used noninvasive method due to its excellent temporal resolution, ease of use, and low cost. EEG signals are categorized by their frequency bands including delta, theta, alpha, beta, and gamma. Each frequency band can be used as a discriminating feature for different brain states [19]. There are methods reported in literature to quantify human acute stress in response to induced stressors using EEG signal recordings. In comparison, the classification of long-term or chronic stress using EEG has not been widely assessed.

Our Contributions

In this study, the problem of long-term human stress recognition is addressed by using PSS labels and expert evaluation, which has not been explored before. We have hypothesized that wearable sensors (such as those for recording brain activity using EEG electrodes) can be used for identifying chronic stress, without inducing stress using a stimulus. To this end, our experiments have shown that involving a psychology expert for labeling stressed and control subjects is beneficial for such a classification. It is important to note here that we did not use any stimulus in our study to induce stress so that this system can be administered for detecting stress in daily life routine. Two groups of participants were considered including the stressed group and the control group. A total of forty five different features were extracted from EEG signals in frequency domain to classify these two groups. Discriminating features were selected using a statistical significance test. Five different machine learning classifiers including support vector machine (SVM), Naive Bayes (NB), K-nearest neighbor (KNN), logistic regression (LR), and multi-layer perceptron (MLP) were used to classify human stress using the selected features. Due to limitations of the data size and the noisy nature of the signals, deep-learning-based systems were not suitable for the task at hand. Therefore, we concentrated on machine learning classifiers that are more suitable for the task that we target to solve. The summary of our findings in this study is as follows:

1. We used EEG signals acquired from 33 participants in closed eye conditions using a five-channel EEG headset for long term stress classification (no stimuli used to induce stress) and found that among different feature, three frequency domain features were statistically significant in stress and control groups.
2. To the best of our knowledge, this is the first that the stress level of participants was labeled by a psychology expert in an EEG-based study. We showed its feasibility with a validated set of experiments.
3. The conventional machine learning classifiers suite well to long-term human stress classification and give better performance using psychological expert labeling.

The rest of the paper is organized as follows: Section 2, describes the related work. Section 3 presents the proposed methodology including data collection, feature extraction, and classification algorithms. Section 4 presents the results and a comparison with previously reported studies. Finally, the conclusion of the study is given in Section 5.

2. Related Work

Hemispheric specialization is a major concern in neuro-physiological research. Generally, a healthy brain at rest has a fairly balanced level of activity in both hemispheres of brain [20]. The left hemisphere is associated with the processing of positive emotions, while the right hemisphere is associated with the processing of negative emotions [21]. The extent of asymmetry has been suggested to vary under conditions of chronic stress [22]. Frontal asymmetry is highly related to post-traumatic stress disorder (PTSD) [23]. The results in [24], have shown that major depression disorder (MDD) group is significantly right lateralized relative to controls, and both MDD and PTSD displayed more right- than left-frontal activity.

Recently, the feasibility of using EEG in classifying multilevel mental stress has been demonstrated [19], where alpha rhythm at the right pre-frontal cortex was suggested as a suitable bio-marker. A machine learning framework using EEG signals was proposed in [25], where stress was induced by using the Montreal imaging stress task (MIST), and SVM, NB, and LR classifiers were used to classify the stress level of participants. The EEG of participants in resting-state was recorded under negative, positive, and neutral stimulus using soundtracks from the international affective digitized sounds (IADS-2) dataset [26]. Stress detection based on frontal alpha asymmetry was performed using the DEAP dataset, and classification was performed using SVM, KNN, and fuzzy KNN [27]. In [28], a mobile EEG was used to assess stress in humans using EMOTIV EPOC headset in an out-of-lab environment. In an EEG based study, 11 participants were analyzed for the identification of long-term stress [10], including seven mothers of children with mental disability (stress group) and four mothers of healthy children (control group).

A variant of the trier social stress task (TSST) was used to assess stress in 49 participants [29]. Samples of the salivary cortisol and resting state EEG based alpha asymmetry were assessed before and after performing TSST. The frontal and parietal alpha asymmetry was used to classify depression in elderly people [30]. The correlation between frontal and parietal alpha asymmetry, the geriatric depression scale, and the mini mental state examination were analyzed. A high beta activity at the frontal and occipital lobes was observed on the visual input of negative images [31]. The frontal theta activity was shown to decrease due to a stressful mental arithmetic task [32]. In [33], low beta waves in closed eye condition were found to be a strong predictor of perceived stress, where PSS score was predicted by using multiple linear regression. The pre-frontal relative gamma power i.e., the ratio of gamma band and slow brain rhythms, was proposed as a bio marker for identification of stress [34,35].

The related studies presented here can be grouped as either short-term or long-term stress assessment. Short-term stress is measured using a stress eliciting task, while long-term stress is measured without performing any additional mental task. Different techniques have been adopted to measure stress, but most of these techniques require human intervention. Among different physiological measures, EEG has the potential to be used as a measure of stress in daily life. This is due to the fact that EEG headsets are becoming commercially available for observing brain activity in an easy to wear and cost effective manner. The proposed study uses EEG signals acquired with a commercially available EEG headset to identify baseline or long-term stress without relying on stress-inducing tasks.

3. Methodology

We devised a supervised machine learning model for the classification of human stress (Figure 1). A total of 33 volunteers participated in this study. The resting state EEG data for each participant were acquired using an EMOITV Insight headset (https://www.emotiv.com/insight/) in a closed eye condition for three minutes. After EEG signal recording, participants were asked to fill in the PSS-10 questionnaire followed by an interview with the psychology expert. The average time for the interview was 25 min. Based on the PSS scores and interview, the psychology expert grouped each participant in either the stress or the control group.

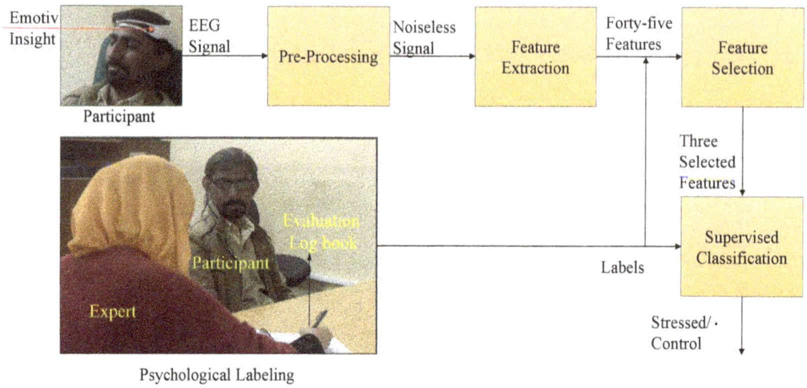

Figure 1. The proposed methodology for long-term human stress classification.

The recorded EEG signals were made noise free in the pre-processing stage. Neuro-physiological features including alpha (α), low beta (β_l), beta (β), gamma (γ), delta (δ), theta (θ), and relative gamma (RG) power were extracted from the signals at each electrode. Frontal and temporal alpha and beta asymmetries, and alpha asymmetry was calculated from these features. Five supervised machine learning algorithms (SVM, NB, KNN, LR, and MLP) were used to classify human stress. Two different labeling methods were used, including the perceived stress scale and expert evaluation, where the PSS and interview scores were simultaneously used. A detailed description of these methods is presented in the following subsections. The flow of events during the data acquisition process is shown in Figure 2.

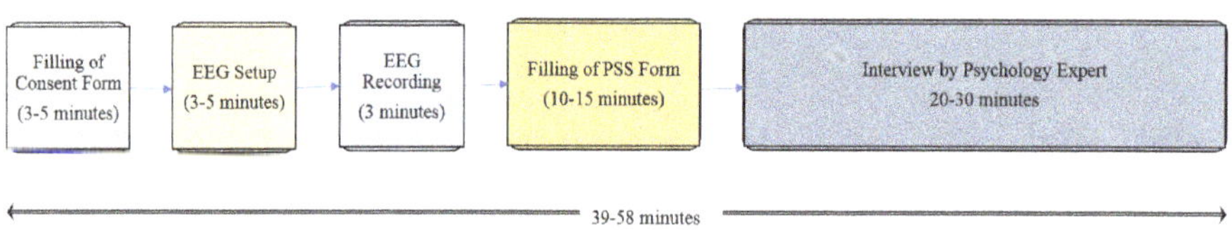

Figure 2. Experimental sequence and the data acquisition process.

3.1. Data Acquisition

All EEG recordings were performed in a noise free lab using the EMOTIV Insight headset, which records brainwaves and provides advanced electronics that are optimized to produce clean and robust signals. Its data transmission rate is 128 samples per second, which provides the ability to perform an in-depth analysis on the brain activity. It has a minimum voltage resolution of 0.51 volts least significant bit (LSB) with 5 EEG electrodes at $AF3, AF4, T7, T8, Pz$ locations and 2 reference electrodes. The headset is shown in Figure 3 with the five electrodes highlighted for reference. The device uses 14 bits for quantization, where 1 LSB = 0.51 μV. A 16-bit analog to digital conversion (ADC) is used, where 2 bits of instrumental noise floor are discarded. The reference electrodes CMS/DRL were located on left mastoid bone. The participants were asked to close their eyes for a duration of three minutes and were instructed to keep their head still to reduce movement artifacts. This also helped in minimizing the muscular motion and reduce these artifacts, since we recorded data at the frontal electrodes. A closed-eye condition was used, since correlates of long-term stress have been found in this condition in previous studies [10,33]. Another advantage of using the closed eye condition is the minimization of eye blink artifact. EEG signal acquisition was performed using the EMOTIV Xavier TestBench v.3.1.21. EEG signals were recorded from the scalp of participants while they were seated in a comfortable chair. Our experiments were specifically carried out in the afternoon (between 3–5 pm) to comply with similar studies where the circadian rhythm was assumed to be similar at this time period for the participants.

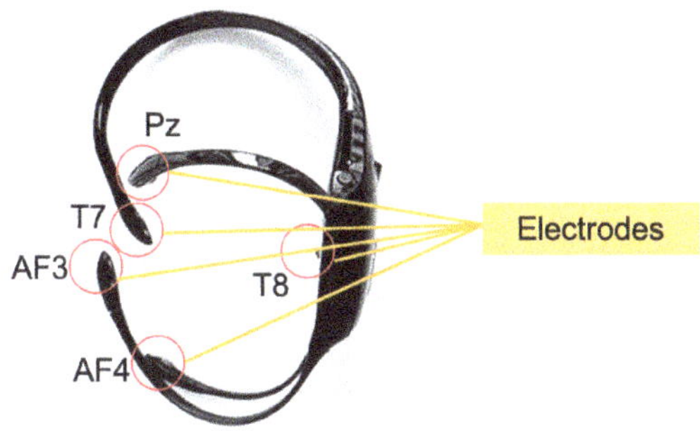

Figure 3. The Emotiv headset with five electrodes marked at positions $AF3, AF4, T7, T8, and Pz$.

3.2. Pre-Processing

The EEG signals recorded from the scalp contained noise due to external interference. Before feature extraction, noise was removed from the signals for better classification results. The data recorded using the EEG electrodes provided by Emotiv have a DC offset in their value that should be removed before doing analysis based on fast Fourier transform. The average value of data from each channel was subtracted from the sample values to remove the DC offset. For reducing muscular artifacts, participants were instructed to minimize their head movements during the EEG acquisition. In a closed eye condition, blink artifacts were also found to be minimal. EMOTIV Insight has a frequency response of 1–43 Hz, which makes the signal noise free from AC line interference at 50 Hz.

3.3. Feature Extraction and Selection

Neural oscillatory features are widely used in literature for EEG-based classification systems. EEG signals are decomposed into different frequency bands. The Welch method was used to extract power spectral densities with a window length of 128 samples with 50 percent overlap. For feature extraction, power spectral densities of different neural oscillations namely, delta (1–3 Hz), theta (4–7 Hz), alpha (8–12 Hz), beta (13–30 Hz), gamma (25–43 Hz), slow (4–13 Hz), and low beta (13–17 Hz) were computed from each channel. Relative gamma waves were computed by taking the ratio of slow and gamma waves. Eight features from each of the five channels adds up to forty features. Moreover, five alpha and beta asymmetries were calculated, giving a total of forty-five neural oscillatory features. The alpha asymmetries were calculated using the following equations,

$$\alpha_f = \frac{\alpha_{AF4} - \alpha_{AF3}}{\alpha_{AF3} + \alpha_{AF4}}, \quad (1)$$

$$\alpha_t = \frac{\alpha_{T8} - \alpha_{T7}}{\alpha_{T8} + \alpha_{T7}}, \quad (2)$$

$$\alpha_a = \alpha_f + \alpha_t, \tag{3}$$

where α_f, α_t and α_a represents the frontal alpha, temporal alpha, and alpha asymmetry respectively and $\alpha_{channel}$ represents the alpha power spectral density of the frontal and temporal EEG channels. Similarly, the frontal and temporal beta asymmetries were calculated using,

$$\beta_f = \frac{\beta_{AF4} - \beta_{AF3}}{\beta_{AF3} + \beta_{AF4}}, \tag{4}$$

$$\beta_t = \frac{\beta_{T8} - \beta_{T7}}{\beta_{T8} + \beta_{T7}}, \tag{5}$$

where β_f, and β_t represents the frontal and temporal beta asymmetries and $\beta_{channel}$ represents the beta power spectral densities for the frontal and temporal EEG channels. Features were selected using a t-test for determining the statistical significance of features in stress and control group. A lower p-value returned by the t-test shows that the feature was significantly discriminating in stress and control group.

3.4. Subject Labeling

The proposed method uses two types of labeling for supervised classification. PSS-10 was used for the questionnaire-based labeling method to subjectively evaluate the stress of participants. This questionnaire consists of ten questions. Each question asks the subject about the frequency of stressful events that have occurred during a period covering the last thirty days. The response for each question is on a scale of 0 to 4, where 0 represents that the event never occurred and 4 represents a frequent occurrence. The total PSS-10 score for each participant has a range between 0 and 40. The participants are divided in two groups i.e., the control and stress group, using the PSS score. A threshold was selected for this purpose, which was given by the following equation,

$$T_p = \mu \pm \frac{\sigma}{2}, \tag{6}$$

where T_p is threshold of PSS score, μ is the mean, and σ is standard deviation of the PSS scores.

The psychologist assigned labels for the stress and control groups after an expert evaluation based on the interview and PSS scores. During the interview, the expert investigated the physical, emotional, behavioral, and cognitive symptoms of stress. Physical symptoms included aches or pain, diarrhea or constipation, nausea, dizziness, chest pain, and rapid heart rate. Emotional symptoms of stress included depression, anxiety, moodiness, irritability, overwhelming feelings, and loneliness. Behavioral and cognitive symptoms included memory problems, inability to concentrate, poor judgment, negativity, racing thoughts, and constant worrying. The interviews were conducted by the psychologist who was affiliated with a public sector hospital. The labels (control/subject) were assigned to participants by the expert based on the responses and the PSS score for each participant. The eighteen symptoms evaluated by the expert are presented in Table 1. The assigned labels were used as ground truth for training the system using the corresponding EEG recordings for each subject.

Table 1. The symptoms evaluated by expert psychologist during the interview process.

Sr. No.	Symptom	Type of Symptom
1	Aches and pains	Physical
2	Diarrhea or constipation	Physical
3	Nausea & Physical pain	Physical
4	Dizziness	Physical
5	Chest pain	Physical
6	Rapid heart beat	Physical
7	Depression or general happiness	Emotional
8	Anxiety or Agitation	Emotional
9	Moodiness	Emotional
10	Irritability	Emotional
11	Feeling overwhelmed	Emotional
12	Loneliness and isolation	Emotional
13	Memory problems	Behavioral and Cognitive
14	Inability to concentrate	Behavioral and Cognitive
15	Poor judgment	Behavioral and Cognitive
16	Seeing only the negative	Behavioral and Cognitive
17	Anxious or racing thoughts	Behavioral and Cognitive
18	Constant worrying	Behavioral and Cognitive

3.5. Stress Classification

In this study, five different types of classifiers were used for classification, which are described in the following subsections very briefly to make the manuscript self contained.

3.5.1. Support Vector Machine

A support vector machine uses the statistical learning theory based on the principle of structural risk minimization. An SVM selects a hyper-plane, which separates the feature space in to control and stress group according to the labels provided. The SVM is a highly efficient classifier and is used widely for stress classification in EEG based studies [19,25]. The use of SVM reduces the risk of data over-fitting and provides good generalization performance.

3.5.2. The Naive Bayes

Naive Bayes is a probabilistic classifier based on Bayes theorem. It uses the maximum posterior hypothesis of statistics and works well for high dimensional input data. It is a nonlinear classifier and gives good results in real world problems. In addition, the Naive Bayes classifier requires a small amount of training data to approximate the statistical parameters [36].

3.5.3. K-Nearest Neighbors

KNN is an instance-based learning classifier, where training instances are stored in their original form. A distance function is used to determine the member of the training set, which is nearest to a test example and used to predict the class. The distance function is easily determined if the attributes are numeric. Most instance-based classifiers use Euclidean distance for distance calculation. The distance between an instance with attribute values $a_1, a_2, ..., a_n$ (where n is the number of attributes) and $b_1, b_2, ..., b_n$ is defined as,

$$D_g = \sqrt{(a_k - b_k)^2}. \tag{7}$$

3.5.4. Logistic Regression

The logistic regression algorithm guards against over-fitting by penalizing large coefficients. The output is set to one for training instances belonging to the class and zero otherwise. Logistic regression builds a linear model based on a transformed target variable, where a transformation function converts a nonlinear function to a linear function.

3.5.5. Multi-Layer Perceptron

In a multi-layer perceptron structure, transfer functions are used for mapping inputs to the output. These functions include sigmoid function, rectified linear unit, and hyperbolic tangent. The classifier uses back-propagation to classify instances. Multi-layer perceptrons are trained by minimizing the squared error of the network output, essentially treating it as an estimate of the class probability, which is given by the following equation,

$$E = \frac{1}{2}((y - f(x)^2)), \quad (8)$$

where $f(x)$ is the network prediction obtained from the output unit and y is the instance class label.

4. Results and Discussion

4.1. Dataset

A total of 33 participants related to the education field volunteered for this study. The participants reported no history of brain injury and they were not using any medications that could have affected their brain activity at the time of experiment. Among these 33 healthy participants, 20 were male and 13 were females (60.6% male and 39.4% female). The participant's ages ranged from 18 to 40 years (μ = 23.85, SD = 5.48). In line with the Helsinki Declaration [37] and the departmental ethics guidelines, all participants of the study were briefed about the research goals. In addition, a signed informed consent was obtained from each participant. This study was approved by the Directorate of Advanced Studies and Research at the University of Engineering and Technology, Taxila.

4.2. Performance Parameters

The parameters used in this study include average accuracy rate, Kappa statistic, F-measure, mean absolute error (MAE), and root mean absolute error (RMAE). Accuracy is the ratio of truly classified instances over total number of instances in the recorded data. F-measure is calculated by considering the precision and recall values. The Kappa statistic values ranges between 0 and 1, where 0 represents chance level classification and 1 means perfect classification. A value less than zero shows that the classification is worse than chance level. For stress classification, the generalization performance of the proposed system was tested using cross validation to avoid over- and under-fitting as well as to make sure the the proposed system adopts well to unseen data. A 10-fold cross validation technique was used in this study, where the training data was randomly divided into ten equal parts (nine parts for train and one part for test) and the process was repeated 10 times. During the process, every instance was used for testing at a time and the remaining instances were used for training of the classifier.

4.3. Stress and Control Group

The scores acquired from participants using the PSS questionnaire are shown in Figure 4. The green and red bars represent the PSS scores of participants belonging to the control and stress groups respectively. The yellow bars indicate the PSS scores of participants not considered in either the stress or the control group. Overall, for the PSS scores we have $(\mu, \sigma) = (20.4 \pm 6.14)$. A participant with a PSS score below 17.33 was considered to be in control group, whereas a participant with a PSS score higher than 23.47 was categorized in the stress group. These values were calculated using the threshold criteria defined in Equation (6). Hence 12 participants were put into the stress group (red bars) and 9 into the control group (green bars).

In expert (hybrid) evaluation, the psychology expert considered both PSS scores and the symptoms obtained from the interview method. The expert interviewed each participant for an average duration of 25 min. Out of the 33 participants, 10 were assigned to the stress group and 10 were assigned to the control group. The details about each participant regarding gender, age, PSS score, the label assigned by using PSS score, and the label assigned by expert is given in Table 2. There were fifteen differences in the assigned labels between those assigned using PSS scores and the expert (hybrid) evaluation. The experimental results show that expert (hybrid) labeling helps in improving the classification of long-term stress. It is important to note here that in a majority of the cases regarding label mismatch (13 out of 15), the PSS score ranges between 17 and 25, which covers the neutral range. Since we hypothesize that the expert (hybrid) labeling is better suited for the classification task, we have used these labels as ground truth.

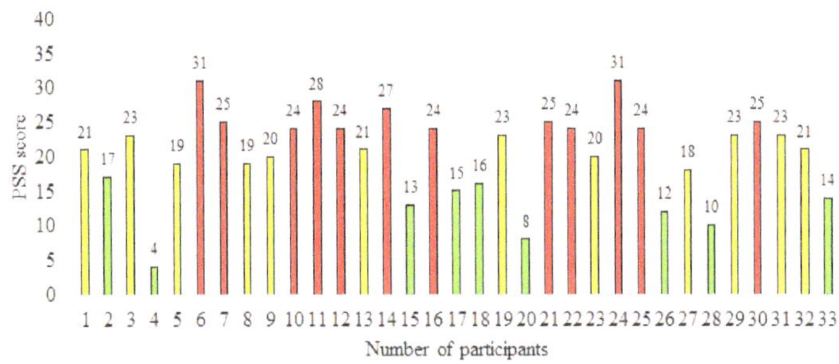

Figure 4. A graphical representation of Perceived Stress Scale (PSS) scores for participants showing labels assigned using the PSS based labeling method (green: control group, red: stress group, yellow: neutral).

Table 2. Gender, age, PSS score, and labels for the participants according to PSS and expert-based (hybrid) labeling (A-control group, B-stress group, X-neutral).

Participant No.	Gender	Age	PSS Score	PSS Label	Expert Label
1	M	28	21	X	X
2	M	29	17	A	X
3	M	23	23	X	X
4	M	32	4	A	A
5	F	19	19	X	A
6	F	18	31	B	B
7	M	24	25	B	X
8	M	33	19	X	A
9	M	21	20	X	B
10	M	22	24	B	X
11	F	20	28	B	B
12	M	19	24	B	B
13	M	24	21	X	A
14	F	20	27	B	B
15	M	23	13	A	X
16	M	21	24	B	X
17	F	19	15	A	A
18	M	25	16	A	A
19	F	21	23	X	B
20	M	34	8	A	A
21	M	33	25	B	X
22	F	21	24	B	B
23	M	31	20	X	B
24	F	24	31	B	B
25	F	20	24	B	B
26	M	19	12	A	A
27	M	21	18	X	A
28	M	21	10	A	X
29	F	21	23	X	X
30	F	23	25	B	X
31	M	20	23	X	X
32	M	40	21	X	X
33	F	20	14	A	A

4.4. Feature Selection Using t-Test

We used a two-sided Student's *t*-test with a significance level of 0.05 and results using the *p*-values are shown in Table 3 for different EEG oscillations. For the *t*-test, the degree of freedom was 9 and the null hypothesis was tested for various features for stress and control groups. It is evident that at a confidence level of 0.05, none of the extracted feature were found statistically significant in the stress and control condition when PSS-based labeling was used for the reference standard. It is also revealed that beta and gamma waves from $AF3$ are statistically significant features in the stress and control group, when labels assigned by expert evaluation were used as a reference standard. Five additional features, namely frontal (α_f) and temporal (α_t) alpha asymmetries, frontal (β_f) and temporal (β_t) beta asymmetries, and alpha asymmetry (α_a) were also used (see Equations (1)–(5)). Results of the *t*-test applied over these features in stress and control groups are presented in Table 4. It can be seen that alpha asymmetry is statistically different between the stress group and the control group using expert-based labeling. Three significant features, namely beta (AF3), gamma (AF3), and alpha asymmetry were selected for long-term stress classification based on the results of *t*-test. A *p*-value of 0.04 and 0.03 for beta and gamma oscillations indicated their statistical significance. The *p*-value of alpha asymmetry from frontal and temporal channels was 0.0005, indicating the statistical significance of alpha asymmetry from both temporal and frontal regions.

The box plots are presented in Figure 5, where the first row represents features acquired through PSS labeling including alpha asymmetry, beta, and gamma respectively. The second row shows the same features acquired through expert evaluation. The + indicates an outlier, and the red line within the box represents the median value. A comparatively short box plot suggests that the features are in agreement with each other. A taller box plot suggests features show different distribution within themselves. From box plots (Figure 5b,c,e,f) it is observed that there is not much difference in the beta and relative gamma features to differentiate stress and control groups for both expert- and PSS-based labels. However, Figure 5a,d are candidates for good features as they appear to differentiate the stress and control group. In Figure 5a, the alpha asymmetry for the stressed group does not have a long lower whisker, which shows alpha asymmetry is not varied along the negative quartile, while in Figure 5d, the stressed group has varied alpha asymmetry as shown by the lower and upper whiskers. Also, in Figure 5d the median is comparatively at the center of the distribution. This suggests that alpha asymmetry is a good candidate to be used in the stress classification task.

Table 3. Results for the *t*-test on various neural oscillations including PSS and expert-based labeling methods.

labeling Method	Channel	delta (δ)	theta (θ)	slow	Neural Oscillations alpha (α)	β_l	beta (β)	gamma (γ)	RG
PSS	AF3	0.12	0.09	0.13	0.28	0.30	0.21	0.32	0.53
	T7	0.89	0.81	0.61	0.21	0.58	0.85	0.52	0.36
	Pz	0.15	0.16	0.16	0.19	0.29	0.46	0.64	0.30
	T8	0.89	0.97	0.95	0.87	0.49	0.97	0.90	0.26
	AF4	0.14	0.12	0.13	0.22	0.20	0.15	0.23	0.79
Expert	AF3	0.65	0.50	0.51	0.08	0.95	**0.04**	**0.03**	0.23
	T7	0.92	0.60	0.51	0.15	0.99	0.42	0.54	0.99
	Pz	0.91	0.89	0.90	0.90	0.93	0.69	0.34	0.40
	T8	0.54	0.51	0.55	0.48	0.85	0.96	0.85	0.56
	AF4	0.11	0.12	0.12	0.35	0.25	0.21	0.28	0.61

Table 4. Asymmetries in PSS and expert evaluation.

Features	α_t	α_f	β_t	β_f	α_a
PSS	0.23	0.39	0.91	0.45	0.11
Expert	0.21	0.07	0.49	0.73	0.0005

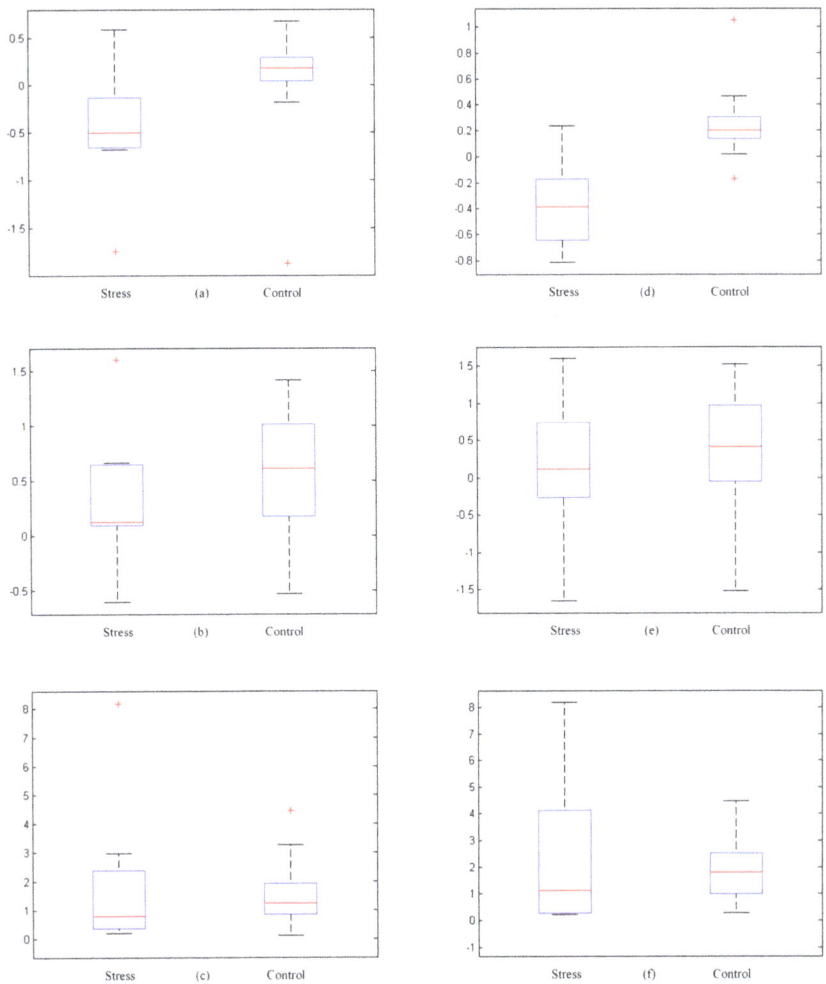

Figure 5. Box plots of features. (**a**) Alpha asymmetry; (**b**) beta; (**c**) gamma; (**d**) alpha asymmetry (EE); (**e**) beta (EE); (**f**) gamma (EE); EE represents the labeling method of expert evaluation.

4.5. Classification

We performed a comprehensive set of experiments to test and validate our proposed model using five classifiers, namely, KNN, NB, SVM, LR, and MLP. These classifiers were used with alpha asymmetry, beta, and gamma waves from channel *AF*3 as features to classify long-term stress. Each combination of the selected features was analyzed with each of the classifiers. The results of these classifiers in terms of average accuracy are shown in Table 5. We used 10-fold cross validation in these experiments since our dataset was limited. We used 10 folds, where in each fold 90% of the data were used for training and 10% for testing and reporting the average values of parameters across all 10 folds. The hyper parameters for classifiers used in our experiment were chosen using a grid search.

We observed that the classifier accuracy was high whenever alpha asymmetry was either used as a single feature or in combination with other features. The SVM- and LR-based classifiers give the highest accuracy when alpha asymmetry was used as a feature. The performance evaluation parameters for these classifiers are given in Table 6. We also observed that both SVM and LR show very similar values for kappa statistic and F-measure. SVM may have a slightly lesser mean absolute error of 0.15 than that of logistic regression with a value of 0.22, whereas LR has a lesser RMAE of 0.36 than that of SVM i.e., 0.38. The overall classification accuracy of both these classifier is similar. Overall, we concluded that SVM may be a better choice for an assisting system for stress recognition.

Table 5. Accuracy of classifiers for various combinations of statistically significant features.

Features	SVM	NB	KNN	LR	MLP
α_a	85.20	80.11	65.32	85.15	80.12
γ	70.32	50.21	50.43	50.33	50.17
β	55.07	50.01	50.51	50.48	50.70
β, γ	70.45	50.65	50.09	50.65	50.02
α_a, β	85.15	80.02	65.38	85.04	85.01
α_a, γ	80.91	80.79	65.55	85.08	85.05
α_a, β, γ	80.83	80.77	65.96	85.09	85.13

Table 6. Evaluation parameters for the best performing classifiers with α_a as a feature.

Classifier	Average Accuracy	Kappa	F-Measure	MAE	RMAE
LR	85.15	0.70	0.85	0.22	0.36
SVM	85.20	0.71	0.87	0.15	0.39

4.6. Discussion

Numerous studies have analyzed brain activities under stressful conditions, which are induced by a task such as impromptu speech, examination, mental task, public speaking, and the cold pressor test [38–43]. These studies evaluate short-term induced stress, whereas the classification of long-term stress using EEG has not been widely investigated. In Table 7, studies involving EEG to classify human stress are presented for comparison. It is observed that different stress-inducing tasks were used such as driving simulation, examination, and mental arithmetic tasks. Specialized instruments like MIST and Stroop tests were also used to induce stress. For chronic stress there could be several stressors that affect the physical, emotional, cognitive, or behavioral well being of a human being. Therefore, it is proposed that recording resting state EEG for stress classification is a better choice without involving stress induction. The number of participants involved in such studies vary from 5 to 42. The SVM and NB were used as classifiers in most of the studies. SVM was found to be the most efficient classifier, giving a maximum accuracy of 96%, when stress was induced by mental arithmetic test. In [10], the resting state EEG was recorded for two minutes and a nonlinear analysis was performed but no classification algorithm was used. In [44], chronic stress has been classified with an accuracy of 90%, using EEG recordings from eight electrodes and a stress-inducing condition.

Despite the difficulties of EEG in stress studies, there are cases where the use of EEG is vital and it has a clinical meaning in various conditions. For instance, ECG is not a direct stress measurement system, especially when mental stress originates in the brain. Furthermore, we studied long-term stress, and we did not have any stress inducer in our study (unlike other ECG- and HRV-based studies); hence, EEG can be a modality of choice for our experiments and we show its effectiveness with our experimental results. Although EEG has not been widely used for long term stress classification in clinical practice, our proposed method attempts to establish this approach. It has been shown that conditions such as anxiety, tension, and depression decrease as the frontal asymmetry shifts to the right hemisphere of the brain giving significance to EEG laterality [22]. It was demonstrated that variations in the beta activity [31] and pre-frontal gamma [34] contribute towards stress assessment. Hence there is evidence suggesting that these oscillations in the pre-frontal brain region can be used for assessment of stress using EEG recordings.

It is shown in this study that the alpha asymmetry of the brain can be considered as a potential marker for the recognition of chronic stress in humans. We observed (Table 5) that the classification accuracy using beta and gamma oscillations was lower when compared to alpha asymmetry. Whenever a combination of alpha asymmetry from beta and gamma oscillations was used, the decision boundaries were changed. Due to this, the classification accuracy was lower when compared to the case when alpha asymmetry was individually used as a feature. The labeling should be performed by using a hybrid method (psychology expert and PSS scores) for training the system in a supervised manner. Due to the limited size of the data, we have shown that MLP is the only class of neural network based classifier that can fit to the task of stress classification. For more deeper networks, we would need more instances of EEG recordings.

Table 7. Comparison of results with previously related EEG-based studies.

Related Work	Stress Inducer	Participants	Classifier	Accuracy
Lin et. al. [45]	Driving simulator	6	KNN and **NBC**	71.77
Vijean et. al. [46]	Mental arithmetic task	5	NN	91.17
Khosrowabadi et. al. [44]	Examination	26	KNN and SVM	90.00
Jun et. al. [47]	Arithmetic task and stroop test	10	SVM	96.00
Al-Shargie et. al. [19]	Mental arithmetic task	18	SVM and **ECoC**	95.37
Subhani et. al. [25]	MIST	42	LR, SVM and NB	94.60
Saeed et. al. [33]	None	28	NB	71.43
Proposed	**None**	**33**	**SVM**	**85.20**

5. Conclusions

In this paper, two different labeling methods were used for the classification of long-term stress in humans using EEG signals. Forty-five signal features were analyzed for the classification of chronic stress, and alpha asymmetry was found to be a discriminating feature when using expert's evaluation as ground truth. The PSS scores, when used solely for labeling, returned no significant features. Furthermore, it is evident from our experimental results that SVM and LR give the highest accuracy (85.20%) for classification. We also observed that the stress group was better classified when compared to the control group irrespective of the classifiers used. Finally, we established that alpha asymmetry can be used a potential bio-marker for the classification of long-term stress with SVM. To the best of our knowledge, no previous EEG-based studies have involved a psychology expert for labeling of groups for long-term stress assessment. In the future, more features and participants will be considered for the analysis. With the availability of more data, deep learning based strategies can be applied for potentially improved methods.

Author Contributions: Conceptualization, S.M.U.S.; data curation, S.M.U.S. and H.K.; formal analysis, H.K., M.M., and U.B.; methodology, S.M.U.S. and S.M.A.; writing—original draft, S.M.U.S. and S.M.A.; writing—review and editing, S.M.A., M.M., and U.B. All authors have read and agreed to the published version of the manuscript.

Funding: This research received no external funding.

Conflicts of Interest: The authors declare that the research was conducted in the absence of any commercial or financial relationships that could be construed as a potential conflict of interest.

References

1. Selye, H. The stress syndrome. *Am. J. Nurs.* **1965**, *65*, 97–99.
2. Heim, C.; Nemeroff, C.B. Neurobiology of early life stress: Clinical studies. *Semin. Clin. Neuropsychiatry* **2002**, *7*, 147–159. [CrossRef] [PubMed]
3. McGonagle, K.A.; Kessler, R.C. Chronic stress, acute stress, and depressive symptoms. *Am. J. Commun. Psychol.* **1990**, *18*, 681–706. [CrossRef] [PubMed]
4. Cohen, S.; Janicki-Deverts, D.; Miller, G.E. Psychological stress and disease. *JAMA* **2007**, *298*, 1685–1687. [CrossRef]

5. Steptoe, A.; Kivimäki, M. Stress and cardiovascular disease. *Nat. Rev. Cardiol.* **2012**, *9*, 360. [CrossRef]
6. Van Praag, H. Can stress cause depression? *Prog. Neuro-Psychopharmacol. Biol. Psychiatry* **2004**, *28*, 891–907. [CrossRef]
7. Hammen, C.; Dalton, E.D.; Thompson, S.M. Measurement of chronic stress. *Encycl. Clin. Psychol.* **2014**, 1–7. [CrossRef]
8. Sobell, L.C.; Toneatto, T.; Sobell, M.B.; Schuller, R.; Maxwell, M. A procedure for reducing errors in reports of life events. *J. Psychosom. Res.* **1990**, *34*, 163–170. [CrossRef]
9. McQuaid, J.R.; Monroe, S.M.; Roberts, J.R.; Johnson, S.L.; Garamoni, G.L.; Kupfer, D.J.; Frank, E. Toward the standardization of life stress assessment: Definitional discrepancies and inconsistencies in methods. *Stress Med.* **1992**, *8*, 47–56. [CrossRef]
10. Peng, H.; Hu, B.; Zheng, F.; Fan, D.; Zhao, W.; Chen, X.; Yang, Y.; Cai, Q. A method of identifying chronic stress by EEG. *Pers. Ubiquitous Comput.* **2013**, *17*, 1341–1347. [CrossRef]
11. Zheng, R.; Yamabe, S.; Nakano, K.; Suda, Y. Biosignal analysis to assess mental stress in automatic driving of trucks: Palmar perspiration and masseter electromyography. *Sensors* **2015**, *15*, 5136–5150. [CrossRef] [PubMed]
12. Ahn, J.W.; Ku, Y.; Kim, H.C. A Novel Wearable EEG and ECG Recording System for Stress Assessment. *Sensors* **2019**, *19*, 1991. [CrossRef] [PubMed]
13. Mehreen, A.; Anwar, S.M.; Haseeb, M.; Majid, M.; Ullah, M.O. A Hybrid Scheme for Drowsiness Detection using Wearable Sensors. *IEEE Sens. J.* **2019**, *19*, 5119–5126. doi:10.1109/JSEN.2019.2904222. [CrossRef]
14. Asif, A.; Majid, M.; Anwar, S.M. Human stress classification using EEG signals in response to music tracks. *Comput. Biol. Med.* **2019**. [CrossRef] [PubMed]
15. Saeed, U.; Muhammad, S.; Anwar, S.M.; Majid, M.; Awais, M.; Alnowami, M. Selection of Neural Oscillatory Features for Human Stress Classification with Single Channel EEG Headset. *BioMed Res. Int.* **2018**, *2018*, 1049257.
16. Raheel, A.; Anwar, S.M.; Majid, M. Emotion recognition in response to traditional and tactile enhanced multimedia using electroencephalography. *Mult. Tools Appl.* **2018**, *78*, 1–15. [CrossRef]
17. Anwar, S.; Saeed, S.; Majid, M.; Usman, S.; Mehmood, C.; Liu, W. A Game Player Expertise Level Classification System Using Electroencephalography (EEG). *Appl. Sci.* **2018**, *8*, 18. [CrossRef]
18. Sanei, S.; Chambers, J.A. *EEG Signal Processing*; Wiely: Hoboken, NJ, USA, 2007.
19. Al-shargie, F.; Tang, T.B.; Badruddin, N.; Kiguchi, M. Towards multilevel mental stress assessment using SVM with ECOC: An EEG approach. *Med. Biol. Eng. Comput.* **2018**, *56*, 125–136. [CrossRef]
20. Fisch, B. *Fisch and Spehlmann's EEG Primer: Basic Principles of Digital and Analog EEG*, 3rd ed.; Elsevier: Amsterdam, The Netherlands, 1999; p. 642.
21. Davidson, R.J. What does the prefrontal cortex "do" in affect: Perspectives on frontal EEG asymmetry research. *Biol. Psychiatry* **2004**, *67*, 219–234. [CrossRef]
22. Papousek, I.; Schulter, G. Covariations of EEG asymmetries and emotional states indicate that activity at frontopolar locations is particularly affected by state factors. *Psychophysiology* **2002**, *39*, 350–360. [CrossRef]
23. Lobo, I.; Portugal, L.C.; Figueira, I.; Volchan, E.; David, I.; Pereira, M.G.; de Oliveira, L. EEG correlates of the severity of posttraumatic stress symptoms: A systematic review of the dimensional PTSD literature. *J. Affect. Disord.* **2015**, *183*, 210–220. [CrossRef] [PubMed]
24. Goncharova, I.I.; Barlow, J.S. Changes in EEG mean frequency and spectral purity during spontaneous alpha blocking. *Electroencephalogr. Clin. Neurophysiol.* **1990**, *76*, 197–204. [CrossRef]
25. Subhani, A.R.; Mumtaz, W.; Saad, M.N.B.M.; Kamel, N.; Malik, A.S. Machine learning framework for the detection of mental stress at multiple levels. *IEEE Access* **2017**, *5*, 13545–13556. [CrossRef]
26. Cai, H.; Han, J.; Chen, Y.; Sha, X.; Wang, Z.; Hu, B.; Yang, J.; Feng, L.; Ding, Z.; Chen, Y.; et al. A Pervasive Approach to EEG-Based Depression Detection. *Complexity* **2018**, *2018*, 5238028. [CrossRef]
27. Baghdadi, A.; Aribi, Y.; Alimi, A.M. Efficient Human Stress Detection System Based on Frontal Alpha Asymmetry. In Proceedings of the 24th International Conference, ICONIP 2017, Guangzhou, China, 14–18 November 2017; pp. 858–867.
28. Aspinall, P.; Mavros, P.; Coyne, R.; Roe, J. The urban brain: analysing outdoor physical activity with mobile EEG. *Br. J. Sports Med.* **2015**, *49*, 272–276. [CrossRef]
29. Düsing, R.; Tops, M.; Radtke, E.L.; Kuhl, J.; Quirin, M. Relative frontal brain asymmetry and cortisol release after social stress: The role of action orientation. *Biol. Psychiatry* **2016**, *115*, 86–93. [CrossRef]

30. Kaiser, A.K.; Doppelmayr, M.; Iglseder, B. Electroencephalogram alpha asymmetry in geriatric depression. *Zeit. Für Geront. Und Ger.* **2018**, *51*, 200–205. [CrossRef]
31. Seo, S.H.; Lee, J.T. Stress and EEG. In *Convergence and Hybrid Information Technologies*; InTech: Rijeka, Croatia, 2010.
32. Gärtner, M.; Grimm, S.; Bajbouj, M. Frontal midline theta oscillations during mental arithmetic: Effects of stress. *Front. Behav. Neurosci.* **2015**, *9*, 96. [CrossRef]
33. Saeed, S.M.U.; Anwar, S.M.; Majid, M. Quantification of human stress using commercially available single channel EEG Headset. *IEICE Trans. Inf. Syst.* **2017**, *100*, 2241–2244. [CrossRef]
34. Minguillon, J.; Lopez-Gordo, M.A.; Pelayo, F. Stress assessment by prefrontal relative gamma. *Front. Comput. Neurosci.* **2016**, *10*, 101. [CrossRef]
35. Arsalan, A.; Majid, M.; Butt, A.R.; Anwar, S.M. Classification of Perceived Mental Stress Using a Commercially Available EEG Headband. *IEEE J. Biomed. Health Inform.* **2019**, *23*, 2257–2264. [CrossRef] [PubMed]
36. Kotsiantis, S.B.; Zaharakis, I.; Pintelas, P. Supervised machine learning: A review of classification techniques. *Emerg. Artif. Intell. Appl. Comput. Eng.* **2007**, *160*, 3–24.
37. Association, W.M. World Medical Association Declaration of Helsinki: Ethical principles for medical research involving human subjects. *J. Am. Coll. Dent.* **2014**, *81*, 14.
38. Knaus, J.; Wiese, R.; Janßen, U. The processing of word stress: EEG studies on task-related components. In Proceedings of the 16th International Congress of Phonetic Sciences, Saarbrücken, Germany, 6–10 August 2007; pp. 709–712.
39. Matsunami, K.; Homma, S.; Han, X.Y.; Jiang, Y.F. Generator sources of EEG large waves elicited by mental stress of memory recall or mental calculation. *Jpn. J. Phys.* **2001**, *51*, 621–624. [CrossRef]
40. Lewis, R.S.; Weekes, N.Y.; Wang, T.H. The effect of a naturalistic stressor on frontal EEG asymmetry, stress, and health. *Biol. Psychiatry* **2007**, *75*, 239–247. [CrossRef]
41. Seo, S.; Gil, Y.; Lee, J. The relation between affective style of stressor on EEG asymmetry and stress scale during multimodal task. In Proceedings of the Third International Conference on Convergence and Hybrid Information Technology, ICCIT'08, Busan, Korea, 11–13 November 2008; Volume 1, pp. 461–466.
42. Miller, P.F.; Light, K.C.; Bragdon, E.E.; Ballenger, M.N.; Herbst, M.C.; Maixner, W.; Hinderliter, A.L.; Atkinson, S.S.; Koch, G.G.; Sheps, D.S. Beta-endorphin response to exercise and mental stress in patients with ischemic heart disease. *J. Psychiatr. Res.* **1993**, *37*, 455–465. [CrossRef]
43. Hassellund, S.S.; Flaa, A.; Sandvik, L.; Kjeldsen, S.E.; Rostrup, M. Long-term stability of cardiovascular and catecholamine responses to stress tests: An 18-year follow-up study. *Hypertension* **2010**, *55*, 131–136. [CrossRef]
44. Khosrowabadi, R.; Quek, C.; Ang, K.K.; Tung, S.W.; Heijnen, M. A Brain-Computer Interface for classifying EEG correlates of chronic mental stress. In Proceedings of the 2011 International Joint Conference on Neural Networks, San Jose, CA, USA, 31 July–5 August 2011; pp. 757–762.
45. Lin, C.T.; Ko, L.W.; Chiou, J.C.; Duann, J.R.; Huang, R.S.; Liang, S.F.; Chiu, T.W.; Jung, T.P. Noninvasive neural prostheses using mobile and wireless EEG. *IEEE* **2008**, *96*, 1167–1183.
46. Vijean, V.; Hariharan, M.; Saidatul, A.; Yaacob, S. Mental tasks classifications using S-transform for BCI applications. In Proceedings of the 2011 IEEE Conference on Sustainable Utilization and Development in Engineering and Technology (STUDENT), Selangor Darul Ehsan, Malaysia, 20–21 October 2011; pp. 69–73.
47. Jun, G.; Smitha, K. EEG based stress level identification. In Proceedings of the 2016 IEEE International Conference on Systems, Man, and Cybernetics (SMC), Budapest, Hungary, 9–12 October 2016; pp. 3270–3274.

sensors

MDPI

Article

The uulmMAC Database—A Multimodal Affective Corpus for Affective Computing in Human-Computer Interaction

Dilana Hazer-Rau [1,*], Sascha Meudt [2], Andreas Daucher [1], Jennifer Spohrs [1], Holger Hoffmann [1,†], Friedhelm Schwenker [2,†] and Harald C. Traue [1,†]

[1] Section Medical Psychology, University of Ulm, Frauensteige 6, 89075 Ulm, Germany
[2] Institute of Neural Information Processing, University of Ulm, James-Frank-Ring, 89081 Ulm, Germany
* Correspondence: dilana.hazer@uni-ulm.de
† These authors contributed equally to this work.

Received: 11 March 2020; Accepted: 14 April 2020; Published: 17 April 2020

Abstract: In this paper, we present a multimodal dataset for affective computing research acquired in a human-computer interaction (HCI) setting. An experimental mobile and interactive scenario was designed and implemented based on a gamified generic paradigm for the induction of dialog-based HCI relevant emotional and cognitive load states. It consists of six experimental sequences, inducing *Interest, Overload, Normal, Easy, Underload*, and *Frustration*. Each sequence is followed by subjective feedbacks to validate the induction, a respiration baseline to level off the physiological reactions, and a summary of results. Further, prior to the experiment, three questionnaires related to emotion regulation (ERQ), emotional control (TEIQue-SF), and personality traits (TIPI) were collected from each subject to evaluate the stability of the induction paradigm. Based on this HCI scenario, the *University of Ulm Multimodal Affective Corpus (uulmMAC)*, consisting of two homogenous samples of 60 participants and 100 recording sessions was generated. We recorded 16 sensor modalities including 4 × video, 3 × audio, and 7 × biophysiological, depth, and pose streams. Further, additional labels and annotations were also collected. After recording, all data were post-processed and checked for technical and signal quality, resulting in the final *uulmMAC* dataset of 57 subjects and 95 recording sessions. The evaluation of the reported subjective feedbacks shows significant differences between the sequences, well consistent with the induced states, and the analysis of the questionnaires shows stable results. In summary, our *uulmMAC* database is a valuable contribution for the field of affective computing and multimodal data analysis: Acquired in a mobile interactive scenario close to real HCI, it consists of a large number of subjects and allows transtemporal investigations. Validated via subjective feedbacks and checked for quality issues, it can be used for affective computing and machine learning applications.

Keywords: affective corpus; multimodal sensors; overload; underload; interest; frustration; cognitive load; emotion recognition; stress research; affective computing; machine learning; human-computer interaction

---

1. Introduction

The rapid technological advancements and the expectations for fast adaptation impose high pressure on humans to deliver maximum effort in stressful constraints and multitasking situations of HCI. Among the variety of emotional and cognitive states in HCI, cognitive load is a prominent "multi-dimensional construct representing the load imposed on the working memory during performance of a cognitive task" [1]. It is highly associated with human effort and with the efficiency of cognitive technical systems during Human-Computer Interaction (HCI) [2]. Following Sweller [3], which focuses on human learning, the intensity of cognitive load experienced for a specific mental task varies between individuals depending on their working memory capacity. Individuals can raise their cognitive effort to adapt to increasing difficulties until mental limit capacities are reached. Above this limit, human performance decreases, involving increase in errors, emergence of stress, and negative affects [2]. An adequate level of cognitive load for an individual is desirable, in order to perform a task in an optimal manner. Results from our transsituational study show indeed the existence of a biological basis for success in human-computer interaction [4]. Therefore, particularly in the context of HCI, knowledge about cognitive load is essential in order to intelligently match the level and nature of the interaction in such systems. The recognition of cognitive load in HCI can enable real-time user's state monitoring and adaptation to the individual users. Individual content generation for distant learning and adaptive learning systems [5], practical training sessions [6], monitoring pilots [7], and truck drivers [8], usability testing and evaluation of user-interface and mobile applications [9], or digital assistance providing personalized advises for stress reduction and health risk prevention strategies [10] are some relevant fields of useful applications.

Estimation of cognitive load can be achieved via various measuring approaches, including subjective measures, performance measures, physiological measures, and behavioral measures [11–15]. Traditional simple measures are based on subjective ratings, asking the users to perform a self-assessment of their mental state. These measures lack objectivity and are not reliable for computational recognition techniques. They are generally used as ground truths in experiments, however with the disadvantage of being acquired after the event. Performance measures can be measured in parallel, but are difficult to evaluate in real-life applications and generally insensitive to load capacity variations. Physiological and behavioral procedures are non-intrusive methods providing a more reliable and direct access to cognitive load in an objective way. Cognitive load recognition using multimodal sensors has the potential to increase the robustness and accuracy compared to estimation from single modality data. Unlike subjective measurements prevalent in psychological research, cognitive load estimation based on human responses is necessary for advanced computational techniques. Further, real-life investigation requires the implementation of mobile measurements "in-the-wild" [16]. Despite all technological advancements, mobile measurements still represent a challenge and can only be realistic if the measuring devices and sensor techniques are reliable and sensitive to wild movements, are at low cost, and easy to wear.

Various datasets were specifically collected for the study of cognitive load. While most of the studies are based on statistical approaches or functional magnetic resonance imaging (fMRI) [17,18], alternative methods including physiological [19,20], text [21,22] speech [23,24], brain [25,26], and pupil change [27,28] analyses, are used to detect cognitive load. The relationship between cognitive load and writing behavior was examined using the CLTex (Cognitive Load via Text), CLSkt (Cognitive Load Sketching) and CLDgt (Cognitive Load via Digits) datasets [29]. The datasets are composed of writing samples of 20 subjects under three cognitive load levels, induced from a writing task experiment. Speech-based cognitive load examination is supported by the Cognitive Load with Speech and Electroglottography (CLSE) dataset [30]. It includes recordings of 26 subjects for the determination of a speaker's cognitive load during speech based on acoustic features. Mattys et al. developed an experiment to induce cognitive load based on a concurrent visual search task for the investigation of the impact of cognitive load on the Ganong effect [24]. The effect of visual presentation was also investigated for the detection of cognitive load: Liu et al. present a contact-free method to improve cognitive load recognition from eye movement signals and for this purpose designed an experiment to induce cognitive load [31]. In their final project report for AOARD Grant, Chen et al. summarize research activities and issues related to multimodal cognitive load recognition in the real world. They examine the use of various electroencephalography (EEG) features, eye activities, linguistic features, skin conductance response, facial activities and writing behavior. An extended version of the report is their book "Robust multimodal cognitive load measurement" presenting all the related issues in details [29].

As for the induction of emotional states, many studies exist focusing on basic emotions in both discrete (i.e., fear, anger, joy, sadness, surprise or disgust) or dimensional (i.e., valence, arousal, dominance) models. These emotional states are especially induced using standardized pictures [32,33] for instance from the International Affective Picture System (IAPS) [34] or relying on audiovisual stimuli [35] used as movie clips [36,37] or as music clips [38]. Emotional states can be also induced using game scenarios by asking the user to perform a certain task [39]. This elicitation method is especially useful for the induction of HCI relevant emotional states such as *Frustration* and *Interest* [40]. These states are relevant in designing efficient and easy-to-use interactive systems [41], in interactive educational and social applications [42], or in therapeutic settings by providing tailored feedback for instance to reduce *Frustration* states [43].

Taylor et al. conducted a study to induce *Frustration* in subjects based on the inclusion of latency between the user's touch and the reaction of the breakout engine [44]. A more recent study on *Frustration* is given by Aslam et al. examining the effects of annoying factors in HCI on feelings of *Frustration* and disappointment [45]. For the induction of *Frustration*, they asked the subjects to fill in a registration form, which fails twice based on intended system errors, before it succeeds in the third time. Additionally, Lisetti et al. designed an experiment for the elicitation of six emotions including *Frustration* in the context of HCI [46]. They collected physiological data via wearable computers and included classification results of three different supervised learning algorithms. In their paper on human-robot interaction, Liu et al. present a comparative study of four machine learning methods using physiological signals for the recognition of five different emotions including *Frustration* [47].

In her article "Interest—the curious emotion", Silvia focuses on the role of *Interest* in learning and motivation and describes its central role in cultivating knowledge and expertise [48]. Additionally, Reeve et al. present a concept of *Interest* in three ways: as a basic emotion, as an affect, and as an emotion schema [49]. They explain the importance of *Interest* in educational settings as a mean to motivate high-quality engagement that leads to positive learning outcomes and as an enrichment of motivational and cognitive resources that leads to high-vitality experience rather than exhaustion. According to Ellsworth, *Interest* can be related to the uncertainty of a positive event which may also lead to curiosity and hope, while lack of control often results in *Frustration*, which if sustained can lead to desperation and resignation [50]. Thus, in a HCI context, providing excitement through an appropriate degree of uncertainty might increase *Interest*, while providing a certain level of controllability, by preventing inexplicit system errors can reduce *Frustration*. The recognition of *Frustration* and the system reaction to turn it into a positive *Interest* state are critical aspects for avoiding negative affective consequences and valuable for enhancing positive interaction effects.

Despite the many studies investigating emotional and cognitive states, particularly *Overload*, *Underload*, *Frustration* and *Interest*, their measurement still poses many challenging issues especially with respect to multimodal, mobile and transtemporal acquisition. Additionally, regarding the validation of the experimental induction, most of the studies limit their validation to one subjective modality. Further, previous studies restrict their induction to either cognitive or emotional elicitation and rarely include both states into one single dataset. In this paper, we focus on these issues and present a database for affective computing research, based on systematic induction of cognitive load (*Overload*, *Underload*) and specific emotions relevant to HCI (*Interest*, *Frustration*) as well as a neutral and a transition state (*Normal*, *Easy*) (see Section 2.2). The database is (1) designed and acquired in a mobile interactive HCI setting, (2) based on multimodal sensor data, (3) involving transtemporal acquisition including different recording times, and (4) validated via three different subjective modalities. Combining these challenging issues related to mobile, interactive, multimodal, transtemporal, and validated acquisition into one large dataset for both cognitive and emotional states are the main contributions of this work.

In the next section (Section 2), the methods are described including a description of the participants and cohorts, interaction scheme, experiment structure, technical implementation and multimodal sensors infrastructure. Following (Section 3), the results are presented including the generated *uulmMAC* database, the validation via questionnaires and subjective feedback, as well as the data annotation. Finally (Section 4), we conclude with a discussion and a summary of the results.

2. Materials and Methods

An experimental mobile interactive and multimodal emotional-cognitive load scenario was designed and implemented for the induction of various cognitive and emotional states in an HCI setting. Based on this mobile and interactive scenario, multimodal data were acquired generating the *University of Ulm Multimodal Affective Corpus (uulmMAC)*. The basic concept of our cognitive load scenario follows a generic scheme from Schüssel et al. who proposed a gamified setup for the exploration of various aspects with potential influence on users' way of interaction [51]. The generic scheme is, however, an abstract fundament for HCI exploration with no specific application field. The induction of emotional and cognitive states depends on various factors related to the specific nature of human reactions [52]. Therefore, for our research question focusing on emotional and cognitive states induction in real-life HCI, the development of the current experiment required further developments with an in-depth adjustment and re-implementation of the original generic paradigm such that to comply with the induction requirements of cognitive load and affective states. The main development contributions include the design of the interaction sequences scheme inducing cognitive, emotional and neutral states (Section 2.2), the development of the experimental structure (Section 2.3) and the software implementation and platform embedment (Section 2.4). Furthermore, for the experimental data acquisition, we developed and implemented a technical infrastructure with multimodal sensors system for the distributed experimental and recording setup (Section 2.5).

2.1. Participants and Cohort Description

The *uulmMAC* dataset consists of two homogenous samples of 60 participants (30 females, 30 males; 17–27 years; mean age = 21.65 years, SD = 2.65) with a total of 100 recording sessions (N = 100) of about 45 minutes each. The 60 subjects are medical students and were recruited through bulletin notices distributed at the campus of the Ulm University. The first sample includes 40 subjects who underwent one measurement each, while the second sample consists of 20 subjects who underwent three measurements each. The three different measurements were acquired at three different times with one week of time-interval in-between. The second sample allows for instance the investigation of additional transtemporal research questions. While both samples underwent exactly the same experiment, they slightly differ in one modality acquisition: The first sample does not include facial electromyography (EMG) measurements, allowing better conditions for the analysis of facial expressions via video data. Both samples are evenly balanced between male and female. All subjects gave their informed consent for inclusion before they participated in the experiment and the study was approved by the Ethics Committee of the Ulm University (Project: C4 - SFB TRR62).

In summary, the original dataset of *uulmMAC* consists of 100 individual recording sessions: The first sample with 40 recording sessions (40 subjects × 1 measurement) and the second sample with 60 recording sessions (20 subjects × 3 measurements).

2.2. The Interaction Scheme

The goal of the experiment was the induction of various dialog-based cognitive and emotional states in a real HCI environment. Therefore, the participants were asked by the system to solve a series of cognitive games in order to investigate their reaction to various cognitive tasks difficulties, varying from high interest and overwhelming to boring and frustrating levels. The aim of each game task was to identify the single one item that is unique in shape and color (i.e., the number 36 and the number 2 in Figure 1), based on a visual search task. The difficulty was set by adjusting the number of objects, shapes and colors shown per task as well as the available time given to solve that task. Thus, cognitive *Overload* was induced by increasing the task field objects and decreasing the available time, while cognitive *Underload* was induced by decreasing the task field objects and increasing the available time. Further, for each individual task, the subject could earn a certain amount of money (up to ten cents) according to the individual speed of the given response. The amount of reward money earned for solving a task was increasingly reduced, the longer the subject needed to answer. If the given answer was incorrect, the participant received no reward at all for that particular task. Figure 1 shows screenshots of the visual search task.

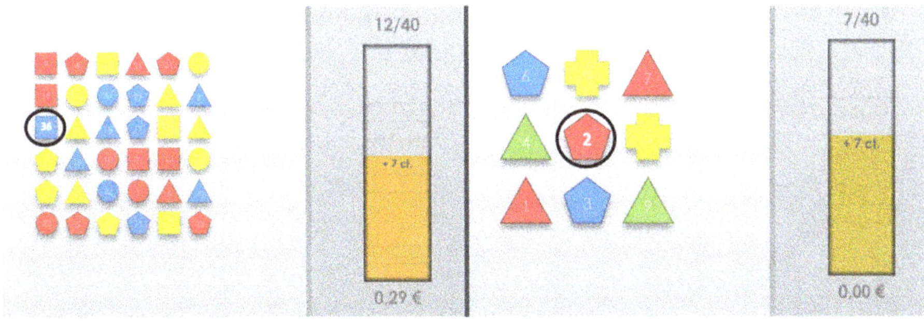

Figure 1. The visual search task on the example of *Overload* (**left**) and *Underload* (**right**) induction scheme. The user has to spot the single unique object. The correct answers are 36 for *Overload* (unique blue and square object) and 2 for *Underload* (unique red and pentagon object).

2.3. Experiment Structure

The experiment structure consists of six *induction sequences*, separated by *subjective feedback* related to the actual sequence, and followed by a respiration *baseline* and a *summary* of the achieved results in that sequence. While the experimental sequences are used to induce various cognitive and emotional states, the subjective feedback is used for the validation of the induction. These are described in details in the following subsections. Further, prior to the experiment, each subject received an introduction and instructions to the experimental steps in form of a short PowerPoint presentation and was afterwards asked to fill in three questionnaires related to: (1) emotion regulation based on the Emotion Regulation Questionnaire (ERQ) [53,54]; (2) emotional control based on the Trait Emotional Intelligence Questionnaire Short Form (TEIQue-SF) [55,56]; and (3) personality traits based on the Ten Item Personality Measure (TIPI) [57,58]. These questionnaires are also used as further subjective evaluation of the stability of the induction paradigm.

2.3.1. Induction Sequences

Six consecutive sequences of different difficulties, with 40 single tasks each, are implemented for the induction of six different emotional and cognitive load states. All tasks within a sequence have thereby the same or comparable difficulty levels. The first introductory sequence is designed to induce *Interest* and is of moderate difficulty to gain the users' interest and familiarize them with the visual search task procedure. The *Interest* sequence has 40 tasks and is designed with a mix of 3×3 and 4×4 matrices, and 10 s time per task to give the right answer. The second sequence is designed to induce *Overload* and consists of 40 difficult tasks with a 6×6 matrix each and with short time of 6 s per task to provide an answer. The third sequence has a moderate *Normal* difficulty and is defined with 40 tasks with 4×4 matrices and moderate time of 10 s to respond per task. This *Normal* sequence is the neutral (cognitive and emotional) state to be considered as baseline between the sequences. The fourth sequence is implemented as an *Easy* sequence with 40 tasks with 3×3 matrices and very long time of 100 s for responding. In order to induce *Underload*, the fifth sequence is defined as a repetition of the previous *Easy* scheme of low difficulty, with again 40 tasks with 3×3 matrices and 100 s to provide an answer. This originates from the trivial idea that repeating an easy well-known task in the same way two times in a row, generates a state of boredom and leads to *Underload*. Based on this idea, the *Easy* sequence is considered as a transition state used as a mean to induce *Underload*. Finally, the last sixth sequence is intended to induce *Frustration* by purposely logging in a wrong answer at randomly distributed tasks (eight wrong out of 40), even when the subject provides a right answer. This *Frustration* sequence has 40 tasks with a mix of 3 × 3 and 4 × 4 matrices each and 10 s time to provide an answer. Table 1 illustrates a summary of the experimental procedure.

Table 1. Illustration of the experimental procedure and sequences description.

Sequence Nr.	Induction of	Number of Tasks	Search Field	Time (s) to Answer
*Sequence 1	Interest	40 tasks	3 × 3 and 4 × 4 items	10 s/task
*Sequence 2	Overload	40 tasks	6 × 6 items	6 s/task
*Sequence 3	Normal	40 tasks	4 × 4 items	10 s/task
*Sequence 4	Easy	40 tasks	3 × 3 items	100 s/task
*Sequence 5	Underload	40 tasks	3 × 3 items	100 s/task
*Sequence 6	Frustration	40 tasks	3 × 3 and 4 × 4 items	10 s/task

*(Every) Sequence is followed by subjective feedback, respiration baseline, and results' summary.

The user-system interaction during all the tasks is a mobile interaction conducted via natural speech while the participants could freely move and walk in the room (standing position). The walking area is limited to a field of 1 m × 3 m, represented by an electrostatic floor mat to prevent any signal disturbance caused by any electrostatic charge influence.

2.3.2. Subjective Feedback

In order to evaluate the validity of the induction paradigm, various kinds of subjective feedback are implemented, including *Free Speech*, *SAM Ratings*, and *Direct Questions* parts. These are presented to the subjects on the screen as illustrated in Figure 2. After each of the six accomplished sequences, the participants provided a series of information about their current emotional state in three different ways, including: (1) expressing in own words via *Free Speech* feedback of 12 s duration, how they felt during that particular sequence, (2) rating their emotions via Self-Assessment-Manikin *SAM Ratings* on the Valence-Arousal-Dominance (VAD) scale, and (3) answering *Direct Questions* related to the assessment of their own performance. The aim of this subjective feedback is to determine the current subjective emotional state experienced in that particular sequence, which, in turn, can be used as ground truth to evaluate and validate the induction paradigm. While the Free Speech feedbacks are given via natural speech, logging of the SAM Ratings and Direct Questions was carried out per mouse-click to ensure correct logging documentation. The user was thereby guided and instructed by the system via speech output. The user-system interaction modality (mouse, speech or both) within the experiment is part of the technical implementation as described in Section 2.4.

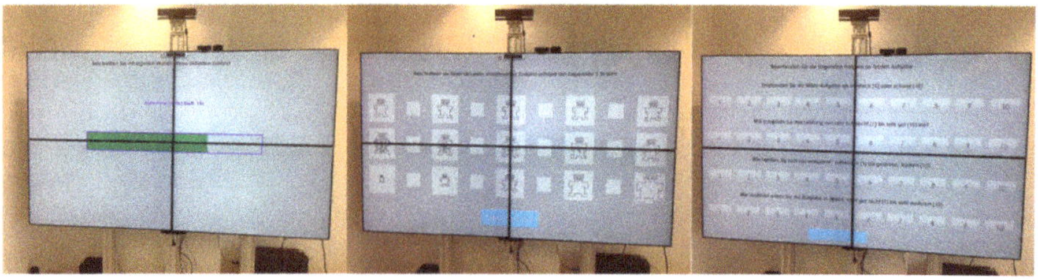

Figure 2. Illustration of the subjective feedback screens including Free Speech (**left**), SAM Ratings (**middle**) and Direct Questions (**right**) parts. Note that the SAM Ratings are scored on a nine-point Likert scale (represented by both the big labeled fields and the small empty fields).

2.3.3. Respiration Baseline and Results' Summary

Following the subjective feedback, a baseline phase consisting of a breathing exercise to level off the physiological reactions related to that particular sequence is conducted by the subjects. Additionally, here, the users are thereby guided by the system via speech to first deeply breathe, then hold their breath for few seconds, and finally breathe out. The exercise was repeated three times subsequently. Finally, after the baseline phase, the system informs the user via speech about his performance during the last sequence and the related results achieved, including the earned money, are presented on the screen.

2.4. *Technical Implementation*

The further developments of the generic paradigm and software implementation of the interaction scheme and experimental structure for the induction of various cognitive, stress, and affective states are realized using C# programming and integrated within the Semaine platform [59].

The workflow of the experiment including the structure, order and content of the different sequences as well as the subjective feedback and baseline sections in between are defined in an external *taskset* file which can be imported at the beginning of the experiment. Within a *taskset*, the course setting of the sequences can be defined individually for every task and every subject, allowing a high flexibility and an easy-to-handle workflow setup. Additionally, the user-system interaction modality (mouse, speech, or both) for every part within the experiment is predefined in this file. This also includes the text content (spoken and written) given by the system. The *taskset* describes the course of events of the entire experiment and is consistent for all the participants, except for the second sample who underwent three repeated measurements at three different times. For this group, the content of speech output given by the system is slightly modified for the second and third measurements by using alternative synonyms while keeping the content the same. The intention here is to keep the interaction as natural as possible by preventing a repetition of exactly the same words every time.

During the visual search task, the user is instructed to give his answer by speech command. To recognize the speech content, our experimental implementation includes an integrated automated speech recognition algorithm. If well trained in advance, the speech recognition works properly in most of the cases. Nevertheless, in order to ensure a smooth interaction between the user and the system, a "Wizard of Oz" (WOZ) scenario was also implemented and used to support the integrated automated speech recognition algorithm. This was especially useful if the automated recognition fails for instance because of language dialect disparity of specific subjects that strongly diverge from the norm language on which the recognition algorithm was trained. Within the WOZ scenario, the experiment was observed on an external monitor in a separate room by the experimenter, who controlled and adjusted the (correct) login of the given answers, if necessary.

Finally, the behavior of the subjects and all their conducted actions as well as the whole course of the experiment are triggered after the events. As a result, for every individual subject, a .log file is generated after every experiment including all the course details of the experiment and can be used for the later processing and analysis of the signal data.

2.5. *Multimodal Sensors for Data Acquisition*

In order to collect high quality data for a wide kind of multimodal analysis there are mainly two important issues regarding the technical data acquisition. First, a wide set of different modalities with a maximum of data quality in each sensor needs to be ensured. Second, the synchronization between all sensors and the user interface components has to be as congruent as possible. The sensors used here can be divided into two kinds. Sensors attached to the participant and sensors mounted to the environment. To ensure a high mobility of the participant, and, therefore, less influence on the participant's natural behavior, wireless sensors were used.

In particular, they include a small theatre stereo headset microphone with a frequency range of 20 to 20.000 Hz, sampled at 48 kHz, transmitted via digital radio and a g.tec g.MOBIlab+ Bluetooth amplifier for biophysiological sensors. The bioamplifier was equipped with sensors for electromyography (EMG), electrocardiography (ECG), skin conductance level (SCL), respiration, and body temperature at a sampling rate of 256 Hz. To ensure accurate recordings free of motion artifacts, the signals from the physiological sensors underwent an online monitoring check adapted for our experiment using Simulink®software. This online signal quality check was conducted during an initial baseline record at rest in sitting position and prior to the first sequence of the experiment.

A stationary mounted frontal webcam with HD resolution of 1920 × 1080 pixels at 30 frames per second was used. Further a Microsoft Kinect v2 also was mounted in the front. The Kinect includes a full HD RGB color video stream (1080p @ 30 Hz), an infrared (IR) video stream (512 × 424 @ 30 Hz), a depth stream (512 × 424 @ 30 Hz), a directed audio stream (virtual beam forming by a microphone array) and pose estimation stream including skeleton information containing 25 joints. Kinect and primary webcam were placed on top of the interaction screen in front of the scenery looking towards the participants face. Finally, a second webcam with a resolution of 1280 × 720 @ 30 fps was placed in the rear of the experimental setting in order to monitor the scenery overview and sample the atmosphere sounds. Figure 3 shows the views from the frontal and rear cameras and the acquired depth information.

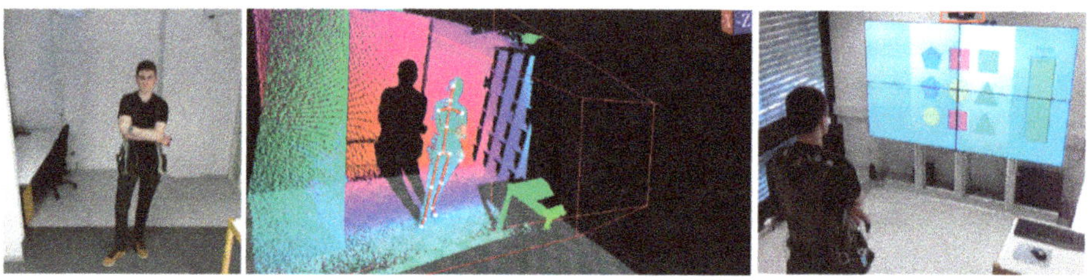

Figure 3. View of the frontal camera (**left**), depth information (**middle**) and scenery overview of the rear camera (**right**).

Summarized, we recorded 16 sensor modalities, including four video streams (front/rear/Kinect RGB/Kinect IR), three audio streams (headset/directed array/atmosphere), seven biophysiological streams (3 × EMG/ECG/SCL/respiration/temperature), depth, and pose stream. Further, several label information streams extracted from an application log file, described later, were also recorded. After recording, all data were post-processed in order to prove a high quality towards technical and signal quality issues. As visualization tool we used ATLAS [60,61] to present (and playback) all recorded data to the experts. Only sessions which passed all technical and manual quality checks belong to the final dataset of 100 (40 × 1 + 20 × 3) sessions. These are described in Section 3. In addition to the annotation extracted from the log file entries and experimental design structural issues, some additional labels are achieved by a semi-automatic active learning procedure as described in the Annotation section (Section 3.4) of this work.

Figure 4 shows an overview of the collected data of a single session displayed in the visualization tool ATLAS. All video streams, time series type data and some label information are illustrated. The timescale is at minimum zoom, so the structure of experimental phases can be seen in the upper annotation line. It is not possible to record this massive amount of data on a single PC, so we developed a modular network-based recording infrastructure called MAR^{2}S (Multimodal Activity Recognition and Recording System). This contains a specific recording module which on the one hand controls each specific sensor according to its specific API. This can include preparation and initialization commands, trigger and timing control, data format transformations, disk read/write control of the streams, etc. On the other hand, each module accomplishes the defined network commands and synchronization protocols. The modules are mostly written in C#, but due to the inter-module communication by network, there is no technical limitation to a specific programming language, operating system or hardware type. Depending on the sensors, hard and software requirements, in most cases more than one sensor can be grouped on a PC without influencing each other.

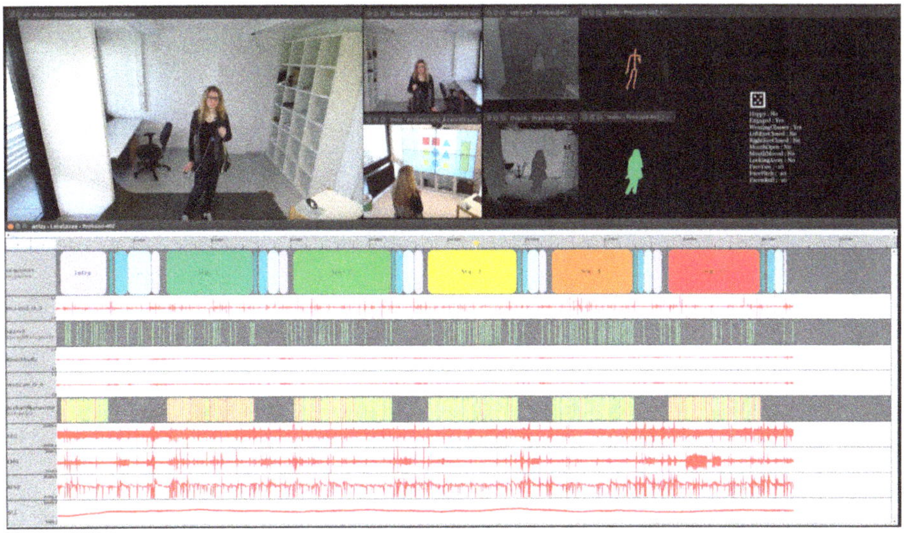

Figure 4. Overview of a whole recording session displayed in the multimodal annotation tool ATLAS: Kinect video (**top left**); Front webcam view, infrared and pose (first video row); rear camera, depth images (second video row); face position and simple facial estimations (**top right**). The data window contains from top to bottom: Sequence start and end information from logfile, audio, speech recognition information from logfile, stereo audio, front webcam, search, and answer phases including hit and miss from logfile, ECG, EMG, respiration, SCL.

In addition to the sensor modules, the user interface (UI) and WOZ module were also encapsulated in such a network module in order to control and monitor their behavior in the same synchronous manner. Finally, a logging module was established acting like a sensor, not recording physical data, but recording the whole system behavior. This includes exact time stamps on all participants and WOZ inputs, global information on the internal and external systems states, information about the sensors states, any network communication, etc. With this log file and the recorded sensor streams it is possible to reconstruct the whole experimental procedure in detail up to a virtual playback without a real participant. Therefore, the data can not only be used for numerous offline analyses but also for the development of real time capable online recognition systems.

Finally, each involved PC had a network monitoring module, measuring the current network latency to ensure synchronous recording. Due to the usage of "of the shelf" sensors like the Kinect sensor and webcams, which do not include physical trigger input capabilities, and the complex multi-PC network environment, we are not able to ensure synchronicity on a nanosecond level, like highly-specialized, expensive, hardware-triggered setups do. Hence, our setup is much more flexible and a great deal more realistic towards future end user implementations on custom hardware and smart devices. The recorded emotions and mental states occur typically in a longer range and all multimodal recognition approaches typically use time windows from 50 ms up to several seconds. Thus, inter-modality delays from under one millisecond are acceptable. To ensure this, each involved PC was directly attached to a separate recording control sub network containing just one switch transmitting only record timing and control information (no sensor streams, they are processed locally). Figure 5 shows the technical infrastructure of the distributed experimental and recording setup.

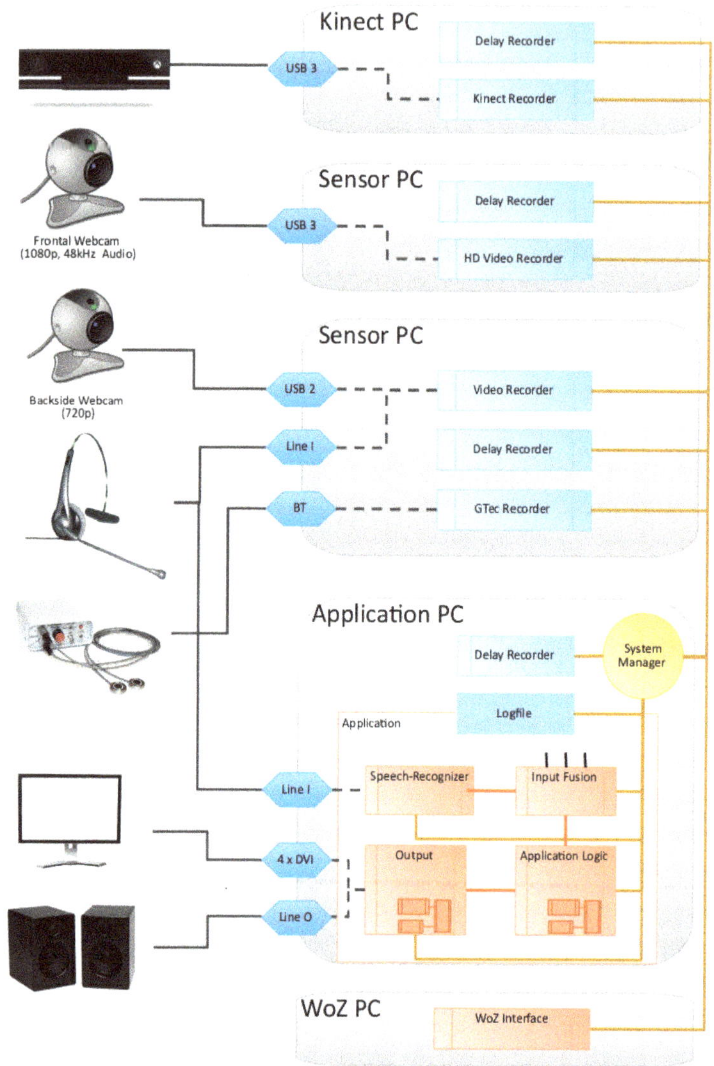

Figure 5. Technical infrastructure of the distributed experimental and recording setup.

The module which initiates the recording start also listens to its own send "start" message, and starts recording after the message returns back to itself to prevent time leading of the initiator module. Each module further sends a roundtrip message to itself to measure the network latency at the beginning of each recording session. The round trip times can be seen in Table 2. Thus, we can assume that the average delay or desynchronization is within an acceptable range. Additionally, the synchronicity can be improved by taking the individual delays into account and shifting the timestamps after recording in the post-processing step. This is not done in the raw data.

Table 2. Average network latency between recording modules and estimated maximum delay between modalities.

Modality	Average Network Latency (ms)	Estimated Max. Delay (ms)
Kinect (Video / Depth / IR / Audio / Pose)	5.61	1.35
Front video	4.95	1.27
Rear video, atmosphere audio, headset audio, biosignals	5.37	0.83
User interface	5.17	1.17
WOZ interface	5.52	1.17
Average estimated synchronization error		1.35 − 0.83 = **0.52**

3. Results

In the following, the resulting database, the validation of the induction via questionnaires and via subjective feedback, as well as the data annotation results are presented.

3.1. *The Database*

In total, three subjects were excluded from the analysis: two subjects from the first sample because of missing biosignal data (ID-04) and an absent logger data (ID-40) as well as one subject from the second sample because of missing sequences due to a technical error (*Underload* and *Frustration* for ID-90). Because ID-90 represents the second measurement of a participant from the second sample, the first and third measurement data of that subject (ID-80 and ID-100) were also excluded from the analysis. Consequently, the final dataset *uulmMAC* consists of 95 recording sessions from 57 subjects, presented for the following groups and subgroups:

- Group A involves 38 subjects from the first sample who underwent one single measurement. It consists of 38 recording sessions.
- Group B involves 19 subjects from the second sample who underwent three different measurements. It consists of 57 recording sessions. Group B includes three different subgroups: Group B1, Group B2, and Group B3 consisting of 19 recording sessions each and representing the first, second and third measurement time of the 19 subjects, respectively.

While both groups underwent exactly the same experiment, they slightly differ in one modality acquisition: The EMG data of Group A include only musculus trapezius activity measurements (thus, without facial electrodes, which allows a better analysis of facial expressions from the video data). As for Group B, the EMG data include activity measurements of three muscles: musculus trapezius, musculus currogator and musculus cygomaticus. In the following, the results of Group A, Group B1, Group B2, and Group B3 are separately analyzed and presented.

3.2. *Evaluation via Questionnaires*

The three questionnaires TEIQue-SF, ERQ, and TIPI collected from all the participants prior to the experiment are first evaluated for Group A, Group B1, Group B2, and Group B3. For all questionnaires items, the possible score values range between 1 (minimum) and 7 (maximum).

3.2.1. TEIQue-SF Questionnaire

In Figure 6 the four dimensions of the TEIQue-SF, consisting of Well-Being, Self-Control, Emotionality, and Sociability factors, are presented for the different groups. The mean values vary between 5.61 and 5.82 for the Well-Being factor, between 5.04 and 5.26 for the Self-Control factor, between 4.80 and 5.06 for the Emotionality factor and between 5.02 and 5.34 for the Sociability factor. The standard deviations (SD) range between 0.55 and 1.28 for all groups and all factors. The first three factors of Well-Being, Self-Control and Emotionality have a small decreasing tendency within Group B1, Group B2 and Group B3, with the highest value obtained for Group B1. Only for Sociability the mean value of Group B3 slightly increases compared to the value of Group B2. Nevertheless, in total the mean values of all factors present a homogenous distribution within all the groups, showing minimal deviations and, thus, stable results.

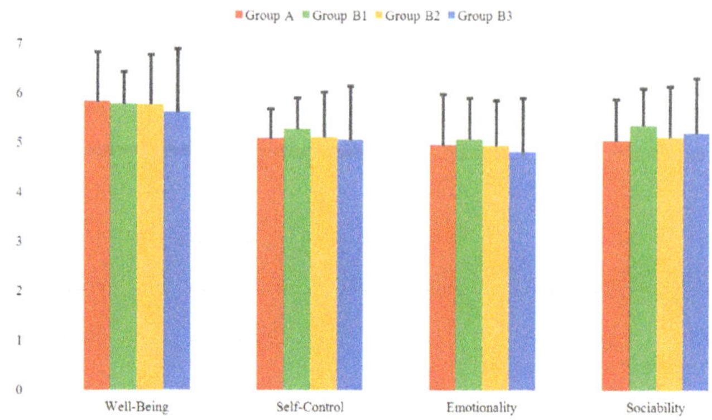

Figure 6. Mean values of the four dimensions of the TEIQue-SF questionnaire for the different groups. The error bars represent the corresponding standard deviations.

3.2.2. ERQ and TIPI Questionnaires

In Figure 7 the results of the ERQ and TIPI questionnaires are presented for Group A, Group B1, Group B2 and Group B3. Reappraisal and Suppression are the factors related to the ERQ, while Extraversion, Agreeableness, Conscientiousness, Emotional Stability and Openness to Experience are the factors related to the TIPI questionnaire. The range of the Reappraisal mean values varies between 2.94 and 3.05 (range of SD: 0.81 to 1.19), while the range of the Suppression values varies between 4.35 and 4.46 (range of SD: 1.14 to 1.43). For the TIPI questionnaire, the Extraversion has values between 4.61 and 4.76, Agreeableness between 5.11 and 5.29, Conscientiousness between 5.53 and 5.79, Emotional Stability between 5.39 and 5.63 and Openness to Experience between 5.13 and 5.37. The standard deviations of the five factors of the TIPI have values from 0.74 to 1.64. Similar to the TEIQue-SF questionnaire, in summary, the mean values of all factors present homogenous distribution within all the groups/subgroups, showing minimal deviations and, thus, stable results.

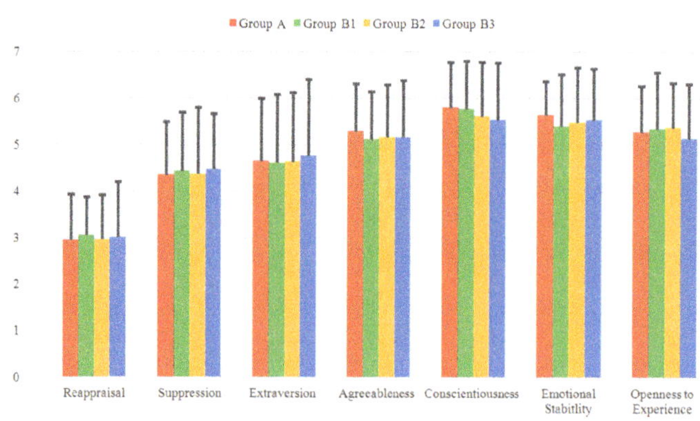

Figure 7. Mean values of the ERQ and TIPI questionnaires for the different groups: Reappraisal and Suppression are the ERQ dimensions, while Extraversion, Agreeableness, Conscientiousness, Emotional Stability, and Openness to Experience are the TIPI dimensions.

3.3. Validation via Subjective Feedback

Following, the evaluations obtained from the subjective feedback of the participants are presented for Group A, Group B1, Group B2, and Group B3. They include the analysis of the SAM Ratings and the Direct Questions. The evaluation of the subjective feedback is necessary to provide ground truth and validation of the dataset, which is in turn essential for further analysis and applications. The Free Speech data are not analyzed here but are part of the dataset in their raw state.

3.3.1. SAM Ratings

The SAM Ratings were collected from every subject after each accomplished sequence during the experiment. With the help of the three dimensions, Valence, Arousal, and Dominance, the induction of the different sequence levels of cognitive load and affective states is evaluated. First, the evaluation of the ratings of Group A is presented. Then, the ratings of the three different measurements of Group B are separately analyzed (Group B1, Group B2, and Group B3). Finally, repeated measures ANOVA and post-hoc corrections were performed to examine the significance of the variations between the different sequences.

In Figure 8, the mean SAM Ratings for Group A are presented. The highest valence values are found for the sequences *Easy* (7.32) and *Interest* (6.92) and *Underload* (6.84), while the lowest values are found for the *Overload* (5.13) and the *Frustration* (5.68) sequences. On the other hand, the highest Arousal was perceived for these two latter sequences, *Overload* (5.18) and *Frustration* (4.37), while the lowest Arousal was registered for *Underload* (2.11) and *Easy* (2.39). As for the Dominance values, the highest mean values were also obtained for these two sequences, *Underload* and *Easy* (7.03 each), while the lowest values were registered for *Overload* (3.66) and *Frustration* (4.26).

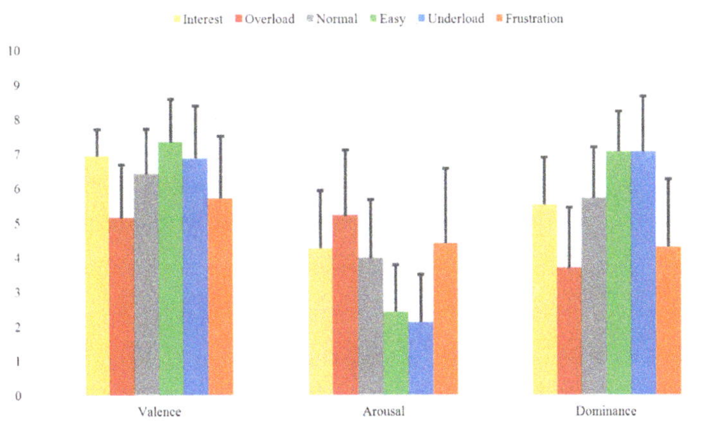

Figure 8. SAM Ratings of Group A for all sequences with mean Valence, Arousal, and Dominance.

Figures 9–11 illustrate the SAM Ratings results for the first (Group B1), second (Group B2), and third (Group B3) measurement time of Group B, respectively. Additionally, here, the mean SAM Ratings values are consistent with each other for all three measurement times showing transtemporal stability in the subjective evaluation. Additionally, compared to the rating results of the one-measurement group (Group A) illustrated in Figure 8, the mean Valence, Arousal, and Dominance values show similar rating tendencies.

In summary, the SAM Ratings show overall stable course with highest Valence, lowest Arousal, and highest Dominance values for the *Easy* and *Underload* sequences and with lowest Valence, highest Arousal and lowest Dominance values for the *Overload* and *Frustration* sequences.

Both transtemporal stability and the similar rating tendencies between the subjects of the first and second samples prove the robust quality of the induction as evaluated by the subjects.

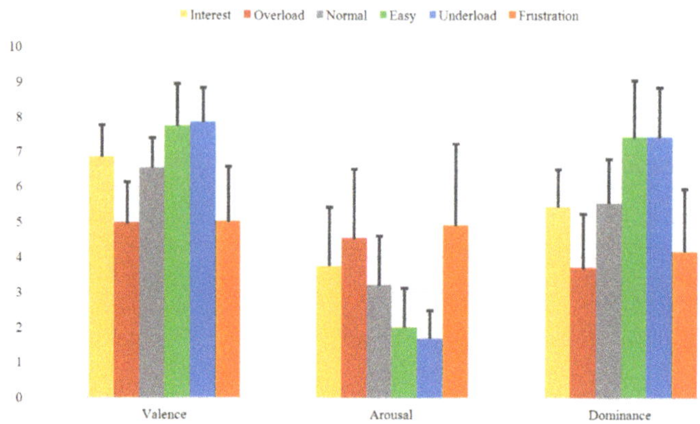

Figure 9. SAM Ratings of Group B1 for all sequences with mean Valence, Arousal, and Dominance.

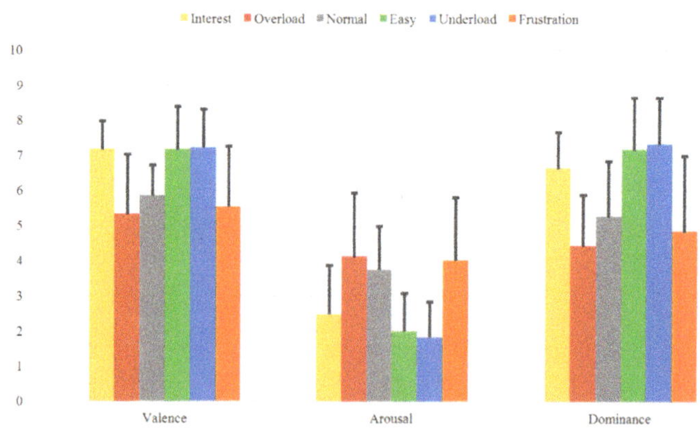

Figure 10. SAM Ratings of Group B2 for all sequences with mean Valence, Arousal, and Dominance.

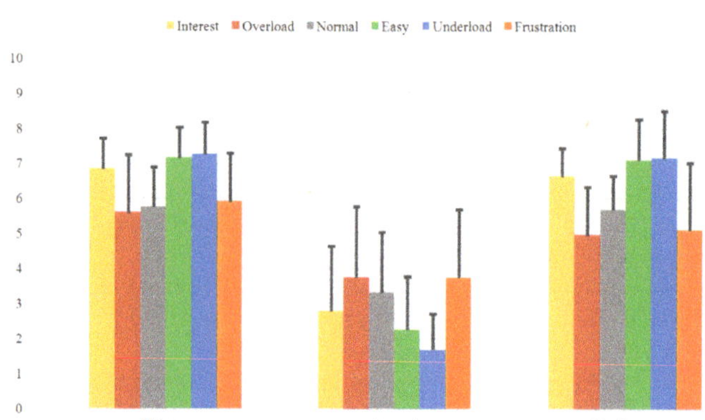

Figure 11. SAM Ratings of Group B3 for all sequences with mean Valence, Arousal, and Dominance.

The distribution of the SAM Ratings for all the measurements are presented as scatter-plots in the Appendix A in Figure A1a (Valence), Figure A1b (Arousal), and Figure A1c (Dominance).

In order to examine if the differences of the SAM Ratings evaluations are statistically significant in the VAD-space between the different sequences, further statistical analysis was carried out. To analyze the ratings for Group A and Group A + Group B1, we conducted separate repeated measures ANOVA with the factors Sequence and VAD (for Valence, Arousal, and Dominance, respectively). Post-hoc, Newman-Keuls corrections were carried out to compare the mean differences between the sequences. For Group A, the repeated measures ANOVA revealed a significant effect of Sequence (F(5.185) = 16.866, $p < 0.001$, ηp2 = 0.313), VAD (F(2.74) = 57.996, $p < 0.001$, ηp2 = 0.611) and the interaction (F(10.370 = 30.748, $p < 0.001$, ηp2 = 0.454). Additionally, post-hoc tests using Newman-Keuls correction revealed significant differences (see Table A1a in the Appendix A). For the combined Group A + Group B1, the repeated measures ANOVA revealed a significant effect of Sequence (F(5.280) = 30.190, $p < 0.001$, ηp2 = 0.353), VAD (F(2.112) = 106.429, $p < 0.001$, ηp2 = 0.774) and the interaction (F(10.560) = 51.405, $p < 0.001$, ηp2 = 0.447). Again, post-hoc tests using Newman-Keuls correction revealed significant differences (see entire Table A1b in the Appendix A).

A direct analysis of the SAM Ratings of all sequences in comparison to the *Normal* sequence as baseline is presented in Table 3, while the results of the SAM Ratings between the *Overload* vs. *Underload* sequences and between the *Interest* vs. *Frustration* sequences are presented in Table 4 (the entire results can be found in the Appendix A as Tables A1a and A1b).

Table 3. Post-hoc Newman-Keuls corrections for the Valence (V), Arousal (A), and Dominance (D) ratings between all sequences compared to *Normal*. Mean-Differences (Mean-Diff.) and *p*-values are presented.

SAM Ratings	Mean-Diff. Group A	*p*-Value Group A	Mean-Diff. Group A + Group B1	*p*-Value Group A + Group B1
V_Normal – V_Interest	−0.526	0.272	−0.509	0.100
V_Normal – V_Overload	1.263	0.003**	1.351	0.000***
V_Normal – V_Easy	−0.921	0.076	−1.053	0.003**
V_Normal – V_Underload	−0.447	0.190	−0.772	0.061
V_Normal – V_Frustration	0.711	0.094	0.930	0.002**
A_Normal – A_Interest	−0.289	0.397	−0.386	0.184
A_Normal – A_Overload	−1.237	0.004**	−1.246	0.000***
A_Normal – A_Easy	1.553	0.000***	1.491	0.000***
A_Normal – A_Underload	1.842	0.000***	1.754	0.000***
A_Normal – A_Frustration	−0.421	0.606	−0.807	0.013*
D_Normal – D_Interest	0.184	0.852	0.175	0.569
D_Normal – D_Overload	2.026	0.000***	2.018	0.000***
D_Normal – D_Easy	−1.342	0.001**	−1.561	0.000***
D_Normal – D_Underload	−1.342	0.001**	-1.579	0.000***
D_Normal – D_Frustration	1.421	0.001**	1.404	0.000***

*$p < 0.05$, **$p < 0.01$, ***$p < 0.001$.

Table 4. Post-hoc Newman-Keuls corrections for the Valence (V), Arousal (A), and Dominance (D) ratings between *Overload* vs. *Underload* and between *Interest* vs. *Frustration*. Mean-Differences (Mean-Diff.) and *p*-values are presented.

SAM Ratings	Mean-Diff. Group A	*p*-Value Group A	Mean-Diff. Group A + Group B1	*p*-Value Group A + Group B1
V_Overload – V_Underload	−1.711	0.000***	−2.123	0.000***
V_Interest – V_Frustration	1.237	0.003**	1.439	0.000***
A_Overload – A_Underload	3.079	0.000***	3.000	0.000***
A_Interest – A_Frustration	−0.132	0.921	−0.421	0.202
D_Overload – D_Underload	−3.368	0.000***	−3.596	0.000***
D_Interest – D_Frustration	1.237	0.003**	1.228	0.000***

*$p < 0.05$, **$p < 0.01$, ***$p < 0.001$.

Further, in order to justify the combination of Group A and Group B1 (all first measurements) in the statistical analysis, an ANOVA was additionally computed for the Valence, Arousal, and Dominance scores of the SAM Ratings between Group A and Group B1. Based on a one-way ANOVA, we found no statistically significant difference in the Valence scores ($F(2.6) = 1.650$, $p = 0.153$), nor in the Arousal scores ($F(2.6) = 0.978$, $p = 0.450$) nor in the Dominance scores ($F(2.6) = 0.376$, $p = 0.891$) between Group A and Group B1.

According to Table 3, most of the Valence, Arousal, and Dominance values of the SAM Ratings can be significantly distinguished from each other for all the sequences compared to *Normal*. Exceptions for Group A + Group B1 are the Valence, Arousal, and Dominance of *Interest* and the Valence of *Underload*. For Group A, more exceptions could be observed especially on the Valence dimension. More context-relevant results are the implications in Table 4, showing that the states *Overload* vs. *Underload* and *Interest* vs. *Frustration* can be significantly distinguished from each other on all SAM dimensions for both the Group A and the Group A + Group B1 except for Arousal between *Interest* and *Frustration*.

3.3.2. Direct Questions

A further subjective feedback evaluation was carried out in terms of Direct Questions. Therefore, after each sequence, the subjects were asked to answer Direct Questions related to the assessment of their own perception. Four questions related to "Difficulty", "Performance", "Stress", and "Motivation" were processed: With the help of the first question, the subjects described how difficult the sequence was (very easy = 1; very difficult = 10). The second question is a personal performance assessment (performed very bad = 1; performed very well = 10). For the first sequence of *Interest*, this "Performance" question was adapted to answer the subjects' interest. The third question describes the individually experienced stress level (very relaxed = 1; very stressed = 10), and the fourth question reflects the motivation of the participant (not motivated = 1; very motivated = 10).

In Figure 12 the results of the Direct Questions are shown for Group A. It can be seen, that for the first question "Difficulty", *Overload* has the highest rating (9.18), while *Easy* and *Underload* have the lowest ratings (1.66 and 1.68, respectively). As expected, the sequences *Interest*, *Normal*, and *Frustration* present middle ratings (5.24, 5.50, and 4.58, respectively). As for the second question "Performance", the lowest rating is observed for *Overload* (2.13), while the highest ratings were obtained for *Easy* and *Underload* (8.34 and 8.50, respectively). The "Interest" rating for the first sequence *Interest* was 7.63.

Further, the third "Stress" question shows similar course as the first "Difficulty" question. The "Stress" ratings for the sequences *Interest*, *Normal* and *Frustration* are in the same range (5.13, 5.03 and 5.34, respectively), while *Overload* has the highest rating (7.00) and *Easy* and *Underload* the lowest ones (2.32 and 2.58, respectively). An interesting observation here, is the slightly increasing stress from *Easy* to *Underload*. The last "Motivation" question shows the highest rating for *Interest* (9.13), and the lowest ratings for *Overload* (7.13), *Underload* (7.55), and *Frustration* (7.11).

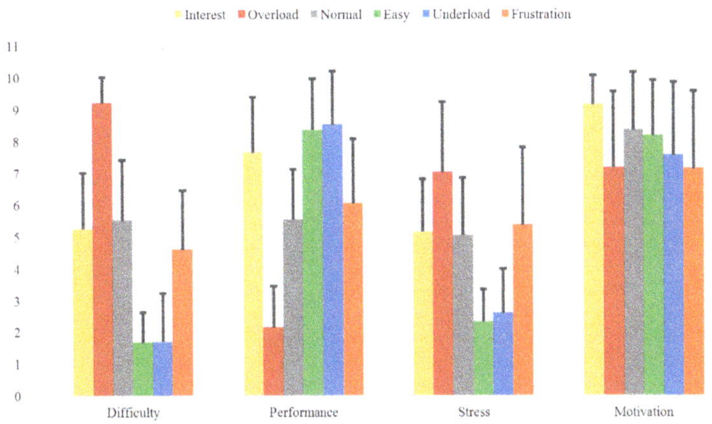

Figure 12. Direct Questions of Group A for all the sequences with mean values.

With regard to Group B with the subjects who underwent three measurements each, some changes over time can be observed. Figures 13–15 illustrate the Direct Questions ratings for the first (Group B1), second (Group B2), and third (Group B3) measurement, respectively. The mean rating distributions for each sequence for Group B1 (first measurement) presented in Figure 13 are comparable to the results obtained for Group A (single measurement) presented in Figure 12.

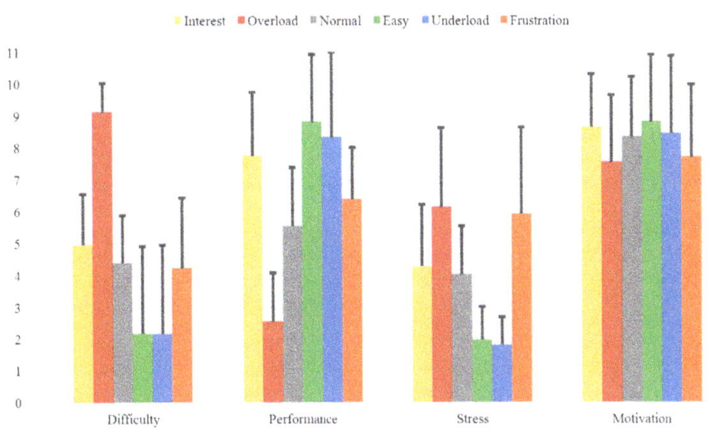

Figure 13. Direct Questions of Group B1 for all the sequences with mean values.

Comparing Group B1 and Group B2, the mean rating values of the first question "Difficulty" for the sequences *Interest*, *Easy* and *Underload* decrease from the first to the second measurement. On the other hand, the mean rating values for *Normal* increase from 4.37 to 5.16. As for the second "Performance" question, the mean rating values for *Overload* (2.53 vs. 3.58) and *Underload* (8.32 vs. 8.68) increase from the first to the second measurement, while the rating related to *Normal* decreases (5.53 vs. 4.89). As for the third "Stress" question, higher differences are observed for the *Interest* (4.26 vs. 3.00), *Normal* (4.00 vs. 4.89) and *Frustration* (5.89 vs. 5.32) sequences. Finally, the last "Motivation" question has comparable tendencies and values for the first and second measurements.

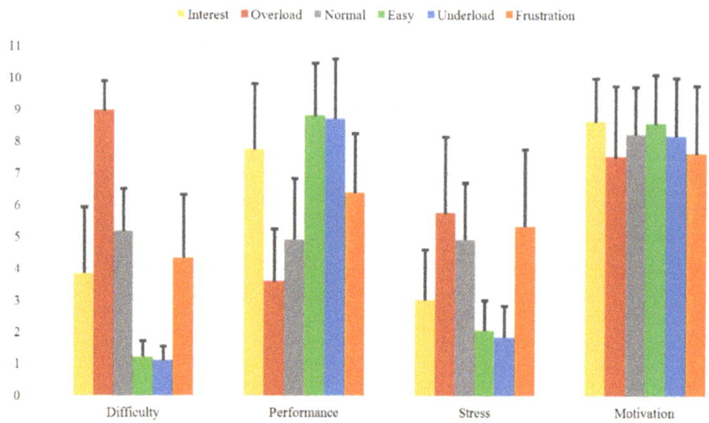

Figure 14. Direct Questions of Group B2 for all the sequences with mean values.

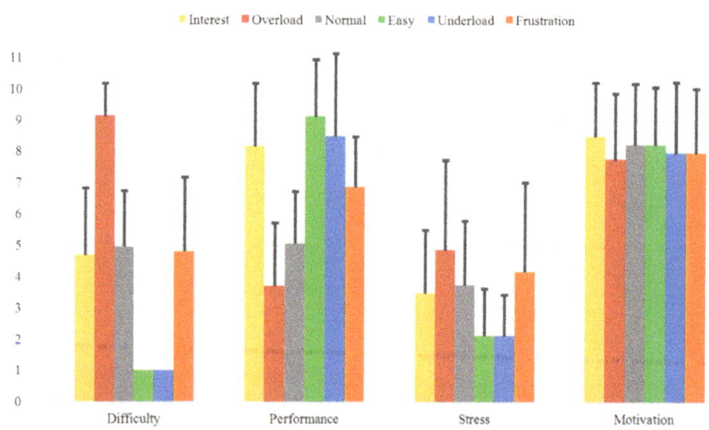

Figure 15. Direct Questions of Group B3 for all the sequences with mean values.

Finally, comparing Group B2 and Group B3 illustrated in Figures 14 and 15, the mean values of the "Difficulty" question for the sequences *Interest*, *Overload* and *Frustration* increase (3.84 vs. 4.68, 8.95 vs. 9.11 and 4.32 vs. 4.79, respectively), while the values for *Normal*, *Easy* and *Underload* decrease (5.16 vs. 4.95, 1.21 vs. 1.00 and 1.11 vs. 1.00, respectively). The highest rating in the third measurement is again obtained for the sequence *Overload* (9.11), while the lowest values are observed for the sequences *Easy* and *Underload* with values of 1.00 each. Furthermore, the "Performance" question shows increasing values for *Interest* from the second to the third measurement time (7.74 vs. 8.16), while the related mean ratings of the remaining sequences have nearly the same values with small variations. As for the "Stress" question, the highest mean values are also observed for the *Overload* and *Frustration* sequences and show a decreasing tendency compared to the second measurement (5.74 vs. 4.84 and 5.32 vs. 4.16, respectively). The "Motivation" question for all sequences results in mean ratings ranging around the value of 8, with the lowest values obtained for the sequences *Overload* (7.50 vs. 7.74) and *Frustration* (7.60 vs. 7.95).

The ratings distribution of the Direct Questions for all the measurements are presented as scatter-plots in the Appendix A in Figure A2a ("Difficulty"), Figure A2b ("Performance"), Figure A2c ("Stress"), and Figure A2d ("Motivation").

In order to examine if the differences of the Direct Questions evaluations are statistically significant between the different sequences, further statistical analysis was carried out. Similar to the SAM Ratings, we conducted separate repeated measures ANOVA with a post-hoc Newman-Keuls correction to analyze differences between the respective ratings of the individual questions "Difficulty" (Dif), "Performance" (Per), "Stress" (Str) and "Motivation" (Mot) for Group A and Group A + Group B1. For Group A, the repeated measures ANOVA revealed a significant effect of Sequence (F(5.185) = 43.379, $p < 0.001$, ηp2 = 0.540), Question (F(3.111) = 74.360, $p < 0.001$, ηp2 = 0.668) and the interaction (F(15.555 = 81.485, $p < 0.001$, ηp2 = 0.688). Additionally, post-hoc tests using Newman-Keuls correction revealed significant differences (see Table A2a in the Appendix A). For the combined Group A + Group B1, the repeated measures ANOVA revealed a significant effect of Sequence (F(5.280) = 35.164, $p < 0.001$, ηp2 = 0.386), Question (F(3.168) = 126.204, $p < 0.001$, ηp2 = 0.693) and the interaction (F(15.840) = 111.873, $p < 0.001$, ηp2 = 0.666). Again, post-hoc tests using Newman-Keuls correction revealed significant differences (see Table A2b in the Appendix A).

A direct analysis of the Direct Questions of all sequences in comparison to the *Normal* sequence as baseline is presented in Table 5, while the results of the Direct Questions between the *Overload* vs. *Underload* sequences and between the *Interest* vs. *Frustration* sequences are presented in Table 6 (the entire results can be found in the Appendix A as Tables A2a and A2b).

Further, in order to justify the combination of Group A and Group B1 (all first measurements) in the statistical analysis, an ANOVA was additionally computed for the individual ratings of the Direct Questions between Group A and Group B1. Based on a one-way ANOVA, we did not find any statistically significant difference in the "Difficulty" scores (F(2.6) = 1.333, $p = 0.260$), nor in the "Performance" (F(2.6) = 0.6778, $p = 0.668$), nor in the "Stress" (F(2.6) = 1.740, $p = 0.131$), nor in the "Motivation" (F(2.6) = 1.072, $p = 0.392$) scores between Group A and Group B1.

Table 5. Post-hoc Newman-Keuls corrections for the "Difficulty" (Dif), "Performance" (Per), "Stress" (Str) and "Motivation" (Mot) questions between all sequences compared to *Normal*. Mean-Differences (Mean-Diff.) and *p*-values are presented.

Direct Questions	Mean-Diff. Group A	*p*-Value Group A	Mean-Diff. Group A + Group B1	*p*-Value Group A + Group B1
Dif_Normal – Dif_Interest	0.263	0.742	−0.035	0.907
Dif_Normal – Dif_Overload	−3.684	0.000***	−3.965	0.000***
Dif_Normal – Dif_Easy	3.842	0.000***	3.351	0.000***
Dif_Normal – Dif_Underload	3.816	0.000***	3.333	0.000***
Dif_Normal – Dif_Frustration	0.921	0.103	0.719	0.080
Per_Normal – Per_Interest	−2.105	0.000***	−2.123	0.000***
Per_Normal – Per_Overload	3.395	0.000***	3.316	0.000***
Per_Normal – Per_Easy	−2.816	0.000***	−3.000	0.000***
Per_Normal – Per_Underload	−2.974	0.000***	−2.947	0.000***
Per_Normal – Per_Frustration	−0.500	0.162	−0.632	0.036*

Table 5. *Cont.*

Direct Questions	Mean-Diff. Group A	p-Value Group A	Mean-Diff. Group A + Group B1	p-Value Group A + Group B1
Str_Normal – Str_Interest	–0.105	0.768	–0.123	0.684
Str_Normal – Str_Overload	–1.974	0.000***	–2.053	0.000***
Str_Normal – Str_Easy	2.711	0.000***	2.544	0.000***
Str_Normal – Str_Underload	2.447	0.000***	2.421	0.000***
Str_Normal – Str_Frustration	–0.316	0.813	–0.737	0.104
Mot_Normal – Mot_Interest	–0.789	0.070	–0.667	0.176
Mot_Normal – Mot_Overload	1.211	0.009**	1.035	0.005**
Mot_Normal – Mot_Easy	0.184	0.864	0.000	1.000
Mot_Normal – Mot_Underload	0.789	0.176	0.544	0.072
Mot_Normal – Mot_Frustration	1.237	0.010*	1.035	0.003**

*$p < 0.05$, **$p < 0.01$, ***$p < 0.001$.

Table 6. Post-hoc Newman-Keuls corrections for the "Difficulty" (Dif), "Performance" (Per), "Stress" (Str) and "Motivation" (Mot) questions between *Overload* vs. *Underload* and *Interest* vs. *Frustration*. Mean-Differences (Mean-Diff.) and p-values are presented.

Direct Questions	Mean-Diff. Group A	p-Value Group A	Mean-Diff. Group A + Group B1	p-Value Group A + Group B1
Dif_Overload – Dif_Underload	7.500	0.000***	7.298	0.000***
Dif_Interest – Dif_Frustration	0.658	0.254	0.754	0.091
Per_Overload – Per_Underload	–6.368	0.000***	–6.263	0.000***
Per_Interest – Per_Frustration	1.605	0.000***	1.491	0.000***
Str_Overload – Str_Underload	4.421	0.000***	4.474	0.000***
Str_Interest – Str_Frustration	–0.211	0.826	–0.614	0.175
Mot_Overload – Mot_Underload	–0.421	0.239	–0.491	0.363
Mot_ Interest – Mot_Frustration	2.026	0.000***	1.702	0.000***

*$p < 0.05$, **$p < 0.01$, ***$p < 0.001$.

According to Table 5, most of the Direct Questions can be significantly distinguished from each other for all the sequences compared to *Normal*. Exceptions are the "Difficulty" and "Stress" questions of *Interest* and *Frustration* as well as the "Motivation" question of *Interest*, *Easy* and *Underload* for both Group A and Group A + Group B1, in addition to the "Performance" question of *Frustration* for Group A. More context-relevant results are the implications in Table 6, which show that the states *Overload* vs. *Underload* and *Interest* vs. *Frustration* can be significantly distinguished from each other for all the Direct Questions except for the "Difficulty" and "Stress" question between *Interest* vs. *Frustration* and the "Motivation" question between *Overload* vs. *Underload* for both Group A and Group A + Group B1.

3.4. Data Annotation

In addition to the basic annotation, leading from the experimental design and application log files, the dataset is enhanced by various semi-automatic generated labels. The basic annotation contains the exact timing information at millisecond level of the beginning and ending of all sequences: Timestamps when each search item was presented, if and when a subject pronounced the solution, including whether the solution was correct or wrong and all information of the given subjective feedback. These mostly technical annotations do not necessarily contain emotional information, although they can give hints on situations where the probability of emotional reactions raises, for instance in case of timeouts or wrong answers, or during maximum load phases.

The semi-automatic labels are generated by our data driven active learning approach, presented in [62,63]. The basic assumption of this approach is the sparseness of emotional reactions in the audio and video modalities. In several pre-studies we figured out, that in HCI scenarios, users mostly tend to have emotions only in a few situations, or at least show them only in sparseness [64]. This leads to the assumption that most of the recorded data represent neutral emotional content. Based on this assumption, we train different density estimation models, such as One Class SVM, SVDD, or GMM on the whole dataset (ignoring the underling experimental structure) and then compare each feature vector instance with this neutral or background model. If a specific feature vector has a high distance compared to the background model, the probability of having an emotional instance increases. As such, less fitting-points are then presented to experts, which rate the points towards the emotional content. After having the first points labeled (emotion or neutral), these labels are used to improve the background model iteratively, until most of the outlier data points are labeled. Details of this active learning-based process can be found in the cited papers. The main conclusion of our active learning algorithms is the dramatic reduction of annotation effort in case of affective datasets like the one presented here. In most cases, only 10% of a naturalistic HCI dataset has to be annotated in order to achieve the same classification results as the baseline classifier using the full dataset. The active learning based, semi-automatic generated labels are part of the dataset. Further we had to manually label nine participants in order to evaluate our active learning approach. These manual labels are also part of the dataset.

Additionally, we provide some further manual created labels regarding the body pose information. As described in [65], we annotated several body poses based on distance measures of the skeleton provided by the Kinect sensor. Static poses include onsets and offsets of: arms crossed, hands behind back, hands on hips, legs crossed, and legs in step position. Dynamic poses include: sideways moving hands away from body, facial hand touch, and quick movement of feet.

4. Discussion and Summary

The resulting multimodal *uulmMAC* database from our emotional and cognitive load scenario conducted in a mobile interactive HCI setting is a valuable contribution to research fields related to multimodal affective computing and machine learning applications in HCI. Summarized, the main contributions of our work include the following:

- **Dataset for affective computing research:** We designed and implemented a HCI scenario and acquired the *uulmMAC* dataset for emotional and cognitive states recognition. The dataset consists of six different sequences, including the states *Interest, Overload, Normal, Easy, Underload,* and *Frustration*. The emotional-cognitive conditions were thereby induced by increasing the task field objects and colors as well as decreasing the available time of an interactive game paradigm.
- **Multimodal mobile and interactive:** It consists of highly multimodal (biosignals, videos, audios, Kinect), mobile (standing, walking, freely moving positions with wireless physiology sensors), and interactive (HCI via natural speech) emotional-cognitive HCI scenario.
- **Large number of subjects/recording sessions:** The original experiment includes 60 subjects and 100 recording sessions, from which 57 subjects and 95 recording sessions are left as part of the

final dataset. Depending on the focus of the research question or modality, the recording sessions of 38 subjects or 19 subjects or all 57 subjects can be analyzed: For instance, when focusing on physiological reactions with no specific interest in facial EMG, all 57 subjects (Group A + Group B1) can be analyzed; while when focusing on facial expression from video data, 38 subjects (Group A) can be analyzed.

- **Transtemporal analysis:** Our dataset also allows transtemporal research and investigations of the changes and variations of the induction, reactions and recognition over time: This is possible with our second sample of 20 subjects who underwent three different measurements at three different times with one-week interval-time inbetween. The transtemporal part of the final dataset includes 19 subjects (Group B) out of the original 20 subjects, left after quality check. Further, the data analysis and evaluation of the subjective feedback and questionnaires show transtemporally stable and valid induction results.
- **Validated dataset:** The induction of the various emotional and cognitive load states via six different sequences is validated through evaluation of subjective feedback acquired during the experiment. These are used as ground truth for our paradigm. On one side, the reported SAM Ratings vary between the sequences and show significant differences between the relevant induced states (Tables 3 and 4 or Tables A1a and A1b). Additionally, the SAM Ratings results of Group B1 show a consistent course with the results of Group A (first measurements from both samples) with no statistically significant differences based on an ANOVA. On the other side, the results from the Direct Questions are compatible with the related induction state (i.e., the *Overload* and *Frustration* sequences have high "Stress" answers rates, while the *Interest* sequence has high "Motivation" answers rates etc.) and show significant differences between the relevant induced states (Tables 5 and 6 or Tables A2a and A2b). Additionally, the Direct Questions ratings of Group B1 present similar course as the results of Group A for all the four questions with no statistically significant differences based on an ANOVA. Finally, the evaluation and analysis of the various questionnaires, acquired from the subjects prior the experiment, also show stable results.
- **High technical quality:** The technical quality of the data and related signals is also checked and demonstrated via different preliminary classifications conducted on various subsets of the database including: the video data [63], the gesture data [65], the audio data [66], the biophysiological data [67], the speech and the biophysiological data [68], and the multimodal data [69].

Overall, we created a dataset for various applications in the fields of affective computing and machine learning, including classifications, feature analysis, multimodal fusion or transtemporal investigations. The dataset includes multimodal sensor data as well as various annotations and extracted labels. Limitations of this work include the relatively limited number of transtemporal data (57 measurements from 19 subjects) as well as the absence of electroencephalography (EEG) or electrooculography (EOG) data for brain and eye movement analysis, both relevant for cognitive reactions. Finally, the experiment was conducted in a laboratory setting designed to be close to real HCI, and the next step would be to transfer our settings and findings into-the-wild for closer real-life induction and recognition research.

Future work will include numerical evaluations based on classification models using machine learning for the full dataset. Thereby, standardized sets of feature extraction techniques for each recorded modality will be generated and standard features for each emotional and cognitive state will be defined. A multimodal fusion analysis will be conducted to investigate the effect of each modality on the recognition rates of the different states. Further, a transtemporal analysis of the Group B data will be conducted to investigate the changes in time including features and classifications. Further, investigations related to the analysis of human-computer dialogs could be conducted, for instance to investigate the effects of computer feedbacks on human performance and the psychophysiological responses. Similarly, a gender analysis could also be conducted to investigate differences in the elicitation levels, emotional-cognitive psychophysiological responses or in the recognition rates and individual performance.

Finally, considering the relevance of emotional *Frustration* and cognitive *Overload* in the emergence of stress, which was investigated in many studies [70–73], we believe that our *uulmMAC* database on emotional and cognitive load states can also be used for affective computing and machine learning applications in the field of stress research. The well-adapted TSST—Trier Social Stress Test [70] employs a mental arithmetic task to induce high cognitive load (beside a social-evaluative part based on a public speaking task). The Stroop Color Test [71] employs a word-color task to induce high cognitive load and was further adopted by Choi et al. [74] in their experiments to develop a wearable stress monitoring system. Additionally, Wijsman et al. employ computer tasks (calculation, puzzle, memorization) under time pressure to induce stress [72]. In a similar context, a multimodal dataset was recently collected within the SWELL project [73] to induce stress by manipulating the working conditions of the subjects through mail interruptions and time pressure. Based on these studies, we will investigate in our future work the application of our database to the field of stress recognition research. It would be of interest if specialized machine learning techniques like transfer learning and/or deep learning approaches can be applied to transfer features and classifiers created on the *uulmMAC* dataset into the stress classification scenario.

Author Contributions: Conceptualization: D.H.-R., H.H., and H.C.T.; methodology: D.H.-R. and S.M.; software: D.H.-R., S.M., and A.D.; validation: D.H.-R., S.M., A.D., H.H., and H.C.T.; formal analysis: D.H.-R., S.M., A.D., and J.S.; investigation: D.H.-R. and A.D.; resources: H.C.T. and F.S.; data curation: D.H.-R., S.M., and A.D.; writing—original draft preparation: D.H.-R., S.M., and A.D.; writing—review and editing: D.H.-R., F.S., J.S. and H.C.T.; visualization: D.H.-R., S.M., A.D., and J.S.; supervision: D.H.-R., H.H., H.C.T., and F.S.; project administration: D.H.-R., H.H., H.C.T., and F.S.; funding acquisition: D.H.-R.., H.H., H.C.T., and F.S. All authors have read and agreed to the published version of the manuscript.

Funding: This research was supported by grants from the Transregional Collaborative Research Center SFB/TRR 62 Companion Technology for Cognitive Technical Systems funded by the German Research Foundation (DFG). It is also supported by a Margarete von Wrangell (MvW) habilitation scholarship funded by the Ministry of Science, Research and Arts (MWK) of the state of Baden-Württemberg for Dilana Hazer-Rau.

Conflicts of Interest: The authors declare no conflict of interest.

Appendix A

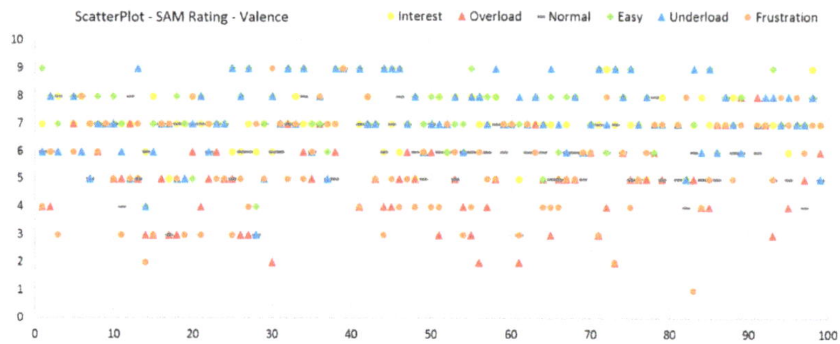

(a)

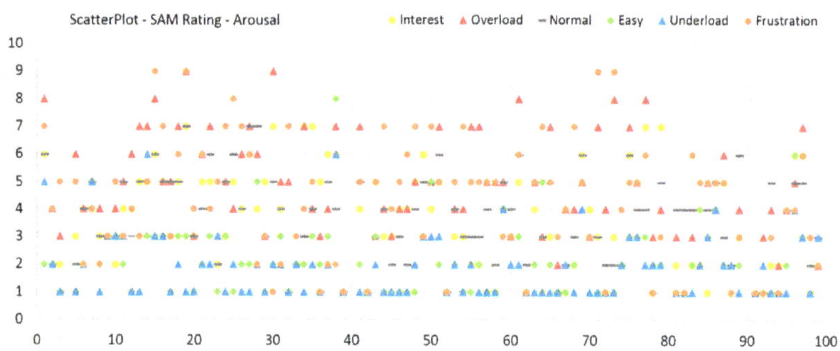

(b)

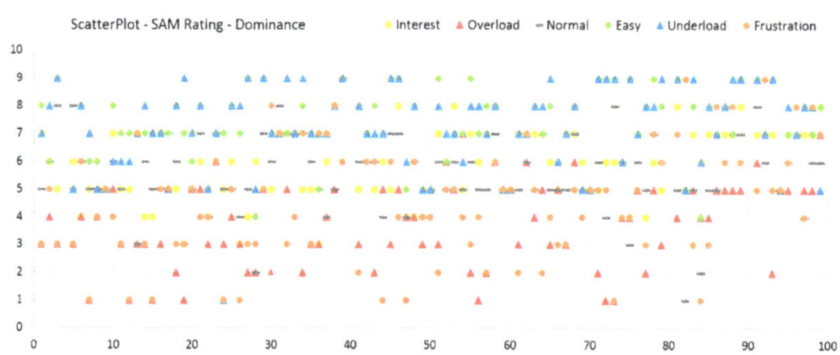

(c)

Figure A1. (**a**) Ratings distribution of the Valence dimension for all the measurements. (**b**) Ratings distribution of the Arousal dimension for all the measurements. (**c**) Ratings distribution of the Dominance dimension for all the measurements.

(**a**)

(**b**)

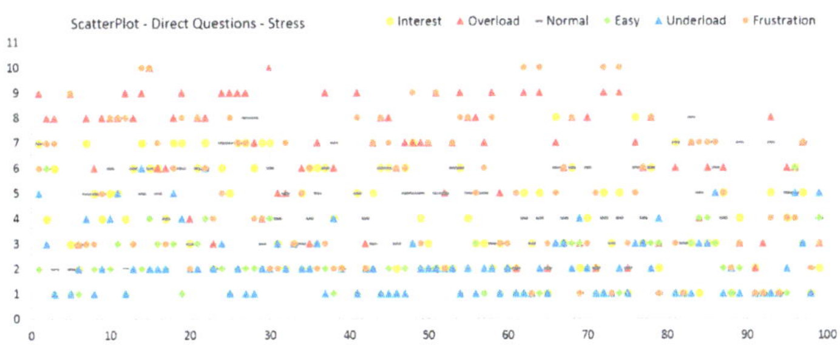

(**c**)

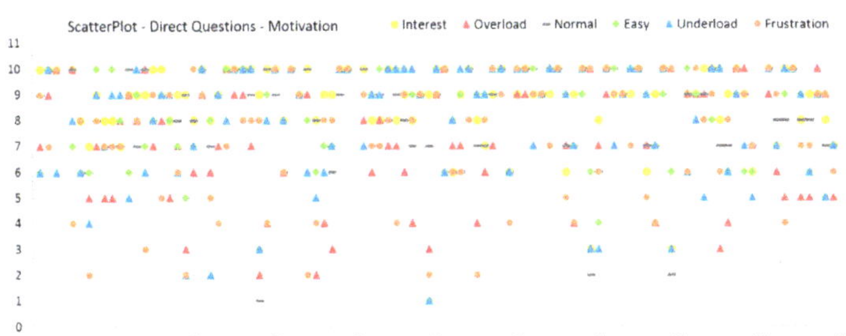

(**d**)

Figure A2. (**a**) Ratings distribution of the "Difficulty" question for all the measurements. (**b**) Ratings distribution of the "Performance" question for all the measurements. (**c**) Ratings distribution of the "Stress" question for all the measurements. (**d**) Ratings distribution of the "Motivation" question for all the measurements.

Table A1. (**a**) Newman-Keuls post-hoc corrections for the VAD dimensions (V = Valence; A = Arousal; D = Dominance) of the SAM Ratings and the six sequences Seq. (I= *Interest*, O= *Overload*, N= *Normal*, E= *Easy*, U= *Underload*, F= *Frustration*) for Group A. Significant effects are highlighted in bold ($p < 0.05$). (**b**) Newman-Keuls post-hoc corrections for the VAD dimensions (V = Valence; A = Arousal; D = Dominance) of the SAM Ratings and the six sequences Seq. (I= *Interest*, O= *Overload*, N= *Normal*, E= *Easy*, U= *Underload*, F= *Frustration*) for Group A + Group B1. Significant effects are highlighted in bold ($p < 0.05$).

VAD		V	V	V	V	V	V	A	A	A	A	A	A	D	D	D	D	D	D
	Seq.	I	O	N	E	U	F	I	O	N	E	U	F	I	O	N	E	U	F
V	I		**0.000**	0.272	0.655	0.817	**0.003**	**0.000**	**0.000**	**0.000**	**0.000**	**0.000**	**0.000**	**0.000**	**0.000**	**0.002**	0.758	0.949	**0.000**
V	O	**0.000**		**0.003**	**0.000**	**0.000**	0.369	**0.044**	0.878	**0.005**	**0.000**	**0.000**	**0.025**	0.527	**0.000**	0.486	**0.000**	**0.000**	**0.030**
V	N	0.272	**0.003**		0.076	0.190	0.094	**0.000**	**0.004**	**0.000**	**0.000**	**0.000**	**0.000**	**0.044**	**0.000**	**0.038**	0.250	0.345	**0.000**
V	E	0.655	**0.000**	0.076		0.636	**0.000**	**0.000**	**0.000**	**0.000**	**0.000**	**0.000**	**0.000**	**0.000**	**0.000**	**0.000**	0.673	0.397	**0.000**
V	U	0.817	**0.000**	0.190	0.636		**0.004**	**0.000**	**0.000**	**0.000**	**0.000**	**0.000**	**0.000**	**0.001**	**0.000**	**0.002**	0.852	0.949	**0.000**
V	F	**0.003**	0.369	0.094	**0.000**	**0.004**		**0.000**	0.309	**0.000**	**0.000**	**0.000**	**0.001**	0.590	**0.000**	1.000	**0.001**	**0.002**	**0.000**
A	I	**0.000**	**0.044**	**0.000**	**0.000**	**0.000**	**0.000**		**0.044**	0.397	**0.000**	**0.000**	0.921	**0.003**	0.207	**0.001**	**0.000**	**0.000**	0.939
A	O	**0.000**	0.878	**0.004**	**0.000**	**0.000**	0.309	**0.044**		**0.004**	**0.000**	**0.000**	**0.045**	0.355	**0.000**	0.460	**0.000**	**0.000**	**0.035**
A	N	**0.000**	**0.005**	**0.000**	**0.000**	**0.000**	**0.000**	0.397	**0.004**		**0.000**	**0.000**	0.606	**0.000**	0.397	**0.000**	**0.000**	**0.000**	0.625
A	E	**0.000**	**0.000**	**0.000**	**0.000**	**0.000**	**0.000**	**0.000**	**0.000**	**0.000**		0.397	**0.000**	**0.000**	**0.000**	**0.000**	**0.000**	**0.000**	**0.000**
A	U	**0.000**	**0.000**	**0.000**	**0.000**	**0.000**	**0.000**	**0.000**	**0.000**	**0.000**	0.397		**0.000**	**0.000**	**0.000**	**0.000**	**0.000**	**0.000**	**0.000**
A	F	**0.000**	**0.025**	**0.000**	**0.000**	**0.000**	**0.001**	0.921	**0.045**	0.606	**0.000**	**0.000**		**0.005**	0.229	**0.002**	**0.000**	**0.000**	0.758
D	I	**0.000**	0.527	**0.044**	**0.000**	**0.001**	0.590	**0.003**	0.355	**0.000**	**0.000**	**0.000**	**0.005**		**0.000**	0.852	**0.000**	**0.000**	**0.003**
D	O	**0.000**	**0.000**	**0.000**	**0.000**	**0.000**	**0.000**	0.207	**0.000**	0.397	**0.000**	**0.000**	0.229	**0.000**		**0.000**	**0.000**	**0.000**	0.287
D	N	**0.002**	0.486	**0.038**	**0.000**	**0.002**	1.000	**0.001**	0.460	**0.000**	**0.000**	**0.000**	**0.002**	0.852	**0.000**		**0.001**	**0.001**	**0.001**
D	E	0.758	**0.000**	0.250	0.673	0.852	**0.001**	**0.000**	**0.000**	**0.000**	**0.000**	**0.000**	**0.000**	**0.000**	**0.000**	**0.001**		1.000	**0.000**
D	U	0.949	**0.000**	0.345	0.397	0.949	**0.002**	**0.000**	**0.000**	**0.000**	**0.000**	**0.000**	**0.000**	**0.000**	**0.000**	**0.001**	1.000		**0.000**
D	F	**0.000**	**0.030**	**0.000**	**0.000**	**0.000**	**0.000**	0.939	**0.035**	0.625	**0.000**	**0.000**	0.758	**0.003**	0.287	**0.001**	**0.000**	**0.000**	

(**a**)

Table A1. *Cont.*

VAD	Seq.	V I	V O	V N	V E	V U	V F	A I	A O	A N	A E	A U	A F	D I	D O	D N	D E	D U	D F
V	I		0.000	0.100	0.254	0.742	0.000	0.000	0.000	0.000	0.000	0.000	0.000	0.000	0.000	0.000	0.343	0.609	0.000
V	O	0.000		0.000	0.000	0.000	0.164	0.003	0.704	0.000	0.000	0.000	0.139	0.313	0.000	0.179	0.000	0.000	0.013
V	N	0.100	0.000		0.003	0.061	0.002	0.000	0.000	0.000	0.000	0.000	0.000	0.001	0.000	0.004	0.026	0.047	0.000
V	E	0.254	0.000	0.003		0.311	0.000	0.000	0.000	0.000	0.000	0.000	0.000	0.000	0.000	0.000	0.705	0.529	0.000
V	U	0.742	0.000	0.061	0.311		0.000	0.000	0.000	0.000	0.000	0.000	0.000	0.000	0.000	0.000	0.998	0.950	0.000
V	F	0.000	0.164	0.002	0.000	0.000		0.000	0.179	0.000	0.000	0.000	0.006	0.950	0.000	0.802	0.000	0.000	0.000
A	I	0.000	0.003	0.000	0.000	0.000	0.000		0.007	0.184	0.000	0.000	0.202	0.000	0.313	0.000	0.000	0.000	0.569
A	O	0.000	0.704	0.000	0.000	0.000	0.179	0.007		0.000	0.000	0.000	0.129	0.257	0.000	0.114	0.000	0.000	0.022
A	N	0.000	0.000	0.000	0.000	0.000	0.000	0.184	0.000		0.000	0.000	0.013	0.000	0.899	0.000	0.000	0.000	0.139
A	E	0.000	0.000	0.000	0.000	0.000	0.000	0.000	0.000	0.000		0.282	0.000	0.000	0.000	0.000	0.000	0.000	0.000
A	U	0.000	0.000	0.000	0.000	0.000	0.000	0.000	0.000	0.000	0.282		0.000	0.000	0.000	0.000	0.000	0.000	0.000
A	F	0.000	0.139	0.000	0.000	0.000	0.006	0.202	0.129	0.013	0.000	0.000		0.007	0.014	0.001	0.000	0.000	0.255
D	I	0.000	0.313	0.001	0.000	0.000	0.950	0.000	0.257	0.000	0.000	0.000	0.007		0.000	0.569	0.000	0.000	0.000
D	O	0.000	0.000	0.000	0.000	0.000	0.000	0.313	0.000	0.899	0.000	0.000	0.014	0.000		0.000	0.000	0.000	0.179
D	N	0.000	0.179	0.004	0.000	0.000	0.802	0.000	0.114	0.000	0.000	0.000	0.001	0.569	0.000		0.000	0.000	0.000
D	E	0.343	0.000	0.026	0.705	0.998	0.000	0.000	0.000	0.000	0.000	0.000	0.000	0.000	0.000	0.000		1.000	0.000
D	U	0.609	0.000	0.047	0.529	0.950	0.000	0.000	0.000	0.000	0.000	0.000	0.000	0.000	0.000	0.000	1.000		0.000
D	F	0.000	0.013	0.000	0.000	0.000	0.000	0.569	0.022	0.139	0.000	0.000	0.255	0.000	0.179	0.000	0.000	0.000	

(b)

Table A2. (**a**) Newman-Keuls post-hoc corrections for the Direct Questions Quest. (Dif = "Difficulty"; Per = "Performance"; Str = "Stress"; Mot = "Motivation") and the sequences Seq. (I= *Interest*, O= *Overload*, N= *Normal*, E= *Easy*, U= *Underload*, F= *Frustration*) for Group A. Significant effects are marked bold ($p < 0.05$). (**b**) Newman-Keuls post-hoc corrections for the Direct Questions Quest. (Dif = "Difficulty"; Per = "Performance"; Str = "Stress"; Mot = "Motivation") and the sequences Seq. (I= *Interest*, O= *Overload*, N= *Normal*, E= *Easy*, U= *Underload*, F= *Frustration*) for Group A + Group B1. Significant effects are marked bold ($p < 0.05$).

Ques	Seq.	Dif I	Dif O	Dif N	Dif E	Dif U	Dif F	Per I	Per O	Per N	Per E	Per U	Per F	Str I	Str O	Str N	Str E	Str U	Str F	Mot I	Mot O	Mot N	Mot E	Mot U	Mot F
Dif	I		0.000	0.742	0.000	0.000	0.254	0.000	0.000	0.850	0.000	0.000	0.176	0.768	0.000	0.826	0.000	0.000	0.768	0.000	0.000	0.000	0.000	0.000	0.000
Dif	O	0.000		0.000	0.000	0.000	0.000	0.000	0.000	0.000	0.128	0.135	0.000	0.000	0.000	0.000	0.000	0.000	0.000	0.883	0.000	0.086	0.047	0.000	0.000
Dif	N	0.742	0.000		0.000	0.000	0.103	0.000	0.000	0.941	0.000	0.000	0.304	0.731	0.000	0.675	0.000	0.000	0.659	0.000	0.000	0.000	0.000	0.000	0.000
Dif	E	0.000	0.000	0.000		0.941	0.000	0.000	0.381	0.000	0.000	0.000	0.000	0.000	0.000	0.000	0.254	0.075	0.000	0.000	0.000	0.000	0.000	0.000	0.000
Dif	U	0.000	0.000	0.000	0.941		0.000	0.000	0.211	0.000	0.000	0.000	0.000	0.000	0.000	0.000	0.181	0.059	0.000	0.000	0.000	0.000	0.000	0.000	0.000
Dif	F	0.254	0.000	0.103	0.000	0.000		0.000	0.000	0.111	0.000	0.000	0.001	0.269	0.000	0.211	0.000	0.000	0.205	0.000	0.000	0.000	0.000	0.000	0.000
Per	I	0.000	0.000	0.000	0.000	0.000	0.000		0.000	0.000	0.115	0.107	0.000	0.000	0.393	0.000	0.000	0.000	0.000	0.000	0.341	0.192	0.141	0.825	0.454
Per	O	0.000	0.000	0.000	0.381	0.211	0.000	0.000		0.000	0.000	0.000	0.000	0.000	0.000	0.000	0.606	0.423	0.000	0.000	0.000	0.000	0.000	0.000	0.000
Per	N	0.850	0.000	0.941	0.000	0.000	0.111	0.000	0.000		0.000	0.000	0.162	0.804	0.000	0.728	0.000	0.000	0.864	0.000	0.000	0.000	0.000	0.000	0.000
Per	E	0.000	0.128	0.000	0.000	0.000	0.000	0.115	0.000	0.000		0.898	0.000	0.000	0.003	0.000	0.000	0.000	0.000	0.121	0.006	1.000	0.606	0.121	0.007
Per	U	0.000	0.135	0.000	0.000	0.000	0.000	0.107	0.000	0.000	0.898		0.000	0.000	0.001	0.000	0.000	0.000	0.000	0.077	0.002	0.659	0.774	0.085	0.002
Per	F	0.176	0.000	0.304	0.000	0.000	0.001	0.000	0.000	0.162	0.000	0.000		0.123	0.006	0.076	0.000	0.000	0.222	0.000	0.011	0.000	0.000	0.000	0.007
Str	I	0.768	0.000	0.731	0.000	0.000	0.269	0.000	0.000	0.804	0.000	0.000	0.123		0.000	0.768	0.000	0.000	0.826	0.000	0.000	0.000	0.000	0.000	0.000
Str	O	0.000	0.000	0.000	0.000	0.000	0.000	0.393	0.000	0.000	0.003	0.001	0.006	0.000		0.000	0.000	0.000	0.000	0.000	0.928	0.004	0.015	0.410	0.768
Str	N	0.826	0.000	0.675	0.000	0.000	0.211	0.000	0.000	0.728	0.000	0.000	0.076	0.768	0.000		0.000	0.000	0.813	0.000	0.000	0.000	0.000	0.000	0.000
Str	E	0.000	0.000	0.000	0.254	0.181	0.000	0.000	0.606	0.000	0.000	0.000	0.000	0.000	0.000	0.000		0.462	0.000	0.000	0.000	0.000	0.000	0.000	0.000
Str	U	0.000	0.000	0.000	0.075	0.059	0.000	0.000	0.423	0.000	0.000	0.000	0.000	0.000	0.000	0.000	0.462		0.000	0.000	0.000	0.000	0.000	0.000	0.000
Str	F	0.768	0.000	0.659	0.000	0.000	0.205	0.000	0.000	0.864	0.000	0.000	0.222	0.826	0.000	0.813	0.000	0.000		0.000	0.000	0.000	0.000	0.000	0.000
Mot	I	0.000	0.883	0.000	0.000	0.000	0.000	0.000	0.000	0.000	0.121	0.077	0.000	0.000	0.000	0.000	0.000	0.000	0.000		0.000	0.070	0.050	0.000	0.000
Mot	O	0.000	0.000	0.000	0.000	0.000	0.000	0.341	0.000	0.000	0.006	0.002	0.011	0.000	0.928	0.000	0.000	0.000	0.000	0.000		0.009	0.021	0.239	0.941
Mot	N	0.000	0.086	0.000	0.000	0.000	0.000	0.192	0.000	0.000	1.000	0.659	0.000	0.000	0.004	0.000	0.000	0.000	0.000	0.070	0.009		0.864	0.176	0.010
Mot	E	0.000	0.047	0.000	0.000	0.000	0.000	0.141	0.000	0.000	0.606	0.774	0.000	0.000	0.015	0.000	0.000	0.000	0.000	0.050	0.021	0.864		0.208	0.027
Mot	U	0.000	0.000	0.000	0.000	0.000	0.000	0.825	0.000	0.000	0.121	0.085	0.000	0.000	0.410	0.000	0.000	0.000	0.000	0.000	0.239	0.176	0.208		0.423
Mot	F	0.000	0.000	0.000	0.000	0.000	0.000	0.454	0.000	0.000	0.007	0.002	0.007	0.000	0.768	0.000	0.000	0.000	0.000	0.000	0.941	0.010	0.027	0.423	

(a)

Table A2. *Cont.*

Ques	Seq.	Dif I	Dif O	Dif N	Dif E	Dif U	Dif F	Per I	Per O	Per N	Per E	Per U	Per F	Str I	Str O	Str N	Str E	Str U	Str F	Mot I	Mot O	Mot N	Mot E	Mot U	Mot F
Dif	I		0.000	0.907	0.000	0.000	0.091	0.000	0.000	0.584	0.000	0.000	0.011	0.511	0.000	0.431	0.000	0.000	0.352	0.000	0.000	0.000	0.000	0.000	0.000
Dif	O	0.000		0.000	0.000	0.000	0.000	0.000	0.000	0.000	0.091	0.106	0.000	0.000	0.000	0.000	0.000	0.000	0.000	0.485	0.000	0.043	0.030	0.000	0.000
Dif	N	0.907	0.000		0.000	0.000	0.080	0.000	0.000	0.687	0.000	0.000	0.012	0.323	0.000	0.343	0.000	0.000	0.548	0.000	0.000	0.000	0.000	0.000	0.000
Dif	E	0.000	0.000	0.000		0.954	0.000	0.000	0.441	0.000	0.000	0.000	0.000	0.000	0.000	0.000	0.576	0.443	0.000	0.000	0.000	0.000	0.000	0.000	0.000
Dif	U	0.000	0.000	0.000	0.954		0.000	0.000	0.245	0.000	0.000	0.000	0.000	0.000	0.000	0.000	0.441	0.363	0.000	0.000	0.000	0.000	0.000	0.000	0.000
Dif	F	0.091	0.000	0.080	0.000	0.000		0.000	0.000	0.009	0.000	0.000	0.000	0.343	0.000	0.323	0.000	0.000	0.008	0.000	0.000	0.000	0.000	0.000	0.000
Per	I	0.000	0.000	0.000	0.000	0.000	0.000		0.000	0.000	0.043	0.049	0.000	0.000	0.032	0.000	0.000	0.000	0.000	0.000	0.375	0.091	0.155	0.771	0.181
Per	O	0.000	0.000	0.000	0.441	0.245	0.000	0.000		0.000	0.000	0.000	0.000	0.000	0.000	0.000	0.954	0.888	0.000	0.000	0.000	0.000	0.000	0.000	0.000
Per	N	0.584	0.000	0.687	0.000	0.000	0.009	0.000	0.000		0.000	0.000	0.036	0.223	0.000	0.124	0.000	0.000	0.954	0.000	0.000	0.000	0.000	0.000	0.000
Per	E	0.000	0.091	0.000	0.000	0.000	0.000	0.043	0.000	0.000		0.862	0.000	0.000	0.000	0.000	0.000	0.000	0.000	0.163	0.001	0.848	0.694	0.067	0.000
Per	U	0.000	0.106	0.000	0.000	0.000	0.000	0.049	0.000	0.000	0.862		0.000	0.000	0.000	0.000	0.000	0.000	0.000	0.259	0.001	0.798	0.523	0.069	0.001
Per	F	0.011	0.000	0.012	0.000	0.000	0.000	0.000	0.000	0.036	0.000	0.000		0.000	0.027	0.000	0.000	0.000	0.080	0.000	0.001	0.000	0.000	0.000	0.002
Str	I	0.511	0.000	0.323	0.000	0.000	0.343	0.000	0.000	0.223	0.000	0.000	0.000		0.000	0.684	0.000	0.000	0.175	0.000	0.000	0.000	0.000	0.000	0.000
Str	O	0.000	0.000	0.000	0.000	0.000	0.000	0.032	0.000	0.000	0.000	0.000	0.027	0.000		0.000	0.000	0.000	0.000	0.000	0.163	0.000	0.000	0.021	0.343
Str	N	0.431	0.000	0.343	0.000	0.000	0.323	0.000	0.000	0.124	0.000	0.000	0.000	0.684	0.000		0.000	0.000	0.104	0.000	0.000	0.000	0.000	0.000	0.000
Str	E	0.000	0.000	0.000	0.576	0.441	0.000	0.000	0.954	0.000	0.000	0.000	0.000	0.000	0.000	0.000		0.684	0.000	0.000	0.000	0.000	0.000	0.000	0.000
Str	U	0.000	0.000	0.000	0.443	0.363	0.000	0.000	0.888	0.000	0.000	0.000	0.000	0.000	0.000	0.000	0.684		0.000	0.000	0.000	0.000	0.000	0.000	0.000
Str	F	0.352	0.000	0.548	0.000	0.000	0.008	0.000	0.000	0.954	0.000	0.000	0.080	0.175	0.000	0.104	0.000	0.000		0.000	0.000	0.000	0.000	0.000	0.000
Mot	I	0.000	0.485	0.000	0.000	0.000	0.000	0.000	0.000	0.000	0.163	0.259	0.000	0.000	0.000	0.000	0.000	0.000	0.000		0.000	0.176	0.121	0.001	0.000
Mot	O	0.000	0.000	0.000	0.000	0.000	0.000	0.375	0.000	0.000	0.001	0.001	0.001	0.000	0.163	0.000	0.000	0.000	0.000	0.000		0.005	0.008	0.363	1.000
Mot	N	0.000	0.043	0.000	0.000	0.000	0.000	0.091	0.000	0.000	0.848	0.798	0.000	0.000	0.000	0.000	0.000	0.000	0.000	0.176	0.005		1.000	0.072	0.003
Mot	E	0.000	0.030	0.000	0.000	0.000	0.000	0.155	0.000	0.000	0.694	0.523	0.000	0.000	0.000	0.000	0.000	0.000	0.000	0.121	0.008	1.000		0.169	0.005
Mot	U	0.000	0.000	0.000	0.000	0.000	0.000	0.771	0.000	0.000	0.067	0.069	0.000	0.000	0.021	0.000	0.000	0.000	0.000	0.001	0.363	0.072	0.169		0.234
Mot	F	0.000	0.000	0.000	0.000	0.000	0.000	0.181	0.000	0.000	0.000	0.001	0.002	0.000	0.343	0.000	0.000	0.000	0.000	0.000	1.000	0.003	0.005	0.234	

(b)

References

1. Paas, F.; Tuovinen, J.E.; Tabbers, H.; Van Gerven, P.W.M. Cognitive load measurement as a means to advance cognitive load theory. *Educ. Psychol.* **2003**, *38*, 63–71. [CrossRef]
2. Sweller, J.; Merrienboer, J.; Paas, F. Cognitive Architecture and Instructional Design. *Educ. Psychol. Rev.* **1998**, *10*, 251–296. [CrossRef]
3. Sweller, J.; Ayres, P.; Kalyuga, S. *Cognitive Load Theory*; Springer: New York, NY, USA, 2011.
4. Rösner, D.; Hazer-Rau, D.; Kohrs, C.; Bauer, T.; Günther, S.; Hoffmann, H.; Zhang, L.; Brechmann, A. Is there a Biological Basis for Success in Human Companion Interaction?—Results from a Transsituational Study. In *Human-Computer Interaction—Theory, Design, Development and Practice*; Kurosu, M., Ed.; Part I; Lecture Notes in Computer Science LNCS; Springer: Cham, Switzerland, 2016; Volume 9731, pp. 77–88. [CrossRef]
5. Moos, D. Examining hypermedia learning: The role of cognitive load and self-regulated learning. *J. Educ. Multimed. Hypermed.* **2013**, *22*, 39–61.
6. Creemers, B.; Kyriakides, L.; Panayiotis, A. Improvement of Teaching by Mastering Specific Competences: The Competency-Based Approach. In *Teacher Professional Development for Improving Quality of Teaching*; Springer: Berlin, Germany, 2013; pp. 13–28.
7. Lini, S.; Bey, C.; Hourlier, S.; Vallespir, B.; Johnston, A.; Favier, P.A. Evaluating ASAP (Anticipation Support for Aeronautical Planning): A user-centered case study. In Proceedings of the 17th International Symposium on Aviation Psychology, Dayton, OH, USA, 6–9 May 2013.
8. Taylor, T.; Pradhan, A.K.; Divekar, G.; Romoser, M.; Muttart, J.; Gomez, R.; Pollatsek, A.; Fisher, D.L. The view from the road: The contribution of on-road glance-monitoring technologies to understanding driver behavior. *Accid. Anal. Prev.* **2013**, *58*, 175–186. [CrossRef]
9. Hirshfield, L.M.; Solovey, E.T.; Girouard, A.; Kebinger, J.; Jacob, R.J.K.; Sassaroli, A.; Fantini, S. Brain measurement for usability testing and adaptive interfaces: An example of uncovering syntactic workload with functional near infrared spectroscopy. In Proceedings of the 27th SIGCHI Conference on Human Factors in Computing Systems (CHI '09), Boston, MA, USA, 4–9 April 2009; pp. 2185–2194. [CrossRef]
10. Greene, S.; Thapliyal, H.; Caban-Holt, A. A Survey of Affective Computing for Stress Detection: Evaluating technologies in stress detection for better health. *IEEE Consum. Electron. Mag.* **2016**, *5*, 44–56. [CrossRef]
11. Wilson, G.F.; Schlegel, R.E. *Operator Functional State Assessment*; NATO RTO Publication RTO-TR-HF M-104; NATO Research and Technology Organization: Neuilly sur Seine, France, 2004.
12. Hart, S.; Staveland, L. Development of the NASA-TLX (Task Load Index): Results of empirical and theoretical research. In *Human Mental Workload*; Hancock, P.A., Meshkati, N., Eds.; Elsevier Science Publishers, North Holland Press: Amsterdam, The Netherlands, 1988; pp. 139–183.
13. O'Donnell, R.; Eggemeier, F. Workload assessment methodology. In *Handbook of Perception and Human Performance. Cognitive Processes and Performance*; Boff, K., Kaufman, L., Thomas, J., Eds.; Wiley: New York, NY, USA, 1986; pp. 1–49.
14. Wierwille, W.W.; Eggemeier, F.T. Recommendations for mental workload measurement in a test and evaluation environment. *Hum. Factors J. Hum. Factors Erg.Soc.* **1993**, *35*, 263–281. [CrossRef]
15. Gingell, R. Review of Workload Measurement, Analysis and Interpretation Methods. *Eur. Organ. Saf. Air Navig.* **2003**, *33*, 1–33.
16. Fridman, L.; Reimer, B.; Mehler, B.; Freeman, W.T. Cognitive Load Estimation in the Wild. In Proceedings of the CHI Conference on Human Factors in Computing Systems (CHI '18), Montreal, QC, Canada, 21–26 April 2018; pp. 1–9. [CrossRef]
17. Van Dille, L.F.; Heslenfeld, D.J.; Koole, S.L. Tuning down the emotional brain: An fMRI study of the effects of cognitive load on the processing of affective images. *NeuroImage* **2009**, *45*, 1212–1219. [CrossRef]
18. Just, M.A.; Carpenter, P.A.; Miyake, A. Neuroindices of cognitive workload: Neuroimaging, pupillometric and event-related potential studies of brain work. *Theor. Issues Ergon. Sci.* **2003**, *4*, 56–88. [CrossRef]
19. Jerčić, P.; Sennersten, C.; Lindley, C. Modeling cognitive load and physiological arousal through pupil diameter and heart rate. *Multimed. Tools Appl.* **2020**, *79*, 3145–3159. [CrossRef]
20. Von Bauer, P. Cognitive State Classification Using Psychophysiological Measures. Master's Thesis, Universität Konstanz, Konstanz, Germany, 2018.

21. Luria, G.; Rosenblum, S. A computerized multidimensional measurement of mental workload via handwriting analysis. *Behav. Res. Methods.* **2012**, *44*, 575–586. [CrossRef] [PubMed]
22. Yu, K. Cognitive load examination via pen interactions. Ph.D. Thesis, The University of New South Wales, Sydney, Australia, 2015. Available online: http://unsworks.unsw.edu.au/fapi/datastream/unsworks:38410/SOURCE02?view=true (accessed on 14 February 2020).
23. Su, J.; Luz, S. Predicting Cognitive Load Levels from Speech Data. In *Recent Advances in Nonlinear Speech Processing*; Esposito, A., Faundez-Zanuy, M., Esposito, A.M., Cordasco, G., Drugman, T., Solé-Casals, J., Morabito, F.C., Eds.; Springer: Berlin Heidelberg, Germany, 2016; pp. 255–263.
24. Mattys, S.; Wiget, L. Effects of cognitive load on speech recognition. *J. Mem. Langg.* **2011**, *65*, 145–160. [CrossRef]
25. Das, D.; Chatterjee, D.; Sinha, A. Unsupervised approach for measurement of cognitive load using EEG signals. In Proceedings of the 13th IEEE Conference on BioInformatics and BioEngineering, Chania, Greece, 10–13 November 2013; pp. 1–6.
26. Bauer, R.; Gharabaghi, A. Estimating cognitive load during self-regulation of brain activity and neurofeedback with therapeutic brain-computer interfaces. *Front. Behav. Neurosci.* **2015**, *9*, 1–9. [CrossRef] [PubMed]
27. Klingner, J.; Tversky, B.; Hanrahan, P. Effects of visual and verbal presentation on cognitive load in vigilance, memory and arithmetic tasks. *J. Psychophys.* **2011**, *48*, 323–332. [CrossRef] [PubMed]
28. Hossain, G.; Yeasin, M. Understanding Effects of Cognitive Load from Pupillary Responses Using Hilbert Analytic Phase. In Proceedings of the IEEE on Computer Vision and Pattern Recognition Workshops 2014, Columbus, OH, USA, 23–28 June 2014; pp. 381–386.
29. Chen, F.; Zhou, J.; Wang, Y.; Yu, K.; Arshad, S.Z.; Khawaji, A.; Conway, D. *Robust Multimodal Cognitive Load Measurement*; Human-Computer Interaction Series; Springer: Berlin, Germany, 2016.
30. Schuller, B.; Steidl, S.; Batliner, A.; Epps, J.; Eyben, F.; Ringeval, F.; Marchi, E.; Zhang, Y. The Interspeech 2014 Computational Paralinguistics Challenge: Cognitive & Physical Load. In Proceedings of the 15th Annual Conference of the International Speech Communication Association (ISCA), Singapore, 14–18 September 2014; pp. 427–431.
31. Liu, X.; Chen, T.; Xie, G.; Liu, G. Contact-Free Cognitive Load Recognition Based on Eye Movement. *J. Electr. Comput. Eng.* **2016**, *2016*. [CrossRef]
32. Zhang, L.; Rukavina, S.; Gruss, S.; Traue, H.C.; Hazer, D. Classification analysis for the emotion recognition from psychobiological data. In Proceedings of the International Symposium on Companion-Technologies (ISCT'15), Ulm, Germany, 23–25 September 2015.
33. Hamdi, H.; Richard, P.; Suteau, A.; Allain, P. Emotion assessment for affective computing based on physiological responses. In Proceedings of the IEEE International Conference on Fuzzy Systems (FUZZ-IEEE), Brisbane, QLD, Australia, 10–15 June 2012; pp. 1–8.
34. Lang, P.; Bradley, M.; Cuthbert, B. *The International Affective Picture System (IAPS): Technical Manual and Affective Ratings*; The Center for Research in Psychophysiology, University of Florida: Gainesville, FL, USA, 1999.
35. Gross, J.; Levenson, R. Emotion elicitation using films. *Cogn. Emot.* **1995**, *9*, 87–108. [CrossRef]
36. Soleymani, M.; Pantic, M.; Pun, T. Multimodal emotion recognition in response to videos. *IEEE Trans. Affect. Comput.* **2012**, *3*, 211–223. [CrossRef]
37. Hazer, D.; Ma, X.Y.; Rukavina, S.; Gruss, S.; Walter, S.; Traue, H.C. Emotion Elicitation Using Film Clips: Effect of Age Groups on Movie Choice and Emotion Rating. In *Human-Computer Interaction*; Stephanidis, C., Ed.; HCI 2015, Communications in Computer and Information Science; Springer: Cham, Switzerland, 2015; Volume 528, pp. 110–116. [CrossRef]
38. Koelstra, S.; Muhl, C.; Soleymani, M.; Lee, J.; Yazdani, A.; Ebrahimi, T.; Pun, T.; Nijholt, A.; Patras, I. Deap: A database for emotion analysis using physiological signals. *IEEE Trans. Affect. Comput.* **2012**, *3*, 18–31. [CrossRef]
39. Chanel, G.; Rebetez, C.; Bétrancourt, M.; Pun, T. Emotion Assessment From Physiological Signals for Adaptation of Game Difficulty. *IEEE Trans. Syst. Man Cybern. Part. A Syst. Hum.* **2011**, *41*, 1052–1063. [CrossRef]
40. Hudlicka, E. To feel or not to feel: The role of affect in human–computer interaction. *Int. J. Hum.-Comput. Stud.* **2003**, *59*, 1–32. [CrossRef]

41. Brave, S.; Nass, C. Emotion in Human–Computer Interaction. In *The Human-Computer Interaction Handbook: Fundamentals, Evolving Technologies and Emerging Applications*; L. Erlbaum Associates Inc.: Mahwah, NJ, USA, 2003; pp. 81–96.
42. Conati, C. Probabilistic assessment of user's emotions in educational games. *Appl. Ai.* **2002**, *16*, 555–576. [CrossRef]
43. Klein, J.; Moon, Y.; Picard, R.W. This computer responds to user frustration: Theory, design and results. *Interact. Comput.* **2002**, *14*, 119–140. [CrossRef]
44. Taylor, B.; Dey, A.; Siewiorek, D.; Smailagic, A. Using Physiological Sensors to Detect Levels of User Frustration Induced by System Delays. In Proceedings of the International Joint Conference on Pervasive and Ubiquitous Computing (UbiComp'15), Osaka, Japan, 7–11 September 2015; pp. 517–528. [CrossRef]
45. Suhaib, A.; Zwart, N.; Gouweleeuw, K.; Verhoeven, G. Classification of Disappointment and Frustration Elicited by Human-Computer Interaction: Towards Affective HCI, 2019. Available online: https://www.researchgate.net/publication/335158621_Classification_of_Disappointment_and_Frustration_Elicited_by_Human-Computer_Interaction_Towards_Affective_HCI (accessed on 7 April 2020).
46. Lisetti, C.; Nasoz, F. Using Noninvasive Wearable Computers to Recognize Human Emotions from Physiological Signals. *Eurasip J. Adv. Signal. Process.* **2004**, *11*, 1672–1687. [CrossRef]
47. Liu, C.; Rani, P.; Sarkar, N. An empirical study of machine learning techniques for affect recognition in human-robot interaction. In Proceedings of the IEEE International Conference on Intelligent Robots and Systems (IEEE'RS), Edmonton, Canada, 2–6 August 2005; pp. 2662–2667. [CrossRef]
48. Silvia, P.J. Interest—The curious emotion. *Curr. Dir. Psychol. Sci.* **2008**, *17*, 57–60. [CrossRef]
49. Reeve, J.; Lee, W.; Won, S. Interest as Emotion, as Affect, and as Schema. In *Interest in Mathematics and Science Learning*; American Educational Research Association: Washington, DC, USA, 2015; Chapter 5; pp. 79–92. [CrossRef]
50. Ellsworth, P.C. Some reasons to expect universal antecedents of emotion. In *The Nature of Emotion: Fundamental Questions*; Ekman, P., Davidson, R.J., Eds.; Oxford University Press: New York, NY, USA, 1994; pp. 150–154.
51. Schüssel, F.; Honold, F.; Bubalo, N.; Huckauf, A.; Traue, H.C.; Hazer-Rau, D. In-depth analysis of multimodal interaction: An explorative paradigm. In *Human-Computer Interaction—Interaction Platforms and Techniques*; Kurosu, M., Ed.; Lecture Notes in Computer Science LNCS; Springer: Cham, Switzerland, 2016; Volume 9732, pp. 233–240. [CrossRef]
52. Traue, H.C.; Ohl, F.; Limbrecht, K.; Scherer, S.; Kessler, H.; Schwenker, F.; Kozyba, M.; Brechmann, A.; Hoffmann, H.; Scheck, A.; et al. Framework for Emotions and Dispositions in Man-Companion Interaction. In *Coverbal Synchrony in Human-Machine Interaction*; Campbell, N., Rojc, M., Eds.; Science Publishers: Rawalpindi, Pakistan, 2013; pp. 99–140. [CrossRef]
53. Gross, J.J.; John, O.P. Individual differences in two emotion regulation processes: Implications for affect, relationships, and well-being. *J. Personal. Soc. Psychol.* **2003**, *85*, 348–362. [CrossRef] [PubMed]
54. Abler, B.; Kessler, H. Emotion Regulation Questionnaire—Eine deutschsprachige Fassung des ERQ von Gross und John (2003). *Diagnostica* **2009**, *55*, 144–152. [CrossRef]
55. Cooper, A.; Petrides, K.V. A psychometric analysis of the Trait Emotional Intelligence Questionnaire-Short Form (TEIQue-SF) using item response theory. *J. Personal. Assess.* **2010**, *92*, 449–457. [CrossRef]
56. Jacobs, I.; Sim, C.W.; Zimmermann, J. The German TEIQue-SF: Factorial structure and relations to agentic and communal traits and mental health. *Pers. Individ. Differ.* **2015**, *72*, 72–189. [CrossRef]
57. Gosling, S.D.; Rentfrow, P.J.; Swann, W.B. A very brief measure of the big five personality domains. *J. Res. Personal.* **2003**, *37*, 504–528. [CrossRef]
58. Muck, P.M.; Hell, B.; Gosling, S.D. Construct validation of a short five-factor model instrument: A self-peer study on the German adaptation of the Ten-Item Personality Inventory (TIPI-G). *Eur. J. Psychol. Assess.* **2007**, *23*, 23–166. [CrossRef]
59. The Semaine Developer Website. Available online: http://semaine.opendfki.de (accessed on 27 September 2015).
60. Meudt, S.; Bigalke, L.; Schwenker, F. ATLAS—Annotation tool using partially supervised learning and Multi-view Co-learning in Human-Computer-Interaction Scenarios. In Proceedings of the 11th International Conference on Information Science, Signal Processing and their Applications (ISSPA'12), Montreal, QC, Canada, 2–5 July 2012; pp. 1309–1312.

61. Meudt, S.; Schwenker, F. ATLAS—Machine learning based annotation of multimodal data recorded in human-computer interaction scenarios. In Proceedings of the International Symposium on Companion Technology (ISCT'15), Ulm, Germany, 23–25 September 2015; pp. 181–186.
62. Thiam, P.; Kächele, M.; Schwenker, F.; Palm, G. Ensembles of Support Vector Data Description for Active Learning Based Annotation of Affective Corpora. In Proceedings of the IEEE Symposium Series on Computational Intelligence, Cape Town, South Africa, 7–10 December 2015; pp. 1801–1807.
63. Thiam, P.; Meudt, S.; Palm, G.; Schwenker, F. Temporal Dependency Based Multi-modal Active Learning Approach for Audiovisual Event Detection. *Neural Process. Lett.* **2018**, *48*, 709–732. [CrossRef]
64. Thiam, P.; Meudt, S.; Kächele, M.; Schwenker, F.; Palm, G. Detection of Emotional Events Utilizing Support Vector Methods in an Active Learning HCI Scenario. In Proceedings of the Workshop on Emotion Representation and Modelling in Human-Computer-Interaction-Systems (ERM4HCI'14), Istanbul, Turkey, 16 November 2014; pp. 31–36. [CrossRef]
65. Hihn, H.; Meudt, S.; Schwenker, F. Inferring Mental Overload based on Postural Behavior and Gestures. In Proceedings of the Emotion Recognition and Modeling for Companion Technology (ERM4CT'16), Tokyo, Japan, 16 November 201; pp. 1–8. [CrossRef]
66. Thiam, P.; Meudt, S.; Schwenker, F.; Palm, G. Active Learning for Speech Event Detection in HCI. In Proceedings of the Artificial Neural Networks in Pattern Recognition (ANNPR'16), Ulm, Germany, 28–30 September 2016; pp. 285–297.
67. Daucher, A.; Gruss, S.; Jerg-Bretzke, L.; Walter, S.; Hazer-Rau, D. Preliminary classification of cognitive load states in a human machine interaction scenario. In Proceedings of the International Conference on Companion Technology (ICCT'17), Ulm, Germany, 11–13 September 2017; pp. 1–5.
68. Held, D.; Meudt, S.; Schwenker, F. Bimodal Recognition of Cognitive Load Based on Speech and Physiological Changes. In *Multimodal Pattern Recognition of Social Signals in Human-Computer Interaction (MPRSS 2016)*; Schwenker, F., Scherer, S., Eds.; Lecture Notes in Computer Science LNCS; Springer: Cham, Switzerland, 2017; Volume 10183, pp. 12–23. [CrossRef]
69. Kindsvater, D.; Meudt, S.; Schwenker, F. Fusion Architectures for Multimodal Cognitive Load Recognition. In *Multimodal Pattern Recognition of Social Signals in Human-Computer-Interaction (MPRSS 2016)*; Schwenker, F., Scherer, S., Eds.; Lecture Notes in Computer Science LNCS; Springer: Cham, Switzerland, 2017; Volume 10183, pp. 12–23. [CrossRef]
70. Kirschbaum, C.; Pirke, K.; Hellhammer, D. The Trier Social Stress Test—A Tool for Investigating Psychobiological Stress Responses in a Laboratory Setting. *Neuropsychobiology* **1993**, *28*, 76–81. [CrossRef] [PubMed]
71. Stroop, R. Studies of interference in serial verbal reactions. *J. Exp. Psychol.* **1935**, *18*, 643–662. [CrossRef]
72. Wijsman, J.; Grundlehner, B.; Liu, H.; Hermens, H. Wearable Physiological Sensors Reflect Mental Stress State in Office-Like Situations. In Proceedings of the Humaine Association Conference on Affective Computing and Intelligent Interaction, Geneva, Switzerland, 2–5 September 2013; pp. 600–605.
73. Koldijk, S.; Sappelli, M.; Verberne, S.; Neerincx, M.; Kraaij, W. The SWELL Knowledge Work Dataset for Stress and User Modeling Research. In Proceedings of the 16th International Conference on Multimodal Interaction (ICMI'14), Istanbul, Turkey, 12–16 November 2014; pp. 291–298.
74. Choi, J.; Ahmed, B.; Gutierrez-Osuna, R. Development and Evaluation of an Ambulatory Stress Monitor based on wearable sensors. *IEEE Trans. Inf. Technol. Biomed.* **2012**, *16*, 279–286. [CrossRef] [PubMed]

sensors

MDPI

Article

Facial Expression Recognition Based on Weighted-Cluster Loss and Deep Transfer Learning Using a Highly Imbalanced Dataset

Quan T. Ngo and Seokhoon Yoon *

Department of Electrical and Computer Engineering, University of Ulsan, Ulsan 44610, Korea; tanquan.dn@gmail.com

* Correspondence: seokhoonyoon@ulsan.ac.kr

Received: 7 April 2020; Accepted: 3 May 2020; Published: 5 May 2020

Abstract: Facial expression recognition (FER) is a challenging problem in the fields of pattern recognition and computer vision. The recent success of convolutional neural networks (CNNs) in object detection and object segmentation tasks has shown promise in building an automatic deep CNN-based FER model. However, in real-world scenarios, performance degrades dramatically owing to the great diversity of factors unrelated to facial expressions, and due to a lack of training data and an intrinsic imbalance in the existing facial emotion datasets. To tackle these problems, this paper not only applies deep transfer learning techniques, but also proposes a novel loss function called weighted-cluster loss, which is used during the fine-tuning phase. Specifically, the weighted-cluster loss function simultaneously improves the intra-class compactness and the inter-class separability by learning a class center for each emotion class. It also takes the imbalance in a facial expression dataset into account by giving each emotion class a weight based on its proportion of the total number of images. In addition, a recent, successful deep CNN architecture, pre-trained in the task of face identification with the VGGFace2 database from the Visual Geometry Group at Oxford University, is employed and fine-tuned using the proposed loss function to recognize eight basic facial emotions from the AffectNet database of facial expression, valence, and arousal computing in the wild. Experiments on an AffectNet real-world facial dataset demonstrate that our method outperforms the baseline CNN models that use either weighted-softmax loss or center loss.

Keywords: facial expression recognition; deep convolutional neural network; transfer learning; auxiliary loss; weighted loss; class center

1. Introduction

Facial expressions are undoubtedly a dominant, natural, and effective channel used by people to convey their emotions and intentions during communication. Over the last few decades, automatic facial expression analysis has attracted significant attention, and has become one of the most challenging problems in computer vision and artificial intelligence fields. Numerous studies have been conducted on developing reliable automated facial expression recognition (FER) systems for use over a wide range of applications, such as human–computer interaction, social robotics, medical treatment, virtual reality, augmented reality, and games [1–4]. In this work, we constructed deep convolutional neural network (CNN)-based FER models to recognize eight common facial expressions (happy, sad, surprise, fear, contempt, anger, disgust, and neutral) from the AffectNet database of facial expression, valence, and arousal computing in the wild [5]. This was motivated by a upcoming challenge on AffectNet's website [6].

The FER model is usually composed of three main stages: face detection, feature extraction, and emotion classification. First, the face and its components (e.g., eyes, mouth, nose, and eyebrows) are detected from images or video sequences. Then, features that are the most effective at distinguishing one expression from another are extracted from the face region. Finally, a classifier is constructed, given the extracted feature set for each target facial expression. The literature is rich with handcrafted face detection and feature extraction methods for FER that have achieved satisfactory results in laboratory-controlled settings [7–11]. However, these traditional methods have been reported to be incapable of discriminating a great diversity of unrelated factors in facial expressions (e.g., subtle facial appearances, head poses, illumination intensity, and occlusions) with FER tasks for in-the-wild settings [12,13].

Recently, the success of convolutional neural networks in both computer vision and pattern recognition has promoted a transition in FER from using handcrafted feature-learning methods to using deep learning technologies. A deep learning-based FER system commonly uses a CNN model to extract and learn high-level features directly from input images. Then, an output layer (which usually uses softmax as an activation function) is attached on top of the CNN model to distinguish the emotion to be detected. This allows a faster emotion recognition system with higher accuracy in challenging, real-world environments [14–16]. In this paper, we make use of the powerful feature-learning capacity of the deep CNN to build FER models. However, a few significant problems arise when applying deep learning to FER systems.

First, deep learning-based FER models require a large amount of data for training to acquire suitable values for model parameters. Directly training the FER model on small-scale datasets is prone to overfitting [17], which leads the model to be less generalized and incapable of handling FER tasks in real-world environments. Although a great effort has been made to collect facial expression training datasets, large-scale, annotated, facial expression databases are still limited [13]. Therefore, overfitting caused by a shortage of data remains a challenging issue for most FER systems.

Second, imbalances in the distribution of facial expression samples from real-world FER datasets may degrade the overall performance of the system [18]. Due to the nature of emotions, the number of collected facial images for the major classes (e.g., happiness, sadness, and anger) is much larger than for the minor classes (e.g., contempt, disgust, and fear). In the AffectNet dataset, happy category comprises about 47% of all the images, whereas contempt category comprises only 1.2%. FER systems being trained on an imbalanced dataset may perform well on dominant emotions, but they perform poorly on the under-represented ones. Usually, the weighted-softmax loss approach [5] is used to handle this problem by weighting the loss term for each emotion class based on its relative proportion in the training set. However, this weighted-loss approach is based on the softmax loss function, which is reported to simply force features of different classes to remain apart without paying attention to intra-class compactness. One effective strategy to address the problem of softmax loss is to use auxiliary loss to train the neural network. For instance, triplet loss [19] and center loss [20] introduce multi-loss learning to enhance the discriminating ability of CNN models. Although these loss functions do boost the discriminative ability of the conventional softmax loss, they usually come with limitations. Triplet loss requires a comprehensive process of choosing image pairs or triplet samples, which is impractical and extremely time-consuming owing to the huge number of pairs and samples in the training phase. Center loss does not consider inter-class similarity, which may lead to poor performance by the FER system. In addition, none of these auxiliary loss functions is able to address data-imbalance problems.

To address the first problem (the shortage of data), in this work, the transfer learning technique is applied to build the FER system. Transfer learning is a machine learning technique by which a model trained on one task is repurposed for another related task. It not only helps to handle the shortage of data but also speeds up training and improves the performance of the prediction model. In this paper, we take a transfer learning approach by employing two recent CNN architectures in two-stage, supervised pre-training and fine-tuning. Specifically, a squeeze-and-excitation network (SENet) model [21] which is pre-trained for the face identification task on the VGGFace2 [22] database from the Visual Geometry Group at Oxford University, was fine-tuned on the AffectNet dataset [5] to recognize the above-mentioned eight common facial expressions.

Tackling the second problem of imbalanced data distribution in existing FER datasets, we propose a new loss function called the weighted-cluster loss, which integrates the advantages of the weighted-softmax approach and the auxiliary loss approach. First, weighted-cluster loss learns a class center for each emotion, which simultaneously reduces the intra-class variations and increases the inter-class differences. Next, the proposed loss gives weights to each emotion class's loss terms based on their relative proportion of the total number of samples in the training dataset. In other words, weighted-cluster loss penalizes networks more for misclassifying samples from minor classes while penalizing those networks less for misclassifying examples from major classes. Furthermore, the training process is simple because weighted-cluster loss does not require preselected sample pairs or triples.

Experiments were conducted to show the effectiveness of the proposed method. In addition to widely used metrics for classification (accuracy, F1-score [23], area under the receiver operating characteristic [ROC] curve [AUC] [24], and area under the precision-recall curve [AUC-PR] [25]), two measures of inter-annotator agreement (Cohen's kappa [26] and Krippendorff's alpha [27]) are used to evaluate the models. The experimental results with the AffectNet dataset [5] show that our transfer learning-based model with weighted-cluster loss outperforms other models that use either weighted softmax-loss or center loss.

In summary, the main contributions of this paper are listed as follows:

- First, the objective of this paper is to mitigate the overfitting problem caused by an insufficient amount of data and imbalanced data problem when building a FER model which recognizes eight common facial expressions on AffectNet dataset [5].
- Second, to address the overfitting problem, a deep transfer-based framework is proposed in which we utilize an SE-ResNet-50 model [21] (which was pre-trained on VGGFace2 data [22]) for fine-tuning on AffectNet dataset.
- Third, to alleviate the imbalanced-data problem, we propose a new loss function, named weighted-cluster loss, which gives weights to each emotion class's loss terms based on their relative proportion of the total number of samples in the training dataset.
- Last, experiments are conducted to validate the effectiveness of the proposed method. The experimental results on AffectNet data show that our FER model outperforms its counterpart models in term of various evaluation metrics such as accuracy, F1-score, Cohen's kappa score [26], Krippendorff's alpha score [27], area under the receiver operating characteristic curve (AUC), and area under the precision-recall curve (AUC-PR).

The rest of this manuscript is organized as follows. Section 2 summarizes the existing literature related to facial expression recognition. Then, our proposed method is presented in detail in Section 3. The experiments with results discussion are presented in Section 4. Conclusions drawn from this work, in addition to possible future work are discussed in Section 5.

2. Related Works

This section summarizes recent studies in the literature that are related to facial expression recognition, deep transfer learning techniques used to solve FER tasks, and recent successful loss functions for training deep models.

2.1. Facial Expression Recognition Approaches

Over the last few decades, several approaches have been proposed to build FER models. Traditional methods mostly detect the face region and extract the geometry, appearance, texture, or other highlighted facial characteristics using handcrafted features and shallow learning, such as Gabor wavelet coefficients [7], Haar features [8], histograms of local binary patterns (LBPs) [9], LBPs on three orthogonal planes (LBP-TOP) [10], histograms of oriented gradients (HOG) [11], non-negative matrix factorization (NMF) [28], scale-invariant feature transform (SIFT) descriptors [29], and sparse learning [30]. Overall, these methods can solve the FER tasks in laboratory settings, where emotion images are produced in a controlled manner. However, with the introduction of relatively large real-world databases from emotion recognition competitions such as FER2013 [31] and emotion recognition in the wild [32–36], FER tasks have observed a big transition from laboratory-controlled setups to more challenging real-world scenarios. Traditional methods face many difficulties in solving inter-class similarity and intra-class differences, and are reported to be incapable of addressing the great diversity of factors unrelated to facial expressions.

Recently, the development of deep learning and the increase in the number of powerful CNN architectures [37–39] in the computer vision field implicitly promote using deep learning methods when building FER models. For example, the winner of the 2013 International Conference on Machine Learning [40] combined a deep CNN with a support vector machine (SVM) classifier to distinguish common emotions. However, despite the impressive feature learning ability of deep learning, difficulties remain when applied to FER such as the shortage of training data, the inter-class similarity, the intra-class differences, and the high imbalance of emotion classes appearing in most existing real-world facial expression databases. This work is taking advantage of deep models to extract robust facial features and translate them to recognize facial emotions. Table 1 summarizes the comparison between our approaches with existing studies. In this table, human resource refers to how much human labor needed to construct features learning/extraction model, computing resource refers to hardware resources needed to operate model, and computational complexity refers to operations per pixel.

Table 1. Comparison between FER approaches.

Approach	Method	Human Resource	Computing Resource	Computational Complexity	Accuracy
Conventional	Gabor wavelets coefficients [7]	High	Low	Low	Medium
	Haar features [8]	High	Low	Low	Medium
	Local binary pattern (LBP) [9]	High	Low	Low	Medium
	LBP on three orthogonal planes (LBP-TOP) [10]	High	Low	Low	Medium
	Scale-invariant feature transform (SIFT) [29]	High	Low	Low	Medium
	Histogram of gradient (HOG) [11]	High	Low	Low	Medium
Deep learning-based	Convolutional neural network [40]	Low	High	High	High
	Transfer learning-based CNN (Our approaches)	Low	Medium	Medium	High

2.2. Transfer Learning for Facial Expression Recognition

Studies in FER have suffered from the lack of data for training deep CNN models, which may have resulted in overfitting. To work around this problem, transfer learning has been widely used for facial recognition tasks. In fact, the use of auxiliary data can help FER models to obtain a high capacity without overfitting, thus improving the overall performance of the system. Usually, the weights of the CNN are initialized and pre-trained on additional data from other relative tasks (e.g., object detection and face recognition) before being fine-tuned using the target dataset. Clearly, applying transfer learning to the FER task has consistently achieved better results, compared to directly training the network on a small-scale FER dataset. Some popular datasets can be used as auxiliary data, such as ImageNet [41] for object recognition tasks and VGGFace from the Visual Geometry Group [42] for the face recognition task. In the work of Ly et al. [43], a Inception-ResNetV1 model was pre-trained on VGGFace2 [22] and AffectNet [5] then was used to develop a multi-modal 2D and 3D for real-word FER task. However, the number of existing 3D data for FER is limited thus make it difficult to construct a multi-modal FER model. Do et al. [44] used a ResNet-50 model pre-trained on VGGFace2 [22] as a feature extraction model. The model was then integrated with a LSTM [45] model to analyses facial expression on the video data.

It is worthwhile to note that in preliminary experiments, Ngo and Yoon showed the improvement in recognition performance when using ImageNet data [41] as auxiliary data for building a transfer learning-based FER model [46]. The authors fine-tuned a ResNet-50 [47] model, which was pre-trained with ImageNet data. However, the ImageNet dataset was developed for object detection task, which may not be sufficiently related to the FER task. In this paper, to build the transfer learning-based

FER model, we employ the more advanced CNN architecture (i.e., SENet [21]) and then fine-tune it for FER task. Note that this transferred model is pre-trained with VGGFace2 data [22] for the face identification task which is more related to the FER task than object detection tasks. This may help improve the performance of the transfer learning-based FER system. Furthermore, in [46], authors simply fine-tuned their models using softmax loss and did not consider the imbalanced data problem, which degraded the recognition performance of FER system. In this paper, we focus on handling the shortage of data as well as tackling the imbalanced data problem by applying weighted loss and auxiliary loss approaches.

2.3. Data Re-Sampling and Augmentation

As the nature of emotions, facial expression data collected in real-world settings are highly imbalanced in the number of samples in each class (e.g., the number of images in the happy class is much greater than the number of images that show disgust). Using imbalanced data for training may degrade the performance of FER models [18]. Data re-sampling and generating samples using generative adversarial networks (GAN) [48] are usually considered as solutions to mitigate the imbalanced data problem. Mollahoseini et al. [5] used down-sampling and up-sampling methods to balance the distribution of data in the training set, which alleviates the imbalanced data problem. However, the two data re-sampling methods simply randomly reduce the number of samples in major classes or duplicate samples in minor classes without actually collecting further data. Under-sampling may discards samples that could be important for the model while over-sampling significantly increases the model training time [49]. Lai et al. [50] proposed a GAN that generates frontal face images from input non-frontal face images. This model can be employed to augment more data for training FER models. Nonetheless, the reliability of the new data needs to be carefully verified before being used to train FER models, otherwise, the data may degrade the performance of FER systems. Our solution is based on a weighted loss approach that tackles the imbalanced data problem without the need of data re-sampling or augmentation step.

2.4. Weighted Loss and Auxiliary Loss

Weighted-softmax loss gives weights to the loss terms of each emotion class based on its relative numbers of samples in the training set. In this way, the loss function heavily penalizes the FER model for misclassifying examples from minor classes, while lightly penalizing the model for misclassifying examples from major classes. Mollahoseini et al. [5] showed that the weighted-loss methods achieve the highest performance and outperform the re-sampling methods. However, since the weighted-softmax loss function is built based on conventional softmax loss, it inherits the limitations of softmax loss. For example, weighted-softmax loss simply forces features of different classes to remain apart, without paying attention to intra-class compactness.

To tackle the limitations of softmax loss, several auxiliary loss approaches have been proposed, and they can be used to improve the discriminative ability of FER models. Contrastive loss [51] inputs the CNN model with a pair of training samples, which forces the features of same-class pairs to be as similar as possible while requiring a pairwise distance longer than a predefined margin if the input samples belong to different classes. Similar to contrastive loss, triplet loss [19] encourages a distance constraint between samples that come from different classes. Specifically, triplet loss requires one positive sample to be closer to a preselected anchor than one negative sample with a fixed margin. Based on triplet loss, there are two variations proposed to support softmax loss. (N + M)-tuples cluster loss [52] mitigates the difficulty of anchor selection and threshold validation, whereas exponential triplet-based loss [53] gives more weight to difficult samples during an update of network parameters. In spite of some success, these loss functions still need a careful pre-selection process. However, this becomes impossible when the number of training samples is relatively large. Recently, center loss [20] has been proposed for the face recognition task; it penalizes the distance between deep features and their corresponding class clusters in order to reduce intra-class differences. However, center loss does not take inter-class similarity into account. Motivated by this, island loss [54] was proposed to simultaneously force deep features of a sample close to its cluster, and to push class cluster centers far away from each other to enlarge the distance between samples of different expressions. Nonetheless, adding more loss terms usually comes with difficulty in the training phase caused by extra hyper-parameters that need to be adjusted. In addition, none of these aforementioned loss functions take into consideration the imbalance in the number of training samples from each emotion class in the training dataset. This may lead FER models to perform poorly with minority classes, even though they perform well with majority classes.

By analyzing the complementary nature of weighted loss and center loss, we propose a new loss function, named weighted-cluster loss, which not only takes highly skewed facial emotion data into consideration, but it also uses multiple loss terms to improve the performance of the FER models.

3. Methods

In this section, we first describe our deep transfer–learning framework and then propose weighted-cluster loss. The framework of the proposed method is shown in Figure 1.

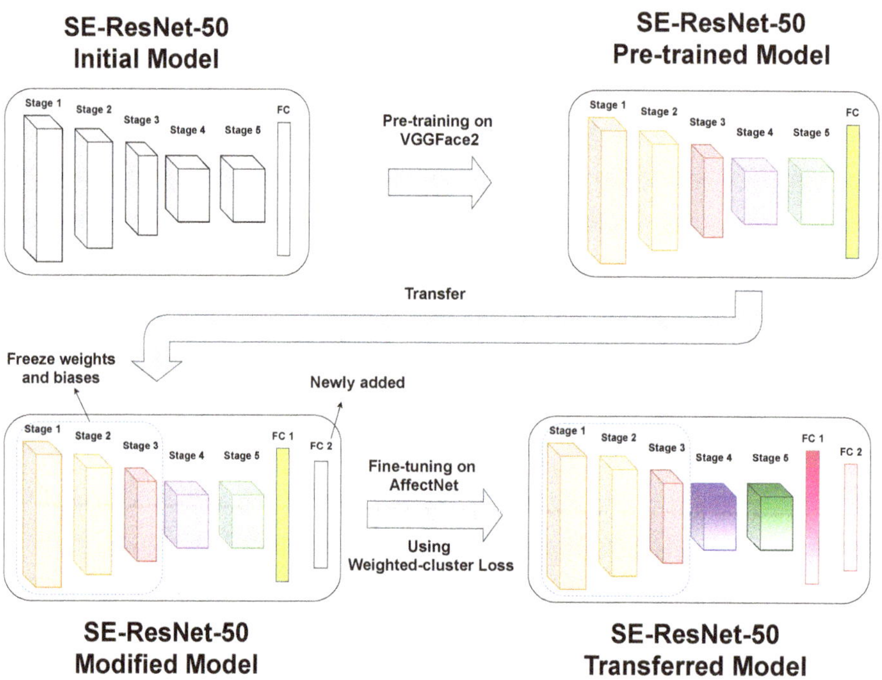

Figure 1. Framework of the proposed method. A SE-ResNet-50 model [21], which was pre-trained on VGGFace2 data [22] for face identification, is fine-tuned with AffectNet data [5] for facial recognition using weighted-cluster loss. Before the fine-tuning phase, we add one more fully connected layer to the model while froze the three first stages of the pre-trained model to save computing power. The weighted-cluster loss is used at the output layer to update model parameters. Best view in color.

3.1. Base Model and Pre-Training

3.1.1. Base Models

Convolutional neural networks have achieved great success in the fields of pattern recognition and computer vision. This motivated us to base our FER models on a recent representative CNN architecture in the computer vision field. In this work, we employed the SE-ResNet-50 model [21], which is the ResNet-50 model [47] integrated with SE-ResNet modules as our base model. This CNN architecture uses SE modules that integrated with ResNet CNN architectures and improved the feature learning capacity of the integrated model. A wide range of experiments showed the effectiveness of SENets that achieve state-of-the-art performance across multiple datasets and tasks. This was demonstrated for object and scene classification, with a squeeze-and-excitation network winning the ILSVRC 2017 competition.

3.1.2. Pre-Training

This work uses publicly available pre-trained model of the base model (i.e., transfer learning-based model) instead of directly pre-training it. Specifically, the SE-ResNet-50 pre-trained model, which is published on the official website of the VGGFace2 database [55], was trained with the VGGFace2 dataset [22] for the task of recognizing 8631 faces. As a results, the network learned rich feature represent for face factors.

3.2. Fine-Tuning

In this stage, we add one more fully connected layer to the pre-trained SE-ResNet-50 model (before the output layer; see Table 2) and we then changed the output layer to eight-class classification. In addition, we froze the weights and biases of Stage 1 to Stage 3 on the pre-trained model. This process is usually found in many transfer learning-based systems when computing power is limited. Last we fine-tuned the adjusted model with the AffectNet dataset [5] to recognize the eight facial expressions (happy, sad, surprise, fear, contempt, anger, disgust, and neutral). Specifically,

Table 2. Detailed architectures of SE-ResNet-50 in the fine-tuning phase. Convolution blocks (SE-ResNet-50 uses the SE-ResNet block [21]) are shown in brackets, with the numbers of blocks stacked.

	Output Size	**Kernel**	**Repeat**
Stage 1 (Freeze)	112×112	convolution 7×7	1
Stage 2 (Freeze)	56×56	convolution block	3
Stage 3 (Freeze)	28×28	convolution block	4
Stage 4	14×14	convolution block	6
Stage 5	7×7	convolution block	3
	1×1	global average pooling	
Fully connected	[2048×512] Fully connected		
Fully connected	[512×8] Fully connected		
Output layer	8-d softmax		

3.3. Weighted-Cluster Loss

3.3.1. Review of Weighted-Softmax Loss

Weighted-softmax [5] often comes as a solution for tackling the imbalanced dataset problem. Weighted-softmax loss weighs the entropy loss of each of the emotion classes by their relative proportion of the total number of samples in the training dataset. The entropy weighted-softmax loss for the ith training sample is defined as:

$$L_{\text{Weighted-softmax}} = -\sum_{i=1}^{m} W_{y_i} \log(\hat{p}_i) \tag{1}$$

where m is the number of training samples in the mini batch (i.e., the batch size), y_i is the class label of the ith training sample, W_{y_i} denotes the weight assigned to the class where the label y_i, and $\hat{p}_i$ is the softmax-predicted probability of the ith sample.

Although weighted-softmax loss is able to tackle the imbalanced dataset problem, it still has some limitations. Since weighted loss simply adds weight to the conventional softmax loss, it is incapable of handling not only high inter-class similarity but also high intra-class variations.

3.3.2. Review of Center Loss

Center loss [20] can be incorporated with conventional softmax loss to reduces the intra-class variations by compressing samples towards their corresponding class center in the feature space during training. Center loss for the ith training sample is defined as:

$$L_{\text{Center}} = \frac{1}{2}\sum_{i=1}^{m} ||\mathbf{x}_i - \mathbf{c}_{y_i}||_2^2, \tag{2}$$

where L_{Center} denotes the center loss, $\mathbf{x}_i \in \mathbb{R}^d$ denotes the deep feature of the ith training sample (i.e., taken from the last fully connected layer, before the output layer), and $\mathbf{c}_{y_i} \in \mathbb{R}^d$ denotes the center of class label y_i; d is the feature dimension.

To train the CNN model, joint supervision of softmax loss and center loss is employed as follows:

$$L = L_{\text{Softmax}} + \lambda L_{\text{Center}} \tag{3}$$

where L denotes the total loss, $L_{\text{Softmax}} = -\sum_{i=1}^{m} \log(\hat{p}_i)$ is the the entropy softmax loss and λ is a scalar used for balancing the two losses.

Since inter-class similarities are ignored by the center loss, the class clusters may move closer to, or even overlap, each other. Furthermore, center loss was not proposed to deal with the imbalanced dataset problem. Due to dataset imbalances, the centers of major emotion classes are more frequently updated than those of minor classes, which leads to poor performance of FER model on the minor classes.

3.3.3. The Proposed Weighted-Cluster Loss

To address the limitations of existing loss functions (e.g., weighted-softmax loss and center loss), we propose a new loss function called weighted-cluster loss. It effectively tackles the imbalanced data problem by taking into consideration the imbalanced proportion in the number of samples of each emotion class in the training set. Furthermore, weighted-cluster loss adds a new term to center loss, which simultaneously pulls the centers of each class apart. This may allow the model to simultaneously handle the high intra-variation and the inter-class similarity in the FER dataset. The weighted-cluster loss for the ith training sample is given by Equation (4):

$$L_{\text{Weighted-cluster}} = \frac{1}{2}\sum_{i=1}^{m} W_{y_i} \frac{||\mathbf{x}_i - \mathbf{c}_{y_i}||_2^2}{\left(\sum_{j=1, j\neq i}^{k} ||\mathbf{c}_j - \mathbf{c}_{y_i}||_2^2\right) + \alpha} \tag{4}$$

where W_{y_i} denotes the weight assigned for the class where the label is y_i, k denotes the number of emotion classes, and the constant, α, prevents the denominator from equaling 0. In this paper, we set $\alpha = 1$ by default.

In Equation (4), the numerator penalizes the distance between the deep feature of the training sample (i.e., taken from the last fully connected layer before the output layer) and its corresponding center, and the denominator penalizes the distance between the corresponding center and all other class centers. By minimizing weighted-cluster loss, the deep features of training samples from the same class (i.e., the cluster) will be compacted in the feature space, whereas the distance between different classes of clusters will be enlarged. In addition, weight W_{y_i} is used to tackle the imbalanced dataset problem by penalizing the models less for misclassifying samples from majority classes while heavily penalizing the model for misclassifying samples from minority classes.

The overall loss function of FER training is given by:

$$\begin{aligned} L &= L_{\text{Weighted-softmax}} + \lambda L_{\text{Weighted-cluster}} \\ &= -\sum_{i=1}^{m} W_{y_i}\left(\log(\hat{p}_i) - \frac{\lambda}{2}\frac{\|\mathbf{x}_i - \mathbf{c}_{y_i}\|_2^2}{\sum_{j=1,j\neq i}^{k}\|\mathbf{c}_j - \mathbf{c}_{y_i}\|_2^2 + \alpha}\right) \end{aligned} \tag{5}$$

where λ is used to balance the two losses.

In this work, we define W_{y_i} as:

$$W_{y_i} = \frac{N_{\text{min}}}{N_{y_i}} \tag{6}$$

where N_{min} denotes the number of samples from the smallest class (i.e., disgust), and N_{y_i} denotes the number of samples from the class where the label is y_i.

The joint weighted-cluster loss and weighted-softmax loss can be directly used for training deep neural networks. The class centers, $\mathbf{c}_{y_i}$, are updated each iteration through the training process. Specifically, the class centers are updated based on mini-batches, and use a scalar γ to control the learning rate of class centers (and in our experiments, we set $\gamma = 1$). The partial derivative of the weighted cluster loss, $L_{\text{Weighted-cluster}}$, with respect to the sample's feature, $\mathbf{x}_i$, can be calculated as follows:

$$\frac{\partial L_{\text{Weighted-cluster}}}{\partial \mathbf{x}_i} = W_{y_i}\frac{\mathbf{x}_i - \mathbf{c}_{y_i}}{\left(\sum_{l=1,l\neq i}^{k}\|\mathbf{c}_l - \mathbf{c}_{y_i}\|_2^2\right) + \alpha} \tag{7}$$

The update of the jth class center can be calculated with Equation (8):

$$\Delta\mathbf{c}_j^t = \sum_{i=1}^{m} W_{y_i}\frac{\delta(y_i = j)(\mathbf{x}_i - \mathbf{c}_j)}{\left(\sum_{l=1,l\neq i}^{k}\|\mathbf{c}_l - \mathbf{c}_{y_i}\|_2^2\right) + \alpha} \tag{8}$$

where $\delta(y_i = j) = 1$ if $y_i = j$, and $\delta(y_i = j) = 0$ if $y_i \neq j$.

Then, the class centers can be updated in each mini-batch with a learning rate γ:

$$\mathbf{c}_j^{t+1} = \mathbf{c}_j^t - \gamma\Delta\mathbf{c}_j^t \tag{9}$$

In Algorithm 1, we summarize the learning process of the FER model with the joint loss functions.

Algorithm 1 Learning algorithm of the FER model with the jointly loss functions

---

Input: Training data $\{\mathbf{x}_i\}$, mini-batch size m, number of iterations T, learning rates μ and γ, and hyper-parameters λ

Initialize: Network layer parameters W, weighted-softmax loss parameters θ, and weighted-cluster loss parameters (i.e., centers) $\{\mathbf{c}_j | j = 1, 2, ..., n\}$.

for $t = 1$ to T **do**

  Calculate the joint loss

  $L^t = L^t_{\text{Weighted-softmax}} + \lambda L^t_{\text{Weighted-cluster}}$

  Compute the backpropagation error for each i

  $\frac{\partial L^t}{\partial \mathbf{x}^t_i} = \frac{\partial L^t_{\text{Weighed-softmax}}}{\partial \mathbf{x}^t_i} + \lambda \frac{\partial L^t_{\text{Weighted-cluster}}}{\partial \mathbf{x}^t_i}$

  Update the weighted-softmax loss parameters θ

  $\theta^{t+1} = \theta^t - \mu \frac{\partial L^t_{\text{Weighed-softmax}}}{\partial \theta^t_i}$

  Update the weighted-cluster loss parameters $\mathbf{c}_j$ for each j as Equation (9)

  $\mathbf{c}^{t+1}_j = \mathbf{c}^t_j - \gamma \Delta \mathbf{c}^t_j$

  Update the network parameters W

  $W^{t+1} = W^t - \mu \frac{\partial L^t}{\partial \mathbf{x}^t_i} \frac{\partial \mathbf{x}^t_i}{\partial W^t}$

end for

Output: Network layer parameters W

---

4. Experiments

4.1. Experimental Datasets

4.1.1. AffectNet Dataset

In this work, we consider the problem in recognizing eight common facial expressions from the AffectNet dataset [5]. The AffectNet dataset contains more than 1,000,000 images from the Internet that were obtained by querying different search engines using emotion-related tags. AffectNet is by far the largest database that provides facial expressions in two different emotion models (a categorical model and a dimensional model), which can be used for studies in automated recognition of facial expressions, valences, and arousal in real-world scenarios. About 450,000 images already have manually annotated labels for eight basic expressions which are neutral, happy, sad, surprise, fear, disgust, anger, and contempt, as well as some non-emotion–related image classes such as none, uncertain, and non-face. Figure 2 shows some sample images from the dataset.

Figure 2. Sample images from the AffectNet dataset (0: neutral; 1: happy; 2: sad; 3: surprise; 4: fear; 5: disgust; 6: anger; 7: contempt).

In this work, only images of the eight common emotion classes (manually annotated) were used to train the FER model. From each of the eight emotion classes, we randomly selected 500 samples for a validation set and another 500 samples were selected for a test set. The remainder were used for fine-tuning the FER models. The numbers of samples in the training, validation, and test sets are shown in Table 3.

Table 3. Numbers of samples in training, validation, and test sets.

Emotion	Training	Validation	Test
Neutral	74,374	500	500
Happy	133,915	500	500
Sad	24,959	500	500
Surprise	13,590	500	500
Fear	5878	500	500
Disgust	3303	500	500
Anger	24,382	500	500
Contempt	3250	500	500
Total	283,651	4000	4000

Although AffectNet is considered one of the largest facial expression databases, it still has shortcomings when used for training FER models. This database is highly imbalanced, as can be seen in Figure 3. Specifically, as seen in Table 3, the number of images in the largest category (happy with 134,915 images) is approximately 30 times larger than the smallest category (contempt, with 4250 images). Furthermore, images were manually annotated, which may result in a low-reliability dataset. Therefore, transfer learning is still needed to mitigate these drawbacks.

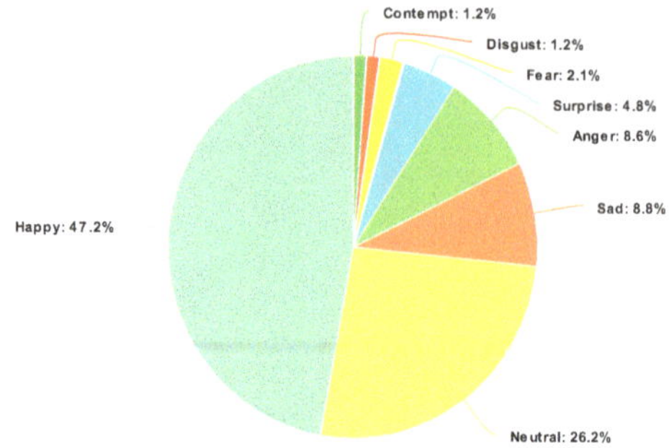

Figure 3. Distribution of the eight classes in the training set.

4.1.2. VGGFace2 Dataset

VGGFace2 [22] is a new large-scale face dataset that contains 3.31 million images of 9131 subjects, with an average of 362 images for each subject. Images were downloaded from Google's image search function, and they have large variations in pose, age, illumination, ethnicity, and profession (e.g., actors, athletes, politicians). Figure 4 shows some sample images from the VGGFace2 dataset.

Figure 4. Sample images from the VGGFace2 dataset.

4.2. Evaluation Metrics

There are various evaluation metrics in the literature to measure discriminative performance of the FER model. In addition to several widely used metrics for classification, such as accuracy, F1-score [23], area under the ROC curve (AUC) [24], and area under the precision–recall curve (AUC-PR) [25], two measures of inter-annotator agreement (Cohen's kappa [26] and Krippendorff's alpha [27]) are used in our work. In statistics, Cohen's kappa measures inter-rater reliability, which is the degree of agreement among raters, given the same data. Krippendorff's alpha (also called Krippendorff's coefficient) is an alternative to Cohen's kappa for determining inter-rater reliability. Table 4 lists acronyms used in this paper.

Table 4. List of acronyms.

Abbreviation	Meaning
ResNet	Residual neural network
SE-ResNet	ResNet-based squeeze and excitation neural network
Alpha	Krippendorff's alpha score
Kappa	Cohen's kappa score
AUC	Area under the receiver operating characteristic curve
AUC-PR	Area under precision recall curve

4.3. Experiment Setups and Implementation Details

We experimented with different schemes to point out the effectiveness of using transfer learning as well as the proposed loss function for FER tasks. In detail, we fine-tune the transfer learning-based model (e.g., SE-ResNet-50 pre-trained with VGGFace2 dataset) with different loss settings, such as the conventional softmax loss, center loss with softmax loss [20], weighted-softmax loss [5], center loss with weighted-softmax loss (i.e., replace softmax loss by weighted-softmax loss), and the proposed weighted-cluster loss with weighted softmax loss. In addition, to evaluate the effectiveness of transfer learning, we also trained the base model using only the AffectNet training dataset with the same loss settings as those used for transfer learning-based model. In all experiments, we set the λ value (i.e., the scalar used to balance the loss terms) to 0.5 when using center loss, and to 1.0 when using weighted-cluster loss.

The pre-trained model was fine-tuned using the stochastic gradient descent algorithm with hyper-parameters (momentum = 0.9, weight decay = 0.0005). Note that we fine-tuned the pre-trained CNN model using a much smaller dataset (AffectNet compared with VGGFace2), and thus, the initial learning rate was set to 0.001, which is lower than the typical value of 0.01, in order to not drastically alter the pre-trained weights. The learning rate was dropped by a factor of 2 following every 10 epochs of training. For the base models that were trained using only AffectNet data (i.e., no pre-training), the initial learning rate was set to 0.01, and all other settings were kept the same. All experiments were implemented using the PyTorch library and were trained on a four-core Xeon CPU with a single Titan-XP GPU. Batch size for fine-tuning the transfer learning-based models was set to 36, while the batch size for training the base models from scratch was set to 30.

To enrich the scale of the dataset and mitigate the overfitting problem, it is necessary to conduct data augmentation. During the fine-tuning phase, input images were randomly cropped and resized to 224×224 pixels; the horizontal flip was randomly extracted from the cropped images. In addition, before being input into the FER model, all input samples were normalized by using the ImageNet mean and standard deviation (std) (mean = [0.485, 0.456, 0.406], std = [0.229, 0.224, 0.225]), which is a common practice for deep CNN models when working with RGB images.

4.4. Results and Discussions

Table 5 shows the accuracy, F1-score, Cohen's kappa, Krippendorff's alpha, AUC, and AUC-PR of different FER models on the test set. These values are averages over the eight classes. All metrics except for accuracy were calculated in a binary-class manner, whereas accuracy is defined in a multi-class manner. From the results in Table 5, we have the following observations.

Table 5. Recognition performance of different FER models on the test set.

Base Model	Model No.	Pre-Trained	Loss Function	Accuracy	F1-Score	Kappa	Alpha	AUCPR	AUC
SE-ResNet-50	1	No	Softmax	50.65	46.87	43.60	42.60	62.87	90.67
	2	No	Center with softmax	46.07	39.24	38.37	36.89	55.96	87.92
	3	No	Weighted-softmax	56.37	56.41	50.14	50.09	62.27	90.49
	4	No	Center with weighted-softmax	56.90	57.06	50.74	50.71	62.24	90.33
	5	No	Weighted-cluster with weighted-softmax	56.27	56.42	50.03	49.97	61.57	90.10
	6	Yes	Softmax	52.22	49.51	45.4	44.54	63.27	90.75
	7	Yes	Center with softmax	47.08	40.02	39.51	38.27	54.91	86.75
	8	Yes	Weighted-softmax	59.72	59.72	53.97	53.93	66.47	91.85
	9	Yes	Center with weighted-softmax	59.60	59.50	53.83	53.82	65.35	91.21
	10	Yes	Weighted-cluster with weighted-softmax (proposed)	60.70	60.49	55.09	55.06	66.55	91.82

First, the proposed FER model achieved the highest performance in terms of all evaluation metrics, outperforming their counterparts. In detail, model 10 (the transferred SE-ResNet-50 model fine-tuned using the proposed joint weighted-cluster and weighted-softmax) achieved recognition accuracy of 60.70%. They led the second-best model (i.e., model 8) by approximately 1% in terms of recognition accuracy. As reported in [5], the average agreement over eight emotion categories between two human annotators (randomly chosen out of a total of 12 annotators) on only a part of the AffectNet data was only 65.56%, which may be considered the maximum achievable recognition accuracy. This emphasizes the fact that recognizing human emotions from facial expressions in a real-world scenario is a challenging task, and the newly proposed weighted-cluster loss function is capable of addressing challenging factors such as subtle facial appearance, head pose, illumination intensity, and occlusions.

In terms of F1-score, the proposed model (model 10) surpassed its counterparts. In most imbalanced classification problems (e.g., facial expression recognition tasks on an imbalanced dataset like AffectNet), F1-score, which is the weighted average of precision and recall, gives a better measure of incorrectly classified cases than the accuracy metric. In addition, model 10 also achieved the best kappa, alpha, and AUC-PR values, which outperforms models that use other loss functions. Similar to F1-score, these values give us alternatives to accuracy when measuring the reliability of automated FER systems. This shows that the proposed FER model is dependable when it comes to solving facial expression recognition problems in a real-world scenario. It should be noted that model 8 achieves the best AUC value. This is because model 8 achieves good performance on the positive class (high AUC) at the cost of a high false negatives rate. On the other hand, the proposed model (model 10) tries to reduce the false negatives rate while maintaining good performance on the positive class.

Second, under the supervision of the same loss functions, transferred FER models performed better than FER models that are trained from scratch (i.e., models 1, 2, 3, 4, and 5, vs. models 6, 7, 8, 9, and 10, respectively). This shows the benefit of using transfer learning where the pre-trained model, which has the ability to learn features for face identification, can be transferred to recognize facial expressions to improve the recognition performance. The proposed model (model 10) which integrated both transfer learning and weighted loss approaches thus achieved the highest performance.

Next, the performance of model 1 and model 6 (models using conventional softmax loss) are lower than the performance model 5 and model 10 (models using weighted-cluster loss) by large margin (50.65% and 52.22% vs. 56.27% vs. 60.70%). This is because the softmax loss is incapable to handle the imbalanced data problem which leads to the poor performance of FER model on minor classes (e.g., contempt, disgust). In contrast, the proposed weighted-cluster alleviates the imbalanced data problem by giving weight to the loss term of each class. This help improve performance of model on small classes. In addition, center loss-based models (models that were fine-tuned using joint center loss with either softmax loss or weighted-softmax loss) performed worse than the models fine-tuned using only softmax loss or weighted-softmax loss. The recognition accuracy of model 7 which uses center with softmax loss function is only 46.07%, which is even lower than that of model 6 (using softmax loss). Similar phenomena can be found in the case of model 9. Model 9 that using center loss with weighted-softmax loss achieves lower accuracy than model 8 that using only weighted-softmax. This shows that the existing center loss is not suitable for tackling data imbalance problems where the centers of the major emotion classes are updated more frequently than the centers of minor emotion classes.

Last, model 10 (transferred model that was fine-tuned using the proposed joint weighted-cluster loss and weighted-softmax loss) achieved better performance compared to model 9 (transferred model that was fine-tuned using joint center loss and weighted-softmax loss) in terms of all evaluation metrics (e.g., in terms of accuracy: 60.70% vs. 59.60%, respectively). This shows that the proposed cluster loss effectively handles the limitations of center loss and weighted-softmax loss by not only taking the imbalanced data into consideration but by also simultaneously improving intra-class compactness and enlarging inter-class differences.

Figure 5 shows the training loss and the validation accuracy of different models during the training phase. We can see that the training loss of auxiliary loss-based models (i.e., models 8, 9, and 10) is higher than the model using only softmax loss or weighted-softmax loss. This is because we added more loss terms to the total loss function. During the training phase, we can observe that all models are convergent as epochs are trained.

To compare our model with existing models, we also conducted experiments in [46] where authors used a ResNet-50 model [47] which is pre-trained with ImageNet data [41] for object detection task as their base model and then fine-tuned it with AffectNet data for FER task. It is worthwhile to note that, in [46], authors fine-tuned their model using only the conventional softmax loss. In our experiments, we further fine-tuned the ResNet-50 model (pre-trained with ImageNet) using the same loss settings that were used to fine-tune SE-ResNet-50 model (e.g., weighted-softmax loss, center with softmax loss, and so forth). Recognition performance of ResNet-50-based models on are shown in Table 6.

Table 6. Recognition performance of ResNet-50-based models on the test set.

Base Model	Model No.	Pre-Trained	Loss Function	Accuracy	F1-score	Kappa	Alpha	AUCPR	AUC
ResNet-50	11	No	Softmax	49.85	46.67	42.69	41.46	62.80	90.63
	12	No	Center with softmax	46.62	39.96	39.00	37.67	54.08	86.26
	13	No	Weighted-softmax	57.40	57.33	51.31	51.23	63.05	90.82
	14	No	Center with weighted-softmax	57.20	57.08	51.09	51.06	62.61	90.50
	15	No	Weighted-cluster with weighted-softmax	57.37	57.40	51.29	51.24	62.79	90.52
	16	Yes	Softmax	51.88	48.89	45.00	44.10	61.6	90.22
	17	Yes	Center with softmax	48.33	44.00	40.94	39.8	56.96	87.89
	18	Yes	Weighted-softmax	58.65	58.59	52.74	52.71	64.58	91.17
	19	Yes	Center with weighted-softmax	58.27	58.07	52.31	52.24	63.59	90.54
	20	Yes	Weighted-cluster with weighted-softmax	59.45	59.42	53.66	53.66	65.26	91.51

As can be seen in Table 6, the transferred ResNet-50 model that was fine-tuned using the proposed weighted-cluster loss (i.e., model 20) outperforms its counterpart models that were fine-tuned using other loss functions. This strengthens the point that the proposed weighted-cluster loss function is capable to handle the imbalanced data problem in the AffectNet dataset.

By comparing the performance of our models (Table 5) with the performance of the existing models in [46] (Table 6), we have following observation. Using the same loss function, our models (i.e., model 6, 7, 8, 9, 10) that used the SE-ResNet-50 model achieve better recognition performance than the models proposed in [46] (i.e., model 16, 17, 18, 19, 20, respectively) that used the ResNet-50 model pre-trained for object detection. For example, in terms of recognition accuracy, our best fine-tuned model (i.e., model 10) leads the best fine-tuned model that used the ResNet-50 model (i.e., model 20) about 1.25% (e.g., 60.70% vs 59.45%). This is because we used a more advanced CNN architecture (i.e., SE-ResNet-50) that was then pre-trained for face identification, instead of object detection.

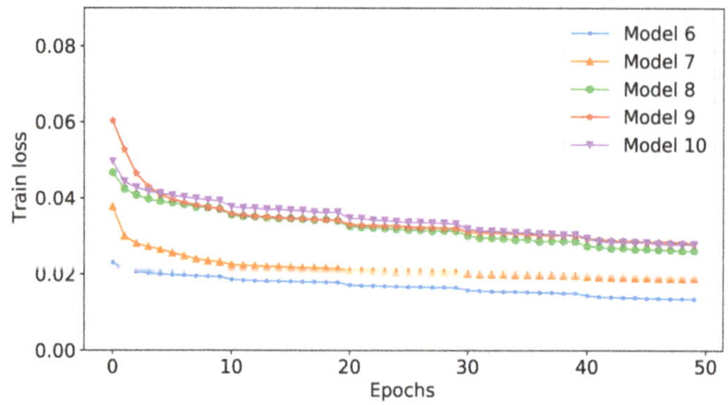

(a)

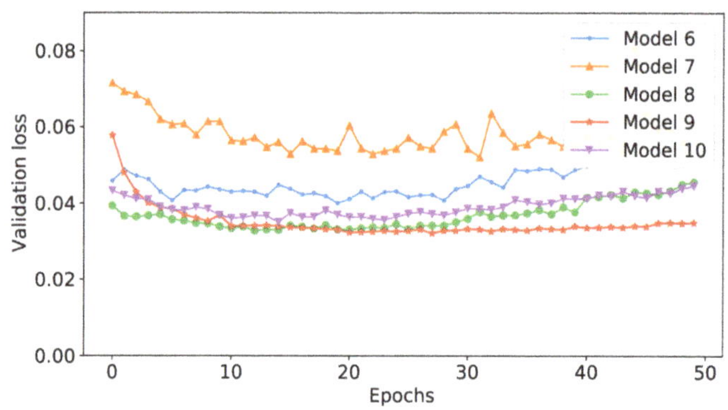

(b)

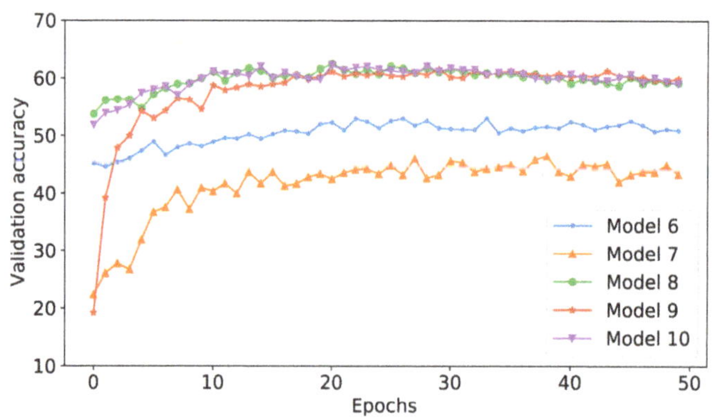

(c)

Figure 5. Learning curves of different models over number of epoch trained. (**a**) Train loss during the training. (**b**) Validation loss during the training. (**c**) Validation accuracy during the training. Best view in color.

To investigate the effectiveness of the proposed weighted-cluster loss function when handling the imbalanced dataset problem, we further plotted in Figure 6 the confusion matrices of the transfer learning-based models that were fine-tuned using the conventional softmax loss (model 1 and model 6) and the proposed joint weighted-cluster loss and weighted-softmax loss (model 10 and model 20).

We can see that weighted-cluster loss effectively solved the high imbalanced data problem with the AffectNet dataset. In particular, the FER performance with minor emotion classes dramatically improved. For example, recognition accuracy of the transferred SE-ResNet-50 model for the contempt class improved by approximately 1000%—from 6% when fine-tuned using the softmax loss, to 59% when fine-tuned using the proposed weighted-cluster loss. For the disgust class, it improved by about 150%—from 36% when fine-tuned using the softmax loss to 54% when fine-tuned using the proposed weighted-cluster loss. This is because the proposed loss function penalizes misclassifying samples from these classes more. However, the FER performance with major emotion classes (e.g., happy and neutral) slightly decreased, because the loss function may not sufficiently penalize misclassifying samples from these classes. The same trend can be found in case of the transferred ResNet-50 model where the recognition accuracy of class contempt increases about 500% (from 10% to 53%) and that of class disgust increases about 150% (from 30% to 47%).

It is worthwhile to note that although the weighted-cluster loss tries to increase the inter-class difference, the similarities between some emotion classes such as happy vs. contempt, neutral vs. contempt, surprise vs. fear, and disgust vs. anger still remain to some extent. This is due to the natural similarity between these emotion classes. For this reason, there were up to 16% samples of the neutral class and and 15% samples of happy class that was misclassified as the contempt class by model 10. In the opposite direction, there was also a high percentage (e.g., 11% and 12%) samples of the contempt class misclassified as the neutral and happy classes, respectively. Similarly, in model 10, 18% samples of surprise class was misclassified as the fear class, and 13% samples of the fear class was misclassified as the surprise class while the disgust and anger classes were falsely recognized as the other at the rate of 15% and 12%, respectively.

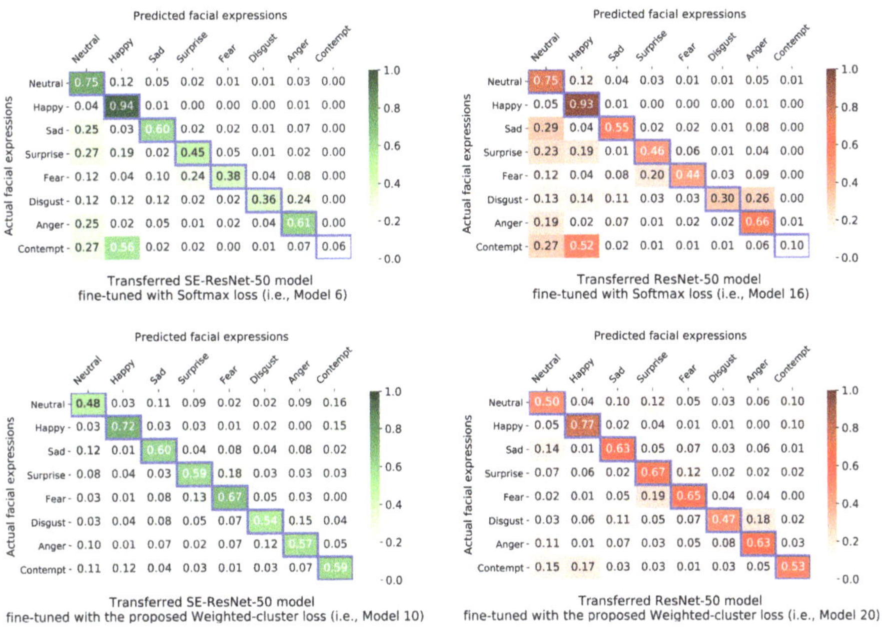

Figure 6. Confusion matrices of the transfer learning-based models on the test set.

4.5. Threats to Validity

- Threat to internal validity: Threat to internal validity include errors when implementing the codes and conducting experiments. Although the implementations and experiments were carefully verified, errors are possible.

- Threat to external validity: Threats to external validity include the generalization to other datasets of the results in this paper. Applying the proposed methods in this paper to additional datasets will reduce this threat.
- Threats to construct validity: Threats to construct validity include the appropriateness of the benchmark algorithms and evaluation metrics. For the most part, the proposed FER model outperformed their counterparts on the same test dataset as expected. The evaluation metrics used for validating models (i.e., accuracy, F1-Score, Kappa score, AUC, and AUC-PR) are common in machine learning studies to evaluate the performance of classification models.

5. Conclusions

This paper proposed a facial expression recognition model based on deep transferred learning and a novel weighted-cluster loss to mitigate the shortage an imbalanced data problems. An SE-ResNet-50 model which is pre-trained for face identification task is fine-tuned to recognize eight common facial expressions in AffectNet data. This not only helps to save computing resource but also alleviate the shortage of training data problem. Then, the proposed weighted-cluster loss was used in the fine-tuning phase to tackle the high imbalance in data distribution of AffectNet data. Multiple metrics have been used to evaluate the effectiveness of the proposed model. Experimental results on the test set indicate that the proposed FER model can outperform its counterpart models which uses either weighted-softmax loss or center loss. However, the proposed model is built to recognize facial expressions on static image data, which may limit its applicability. Moreover, using generative adversarial networks (GAN) to generate more data for training FER models is considered as our other future work.

Author Contributions: Conceptualization, Q.T.N. and S.Y.; Data curation, Q.T.N. and S.Y.; Formal analysis, Q.T.N. and S.Y.; Funding acquisition, S.Y.; Investigation, S.Y.; Methodology, Q.T.N. and S.Y.; Project administration, S.Y.; Resources, S.Y.; Software, Q.T.N.; Supervision, S.Y.; Validation, Q.T.N. and S.Y.; Visualization, Q.T.N. and S.Y.; Writing—original draft, Q.T.N. and S.Y.; Writing—review and editing, Q.T.N. and S.Y. All authors have read and agreed to the published version of the manuscript.

Funding: This research was funded by the 2020 Research Fund of University of Ulsan.

Conflicts of Interest: The authors declare no conflict of interest.

References

1. Hickson, S.; Dufour, N.; Sud, A.; Kwatra, V.; Essa, I. Eyemotion: Classifying facial expressions in VR using eye-tracking cameras. In Proceedings of the 2019 IEEE Winter Conference on Applications of Computer Vision (WACV), Waikoloa Village, HI, USA, 7–11 January 2019; pp. 1626–1635.
2. Chen, C.H.; Lee, I.J.; Lin, L.Y. Augmented reality-based self-facial modeling to promote the emotional expression and social skills of adolescents with autism spectrum disorders. *Res. Dev. Disabilities* **2015**, *36*, 396–403. [CrossRef] [PubMed]
3. Dornaika, F.; Raducanu, B. Efficient facial expression recognition for human robot interaction. In Proceedings of the International Work-Conference on Artificial Neural Networks, San Sebastián, Spain, 20–22 June 2007; pp. 700–708.
4. Zhan, C.; Li, W.; Ogunbona, P.; Safaei, F. A real-time facial expression recognition system for online games. *Int. J. Comput. Games Technol.* **2008**, *2008*, 10. [CrossRef]
5. Mollahosseini, A.; Hasani, B.; Mahoor, M. AffectNet: A Database for Facial Expression, Valence, and Arousal Computing in the Wild. *IEEE Trans. Affective Comput.* **2017**. [CrossRef]
6. AffectNet Database. Available online: http://mohammadmahoor.com/affectnet/ (accessed on 15 August 2019).
7. Tian, Y.l.; Kanade, T.; Cohn, J.F. Evaluation of Gabor-wavelet-based facial action unit recognition in image sequences of increasing complexity. In Proceedings of the Fifth IEEE International Conference on Automatic Face Gesture Recognition, Washington, DC, USA, 21 May 2002; pp. 229–234.
8. Whitehill, J.; Omlin, C.W. Haar features for facs au recognition. In Proceedings of the 7th International Conference on Automatic Face and Gesture Recognition (FGR06), Southampton, UK, 10–12 April 2006.

9. Shan, C.; Gong, S.; McOwan, P.W. Facial expression recognition based on local binary patterns: A comprehensive study. *Image Vision Comput.* **2009**, *27*, 803–816. [CrossRef]
10. Zhao, G.; Pietikainen, M. Dynamic texture recognition using local binary patterns with an application to facial expressions. *IEEE Trans. Pattern Anal. Mach. Intell.* **2007**, *29*, 915–928. [CrossRef] [PubMed]
11. Dahmane, M.; Meunier, J. Emotion recognition using dynamic grid-based HoG features. In Proceedings of the Face and Gesture 2011, Santa Barbara, CA, USA, 21–25 March 2011.
12. Li, S.; Deng, W. Deep facial expression recognition: A survey. *arXiv* **2018**, arXiv:1804.08348.
13. Ko, B. A brief review of facial emotion recognition based on visual information. *Sensors* **2018**, *18*, 401. [CrossRef] [PubMed]
14. Ebrahimi Kahou, S.; Michalski, V.; Konda, K.; Memisevic, R.; Pal, C. Recurrent neural networks for emotion recognition in video. In Proceedings of the 17th ACM International Conference on Multimodal Interaction, Seattle, DC, USA, 9–13 November 2015; pp. 467–474.
15. Walecki, R.; Rudovic, O.; Pavlovic, V.; Schuller, B.; Pantic, M. Deep structured learning for facial expression intensity estimation. *Image Vis. Comput.* **2017**, *259*, 143–154.
16. Kim, D.H.; Baddar, W.J.; Jang, J.; Ro, Y.M. Multi-objective based spatio-temporal feature representation learning robust to expression intensity variations for facial expression recognition. *IEEE Trans. Affective Comput.* **2017**, *10*, 223–236. [CrossRef]
17. Hawkins, D.M. The problem of overfitting. *J. Chem. Inf. Comput. Sci.* **2004**, *44*, 1–12. [CrossRef] [PubMed]
18. Visa, S.; Ralescu, A. Issues in mining imbalanced data sets-a review paper. In Proceedings of the Sixteen midwest artificial intelligence and cognitive science conference, Dayton, OH, USA, 22 February 2005.
19. Schroff, F.; Kalenichenko, D.; Philbin, J. Facenet: A unified embedding for face recognition and clustering. In Proceedings of the IEEE conference on computer vision and pattern recognition, Boston, MA, USA, 8–10 June 2015; pp. 815–823.
20. Wen, Y.; Zhang, K.; Li, Z.; Qiao, Y. A discriminative feature learning approach for deep face recognition. In Proceedings of the European conference on computer vision, Amsterdam, The Netherlands, 8–16 October 2016; pp. 499–515.
21. Hu, J.; Shen, L.; Sun, G. Squeeze-and-excitation networks. In Proceedings of the IEEE Conference on Computer Vision and Pattern Recognition, Salt Lake City, UT, USA, 19–21 June 2018; pp. 7132–7141.
22. Cao, Q.; Shen, L.; Xie, W.; Parkhi, O.M.; Zisserman, A. Vggface2: A dataset for recognising faces across pose and age. In Proceedings of the 13th IEEE International Conference on Automatic Face & Gesture Recognition (FG 2018), Xi'an, China, 15–19 May 2018; pp. 67–74.
23. Sokolova, M.; Japkowicz, N.; Szpakowicz, S. Beyond accuracy, F-score and ROC: A family of discriminant measures for performance evaluation. In Proceedings of the Australasian joint conference on artificial intelligence, Hobart, Australia, 4–8 December 2006; pp. 1015–1021.
24. Fawcett, T. An introduction to ROC analysis. *Pattern Recognit. Lett.* **2006**, *27*, 861–874. [CrossRef]
25. Davis, J.; Goadrich, M. The relationship between Precision-Recall and ROC curves. In Proceedings of the 23rd international conference on Machine learning, Pittsburgh, PA, USA, 25–29 June 2006; pp. 233–240.
26. Cohen, J. A coefficient of agreement for nominal scales. *Educ. Psychol. Meas.* **1960**, *20*, 37–46. [CrossRef]
27. Krippendorff, K. Estimating the reliability, systematic error and random error of interval data. *Educ. Psychol. Meas.* **1970**, *30*, 61–70. [CrossRef]
28. Zhi, R.; Flierl, M.; Ruan, Q.; Kleijn, W.B. Graph-preserving sparse nonnegative matrix factorization with application to facial expression recognition. *IEEE Trans. Syst. Man Cybern. Part B Cybern.* **2011**, *41*, 38–52.
29. Hu, Y.; Zeng, Z.; Yin, L.; Wei, X.; Zhou, X.; Huang, T.S. Multi-view facial expression recognition. In Proceedings of the 8th IEEE International Conference on Automatic Face & Gesture Recognition, Amsterdam, The Netherlands, 17–19 September 2008; pp. 1–6.
30. Zhong, L.; Liu, Q.; Yang, P.; Liu, B.; Huang, J.; Metaxas, D.N. Learning active facial patches for expression analysis. In Proceedings of the 2012 IEEE Conference on Computer Vision and Pattern Recognition, Providence, RI, USA, 16–21 June 2012; pp. 2562–2569.
31. Goodfellow, I.J.; Erhan, D.; Carrier, P.L.; Courville, A.; Mirza, M.; Hamner, B.; Cukierski, W.; Tang, Y.; Thaler, D.; Lee, D.H.; et al. Challenges in representation learning: A report on three machine learning contests. In Proceedings of the International Conference on Neural Information Processing, Daegu, Korea, 3–7 November 2013; pp. 117–124.

32. Dhall, A.; Goecke, R.; Joshi, J.; Wagner, M.; Gedeon, T. Emotion recognition in the wild challenge 2013. In Proceedings of the 15th ACM International Conference on Multimodal Interaction, Sydney, Australia, 24–31 July 2013; pp. 509–516.
33. Dhall, A.; Goecke, R.; Joshi, J.; Sikka, K.; Gedeon, T. Emotion recognition in the wild challenge 2014: Baseline, data and protocol. In Proceedings of the 16th ACM International Conference on Multimodal Interaction, Istanbul, Turkey, 12–16 November 2014; pp. 461–466.
34. Dhall, A.; Ramana Murthy, O.; Goecke, R.; Joshi, J.; Gedeon, T. Video and image based emotion recognition challenges in the wild: Emotiw 2015. In Proceedings of the 17th ACM International Conference on Multimodal Interactionn, Seattle, DC, USA, 9–13 November 2015; pp. 423–426.
35. Dhall, A.; Goecke, R.; Joshi, J.; Hoey, J.; Gedeon, T. Emotiw 2016: Video and group-level emotion recognition challenges. In Proceedings of the 18th ACM International Conference on Multimodal Interaction, Tokyo, Japan, 12–16 November 2016; pp. 427–432.
36. Dhall, A.; Goecke, R.; Ghosh, S.; Joshi, J.; Hoey, J.; Gedeon, T. From individual to group-level emotion recognition: EmotiW 5.0. In Proceedings of the 19th ACM International Conference on Multimodal Interaction, Glasgow, Scotland, 13–17 November 2017; pp. 524–528.
37. Krizhevsky, A.; Sutskever, I.; Hinton, G.E. Imagenet classification with deep convolutional neural networks. In Proceedings of the Advances in neural information processing systems, Lake Tahoe, Nevada, USA, 3–6 December 2012; pp. 1097–1105.
38. Simonyan, K.; Zisserman, A. Very Deep Convolutional Networks for Large-Scale Image Recognition. *arXiv* **2014**, arXiv:1409.1556.
39. Szegedy, C.; Liu, W.; Jia, Y.; Sermanet, P.; Reed, S.; Anguelov, D.; Erhan, D.; Vanhoucke, V.; Rabinovich, A. Going deeper with convolutions. In Proceedings of the IEEE conference on computer vision and pattern recognition, Boston, MA, USA, 8–10 June 2015; pp. 1–9.
40. Tang, Y. Deep Learning Using Linear Support Vector Machines. *arXiv* **2013**, arXiv:1306.0239.
41. Deng, J.; Dong, W.; Socher, R.; Li, L.J.; Li, K.; Li, F.-F. Imagenet: A large-scale hierarchical image database. In Proceedings of the 2009 IEEE Conference on Computer Vision and Pattern Recognition, Miami, FL, USA, 20–25 June 2009; pp. 248–255.
42. Parkhi, O.M.; Vedaldi, A.; Zisserman, A. Deep face recognition. In Proceedings of the 26th British Machine Vision Conference (BMVC), Swansea, UK, 7–10 September 2015; pp. 1–12.
43. Do, N.T.; Kim, S.H. Affective Expression Analysis in-the-Wild Using Multi-Task Temporal Statistical Deep Learning Model. *arXiv* **2020**, arXiv:2002.09120.
44. Thai Ly, S.; Do, N.T.; Lee, G.S.; Kim, S.H.; Yang, H.J. Multimodal 2D and 3D for In-the-wild Facial Expression Recognition. In Proceedings of the IEEE Conference on Computer Vision and Pattern Recognition Workshops, Long Beach, CA, USA, 16–20 June 2019.
45. Hochreiter, S.; Schmidhuber, J. Long short-term memory. *Neural Comput.* **1997**, *9*, 1735–1780. [CrossRef] [PubMed]
46. Ngo, T.Q.; Yoon, S. Facial Expression Recognition on Static Images. In Proceedings of the International Conference on Future Data and Security Engineering, FDSE2019, Nha Trang, Vietnam, 27–29 November 2019; pp. 640–647.
47. He, K.; Zhang, X.; Ren, S.; Sun, J. Deep residual learning for image recognition. In Proceedings of the IEEE conference on computer vision and pattern recognition, Las Vegas, NV, USA, 27–30 June 2016; pp. 770–778.
48. Creswell, A.; White, T.; Dumoulin, V.; Arulkumaran, K.; Sengupta, B.; Bharath, A.A. Generative adversarial networks: An overview. *IEEE Signal Process Mag.* **2018**, *35*, 53–65. [CrossRef]
49. Haixiang, G.; Yijing, L.; Shang, J.; Mingyun, G.; Yuanyue, H.; Bing, G. Learning from class-imbalanced data: Review of methods and applications. *Expert Syst. Appl.* **2017**, *73*, 220–239. [CrossRef]
50. Lai, Y.H.; Lai, S.H. Emotion-preserving representation learning via generative adversarial network for multi-view facial expression recognition. In Proceedings of the 13th IEEE International Conference on Automatic Face & Gesture Recognition (FG 2018), Xi'an, China, 15–19 May 2018; pp. 263–270.
51. Sun, Y.; Chen, Y.; Wang, X.; Tang, X. Deep learning face representation by joint identification-verification. In Proceedings of the Advances in Neural Information Processing Systems, Montreal, QC, Canada, 8–13 December 2014; pp. 1988–1996.

52. Liu, X.; Vijaya Kumar, B.; You, J.; Jia, P. Adaptive deep metric learning for identity-aware facial expression recognition. In Proceedings of the IEEE Conference on Computer Vision and Pattern Recognition Workshops, Honolulu, HI, USA, 21–26 July 2017; pp. 20–29.
53. Guo, Y.; Tao, D.; Yu, J.; Xiong, H.; Li, Y.; Tao, D. Deep neural networks with relativity learning for facial expression recognition. In Proceedings of the 2016 IEEE International Conference on Multimedia & Expo Workshops (ICMEW), Seattle, WA, USA, 11–15 July 2016; pp. 1–6.
54. Cai, J.; Meng, Z.; Khan, A.S.; Li, Z.; O'Reilly, J.; Tong, Y. Island loss for learning discriminative features in facial expression recognition. In Proceedings of the 13th IEEE International Conference on Automatic Face & Gesture Recognition (FG 2018), Xi'an, China, 15–19 May 2018; pp. 302–309.
55. VGGFace2—A Large Scale Image Dataset for Face Recognition. Available online: http://www.robots.ox.ac.uk/~vgg/data/vgg_face2/ (accessed on 15 August 2019).

sensors

MDPI

Article

StressFoot: Uncovering the Potential of the Foot for Acute Stress Sensing in Sitting Posture

Don Samitha Elvitigala [1,*], Denys J. C. Matthies [1,2] and Suranga Nanayakkara [1]

1 Augmented Human Lab, Auckland Bioengineering Institute, The University of Auckland, Auckland 1010, New Zealand; denys.matthies@th-luebeck.de (D.J.C.M.); suranga@ahlab.org (S.N.)

2 Department of Electrical Engineering and Computer Science, Technical University of Applied Sciences Lübeck, 23562 Lübeck, Germany

* Correspondence: samitha@ahlab.org

Received: 16 April 2020; Accepted: 16 May 2020; Published: 19 May 2020

Abstract: Stress is a naturally occurring psychological response and identifiable by several body signs. We propose a novel way to discriminate acute stress and relaxation, using movement and posture characteristics of the foot. Based on data collected from 23 participants performing tasks that induced stress and relaxation, we developed several machine learning models to construct the validity of our method. We tested our models in another study with 11 additional participants. The results demonstrated replicability with an overall accuracy of 87%. To also demonstrate external validity, we conducted a field study with 10 participants, performing their usual everyday office tasks over a working day. The results showed substantial robustness. We describe ten significant features in detail to enable an easy replication of our models.

Keywords: stress sensing; smart insoles; smart shoes; unobtrusive sensing; stress; center of pressure

---

1. Introduction

Stress has a direct impact on our well-being [1]. Although stress is often perceived negatively, it can have positive aspects. For instance, acute stress aims to mentally and physically prepare our body to accomplish a demanding task. In contrast, episodic occurring acute stress can cause a variety of negative symptoms, such as sleep disorders, headache, stomach pain and exhaustion. A constant elevated stress level may even result in chronic stress [2], which can lead to severe health conditions, such as depression, anxiety, hypertension and cardiovascular diseases [3]. Therefore, it is imperative to identify stressful situations to prevent resulting illnesses.

Previous studies have demonstrated the capability of identifying stress based on physiological parameters [4], such as electrodermal activity (EDA) [5], heart rate variability (HRV), brainwaves via electroencephalography (EEG) [6–8], muscle tension [9,10], facial expressions [11] and body language [12,13], as well as self-reporting [14,15]. Although these methods provide reliable stress indicators, there are drawbacks. For instance, sensing physiological parameters are sensitive to movement artifacts. The sensing device also needs to be instrumented tightly on the user's body, resulting in low comfort. An alternative method is contact free sensing, such as using cameras [16–18]. However, these systems suffer from varying lighting conditions, require line of sight, and typically create privacy concerns.

Manual approaches, such as having an experimenter interpret facial expressions and body language or relying on self-reporting, are prone to issues such as scalability and subjective bias. Several other studies explore stress detection based on smartphone usage [19–21] by correlating screen-time with daytime, and utilising the phone's sensor data. These studies show the capability of detecting stress over long periods of time. Identifying and predicting acute stress in the short-term may also be possible, although still problematic [21].

Smart shoes, in particular insoles, have been used in a variety of scenarios, such as to analyse gait [22], identify postures [23], calculate walking speeds [24], determine the ground surface [25] and recognising foot tapping gestures for an interaction control [26]. These smart insoles provide an unobtrusive way of collecting data. However, to our knowledge, insole-based tracking has not yet been proposed to identify stress.

Motivated by the fact that feet and legs may carry essential information about a person's stress level [12,27] and the widespread availability of shoes, we present StressFoot. Our prototype encapsulates a smart shoe system that incorporates a pressure-sensitive insole based on force-sensitive resistor (FSR) technology and an inertial measurement unit (IMU). We drive a machine learning approach to reliably detect acute stress situations in sitting postures, such as sedentary office work. To scientifically validate this, we followed the standard research design process [28] by Constructing Validity—Study 1 developing a machine learning model based on the data of 23 participants, evidencing Empirical Replicability—Study 2 with 11 participants showing a distinguishability between stressed and relaxed conditions and finally, testing for External Validity—Study 3 showing the generalisability in terms of robustness of our model for office workers during a typical 8 h work day with 10 participants. In summary, we contribute:

- a novel way to discriminate acute stress and relaxation by four distinct foot movements and posture characteristics,
- ten mathematical features to train machine learning models,
- and design implications for future applications in ubiquitous computing.

1.1. Background

1.1.1. Physiological Responses

The most common approach in stress sensing is interpreting physiological responses, such as EDA [29], heart rate [30] (HR), HRV [31], pupil dilation [32] (PD), skin temperature [33,34] (ST) or EEG [6,8,35]. In addition, muscle activity, such as microvibrations [9] and muscle tension [10,36], is affected by stress. The Sympathetic Nervous System (SNS) of the Autonomous Nervous System unconsciously controls these vital signs and are thus considered reliable sources [37]. Prior work demonstrated EDA as being linearly related to arousal and widely used in the context of stress sensing [38–40]. More specifically, EDA has been used to measure stress in applications, such as measuring the stress of call centre agents [39] and the discrimination of stress from the cognitive load [40]. In addition, in many laboratory and field studies, EDA is considered one of the gold-standard methods to sense stress [41]. As the SNS mainly controls EDA, it is regarded as a reliable physiological sign for acute stress [42].

Furthermore, prior work revealed that mental stress related to cognitive load has an impact on HRV [6,8], in particular towards reduced HF components [43–45]. Measurable changes in HR has also been observed during high attention tasks [44]. Electrocardiography (ECG) and Photoplethysmogram (PPG) [46,47] are the earliest technologies used in literature to measure HR and HRV. High power consumption and tight sensor placements are the major draw back of these technologies. Meanwhile, various studies, such as BioWatch [48], SenseGlass [49], SeismoTracker [50], investigated smart devices capable of measuring HR and HRV based on ballistorcardiography (BCG). Although BCG can compete with state-of-the-art techniques [51], it is susceptible to unwanted motion artifacts and thus only reliable in resting states, such as sleeping, standing or sitting. In summary, wearable sensing using wristbands, nail clips, vests and headbands are often uncomfortable, given the bulkiness and tight sensor mounting.

Another option is contact-free sensing of physiological data [52], such as using thermal cameras [53,54] or Doppler radar [55]. As cameras are typically expensive, webcams are a low-cost alternative to measuring HR and HRV from the human face [17,18]. Inspired by previous studies, COGCAM [16] measured cognitive stress with digital cameras, placed 3m away from the user.

In these studies, participants are instructed not to move their head, which restricts natural behaviours. An ambient light source ensuring constant light conditions is essential. Privacy concerns may arise when using cameras for tracking.

An increased muscle activity can also indicate stress. For instance, an increased amplitude of the muscles' microvibrations [9] and an increased muscle tension can provide stress indicators [10]. Other researchers [41,56,57] have detected some types of muscle tension implicitly by analysing keyboard and mouse control. An increased typing pressure and a greater contact with the surface of the mouse has been observed in stressed conditions [41]. We believe such implicit sensing is an unobtrusive and thus desirable approach we would also like to explore.

1.1.2. Facial Expressions

In recent years, identifying affective states based on facial expressions has been widely explored in the area of affective computing [58]. Technology-wise, vision-based camera tracking, including depth cameras [59], are the most commonly used technologies in identifying facial expressions [60]. A recent work [61] investigated stress and anxiety sensing using facial cues, in which the authors' induced acute stress by internal and external stressors. All videos recorded were analysed posterior and a machine learning model scored between 80–90% accuracy in recognizing stress tasks. Although vision-based identification is demonstrably effective, unfavourable light conditions and movement artifacts easily affect detection accuracy.

Other researchers identify facial expressions based on skin-contact electrodes, such as piezoelectric sensing [62], EMG [63,64], capacitive sensing [65,66] and electric field sensing [67]. These researches demonstrate the detection of facial gestures, such as frowning, eye wink, eye-down, mouth movements, etc. and infer on emotions such as frustration, confusion and interest engagement. Although the identification of emotions and facial expressions seem reliable, the outcome may differ in real-life scenarios given that facial expressions are known to deceive easily [68].

2. StressFoot

In this paper, we explore the feasibility of utilising foot movements and posture characteristics to identify acute stress.

2.1. Concept

Previous studies demonstrate that body expressions are as powerful as facial expressions in conveying emotions [13,69,70]. Body expressions can play an important role in non-verbal communication than previously thought [69,71]. Studies have also utilized body expressions to detect stress [72]. According to Wallbott et al. [73], changes in body posture provides a strong indication for the changes in affective states. Body expressions are stated as being greater in revealing a deception than relying on facial expressions [68]. Some affective expressions may even be better pronounced using body posture than facial expressions [74]. Recent advances in ubiquitous computing enable the recognition of these body expressions. For instance, computer games such as Nintendo Wii and Microsoft Kinect [13,75] not only utilise body movements as a means to control the game but also to capture the emotional and cognitive performance of the player. The majority of previous works on affective recognition systems utilising body postures and body expressions rely on vision-based techniques that analyse motion data by rgb and depth cameras [75,76]. For instance, Kleinsmith et al. [75] classified four affective states (concentration, defeat, frustration and triumph) with people playing a sports game on Nintendo Wii, and achieved an accuracy of 50%. Another work demonstrates the detection of sadness, joy, anger and fear with a 84–94% accuracy using a 6-camera Vicon motion tracking system. Although vision-based systems seem promising in identifying body postures, these are impractical to integrate into one's daily routine when aiming to identify emotions, such as those induced by acute stress. As an alternative for vision-based approaches, inferring stress based on posture has been explored previously by embedding pressure sensor in chairs [77]. In this work, authors identified fast movements of Centre of Pressure as a valuable feature in detecting stress. A similar approach was used to detect interest levels of students [78]. This approach required instrumenting chairs with high-resolution pressure sensors, which may not be scalable. However, we believe the sitting posture variations should reflect from the foot posture and motion variations. In literature, smart insoles has been used to detect postures unobtrusively [23,79]. Hence, a smart insole-based solution could be a practical and unobtrusive solution to identify stress that is induced by body language, such as through tracking leg movements and posture characteristics.

2.2. Prototype

The base of the StressFoot prototype is a pair of sport shoes (Figure 1A). To sense the pressure distribution of the foot, we insert a pair of pressure sensitive insoles (UK size 9–10). These insoles are commercially available from sensing.tex [80] and offer 16 pressure sensors that rely on a force sensitive resistance technology (Figure 1B). In addition, the sensor placement is aligned with the critical pressure points discovered in previous work [81,82]. To drive the insoles, we developed a voltage divider circuit interfacing the insole with the Sparkfun Razor board that consists of an SAMD21 microprocessor manufactured by Atmel Corporation in San Jose, CA, USA. To greater understand the angle and motion of the leg and feet, we also queried the accelerometer of 9 DOF IMU (MPU-9250) manufactured by InvenSense in San Jose, CA, USA that comes with the microcontroller board, which was tied to the user's ankle. We decided to only utilise the Accelerometer (X,Y,Z) data, since it provides sufficient information. The prototype has a 3.7 V, 400 mAh LiPo battery and an SD card, to allow a fully wireless operation, without constraining the user's movements. The prototype can be used for approximately 16 h when it is fully charged.

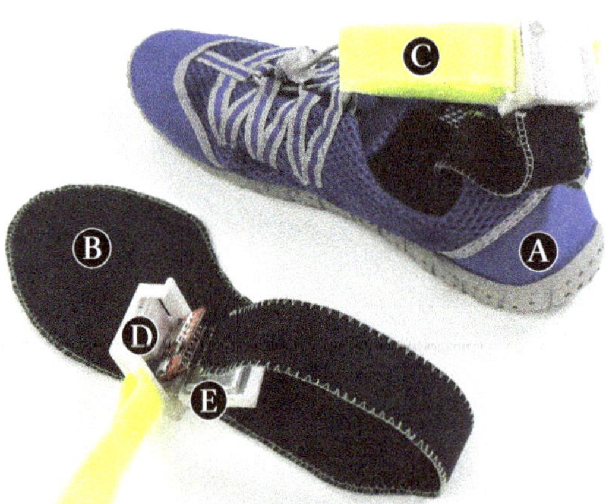

Figure 1. Prototype: (**A**) shoe, (**B**) sensing.tex insole, (**C**) ankle strip, (**D**) SAMD21 microprocessor board with inertial measurement unit (IMU), secure digital (SD) card, voltage divider circuit and (**E**) 400 mAh Battery.

2.3. Features

We ran a pilot study with 5 participants, in which we asked the users to complete the minesweeper game in a limited time. We presented a clock to the user, and an observer was also present to make negative comments on the user's performance. Among all participants in the stress situation, we observed a change in leg posture and a different foot contact with the floor. Two participants demonstrated a nervous leg shaking and foot tapping. Our prototype is designed to identify these unique characteristics by sensing plantar pressure distributions of the foot, as well as tracking the foot angle and motions. We defined 10 low-level features reflecting these behaviours, which are:

2.3.1. A: Foot Pressure

The forefoot pressure was calculated by adding the pressure of 12 sensor points located at the forefoot area of the insole. To calculate rear-foot pressure, the sensor readings of 4 sensors located at the heel area were added together. The sensor layout is depicted in Figure 2.

If the ith sensor is P_i:

$$P_F = \sum_{i=1}^{12} P_i \tag{1}$$

$$P_R = \sum_{i=13}^{16} P_i \tag{2}$$

Before summing the raw data, we applied a moving average filter of a 1 s window size to filter out high frequency noises. Finally, we calculated the total foot pressure by summing up all pressure points.

$$P_T = \sum_{i=1}^{16} P_i \tag{3}$$

Since the features are independent from previous data points, all pressure features were segmented into 10 s non-overlapping windows. Then, the mean of each window is used for the model training.

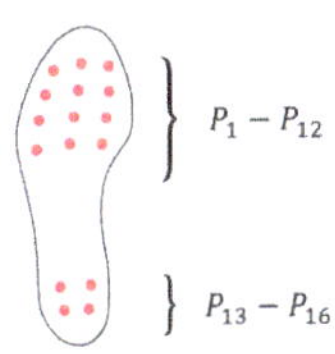

Figure 2. The sensor layout of the insole.

2.3.2. B: Centre of Pressure

Literature has shown the Centre of Pressure (CoP) as an interesting feature to understand different postures and motion while standing [23], as well as sitting [77]. Therefore, we calculated the CoP for both dimensions (X-axis and Y-axis—see also Figure 3B) by using a weighted average calculation as mentioned in prior work [23]. The CoP was also calculated over 10 s of non-overlapping windows since it is independent from the previous window.

$$Y_{CoP} = \frac{\sum_{i=1}^{12} P_i y_i + 3\sum_{i=13}^{16} P_i y_i}{P_F + 3P_R} \tag{4}$$

$$X_{CoP} = \frac{\sum_{i=1}^{12} P_i x_i + 3\sum_{i=13}^{16} P_i x_i}{P_F + 3P_R} \tag{5}$$

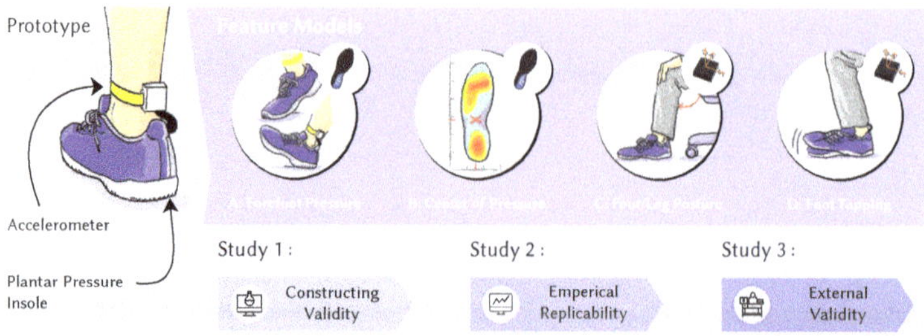

Figure 3. To detect acute stress, we present a shoe prototype incorporating a pressure sensitive insole and an IMU worn around the ankle. Based on literature research and observations, we propose four features that can reveal the user's stress level. We show that a significant forefoot pressure, deviations in the centre of pressure, different foot/leg postures, and tapping of the foot, can indicate an elevated stress level. To validate generalisability, we conducted three studies: Study 1—Constructing Validity, Study 2—evidencing Empirical Replicability and Study 3—testing for External Validity. Our proposed models score reasonable accuracy and show robustness across different users.

2.3.3. C: Foot/Leg Posture

The acceleration data of each axis (aX, aY, aZ) primarily indicates the leg/foot posture defined by the knee angle as seen in Figure 3C. Similar to the above mentioned features, the raw data stream of each axis was segmented into non-overlapping 10 s windows and the mean of the each window was used in model training. Since the sampling rate was 50 Hz, the mean of 500 data points were taken in each 10 s window for all 3 axes. If the number of data points per window is W where $W = 500$:

$$aX_{mean} = \frac{1}{W}\sum_{i=1}^{500} aX \tag{6}$$

$$aY_{mean} = \frac{1}{W}\sum_{i=1}^{500} aY \tag{7}$$

$$aZ_{mean} = \frac{1}{W}\sum_{i=1}^{500} aZ \tag{8}$$

2.3.4. D: Foot Tapping

The dominant and the median frequency of each axis are used to describe foot motions, such as foot tapping, waving and leg shaking. We recognised that some motions may be very minimal. To avoid cancelling out these low-amplitude motions when calculating a 3D vector-norm, we deployed the axis-inversion algorithm Matthies et al. introduced [83]. For instance, if we inverted X axis of the accelerometer the inverted axis aX_{inv} would be: $aX_{inv} = 2MAX(aX) - aX$. After segmenting the signal into 10 s non-overlapping windows, we calculated three 3D norms, each with a different inverted axis. From the resulting signals, the 3D-norm with the highest peak-to-peak, V_{3D} was selected for further processing. Then, FFT was used to identify the frequency components of the signal before we extract the dominant and median frequency [84].

$$F_D = F_{MaxEnergy}(FFT(V_{3D})) \tag{9}$$

$$F_M = F_{MedianEnergy}(FFT(V_{3D})) \tag{10}$$

3. Construct Validity—Study 1

As elaborated before, conceptual-wise leg/foot postures and movements should yield information that could indicate stress. Therefore, in our first study, we aim to construct the validity of this concept by developing a machine learning model that aims to distinguish acute stress and relaxation. The study protocol was approved by the University of Auckland Human Participants Ethics Committee.

3.1. *Participants*

The participants had to be at least 18 years of age, English speaking and able to provide written consent. In addition, participants needed to have a foot size of UK 9–10 to match the size of the prototype. Participants with prolonged foot-related injuries were excluded. We recruited 23 healthy participants (12 Males and 11 Females) aged between 21–32 years old (M = 26.4, SD = 3.17).

3.2. *Apparatus*

To collect the data, we used the StressFoot prototypes. In addition, the Empatica E4 wristband [85] was used to collect the participant's EDA. All the study tasks were executed and visualised on a 13" Macbook Pro, which was connected to an extended DELL LED monitor. (see Figure 4).

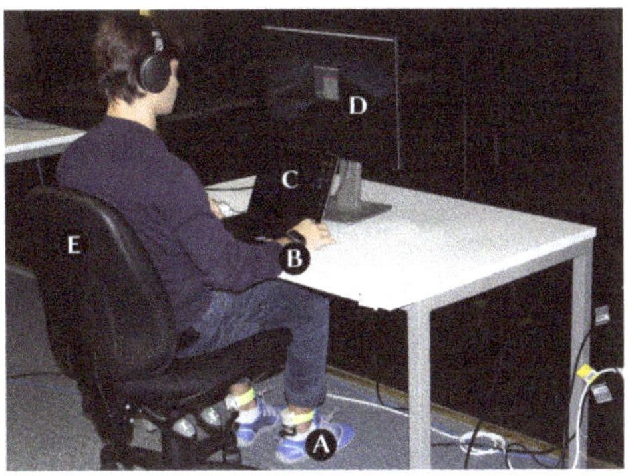

Figure 4. Experimental setup: (**A**) StressFoot prototype, (**B**) Empatica E4 wristband, (**C**) Macbook Pro, (**D**) Dell LED monitor and (**E**) adjustable chair.

3.3. *Procedure & Tasks*

After explaining the aim of the study, participants were required to fill out consent forms. Participants were informed that the study aims to explore behavioural changes while performing tasks which induce stress and relaxation. We then collected demographic data and asked the participant to wear the apparatuses. The participants were then asked to sit in-front of the Macbook Pro with an extended display as shown in Figure 4. We allowed the user to adjust the chair and the distance of the table according to their personal preference. During the study, they were also allowed to cross their legs. However, we asked them not to take the foot up to the chair. Once the setup was complete and the user was comfortable, we asked them to complete four tasks. The first, as well as the third task, were stress-inducing tasks, while the second and fourth tasks intended to create relaxation. The task order remained consistent for all participants. After the completion of each task, the participant had to fill a questionnaire. On a 7-point Likert scale, the users had to rate their perceived stress level, energy level and how pleasant they found the task. A NASA Task Load Index (NASA TLX) [86] was conducted to calculate the overall workload per task.

3.3.1. Task 1 [Stress]: Stroop Color and Word Test

The Stroop color and word test is a common stress test (Stroop Test) [87] used in many previous works [10,88,89]. For our study, we customised an open source MatLab-based Stroop Test tool to display four words ("Red", "Magenta", "Green" and "Blue") on the screen. Participants were then asked to type the first letter of the colour ('R', 'M', 'G', 'B') the displayed word was coloured with. After each answer, the program provides the result by displaying the word "correct" or "incorrect".

The participant has to perform 5 rounds. Each round has 20 words with a mismatch of 80% between the word and the colour it is printed with. We use several tactics to induce stress among participants. As participants were required to provide answers as quickly as possible, we first induced stress by providing performance feedback on their speed and accuracy. We further elevated stress levels by exposing the participant to loud traffic noises through a headset. Moreover, an experimenter continuously observed the participant, and verbally commented on the participant's performance. Completing the stress task took 5 min.

3.3.2. Task 2 [Relaxation]: Minesweeper Introduction Video

The second task aimed to induce relaxation among the user after the stress task. It was also intended to provide an introduction to the game, Minesweeper. For this, we selected an entertaining Minesweeper Let's Play Video from the platform youtube.com. The selection of this video was completed beforehand with 3 pilot study participants. We explicitly informed the user that the video is purely for relaxation purposes and also encouraged the user to relax and enjoy the video. The video took 5 min.

3.3.3. Task 3 [Stress]: Minesweeper

Previous studies already utilized Minesweeper to induce a stress condition [90,91]. Therefore, we used an implementation, in which the user had to find 10 mines in an area of 10×10 fields within a time frame of 100 s. The user had to complete 10 rounds. To increase the stress level, the given time was reduced by 10 s every round the user successfully completed the task. The score was indicated at all times, and the user was challenged to complete at least 7 rounds successfully, which was deemed to increase stress. Moreover, similar to task 1, the experimenter provided verbal comments aiming to further elevate the stress level.

3.3.4. Task 4 [Relaxation]: Nature Video

This is the second relaxation task, in which we presented a relaxing nature video for 5 min. Similar to the previous relaxation task, we selected this video based on user feedback from a previous pilot study.

3.4. Data Gathering

The participant's plantar pressure data, as well as the accelerometer data, were collected as a time series with a sampling rate of 50 Hz and stored with the timestamp on an SD card. The EDA data was collected with a sampling rate of 4 Hz at the Empatica E4 wristband. From this data, the mean EDA and the average slope over a non-overlapping windows of 10 s were calculated. In addition to collecting sensor data, the questionnaire collected the user's perceived stress level, energy level and the level of pleasantness: "How much did you feel stressed during the task? How energetic did you feel during the task? How pleasant did you find the task?" We quantified the answers by letting participants rate their answer on a 7-pnt Likert scale (1: low, 7: high). Moreover, we conducted the NASA TLX to account for the induced task load. It consists of 6 questions which are equally weighted.

3.5. Data Analysis

To identify whether our task induced stress or relaxation as we expected, we conducted a pairwise *t*-Test on user's perceived stress level, energy level, the level of pleasantness and the task load. In addition, to detect significant difference of average EDA slope we conducted another pairwise *t*-Test. Since the data was normally distributed according to Shapiro Wilk's test, parametric tests such as pairwise *t*-Test and oneway-Anova were used. The participant's plantar pressure data and IMU data were used to calculate features as mentioned above and used in model training.

3.6. Results

As shown in Figure 5, both stress tests, Task 1: Stroop Colour and Word Test, as well as Task 3: Minesweeper, indeed induced an elevated stress level based on the users' subjective rating. Task 1 scored $M = 5.06$ ($SD = 0.68$) and Task 3 scored $M = 6.05$ ($SD = 0.76$), respectively. In both relaxation tasks, Task 2: Minesweeper Introduction Video, as well as Task 4: Nature Video, the average rated stress level significantly reduced to $M = 2$ ($SD = 0.76$) and $M = 1.39$ ($SD = 0.58$). A pairwise t-Test ($t = 19.33$; $p < 0.05$), demonstrating significance and provides proven evidence of our success in inducing stress and relaxation among the users.

	Task 1 [Stress]	Task 2 [Relaxation]	Task 3 [Stress]	Task 4 [Relaxation]
How much did you feel stressed during the task?	5.06 /7	2 /7	6.05 /7	1.39 /7
How energetic did you feel during the task?	5.38 /7	3.67 /7	5.85 /7	2.91 /7
How pleasant did you find the task?	4.5 /7	4.47 /7	4.95 /7	6.43 /7
Task Load	64.03 /100	33.55 /100	74.67 / 100	21.3 /100

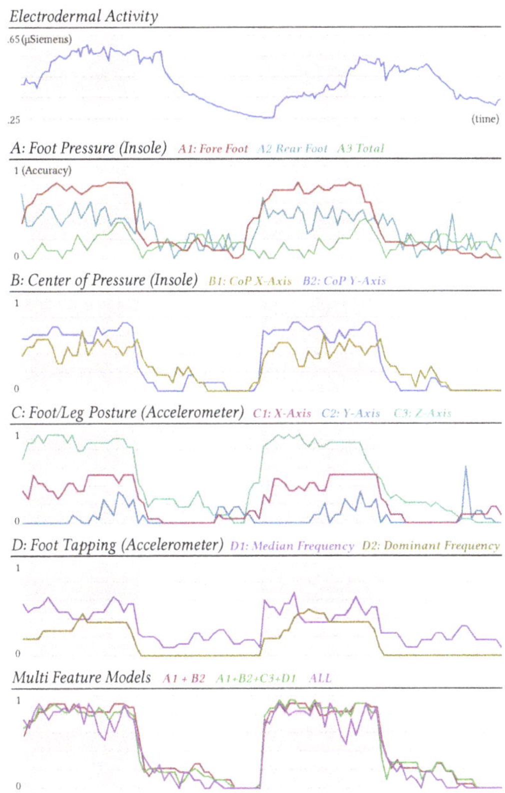

Figure 5. Summary of the results for each task in study 1. The graph of electrodermal activity was computed using the average of the mean electrodermal activity (EDA) over non-overlapping windows of 10 s of all the participants. Accuracy graphs were computed by calculating the average stress detection accuracy of all the participants over non-overlapping windows of 10 s.

Acute stress raises adrenaline levels, making participants feel energetic. In accordance with the answers, all participants stated they felt significantly more energetic during the stress task 1 ($M = 5.38$; $SD = 0.89$) and 3 ($M = 5.85$; $SD = 1.04$), than at the relaxation task 2 ($M = 3.67$; $SD = 1.67$) and 4 ($M = 2.91$; $SD = 1.78$). A pairwise comparison between the stress and relaxation task demonstrated a statistical difference ($t = 7.09$; $p < 0.05$). We hypothesised that a relaxation task will also be more pleasant than a stress task. The perceived pleasure was positive throughout all tasks (T1: $M = 4.5$; $SD = 1.51$, T2: $M = 4.47$; $SD = 1.6$, T3: $M = 4.95$; $SD = 1.9$, T4: $M = 6.43$; $SD = 0.79$). Grouping both relaxation tasks together demonstrated significance to the stress tasks ($t = 2.84$; $p < 0.05$). However, this effect is only significant because the last relaxation task, watching a nature video, was rated as extraordinarily pleasurable. A oneway-ANOVA ($F_{3,71} = 7.88$, $p < 0.05$) confirmed this task as significantly more pleasurable than all other tasks. In addition, The NASA TLX seems to be correlated with the actual stress level. A pairwise t-Test ($t = 13.24$; $p < 0.05$) indicated two groups as significantly different from each other. Both stress tasks (T1: $M = 64.03$; $SD = 17.46$ and T3: $M = 74.67$; $SD = 11.32$) showed a significantly increased task load in comparison to the relaxation tasks (T2: $M = 33.55$; $SD = 11.78$ and T4 $M = 21.3$; $SD = 8.08$). The data coincides with the self-perceived stress level.

3.6.1. Electrodermal Activity (EDA)

As displayed in Figure 5, the average EDA profile of the participants show a significant increase in both stress tasks (T1 and T3) in comparison to the relaxation task (T2 and T4), which is evidenced by a pairwise t-Test ($t = 2.05$; $p < 0.05$). These findings show that the physiological response also correlates to the self-reported stress level and with the Task Load of the participants.

3.6.2. Model Training

As previously stated, we identified four general characteristics that occurs at our foot when placed under pressure. Since our data is low-dimensional to identify the quality of these features, we developed a model for each individual feature using a machine learning classifier. We tested 5 different classifiers with our data, which were: Random Forest (RF) (ntree = 3, mtry = 2), K-Nearest-Neighbour (KNN) (k = 9), Support Vector Machines (SVM) (sigma = 13.9779 C = 0.25, kernel = radial), Decision Trees (CART), cp = 0.02294894 and Linear Discriminant Analysis (LDA). A one-way ANOVA for correlated samples ($F_{4,60} = 4.82$, $p < 0.05$) showed significant differences. A Tukey's post-hoc analysis suggested that LDA-KNN, LDA-SVM, LDA-RF pairs yielded a significant difference. We selected the *LDA* classifier because it showed a consistent and higher mean performance with our data.

Using a supervised learning approach, we trained 10 single-feature models (A1, A2, ..., D2) based on the annotation of our ground truth data. For the ground truth data, we considered the data gathered from all participants, whose stress rating was $M > 4$ for the stress tasks and whose stress rating was $M < 4$ for the relaxation tasks. Hence, for training our model, we excluded ambiguous data, which showed a low stress level at stress tasks (T1: 7/23p., T3: 3/23p.) and an elevated stress level at a relaxation task (T2: 8/23p., T4 0/23p.). Moreover, we excluded the first 60 s and the last 60 s of each task when training our model. Exclusion was necessary to reduce possible noise created by a task accustomisation at the beginning and task exhaustinction at the end of the task.

3.6.3. Model Validation

Each model (*A: Foot Pressure, A1: Fore Foot, A2: Rear Foot, A3: Total Foot, B: Center of Pressure B1: CoP X-Axis, B2: CoP Y-Axis, C: Accelerometer C1: X-Axis, C2: Y-Axis, C3: Z-Axis, D: Foot Tapping, D1: Median Frequency, D2: Dominant Frequency*) was trained and validated using a leave-one$_{User}$-out method. Meaning, we built a user-specific model, which was trained by all other users, but does not include the one we tested the model with. Figure 5 depicts the accuracy rates for a stress detection. Instead of calculating an overall accuracy for each task, we calculated the accuracy for non-overlapping windows of 10 s to allows us to observe the progressing confidence throughout the task. Table 1 summarises the overall accuracy rates across all single feature and multi-feature Models. Creating a model using all features provides a reasonable accuracy (M = 83.9%). However, the standard deviation (SD = 12.01) is higher than any other model. The highest accuracy (M = 85.32; SD = 8.1) was from the combination of all four high performers (A1+B2+C3+D1). Although the standard deviation is relatively high as a single feature model, C3 showed the highest accuracy (M = 83.1; SD = 11.9) compared to other single feature models. In addition, a pair-wise t-Test was conducted to identify the separation sharpness between the stress and relaxation tasks. Except for model A3, all other models showed a high distinguishability (See Table 1).

Table 1. Model performance (selected classifier: Linear Discriminant Analysis (LDA)).

Feature Model	A1+B2+C3+D1	C3+A1	ALL	C3	A1	B2	D1	C1	B1	D2	A2	C2	A3
Accuracy [%]	85.32	85.0	83.9	83.1	79.8	78.8	69.3	67.3	66.6	65.3	62.5	52.8	50.7
SD [%]	8.1	9.7	12.01	11.9	11.1	6.7	7.9	8.0	11.3	6.7	11.4	10.6	9.25
Distinguishablity [p]	<0.0001	<0.0001	<0.0001	<0.0001	<0.0001	<0.0001	<0.0001	<0.0001	<0.0001	<0.0001	<0.0001	<0.001	=0.2

4. Empirical Replicability—Study 2

To validate the generalisability of our models, we replicated the previous study with different parameters, such as using different users and different tasks. The study protocol was approved by the University of Auckland Human Participants Ethics Committee.

4.1. Study Design

The apparatus and data gathering remained the same. The procedure was very similar, with the only difference being the deployment of a single stress and relaxation task. After each task, the participants were asked to answer the same questionnaire from the previous study and to complete a NASA TLX. The data analysis remained the same. We recruited 11 new participants (7 males and 4 females), aged between 22–34 (M = 26.4, SD = 3.17) with different ethnicity. The inclusion/exclusion criteria were similar to previous study. We also utilised a different stress (Task 5: Mental Arithmetic Test) and relaxation (Task 6: Nature Video) task.

4.1.1. Task 5 [Stress]: Mental Arithmetic Test

Participants were asked to complete 20 challenging maths questions based on fundamental mathematics within 5 min. We created additional pressure by informing participant's their performance will be graded. Similar to previous tasks, an experimenter observed their performance and commented on their performance.

4.1.2. Task 6 [Relaxation]: Nature Video

To relax the user, we showed a 5 min nature video with soothing music. The video was different to the one in Task 4.

4.2. Results

In accordance with the questionnaire, all the participants agreed they felt stressed (M = 5.36; SD = 0.92) during the mental arithmetic task and relaxed (M = 1.72; SD = 0.64) during the relaxation task (see Figure 6). A pairwise t-Test (t = 11.74; $p < 0.05$) confirmed that all participants were significantly more stressed during stress Task 5 compared to the relaxation Task 6. All participants stated they felt more energetic during stress task 5 (M = 5.18; SD = 1.47), than at the relaxation task 6 (M = 3.45; SD = 2.01), which was confirmed as statistically different following a pairwise t-Test (t = 2.55; $p < 0.05$). Additionally, participants rated the relaxed task (M = 6.36; SD = 0.67) as more pleasant than the stress task (M = 4.18; SD = 2.13), which was evidenced as significant by a pairwise t-Test (t = 3.54; $p < 0.05$). Most crucially, in terms of Task Load (equally weighted), a pairwise comparison (t = 9.49; $p < 0.05$) between both tasks clearly indicated that stress Task 5 significantly induced a higher task load than relaxation Task 6.

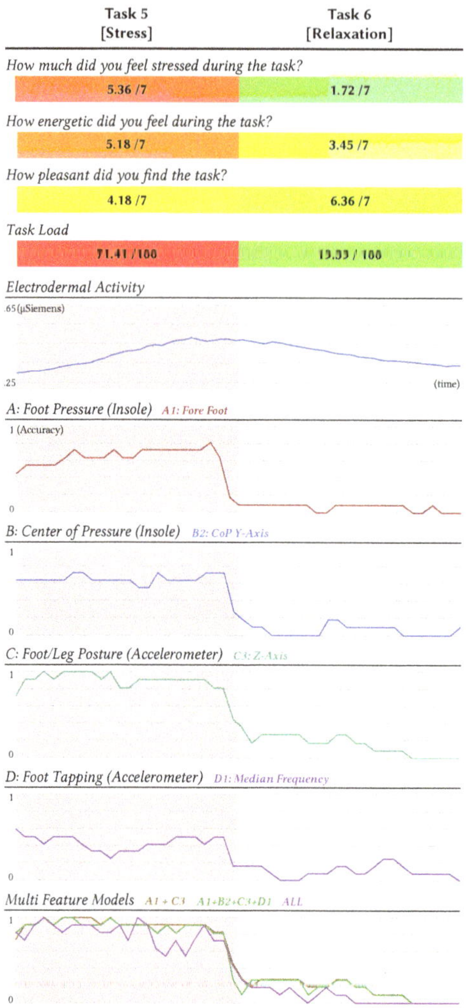

Figure 6. Results of generalisability. The graph of electrodermal activity and accuracy shows the averages of the 11 participants over non-overlapping windows of 10 s.

4.2.1. Electrodermal Activity (EDA)

These subjective ratings are also consistent with the physiological data gathered—the average EDA response. It showed an overall positively increasing slope at the stress task 5 and negatively decreasing slope for relaxation task 6. Comparing the trends of both tasks by a *t*-Test confirmed the average EDA slopes to be significantly different ($t = 6.9$; $p < 0.05$).

4.2.2. Overall Model Validation

The features were extracted as mentioned before and classified (using *LDA*) with seven previously built models. We chose four single feature models based on the best performer for each observed characteristic, which are: A1, B2, C3 and D1. Furthermore, we used a multi-feature model combining all four. The sixth model was generated based on previous data and by combining the most meaningful features of A1 and C3. The final model was a combination of all computed features. Again, we windowed the accuracy of models over 10 s to observe any periodic trends (see Figure 6).

The summary of the overall accuracy rates are shown in the Table 2. A one-way ANOVA for correlated samples ($F_{6,339} = 12.75$, $p < 0.05$) showed significant differences between the accuracy of the models. A Tukey's post-hoc analysis revealed that the significant difference occurred due to the low mean accuracy of model D1. Since model A1+B2+C3+D1 showed slightly higher accuracy than model C3+A1 and a lower standard deviation ($M = 87.45$; $SD = 8.5$), we suggest it is the best performing model. Furthermore, it will work for most of the users, since it considers a feature related to foot tapping such as D1. The model incorporating all features showed an overall accuracy of 85.6%, but with a comparably high standard deviation ($SD = 12.0$). Model C3 showed the highest accuracy as an individual feature model ($M = 86.7$; $SD = 10.0$). In a pair wise t-Test, all models indicated high separation sharpness between stress and relaxation ($p < 0.0001$).

Table 2. Model Performance for second set of participants for different tasks (Selected classifier: LDA).

Feature Model	A1+B2 +C3+D1	C3+A1	ALL	C3	A1	B2	D1
Accuracy [%]	87.45	87.3	85.6	86.7	79.3	79.6	66.5
SD [%]	8.5	9.62	12.0	10.0	8.5	6.5	7.6

5. External Validity—Study 3

In this study, the goal is to demonstrate the robustness of our model. Therefore, we conducted an in-field study at which we recorded data over a working day from users performing their usual everyday office tasks. Similar to the previous studies, the University of Auckland Human Participants Ethics Committee approved the protocol.

5.1. Participants

We recruited 10 participants (7 males and 3 females) aged between 25 and 34 ($M = 29.9$, $SD = 3.4$). Among these participants, four were from the first study, another four were from the second study and two were newly recruited. A requirement for our participant selection was a foot size that matched the prototype. In addition, the participant had to work in an office and spent the majority of the time in a sitting posture (>70%). Apart from this, the inclusion/exclusion criteria remained similar to previous studies.

5.2. Task and Procedure

The study was conducted on a particular day that the participants expected to have some periods of acute stress. Some of the stress tasks included working for a deadline, debugging a firmware/software, having a meeting with their supervisor, writing a paper for an upcoming submission, etc. Having lunch/coffee with friends and having casual chats with friends were some activities that would supposedly relax the participants.

The study began at 09:45 a.m. local time at the participants' office space. After elaborating on the study procedures, filling consent forms and collecting demographic data, the experimenter asked participants to wear the StressFoot and E4 wristband. At 10:00 a.m., the experimenter initialised apparatuses for data collection and asked participants to fill a questionnaire, asking them to rate their current stress level, energy level and how pleasant they felt on a 7-point Likert scale. Then, participants were asked to continue their work as usual. We asked participants to fill out two forms, a calendar application reporting the type and duration of tasks performed, as well as a questionnaire. These were the only tasks we asked the participants to perform hourly over 8 h from 10:00 a.m. to 6:00 p.m. To ensure the participants would remember to report and to fill out the calendar and the form, they received reminders via visual and audio pop-ups at the end of each hour. The questionnaire asked the participants to rate their perceived stress level, energy level, as well as how pleasant they felt during last hour. Part of the questionnaire was to fill a NASA TLX, which we used to calculate the overall workload.

5.3. Apparatus and Data Gathering

Similar to the previous studies, the StressFoot prototype was used to collect foot motion and pressure data. The Empatica E4 wristband was used to collect EDA, as well as motion data from the participants.

The data gathering was thus similar to the previous study. Additionally, we collected accelerometer data from the E4 wristband to identify motion (walking) and posture (sitting/standing). We used the E4's preset sampling rate of 32 Hz. The data analysis also remained similar to the previous study. Next, we segmented the accelerometer data into non-overlapping windows. Literature suggests using a window size less than 10 s, such as 2.5 s [92] or even 1 s [93]. Through an experimentation with three users, we determined 5 s as a suitable window size for our use case. To identify motion and posture, we relied on a threshold analysis with a single feature, as suggested by Gjoreski et al. [94]. The feature used is the first derivative over the entire window, also known as an "Acceleration Vector Change" (AVC) [94]. This feature can identify: walking (AVC >= 2 ms^{-3}), standing (AVC < 0.1 ms^{-3}) and sitting (0.1 ms^{-3} =< AVC < 0.2 ms^{-3}). We applied this to extract the timestamps to identify the sitting time. Using these timestamps, we selected the sitting data from the corresponding foot motion and foot pressure data from the StressFoot prototype.

Finally, the sitting data was segmented into 10 s of non-overlapping windows. Then, all features were extracted, as mentioned before, and classified using the best performing model, which was a multi-feature model (A1+B2+C3+D1). At every 10 s window, the model classifies whether a participant is stressed or relaxed. We then calculated the percentage of the number of windows that classified stressed. This provides a measurement of the duration that a participant might have been stressed. Therefore, for each hour, we calculated a ratio (R_S), where:

$$R_S = \frac{nWindows_{\text{sitting}}(Stressed)}{nWindows_{\text{sitting}}(Total)} \tag{11}$$

5.4. Results

5.4.1. Activities

As depicted in the Table 3, all the participants spent the majority of their time in a sitting posture ($M = 79.5$, $SD = 6.1$). Designing PCBs, coding, soldering, debugging firmware/software, reading/writing papers, meetings, writing emails, having coffee/lunch and watching YouTube, were the major activities that participants reported doing during the study.

Table 3. Percentage of being in sitting posture of each participant during the 8 h field study.

Participant No:	P1	P2	P3	P4	P5	P6	P7	P8	P9	P10
Sitting time [%]	85.7	76.5	76.3	73.9	88.4	86.7	70.6	75.1	83.8	78.2

Most of the participants marked higher stress levels (>4 on a 7-pnt Likert scale) when they performed office work-related activities, such as coding, debugging and writing papers. A summary of the self-reported stress level, level of energy and level of pleasantness across all participants is depicted in Table 4. According to the table, when participants felt stressed (Stress Level > 4), a lower level of pleasantness ($M = 3.8$, $SD = 1.4$) was reported compared to when participants were relaxed (Stress Level < 4). Welch's t-test further confirmed that there is a significant difference ($t = -4.10$; $p < 0.05$). Moreover, Welch's t-test showed a significantly higher task load when participants were stressed, compared to being relaxed ($t = 7.02$; $p < 0.05$). These findings confirm that we have collected two significantly different levels of stress (stress and relaxation).

Table 4. Summary of level of pleasantness, energy level and task load.

	Level of Stress > 4	Level of Stress < 4	p	t
Level of Pleasantness	M = 3.8, SD = 1.4	M = 5.2, SD = 0.9	<0.05	4.10
Level of Energy	M = 4.9, SD = 1.1	M = 4.3, SD = 1.4	>0.05	1.64
Task Load	M = 62.3, SD = 8.6	M = 42.4, SD = 11.4	<0.05	7.02

Moreover, the rated "level of energy" was not statistically different when dividing the gathered data into two groups (stressed and relaxed) following the Welch's t-test ($t = 1.64$; $p > 0.05$). This means the perceived level of energy does not relate to stress since energy can be both positive and negative.

5.4.2. Electrodermal Activity (EDA)

Table 5 summarises the average EDA slopes of the participants in stressed and relaxed conditions. When a participant marked the previous hour with a stress level > 4, we considered that hour as a high stress duration. We considered ratings below 4 as a relaxed period. When stressed, the majority of participants (5 out of 6) demonstrated an average positive EDA slope. Comparatively, when participants were relaxed, 8 out of 10 participants showed an overall negative EDA slope. However, none of the participants showed a significantly different mean EDA slope according to Welch's t-test (see Table 5). The limited difference is due to high variances of EDA readings caused by external factors, such as motion artefacts, ambient temperature variances and loose contact of electrodes on the skin.

Table 5. Summary of EDA slopes of each participants while being stressed and relaxed.

Parti. No:	Stress	Relax	p	t
P1	1.99×10^{-6}	-1.61×10^{-5}	>0.05	0.08
P2	-	-3.58×10^{-6}	-	-
P3	5.16×10^{-6}	-3.93×10^{-5}	>0.05	0.02
P4	2.91×10^{-6}	-9.69×10^{-5}	>0.05	0.17
P5	7.04×10^{-5}	-2.38×10^{-4}	>0.05	0.15
P6	2.95×10^{-5}	-6.81×10^{-5}	>0.05	0.38
P7	-	9.03×10^{-6}	-	-
P8	-	-5.77×10^{-5}	-	-
P9	-	4.48×10^{-6}	-	-
P10	-8.08×10^{-5}	-3.52×10^{-5}	>0.05	0.14

5.4.3. Overall in Field Validation

We now analyse how our model compares to the users' Self-Reported Stress Level (SRSL). Figure 7 depicts the SRSL and R_S of each hour for each participant. The graphs already demonstrate a positive correlation. To identify the significance of this relationship, we calculated the Pearson's Correlation Coefficients (r) for each participant. All participants showed a positive correlation coefficient, with an average of $r = 0.79$ ($SD = 0.10$). Eight out of 10 participants demonstrated a statistically significant positive relationship ($r > 0.7$, $p < 0.05$) between SRSL and R_S. Among these participants, three had a high correlation coefficient ($r > 0.9$) which shows strong relationship between SRSL and R_S. Only two participants showed a moderate Pearson's Correlation Coefficient ($r = 0.67$ and $r = 0.65$). However, this result was not statistically significant ($p > 0.05$). Even from those two participants, P8 reported lower stress levels (<4) for the entire study duration, which is also confirmed by our model showing a lower R_S (<0.5) value during the entire study period. Overall, we can conclude that all the participants showed a positive correlation, given the majority demonstrated a significantly high correlation between SRSL and R_S. Such findings evidences the robustness of our model in-field, beyond a controlled laboratory setting.

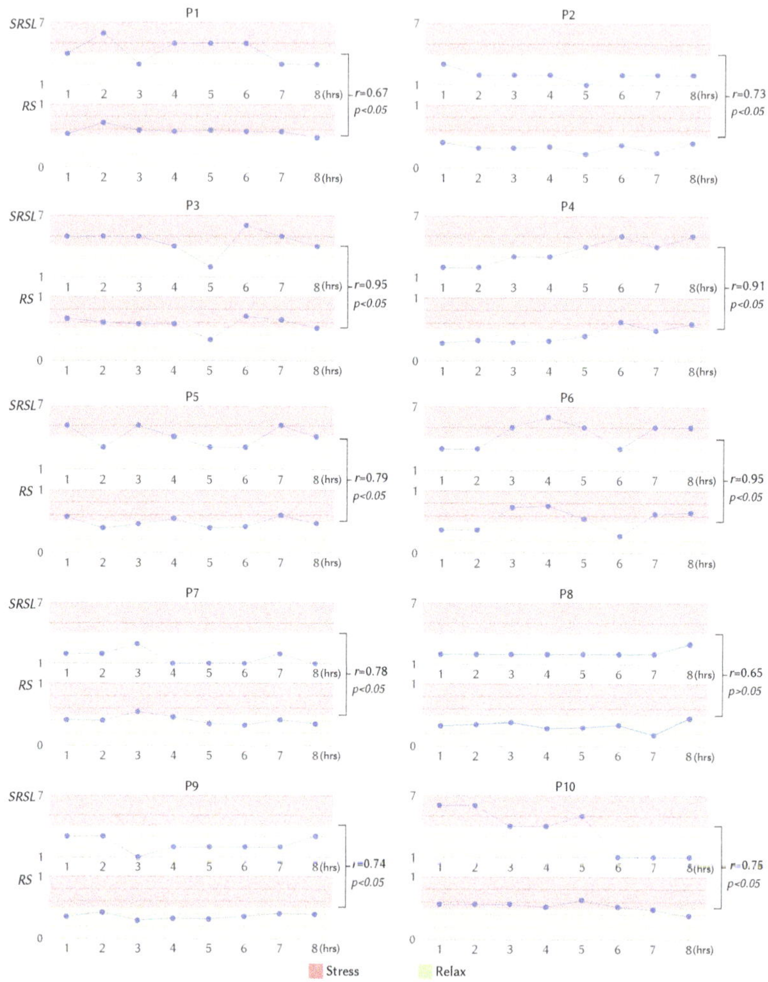

Figure 7. The figure shows the Self-Reported Stress Level (*SRSL*) and R_S for each participant for each hour. Where $R_S = (nWindows_{\text{sitting}}(Stressed)/nWindows_{\text{sitting}}(Total))$. The Pearson's Correlation Coefficient r and the level of significance p for each participant's *SRSL* and R_S is also depicted.

6. Discussion

6.1. Accuracy

Our highest performing model is an LDA classifier with an accuracy of 86% in laboratory conditions. In literature, the models based on physiological parameters such as EDA, ECG, HRV, showed similar or higher accuracy in laboratory validations [32,39]. For example, an SVM model, based on features from a combination of physiological signals such as EDA, Blood Volume Pulse, ST and PD achieved an accuracy of 90.1% [32]. Although the model has achieved a higher accuracy, it has some limitations in field deployment. PD requires line of cite and may generate privacy concerns.

In addition, according to authors, removing PD may drop the accuracy closer to 60%. In another study, authors were able to discriminate stress from cognitive load by using a LDA classifier based on EDA [40]. In a laboratory setting, they achieved an accuracy of 82.8%, which is slightly lower than the results we achieved in our study. In addition, an SVM model based on facial EMG, respiration, EDA and ECG was used to recognise 5 emotional states such as high stress, low stress, disappointment, euphoria and neutral. The paper reports an accuracy of 86% in a laboratory study [95]. In addition, another study reported a system which can classify stress with 86% accuracy based on 15 features extracted from EEG, ECG and EDA signal [96]. All these systems may be inconvenient, given the need to tightly attach multiple wearable sensors onto the body. In our approach, we can recognise stress and relaxation with a similar accuracy by using a simple accelerometer model, such as model C3. However, our current model cannot detect different levels of stress.

Some of the prior studies related to body language [76] study and facial [61] expression also achieved similar or slightly higher accuracy. However, many of these methods used camera-based systems, which may have limitations in real-life implementations. Some of the higher accuracy methods use multiple motion capture cameras, which is impractical to deploy in real-life, specifically in an office environment. For example, a work which detected emotions related to negative stress such as sadness, joy, anger and fear showed an accuracy of 93% [76]. However, the method uses 6 camera vicon motion capture system. Although the accuracy is slightly lower than vision-based methods, our approach captures certain body language related to stress while sitting by using a more practical method, which can be easily used in real-life applications.

On the other hand, there are several real life validations reported in literature. Healey and Picard proposed an LDA classifier to recognise stress of drivers using ECG, EMG, EDA and respiration [4]. Regardless of the cumbersome setup which consists of many on body sensors, they were able to recognise three levels of stress (low, medium and high) in 97% of accuracy. In another study, Hernandez et al. achieved an accuracy of 74% in call-center stress detection [39]. They proposed a person specific SVM model, which uses EDA response for stress detection. In our field study, by using four features related to foot motion and posture, we identified significantly high correlations between self-rated stress levels and model-derived stress levels across users, ultimately showing the robustness of the method.

Overall, methods which utilise a combination of physiological parameters seem to achieve higher accuracy than our proposed method [4,32]. However, sensing multiple physiological parameters requires attaching multiple sensors onto the user, compromising the comfort. Contrarily, previous methods based on single physiological parameter demonstrated either similar accuracy or lower accuracy [39,40]. The lower accuracy could be due to data losses and subjective differences. Some methods based on body language and facial expressions have shown higher accuracy [61,76] due to high sensing accuracy in visual-based sensing. However, those methods seem highly obtrusive in real-life.

In addition, a recent study compared unobtrusive sensors for stress detection at sedentary computer work [97]. In their analysis, they considered wrist worn, chest worn and thermal imaging based sensors. They identified that wrist worn sensors, such as EDA and PPG, may not capture stress accurately due to frequent data losses. This is mainly due to motion artefacts, such as electrode movements, detaching from the skin and a change in pressure on the skin. In addition, chest worn sensors which sense HR showed similar issues due to posture changes generating high noise and thus failing to maintain proper contact with the skin. This is highly problematic in 6–8 h of sensing in a typical working day. However, our approach does not result in such issues, specifically in an office working environment. We have proven that risks of data losses are not present with our method, given we sense foot motion and posture characteristics while sitting. While the study identified that thermal imaging resulted in a greater identification of stress during computer work, this is not always a practical method. Thermal imaging requires a consistent line of sight, which is problematic when the individual needs to attend to tasks away from their typical working desk. Our method poses no such issues.

6.2. Applications

6.2.1. Professional Environment

In our studies, the recruited participants mainly performed computer-aided tasks in an office environment while in a sitting posture. Other occupations, such as cashiers, emergency/non emergency call centre workers, also perform their tasks while in sitting posture for prolonged periods. In addition, the working environment of these occupations are similar to an office environment. Hence, our method may work for these occupations as well. In such a scenario, stress sensing could be used to improve mental health. However, further studies need to identify the viability of foot-based stress sensing method for such occupations. In addition, remote stress monitoring of adults under home care is a another potential application. Simple accelerometer can be embedded to a sock and thus monitor both stress and activity level using a single sensor. However, to accomplish this further studies with better classification algorithms stated in literature are required [98,99].

6.2.2. The Quantified Self

Enriching self-tracking with a stress detection is another application. In this community, stress is being triangulated with several data of wearable sensors [100]. A simple feature could substantially increase accuracy. Moreover, using an already available wearable, like a shoe, can address users who prefer not to wear additional garment accessories.

6.3. Limitations and Future Work

6.3.1. Quantifying Multiple Stress Levels

Both lab studies only investigated discriminating stress from relaxation without aiming to identify different levels of stress. However, the results of study 3 indicates that the frequency of demonstrating stress related postures could reveal the extent of stress. However, it requires further research to infer on different levels of stress that could be based on the frequency of stress-related foot movements and foot posture characteristics.

6.3.2. Stress Detection in Sitting Posture and Other Activities

In our society, sitting is the most common posture demonstrated during both the working week and weekends [101], which our findings also confirm (see Table 3). Therefore, our proposed models would principally work for a majority of the time, whenever an individual is doing some sort of intellectual work while seated. Detecting stress in other postures exceeds the scope of our current research. Identifying stressful situations when performing regular activities, such as standing, walking and running are thus considered future work. For this, we would first need to identify the current posture and activity, such as by using an insole [24,25] or IMU [92,94,102], as shown in the literature.

6.3.3. Accuracy Boost with Personalised Models

We aimed to develop a generalised model capable of working across different users. We observed that the accuracy of our generalised models decreases for features, such as Median Frequency and Dominant Frequency. This is because not all users demonstrate an elevated foot tapping/shaking when stressed. In addition, we observed that individuals may have slightly different foot posture and motion characteristics depending on the BMI and other personal habits. Relying on a personalised model will further boost accuracy, particularly for features like foot-tapping. Further, a Neural Network approach can be advantageous, particularly when building a personalised model with a high volume of data gathered in the wild.

6.3.4. Improved Hardware

In our current prototype, the IMU (in conjunction with the control unit) was attached to the ankle. Our long-term goal is to integrate all these components into a smart footwear. This may change the orientation of the IMU. Thus, the main signal could spread to different axis, resulting in another axis to provide higher separation sharpness.

7. Conclusions

We presented StressFoot, a pair of smart shoes that can sense stress unobtrusively by using a pressure sensitive insole and an IMU. Our prototype is capable of identifying acute stress and relaxation while sitting, such as performing office tasks. We identified four characteristics, which reflect foot pressure distributions, foot posture variations and foot tapping. Based on these features, we trained several machine learning models with 23 participants by using a leave-one$_{user}$-out and validated this as a method to detect stress with an average accuracy of $\sim$85%. Then, with 11 additional participants, we demonstrated the replicability of our model with a similar overall accuracy of $\sim$87%. Finally, to evidence external validity, we conducted a field study with 10 participants, and evaluated the robustness of our models in an actual office setting. The outcome was that the computed stress level provided by our machine learning model correlates with the self-reported stress level with a coefficient of $r = 0.79$. We envision StressFoot to be an unobtrusive system capable of detecting the user's stress level on a daily basis. By drawing attention to the user's mental stress condition, such a system may already be able to contribute to an improvement in overall mental well-being in the future.

Author Contributions: D.S.E. conceptualised and Investigation as part of his PhD research. D.J.C.M. and S.N. contributed with overall supervision, review editing. All authors have read and agreed to the published version of the manuscript.

Funding: This work was supported by *Assistive Augmentation* research grant under the Entrepreneurial Universities (EU) initiative of New Zealand.

Acknowledgments: We acknowledge Ridmi Nimeshani Induruwa Bandarage for helping us in technical illustrations.

Conflicts of Interest: The authors declare no conflict of interest.

Abbreviations

The following abbreviations are used in this manuscript:

IMU	Inertial Measurement Units
CoP	Centre of Pressure
EDA	Electrodermal Activity
SRSL	Self-Reported Stress Level

References

1. Pugliesi, K. The consequences of emotional labor: Effects on work stress, job satisfaction, and well-being. *Motiv. Emot.* **1999**, *23*, 125–154. [CrossRef]
2. Renneberg, B.; Hammelstein, P. *Gesundheitspsychologie*; Springer: Cham, Switzerland, 2006.
3. Quick, J.D.; Horn, R.S.; Quick, J.C. Health consequences of stress. *J. Organ. Behav. Manag.* **1987**, *8*, 19–36. [CrossRef]
4. Healey, J.; Picard, R.W. Detecting stress during real-world driving tasks using physiological sensors. *IEEE Trans. Intell. Transp. Syst.* **2005**, *6*, 156–166. [CrossRef]
5. Villarejo, M.V.; Zapirain, B.G.; Zorrilla, A.M. A stress sensor based on Galvanic Skin Response (GSR) controlled by ZigBee. *Sensors* **2012**, *12*, 6075–6101. [PubMed]

6. Hamid, N.H.A.; Sulaiman, N.; Aris, S.A.M.; Murat, Z.H.; Taib, M.N. Evaluation of human stress using EEG power spectrum. In Proceedings of the 2010 6th International Colloquium on Signal Processing & Its Applications, Mallaca City, Malaysia, 21–23 May 2010; pp. 1–4.
7. Nagae, D.; Mase, A. Measurement of heart rate variability and stress evaluation by using microwave reflectometric vital signal sensing. *Rev. Sci. Instrum.* **2010**, *81*, 094301. [CrossRef]
8. Subhani, A.R.; Xia, L.; Malik, A.S. EEG signals to measure mental stress. In Proceedings of the 2nd International Conference on Behavioral, Cognitive and Psychological Sciences, Maldives, 25–27 November 2011; pp. 84–88.
9. Rohracher, H. Microvibration, permanent muscle-activity and constancy of body-temperature. *Percept. Mot. Ski.* **1964**, *19*, 198. [CrossRef]
10. Lundberg, U.; Kadefors, R.; Melin, B.; Palmerud, G.; Hassmén, P.; Engström, M.; Dohns, I.E. Psychophysiological stress and EMG activity of the trapezius muscle. *Int. J. Behav. Med.* **1994**, *1*, 354–370. [CrossRef]
11. Lerner, J.S.; Dahl, R.E.; Hariri, A.R.; Taylor, S.E. Facial expressions of emotion reveal neuroendocrine and cardiovascular stress responses. *Biol. Psychiatry* **2007**, *61*, 253–260. [CrossRef]
12. Navarro, J.; Karlins, M. *What Every Body is Saying*; HarperCollins Publishers: New York, NY, USA, 2008.
13. Kleinsmith, A.; Bianchi-Berthouze, N. Affective body expression perception and recognition: A survey. *IEEE Trans. Affect. Comput.* **2012**, *4*, 15–33. [CrossRef]
14. Kim, D.; Seo, Y.; Cho, J.; Cho, C.H. Detection of subjects with higher self-reporting stress scores using heart rate variability patterns during the day. In Proceedings of the 2008 30th Annual International Conference of the IEEE Engineering in Medicine and Biology Society, Vancouver, BC, Canada, 20–25 August 2008; pp. 682–685.
15. Macleod, J.; Smith, G.D.; Heslop, P.; Metcalfe, C.; Carroll, D.; Hart, C. Limitations of adjustment for reporting tendency in observational studies of stress and self reported coronary heart disease. *J. Epidemiol. Community Health* **2002**, *56*, 76–77.
16. McDuff, D.J.; Hernandez, J.; Gontarek, S.; Picard, R.W. COGCAM: Contact-free Measurement of Cognitive Stress During Computer Tasks with a Digital Camera. In *Proceedings of the 2016 CHI Conference on Human Factors in Computing Systems*; ACM: New York, NY, USA, 2016; pp. 4000–4004. [CrossRef]
17. McDuff, D.; Gontarek, S.; Picard, R.W. Improvements in remote cardiopulmonary measurement using a five band digital camera. *IEEE Trans. Biomed. Eng.* **2014**, *61*, 2593–2601. [CrossRef] [PubMed]
18. Poh, M.Z.; McDuff, D.J.; Picard, R.W. Advancements in noncontact, multiparameter physiological measurements using a webcam. *IEEE Trans. Biomed. Eng.* **2010**, *58*, 7–11. [CrossRef] [PubMed]
19. Gimpel, H.; Regal, C.; Schmidt, M. myStress: Unobtrusive Smartphone-Based Stress Detection. Available online: https://www.fim-rc.de/Paperbibliothek/Veroeffentlicht/494/wi-494.pdf (accessed on 19 May 2020).
20. Sano, A.; Picard, R.W. Stress recognition using wearable sensors and mobile phones. In Proceedings of the 2013 Humaine Association Conference on Affective Computing and Intelligent Interaction, Geneva, Switzerland, 2–5 September 2013; pp. 671–676.
21. Stütz, T.; Kowar, T.; Kager, M.; Tiefengrabner, M.; Stuppner, M.; Blechert, J.; Wilhelm, F.H.; Ginzinger, S. Smartphone based stress prediction. In *International Conference on User Modeling, Adaptation, and Personalization*; Springer: Cham, Switzerland, 2015; pp. 240–251.
22. Martínez-Nova, A.; Cuevas-García, J.C.; Pascual-Huerta, J.; Sánchez-Rodríguez, R. BioFoot® in-shoe system: Normal values and assessment of the reliability and repeatability. *Foot* **2007**, *17*, 190–196. [CrossRef]
23. Elvitigala, D.S.; Matthies, D.J.; David, L.; Weerasinghe, C.; Nanayakkara, S. GymSoles: Improving Squats and Dead-Lifts by Visualizing the User's Center of Pressure. In *Proceedings of the 2019 CHI Conference on Human Factors in Computing Systems*; ACM: New York, NY, USA, 2019; pp. 174:1–174:12. [CrossRef]
24. Haescher, M.; Matthies, D.J.; Bieber, G.; Urban, B. Capwalk: A capacitive recognition of walking-based activities as a wearable assistive technology. In Proceedings of the 8th ACM International Conference on PErvasive Technologies Related to Assistive Environments, Corfu, Greece, 1–3 July 2015; p. 35.
25. Matthies, D.J.; Roumen, T.; Kuijper, A.; Urban, B. CapSoles: Who is walking on what kind of floor? In Proceedings of the 19th International Conference on Human-Computer Interaction with Mobile Devices and Services, Vienna, Austria, 4–7 September 2017; p. 9.
26. Saunders, W.; Vogel, D. Tap-kick-click: Foot interaction for a standing desk. In Proceedings of the 2016 ACM Conference on Designing Interactive Systems, Brisbane, Australia, 4–8 June 2016; pp. 323–333.

27. What the Feet and Legs Say about Us. *Psychology Today*. Available online: https://www.psychologytoday.com/nz/blog/spycatcher/200911/what-the-feet-and-legs-say-about-us?fbclid=IwAR2MVdwmirSc9GpOOAzl7wNwrkX5dHCxgfcvO0xky7uR46yANdAI-OLnaQY (accessed on 1 August 2019).
28. Brewer, M.B.; Crano, W.D. Research design and issues of validity. In *Handbook of Research Methods in Social and Personality Psychology*; Reis, H.T., Judd, C.M., Eds.; Cambridge University Press: Cambridge, UK, 2014.
29. Blain, S.; Mihailidis, A.; Chau, T. Assessing the potential of electrodermal activity as an alternative access pathway. *Med. Eng. Phys.* **2008**, *30*, 498–505. [CrossRef]
30. Jerritta, S.; Murugappan, M.; Nagarajan, R.; Wan, K. Physiological signals based human emotion recognition: A review. In Proceedings of the 2011 IEEE 7th International Colloquium on Signal Processing and Its Applications, Penang, Malaysia, 4–6 March 2011; pp. 410–415.
31. Bousefsaf, F.; Maaoui, C.; Pruski, A. Remote Assessment of the Heart Rate Variability to Detect Mental Stress. In Proceedings of the (Institute for Computer Sciences, Social-Informatics and Telecommunications Engineering): ICST, Brussels, Belgium, 5 May 2013; pp. 348–351. [CrossRef]
32. Barreto, A.; Zhai, J.; Adjouadi, M. Non-intrusive physiological monitoring for automated stress detection in human-computer interaction. In *International Workshop on Human-Computer Interaction*; Springer: Berlin/Heidelberg, Germany, 2007; pp. 29–38.
33. Herborn, K.A.; Graves, J.L.; Jerem, P.; Evans, N.P.; Nager, R.; McCafferty, D.J.; McKeegan, D.E. Skin temperature reveals the intensity of acute stress. *Physiol. Behav.* **2015**, *152*, 225–230. [CrossRef]
34. Kataoka, H.; Kano, H.; Yoshida, H.; Saijo, A.; Yasuda, M.; Osumi, M. Development of a skin temperature measuring system for non-contact stress evaluation. In Proceedings of the 20th Annual International Conference of the IEEE Engineering in Medicine and Biology Society. Vol. 20 Biomedical Engineering Towards the Year 2000 and Beyond (Cat. No. 98CH36286), Hong Kong, China, 1 November 1998; Volume 2, pp. 940–943.
35. Seo, S.H.; Lee, J.T. Stress and EEG. In *Convergence and Hybrid Information Technologies*; IntechOpen: London, UK, 2010; pp. 413–426.
36. Ekberg, K.; Eklund, J.; Tuvesson, M.A.; Örtengren, R.; Odenrick, P.; Ericson, M. Psychological stress and muscle activity during data entry at visual display units. *Work Stress* **1995**, *9*, 475–490. [CrossRef]
37. Andreassi, J.L. *Psychophysiology: Human Behavior and Physiological Response*; Psychology Press: London, UK, 2010.
38. Choi, J.; Ahmed, B.; Gutierrez-Osuna, R. Development and evaluation of an ambulatory stress monitor based on wearable sensors. *IEEE Trans. Inf. Technol. Biomed.* **2011**, *16*, 279–286. [CrossRef]
39. Hernandez, J.; Morris, R.R.; Picard, R.W. Call Center Stress Recognition with Person-specific Models. In *International Conference on Affective Computing and Intelligent Interaction*; Springer: Berlin/Heidelberg, Germany, 2011; pp. 125–134.
40. Setz, C.; Arnrich, B.; Schumm, J.; La Marca, R.; Tröster, G.; Ehlert, U. Discriminating stress from cognitive load using a wearable EDA device. *IEEE Trans. Inf. Technol. Biomed.* **2009**, *14*, 410–417. [CrossRef]
41. Hernandez, J.; Paredes, P.; Roseway, A.; Czerwinski, M. Under Pressure: Sensing Stress of Computer Users. In *Proceedings of the SIGCHI Conference on Human Factors in Computing Systems*; ACM: New York, NY, USA, 2014; pp. 51–60. [CrossRef]
42. Boucsein, W. *Electrodermal Activity*; Springer Science & Business Media: Berlin/Heidelberg, Germany, 2012.
43. Hjortskov, N.; Rissén, D.; Blangsted, A.K.; Fallentin, N.; Lundberg, U.; Søgaard, K. The effect of mental stress on heart rate variability and blood pressure during computer work. *Eur. J. Appl. Physiol.* **2004**, *92*, 84–89. [CrossRef]
44. Moses, Z.B.; Luecken, L.J.; Eason, J.C. Measuring task-related changes in heart rate variability. In Proceedings of the 2007 29th Annual International Conference of the IEEE Engineering in Medicine and Biology Society, Lyon, France, 22–26 August 2007; pp. 644–647.
45. Lee, H.B.; Kim, J.S.; Kim, Y.S.; Baek, H.J.; Ryu, M.S.; Park, K.S. The relationship between HRV parameters and stressful driving situation in the real road. In Proceedings of the 2007 6th International Special Topic Conference on Information Technology Applications in Biomedicine, Tokyo, Japan, 8–11 November 2007; pp. 198–200.
46. Allen, J. Photoplethysmography and its application in clinical physiological measurement. *Physiol. Meas.* **2007**, *28*, R1. [CrossRef]

47. Lyu, Y.; Luo, X.; Zhou, J.; Yu, C.; Miao, C.; Wang, T.; Shi, Y.; Kameyama, K.I. Measuring Photoplethysmogram-Based Stress-Induced Vascular Response Index to Assess Cognitive Load and Stress. In *Proceedings of the 33rd Annual ACM Conference on Human Factors in Computing Systems*; ACM: New York, NY, USA, 2015; pp. 857–866. [CrossRef]
48. Hernandez, J.; McDuff, D.; Picard, R.W. BioWatch: Estimation of Heart and Breathing Rates from Wrist Motions. In Proceedings of the 2015 9th International Conference on Pervasive Computing Technologies for Healthcare (PervasiveHealth), Istanbul, Turkey, 20–23 May 2015; pp. 169–176.
49. Hernandez, J.; Picard, R.W. SenseGlass: Using Google Glass to Sense Daily Emotions. In *Proceedings of the Adjunct Publication of the 27th Annual ACM Symposium on User Interface Software and Technology*; ACM: New York, NY, USA, 2014; pp. 77–78. [CrossRef]
50. Haescher, M.; Matthies, D.J.; Trimpop, J.; Urban, B. SeismoTracker: Upgrade Any Smart Wearable to Enable a Sensing of Heart Rate, Respiration Rate, and Microvibrations. In *Proceedings of the 2016 CHI Conference Extended Abstracts on Human Factors in Computing Systems*; ACM: New York, NY, USA, 2016; pp. 2209–2216. [CrossRef]
51. Haescher, M.; Matthies, D.J.; Trimpop, J.; Urban, B. A study on measuring heart-and respiration-rate via wrist-worn accelerometer-based seismocardiography (SCG) in comparison to commonly applied technologies. In Proceedings of the 2nd international Workshop on Sensor-Based Activity Recognition and Interaction, Rostock, Germany, 25–26 June 2015; p. 2.
52. Garbey, M.; Sun, N.; Merla, A.; Pavlidis, I. Contact-free measurement of cardiac pulse based on the analysis of thermal imagery. *IEEE Trans. Biomed. Eng.* **2007**, *54*, 1418–1426. [CrossRef] [PubMed]
53. Puri, C.; Olson, L.; Pavlidis, I.; Levine, J.; Starren, J. StressCam: Non-contact measurement of users' emotional states through thermal imaging. In *CHI'05 Extended Abstracts on Human Factors in Computing Systems*; ACM: New York, NY, USA, 2005.
54. Yun, C.; Shastri, D.; Pavlidis, I.; Deng, Z. O'game, can you feel my frustration?: Improving user's gaming experience via stresscam. In Proceedings of the SIGCHI Conference on Human Factors in Computing Systems, Boston, MA, USA, 4–9 April 2009; pp. 2195–2204.
55. Ulyanov, S.S.; Tuchin, V.V. Pulse-wave monitoring by means of focused laser beams scattered by skin surface and membranes. In *Static and Dynamic Light Scattering in Medicine and Biology*; International Society for Optics and Photonics: Los Angeles, CA, USA, 1993; Volume 1884, pp. 160–167.
56. Zimmermann, P.; Guttormsen, S.; Danuser, B.; Gomez, P. Affective computing—A rationale for measuring mood with mouse and keyboard. *Int. J. Occup. Saf. Ergon.* **2003**, *9*, 539–551. [CrossRef]
57. Kołakowska, A. A review of emotion recognition methods based on keystroke dynamics and mouse movements. In Proceedings of the 2013 6th International Conference on Human System Interactions (HSI), Sopot, Poland, 6–8 June 2013; pp. 548–555.
58. Picard, R.W. *Affective Computing*; MIT Press: Cambridge, MA, USA, 2000.
59. Grafsgaard, J.F.; Wiggins, J.B.; Boyer, K.E.; Wiebe, E.N.; Lester, J.C. Automatically Recognizing Facial Expression: Predicting Engagement and Frustration. Available online: http://educationaldatamining.org/EDM2013/proceedings/paper_95.pdf (accessed on 18 May 2020).
60. Fasel, B.; Luettin, J. Automatic facial expression analysis: A survey. *Pattern Recognit.* **2003**, *36*, 259–275. [CrossRef]
61. Giannakakis, G.; Pediaditis, M.; Manousos, D.; Kazantzaki, E.; Chiarugi, F.; Simos, P.G.; Marias, K.; Tsiknakis, M. Stress and anxiety detection using facial cues from videos. *Biomed. Signal Process. Control* **2017**, *31*, 89–101. [CrossRef]
62. Scheirer, J.; Fernandez, R.; Picard, R.W. Expression Glasses: A Wearable Device for Facial Expression Recognition. In *CHI'99 Extended Abstracts on Human Factors in Computing Systems*; ACM: New York, NY, USA, 1999; pp. 262–263. [CrossRef]
63. Hazlett, R. Measurement of User Frustration: A Biologic Approach. In *Proceedings of the CHI'03 Extended Abstracts on Human Factors in Computing Systems*; ACM: New York, NY, USA, 2003; pp. 734–735. [CrossRef]
64. San Agustin, J.; Hansen, J.P.; Hansen, D.W.; Skovsgaard, H. Low-cost gaze pointing and EMG clicking. In *CHI'09 Extended Abstracts on Human Factors in Computing Systems*; ACM: New York, NY, USA, 2009; pp. 3247–3252.

65. Rantanen, V.; Niemenlehto, P.H.; Verho, J.; Lekkala, J. Capacitive facial movement detection for human–computer interaction to click by frowning and lifting eyebrows. *Med. Biol. Eng. Comput.* **2010**, *48*, 39–47. [CrossRef] [PubMed]
66. Rantanen, V.; Venesvirta, H.; Spakov, O.; Verho, J.; Vetek, A.; Surakka, V.; Lekkala, J. Capacitive measurement of facial activity intensity. *IEEE Sens. J.* **2013**, *13*, 4329–4338. [CrossRef]
67. Matthies, D.J.; Strecker, B.A.; Urban, B. EarFieldSensing: A Novel In-Ear Electric Field Sensing to Enrich Wearable Gesture Input Through Facial Expressions. In *Proceedings of the 2017 CHI Conference on Human Factors in Computing Systems*; ACM: New York, NY, USA, 2017; pp. 1911–1922. [CrossRef]
68. Ekman, P.; Friesen, W.V. Detecting deception from the body or face. *J. Personal. Soc. Psychol.* **1974**, *29*, 288. [CrossRef]
69. Argyle, M. *Bodily Communication*; Routledge: Abingdon, UK, 2013.
70. Bull, P.E. *Posture & Gesture*; Elsevier: Amsterdam, The Netherlands, 2016; Volume 16.
71. Mehrabian, A.; Friar, J.T. Encoding of attitude by a seated communicator via posture and position cues. *J. Consult. Clin. Psychol.* **1969**, *33*, 330. [CrossRef]
72. Greene, S.; Thapliyal, H.; Caban-Holt, A. A Survey of Affective Computing for Stress Detection: Evaluating technologies in stress detection for better health. *IEEE Consum. Electron. Mag.* **2016**, *5*, 44–56. [CrossRef]
73. Wallbott, H.G. Bodily expression of emotion. *Eur. J. Soc. Psychol.* **1998**, *28*, 879–896. [CrossRef]
74. Ekman, P.; Friesen, W.V. Nonverbal leakage and clues to deception. *Psychiatry* **1969**, *32*, 88–106. [CrossRef] [PubMed]
75. Kleinsmith, A.; Bianchi-Berthouze, N.; Steed, A. Automatic recognition of non-acted affective postures. *IEEE Trans. Syst. Man Cybern. Part B (Cybernetics)* **2011**, *41*, 1027–1038. [CrossRef] [PubMed]
76. Kapur, A.; Kapur, A.; Virji-Babul, N.; Tzanetakis, G.; Driessen, P.F. Gesture-based affective computing on motion capture data. In *International Conference on Affective Computing and Intelligent Interaction*; Springer: Berlin/Heidelberg, Germany, 2005; pp. 1–7.
77. Arnrich, B.; Setz, C.; La Marca, R.; Tröster, G.; Ehlert, U. What does your chair know about your stress level? *IEEE Trans. Inf. Technol. Biomed.* **2009**, *14*, 207–214. [CrossRef] [PubMed]
78. Mota, S.; Picard, R.W. Automated posture analysis for detecting learner's interest level. In Proceedings of the 2003 Conference on Computer Vision and Pattern Recognition Workshop, Madison, WI, USA, 16–22 June 2003; Volume 5, p. 49.
79. Ohnishi, A.; Terada, T.; Tsukamoto, M. A Motion Recognition Method Using Foot Pressure Sensors. In Proceedings of the 9th Augmented Human International Conference, Seoul, Korea, 7–9 February 2018; p. 10.
80. Available online: http://www.sensingtex.com/ (accessed on 13 August 2019).
81. Shu, L.; Hua, T.; Wang, Y.; Li, Q.; Feng, D.D.; Tao, X. In-shoe plantar pressure measurement and analysis system based on fabric pressure sensing array. *IEEE Trans. Inf. Technol. Biomed.* **2010**, *14*, 767–775.
82. Kellis, E. Plantar pressure distribution during barefoot standing, walking and landing in preschool boys. *Gait Posture* **2001**, *14*, 92–97. [CrossRef]
83. Matthies, D.J.; Haescher, M.; Nanayakkara, S.; Bieber, G. Step Detection for Rollator Users with Smartwatches. In Proceedings of the Symposium on Spatial User Interaction, Berlin, Germany, 13–14 October 2018; pp. 163–167.
84. Telgarsky, R. Dominant frequency extraction. *arXiv* **2013**, arXiv:1306.0103.
85. The Most Comfortable and Accurate Wristband to Monitor Physiological Signals in Real-Time. 2019. Available online: https://e4.empatica.com/e4-wristband (accessed on 15 September 2019).
86. Hart, S.G.; Staveland, L.E. Development of NASA-TLX (Task Load Index): Results of empirical and theoretical research. In *Advances in Psychology*; Elsevier: Amsterdam, The Netherlands, 1988; Volume 52, pp. 139–183.
87. Stroop, J.R. Studies of interference in serial verbal reactions. *J. Exp. Psychol.* **1935**, *18*, 643. [CrossRef]
88. Jun, G.; Smitha, K.G. EEG based stress level identification. In Proceedings of the 2016 IEEE International Conference on Systems, Man, and Cybernetics (SMC), Budapest, Hungary, 9–12 October 2016; pp. 003270–003274.
89. Zhai, J.; Barreto, A. Stress Recognition Using Non-Invasive Technology. Available online: https://www.aaai.org/Papers/FLAIRS/2006/Flairs06-077.pdf (accessed on 18 May 2020).

90. Jo, N.Y.; Lee, K.C.; Lee, D.S. Computer-mediated task performance under stress and non-stress conditions: Emphasis on physiological approaches. In *Digital Creativity*; Springer: Cham, Switzerland, 2013; pp. 15–27.
91. Smeets, T.; Jelicic, M.; Merckelbach, H. Stress-induced cortisol responses, sex differences, and false recollections in a DRM paradigm. *Biol. Psychol.* **2006**, *72*, 164–172. [CrossRef]
92. Haescher, M.; Matthies, D.J.; Srinivasan, K.; Bieber, G. Mobile assisted living: Smartwatch-based fall risk assessment for elderly people. In *Proceedings of the 5th international Workshop on Sensor-Based Activity Recognition and Interaction*; ACM: New York, NY, USA, 2018; p. 6.
93. Gjoreski, H.; Lustrek, M.; Gams, M. Accelerometer placement for posture recognition and fall detection. In Proceedings of the 2011 Seventh International Conference on Intelligent Environments, Nottingham, UK, 25–28 July 2011; pp. 47–54.
94. Gjoreski, M.; Gjoreski, H.; Luštrek, M.; Gams, M. How accurately can your wrist device recognize daily activities and detect falls? *Sensors* **2016**, *16*, 800. [CrossRef]
95. Katsis, C.D.; Ganiatsas, G.; Fotiadis, D.I. An integrated telemedicine platform for the assessment of affective physiological states. *Diagn. Pathol.* **2006**, *1*, 16. [CrossRef]
96. Betti, S.; Lova, R.M.; Rovini, E.; Acerbi, G.; Santarelli, L.; Cabiati, M.; Del Ry, S.; Cavallo, F. Evaluation of an integrated system of wearable physiological sensors for stress monitoring in working environments by using biological markers. *IEEE Trans. Biomed. Eng.* **2017**, *65*, 1748–1758. [PubMed]
97. Akbar, F.; Mark, G.; Pavlidis, I.; Gutierrez-Osuna, R. An empirical study comparing unobtrusive physiological sensors for stress detection in computer work. *Sensors* **2019**, *19*, 3766. [CrossRef] [PubMed]
98. Zhang, H.; Zhang, H.; Pirbhulal, S.; Wu, W.; Albuquerque, V.H.C.D. Active Balancing Mechanism for Imbalanced Medical Data in Deep Learning–Based Classification Models. *ACM Trans. Multimed. Comput. Commun. Appl.* **2020**, *16*, 1–15. [CrossRef]
99. Sun, X.; Wang, S.; Xia, Y.; Zheng, W. Predictive-Trend-Aware Composition of Web Services with Time-Varying Quality-of-Service. *IEEE Access* **2019**, *8*, 1910–1921. [CrossRef]
100. Wu, W.; Pirbhulal, S.; Zhang, H.; Mukhopadhyay, S.C. Quantitative assessment for self-tracking of acute stress based on triangulation principle in a wearable sensor system. *IEEE J. Biomed. Health Inform.* **2018**, *23*, 703–713. [CrossRef] [PubMed]
101. Smith, L.; Hamer, M.; Ucci, M.; Marmot, A.; Gardner, B.; Sawyer, A.; Wardle, J.; Fisher, A. Weekday and weekend patterns of objectively measured sitting, standing, and stepping in a sample of office-based workers: The active buildings study. *BMC Public Health* **2015**, *15*, 9. [CrossRef]
102. Haescher, M.; Trimpop, J.; Matthies, D.J.; Bieber, G.; Urban, B.; Kirste, T. aHead: Considering the head position in a multi-sensory setup of wearables to recognize everyday activities with intelligent sensor fusions. In *International Conference on Human-Computer Interaction*; Springer: Cham, Switzerland, 2015; pp. 741–752.

sensors

MDPI

Article

Visual and Thermal Image Processing for Facial Specific Landmark Detection to Infer Emotions in a Child-Robot Interaction

Christiane Goulart [1,*], Carlos Valadão [2,*], Denis Delisle-Rodriguez [2,3], Douglas Funayama [4], Alvaro Favarato [5], Guilherme Baldo [6], Vinícius Binotte [2], Eliete Caldeira [6] and Teodiano Bastos-Filho [1,2,6]

1 Northeast Network of Biotechnology (RENORBIO), Postgraduate Program in Biotechnology, Health Sciences Center, Federal University of Espirito Santo (UFES), Av. Marechal Campos, 1468, Vitoria-ES 29043-900, Brazil
2 Postgraduate Program in Electrical Engineering, UFES, Av. Fernando Ferrari, 514, Vitoria-ES 29075-910, Brazil
3 Center of Medical Biophysics, University of Oriente, Patricio Lumumba s/n, Santiago de Cuba 90500, Cuba
4 Computer Engineering Department, UFES, Av. Fernando Ferrari, 514, Vitoria-ES 29075-910, Brazil
5 Mechanical Engineering Department, UFES, Av. Fernando Ferrari, 514, Vitoria-ES 29075-910, Brazil
6 Electrical Engineering Department, UFES, Av. Fernando Ferrari, 514, Vitoria-ES 29075-910, Brazil
* Correspondence: christiane.ufes@gmail.com (C.G.); carlostvaladao@gmail.com (C.V.); Tel.: +55-27-4009-2661 (C.G. & C.V.)

Received: 20 May 2019; Accepted: 22 June 2019; Published: 26 June 2019

Abstract: Child-Robot Interaction (CRI) has become increasingly addressed in research and applications. This work proposes a system for emotion recognition in children, recording facial images by both visual (RGB—red, green and blue) and Infrared Thermal Imaging (IRTI) cameras. For this purpose, the Viola-Jones algorithm is used on color images to detect facial regions of interest (ROIs), which are transferred to the thermal camera plane by multiplying a homography matrix obtained through the calibration process of the camera system. As a novelty, we propose to compute the error probability for each ROI located over thermal images, using a reference frame manually marked by a trained expert, in order to choose that ROI better placed according to the expert criteria. Then, this selected ROI is used to relocate the other ROIs, increasing the concordance with respect to the reference manual annotations. Afterwards, other methods for feature extraction, dimensionality reduction through Principal Component Analysis (PCA) and pattern classification by Linear Discriminant Analysis (LDA) are applied to infer emotions. The results show that our approach for ROI locations may track facial landmarks with significant low errors with respect to the traditional Viola-Jones algorithm. These ROIs have shown to be relevant for recognition of five emotions, specifically disgust, fear, happiness, sadness, and surprise, with our recognition system based on PCA and LDA achieving mean accuracy (ACC) and Kappa values of 85.75% and 81.84%, respectively. As a second stage, the proposed recognition system was trained with a dataset of thermal images, collected on 28 typically developing children, in order to infer one of five basic emotions (disgust, fear, happiness, sadness, and surprise) during a child-robot interaction. The results show that our system can be integrated to a social robot to infer child emotions during a child-robot interaction.

Keywords: Viola-Jones; facial emotion recognition; facial expression recognition; facial detection; facial landmarks; infrared thermal imaging; homography matrix; socially assistive robot

1. Introduction

Child-Robot Interaction (CRI) is a subfield of Human-Robot Interaction (HRI) [1], which is defined as the interaction between humans and robotic systems. Inside the several possibilities of HRI and CRI, socially assistive robots are being used as a therapy-aid tool for children with Autism [2,3]. One feature that could improve this interaction is the ability of recognizing emotions, which can be used to provide a better CRI. For instance, children with autism spectrum disorder (ASD) tend to lack the ability of emotion display, thus, the robots should rely on involuntary biological signals measurements, such as skin thermography [4–6].

The face is a region of the body that has a high response to emotions, and the facial thermal print changes may be linked to the child emotion. Thus, this feature can be a useful parameter to be applied in a CRI, since this biological signal is not voluntary and not easily mutable [7]. Due to this feature, recent studies are focused on facial detection and thermography to evaluate emotion expressions in affective computing [8–10]. Moreover, it is a more comfortable and unobtrusive technique to evaluate emotions, since no sensor touching the child is needed, such as electrodes used in electroencephalography and electrocardiography [8,11,12].

A conventional system for facial emotion recognition is composed of the following three main stages: face and facial component detection, computation of various spatial and temporal features, and emotion classification [10]. Then, the first stage for face detection over an input image, and consequently to locate facial components (such as eyes, nose, and mouth) or landmarks of interest, is a crucial task and still a challenge. In fact, to accurately discriminate emotions, it is necessary to apply geometric or appearance features based methods [10,13–15], being the latter the most popular, due to its superior performance [16]. On the other hand, Facial Landmarks (FL) should be used to locate salient points of facial regions, such as the end of the nose, ends of the eye brows, and the mouth [10,16].

Many studies have demonstrated that dividing the face into specific regions for facial feature extraction can improve the performance during emotion recognition [17–29]. However, this strategy may be affected by improper face alignment. Moreover, other works based on learning [30,31] for feature extraction from specific face regions have been proposed to locate those facial regions with higher contribution for emotion recognition. Nevertheless, these approaches are difficult to be extended as a generic system, due to the fact that positions and sizes of the facial patches vary according to the training data.

It is worth commenting that studies using thermal cameras for emotion recognition have shown promising results, but low-cost thermal cameras typically present a poor resolution, making it difficult to accurately detect facial regions by applying conventional methods as the Viola-Jones algorithm, widely used on visual images [15,32].

Then, we hypothesized that a low-cost system for simultaneous capture of both visual and thermal cameras may increase the accuracy for locations of specific facial regions of interest (ROIs) over faces, and consequently improve the feature extraction, increasing the emotion discrimination. This way, we consider an alternative few-explored, which is to firstly apply Viola-Jones algorithm on the visual image to locate desired ROIs, and after transferring it for its corresponding thermal image, but including as a last stage a method for ROI location correction based on error probability, taking into account manual annotations of a trained expert over a reference frame.

Thus, the goal of this work is to propose a system able to detect facial ROIs for five emotions (disgust, fear, happiness, sadness, and surprise) in typically developing children during an interaction with a social robot (as an affective stimulus). In this study, our low-cost camera system allows obtaining pairs of synchronized images for detecting ROIs in the visual image using the Viola-Jones algorithm as a first stage, and then, transferring these ROIs to the corresponding thermal camera frame through a homography matrix. As the main novelty, we introduced here a new way to accurately improve the ROI locations after applying both Viola-Jones and homography transform. This approach computes the error probability to automatically find that ROI located over thermal images, which is better placed according to manual annotations of a trained expert. This ROI of highest probability (with lowest location error) is latter used to relocate other ROIs, improving the overall accuracy. Then, better appearance features can be extracted, in order to increase the emotion discrimination by our proposed recognition system. Similarly, this method may be extended to other studies aiming to accurately locate ROIs over facial thermal images, which are physiologically relevant, such as described in [17,21], allowing to understand phenomena linked to behaviours, emotions, stress, human interactions, among others. As a relevance of this work, our system is capable of detecting ROIs on the child's face, which has neurophysiological importance for emotion recognition through thermal images recorded in an unobtrusive way. Additionally, methods for feature extraction and dimensionality reduction are applied on specific ROIs for emotion recognition using Linear Discriminant Analysis (LDA). As another highlight, a set of visual and thermal images is acquired in an atypical context in which a social robot is used as an emotional stimulus in an interaction with children, in order to test the proposed system for specific ROIs detection and emotion recognition. For our knowledge, this type of approach has not been explored in other studies.

This work is structured as follows. Section 2 presents a description of several works of the state-of-the-art. Section 3 presents a system for image acquisition, in addition to a proposal based on the Viola-Jones algorithm and error probability for facial ROI location. Moreover, the experimental protocol and methods for feature extraction, dimensionality reduction, and classification are described. Section 4 presents the experimental findings about the automatic method for ROI placement and children's emotion recognition during the interaction with the robot. Afterwards, Section 5 presents the findings of this work and compare them to previous studies, summarizing also its main contributions and limitations. Finally, Section 6 presents the Conclusion and Future Works.

2. Related Works

Many research to recognize facial emotion by contact-free strategies have proposed automatic methods for both face and facial ROI detection over visual and thermal images, as constructing an effective face representation from images is a crucial step for successful automatic facial action analysis, in order to recognize facial emotions. In this field, there is the Facial Action Coding System (FACS), which is a taxonomy of human facial expressions designed to facilitate human annotation of facial behaviour [9,14,33]. For instance, a total of 32 atomic facial muscle actions, termed Action Units (AUs), and 14 additional descriptors related to miscellaneous actions are specified, which are widely used by automatic methods to locate facial landmarks and ROIs. These regions are used by methods based on geometric [9,10,15] and appearance features to discriminate emotions [9,14]. Appearance representations use textural information by considering the intensity value of the pixels, whereas geometric representations ignore texture and describe shape explicitly [9,14,15]. Here, we focused our revision of the state-of-the-art on approaches using only appearance features on the target face, which are generally computed by dividing the face region into regular grid (holistic representation). Appearance features can be obtained to encode low or high-level information. For example, low-level information can be encoded through low-level histograms that are computationally simple and ideal for real-time applications, Gabor representations, data-driven representations by applying bag-of-words, among others. Furthermore, higher level of information can be encoded through Non-Negative Matrix Factorization (NMF) [9]. However, the effectiveness of the feature extraction to increase the emotion discrimination may be affected by several factors, such as head-pose variations, illumination variations, face registration, occlusions, among others [9].

In [16] the authors used the Haar classifier for face detection, which is widely applied, due to its high detection accuracy and real time performance [32]. They extracted appearance features from the global face region by applying Local Binary Pattern (LBP) histogram that take care of minor changes of facial expression for different emotions [9,34], followed by Principal Component Analysis (PCA) for dimensionality reduction, to improve the speed of computation in real time during six emotions (anger, disgust, fear, happiness, sadness, and surprise). This approach is customizable person to person, and achieved an accuracy (ACC) of 97%. It is worth mentioning that unlike a global-feature-based approach, different face regions have different levels of importance for emotion recognition [17]. For example, the eyes and mouth contain more information than the forehead and cheek. Notice that LBP has been widely used in many research of emotion recognition. Refer to Ref. [34] for a comprehensive study about methods based on LBP for emotion recognition.

Another study [14] used specific regions for appearance feature extraction by dividing the entire face region into domain-specific local regions, using the landmark detection method presented in Ref. [35] that uses ensemble of regression trees. These authors used facial point locations to define a set of 29 face regions covering the whole face, which was based on expert knowledge regarding face geometry and AU-specific facial muscle contractions, such as shown in Ref. [33]. Ensemble of regression trees are used to estimate the face landmark locations directly from a sparse subset of pixel intensities, achieving super-real-time performance with high quality predictions. Similarly, they used LPB descriptor for appearance feature extraction, achieving an ACC of 93.60% after applying Support Vector Machine (SVM) with Radial Basic Function (RBF) kernel.

In Ref. [36], a comparative study of methods for feature extraction, such as Kernel Discriminant Isometric Mapping (KDIsomap), PCA, Linear Discriminant Analysis (LDA), Kernel Principal Component Analysis (KPCA), Kernel Linear Discriminant Analysis (KLDA), and Kernel Isometric Mapping (KIsomap) was conducted, achieving the best performance (ACC of 81.59% on the JAFFE database, and 94.88% on the Cohn-Kanade database) for KDIsomap during seven emotions (anger, joy, sadness, neutral, surprise, disgust and fear), but without significant difference compared with other approaches. Here, the authors used the well-known Viola-Jones algorithm to detect the face [32], which is suitable for real-time applications. This method uses a cascade of classifiers by employing Haar-wavelet features, which usually use the eye position detected in the face region to align the other detected face regions.

In Ref. [37] the authors propose the Central Symmetric Local Gradient Coding (CS-LGC) algorithm to define the neighborhood as a 5×5 grid, using the concept of center symmetry to extract the gradient information in four directions (horizontal, vertical, and two diagonals) for feature extraction over target pixels more representative. Afterwards, they also applied PCA for dimensionality reduction, followed by the Extreme Learning Machine (ELM) algorithm. The evaluation of this approach was conducted through JAFFE and Cohn-Kanade databases, which contain grayscale visual images related to the following emotions: anger, disgust, fear, happiness, neutral, sadness and surprise. Accuracies of 98.33% and 95.24% for Cohn-Kanade and JAFFE were obtained, respectively, being relatively better compared with other operators for feature extraction, such as LBP.

Several studies for emotion recognition have been conducted with two kinds of camera (one visual and another infrared), such as in Ref. [20]. Those authors proposed a fusion scheme by applying PCA over both thermal and visual faces for feature extraction, and k-nearest neighbors to recognize two classes (surprised and laughing) with mean ACC of 75%. Additionally, in Ref. [23] a comparison for emotion recognition using visual and infrared cameras was carried out and four typical methods, including PCA, PCA plus LDA, Active Appearance Model (AAM), and AAM-based plus LDA were used on visual images for feature extraction, whereas PCA and PCA plus LDA were applied on infrared thermal images using four ROIs (forehead, nose, mouth, and cheeks). These authors used k-nearest neighbors to recognize six emotions (sadness, anger, surprise, fear, happiness, and disgust). It is worth mentioning that the eye locations over the thermal were manually performed for those authors, which latter were used to locate the aforementioned four ROIs. In Ref. [22] an interesting approach using both kinds of camera was addressed, including the use of eyeglasses, which are opaque to the thermal camera, but visible to the visual camera.

Another interesting work shows that infrared and visual cameras can be combined into a multi-modal sensor system to recognize fear [24], through electroencephalogram (EEG) signals, eye blinking rate, and facial temperature while the user watched a horror movie. An Adaptive Boosting (AdaBoost) algorithm was used to detect the face region, and a geometric transform to make the coordinates of the two images (visible-light and thermal) coincident was used. Similarly, other study was conducted on Post traumatic Stress disorder (PTSD) patients to infer fear through visual and thermal images [38]. In Ref. [39], an algorithm for automatic determination of the head center in thermograms was proposed, which has demonstrated to be sensitive to the head rotation or position. In Ref. [40], the authors proposed an unsupervised Local and Global feature extraction for facial emotion recognition through thermal images. For this purpose, they used a bimodal threshold to locate the face for feature extraction by PCA, after applying a method based on clustering to detect points of interest; for facial expression classification, a Support Vector Machine Committee was used. In Ref. [17], the face was extracted on thermal images after applying both median and Gaussian filters with further binarization to convert the gray scale image into pure black and white, and removing small sets of non-connected pixels to enhance the image quality. Afterwards, appearance features were extracted on defined ROIs over the thermal images, followed by Fast Neighbourhood Component Analysis (FNCA) and LDA for feature selection and recognition of five emotions, respectively.

More details about different methods for feature extraction, dimensionality reduction, feature selection, and classification can be reviewed in some studies [37] and also in extensive reviews, such as in Refs. [9,10].

The next section presents our proposed system for five emotions recognition, which allows accurately locating facial ROIs over thermal images, improving the appearance feature extraction.

3. Materials and Methods

3.1. Experimental Procedure

Seventeen typically developing children, 9 boys and 8 girls (aged between 8 and 12 years) participated in this study, who were recruited from elementary schools in Vitoria-Brazil. All had their parents' permission, through signatures of Terms of Free and Informed Consent. In addition, children signed a Term of Assent, informing their wish in participating. This study was approved by the Ethics Committee of Federal University of Espirito Santo (UFES)/Brazil, under number 1,121,638. The experiments were conducted in a room within the children's school environment, where the room temperature was kept between 20 °C and 24 °C, using a constant luminous intensity, such as done by Ref. [39].

A mobile social robot (see Figure 1b), called N-MARIA (New-Mobile Autonomous Robot for Interaction with Autistics), built at UFES/Brazil to assist children during social relationship rehabilitation, was used in our research. This robot has attached a camera system to record facial images during interaction with children. More details about N-MARIA are given in Section 3.2.1.

The experiment was conducted in three phases, as follows. First, N-MARIA was initially covered at the room with a black sheet, except its attached camera system, that was turned on to record visual and thermal images of the frontal view with sampling rate at 2 fps, for further processing. Afterwards, the child was invited to enter the room, and sit comfortably for explanations about the general activities related to the experiment, being conditioned to a relaxed state for a period of time minimum of 10 min, in order to adapt her/his body to the temperature of the room, allowing her/his skin temperature to stabilize for baseline recordings, according to similar studies carried out in Refs. [21,41]. Once they had completed the relaxation period, the child was placed in front of the covered robot about 70 cm away from it, remaining in standing position. Immediately, recordings of the child face by the camera system were carried out for a period of one minute with the robot covered, one minute with the robot uncovered, and three minutes of interaction with the robot. After, the child spent two minutes answering a questionnaire about the experiment.

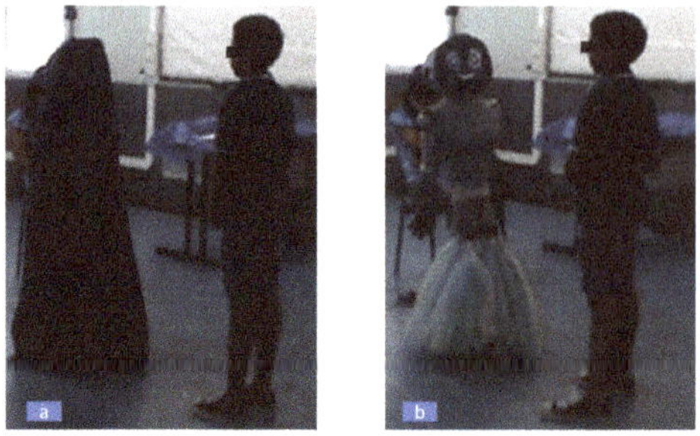

Figure 1. Experimental setup showing the child-robot interaction. (**a**) Before showing the robot; (**b**) After presenting it.

The first part of the recording (robot covered) corresponds to the experimental stage, called Baseline, whereas the next stage, presenting the uncovered robot is called Test. Before the robot is uncovered, the child was asked to permanently look forward without sudden facial movements or touching the face, avoiding any facial obstruction during video recordings.

After the removal of the black sheet that covered the robot, the first dialogue (self-presentation) of the robot was started. In addition to the self-introduction to the child, prompt dialogues during the experiment were related to questions, positive reinforcement and invitations. In the interaction with the child, that lasted two minutes, the child was encouraged to make communication and tactile interaction with the robot. At the end of the experiment, the child was again invited to sit and answer a structured interview about her/his feelings before and after seeing the robot, and also about the robot structure (if the child liked it, what she/he liked more, and what the child would change about it).

3.2. Contact-Free Emotion Recognition

Figure 2 shows the proposed contact-free system for emotion recognition, which is composed of the following four steps: (a) camera calibration; (b) image acquisition and automatic ROI detection; (c) ROI replacement; (d) feature extraction followed by the dimensionality reduction and emotion classification.

Figure 2a shows a first stage to calibrate the camera system by obtaining a homography matrix to map the pixels of the visual camera image into the thermal camera image, considering the relative fixed position between the two cameras. Also, another process is performed to obtain a frame that contains intrinsic noise of the infrared sensor, which is latter used in a second stage (Figure 2b) to remove the sensor noise (inherent to the camera) over the current thermal image captured. In this second stage, the image acquisition process is carried out taking synchronous images from both visual and infrared cameras, which are pre-processed to enhance the automatic facial ROIs detection by applying the Viola-Jones algorithm on the visual image. Then, the ROIs placed on the visual image are projected into the thermal image using the homography matrix. As a third stage, manual annotations by a trained expert over a reference frame are used to accurately relocate the ROIs by applying our approach based on errors of probability, such as shown in Figure 2c. Afterwards, feature vectors related to thermal variations are computed on the detected ROIs, and after reduced by applying PCA for dimensionality reduction for five emotion recognition in a last stage by LDA. More details about the proposed recognition system are given in the next subsections.

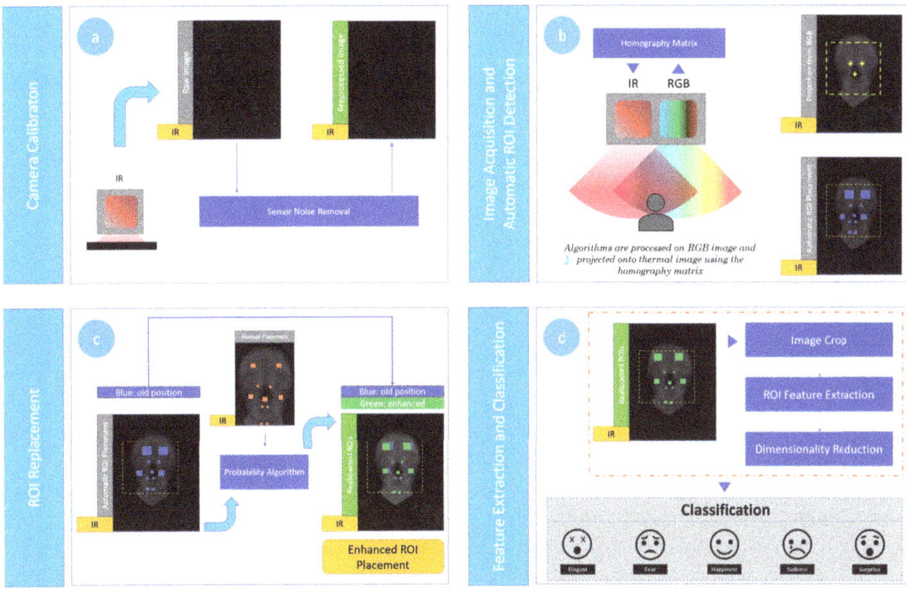

Figure 2. Overview of the proposed system for emotion recognition during a child-robot interaction. (**a**) Camera calibration; (**b**) image acquisition and automatic region of interest (ROI) detection; (**c**) ROI replacement; (**d**) feature extraction followed by the dimensionality reduction and emotion classification.

3.2.1. Camera System and N-MARIA

The camera system composed of both visual and thermal cameras was attached to the head of the social robot, such as shown in Figure 3. These two cameras were fixed so that both had approximately the same visual field. To capture thermal variations, a low-cost camera (Therm-App®) was used, which has spatial resolution of 384 × 288 ppi, frame rate at 8.7 Hz and temperature sensitivity <0.07 °C. The normalization of the thermal images acquired in gray scale consisted of a brightness rate ranging from 0 to 255, where darker pixels correspond to lower temperatures, and lighter pixels correspond to higher temperatures. Moreover, a C270 HD Webcam (Logitech) was used to obtain visual images in RGB format, with a resolution of 1.2 MP.

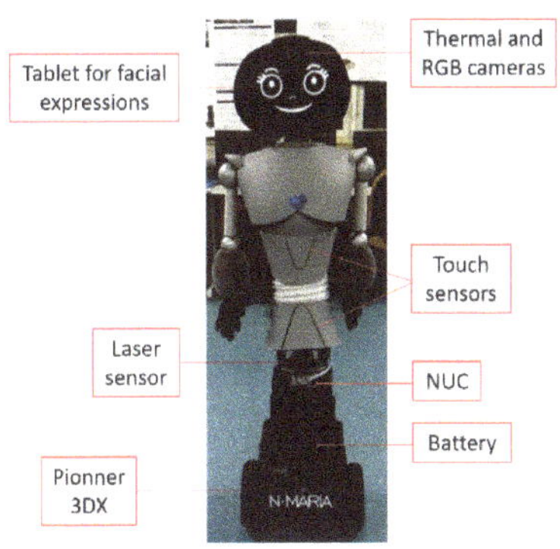

Figure 3. N-MARIA (New-Mobile Robot for Interaction with Autistics) developed at Federal University of Espirito Santo (UFES)/Brazil.

The robot was built 1.41 m tall, considering the standard height of 9–10-year-old children. Additionally, soft malleable materials were used on the robot's structure for protection of both children and internal robot devices. The Pioneer 3-DX mobile platform was responsible by locomotion, a 360° laser sensor was used to locate the child in the environment, and a tablet was used as the robot face to display seven dynamic facial expressions during the robot-child interaction. Those expressions could also be remotely controlled through another tablet by the therapist, who could also control the robot behavior, expressions and dialogues emitted by the speakers.

3.2.2. Camera Calibration

The camera calibration is done through a synchronous acquisition between visual and thermal images, and using a chessboard built with aluminum and electrical tape positioned in several possible angles. Then, the images obtained are processed with OpenCV calibration software [42], which uses Direct Linear Transform (DLT) to return a homography matrix [43], allowing transformation of points from the visual image to the thermal image in a robust way [32]. It is worth mentioning that there is not a homography matrix matching exactly points in all regions of the face (as they are not in the same plane), but the matrix obtained by DLT is used as an efficient approximation.

Also, other procedure to remove the intrinsic thermal noise of infrared sensors is carried out [44], which increases the quality of thermal images by correcting undesirable offsets. For that, a reference of an object with uniform body temperature covering the visual field of the thermal camera is recorded, which contains the intrinsic noise of the infrared sensor. Thus, it was expected to have a frame with the same brightness for all pixels, however, that did not occur. Thus, the frame with the intrinsic sensor noise was used in the pre-processing stage to eliminate the thermal noise, such as described in the next section.

3.2.3. Image Acquisition and Pre-Processing

The thermal camera has maximum acquisition capacity of 8.7 fps whereas the visual camera has maximum capacity of 30 fps. Thus, to obtain temporal consistency, both visual and thermal images were simultaneously recorded with a sampling rate of 2 fps, which was suitable for our purposes.

During acquisition, the frame with the intrinsic sensor noise obtained in the Calibration stage was used to remove the intrinsic thermal noise of the current image acquired by subtracting pixel to pixel. Finally, a median filter was used to reduce salt and pepper noise from the thermal image.

3.2.4. Face Landmark Detection

An automatic method was proposed here for face landmark detection over a given set of frames ($\mathbf{I} = \{\mathbf{i}_1, \mathbf{i}_2, \ldots, \mathbf{i}_b, \ldots, \mathbf{i}_B\}$), taking as reference annotated ROIs by a trained expert on the frame $\mathbf{i}_A (b = A)$, such as shown in Figure 4. It is possible to observe that these manual annotations were located on eleven ROIs ($\mathbf{R}_A = \{\mathbf{R}_A^1, \mathbf{R}_A^2, \ldots, \mathbf{R}_A^k, \ldots, \mathbf{R}_A^{11}\}$) of thermal images, taking into account the relevance of these ROIs in other studies for facial emotion recognition [17,21]. Here, the facial ROI sizes were computed in the same way as in Refs. [17,18,21], using the width of the head ROI and the following defined proportions [18]: 6.49% for nose, 14.28% for forehead, 3.24% for periorbital region, 9.74% for cheek, 3.24% for perinasal region, and 5.19% for chin [17].

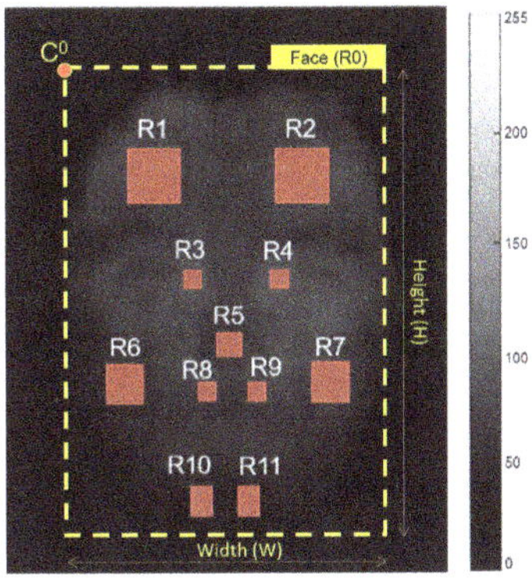

Figure 4. Facial ROIs. $\mathbf{R}^1$, right forehead side; $\mathbf{R}^2$, left forehead side; $\mathbf{R}^3$, right periorbital side; $\mathbf{R}^4$, left periorbital side; $\mathbf{R}^5$, tip of nose; $\mathbf{R}^6$, right cheek; $\mathbf{R}^7$, left cheek; $\mathbf{R}^8$, right perinasal side; $\mathbf{R}^9$, left perinasal side; $\mathbf{R}^{10}$, right chin side; $\mathbf{R}^{11}$, left chin side.

3.2.5. Automatic ROI Detection

Infrared images are more blurred than color images [23], therefore the ROI detection over thermal images of low-cost cameras is a challenge. For this reason, the well-known Viola-Jones algorithm [32] was used on color images for head detection and other facial regions, such as nose and eyes [17,21,32]. Then, these initial detected regions were used as references to automatically locate eleven ROIs within the face (see Table 1), namely the nose, both sides of forehead, cheeks, chin, periorbital area (close to the eyes) and perinasal area (bottom of the nose).

Table 1. Reference ROIs used to locate face landmarks over a frame $\mathbf{i}_b$.

Reference ROIs	Located ROIs
Head	$\mathbf{R}_b^1$, $\mathbf{R}_b^2$, $\mathbf{R}_b^{10}$, $\mathbf{R}_b^{11}$
Eyes	$\mathbf{R}_b^1$, $\mathbf{R}_b^2$, $\mathbf{R}_b^6$, $\mathbf{R}_b^7$, $\mathbf{R}_b^8$, $\mathbf{R}_b^9$
Nose	$\mathbf{R}_b^3$, $\mathbf{R}_b^4$, $\mathbf{R}_b^5$, $\mathbf{R}_b^{10}$, $\mathbf{R}_b^{11}$

$\mathbf{R}_b^1$, right forehead side; $\mathbf{R}_b^2$, left forehead side; $\mathbf{R}_b^3$, right periorbital side; $\mathbf{R}_b^4$, left periorbital side; $\mathbf{R}_b^5$, tip of nose; $\mathbf{R}_b^6$, right cheek; $\mathbf{R}_b^7$, left cheek; $\mathbf{R}_b^8$, right perinasal side; $\mathbf{R}_b^9$, left perinasal side; $\mathbf{R}_b^{10}$, right chin side; $\mathbf{R}_b^{11}$, left chin side.

In our study, the facial ROI sizes were also computed using the width of the head, and the aforementioned proportions [17,18,21]. Additionally, the facial ROIs were spatially placed, taking as reference the expert annotation. Afterwards, the corresponding facial ROIs were projected on the thermal image through the aforementioned transformation using a homography matrix (see Section 3.2.2), such as shown in Figure 4. Here, the ROI set of a thermal frame b is defined by $\mathbf{R}_b = \{\mathbf{R}_b^0, \mathbf{R}_b^1, \ldots, \mathbf{R}_b^k, \ldots, \mathbf{R}_b^{11}\}$, being $\mathbf{R}_b^0$ the head ROI, and $\mathbf{R}_b^k$ for $k = 1$ to 11 the facial ROIs. Notice that $\mathbf{R}_b^k$ is described by several pixels R_{ij} for a range from 0 to 255 (gray scale of 8 bits).

Figure 2b,c show our proposal to accurately locate facial ROIs, formed by the following two-stages: (1) automatic ROI detection and (2) ROI placement correction.

3.2.6. ROI Location Correction

A new method is proposed here to correct with accuracy the detected ROIs by the Viola-Jones algorithm, taking into account all pre-defined ROIs positions, which were manually annotated on a first frame by a trained expert.

Let $\mathbf{R}_b^k$ be an automatic detected ROI over the thermal frame $\mathbf{i}_b$ that presents a coordinate $\mathbf{C}_b^k = (C_{bx}^k, C_{by}^k)$ at the left upper corner, which corresponds to $\mathbf{R}_A^k$ (annotated ROI over $\mathbf{i}_A$) with coordinate $\mathbf{C}_A^k = (C_{Ax}^k, C_{Ay}^k)$ at the left upper corner too. Then, two probabilities p_{bx}^k and p_{by}^k can be calculated for $\mathbf{R}_b^k$, taking into account the expert annotation, such as described in Equations (1) and (2). These p_{bx}^k and p_{by}^k values are computed in relation to **x** and **y**, respectively. They take values closer to 1 if $\mathbf{R}_b^k$ location highly agrees with the manual annotation of the trained expert, which are shown in Equations (1) and (2). Notice that $\mathbf{R}_b^k$ is automatically obtained by applying the Viola-Jones algorithm, fixing defined proportions (see Section 3.2.4).

$$p_{bx}^k = \frac{\exp\left(-\left|\frac{C_{bx}^k}{W_b} - \frac{C_{Ax}^k}{W_A}\right|\right)}{\sum_{i=1}^{11} \exp\left(-\left|\frac{C_{bx}^i}{W_b} - \frac{C_{Ax}^i}{W_A}\right|\right)}, \tag{1}$$

$$p_{by}^k = \frac{\exp\left(-\left|\frac{C_{by}^k}{H_b} - \frac{C_{Ay}^k}{H_A}\right|\right)}{\sum_{i=1}^{11} \exp\left(-\left|\frac{C_{by}^i}{H_b} - \frac{C_{Ay}^i}{H_A}\right|\right)}, \tag{2}$$

$$p_b^k = \min\left\{p_{bx}^k, p_{by}^k\right\}, \tag{3}$$

where k refers to the current facial ROI for analysis, taking values from 1 to 11; W_b and H_b are the width and height of $\mathbf{R}_b^0$ (head ROI for $\mathbf{i}_b$), respectively; W_A and H_A are the width and height of $\mathbf{R}_A^0$ (head ROI for $\mathbf{i}_A$); p_{bx}^k and p_{by}^k are the probabilities that $\mathbf{R}_b^k$ were correctly located on $\mathbf{i}_b$ regarding the trained expert annotation, in relation to x and y axes, respectively.

Finally, $\mathbf{R}_b^k$ (for which $\mathbf{C}_b^k$ is denoted as $\mathbf{C}_b^{ref}$) of lower probability p_b^k is selected as a reference to correct the location for the other ROIs, using Equations (4) and (5). It is worth mentioning that $\mathbf{C}_A^{ref}$ notation is used for the annotated frame $\mathbf{i}_A$.

$$C_{bx}^{k'} = \frac{C_{bx}^{ref} + (C_{Ax}^k - C_{Ax}^{ref})}{W}, \tag{4}$$

$$C_{by}^{k'} = \frac{C_{by}^{ref} + (C_{Ay}^k - C_{Ay}^{ref})}{H}, \tag{5}$$

where $\mathbf{C}_b^{k'} = (C_{bx}^{k'}, C_{by}^{k'})$ is the coordinate of the left upper corner for $\mathbf{R}_b^k$ relocated.

3.2.7. Feature Extraction

Given a thermal frame $\mathbf{i}_b$ formed by a set of ROIs, $\mathbf{R}_b = \{\mathbf{R}_b^1, \mathbf{R}_b^2, \ldots, \mathbf{R}_b^k, \ldots, \mathbf{R}_b^K\}$, being $K = 11$ the total number of ROIs, it is possible to extract from $\mathbf{R}_b$ a feature vector $\mathbf{F}_b = \{\mathbf{f}_b^1, \mathbf{f}_b^2, \ldots, \mathbf{f}_b^k, \ldots, \mathbf{f}_b^K\}$ that describes a pattern related to an emotion, being $\mathbf{f}_b^k = \{f_{b1}^k, f_{b2}^k, \ldots, f_{b14}^k\}$ features of $\mathbf{R}_b^k$. Table 2 presents the features adopted in our study, which agree with [17].

$\mathbf{R}_b^k$ is the current ROI for feature extraction, $\overline{\mathbf{R}_b^k}$ is the average value of $\mathbf{R}_b^k$, σ_b^{2k} is the variance of $\mathbf{R}_b^k$, and $f_{b(c+7)}^k$ for c equal 1 to 7 are other seven features corresponding to the difference of computed features throughout consecutive frames. So we have eleven ROIs (see Figure 4), and 14 features for each of them. This gives a set of 154 features per frame.

Table 2. Features computed in each ROI.

Features	Equations
1. Mean value of the whole ROI	$f_{bc}^{k} = \mathbf{R}_{b}^{k} = \frac{1}{m \cdot n} \sum_{i=1}^{m} \sum_{j=1}^{n} R_{ij}\ , c = 1$
2. Variance form the whole ROI, organized in a vector	$f_{bc}^{k} = \sigma_{b}^{2k} = \frac{1}{(m \cdot n)-1} \sum_{i=1}^{m} \sum_{j=1}^{n} \left(R_{ij} - \overline{\mathbf{R}_{b}^{k}}\right)^{2}, c = 2$
3. Median of the whole ROI, organized as a vector	$f_{bc}^{k} = median(\overline{\mathbf{R}_{b}^{k}})\ , c = 3$
4. Mean of variance values in rows	$f_{bc}^{k} = \frac{1}{m} \sum_{i=1}^{m} \frac{1}{n-1} \sum_{j=1}^{n} \left(R_{ij} - \overline{\mathbf{R}_{bi}^{k}}\right)^{2}, c = 4$
5. Mean of median values in rows	$f_{bc}^{k} = \frac{1}{m} \sum_{i=1}^{m} median(\mathbf{R}_{bi}^{k})\ , c = 5$
6. Mean of variance values in columns	$f_{bc}^{k} = \frac{1}{n} \sum_{j=1}^{n} \frac{1}{m-1} \sum_{i=1}^{m} \left(R_{ij} - \overline{\mathbf{R}_{bj}^{k}}\right)^{2}, c = 6$
7. Mean of median values in columns	$f_{bc}^{k} = \frac{1}{n} \sum_{j=1}^{n} median(\mathbf{R}_{bj}^{k})\ , c = 7$
8. Difference from each item in consecutive frames	$f_{b(c+7)}^{k} = f_{bc}^{k} - f_{(b-1)c}^{k}\ , c = 1, 2, \ldots, 7$

3.2.8. Dimensionality Reduction and Emotion Classification

Let $\mathbf{T} = (\mathbf{F}_1, y_1), (\mathbf{F}_2, y_2), \ldots, (\mathbf{F}_b, y_b), \ldots, (\mathbf{F}_n, y_n)$ be the training set, where n is the number of samples, and $\mathbf{F}_i$ is a d-dimensional feature vector with class label $y_b \in 1, 2, \ldots, 5$. PCA method based on Single Value Decomposition [9,16,20,23,36] is applied on $\mathbf{F}_i$ to obtain the principal component coefficients, which are used in both training and validation sets to reduce the set of 154 features, in order to allow a robust and fast emotion recognition. As advantages, PCA is few sensitive to different training sets, and it can outperform other methods as LDA when the training set is small [45]. This method has been successfully used in many studies to represent, in lower dimensional subspace, the high dimensional feature vectors, which are obtained by applying appearance-based methods [16,23]. It is worth mentioning that before applying PCA in our study, the feature vectors of the training set were normalized using both mean and standard deviation values as reference. Then, the validation was normalized using the same reference values (mean and standard deviation) obtained from the training set.

Some classifiers, such as LDA [17,23,46] and Quadratic Discriminant Analysis (QDA) [12,47] by applying full and diagonal covariance matrices, as well as other three classifiers, such as Mahalanobis discrimination [12,48], Naives Bayes [49], and Linear Support Vector Machine (LSVM) [12,14,50] are used in our study to assign objects to one of several emotion classes based on a feature set.

3.3. Statistical Evaluation

From images recorded during both moments of the experiment (Baseline and Test), a set of 220 thermography frames randomly selected from 11 children was annotated by a trained expert, selecting the ROIs defined in Figure 4. These annotated images were used as reference to evaluate, through Euclidean distances (see Equation (6)), the accuracy and precision of both Viola-Jones without ROI relocation, and Viola-Jones applying our ROI relocation algorithm.

$$D = \sqrt{(A_x - M_x)^2 + (A_y - M_y)^2}, \tag{6}$$

where (A_x, A_y) is the coordinate obtained by the automatic method, and (M_x, M_y) is the coordinate obtained by the manual method. When D is close to zero, that means high accuracy.

The statistical analysis used for comparison between both approaches for each ROI was the Wilcoxon Signed Rank Test for zero median.

In order to evaluate our proposed system for emotion recognition, a published database (available in the supporting information of [17] at the website —Available at https://journals.plos.org/plosone/article?id=10.1371/journal.pone.0212928) was used, which is formed by feature vectors labeled as the following five emotions: disgust, fear, happiness, sadness and surprise. This database was collected on 28 typically developing children (age: 7–11 years) by an infrared thermal camera [17]. It is worth noting that this database was also created with children of similar age range, using the same thermal camera and feature set (a total of 154 features) described in our study (see Sections 3.1, 3.2.2 and 3.2.7), and computing this feature set over the ROIs defined on Figure 4. Notice that the correct locations of these ROIs were visually inspected by a trained expert. For this reason, it was possible to compare the recognition system using one of the following methods: PCA for dimensionality reduction and Fast Neighbor Component Analysis (FNCA) [17] for feature selection. Here, the training and validation sets were chosen for several runs of cross-validation ($kfold = 3$), and metrics such as accuracy (ACC), Kappa, true positive rate (TPR), and false positive rate (FPR) were used [51].

On the other hand, this published database was used to train our proposed system based on PCA, but only using data collected from those children that presented ACC higher than 85% during the emotion recognition [17]. Then, our trained system was used to infer the children emotion during our experimental protocol described in Section 3.1.

4. Results

4.1. Automatic ROI Location

The performance of the proposed method using trained expert criteria and error probability to accurately detect face landmarks was validated on 11 children through two sets of 220 annotated thermal frames by the trained expert, each one obtained for the following two conditions: (1) Baseline, (2) Test (see Section 3.1). Figure 5a,b shows that our proposal significantly improved the ROI placements regarding trained expert criteria. For both conditions, the proposed method closely agreed (errors lower than 10 pixels) with the trained expert, although the children trended to make abrupt face movements for the second condition, as they may have got surprised by the uncovered robot. However, the Viola-Jones algorithm, non-assisted by the error probability, significantly disagreed with the trained expert, such as shown in Figure 5a,b. Notice that Viola-Jones presented the highest error locating both $\mathbf{R}^{10}$ and $\mathbf{R}^{11}$, which corresponds to the right and left chin sides, respectively. This undesirable mistakes may have been caused by mouth movements during talking or by facial expressions, such as happiness and surprise. As a highlight, our proposal takes into account the best located ROI, $\mathbf{R}^1$ or $\mathbf{R}^5$ for Baseline, and $\mathbf{R}^3$ for Test, which reduced the location error as much as possible, such as shown in Figure 5a,b, respectively. It is worth mentioning that our approach for ROI replacements uses the probability error to select (taking into account the manual annotations of a trained expert) that ROI better located during the first stage for automatic ROI locations by applying Viola-Jones. Then, this selected ROI is used to relocate the other neighbor ROIs. Therefore, the first stage using Viola-Jones to detect the three initial ROIs is quite decisive.

Figure 6 shows, for a child, the ROI placement using the automated Viola-Jones algorithm (green), the recalculated algorithm (blue) and the manual placement (yellow). In Ref. [17,21], the authors demonstrated that face regions (see Figure 4) linked to the set of branches and sub-branches of vessels that innervate the face are decisive to study emotions, as the skin temperature variations over these regions can be measured by IRTI. Figure 6 shows that our proposal based on probability error can be used to supervise automatic methods for ROI locations, in order to improve the appearance feature extraction over detected ROIs and, consequently, increase the performance during the emotion recognition.

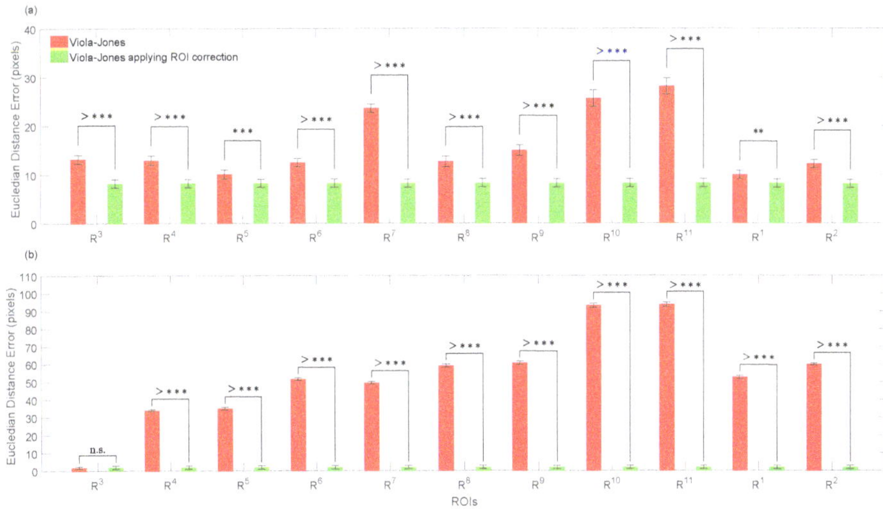

Figure 5. Comparison between Viola-Jones (without applying ROI relocations) and Viola-Jones applying ROI relocations, computing the mean and standard error per ROIs: (**a**) analysis from Baseline; (**b**) analysis from Test. n.s means no significant difference ($p > 0.05$), while * ($p < 0.05$), ** ($p < 0.01$), *** ($p < 0.001$) and >*** ($p < 0.0001$) indicate significant difference. $\mathbf{R}^1$, right forehead side; $\mathbf{R}^2$, left forehead side; $\mathbf{R}^3$, right periorbital side; $\mathbf{R}^4$, left periorbital side; $\mathbf{R}^5$, tip of nose; $\mathbf{R}^6$, right cheek; $\mathbf{R}^7$, left cheek; $\mathbf{R}^8$, right perinasal side; $\mathbf{R}^9$, left perinasal side; $\mathbf{R}^{10}$, right chin side; $\mathbf{R}^{11}$, left chin side.

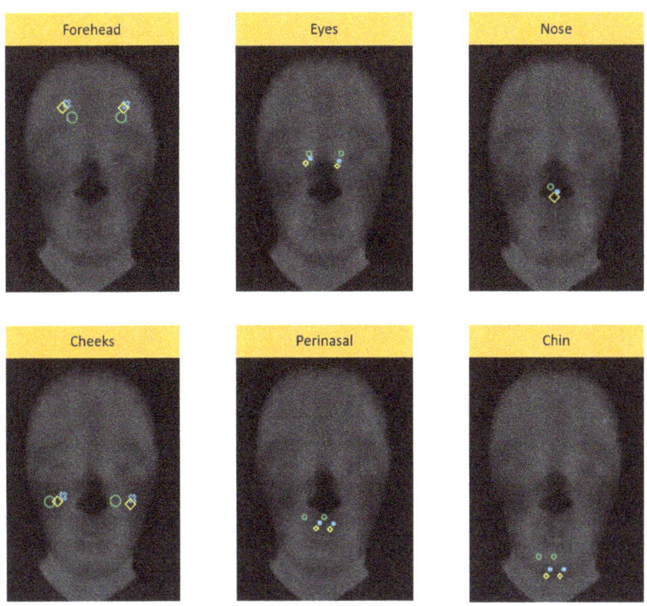

Figure 6. Comparison between manual placement (in yellow, used as reference), Viola-Jones algorithm (green) and Viola-Jones with our replacement algorithm (blue). Such as shown, our replacement algorithm obtained better results than using only Viola-Jones algorithm.

4.2. Emotion Recognition

A database with 28 typically developing children was used to analyze the performance of the proposed system for five emotions recognition [17]. This database contains appearance feature vectors that were computed on eleven face ROIs (see Figure 4), whose correct placements were carefully verified by a trained expert. Then, the effectiveness of PCA plus other classifiers for five emotions classification was evaluated here, being PCA a popular adaptive transform that under controlled head-pose and imaging conditions may be useful to capture features of expressions efficiently [9]. Figure 7a shows the average performance by applying PCA for different principal components plus LDA classifier based on full covariance matrix, being 60 components sufficient for five emotions recognition. Notice that non-significant difference ($p < 0.05$) using more than 60 components was achieved. Similarly, non-significant difference was obtained using PCA with 60 components plus other classifiers, such as LDA using full (Linear) or diagonal (Diag Linear) covariance matrices, and Linear SVM, such as shown in Figure 7b. For instance, LDA using full and diagonal covariance matrices achieved Kappa values of 81.84 ± 1.72% and 77.99 ± 1.74%, respectively, and 78.58 ± 1.83% for Linear SVM. LDA based on full covariance matrix has low-computational cost and can be easily embedded into the N-MARIA hardware for on-line emotion recognition during the child-robot interaction. Moreover, PCA and LDA have been successfully used in other similar studies [16,17,20,23,36], achieving promising results. Then, we selected PCA with 60 principal components plus LDA based on full covariance matrix as the best setup for five emotions recognition.

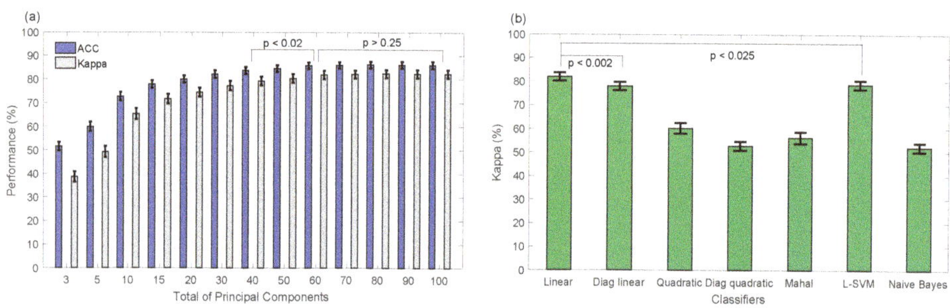

Figure 7. Performance of classification of five emotions applying principal component analysis (PCA) for dimensionality reduction. (**a**) Average accuracy and Kappa obtained by applying PCA for different principal components plus Linear Discriminant Analysis for classification; (**b**) average Kappa achieved for 60 principal components plus other classifiers.

Consequently, a comparison between PCA using 60 components and FNCA was carried out to recognize five emotions of the 28 typically developing children, using LDA based on full covariance matrix. Figure 8 and Tables 3 and 4 show that both PCA and FNCA methods are interchangeable for dimensionality reduction, achieving mean ACC of 85.29% and 85.75%, respectively.

Table 3. Performance of the system for five emotions recognition using FNCA plus LDA based on full covariance matrices.

	Disgust	Fear	Happiness	Sadness	Surprise
TPR (%)	90.02 ± 1.91	87.25 ± 2.37	87.55 ± 2.12	74.69 ± 2.77	86.93 ± 2.12
FPR (%)	3.07 ± 0.49	4.17 ± 0.57	4.61 ± 0.69	3.42 ± 0.78	2.93 ± 0.50
ACC (%)			85.29 ± 1.16		
Kappa (%)			81.26 ± 1.46		

ACC, accuracy; TPR, true positive rate; FPR, false positive rate.

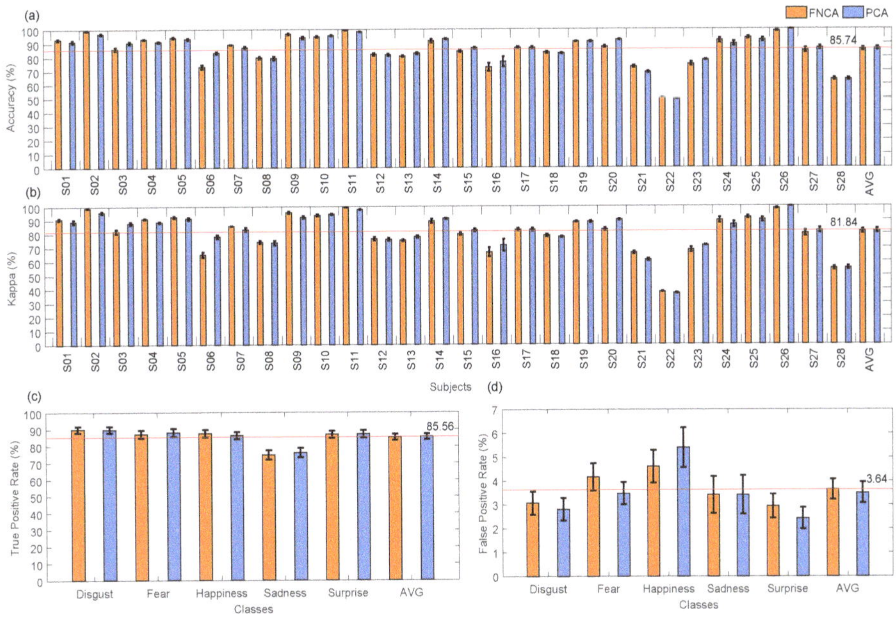

Figure 8. Performance of classification of five emotions applying principal component analysis (PCA) and Fast Neighbourhood Component Analysis (FNCA) for dimensionality reduction.

Table 4. Performance of the proposed system for five emotions recognition using PCA with 60 components plus LDA based on full covariance matrices.

	Disgust	Fear	Happiness	Sadness	Surprise
TPR (%)	89.93 ± 1.88	88.22 ± 2.38	86.36 ± 2.09	76.15 ± 2.80	87.14 ± 2.15
FPR (%)	2.80 ± 0.47	3.48 ± 0.47	5.40 ± 0.83	3.42 ± 0.81	2.42 ± 0.44
ACC (%)			85.75 ± 1.16		
Kappa (%)			81.84 ± 1.46		

ACC, accuracy; TPR, true positive rate; FPR, false positive rate.

It is worth noting that PCA improved the performance for subject S06 (ACC of 83.27% and Kappa of 78.10% for PCA, and ACC of 73.11% and Kappa of 65.09% for FNCA), such as shown in Figure 8a,b. This improvement may be consequence of the PCA robustness for feature extraction under controlled head-pose and imaging conditions [9]. Also, notice that for subject S22 the lowest performance was achieved by applying PCA (ACC of 49.27% and Kappa of 36.75%) and FNCA (ACC of 49.98% and Kappa of 37.78%). However, ACC values higher than 85.74% were achieved for a total of 14 children by applying PCA or FNCA, being the highest ACC of 99.78% for subject S26 using PCA. Labelling feature vectors of emotion patterns is a challenge, mainly when labels are assigned over periods of time, as some delays between both perception and manual label assignment may exist. Similarly, a visual stimulus may produce different facial emotions, thus, the manual labeled procedure is subjective. Then, an automatic method for feature set labelling may be suitable to obtain reliable labels and therefore a classification model for LDA.

On the other hand, Figure 8c shows that four emotions (disgust, fear, happiness, and surprise) were successfully recognized by applying both PCA and FNCA, achieving TPR > 85%, whereas sadness was classified with less sensitivity (TPR of 74.69% and 76.15% for FNCA and PCA, respectively) by the proposed system. As a highlight, low values of FPR (≤5.40%) were obtained to recognize all studied emotions. In general, the results achieved show that PCA and FNCA are interchangeable to obtain the features that increase the emotion discrimination by discovering those regions that are informative in terms of expressions [9]. Similar to other studies [52], the sadness emotion was less recognized (ACC < 77%) than the other emotions, as each child expressed sadness in a different manner, producing a large intra-class variability for sadness.

Many researchers [1] have conducted studies to infer facial emotions over visual images using only three out of the six basic expressions, specifically happiness, surprise and sadness, as these emotions are easier to be recognized for humans, and have been successfully used to generate cohesive expressions across participants. Similarly, other studies using visual images report that the recognition rates for expressions as fear or disgust can be very low, in a range from 40 to 50%. As a highlight, five basic emotions were recognized by our proposed system based on infrared camera, achieving the highest performance for both disgust (TPR of 89.93%) and fear (TPR of 88.22%), followed by happiness (TPR of 86.36%) and surprise (TPR of 87.14%), such as shown in Figure 8c and Table 4. In agree with other studies [17,21], our results show that thermal images are promising for facial emotion analysis and recognition.

Similarly, this database was used as a training set of our recognition system to infer emotions on 17 children, while they firstly remained in front of N-MARIA covered by a period of time, and after it was uncovered. Figures 9a–d and 10a,b show inferred emotions by our system for the seventeen children during the Baseline and Test experiments, being happiness and surprise the most frequent outputs. Emotions such as disgust, fear, and sadness were few inferred by the system, which agree with what the children reported, such as shown in Figure 10c.

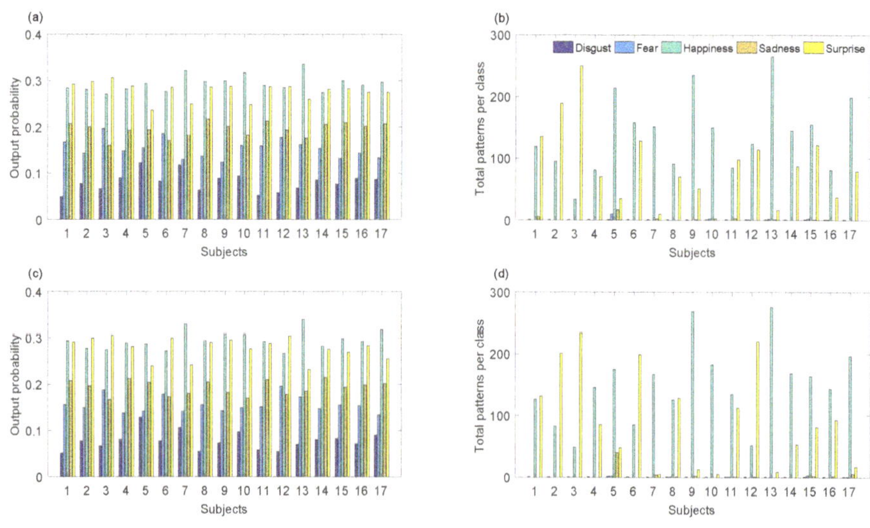

Figure 9. Automatic emotion recognition over unknown patterns. (**a**,**b**) are inferred emotions obtained during the Baseline phase of the experiment; (**c**,**d**) are emotion decisions achieved for the Test phase of the experiment.

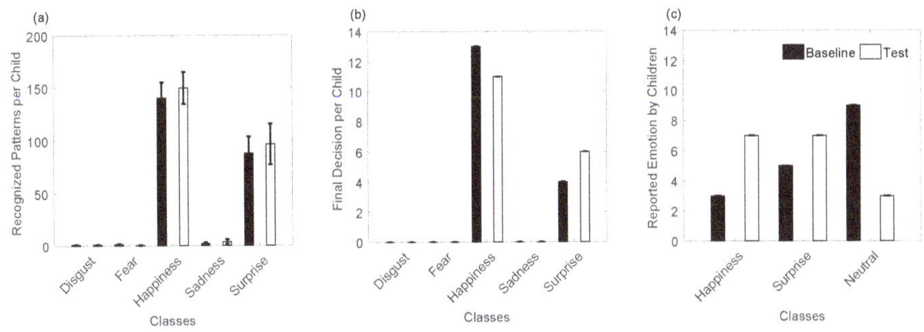

Figure 10. Summary of children emotions before and after they saw the robot for the first time. (**a**) and (**b**) are inferred emotions by the recognition system; (**c**) emotions reported by the children.

Figure 10c shows that a neutral emotion was reported by more children for N-MARIA covered, while they felt surprise and happiness seeing N-MARIA by first time. In contrast, happiness and surprise were slightly more inferred after having uncovered the robot for the children, which also agreed with what the children reported. It is worth noting that emotions may vary among children for a same visual stimulus, such as a social robot. For this reason, a single indicator for evaluation is not sufficient to interpret accurately the responses of children to a robot while they are interacting. Then, a questionnaire was answered by each child, reporting their emotions during both Baseline and Test (see Section 3.1). However, self-assessments have problems with validity and corroboration, as described in Ref. [1], as participants might report differently from how they are actually thinking or feeling. Then, it is not trivial to attribute the children responses to their true behaviours. Similarly, the emotion produced by each child may be conditioned, as they know that they are being observed by the robot, parent, and the specialist conducting the experiment. Also notice that an existent database of 28 typically developing children (age: 7–11 years) was used in our study to train our proposed system to infer one of five basic emotions (disgust, fear, happiness, sadness, surprise) produced by 17 children during the interaction with N-MARIA. However, visual and thermal image processing may be affected due to the quality of the input image, as it depends of the illumination conditions and the distance between child and robot during the interaction.

5. Discussion

Several studies have been addressed to recognize some of the six accepted emotions by psychological theory: surprise, fear, disgust, anger, happiness and sadness [10,53–55]. For instance, facial motion and the tone of the speech have shown in recognition systems their relevant role to infer these aforementioned emotions, achieving accuracy (ACC) from 80% to 98% [56–58] and from 70% to 75% [53,59,60], respectively. Although facial expressions provide important clues about emotions, it is necessary to measure, by optical flow, the movements of specific facial muscles through markers located on the face [53,57,58]. However, this non-contact free technique may not be comfortable to study or infer emotions of children with ASD, as these children present high skin sensitivity.

Following this approach of non-contact techniques, there are some APIs (Application Program Interface) that allow facial detection, such as the "Emotion API" from Microsoft or the Oxford API [61,62]. Some works, alternatively, use other image processing techniques, such as finding the region of the thermogram within a temperature range to detect the face [18]. Another approach is to keep the face in a fixed position using a support for chin or headrest device [41,63]. Furthermore, in Ref. [64] the authors show a solution using neural network and supervised shapes classification methods applied to facial thermogram.

Emotion recognition systems based on IRTI have, in fact, shown promising results. Table 5 shows some studies that use ROI replacement and further emotion analysis, although using other techniques in comparison with our proposal. Unfortunately, the results presented are somewhat disperse, and it is not possible to make a fair comparison among them, due to the different pictures used in the studies. In Ref. [18] the authors proposed techniques for selecting facial ROIs and classify emotions using a FLIR A310 camera. To detect the face, thermogram's temperatures between 32 °C to 36 °C were used to define the face position. The ROIs positions were further calculated by proportions based on the head's width. All other temperature points were considered background. Additionally, for emotion classification, the system was calibrated using a baseline (neutral state) that compensates the induced emotion by applying fuzzy algorithm, and thus, calibrate the induced emotion image. Using the baseline, the temperature is inferred by IF-THEN rules to calibrate the thermal images for the following induced emotions: joy, disgust, anger, fear and sadness. Next, a top-down hierarchical classifier was used to analyze the emotion classification, reaching 89.9% of success rate.

Table 5. Comparison among some strategies used to infer emotions using Infrared Thermal Imaging (IRTI).

Studies	Volunteers	Age	ACC (%)	Summary
Cruz-Albarran et al. [18]	25	19 to 33	89.90	Fuzzy algorithm, IF-THEN rules and top-down hierarquical classifier. The analyzed emotions were joy, disgust, anger, fear and sadness.
Basu et al. [13]	26	11 to 32	87.5	Histogram feature extraction and multiclass support vector machine (SVM). The Kotani Thermal Facial Expression database was used to detect emotions of anger, fear, happiness, and sadness.
Nhan and Chau [41]	12	21 to 27	80.0	Comparison between baseline and affective states. High and low arousal and valence are compared with the baseline.
Wang et al. [65]	38	17 to 31	62.90	Deep Boltzmann Machine to find positive or negative valence.
Bijalwan et al. [66]	1	N/A	97.0	Model for expression recognition in thermal images. Application of PCA for recognizing happiness, anger, disgust, sadness and neutral emotions.
Yoshitomi et al. [67]	1	N/A	90.0	Neural Networks and Backpropagation algorithms that recognize emotions of happiness, surprise and neutral state.
Kosonogov et al. [63]	24	20 to 24	N/A	Studied tip of nose thermal variation in volunteers with images from International Affective Picture System (IAPS) which found that positive or negative images showed more temperature change compared with neutral images
Vukadinovic and Pantic [68]	200	18 to 50	N/A	Algorithm to find facial ROIs. A Viola-Jones adapted algorithm (that applies GentleBoost instead of AdaBoost) was used. For facial feature extraction Gabor wavelet filter was used. No emotion classes were studied, only the facial points for ROI detection.
Bharatharaj et al. [62]	9	6 to 16	N/A	AMRM (indirect teaching method) was studied using a parrot-inspired robot and the Oxford emotion API to recognize and classify emotions in ASD children. Most of them appeared to be happy with the robot.
Mehta et al. [61]	3	N/A	93.8%	Microsoft HoloLens (MHL) system was used to detect emotions achieving a high accuracy to detect happiness, sadness, anger, surprise and neutral emotions.
Our proposal	28	7 to 11	85.75%	PCA and LDA were used on our database, published in [17], to recognize happiness, sadness, fear, surprise and disgust.

ACC, accuracy; N/A means that the age or ACC were not reported.

The functional Infrared Thermal Imaging (fITI) was used in Refs. [21,63], which is considered to be a promising technique to infer emotions through autonomic responses. Similarly, another study was carried out using fITI to compare subjective ratings of displayed pictures for the volunteers [63], where these pictures were categorized into unpleasant, neutral and pleasant. Then, while the volunteers were watching these pictures, the authors collected the nose tip temperature (there was a chin support to keep the face correctly located in the camera image), which is one of the most likely places to change temperature when the person is under some kind of emotion [17]. As a result, they found that pictures that evoke emotions (no matter if it is a positive or a negative emotion) were more susceptible to produce thermal variation, while the difference for the neutral images was not as great as the others. Thus, their findings demonstrate that fITI can be a useful tool to infer emotions in humans.

Another interesting research [68] locates facial points in visual grayscale image using Gabor features based boosted classifiers, in which the authors used an adapted version of Viola-Jones algorithm, using GentleBoost instead of AdaBoost, to detect the face. Also, Gabor wavelet was used for feature extraction, detecting 20 ROIs that represent the facial feature points. All this detection was made automatically and contact-free using the iris and mouth detection. These two parts were detected by dividing the face in two regions, and calculating proportions to find those regions (iris and mouth). From this, all other ROIs were calculated using proportions. Their success rate was high, since the algorithm achieved a 93% of success rate using the Cohn-Kanade database, which has expressionless pictures of 200 people. Although Gabor wavelet transform is a representative method to extract local features, it takes a long time and has a large feature dimension.

Another method was proposed in Ref. [65], where a deep Boltzmann machine (DBM) was applied to recognize emotions from thermal facial images, using a previous database and with the participation of 38 adult volunteers. Their evaluation consisted of finding the emotion valence, which could be positive or negative, and their accuracy rate reached 62.9%. In their study, since the face and the background have different temperatures, they were split by applying the Otsu threshold algorithm in order to binarize images. Then, the projection curves (both vertical and horizontal) were calculated to find the largest gradient and detect the face boundary.

Additionally, a model for expression recognition using thermal images of an adult volunteer was applied in Ref. [66]. These authors used eigenfaces for feature extraction of the volunteer's facial images through PCA to recognize five emotions (happiness, anger, disgust, sad and neutral). As a highlight, that proposal reached an accuracy close to 97%, in which work, they applied eigenvalues and eigenfaces, trained the system with a set of images, used PCA to reduce the dimensionality, and distance classifier to recognize the emotion.

In Ref. [13], authors were able to achieve 81.95% of accuracy using histogram feature extraction combined with multiclass SVM over thermal images of 22 volunteers in the Kotani Thermal Facial Expression (KTFE) database, and four classes were studied: happiness, sadness, fear and anger. They used preprocessing techniques to prepare the image to apply Viola-Jones and for further image enhancement de-noising the image and using the Contrast Limited Adaptive Histogram Equalization. For ROI detection, a ratio-based segmentation was used.

Moreover, the recognition of both baseline and affective states was carried out in Ref. [41], where, to detect the face, they used a headrest to keep it in the correct position, in addition to a reference point (located on the top of the head), which was about 10 °C cooler than the skin temperature. To find the ROIs, the reference point was used, and a radiometric threshold was applied. In case of loss of the reference point, it was manually corrected by the researchers.

Another study [67] applied IRTI in a female adult volunteer, and Neural Networks and Backpropagation algorithms were used to recognize emotions, such as happiness, surprise and neutral state, reaching an ACC of 90%. To find the face they used Otsu segmentation, and the Feret's diameter was found in the binary image together with the center of gravity of the binary image. Then, after segmenting the image, positions of the face based on FACS-AU were used to determine the heat variation and, thus, the emotion.

Another work [61] shows a study about several approaches of emotion recognition and facial detection, such as machine learning and geometric feature-based process, in addition to SVM and a diversity of other classifiers. They also present the use of Microsoft HoloLens (MHL) for detecting human emotions by using an application that has been built to use MHL to detect faces and recognize emotion of people facing it. The set of emotions they worked was composed of happiness, sadness, anger, surprise and neutral. Additionally, they used a webcam to detect emotions and compare with the result using MHL. The system with MHL could achieve much better results than previous works and had remarkable accuracy probably due to the sensors attached to the HoloLens, reaching an accuracy of 93.8% on MHL, using the "Emotion API" from Microsoft.

In Ref. [62], the authors used a parrot-inspired robot (KiliRo) to interact with ASD children by simulating a set of autonomous behaviors. They tested the robot for five consecutive days in a clinical facility. Children's expressions while interacting with the robot were analyzed by the Oxford emotions API, allowing them to make an automated facial detection, emotion recognition and classification system.

Some works, such as Ref. [69], show the use of deep learning to detect the child face and infer the visual attention on a robot during CRI therapy. Authors used the robot NAO from Softbank Robotics, which has two low-resolution cameras that were used to take pictures and record videos. They also used the face detection and tracking system in-built in NAO for the clinical experiments. A total of 6 children participated of the experiment, in which they imitated some robot movements. The children had 14 encounters over a month, and the actual experiments started 7 days after the preliminary encounter, in order to avoid the novelty effect in the results. Different deep learning techniques and classifiers were used, and they could reach an average children attention rate of 59.2%.

Deep-learning-based approaches have shown to be promising for emotion recognition, determining features and classifiers without expert supervisors [10]. However, conventional approaches are still being studied for use in real-time embedded systems because of their low computational complexity and high degree of accuracy [70], although for these systems the methods for feature extraction and classification should be designed by the programmer and cannot be optimized to increase performance [71]. Moreover, it is worth mentioning that conventional approaches require relatively lower computing power and memory than deep learning-based approaches [10]. Similarly, Gabor features are very popular for facial expression classification and face recognition, due to their high discriminative power [72,73], but the computational complexity and memory requirement make them less suitable for a real time implementation.

Our system is composed of low-cost hardware and methods of low-computational cost for visual and thermal image processing, and recognizes five emotions, achieving 85.75% of accuracy. For our system, we proposed a method based on probability error to accurately locate subject-specific landmarks, taking into account the trained expert criteria. As a highlight, our proposal can find frame-to-frame the best located facial ROI using the Viola-Jones algorithm, and adjust the location of its surrounding facial ROIs. As another novelty, our proposal based on probability error showed robustness and good accuracy to locate facial ROIs on thermal images, which were collected while typically developing children interacted with a social robot. As other findings, we extended an existing database of five facial emotions from thermal images, to infer unknown emotion generated while the children interacted with the social robot, using our recognition system based on PCA and LDA, thus, achieving results that agreed with the written reports of children.

As limitation, our system is not able to track head movements, thus, adding a method for facial tracking, such as done by Ref. [74], can make robust our proposal for facial landmarks in uncontrolled scenarios, such as mobile applications for child and social robot interaction. Generally, facial emotion datasets with six basic emotions contain only adult participants, but there are very few databases collected on typically developing children (aged between 7 and 11 years) through infrared camera, containing the basic emotions. Then, it is a challenge to use high quantity of examples during the training stage of a recognition system to infer emotions of children with age range from 7 to 11 years while they interact with a robot for example. In addition, more tests with a higher number of volunteers must be performed, including ASD children.

6. Conclusions

A low-computational cost system for children emotion recognition in an unobtrusive way was proposed in this study, which is composed of low-cost cameras, making possible its extension for research into developing countries. As a first stage, our proposal was tested on visual and thermal images of children interacting with the mobile social robot N-MARIA, achieving promising results (85.75% of accuracy) to locate specific face landmarks, as well as to recognize (or infer) five emotions. All children had a hopeful interaction with the robot, which demonstrated our system is useful to stimulate positive emotions in children, and able to trigger a profitable interaction with them. In future works, this proposal will be integrated in N-MARIA, aiming to know online the children's emotion and making control decisions based on emotions. Additionally, other methods will be explored for facial tracking, in order to reduce the influence of head-pose during the emotion recognition. Furthermore, unsupervised methods for automatic label assignments and classifier learning will be evaluated in our dataset to obtain a robust recognition system for processing patterns of high uncertainty, such as facial expressions and emotions.

Author Contributions: Conceptualization, C.G., C.V., D.D.-R. and D.F.; Methodology, C.G. and D.F.; Software, D.F., V.B., A.F., G.B., C.V. and D.D.-R.; Formal Analysis, C.G. and D.D.-R.; Writing—Original Draft Preparation, C.G., D.D.-R., C.V. and D.F.; Writing—Review and Editing, E.C. and T.B.-F.

Funding: This research was funded by FAPES/Brazil, grant numbers 72982608 and 645/2016.

Acknowledgments: Authors would like to thank the financial support from CAPES, CNPq and FAPES/Brazil (project numbers: 72982608 and 645/2016), and UFES for the technical support.

Conflicts of Interest: The authors declare no conflict of interest. The funders had no role in the design of the study; in the collection, analyses, or interpretation of data; in the writing of the manuscript, or in the decision to publish the results.

References

1. Gunes, H.; Celiktutan, O.; Sariyanidi, E. Live human–robot interactive public demonstrations with automatic emotion and personality prediction. *Philos. Trans. R. Soc. B* **2019**, *374*, 20180026. [CrossRef] [PubMed]
2. Kim, E.S.; Berkovits, L.D.; Bernier, E.P.; Leyzberg, D.; Shic, F.; Paul, R.; Scassellati, B. Social Robots as Embedded Reinforcers of Social Behavior in Children with Autism. *J. Autism Dev. Disord.* **2012**, *43*, 1038–1049. [CrossRef] [PubMed]
3. Valadao, C.; Caldeira, E.; Bastos-Filho, T.; Frizera-Neto, A.; Carelli, R. A New Controller for a Smart Walker Based on Human-Robot Formation. *Sensors* **2016**, *16*, 1116. [CrossRef] [PubMed]
4. Picard, R.; Vyzas, E.; Healey, J. Toward machine emotional intelligence: analysis of affective physiological state. *IEEE Trans. Pattern Anal. Mach. Intell.* **2001**, *23*, 1175–1191. [CrossRef]
5. Conn, K.; Liu, C.; Sarkar, N.; Stone, W.; Warren, Z. Affect-sensitive assistive intervention technologies for children with autism: An individual-specific approach. In Proceedings of the RO-MAN 2008—The 17th IEEE International Symposium on Robot and Human Interactive Communication, Munich, Germany, 1–3 August 2008.
6. Shier, W.A.; Yanushkevich, S.N. Biometrics in human-machine interaction. In Proceedings of the 2015 International Conference on Information and Digital Technologies, Zilina, Slovakia, 7–9 July 2015. doi:10.1109/dt.2015.7222989.
7. Goulart, C.; Valadao, C.; Caldeira, E.; Bastos, T. Brain signal evaluation of children with Autism Spectrum Disorder in the interaction with a social robot. *Biotechnol. Res. Innov.* **2018**. [CrossRef]
8. Latif, M.T.; Yusof, M.; Fatai, S. Emotion Detection from Thermal Facial Imprint based on GLCM Features. *ARPN J. Eng. Appl. Sci.* **2016**, *11*, 345–349.
9. Sariyanidi, E.; Gunes, H.; Cavallaro, A. Automatic analysis of facial affect: A survey of registration, representation, and recognition. *IEEE Trans. Pattern Anal. Mach. Intell.* **2014**, *37*, 1113–1133. [CrossRef] [PubMed]
10. Ko, B. A brief review of facial emotion recognition based on visual information. *Sensors* **2018**, *18*, 401. [CrossRef]
11. Rusli, N.; Sidek, S.N.; Yusof, H.M.; Latif, M.H.A. Non-Invasive Assessment of Affective States on Individual with Autism Spectrum Disorder: A Review. In *IFMBE Proceedings;* Springer: Singapore, 2015; pp. 226–230.
12. Petrantonakis, P.C.; Hadjileontiadis, L.J. Emotion recognition from EEG using higher order crossings. *IEEE Trans. Inf. Technol. Biomed.* **2009**, *14*, 186–197. [CrossRef]

13. Basu, A.; Routray, A.; Shit, S.; Deb, A.K. Human emotion recognition from facial thermal image based on fused statistical feature and multi-class SVM. In Proceedings of the 2015 Annual IEEE India Conference (INDICON), New Delhi, India, 17–20 December 2015.
14. Ghimire, D.; Jeong, S.; Lee, J.; Park, S.H. Facial expression recognition based on local region specific features and support vector machines. *Multimed. Tools Appl.* **2017**, *76*, 7803–7821. [CrossRef]
15. Perikos, I.; Paraskevas, M.; Hatzilygeroudis, I. Facial Expression Recognition Using Adaptive Neuro-fuzzy Inference Systems. In Proceedings of the 2018 IEEE/ACIS 17th International Conference on Computer and Information Science (ICIS), Singapore, 6–8 June 2018; pp. 1–6. [CrossRef]
16. Happy, S.; Routray, A. Automatic facial expression recognition using features of salient facial patches. *IEEE Trans. Affect. Comput.* **2014**, *6*, 1–12. [CrossRef]
17. Goulart, C.; Valadao, C.; Delisle-Rodriguez, D.; Caldeira, E.; Bastos, T. Emotion analysis in children through facial emissivity of infrared thermal imaging. *PLoS ONE* **2019**, *14*, e0212928.. [CrossRef] [PubMed]
18. Cruz-Albarran, I.A.; Benitez-Rangel, J.P.; Osornio-Rios, R.A.; Morales-Hernandez, L.A. Human emotions detection based on a smart-thermal system of thermographic images. *Infrared Phys. Technol.* **2017**, *81*, 250–261. [CrossRef]
19. Wang, S.; Shen, P.; Liu, Z. Facial expression recognition from infrared thermal images using temperature difference by voting. In Proceedings of the 2012 IEEE 2nd International Conference on Cloud Computing and Intelligence Systems, Hangzhou, China, 30 October–1 November 2012; Volume 1, pp. 94–98.
20. Pop, F.M.; Gordan, M.; Florea, C.; Vlaicu, A. Fusion based approach for thermal and visible face recognition under pose and expresivity variation. In Proceedings of the 9th RoEduNet IEEE International Conference, Sibiu, Romania, 24–26 June 2010; pp. 61–66.
21. Ioannou, S.; Gallese, V.; Merla, A. Thermal infrared imaging in psychophysiology: Potentialities and limits. *Psychophysiology* **2014**, *51*, 951–963. [CrossRef] [PubMed]
22. Zheng, Y. Face detection and eyeglasses detection for thermal face recognition. *SPIE Proc.* **2012**, *8300*, 83000C.
23. Wang, S.; Liu, Z.; Lv, S.; Lv, Y.; Wu, G.; Peng, P.; Chen, F.; Wang, X. A Natural Visible and Infrared Facial Expression Database for Expression Recognition and Emotion Inference. *IEEE Trans. Multimed.* **2010**, *12*, 682–691. [CrossRef]
24. Choi, J.S.; Bang, J.; Heo, H.; Park, K. Evaluation of fear using nonintrusive measurement of multimodal sensors. *Sensors* **2015**, *15*, 17507–17533. [CrossRef] [PubMed]
25. Lajevardi, S.M.; Hussain, Z.M. Automatic facial expression recognition: Feature extraction and selection. *Signal Image Video Process.* **2012**, *6*, 159–169. [CrossRef]
26. Jabid, T.; Kabir, M.H.; Chae, O. Robust facial expression recognition based on local directional pattern. *ETRI J.* **2010**, *32*, 784–794. [CrossRef]
27. Kabir, M.H.; Jabid, T.; Chae, O. A local directional pattern variance (LDPv) based face descriptor for human facial expression recognition. In Proceedings of the 2010 7th IEEE International Conference on Advanced Video and Signal Based Surveillance, Boston, MA, USA, 29 August–1 September 2010; pp. 526–532.
28. Shan, C.; Gong, S.; McOwan, P.W. Robust facial expression recognition using local binary patterns. In Proceedings of the IEEE International Conference on Image Processing 2005, Genova, Italy, 14 September 2005; Volume 2, pp. II–370.
29. Shan, C.; Gritti, T. Learning Discriminative LBP-Histogram Bins for Facial Expression Recognition. In Proceedings of the British Machine Vision Conference 2008, Leeds, UK, 1–4 September 2008; pp. 1–10.
30. Song, M.; Tao, D.; Liu, Z.; Li, X.; Zhou, M. Image ratio features for facial expression recognition application. *IEEE Trans. Syst. Man Cybern. Part B Cybern.* **2009**, *40*, 779–788. [CrossRef]
31. Zhang, L.; Tjondronegoro, D. Facial expression recognition using facial movement features. *IEEE Trans. Affect. Comput.* **2011**, *2*, 219–229. [CrossRef]
32. Viola, P.; Jones, M.J. Robust Real-Time Face Detection. *Int. J. Comput. Vis.* **2004**, *57*, 137–154. [CrossRef]
33. Jiang, B.; Martinez, B.; Valstar, M.F.; Pantic, M. Decision level fusion of domain specific regions for facial action recognition. In Proceedings of the 2014 22nd International Conference on Pattern Recognition, Stockholm, Sweden, 24–28 August 2014; pp. 1776–1781.
34. Shan, C.; Gong, S.; McOwan, P.W. Facial expression recognition based on local binary patterns: A comprehensive study. *Image Vis. Comput.* **2009**, *27*, 803–816. [CrossRef]

35. Kazemi, V.; Sullivan, J. One millisecond face alignment with an ensemble of regression trees. In Proceedings of the IEEE Conference on Computer Vision and Pattern Recognition, Columbus, OH, USA, 24–27 June 2014; pp. 1867–1874.
36. Zhao, X.; Zhang, S. Facial expression recognition based on local binary patterns and kernel discriminant isomap. *Sensors* **2011**, *11*, 9573–9588. [CrossRef] [PubMed]
37. Yang, J.; Wang, X.; Han, S.; Wang, J.; Park, D.S.; Wang, Y. Improved Real-Time Facial Expression Recognition Based on a Novel Balanced and Symmetric Local Gradient Coding. *Sensors* **2019**, *19*, 1899. [CrossRef] [PubMed]
38. Giacinto, A.D.; Brunetti, M.; Sepede, G.; Ferretti, A.; Merla, A. Thermal signature of fear conditioning in mild post traumatic stress disorder. *Neuroscience* **2014**, *266*, 216–223. [CrossRef] [PubMed]
39. Marzec, M.; Koprowski, R.; Wróbel, Z. Methods of face localization in thermograms. *Biocybern. Biomed. Eng.* **2015**, *35*, 138–146. [CrossRef]
40. Trujillo, L.; Olague, G.; Hammoud, R.; Hernandez, B. Automatic Feature Localization in Thermal Images for Facial Expression Recognition. In Proceedings of the 2005 IEEE Computer Society Conference on Computer Vision and Pattern Recognition (CVPR'05)—Workshops, San Diego, CA, USA, 21–23 September 2005. doi:10.1109/cvpr.2005.415.
41. Nhan, B.; Chau, T. Classifying Affective States Using Thermal Infrared Imaging of the Human Face. *IEEE Trans. Biomed. Eng.* **2010**, *57*, 979–987. [CrossRef]
42. Bradski, G. The OpenCV library. *Dr Dobb's J. Softw. Tools* **2000**, *25*, 120–125.
43. Malis, E.; Vargas, M. Deeper Understanding of the Homography Decomposition for Vision-Based Control. Ph.D. Thesis, INRIA, Sophia Antipolis Cedex, France, 2007.
44. Budzier, H.; Gerlach, G. Calibration of uncooled thermal infrared cameras. *J. Sens. Sens. Syst.* **2015**, *4*, 187–197. [CrossRef]
45. Martínez, A.M.; Kak, A.C. Pca versus lda. *IEEE Trans. Pattern Anal. Mach. Intell.* **2001**, *23*, 228–233. [CrossRef]
46. Friedman, J.H. Regularized discriminant analysis. *J. Am. Stat. Assoc.* **1989**, *84*, 165–175. [CrossRef]
47. Kwon, O.W.; Chan, K.; Hao, J.; Lee, T.W. Emotion recognition by speech signals. In Proceedings of the Eighth European Conference on Speech Communication and Technology, Geneva, Switzerland, 1–4 September 2003.
48. Bamidis, P.D.; Frantzidis, C.A.; Konstantinidis, E.I.; Luneski, A.; Lithari, C.; Klados, M.A.; Bratsas, C.; Papadelis, C.L.; Pappas, C. An integrated approach to emotion recognition for advanced emotional intelligence. In *International Conference on Human-Computer Interaction*; Springer: Berlin/Heidelberg, Germany, 2009, pp. 565–574.
49. Ververidis, D.; Kotropoulos, C.; Pitas, I. Automatic emotional speech classification. In Proceedings of the 2004 IEEE International Conference on Acoustics, Speech, and Signal Processing, Montreal, QC, Canada, 17–21 May 2004; Volume 1, pp. I–593.
50. Hsu, C.W.; Lin, C.J. A comparison of methods for multiclass support vector machines. *IEEE Trans. Neural Netw.* **2002**, *13*, 415–425. [PubMed]
51. Japkowicz, N.; Shah, M. *Evaluating Learning Algorithms: A Classification Perspective*; Cambridge University Press: Cambridge, UK, 2011.
52. Boucenna, S.; Gaussier, P.; Andry, P.; Hafemeister, L. A robot learns the facial expressions recognition and face/non-face discrimination through an imitation game. *Int. J. Soc. Robot.* **2014**, *6*, 633–652. [CrossRef]
53. Busso, C.; Deng, Z.; Yildirim, S.; Bulut, M.; Lee, C.M.; Kazemzadeh, A.; Lee, S.; Neumann, U.; Narayanan, S. Analysis of emotion recognition using facial expressions, speech and multimodal information. In Proceedings of the 6th International Conference on Multimodal Interfaces, State College, PA, USA, 13–15 October 2004; pp. 205–211.
54. Pantic, M.; Rothkrantz, L.J. Toward an affect-sensitive multimodal human-computer interaction. *Proc. IEEE* **2003**, *91*, 1370–1390. [CrossRef]
55. Cowie, R.; Douglas-Cowie, E.; Tsapatsoulis, N.; Votsis, G.; Kollias, S.; Fellenz, W.; Taylor, J.G. Emotion recognition in human-computer interaction. *IEEE Signal Process. Mag.* **2001**, *18*, 32–80. [CrossRef]
56. Essa, I.A.; Pentland, A.P. Coding, analysis, interpretation, and recognition of facial expressions. *IEEE Trans. Pattern Anal. Mach. Intell.* **1997**, *19*, 757–763. [CrossRef]
57. Mase, K. Recognition of facial expression from optical flow. *IEICE Trans. Inf. Syst.* **1991**, *74*, 3474–3483.
58. Yacoob, Y.; Davis, L. Computing Spatio-Temporal Representations of Human Faces. Ph.D. Thesis, Department of Computer Science, University of Maryland, College Park, MD, USA, 1994.

59. Lee, C.M.; Yildirim, S.; Bulut, M.; Kazemzadeh, A.; Busso, C.; Deng, Z.; Lee, S.; Narayanan, S. Emotion recognition based on phoneme classes. In Proceedings of the Eighth International Conference on Spoken Language Processing, Jeju Island, Korea, 4–8 October 2004.
60. Nwe, T.L.; Wei, F.S.; De Silva, L.C. Speech based emotion classification. In Proceedings of the IEEE Region 10 International Conference on Electrical and Electronic Technology, TENCON 2001 (Cat. No. 01CH37239), Singapore, 19–22 August 2001; Volume 1, pp. 297–301.
61. Mehta, D.; Siddiqui, M.F.H.; Javaid, A.Y. Facial emotion recognition: A survey and real-world user experiences in mixed reality. *Sensors* **2018**, *18*, 416. [CrossRef]
62. Bharatharaj, J.; Huang, L.; Mohan, R.; Al-Jumaily, A.; Krägeloh, C. Robot-Assisted Therapy for Learning and Social Interaction of Children with Autism Spectrum Disorder. *Robotics* **2017**, *6*, 4. [CrossRef]
63. Kosonogov, V.; Zorzi, L.D.; Honoré, J.; Martínez-Velázquez, E.S.; Nandrino, J.L.; Martinez-Selva, J.M.; Sequeira, H. Facial thermal variations: A new marker of emotional arousal. *PLoS ONE* **2017**, *12*, e0183592. [CrossRef] [PubMed]
64. Yoshitomi, Y.; Miyaura, T.; Tomita, S.; Kimura, S. Face identification using thermal image processing. In Proceedings of the 6th IEEE International Workshop on Robot and Human Communication, RO-MAN'97 SENDAI, Sendai, Japan, 29 September–1 October 1997; pp. 374–379. [CrossRef]
65. Wang, S.; He, M.; Gao, Z.; He, S.; Ji, Q. Emotion recognition from thermal infrared images using deep Boltzmann machine. *Front. Comput. Sci.* **2014**, *8*, 609–618. [CrossRef]
66. Bijalwan, V.; Balodhi, M.; Gusain, A. Human emotion recognition using thermal image processing and eigenfaces. *Int. J. Eng. Sci. Res.* **2015**, *5*, 34–40.
67. Yoshitomi, Y.; Miyawaki, N.; Tomita, S.; Kimura, S. Facial expression recognition using thermal image processing and neural network. In Proceedings of the 6th IEEE International Workshop on Robot and Human Communication, RO-MAN'97 SENDAI, Sendai, Japan, 29 September–1 October 1997. [CrossRef]
68. Vukadinovic, D.; Pantic, M. Fully Automatic Facial Feature Point Detection Using Gabor Feature Based Boosted Classifiers. In Proceedings of the 2005 IEEE International Conference on Systems, Man and Cybernetics, Waikoloa, HI, USA, 12 October 2005.
69. Di Nuovo, A.; Conti, D.; Trubia, G.; Buono, S.; Di Nuovo, S. Deep Learning Systems for Estimating Visual Attention in Robot-Assisted Therapy of Children with Autism and Intellectual Disability. *Robotics* **2018**, *7*, 25. [CrossRef]
70. Suk, M.; Prabhakaran, B. Real-time mobile facial expression recognition system—A case study. In Proceedings of the IEEE Conference on Computer Vision and Pattern Recognition Workshops, Columbus, OH, USA, 23–28 June 2014, pp. 132–137.
71. Deshmukh, S.; Patwardhan, M.; Mahajan, A. Survey on real-time facial expression recognition techniques. *IET Biom.* **2016**, *5*, 155–163. [CrossRef]
72. Gu, W.; Xiang, C.; Venkatesh, Y.; Huang, D.; Lin, H. Facial expression recognition using radial encoding of local Gabor features and classifier synthesis. *Pattern Recognit.* **2012**, *45*, 80–91. [CrossRef]
73. Liu, C.; Wechsler, H. Gabor feature based classification using the enhanced fisher linear discriminant model for face recognition. *IEEE Trans. Image Process.* **2002**, *11*, 467–476. [PubMed]
74. Boda, R.; Priyadarsini, M.; Pemeena, J. Face detection and tracking using KLT and Viola Jones. *ARPN J. Eng. Appl. Sci.* **2016**, *11*, 13472–13476.

sensors

MDPI

Article

Identifying Traffic Context Using Driving Stress: A Longitudinal Preliminary Case Study

Olga Vl. Bitkina [1], Jungyoon Kim [2], Jangwoon Park [3], Jaehyun Park [1,*] and Hyun K. Kim [4]

[1] Department of Industrial and Management Engineering, Incheon National University (INU), Incheon 22012, Korea; olgabitkina@inu.ac.kr

[2] Department of Computer Science, Kent State University, Kent, OH 44242, USA; jkim78@kent.edu

[3] Department of Engineering, Texas A&M University—Corpus Christi, Corpus Christi, TX 78412, USA; Jangwoon.Park@tamucc.edu

[4] School of Information Convergence, Kwangwoon University, Seoul 01897, Korea; hyunkkim@kw.ac.kr

* Correspondence: jaehpark@inu.ac.kr; Tel.: +82-32-835-8867

Received: 11 April 2019; Accepted: 7 May 2019; Published: 9 May 2019

Abstract: Many previous studies have identified that physiological responses of a driver are significantly associated with driving stress. However, research is limited to identifying the effects of traffic conditions (low vs. high traffic) and road types (highway vs. city) on driving stress. The objective of this study is to quantify the relationship between driving stress and traffic conditions, and driving stress and road types, respectively. In this study, electrodermal activity (EDA) signals for a male driver were collected in real road driving conditions for 60 min a day for 21 days. To classify the levels of driving stress (low vs. high), two separate models were developed by incorporating the statistical features of the EDA signals, one for traffic conditions and the other for road types. Both models were based on the application of EDA features with the logistic regression analysis. City driving turned out to be more stressful than highway driving. Traffic conditions, defined as traffic jam also significantly affected the stress level of the driver, when using the criteria of the vehicle speed of 40 km/h and standard deviation of the speed of 20 km/h. Relevance to industry: The classification results of the two models indicate that the traffic conditions and the road types are important features for driving stress and its related applications.

Keywords: artificial intelligence; driving stress; electrodermal activity; road traffic; road types

---

1. Introduction

Previous studies have identified that the level of driving stress could be affected by different driving conditions [1–4], such as types of roads [5,6], traffic congestion [7], and weather [8]. Dwight and David [7] identified the relationship between traffic conditions and stress levels based on driver interviews. They found that stress was higher for drivers who have experienced traffic congestions. Therefore, aggressive driving behaviors were observed more in high congestion areas than lower ones. Hill and Boyle [8] studied how different driving tasks and roadway conditions influence the stress perceived by drivers. They conducted a survey to assess drivers' stress under various roads, traffic conditions, and weather-related scenarios. The results of this study showed that driving stress was influenced by not only driver characteristics (age, gender, etc.) but also landscape types and driving distances. Therefore, we can assume that road traffic conditions and road types are associated with driving stress.

One of the most accurate indicators of driver stress is the electrodermal activity (EDA) [9–12], which characterizes the activity of electricity on human skin due to sweat [13]. Zangroniz et al. [14] found that the EDA signal is an accurate measure to distinguish calm/stressful conditions. Healey and Picard [5] presented methods for collecting and analyzing EDA data to detect driver stress during various driving conditions on actual roads. Healey and Picard found that EDA and heart rate metrics are the most significantly correlated with driver stress. Rigas et al. [15] presented a novel methodology, based on a dynamic Bayesian network for the estimation of driver stress in specific driving events, using an electrocardiogram (ECG) and EDA signals. Singh et al. [16] studied the feature extraction method and a few algorithms for detecting stress using EDA, ECG, and photoplethysmography (PPG) signals. Munla et al. [17] used heart rate variability analysis for driving stress detection. Lal and Craig [18] studied the psychophysiological changes that occurred during a driver simulator task based on biological signals.

There has been a series of machine learning algorithms that can be applied to detect different events in various research fields by using physiological signal analysis. Plawiak et al. [19,20] applied deep genetic ensemble of classifiers to detect arrhythmia using the ECG signal and artificial neural network to estimate the state of consumption of a pump, based on dynamic pressure and vibrations. Ksiazec et al. [21] used a machine learning approach to detect Hepatocellular Carcinoma using physiological features. Rzecki et al. [22,23] proposed the computational intelligence methods for person recognition using biometric features and used the same method for the automated identification of paper-ink samples through laser-induced breakdown spectroscopy.

Also, many studies have identified that traffic conditions and road types are associated with the levels of driving stress. For the studies of the relationship between traffic conditions and driving stress, Dwight and David [24] reported that a driver's psychological state depends on the road traffic situation, and driving stress is greater in high congestion areas than in low congestion areas. Neighbors et al. [25] identified that slow traffic was linked to greater feelings of pressure and stress. Meanwhile, traffic congestion occurs when the traffic density is exceeded due to a large number of vehicles. According to the traffic flow theory [26], there are a few important traffic flow parameters, namely speed of vehicles, flow (vehicles per hour), density (number of vehicles occupying a given length of highway or lane), and road capacity. In turn, some studies [27,28] suggest that two of the many important characteristics of road traffic are average speed and standard deviation. Table 1 shows that previous research proved the viability of vehicle speed as a characteristic of traffic congestion.

Table 1. Speed as an indicator of traffic congestion [24–28].

References	Brief Description
Pattara-Aticom et al. [29]	Authors classified three levels of traffic congestion based on GPS speed data using threshold technique. It was shown that vehicle velocity is an important characteristic of traffic congestion.
Palubinskas et al. [30]	Authors introduced the traffic congestion detection approach for image time series and found that average velocity is main the traffic parameter.
Thianniwet et al. [31]	Authors proposed a technique to identify road traffic congestion levels using velocity data from a GPS device. Vehicle moving pattern as an important element was extracted through the sliding window technique.
Xing et al. [32]	Authors studied the road tunnel traffic safety and built up the traffic assessment model contained the parameter of speed variance. It was shown that speed variance is an important element of traffic evaluation.
He et al. [33]	Authors analyzed traffic congestion in urban road networks using speed data. The speed performance index was found as the indicator of road state for congested or smooth traffic.

Based on available research, the current study shows a simplified hypothesis that considers only two traffic congestion parameters, namely mean speed (MS) and standard deviation of speed (STDS). The stress level of the driver was assumed to be high in high traffic conditions and low in low traffic conditions. For the studies of the relationship between road types and driving stress, Liu and Du [10] detected low, medium, and high stress levels using an EDA signal and they found that these levels corresponded to no-driving, highway driving, and city driving conditions, respectively. According to Healey and Picard [5] and Westerink et al. [6], city driving is more stressful than highway driving. Based on the previous studies, we can assume that low and high levels of stress correspond to highway and city driving, respectively. Figure 1 shows an overview of the studied factors including traffic conditions as well as road types in this study, their impact on driving stress in terms of mental and physical, and their consequences, such as increasing risks of accidents and decreasing driving performance.

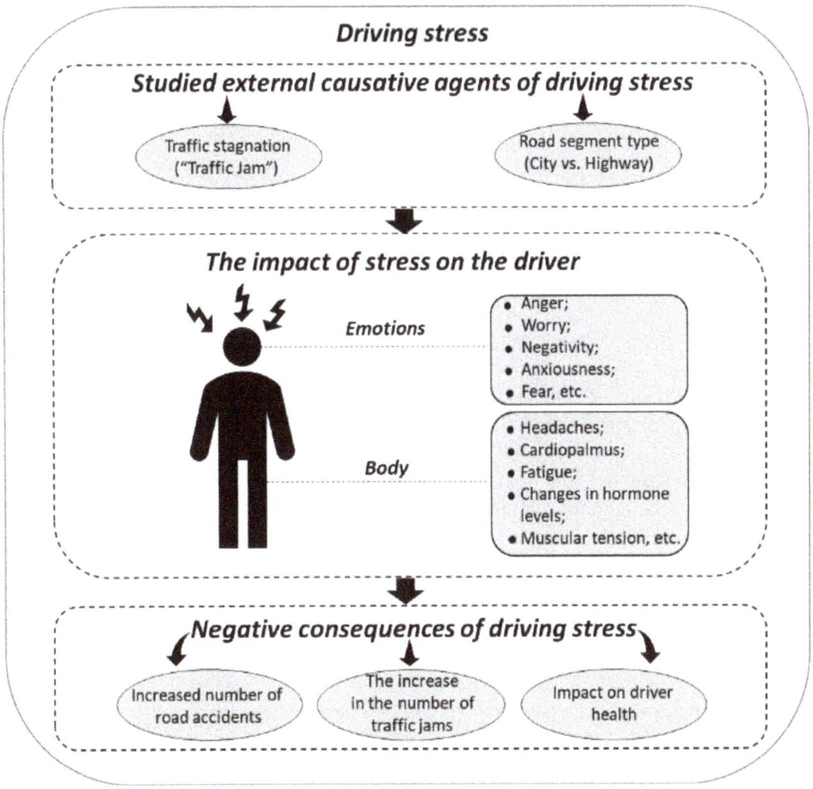

Figure 1. Relationship between driving stress and studied factors.

Summarizing the above information, Table 2 shows a brief compilation of studies which reflects the previous research trends in driving stress detection.

Table 2. Previous studies on driving stress.

References	Collected Mental and Physical Data	Studied Factors	Analysis Methods
Xing et al. [34]	ECG, eye movement, flicker value, face image, self-reported emotional state	Road conditions (three different highways), traffic conditions, driving environment, vehicle behavior	Questionnaire, detection and processing of low/ high-frequency ratio of heart rate variability
Matthews et al. [35]	Self-reported emotional state	Age, type of road (city road, intercity road), frequency of car use, driving conditions (pre-drive, post-drive, weekend), accident involvement, speeding convictions	Questionnaire, factor analysis, ANOVA

Table 2. *Cont.*

References	Collected Mental and Physical Data	Studied Factors	Analysis Methods
Singh et al. [32]	GSR, PPG	Urban driving scenarios (pre-driving, relax driving, busy driving, return driving, rost-driving)	Detection and processing the GSR/PPG signals
Keshan et al. [36]	ECG	Type of road (city road, highway)	Detection and processing the ECG signal
Goel et al. [37]	ECG	Real-time driving in normal road conditions	Detection and processing the ECG signal
Riener [38]	ECG, self-reported emotional state	Specific route, fixed daytime	Post-experiment interview, Detection and processing of low/ high frequency ratio of heart rate variability
Lee et al. [39]	ECG, PPG	Real-time driving in a busy narrow street	Detection and processing the ECG signal
Mundell et al. [40]	GSR	Alternation of rest and driving periods	Detection and processing the GSR signal
Kurniawan et al. [41]	Speech signal, GSR	Real-time driving in usual road conditions	Detection and processing the Speech and GSR signals

The distinctive features of our paper compared with previous research are following conditions and their combinations. The experimental route included city and highway roads without a driver rest time. The driver was influenced by different road types and unstable traffic conditions at the same time. EDA was used as a driving stress measure and for its analysis the special signal features were extracted. Based on Table 1 the vehicle speed signal was used as a traffic conditions indicator and new traffic congestion parameter was extracted by authors. Although many previous studies have identified the relationship between driving stress and traffic conditions, and driving stress and road types, our best knowledge has developed the classification models of driving stress by considering the traffic conditions and road types. In the previous study, most focus has been on the driver's emotional state or stress state. This study was motivated to investigate the relationship between driver stress conditions, road conditions, and road type. We tried to define and predict the traffic jam state itself considering the driver's bio-signals, in addition to the analysis of the road type. The method in this study is expected to contribute to defining a traffic jam. To do this, information was collected on actual roads for a month.

2. Methods

2.1. Participant

The participant of the experiment was a healthy 35-year-old Korean male with a height of 176 cm and a weight of 67 kg. The driver was well acquainted with the driving route and had driving experience of more than 15 years.

2.2. Apparatus

The Empatica E4 wristband (EDA sensor) was used for physiological signal collection in our experiment. The wristband [42] is a wearable and wireless device designed for comfortable, continuous, and real-time data acquisition in daily life. It contains three sensors: (1) PPG sensor that detects cardiovascular features such as blood volume pulse and heart rate variability, (2) EDA sensor that measures an arousal of a sympathetic nervous system and derive features related to stress, engagement, and excitement, and (3) three-axis accelerometer that captures motion-based activity. The wristband operates either in a streaming mode for real-time data viewing on a mobile device, using Bluetooth Low Energy or in recording mode, using its internal memory. In our study, EDA data were collected using the recording mode. The EDA sensor has the following characteristics: sampling frequency of 4 Hz, resolution of one digit –900 pSiemens, range of 0.01–100 μSiemens, and alternating current (8 Hz frequency) with a maximum peak to a peak value of 100 μAmps (at 100 μSiemens) [37].

To record the vehicle movements and on-road situations, a standard on board diagnostic system (OBD-II) was used. OBD-II is a computer-based system built into modern passenger cars, which monitors emission-related controls, and performance of the engine and also detects malfunctions. OBD-II systems provide access to the health information of a vehicle along with numerous parameters and sensors from the engine control unit (ECU). The OBD-II system offers valuable information, including diagnostic trouble codes, when troubleshooting problems [43]. The technical characteristics of OBD-II can be found in the standard signaling protocols for interfaces.

A mid-sized sedan, Hyundai Grandeur (Azera in the U.S.) 6th generation (Hybrid), was used in this study. The humidity values inside the vehicle were maintained between 1016 and 1030 Mbar throughout the experiment.

2.3. Experimental Conditions

In this study, an experiment was conducted during real vehicle driving between the cities of Incheon and Seoul in South Korea, which is a fifty-kilometer route consisting of five main segments: City 1 (Incheon in Korea), Highway 1, Highway 2, Highway 3, and City 2 (Seoul in Korea) with two Tollgate points (TG) (Figure 2). The total time for a one-way drive was approximately 60 min. During this route, two types of recorded data were collected, such as EDA data measured by the Empatica E4 sensor (sampling rate = 4 Hz) and speed signal measured by OBD-II.

All studied data were collected for a month from 13 December 2017 to 13 January 2018. Datasets for 25 weekdays were obtained. Data from four days were omitted because they were inaccurate and unclear, such as unrecorded time data, device failure and zero values of EDA data. The total number of datasets (days) used in the analysis was 21.

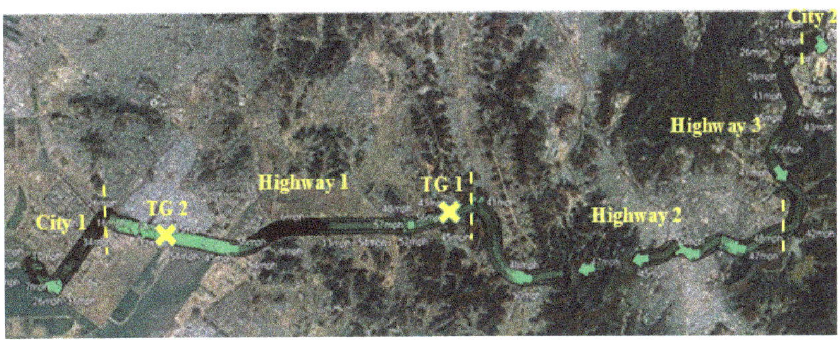

Figure 2. Experiment on the driving route between Incheon and Seoul (captured by Google Earth).

2.4. Measures

Two methods of driving stress evaluation were studied and compared: Dependency between stress and road traffic, and between stress and road type. The vehicle speed and the driver's EDA signals were recorded for different road segments such as city and highway driving as shown in Figure 3.

The road traffic conditions were classified as a low traffic or high traffic state based on the vehicle speed. We determined the high traffic criteria set in Table 3, which characterize potential traffic congestion. The occurrence of high traffic corresponds to high driving stress and its absence to low driving stress. Average speed and standard speed deviation were calculated for every six-minute window of speed data. Speed data for every six-minute window were automatically checked and compared with the adjusted average speed and standard speed deviation sets in Table 3 to determine the potential traffic congestion. If an average and standard deviation of the vehicle speed during a certain period were lower than the criteria set (Table 3), then that period was classified as high traffic conditions, and the other period was classified as low traffic conditions. The driver's stress level was assumed to be low in low traffic conditions and high in high traffic conditions.

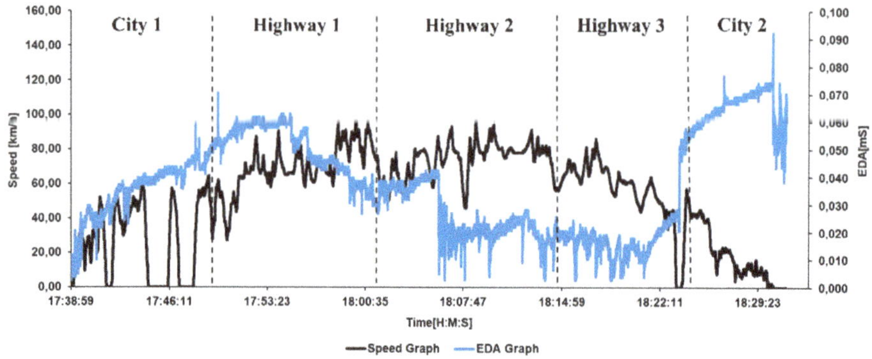

Figure 3. Line plot of the collected speed and electrodermal activity (EDA) data in the same time series for an evening session.

Table 3. Models with various averages and standard deviations of vehicle speed.

Average Speed (km/h)	Standard Deviation (km/h)	Accuracy (%)	Sensitivity (%)	Specificity (%)	Predictive Value (%)
	10	77.3	78	77	64
	15	78.7	68	81	40
20	20	85.8	50	87	15
	25	92.9	0	93	0
	30	97.2	0	97	0
	10	70.9	74	68	67
	15	78	81	76	64
30	20	73	67	75	38
	25	87.1	82	88	36
	30	88.6	80	89	36
	10	71.6	73	69	75
	15	72.3	75	71	65
40	**20**	**80.3**	**85**	**78**	**70**
	25	76.6	70	78	43
	30	79.4	72	81	41
	10	75.9	79	71	79
	15	70.9	73	68	74
50	20	73.8	76	72	66
	25	79.4	82	78	68
	30	75.9	70	78	53

For the second model of dependency between road type and stress state, it was considered that city driving results in a high stress level, and highway driving results in a low stress level. We determined that city driving causes higher stress due to a large number of pedestrians, traffic lights, and traffic congestion. In contrast, such difficult driving conditions are less likely to occur when driving on highways. Therefore, highway driving was classified as low stress driving.

We analyzed the extracted EDA features for each road segment, with a driving time of about 35–40 min on highway-type segments, and 20–25 min on city segments, for a total duration of approximately 60 min. The segments with overlapped periods of high and low levels of stress were excluded. Feature extraction is an important signal processing step for finding more dominating information and for reducing the volume of auxiliary research procedures. It is known that, depending on the type of the signal, different features can be extracted. In this study, for physiological EDA signals, the features proposed by Healey and Picard [5], amplitude (OM) and duration (OD) calculated from signal peaks and valleys, were extracted.

In the current study, the processed EDA signal was resampled at 15.5 Hz. Based on the calculation of mean OD ± 3 × standard deviation (SDOD), it was found that 99.7% of OD falls within the confidence interval. This means that the six-minute window size (approximately equal to 5580 samples) meets the accuracy requirements. In this study, a one-minute sliding window was determined, which was approximately equal to 930 samples, and the moving average window was 60 samples. To find the signal peaks and valleys, the "findpeaks" function in MATLAB (R2017 version) was used. The signal feature extraction process allows us to extract the following EDA characteristics: minimum (min OD and min OM), maximum (max OD and max OM), mean (mean OD and mean OM), standard deviation (SD OD and SD OM), summation (sum of ODs and sum of OMs), and the number of occurrences of duration and amplitude (nOD and nOM). Figure 4 shows the results of applying the OM and OD extraction algorithm for the morning driving session on 11 January 2018.

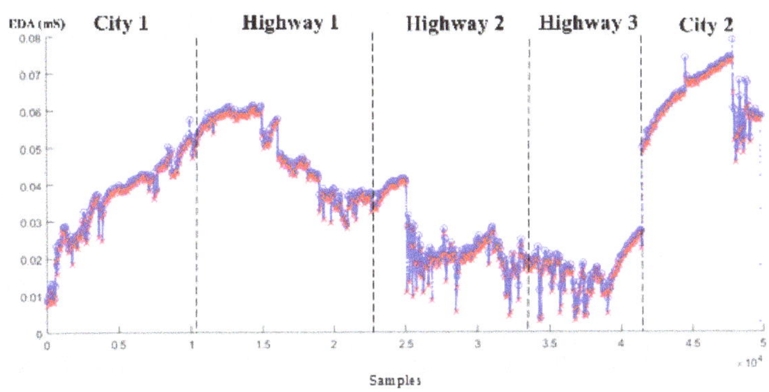

Figure 4. EDA signal processing for January 11, 2018 morning session (peaks and valleys are marked by ○ and ×, respectively).

2.5. Analysis Method

Logistic regression analyses were performed, by using IBM SPSS Statistics Version 23 Software, on every traffic conditions and different road types. Figure 5 contains the schematic description of the development process for both models and their application areas. The physiological features of EDA (OD and OM) and OBD-II data (vehicle speed) were used to construct the same framework in this study. In particular, the model development section of Figure 5 summarizes the study in detail, which is designed to identify the abstract content shown in Figure 1.

The development process in Figure 5 was performed in five steps: Data collection, data pre-processing, analysis, results, and comparison of classifiers. The data collection step shows the period, place, and used sensors during the experiment. Data pre-processing introduces the preliminary processing steps on obtained data for each method. Analysis and results show the analytical methods used, and the main results obtained. Comparison of classifiers provides general comparison for both methods. The model application describes the most applicable areas for the developed methods, such as road traffic management, medicine, and electronic devices.

Based on previous studies in the introductory section, key parameters of physiological signals, road types and traffic situation characteristics have been selected as classifiers. The most important EDA features related to driving stress are minimum, maximum, mean, standard deviation, sum, and the number of occurrences of OD and OM. The most important driving conditions affect the mental state of the driver. Road type [5,6] and traffic jam [24] are representative elements. Based on this, the current study classified the road types and identified traffic jams using the vehicle's speed and the standard deviation of the speed [27,28]. A summary of the models used in this study is shown in Table 4.

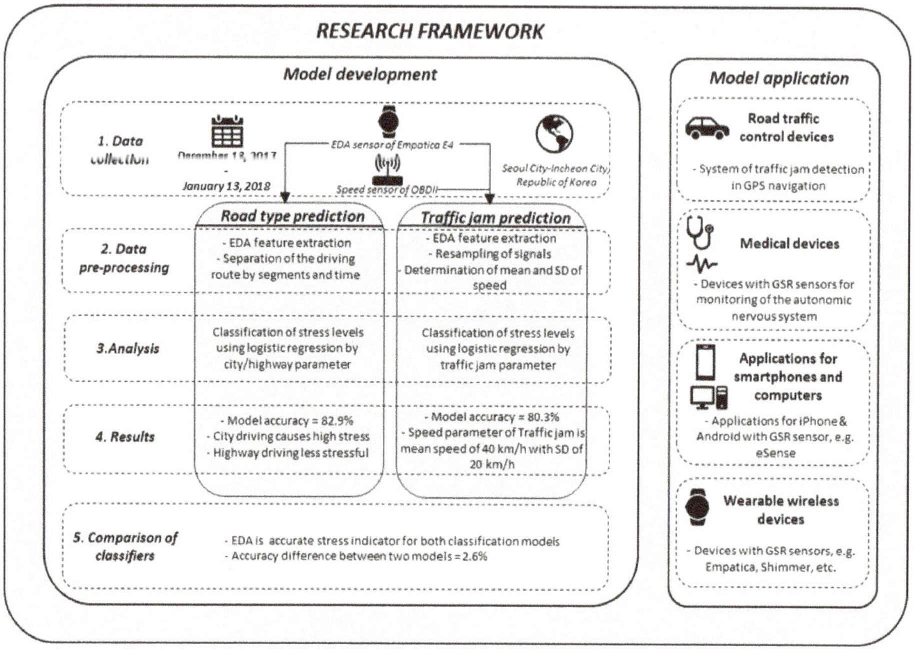

Figure 5. Stepwise development and application of the models.

Table 4. Specification of classification models.

Classification Model	EDA Signal Features	Driving Conditions Features	Analytical Method	Accuracy
Road type prediction	amplitude and duration (min, max, mean, SD, sum, N)	Separation of city and highway section of the path	Logistic regression	82.9%
Traffic jam prediction	amplitude and duration (min, max, mean, SD, sum, N)	Determination of traffic jam criteria using vehicle speed and speed SD	Logistic regression	80.3%

For a complete evaluation of the classification model efficiency, the accuracy (A), sensitivity (Sn), specificity (Sp), and positive predictive value (PPV) were calculated additionally as the specifying characteristics (see Table 3):

$$A = (\text{cases of high stress} + \text{cases of low stress})/(\text{all cases of stress}) = TP + TN/(TP + TN + FP + FN), \quad (1)$$

$$Sn = (\text{cases of high stress})/(\text{all cases of stress}) = TP/(TP + FN), \quad (2)$$

$$Sp = (\text{cases of low stress})/(\text{all cases of low stress}) = TN/(TN + FP), \quad (3)$$

$$PPV = (\text{cases of high stress})/(\text{all cases of high stress}) = TP/(TP+FP), \quad (4)$$

In Equations (1)–(4), FP is the false positive, FN is the false negative, TP is the true positive, and TN is the true negative data. The accuracy determines the ratio of correct predictions to the total analyzed cases. False positives and false negatives are cases when the developed classifier erroneously recognizes low stress as high stress and high stress as low stress, respectively. True positives and true negatives are cases when the developed classifier recognizes stress levels correctly.

3. Results

3.1. Traffic Conditions

A combination of 40 km/h of average speed and 20 km/h of standard deviation, which exhibit relatively higher accuracy, sensitivity, specificity, and positive predictive value, was selected. The mean values of four measures at the combination were the highest. It is highlighted in bold in Table 3. From Table 3, we can see that a few cases have an accuracy of over 80%. However, we cannot select them due to low sensitivity and predictive value, which are under 50%. Low sensitivity shows that a large number of cases with high stress are not recognized, and the low predictive value indicates the low effectiveness of the model in detecting high stress. Thus, considering all the effective parameters of the developed model, the most accurate criterion of low vs. high traffic conditions was found to be with average speed of 40 km/h and standard deviation of 20 km/h: Classification model accuracy = 80.3%, sensitivity = 85%, specificity = 78%, and positive predictivity = 70%. Based on these results, we can conclude that traffic conditions (low vs. high traffic) are an important element to classify driving stress levels (low vs. high stress). In addition, we presumed that traffic conditions (average speed < 40 km/h and SD < 20 km/h) could be a clear threshold of driving stress levels (low vs. high stress), compared to the other 19 conditions in Table 3.

Table 5 shows the developed model for the classification of low stress and high stress levels depending on the high traffic. In the traffic condition model, only NOM was used because the variables NOD and NOM are the same.

Table 5. Model based on traffic conditions (low vs. high traffic).

Predictor	Coefficient	*p*-Value
N	−0.117	0.046
Mean OM	−657.549	0.040
Max OM	66.019	0.047
Min OM	1586.879	0.075
Sum OM	7.747	0.063
SD OM	71.514	0.678
Max OD	0.001	0.727
Min OD	−0.005	0.219
Sum OD	0.000	0.487
Mean OD	−0.001	0.941
Constant	5.444	0.198

It was found that three extracted features (number of occurrences (N), mean amplitude (MeanOM), and maximal amplitude (MaxOM)) of the hand EDA data are significant (all ps < 0.05). The mathematical rule of the developed model for the classification of stress levels depends on the probability concept. High stress was recognized by the model if the probability of case i was greater than 0.5. Otherwise the case was classified as low stress.

The goodness of model fit measure evaluates the ability of developed models to explain variation in the dependent variable (high and low stress). As such measures, pseudo-R-squared are usually used in logistic regression analysis. In the presented study pseudo-R-squared of Cox and Snell and Nagelkerke were obtained using IBM SPSS Statistics Software (Version 23). For traffic condition model Cox and Snell and Nagelkerke R-squares are 0.323 and 0.432 respectively. Figure 6 shows the confusion matrix of the analyzed datasets for the classification of high and low stress levels by using the EDA data, depending on the traffic conditions.

		Actual		
		High traffic	Low traffic	Total
Predicted	High stress	67	8	75
	Low stress	19	45	64
	Total	86	53	139

Figure 6. Confusion matrix of the model using the traffic condition datasets.

3.2. *Road Type*

In the second method for road type prediction, where driving in cities was recognized as a high stress state and highway as a low stress state, the statistical method accuracy reached 82.9% with sensitivity of 81%, specificity of 84%, and positive predictivity of 65%. Table 6 shows the developed model including all used predictors for road type segments.

Table 6. Model based on different types of road segments.

Predictor	Coefficient	*p*-Value
Min OD	0.011	0.031
Max OD	0.000	0.977
Sum OD	0.000	0.378
Mean OD	−0.009	0.240
Mean OM	−128.868	0.604
Max OM	−1.864	0.918
Min OM	18.381	0.976
Sum OM	4.682	0.176
SD OM	94.897	0.383
N	−0.062	0.145
Constant	3.740	0.218

The binary logistic regression model for road type segments shows that Min OD is significant ($p < 0.05$). For road type model goodness of fit was presented by Cox and Snell and Nagelkerke R-square with values of 0.374 and 0.518, respectively. As for the traffic condition model, pseudo-R-squared was obtained through SPSS Statistics Software (Version 23). Figure 7 shows the confusion matrix of the model developed by the road types.

		Actual		
		City driving	Highway driving	Total
Predicted	High stress	71	6	77
	Low stress	14	26	40
	Total	85	32	117

Figure 7. Confusion matrix of the model using the road type datasets.

4. Discussion

4.1. Predictability of Stress State Depending on Traffic Conditions and Road Type

External driving conditions have a decisive influence on the emotional state of the driver and the occurrence of traffic accidents [44,45]. Based on this, two driving conditions of traffic congestion and road type were chosen and compared in this study. Simplified traffic conditions criteria, which characterize traffic congestion, were found to be 40 km/h of average speed with a standard deviation of 20 km/h. If this criterion is met, then we can say that it corresponds to potential traffic congestion and in this moment, the driver is in a high stress state. To confirm this theory, a statistical model was developed. It shows the classification accuracy of 80.3%. From the detected EDA features, the number of occurrences, duration, and amplitude of peaks and valleys were found as significant predictors for the analyzed classification model. The method for determining stress dependency on the road type segments shows the classification accuracy of 82.9% with a significant predictor of Min OD. The goodness of model fit is assessed using various measures and in modern literature there is no consensus which one is better [46]. In our study it was evaluated using Cox and Snell and Nagelkerke pseudo-R-squares, which show improvement of the constant-only model (null model) after applying all predictors (full model with predictors). Cox and Snell pseudo-R^2 is not able to reach "1" even for the perfect model, in turn Nagelkerke R^2 is adjusted Cox and Snell R^2 and its maximal value can be extended to "1" [46]. Obtained results for traffic condition model show that model explain between 32.3% (Cox and Snell) and 43.2% (Nagelkerke) of variance in low and high stress level occurrence, in turn, road type model explain between 37.4% (Cox and Snell) and 51.8% (Nagelkerke) of variance (Table 7). In previous studies, there is no consensus on how to interpret the values of pseudo-R-squares, but some sources [47,48] evaluated the Cox & Snell level higher than 0.3 and Nagelkerke level higher than 0.35 as a satisfactory. Based on this, both our models have a good level of compliance with the observations, but the road type model has higher values and it fits better the observations. The comparison of both methods is shown in Table 7. We also compared the performance of other classifiers using 10-fold cross-validation and separation of training (70%) and testing (30%) data. As shown in Table 8, overall, random forest (RF) has the best performance, the area under the ROC curve (AUC) values of RFs are 85.70% and 79.10% in 10-fold cross-validation and 89.46% and 85.82% in training/testing method, respectively. Surprisingly, multi-layered perceptron which is the neural network-based method shows the overall low performance. Thus, the performance of the tree-based methods is better than neural network-based method in this type of data.

Sensitivity and specificity each reflect how well the model developer is able to grasp high stress and low stress, respectively. A positive predictive value indicates the exactness of high stress. Experimental results show that the overall model of the road type predominates, but it can predict the stress more easily in the traffic-condition-based model. We use the AUC value which reflects the overall performance better as a single value.

Table 7. Comparison of two developed methods.

Method	A (%)	Sn (%)	Sp (%)	PPV (%)	Cox & Snell R^2	Nagelkerke R^2
Traffic conditions	80.3	85	78	70	0.323	0.432
Road Type	82.9	81	84	65	0.374	0.518

Table 8. Comparison of performance (%) with other classifiers.

(a) 10-fold cross-validation.

10-Fold Cross-Validation	ROAD TYPE Sn	Sp	PPV	AUC	Traffic Condition Sn	Sp	PPV	AUC
RF	**64.70**	**88.20**	**76.70**	**85.70**	**60.90**	**86.70**	**79.60**	**79.10**
AB	62.70	90.60	80.00	**86.10**	57.80	88.00	80.40	68.90
NB	52.90	95.30	87.10	84.70	53.10	86.70	77.30	75.60
SVM	56.90	89.40	76.30	73.10	70.30	65.30	63.40	67.80
MLP	54.90	83.30	66.70	75.30	57.80	80.00	71.20	73.80

(b) Training 70% and testing 30%.

Testing (30%) Training (70%)	Road Type Sn	Sp	PPV	AUC	Traffic Condition Sn	Sp	PPV	AUC
RF	**76.47**	**87.50**	**81.25**	**89.46**	**75.00**	**92.31**	**85.71**	**85.82**
AB	76.47	83.33	76.47	83.46	68.75	57.69	50.00	62.74
NB	47.06	91.67	80.00	83.09	68.75	61.54	52.38	73.80
SVM	52.90	91.70	81.80	72.30	12.50	100.00	100.00	59.62
MLP	52.94	62.50	50.00	64.22	52.20	57.90	60.00	54.33

* Random forest (RF), adaBoost (AB), naïve Bayes (NB), support vector machine (SVM), multi-layered perceptron (MLP).

Initial hypothesis about dependency between driving stress, traffic congestion, and road type segments was confirmed. Both methods show accuracies higher than 80% with a small difference of 2–3%, which means that these models are effective and can be used for stress prediction in real driving conditions. Certainly not only these factors have an impact on the predictability of a driving stress state, for example, the internal negative emotions and experiences may affect the driver's state [24,49]. The distinction between these two groups of factors is a difficult scientific task. Based on this, one of the future scientific questions is the improvement of stress detection methods, considering the personal characteristics and emotions of the drivers.

Next factor that can additionally affect the predictability of the stress state is stress measures. Villarejo et al. [50] reported that EDA successfully detects various human emotional states, but it is difficult to distinguish between, for example, the stress situation and the situation of making an effort by a human. EDA sensor may recognize these two different states as the same. Additionally, previous studies devoted to the stress recognition proposed various measures, such as heart rate, blood pressure, muscle tension, etc. [51,52]. In our paper, EDA signal was used as a significant measure of the stress level. It was shown that along with the mentioned characteristics from previous studies, EDA successfully can be used to predict low/high stress during driving in real time. Moreover, EDA sensor is more convenient and useful for long-term monitoring since this type of sensors does not always require the gel-type electrodes and specific contact points on human skin unlike EEG and ECG sensors.

In the presented study, classical statistical models based on logistic regression analysis were developed. In turn, previous studies show that different approaches can be used for stress and emotional state recognition. Rigas et al. [53] proposed a model to predict driving stress (high, medium and low) using EDA and ECG signals based on a dynamic Bayesian network with an accuracy range of 31–94%. Magana et al. [54] introduced the deep learning algorithms based on the heart rate variability (HRV) signal to estimate the stress of drivers and passengers with an accuracy range of 86–92%. Jabon et al. [55] used a few classifiers, including Bayesian nets, decision tables, decision trees, support vector machines, regressions, and LogitBoost to predict unsafe driver behavior by a facial expression based on Kappa values (in most cases, LogitBoost classifier provided the highest Kappa statistic). By comparing these with the methods proposed in our paper, it should be noted that elements of advanced deep learning can be considered in the future to increase the prediction ability of the developed models. From the viewpoint of average accuracy, the presented models show satisfactory results. The proposed models can be extended to include more variables or features, but for other classifiers, there can be limitations due to algorithm inflexibility. Using presented models needs minimal efforts and simple devices. Additionally, it can be simply applied for science and industry. Otherwise, some advanced approaches can be more difficult. In general, the combination of developed and more advanced classifiers can result in offering applications with improved accuracy and predictability.

Our study demonstrates and compares two developed models capable of a satisfactory accuracy to predict the stress state of the drivers based on physiological signals. Obtained results expand previous research and new findings can be used in traffic management, medicine, and design of electronic devices with physiological sensors.

4.2. Limitation of This Study and Future Research

The presented research is the initial result of a long-term study about stress prediction based on physiological signals. Even though both developed classification models showed good results, there are a few limitations, which will need to be addressed in the future. First, a shortcoming of this study is the number of participants, which was limited to one male driver. The authors compensated for it by increasing the experimental period to one month to obtain the optimal number of driving datasets for analysis. Second, the temperature in the car was not regulated during the experiment. It will need to be controlled in future studies because this condition can possibly affect the psychological state and sweating intensity of the driver. Third, the traffic congestion concept is a simplified point of view. The concept of the traffic flow theory [26] will be used in future research. From this theory, the speed of vehicles, traffic flow, density, road capacity, and distance between vehicles will be considered to extend the developed methods. Fourth, the accuracies of models obtained can be improved by increasing their sensitivity to different emotional states of the driver, such as anger, annoyance, stress, etc., as discussed in the previous studies [24,49,50].

Few additional observations will need to be considered in the future. For example, direct EDA data evaluation shows that average EDA data are approximately doubled in the morning route (EDA = 0.03129 mS) than in the evening route (EDA = 0.01543 mS). A higher EDA level was observed in the morning session on 12 January 2018, which was the coldest observed day with a temperature of −14 °C in the experimental period. Potentially, this result indicates that weather conditions and the time of day affect the driver's EDA signal, and these factors will be considered in the future. This and further research can find many applications, including the development of wearable devices containing physiological sensors.

5. Conclusions

This study presented the findings on the levels of driving stress depending on the traffic conditions and road types, using the EDA signals. The study confirmed that EDA is a significant indicator of the psychological stress and an effective tool to determine the stress levels in real driving conditions. Advantages of the developed method are that both developed models have an ability to predict stress levels under actual driving conditions with over 80% accuracy. The developed models can also be used to classify driving stress in different traffic situations and different road types. The definition of a traffic jam state in a driving situation is one of the important contributions of this study. One of the characteristics of the models is the ability to classify the stress level without considering the rest state of the driver. This makes the models universal for use in a variety of situations without the preliminary intervention. The proposed model and the experimental procedure are expected to be easily reconfigured by other researchers. One of the disadvantages of the proposed model is that the traffic congestion concept needs to be defined and improved more specifically based on traffic flow theory. To make a better model in the future, the following points should be considered. First, researchers can collect more road information, such as lane-keeping status, as well as bio-signals, such as ECG. In addition, various kinds of machine-learning techniques can be applied instead of logistic regression. Finally, researchers can incorporate other factors, such as seasonal factors, in-vehicle temperature, and a driver's attention into the model as key features.

Obtained results can be used for practical application. Examples of such uses are sensors for monitoring of the autonomic nervous system, smartphone/computer applications and wearable wireless devices with EDA sensors.

Author Contributions: J.P. proposed the main idea and conducted the experiment; O.V.B. finished the draft manuscript and drew the figures and tables; J.P. and J.K. analyzed the data; H.K.K. revised and finished the manuscript.

Funding: This work was supported by the Incheon National University Research Grant in 2017 (Grant No.: 20170467).

Conflicts of Interest: The authors declare no conflict of interest.

References

1. Antoun, M.; Ding, D.; Bohn-Goldbaum, E.E.; Michael, S.; Edwards, K.M. Driving in an urban environment, the stress response and effects of exercise. *Ergonomics* **2018**, *61*, 1–9. [CrossRef] [PubMed]
2. Rigas, G.; Goletsis, Y.; Bougia, P.; Fotiadis, D.I. Towards driver's state recognition on real driving conditions. *Int. J. Veh. Technol.* **2011**, *2011*, 617210. [CrossRef]
3. Munoz-Organero, M.; Corcoba-Magana, V. Predicting upcoming values of stress while driving. *IEEE Trans. Intell. Transp. Syst.* **2017**, *18*, 1802–1811. [CrossRef]
4. Matthews, G.; Tsuda, A.; Xin, G.; Ozeki, Y. Individual differences in driver stress vulnerability in a Japanese sample. *Ergonomics* **1999**, *42*, 401–415. [CrossRef]
5. Healey, J.A.; Picard, R.W. Detecting stress during real-world driving tasks using physiological sensors. *IEEE Trans. Intell. Transp. Syst.* **2005**, *6*, 156–166. [CrossRef]
6. Westerink, J.H.D.M.; Ouwerkerk, M.; Overbeek, T.J.M.; Pasveer, W.F.; DeRuyter, B. *Probing Experience*; Springer: Berlin, Germany, 2008; p. 14.
7. Dwight, A.H.; David, L.W. The relationship between traffic congestion, driver stress and direct versus indirect coping behaviours. *Ergonomics* **1997**, *40*, 348–361.
8. Hill, J.D.; Boyle, L.N. Driver stress as influenced by driving maneuvers and roadway conditions. *Transp. Res. Part F Traffic Psychol. Behav.* **2007**, *10*, 177–186. [CrossRef]
9. Affanni, A.; Bernardini, R.; Piras, A.; Rinaldo, R.; Zontone, P. Driver's stress detection using skin potential response signals. *Measurement* **2018**, *122*, 264–274. [CrossRef]
10. Liu, Y.; Du, S. Psychological stress level detection based on electrodermal activity. *Behav. Brain Res.* **2018**, *341*, 50–53. [CrossRef]
11. Ollander, S.; Godin, C.; Charbonnier, S.; Campagne, A. Feature and sensor selection for detection of driver stress. In Proceedings of the 3rd International Conference on Physiological Computing Systems, Lisbon, Portugal, 27–28 July 2016; pp. 115–122.
12. Ooi, J.S.K.; Ahmad, S.A.; Chong, Y.Z.; Ali, S.H.M.; Ai, G.; Wagatsuma, H. Driver emotion recognition framework based on electrodermal activity measurements during simulated driving conditions. In Proceedings of the IEEE-EMBS Conference on Biomedical Engineering and Sciences, Orlando, FL, USA, 16–19 February 2017; pp. 365–369.
13. Farnsworth, B. What Is GSR (Galvanic Skin Response) and How Does It Work? Available online: https://imotions.com/blog/gsr/ (accessed on 10 April 2019).
14. Zangroniz, R.; Martinez-Rodrigo, A.; Manuel, P.J.; Lopez, M.T.; Antonio, F.C. Electrodermal activity Sensor for classification of calm/distress condition. *Sensors* **2017**, *17*, 2324. [CrossRef]
15. Rigas, G.; Katsis, C.; Bougia, P.; Fotiadis, D. A reasoning-based framework for car driver's stress prediction. In Proceedings of the 16th Mediterranean Conference on Control and Automation, Ajaccio, France, 25–27 June 2008; pp. 627–632.
16. Singh, R.R.; Conjeti, S.; Banerjee, R. An approach for real-time stress-trend detection using physiological signals in wearable computing systems for automotive drivers. In Proceedings of the 14th International IEEE Conference on Intelligent Transportation Systems, Washington, DC, USA, 5–7 October 2011; pp. 1477–1482.

17. Munla, N.; Khalil, M.; Shahin, A.; Mourad, A. Driver stress level detection using HRV analysis. In Proceedings of the International Conference on Advances in Biomedical Engineering, Beirut, Lebanon, 16–18 September 2015; pp. 61–64.
18. Lal, S.K.; Craig, A. Driver fatigue: Electroencephalography and psychological assessment. *Psychophysiology* **2002**, *39*, 313–321. [CrossRef]
19. Plawiak, P.; Acharya, U.R. Novel deep genetic ensemble of classifiers for arrhythmia detection using ECG signals. *Neural Comput. Appl.* **2019**, 1–25. [CrossRef]
20. Plawiak, P. An estimation of the state of consumption of a positive displacement pump based on dynamic pressure or vibrations using neural networks. *Neurocomputing* **2014**, *144*, 471–483. [CrossRef]
21. Ksiazek, W.; Abdar, M.; Acharya, U.R.; Plawiak, P. A novel machine learning approach for early detection of hepatocellular carcinoma patients. *Cogn. Syst. Res.* **2018**, *54*, 116–127. [CrossRef]
22. Rzecki, K.; Sosnicki, T.; Baran, M.; Nedzwiecki, M.; Krol, M.; Lojewski, T.; Acharya, U.R.; Yildirim, O.; Plawiak, P. Application of computational intelligence methods for the automated identification of paper-ink samples based on LIBS. *Sensors* **2018**, *18*, 3670. [CrossRef]
23. Rzecki, K.; Plawiak, P.; Nedzwiecki, M.; Sosnicki, T.; Leskow, J.; Ciesielski, M. Person recognition based on touch screen gestures using computational intelligence methods. *Inf. Sci.* **2017**, *415–416*, 70–84. [CrossRef]
24. Hennessy, D.A.; Wiesenthal, D.L. Traffic congestion, driver stress, and driver aggression. *Aggress. Behav.* **1999**, *25*, 409–423. [CrossRef]
25. Neighbors, C.; Vietor, N.A.; Knee, C.; Raymond, A. Motivational model of driving anger and aggression. *Personal. Soc. Psychol. Bull.* **2002**, *28*, 324–335. [CrossRef]
26. Mathew, T.V.; Rao, K.V.K. *Introduction to Transportation Engineering*; NPTEL: Mumbai, India, 2006; 6p.
27. Praburam, G.; Koorey, G. Effect of on-street parking on traffic speeds. In Proceedings of the IPENZ Transportation Conference, Christchurch, New Zealand, 14 March 2015; pp. 1–9.
28. Shankar, H.; Raju PL, N.; Rao, K.R.M. Multi model criteria for the estimation of road traffic congestion from traffic flow information based on fuzzy logic. *J. Transp. Technol.* **2012**, *2*, 50–62. [CrossRef]
29. Pattara-Aticom, W.; Pongraibool, P.; Thajchayapong, S. Estimating road traffic congestion using vehicle velocity. In Proceedings of the 6th International Conference on ITS Telecommunications, Chengdu, China, 21–23 June 2006; pp. 1001–1004.
30. Palubinskas, G.; Kurz, F.; Reinartz, P. Traffic congestion parameter estimation in time series of airborne optical remote sensing images. In Proceedings of the SPRS Hannover Workshop 2009—High Resolution Earth Imaging, Hannover, Germany, 2–5 June 2009; p. 6.
31. Thianniwet, T.; Phosaard, S.; Pattara-Atikom, W. Classification of road traffic congestion levels from GPS data using a decision tree algorithm and sliding windows. In Proceedings of the World Congress on Engineering, London, UK, 1–3 July 2009; pp. 1–5.
32. Singh, R.R.; Conjeti, S.; Banerjee, R. Assessment of driver stress from physiological signals collected under real-time semi-urban driving scenarios. *Int. J. Comput. Intell. Syst.* **2014**, *7*, 909–923. [CrossRef]
33. He, F.; Yan, X.; Liu, Y.; Ma, L. A traffic congestion assessment method for urban road networks based on speed performance index. *Procedia Eng.* **2016**, *137*, 425–433. [CrossRef]
34. Xing, J.; Hirai, S. Driving stress of drivers on narrowed line and hard shoulder of motorways. *PEOPLE Int. J. Soc. Sci.* **2017**, *3*, 124–135.
35. Matthews, G.; Dorn, L.; Glendon, I. Personality correlates of driver stress. *Personal. Individ. Differ.* **1991**, *12*, 535–549. [CrossRef]
36. Keshan, N.; Parimi, P.V.; Bichindaritz, I. Machine learning for stress detection from ECG signals in automobile drivers. In Proceedings of the IEEE International Conference on Big Data, Santa Clara, CA, USA, 29 October–1 November 2015; pp. 1–7.
37. Goel, S.; Kau, G.; Toma, P. A novel technique for stress recognition using ECG signal pattern. *Curr. Pediatr. Res.* **2017**, *21*, 674–679.
38. Riener, A. Sitting postures and electrocardiograms: A method for continuous and non-disruptive driver authentication. In *Continuous Authentication Using Biometrics: Data, Models, and Metrics*; IGI Global: Hershey, PA, USA, 2012; pp. 137–168.
39. Lee, H.B.; Kim, J.S.; Kim, Y.S.; Baek, H.J.; Ryu, M.S.; Park, K.S. The relationship between HRV parameters and stressful driving situation in the real road. In Proceedings of the 6th International Special Topic Conference on Information Technology Applications in Biomedicine, Tokyo, Japan, 8–11 November 2007; p. 9827162.

40. Mundell, C.; Vielma, J.P.; Zaman, T. Predicting performance under stressful conditions using galvanic skin response. *arXiv* **2016**, arXiv:160601836.
41. Kurniawan, H.; Maslov, A.V.; Pechenizkiy, M. Stress detection from speech and galvanic skin response signals. In Proceedings of the 26th IEEE International Symposium on Computer-Based Medical Systems, Porto, Portugal, 20–22 June 2013.
42. E4 UM. Empatica, User's Manual 2018. Available online: https://empatica.app.box.com/v/E4-User-Manual (accessed on 10 April 2019).
43. OBD Site. Available online: https://www.obdautodoctor.com/ (accessed on 10 April 2019).
44. Lyu, N.; Xie, L.; Wu, C.; Fu, Q.; Deng, C. Driver's cognitive workload and driving performance under traffic sign information exposure in complex environments: A case study of the highways in China. *Int. J. Environ. Res. Public Health* **2017**, *14*, 203. [CrossRef] [PubMed]
45. Taylor, A.H.; Dorn, L. Stress, fatigue, health, and risk of road traffic accidents among professional drivers: The contribution of physical inactivity. *Annu. Rev. Public Health* **2006**, *27*, 371–391. [CrossRef] [PubMed]
46. Allison, P. What's the Best R-Squared for Logistic Regression 2013? Available online: https://statisticalhorizons.com/r2logistic (accessed on 10 April 2019).
47. CD Site (Comlete Dissertation). Available online: https://www.statisticssolutions.com (accessed on 10 April 2019).
48. IBM Site. Available online: https://www.ibm.com/us-en/?lnk=m (accessed on 10 April 2019).
49. Mesken, J.; Hagenzieker, M.P.; Rothengatter, T.; de Waard, D. Frequency, determinants, and consequences of different drivers' emotions: An on-the-road study using self-reports, (observed) behaviour, and physiology. *Transp. Res. Part F Traffic Psychol. Behav.* **2007**, *10*, 458–475. [CrossRef]
50. Villarejo, M.V.; Zapirain, B.G.; Zorrilla, A.M. A stress sensor based on galvanic skin response (GSR) controlled by zigbee. *Sensors* **2012**, *12*, 6075–6101. [CrossRef]
51. Vrijkotte, T.G.M.; Van Doornen, L.J.P.; De Geus, E.J.C. Effects of work stress on ambulatory blood pressure heart rate and heart rate variability. *Hypertension* **2000**, *35*, 880–888. [CrossRef]
52. Rigas, G.; Goletsis, Y.; Fotiadis, D.I. Real-time drivers stress event detection. *IEEE Trans. Intell. Transp. Syst.* **2012**, *13*, 221–234. [CrossRef]
53. Rigas, G.; Katsis, C.D.; Bougia, P.; Fotiadis, D.I. A reasoning-based framework for car driver's stress prediction. *Control Autom.* **2008**. [CrossRef]
54. Magana, V.C.; Organero, M.M.; Fisteus, J.A.; Fernandez, L.S. Estimating the stress for drivers and passengers using deep learning. *Proc. JARCA* **2016**, *2016*, 1–6.
55. Jabon, M.; Bailenson, J.; Pontikakis, E.; Takayama, L.; Nass, C. Facial expression analysis for predicting unsafe driving behavior. *IEEE Pervasive Comput.* **2011**, *10*, 84–95. [CrossRef]

sensors

MDPI

Article

Recognition of Emotion Intensities Using Machine Learning Algorithms: A Comparative Study

Dhwani Mehta, Mohammad Faridul Haque Siddiqui and Ahmad Y. Javaid *

Electrical Engineering and Computer Science Department, The University of Toledo, 2801 W Bancroft St, MS 308, Toledo, OH 43606, USA; dhwani.mehta@utoledo.edu (D.M.); mohammadfaridulhaque.siddiqui@utoledo.edu (M.F.H.S.)

* Correspondence: ahmad.javaid@utoledo.edu; Tel.: +1-419-530-8161

Received: 24 March 2019; Accepted: 18 April 2019; Published: 21 April 2019

Abstract: Over the past two decades, automatic facial emotion recognition has received enormous attention. This is due to the increase in the need for behavioral biometric systems and human–machine interaction where the facial emotion recognition and the intensity of emotion play vital roles. The existing works usually do not encode the intensity of the observed facial emotion and even less involve modeling the multi-class facial behavior data jointly. Our work involves recognizing the emotion along with the respective intensities of those emotions. The algorithms used in this comparative study are Gabor filters, a Histogram of Oriented Gradients (HOG), and Local Binary Pattern (LBP) for feature extraction. For classification, we have used Support Vector Machine (SVM), Random Forest (RF), and Nearest Neighbor Algorithm (kNN). This attains emotion recognition and intensity estimation of each recognized emotion. This is a comparative study of classifiers used for facial emotion recognition along with the intensity estimation of those emotions for databases. The results verified that the comparative study could be further used in real-time behavioral facial emotion and intensity of emotion recognition.

Keywords: automatic facial emotion recognition; intensity of emotion recognition; behavioral biometrical systems; machine learning

---

1. Introduction

The dual fears of identity theft and password hacking are now becoming a reality, where the only hope of a secure method for preserving data are behavioral systems. Systems which are based on user behavior are usually understood as behavioral systems. Behavioral traits are almost impossible to steal. Multiple commercial, civilian, and government entities have already started using behavioral biometrics to secure sensitive data. One of the major components of behavioral biometrics is the recognition of facial emotion and its intensity [1–3]. In the industry and academic research, physiological traits have been used for identification through biometrics. Any level of biometrics could not be performed without good sensors, and when it comes to facial emotion intensity recognition, apart from high-quality sensors (cameras), there is a need for efficient algorithms to recognize emotional intensity in real time. With the increased use of images over the past decade, the automated facial analytics such as facial detection, recognition, and expression recognition along with its intensity has gained importance and are useful in security and forensics. Components such as behavior, voice, posture, vocal intensity, and emotion intensity of the person depicting the emotion, when combined, help in measuring and recognizing various emotions.

Considering human facial images, as seen in Figure 1, recognizing the emotions and finding their intensity are vital. Primarily, 3D facial images are the most thoroughly researched [4–7] and predictions are made by the available systems based on the features that were extracted from the images for emotion and intensity recognition. The intensity of emotion plays a significant role in behavioral biometrics for future crime prediction systems. The intensity may be referred to as *"the degree of manifestation along the dimension of behavior"* [3]. The intensity of emotion is often directly associated with the intensity of facial muscle movements. This, in turn, indicates that the intensity of muscle movement represents the index of the intensity of emotional state, implying the intensity of the emotion that is being experienced. Such intensities can be measured in both spontaneous and posed expressions. These intensities are affected by the behavior of the person, whether the emotion depicted is voluntary or involuntary. Spontaneous facial expressions of an emotion indicate the behavior of the face that occurs when a person displays involuntary emotion, with no prior planning or intention. Posed facial expressions of emotions, on the other hand, are used on a large-scale for studies involving the intensity of facial emotions. The criterion for the accuracy of intensity detection of the five observed basic emotions (and a neutral expression) is based on the analysis of the facial behavior components that are relevant to emotional intensity communication [8]. This involves detecting the face and recognizing the intensity of emotion depicted, both of which could be achieved using classifiers assisted by a training set. Existing work gives a detailed survey of the algorithms used in this area [9–11]. Also, multiple intensities are measured with a value of rank one recognition and rank five recognition, where rank one is that the intensity is measured at the highest accuracy level and rank five at the lowest. Algorithms that have been used for feature extraction span from classical techniques such as principal component analysis (PCA) and linear discriminant analysis (LDA) to modern techniques such as machine learning (ML) and artificial neural networks (ANN) [10–16].

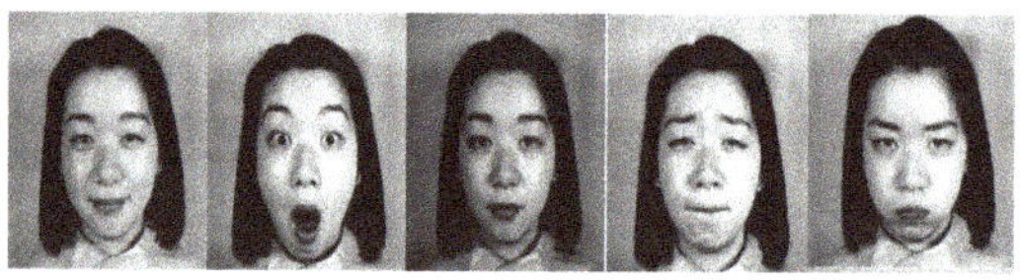

Figure 1. Illustration of 5 Basic Emotions: Happy, Surprise, Neutral, Sad and Angry.

In recent years, major studies were carried out in the field of emotion intensity detection relating to three major areas. The first area relates to the cross-cultural character, where studies concluded that cultures highly agree with each other in facial emotion identification [8,17,18]. These studies also revealed a cross-cultural agreement in the prediction of the intensity of two different expressions for the same emotion [19]. The second area brings forward the research which shows that there exist major differences in gender skills to decode/predict nonverbal signs. In these studies, women have been superior predictors of emotions than men [20–23]. In the third area, several studies have been conducted, based on the five basic emotions, to find major error patterns and effects of the emotional intensity in diseases such as schizophrenia, autism, and borderline personality disorder [24–26].

Emotion recognition is being put to service in diverse real-life applications where a person's emotional state serves as a cue to the successful operation of these systems. Physically and mentally challenged people are impuissant to show their emotions and require an alternate criterion to perceive their emotional state of mind. The Autism Spectrum disorders in an individual have been a core area of research in affective computing, and several research works have laid a tangible emphasis of emotion recognition for such cases [27–29]. Patients seeking therapies [30,31] and counselling for depressions [32], disorder of consciousness [33], and schizophrenia [34] have been a subject of interest in many explorations to unearth their curbed and concealed emotions. The information filtering systems also referred to as recommender systems, find an interest in detecting a person's emotional state. These systems predict the preference of the user for a particular good or service and recommend the best possible option according to its ability. There are health-related recommender systems such as emHealth [35], hospital recommender systems [36], multimedia recommender systems [37,38] and for movies [39,40], web search [41] and e-commerce [42,43]. Prominent use of emotion recognition is in marketing and for getting automatic feedback. The concept, referred to as emotional marketing [44] milk affective computing for decision support [45,46] and for product feedback assessment [47,48].

E-Learning is also setting its feet wet by exploiting emotion recognition [49,50]. An emotional state of the learner may suggest a modification in the presentation style and to be more interactive for effective tutoring. Use of emotion recognition is also getting prevalent in gaming where his emotional state governs a user's interactions. Exergames (or gamercizing) frameworks [51,52], and other interactive and cloud-based gaming are becoming emotional-aware to provide a more immersive gaming experience [53,54]. The use of emotion recognition is also applicable in law for detecting and perceiving the intentions of the suspect [55] and in monitoring for risk prevention [56] and for smart health-care [57].

This work primarily explores the use of popular ML algorithms to recognize the intensity of emotion in combination with popular feature extraction algorithms. Following is a brief explanation of the major contributions of this comparative study:

- Three feature extraction algorithms, Gabor Features, Histogram of Oriented Gradients (HOG) and Local Binary Pattern (LBP), have been used and compared on five popular databases (B DFE, CK, JAFFE, and one of our own).
- Three popular ML algorithms, SVM, RF, and kNN were used for emotion intensity recognition.
- A comparative study and implementation of algorithms for measuring facial emotions and their intensities based on the different AUs (Action Units) are presented.
- The highest accuracy achieved for LBP combined with SVM shows that the intensity of emotions can be further used for real-world applications such as crime prediction systems and drowsy driver detection in vehicles to prevent fatalities [58].

The paper includes an in-depth literature review which discusses recent works in the area of facial emotion intensity recognition and is presented as Section 2. The current challenges and motivation behind this research are also discussed in Section 2. Section 3 gives a brief insight into the experiment along with the techniques that were used. We pay particular attention to the face detection techniques and emotion recognition algorithms in this section. The different correlations between Action Unit (AU) intensities and the description of the method for calculating intensities of the emotions based on facial expressions are described in Section 3.1. AUs can be understood as the specific parts (or units) of a face which come into action while a face depicts emotion. This section also discusses the gold-standard categorization of AUs, popularly used in emotion detection and recognition systems. The accuracy achieved from various experiments and the comparative study is presented in Section 4. Finally, the paper is concluded with a discussion on insights gained from the experiments and future work in Section 5.

2. Literature Review

Facial emotion recognition has been used for a variety of applications such as the identification of Autism and Schizophrenia, detection of a drowsy driver [59], identifying abnormalities in early stages of Alzheimer's disease or schizophrenia, and for crime prediction systems. Before we discuss compare the prevalent works, it is important to understand the datasets that were used to train the recognition algorithms by popular works. Several databases exist that have been used, and are summarized in Table 1 which has been reproduced from [60]. It is also important to understand that several algorithms pose limitations. As an example, the Viola-Jones algorithm can be used along with Haar feature selection and AdaBoost training algorithm for remarkable detection of the eye region and nose bridge region with the limitation that it is only effective for the frontal images and can hardly cope with a 45° face rotation [61].

Table 1. Popular databases for measuring intensity of emotions (Reproduced from [60]). (Type: P: Only Posed, S: Only Spontaneous, SP: Spontaneous and Posed).

Database	Database Description	Emotion Description Intensity	Type
Cohn-Kanade (CK) [62]	100 multi-ethnic subjects, 69% female, 31% male (age: 18–50) with frontal and 30° view	23 series of facial display, available, AU-coded face database (single and combination)	P
Extended CK (CK+) [63]	123 multi-ethnic subjects, extension of CK	66/123 subjects considered, spontaneous smiles, onset to peak coded	SP
JAFFE [64]	Japanese Female Facial Expression, Grayscale images with 10 subjects	6 basic emotions + neutral, 2–4 samples per expression, available	P
Bosphorus	105 subject, 44 female, 61 male	2D/3D AU-coded, pose, and illumination variations, available	P
BU-3DFE [65]	Binghamton University 3D Facial Expression, multi-ethnic, 56 female and 44 male (age: 18–70), 2500 samples	4 intensity levels, 6 basic emotions + neutral, available	P
RU-FACS	Rochester/UCSD Facial Action Coding System, 100 subjects	AU-FACS coded, private database with 33 AUs, unavailable	SP
NVIE	Natural Visible and Infrared (IR) facial Expression, Visible and IR imaging, 215 students (age: 17–31)	temporal analysis for face data, basic 6 expressions, available	SP
MMI	25 multi-ethnic subjects, 12 female, 13 male (age: 20–32)	Onset, offset, apex temporal analysis, single + combined AUs, available	SP
DISFA [60,66,67]	Denver Intensity of Spontaneous Facial Actions, 12 female, 15 male, 130,000 video frames	Intensity of 12 AUs coded, available	S
Belfast Induced Emotion	Three set of tasks, lab-based emotion induction tasks	Intensity plus emotion, trace style rating of intensity and valence	S
PAINFUL DATA	UNBC-McMaster Shoulder Pain Expression Archive Database, 200 video sequences, 66 female, 63 male	Pain related AUs coded, available	S

Furthermore, we studied various works where emotion recognition was reported as an aggregate score of all emotions or expressions and summarized our findings in Table 2. We also studied works that attempt to measure the emotional intensity of AUs, and therefore, facial emotion. Here, we look at some of the works in a little more detail that reported accuracies for individual emotion detection. Jens et al. experimented to find the expression leakage in relation to the emotional intensity with a database created by capturing images of 21 participants [68]. The study was based on three emotions of happiness, sadness, and fear. The results of the study showed that there was facial expression leakage between the posed and genuine emotions. These *leaks* were measured by finding the intensity differences between the emotions. The participants were shown video footage and were told to enact genuine and posed emotions for the experiments. The study was limited to participants from an academic background with an average age of 19.4.

Table 2. Comparative study of accuracy achieved for intensity of emotions.

Work	Technique	Database	Accuracy	Drawbacks
[69]	Iterated Closest Point (ICP), PCA	355 images	92.00%	Low accuracy *
[70]	Geometric-based approach, Haar feature selection technique (HFST)	FRGC v2	97.00%	Unsuitable for real-time applications
[71]	K-means clustering with back-propagation	CK	98%	Unsuitable for real-time applications
[72]	AAM, Lucas-Kanade, BPNN	BU-3DFE	83.80%	Low accuracy
[73]	Feature Distribution Entropy, Euclidean distances between 83 3D facial feature points, Adaboost	BU-3DFE	95.1%	Unsuitable for real-time applications
[74]	Distance vectors, neural network	BU-3DFE	91.30%	Low accuracy
[75]	Principal Component Analysis	FRGC v1	95%	—
[42]	Stochastic Neighbor Embedding, Gabor wavelet, SVM	JAFFE	58.70%	Low accuracy, small database
[76]	Local Binary Patterns (LBP)	10 Volunteers	89.60%	Low accuracy, small database
[77]	Gradient-based ternary texture patterns, SVM	CK	97.10%	—
[78]	Gaussian curvature, Gabor wavelets, shape index, SVM	Bosphorus, (2902 images)	63.10%	Only 25 AUs from still images considered
[79]	Facial data generator, Hercules engine	Webcam data	89.60%	Low accuracy, slow
[80]	Spectral regression, SVM	18 min-video	92%	Limited data
[81]	Dynamic Bayesian Network (DBN), Adaboost (ADB)	DISFA	72.77% (DBN), 69.31% (ADB)	Low accuracy, AU extraction and intensity inference phases independent
[82]	Conditional random Field Model	DISFA, FERA2015	70% (DISFA), 50% (FERA)	Limited data, low accuracy

* Low accuracy indicates it is low for real-world applications.

Hess et al. measured the intensity of emotional facial expression for the emotions of happiness, sadness, anger, and disgust, posed by two men and women [83]. The expressions depicted had six levels of emotional intensities. The accuracy achieved were 85.3% (anger), 79.3% (disgust), 97.9% (sadness), and 91.8% (happiness). The method included collecting and pre-processing the colored images into high-quality grayscale images. This study was conducted using a set of 5 pre-defined intensity of expressions of 20%, 40%, 60%, 80%, and 100%. The authors concluded the study by providing evidence that the perceived intensity of the underlying emotion varied linearly with the physical intensity of expressions.

Biele et al. performed an experiment where the intensity of happiness and anger were measured using static and dynamic images (animations) [23]. Anger was noted to be more intense than the extreme happy emotion. Also, the intensity results for dynamic images were higher than that of the static images. Gender differences also had an impact on the results, e.g., a difference in intensity was observed for anger in males and both emotions among females. They used the 2 (Subject Sex) · 2 (Actor Sex) · 2 (Emotion) · 2 (Stimuli Dynamics) ANOVA method for finding the intensity of emotions. A need for a new methodology for measuring the intensity of emotions to have a better insight into how men and women processed different emotional information was stressed upon in this work.

Another work experimented the perceived emotion by measuring the intensities of the emotions [84]. The basis of this analysis relied entirely on the fact that facial expressions conveyed rich information concerning the state of mind. Researchers have argued that neuromotor programs of facial expression patterns might be able to serve as the building blocks of emotion communication [84–87].

The interaction pattern was observed behaviorally for emotion intensity ratings, and neutrally for functional magnetic resonance imaging activation in the amygdala (a roughly almond-shaped mass of gray matter inside each cerebral hemisphere, involved with the experiencing of emotions), as well as fusiform and medial prefrontal cortices. However, this behavior was observed only for mild-intensity expressions.

Delannoy et al.'s experiments were based on the image-based approach for the three ranks of intensities, low–medium–high [88]. They performed one-against all SVMs for training classifiers. They used only 68 images from 10 subjects from the CK+ dataset and used ten-fold cross-validation. However, the drawback in their approach was that multi-class classifiers approach assume that the rankings of intensities were independent of each other. Hence the relation between labels was not well employed for performance enhancement. Furthermore, even with the classification of expressions in three intensity rankings, the AU must still be extracted from the image sequence as a feature representation.

Several other researchers performed experiments to conduct more accurate expression intensity estimation results [78,89–91]. The authors employed the training data with known intensity labels for these works. These intensity labels were categorized into two sections, discrete rank representations [88,89], and continuous value representations [92]. Once the label intensities were employed, the expression intensity degree was predicted and validated more accurately. The drawback of these approaches was that multiple images or an image sequence were required to estimate the intensity of emotion and. Also, sufficient images were required for estimation in a sequence-based approach.

Littlewort et al. performed their experiment by conducting estimation of the intensity rankings for facial expressions by using SVMs [93]. The authors used the distances to the SVM hyperplanes to estimate the rankings of emotion intensities. Chang et al. performed a manifold learning approach discrimination for recognizing facial expressions and used the distances of the manifold to determine the intensities of the emotions [94]. However, the distances to the classification boundaries in the feature space may not necessarily reflect how neutral/strong was an expression and hence, concluded inaccurate results.

Rudovic et al. proposed a method using intrinsic topology of multidimensional continuous facial affect data. An ordinal manifold first modeled the data. The topology was then used for the (H-CORF) Hidden Conditional Ordinal Random Field. Later, it was used for dynamic ordinal regression [95]. This was done to constrain H-CORF parameters to lie on the ordinal manifold. The resulting model attained simultaneous dynamic recognition and intensity estimation of facial expressions of multiple emotions. This method was used for both posed and spontaneous expressions. They also tested their model on databases such as BU-4DFE, CK, and CK+. All this research on the previously applied techniques and these facts provide the evidence that brings us to the conclusion that more research is required for the accurate detection of emotions with intensities.

Current Challenges and Motivation

Even with a higher accuracy of emotional intensity measurements, practical real-time systems face a lot of problems. These problems are primarily related to time resolution, low-level emotion recognition (facial expressions captured with low peak frequencies), scarcity of the available databases for research-related intensity measurement, face and angle variation, illumination variations, and non-alignment of the faces. The research involving facial emotion recognition is divided into two categories: (i) image-based and (ii) video-based. Image-based recognition uses static frames for the recognition, while the video-based method includes dynamic frames for recognition. A lot of work is needed to address the problems mentioned above. Low-level emotion recognition is one such problem, since measuring the facial intensity of emotion is directly related to human–machine interaction (HMI), where robots can interact more naturally if they know the intensity of the emotion of the person they are communicating with. Security, surveillance, biometrics, and patient monitoring are a few other problems that have not been explored much. Majority of the work concentrates on the emotions depicted at the peak-level of the intensity, neglecting the emotions depicted at lower intensity levels as they are comparatively difficult to recognize. Measuring the low-level frame intensity is a challenge as the available databases lack the discriminating features.

Comprehensive standards were designed by Paul Ekman and Friesen to subdivide each emotion into several special AUs which is popularly known as the Facial Action Coding Units (FACS). These categorizations are considered to be a gold standard for emotion detection and recognition systems. An insight into the relevant literature in the field reveals that the examination of the accuracy of intensity predictions of 5 basic emotions from spontaneous/posed facial expressions has been highly ignored. Besides, the attention has been primarily paid to the examination of the factors that affect the perception of the category of emotion (the one that leads to expression analysis such as happiness/sadness/fear/anger), neglecting the intensity. Naturally, this does not mean that research on the accuracy of intensity prediction is less critical. With advances in technology, providing an input image and classifying it into one of the six emotions (5 basic, and one neutral) is proving to be insufficient. Hence, more attention needs to be given to the intensity of the five basic emotions to judge and understand human emotions for HMI, medical, and biometrical applications. Over the years, psychological research has shown that understanding the dynamics of the expressions is equally essential to understand human emotion. In other words, emotion dynamics is primarily related to emotion intensity variations in spatial and temporal domains. The future where robots successfully understand human emotions through intensity is far away since much work remains to be done. This underlying problem was the primary motivation to follow the AUs and their intensity to solve futuristic issues.

3. Methodology

In this work, a comparison between the feature extraction techniques and the classification algorithms is presented to find the best combination that can be used for emotion intensity recognition. Figure 2 shows an overview of the experiment in the form of a generalized architecture, where training and testing layers are shown in detail. The first layer, called the training layer, has the following stages:

1. Image input and sequencing.
2. Pre-processing such as masking, scaling, converting into grayscale, and noise reduction.
3. Feature extraction algorithms such as LBP, Gabor, HOG are used, and a final feature vector is created using concatenation.
4. To remove the unwanted features, dimensionality reduction was used
5. Classification algorithms such as SVM, RF, and kNN were used to classify the AUs.

The second layer, testing layer, primarily has two stages. First, similar to the training stages, image sequencing, pre-processing, and feature extraction & selection was performed. Second, these features were passed through the trained model and finally an emotion intensity decision was made based on the AUs. Before we further discuss the intricate details of our comparative study, it is important to understand that this comparative study evaluates multiple techniques at both the intermediary step of feature extraction as well as the final step of classification.

HOG and LBP both create histograms to express features. HOG uses gradients to build spatial and orientation cells and assembles histograms of these gradients using overlapping spatial blocks while LBP considers a neighborhood block and computes and normalizes the histograms by converting the binary-threshold code to an integer. On the other hand, Gabor features are extracted using Gabor filters and use frequency patterns of regions of interest to extract features for segmentation and texture analysis. Gabor filter uses functions that relate filter size, oscillation frequency/phase, and orientation. Although technically Gabor filter is closest to the human visual perception system, LBP is known to be computationally simpler and work better in various illuminations. HOG, on the other hand, comes with the advantage of using different block sizes and number of histogram bins, unlike LBP.

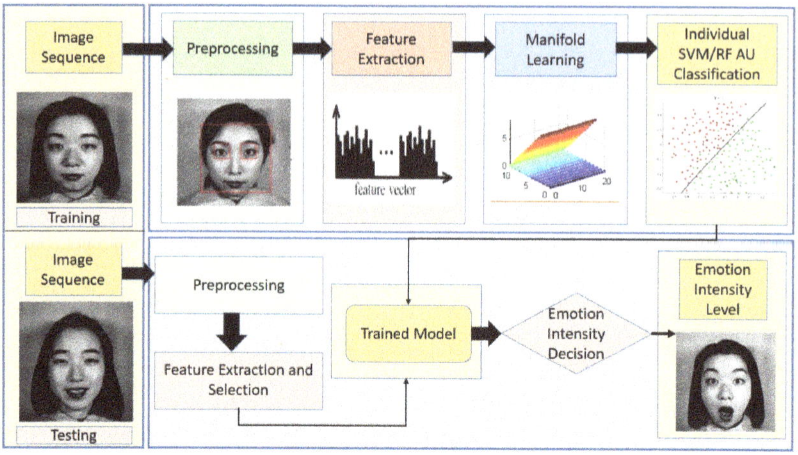

Figure 2. Generalized Architecture for Intensity of Emotion Recognition.

The ML techniques we used in this work include kNN, SVM, and RF. While SVM is known to have a generalization ability by mapping inputs non-linearly to higher dimensional feature spaces through its capability of separating training data with a hyperplane. kNN, a type of instance-based learning, involves the neighbors deciding the class (among k classes) a specific data point belongs to. Closest neighbors are assigned using popular methods such as Euclidean or Hamming distance. RF is a collection of several decision trees which do not need linear features or even features that interact linearly. These three classification algorithms are known to perform well for high-dimensional spaces as well as a large number of training samples. Each of these algorithms works well under specific circumstances, kNN for noisy data, SVM for linearly inseparable data, and RF for categorical features. Due to these specific features, we chose to use these methods. All these techniques have been widely used in the literature, as discussed in Section 2. This section further discusses the essential details of the experiment performed.

3.1. AU Intensity Feature Extraction and Correlation Analysis

This section consists of the description of the observed AU intensity feature extraction model, which consists of facial image registration and representation, dimensionality reduction, feature extractors, and classifications as shown in Figure 3. In this paper, we represent and capture the semantic AUs relations, as well as the correlation between the intensities of the AUs. This is done to measure the intensities of facial emotions more robustly.

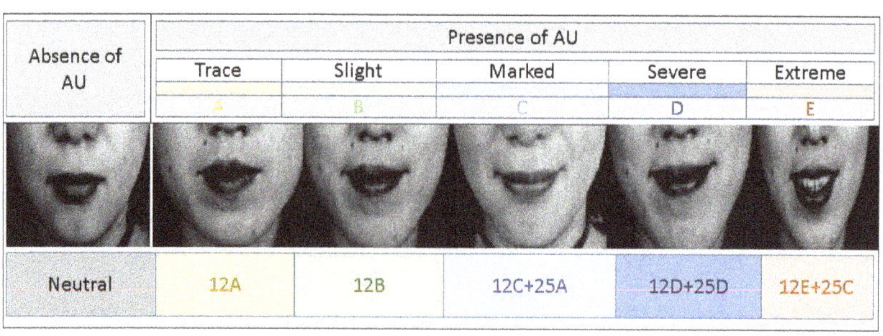

Figure 3. Relation between scale of evidence and intensities of facial action units.

Due to the variety, the dynamics of facial actions, and the ambiguity, it is challenging to measure the intensities of AUs in a single frame. Mostly, databases are created with posed and spontaneous expressions, where it is a challenge to measure the intensity of spontaneous expressions as they occur more randomly. AUs significantly occur in combinations, where they are not always additive. This implies that the occurrence of an AU can be different than its original standalone type. A perfect example is shown in Figure 4, where AU12 occurs alone in Case A and the lip corner are pointing straight-slightly upwards. In Case B, AU15 lip corners appear a bit angled towards the ground, and in Case C, both co-occur, they are non-additive and hence, recognizing that emotion and intensity of the emotion become more difficult.

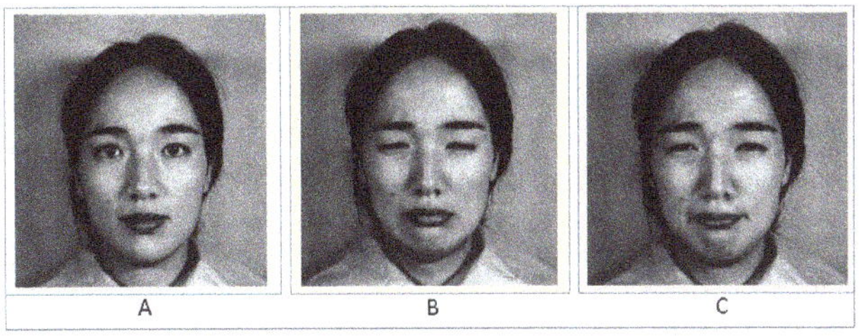

Figure 4. AU Combination: (**A**) AU12 occurs alone; (**B**) AU15 occurs alone; (**C**) AU12 and AU15 occur together—non-additive.

FACS manual gives an insight into the inherent relationship between the AUs that can produce the required information for measuring and analyzing the emotional intensity. The manual mentions that the inherent relationships can be subdivided into two classes—the class of mutual exclusions and the class of co-occurrences. Here, the class of co-occurrences is a class of groups of AUs which generally and most frequently appear together to give meaning to the depicted facial emotions. For example, AU6 + AU12 + AU25 suggest "happy" while AU4 + AU15 + AU17 depict "sad". In the case of mutually exclusive AUs, the FACS manual provides alternative rules. Mutually exclusive AUs rarely occur together in spontaneous emotions in day-to-day life. FACS mentions that it is difficult to demonstrate two AUs such as AU25 (lips apart) and AU24 (pressed lips) together at all. This suggests that the mutually exclusive cases are very much possible, but with very low probability. There are still few limitations to the co-occurrence class in terms of intensity levels of emotion AUs. For example, when AU6 (raised cheeks) occur with AU12 (lip corner puller) as shown in Figure 5, both the AUs present a high/low-intensity level of one another.

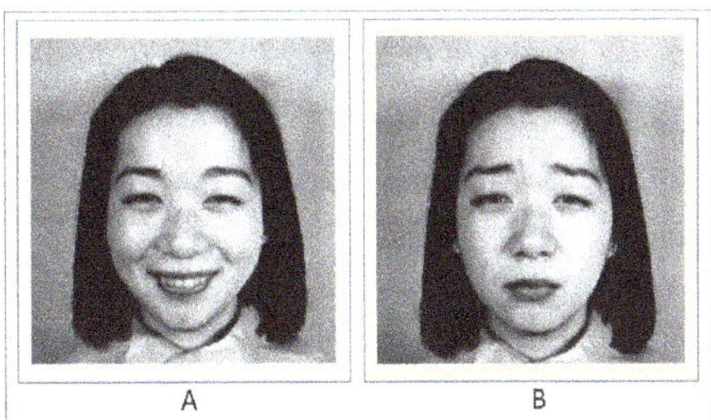

Figure 5. AU Combination: (**Case A**) AU6 + AU12 + AU25, (**Case B**) AU4 + AU15 + AU17.

3.1.1. Face Registration and Representation

This step aligns the data of a similar kind, such as input facial images plus the referenced facial images. Landmark points were used to mark to represent the important location of facial components such as the eyes, nose, and lips. To obtain the landmark points, an averaging solution was used, and this averaging was done over the entire training data set. The images are finally masked for the extraction of important facial regions and re-sized to 128 × 108 pixels. After this step, three well-known algorithms such as Histogram of Oriented Gradient, Gabor Features, and LBP were used for feature extraction for the reason that they are highly capable of representing the appearance-based information accurately.

3.1.2. Feature Extraction through Gabor Features

Gabor features are comparative to the human visual system because of their frequency and orientation representations. A 2D Gabor feature, in the spatial domain, is a Gaussian kernel function which is modulated by a sinusoidal plane wave. These filters can be generated by one major wavelet by rotation and filter. These are the best among the other existing relevant image features such as the edge orientation histograms, box filters. In our experimentations, we extracted magnitudes on 96 × 96 images sizes using the directions of eight wavelets and scales of nine so that the Gabor wavelengths vary from the range of 2 to 32 pixels in half octave intervals. Although the resulting feature vector has 9 × 8 × 96 × 96 = 663,552 components, not all of them are useful. In fact, in our experiment, a very small number of informative components are selected. To perform Gabor analysis, first the eye centers are located, and then the images are aligned accordingly. This alignment is done by performing transformation, rotation, and scaling. This is a typical procedure for 2D images for registration. Normalization is done using manual determination of landmarks. This is done to preclude any misalignment effects from the registration schemes.

3.1.3. Local Binary Pattern Method

The LBP method is based on a texture descriptor that is useful in extracting features from any textured image. We used an LBP for extracting facial features that are used for estimating the intensity of the emotion depicted in the image. The LBP is non-overlapping and uniform when applied to an image. Initially, a user specified number of uniform blocks are used to segment the image. For each patch on the image, the LBP matches the center pixel to its surrounding neighboring pixel to generate an *LBP* value. Equations (1) and (2) mentioned below are used for computation of *LBP*, where N represents the adjacent pixels, *k* is the neighboring size, and C is the center pixel. For this research, we have considered the value of *k* = 8.

$$LBP(N,C) = \sum_{k=0}^{7} P(N_k - C)2^k \tag{1}$$

$$[y] = \begin{cases} 1 & \text{for } y \geq 0 \\ 0 & \text{for } y < 0 \end{cases} \tag{2}$$

The function *LBP* (*N*, *C*) (from Equation (1)) uses the P (Nk-C) as seen in Equation (2), generates a 1 or a 0 depending on the difference between the center pixel and the neighbor. Figure 4 shows an example of a neighboring pixel with their intensity values. Later the differences are calculated considering the center pixel. Equation (1) is used for the transition from difference matrix to Bit String Matrix (is a sequence of 0's and 1's). The most important step in LBP is that the starting position must be arbitrarily chosen for calculation. This is done by unwrapping the bit string and decoding it.

200	190	50
120	150	150
225	40	150

50	40	−100
−30	N/A	0
−75	−110	0

1	1	0
0		1
0	0	1

The number of bit string pattern within a patch is counted to create a feature vector that is used in a distance measure. For an 8-bit string, there are a total of 256 possible bit strings. Furthermore, for simplification of the process, the string is either considered to be uniform or non-uniform. A string is considered to be uniform when its bits, parsed in a circular sequential manner, has a shift of values two or fewer times. Similarly, a string is non-uniform when its bits have changed more than two times. e.g., consider the string 00011110. Here only two shifts occur. One between the third and fourth position and one between the seventh and eighth position. Out of the total 256 patterns, only 58 are uniform. For every patch of an image, a histogram is created which is composed of 59 bins. All the 58 uniform patterns are assigned to those 58 bins in the histogram, where each bin stores the frequencies of the patterns. The one bin which is left (59th bin) keeps an account of all the non-uniform patterns found in the patch. Furthermore, all the histogram vectors from patches are concatenated to represent a histogram representing the features extracted by the LBP.

3.1.4. Histogram of Oriented Gradient Features

This method was initially used in the human detection area, further used as object detectors and finally, they are now used for analyzing and representing the facial emotions. The descriptor HOG can quickly and efficiently describe the local shape and appearance of objects by counting the occurrences of gradient orientations in a localized portion of the images. In this study, the images are divided into small cells, and for every single cell the histogram of the gradient is calculated. This is done to represent the spatial information of the face image. For every image, in our study, 48 cells are constructed out of every image by building a cell with 18 × 16 pixels. A horizontal gradient filter [−1 0 1] was applied with 59 orientation bins in the study. Final step done was the concatenation of all the HOG representations of each cell to form a HOG feature vector (size of 2832 (48 × 59)).

3.1.5. Dimensionality Reduction

High-dimensional features of an image make the analysis of the samples more complicated in the real-world applications where ML and pattern recognition algorithms are used. When extracting and selecting features, several features extracted are redundant and should be removed. e.g., in ML, univariate feature selection is made to avoid the use of redundant features for training. Literature review above has shown that facial expression and intensity of those expressions are embedded along a low dimensional manifold in a high-dimensional space. In our study, we have implemented nonlinear techniques for preserving the local information which is further useful in the classification of the intensity of facial emotions and their representation. Manifold learning is a technique which presumes that the sample data points are collected from a low dimensional manifold and embedded into a high-dimensional space. Quantitatively, Consider, a set of points (6), find a set of points (7), such that y_i represents x_i efficiently.

$$x_1 ... x_n \in R^D \tag{3}$$

$$y_1 ... y_n \in R^D (d \backslash D) \tag{4}$$

The Laplacian eigenmap algorithm was used in our study to reduce the dimensionality of the data. Furthermore, the high-dimensional data was mapped to a 29-dimensional space. The basics of the algorithm are to map the closest points of the high-dimensional space into the close points of the low dimensional space. For problem-solving, the generalized eigenvector problem is applied. Further to describe the embedded d-dimensional Euclidean space the first d eigen vectors in correspondence to the first d eigen values are used. Spectral regression algorithm was used to find a projection function which can map the high-dimensional data, in our study the HOG, Gabor features, and LBPs into low dimensional space.

3.1.6. Classification

AU intensity classification was performed using SVM, RF, and kNN classifiers after reducing the feature vectors. SVMs analyze the data and are used for pattern recognition. They construct a hyperplane or a set of hyperplanes which are further used for regression and classification problems. Discriminative hyper planes are found by the SVM classifiers including the highest margin for dividing the data that belongs to the different classes. However, several kernel types can affect the efficiency of the SVM classifiers. In our study, we used the $C = 6$ AU intensity levels of each of the following AUs. The strategy considered is the one-against-one strategy where $C(C-1)/2$ binary discriminant functions are defined, one for each possible pair of classes. The Gaussian RBF kernel was used. In our study, each AU and each frame were considered individually since it is an appearance-based approach. The results are also largely affected by the face region alignment.

3.2. Databases Considered

Training of ML algorithms depend on the type and size of database used. A low number of images in a database that is being used for training, can cause under-fitting. To counter the issue, we need a large database for training the available ML algorithms. The estimations and results are remarkably affected by the use of larger databases over the smaller ones; hence, for more accurate results a database is required in abundance. Emotion classification and their intensity estimations require a vast and varied dataset for validation and testing. The images used for intensity estimation and recognition of emotion are spontaneous and posed.

- Posed Datasets: Popular for capturing extreme emotions. The disadvantage is the artificial human behavior.
- Spontaneous Datasets: Natural human behavior. However, it is extremely time-consuming for capturing the right emotion.

Hence, a close relationship exists between the models used for the intensity of facial emotions and the databases used. Mainly five databases CK [62,63], JAFFE [64], B-DFE and an in-house dataset of 200 images (20 images of each of the basic emotions considered) taken in an uncontrolled environment through a web camera. Here is a brief description of each database:

1. JAFFE: The database contains 213 images of 7 facial expressions (6 basic facial expressions + 1 neutral) posed by 10 Japanese female models. Each image has been rated on 6 emotion adjectives by 60 Japanese subjects. (Posed – no AUs present)
2. DISFA: In this database the images were acquired using PtGrey stereo imaging system at high resolution (1024 × 768). The intensity of AU's (0–5 scale) for all video frames were manually scored by two human FACS experts. The database also includes 66 facial landmark points of each image in the database. (Spontaneous—AUs present (AU1, AU2, AU4, AU5, AU6, AU9, AU12, AU, 15, AU17, AU20, AU25, AU26))
3. CK: Subjects in the released portion of the Cohn-Kanade AU-Coded Facial Expression Database are 100 university students. They ranged in age from 18 to 30 years. Sixty-five percent were female, 15 percent were African-American, and three percent were Asian or Latino. Subjects were instructed by an experimenter to perform a series of 23 facial displays that included single AUs and combinations of AUs. (Posed: AUs present (AU1, AU2, AU4, AU5, AU27))
4. BU-3DFE: Includes 100 subjects with 2500 facial expression models. The BU-3DFE database is available to the research community (e.g., areas of interest come from as diverse as affective computing, computer vision, human computer interaction, security, biomedicine, law-enforcement, and psychology). The database contains 100 subjects (56% female, 44% male), ranging age from 18 years to 70 years old, with a variety of ethnic/racial ancestries, including White, Black, East Asian, Middle-east Asian, Indian, and Hispanic Latino. (Posed: AUs present (AU1, AU2, AU4, AU5, AU6, AU9, AU12, AU, 15, AU17, AU20, AU25, AU26))
5. Our own: Dataset consists of 200 images (20 images of each of the 5 basic emotions considered) taken in an uncontrolled environment through a web camera. The reason for collecting the dataset in an uncontrolled environment was that the intensity of facial emotions is to be measured in an everyday uncontrolled environment for practical real-time applications. The dataset consists of probe and gallery images which are taken over a month. There are in total 200 probe images and 100 gallery images from the same subjects of different emotions which are used for validation of the algorithm purposes. The images are taken in different angles and lighting conditions to see the effect of these factors on our proposed model. (No AUs present)

It should be noted that the AUs labeled by each database were different. Hence, comparison between posed and spontaneous images was not possible for all databases. The only comparison would be between DISF and B-DFE; however, B-DFE is a 3D image database and features extracted from those images unlike DISF, which had 2D images. Hence, a comparison between those two would not be a fair comparison. The feature vector of a 2D image was in the form [width, height, 3] and the feature vector in the 3D image dataset was of the form [width, height, depth, 3], i.e., both feature representations would be different and hence, not comparable. A possible way was to perform feature extraction by passing 3D volumes through a pre-trained 3D network/algorithm, or to perform 2D feature extraction on each slice of the volume and then combine the features for each slice, using PCA to reduce the dimensionality. However, this would impact the accuracy. Therefore, such a comparison was not presented.

4. Results

4.1. Recognition and Reliability Measures

Recognition rate and the Intra-Class Correlation (ICC) values were exploited to evaluate the proposed automated AU intensity measurement in our study. The statistical index, i.e., the ICC has a range from 0 to 1.

$$ICC = \frac{BMS - EMS}{BMS + (k-1) \times EMS} \tag{5}$$

$$EMS = \frac{ESS}{(k-1) \times (n-1)} \tag{6}$$

$$BMS = \frac{BSS}{n-1} \tag{7}$$

It is also the measure of conformity for our data set since it has multiple targets. This study is basically where n participants are being judged by k number of judges. In our study, we assume $n = 6$ and $k = 2$. The purpose of using ICC is because it is preferred over the Pearson correlation between measurement and judges. The ICC shows the proportion of total variance between the targets. Here, BMS = Between target Mean Squares, EMS = Residual Mean Squares which are defined by the ANOVA (Analysis of Variance).

4.2. Result Analysis Based on Intensity of Emotions

The previous section discusses the feature extraction techniques we implemented for measuring AU intensities in facial emotions. The three techniques implemented include the LBP, Histogram of Oriented Gradient Features, and Gabor Features. These are followed by classification techniques such as Support Vector Machine, Random Forest and Nearest Neighbor Classifiers. Given all the image observations, we implement the network for measuring the intensity of emotion recognition for each AU. The results are presented in Table 3.

As shown in Table 3, the best results were achieved with the *LBP* with the nearest neighbor classifier while using all three features. This is because it models static relationships between the AU intensities. All the values in the table are percentage accuracy in detection. AUs which were not present for certain cases, have been indicated as NA (Not Applicable) while zeros indicate that the AU was present, but accuracy was 0 indicating that it was not recognized at all.

Table 3. AU Intensity Measurement Results—HOG, Gabor and LBP (Acc: Accuracy Percentage).

							HOG							
#		**AU1**	**AU2**	**AU4**	**AU5**	**AU6**	**AU9**	**AU12**	**AU15**	**AU17**	**AU20**	**AU25**	**AU26**	**AVG**
1	ICC	0.71	0.68	0.8	0.55	0.67	0.76	0.82	0.69	0.63	0.61	0.86	0.69	0.70
2	Acc	87.22	71.49	89.48	92.76	89.01	94.18	91.89	77.1	97.15	79.3	96.14	97.74	**88.62**
3	L = 0	88.13	89.67	90.17	87.28	88.97	78.56	93.47	95.17	88.14	95.58	89.66	84.36	89.09
4	L = 1	87.31	77.41	79.36	74.15	68.36	69.48	88.54	87.79	84.58	74.69	78.34	77.14	78.92
5	L = 2	84.26	59.68	36.74	68.45	77.96	85.14	39.62	47.27	53.99	77.4	69.35	65.72	63.79
6	L = 3	77.74	66.03	82.23	91.18	66.89	93.78	86.79	73.8	84.57	78.45	79.88	84.33	80.47
7	L = 4	55.63	59.68	69.74	66.13	64.58	78.69	67.48	61.34	61.79	79.77	78.49	77.41	68.39
8	L = 5	33.19	18.97	92.36	49.89	0	54.36	0	NA	16.89	NA	68.36	69.44	40.34
							Gabor							
#		**AU1**	**AU2**	**AU4**	**AU5**	**AU6**	**AU9**	**AU12**	**AU15**	**AU17**	**AU20**	**AU25**	**AU26**	**AVG**
1	ICC	0.82	0.87	0.89	0.59	0.81	0.87	0.8	0.79	0.76	0.69	0.96	0.87	0.81
2	Acc	88.9	91.74	83.69	96.27	88.74	93.18	96.78	98.36	71.49	68.98	85.47	82.89	**87.20**
3	L = 0	99.1	97.86	97.84	95.36	99.67	97.36	99.87	96.87	98.48	98.36	99.55	99.68	98.33
4	L = 1	68.15	26.68	84.15	79.36	88.36	84.16	91.89	77.98	89.74	85.69	87.48	89.79	79.45
5	L = 2	77.89	74.59	89.61	84.67	80.57	84.36	86.74	75.06	82.71	70.11	78.31	68.03	79.38
6	L = 3	79.36	61.92	56.98	65.69	66.26	69.93	55.53	76.25	73.81	82.11	89.02	61.93	69.89
7	L = 4	64.69	55.81	53.67	59.57	61.26	60.72	48.69	0	70.14	85.97	68.14	60.25	57.40
8	L = 5	33.69	64.56	59.57	55.79	0	15.79	0	NA	0	NA	97.82	93.67	42.08
							LBP							
#		**AU1**	**AU2**	**AU4**	**AU5**	**AU6**	**AU9**	**AU12**	**AU15**	**AU17**	**AU20**	**AU25**	**AU26**	**AVG**
1	ICC	0.89	0.88	0.74	0.85	0.83	0.84	0.77	0.78	0.91	0.74	0.82	0.87	0.82
2	Acc	97.81	98.17	94.09	96.07	96.87	96.74	98.99	98.065	98.77	95.67	95.89	98.88	**97.16**
3	L = 0	85.69	81.02	68.19	79.6	98.24	92.16	77.4	91.8	90.56	96.3	77.49	80.45	84.90
4	L = 1	53.64	27.73	44.69	60.13	42.94	33.02	59.23	67.14	66.98	78.09	74.08	76.31	57.00
5	L = 2	68.42	61.08	78.61	98.07	58.41	60.79	82.66	87.9	61.27	71.08	63.78	55.09	70.59
6	L = 3	59.67	68.07	82.34	57.09	63.87	90.27	83.44	67.07	56.65	77.4	59.77	61.79	68.95
7	L = 4	47.83	80.17	86.49	97.11	84.03	66.08	72.11	77.6	95.17	96.79	93.37	88.27	82.08
8	L = 5	90.81	49.67	56.47	33.96	0	84.17	55.37	NA	0	NA	87.28	66.98	52.47

For observations in images, which are not very accurate, improvements are seen in features of Gabor wavelets using the random forest classifiers. Table 4 shows the performance of individual features when combined with popular ML algorithms. Besides, a correlation analysis for the AUs was done, for which we have listed a correlation matrix especially for the action units 1 and 2 in the Table 5. This matrix is a relation between the AU1 and AU2. The intensity dependency between both the AUs is proportional to each other. High of AU1 results in a high probability of AU2 and vice-versa. When AU2 is at level 0, AU1 probability is 0.982 and when the intensity at AU2 is level "3" AU1 probability at level "3" is 0.88. By calculating such AU dependency relationships between the action units, the ICC and the accuracy for various algorithms improved. Although not shown in the table, but we would like to mention that the accuracy increased from 68.32% to 71.95% for Random Forest when the AU dependency relationship was used while extracting HOG features. Similarly, for Gabor features, the accuracy increased from 79.11% to 82.13% for when the nearest neighbor algorithm was used. Since the AU intensity inference phase and the feature extraction phase are independent of each other, higher accuracy is achieved. Table 4 also shows that LBP performs best when used with SVM.

Table 4. Summary of Accuracies for various Feature Types.

#	Feature Type	kNN	RF	SVM
1.	HOG Feature	68.64	71.95	88.62
2.	Gabor Wavelets	82.13	85.6	87.20
3.	LBP	92.11	96.33	97.16

Table 5. Correlation Matrix for Action Units.

#	AU Intensity	AU1 = 0	AU1 = 1	AU1 = 2	AU1 = 3
1.	AU2 = 0	0.982	0.0097	0.0063	0.0079
2.	AU2 = 1	0.60	0.322	0.078	0.051
3.	AU2 = 2	0.324	0.290	0.313	0.131
4.	AU2 = 3	0.124	0.0837	0.158	0.88

We also performed a comparison of our work with a few other works. As shown in Table 6, we noted that a few recent works [60,81] had used similar features (HOG and Gabor in both; LBP in one) and the DISFA database. Therefore, to have a fair comparison, we applied our approach. It is clear from Table 6 that the proposed method performs far better than other works while using the same feature selection methods and database. Table 7 presents the characteristics of the comparative state-of-the-art methods listing the databases, and feature extraction & ML methods used in each work. It should be noted that one of the primary differences between our work and other recent works is the use of a greater number of databases for training the ML algorithm while using similar feature extraction algorithm. Although the average accuracy and ICC percentages improved for all feature extraction methods, it is noteworthy that the highest improvement was seen with LBP when the SVM model was trained using 5 databases, as seen in Table 6.

Table 6. Comparison with state-of-the-art methods.

#	Works	HOG		Gabor		LBP	
		Acc	ICC	Acc	ICC	Acc	ICC
1.	[60]	79.14%	0.70	85.71%	0.77	81.54%	0.69
2.	[81]	81.08%	0.7016	86.60%	0.7834	—	—
3.	Our Work	88.62%	0.70	87.20%	0.81	**97.16%**	0.82

Table 7. Characteristics of the state-of-the-art comparative methods.

#	Works	Databases for Training	Feature Extraction Techniques	Machine Learning Techniques
1.	[60]	DISFA	HOG, Gabor, LBP	SVM
2.	[81]	DISFA	HOG, Gabor	DBN
3.	Our Work	B-DFE, JAFFE, CK & In-house	HOG, Gabor, LBP	SVM, kNN, RF

Furthermore, we also evaluated the performance of the proposed method for a few other databases including JAFFE, CK, B-DFE, and our dataset of 200 images. These results are presented in Table 8 and clearly shows that LBP performs better for the first three databases while the performance is quite close to the best performance of Gabor for the CK and databases. A few of these results are also presented using bar charts for a better visual comparison in Figures 6–8 for databases, respectively. It is evident from these figures that LBP gives better results for almost all AUs and gives the best results when combined with SVM. Therefore, we conclude that the LBP feature extraction method, when used with SVM, works best for facial emotion intensity recognition.

(**a**) Using kNN, kNN-HOG, kNN-Gabor, and kNN-LBP

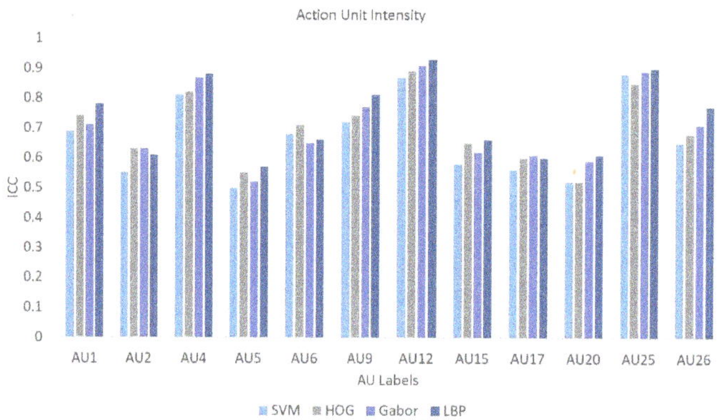

(**b**) Using SVM, SVM-HOG, SVM-Gabor, and SVM-LBP

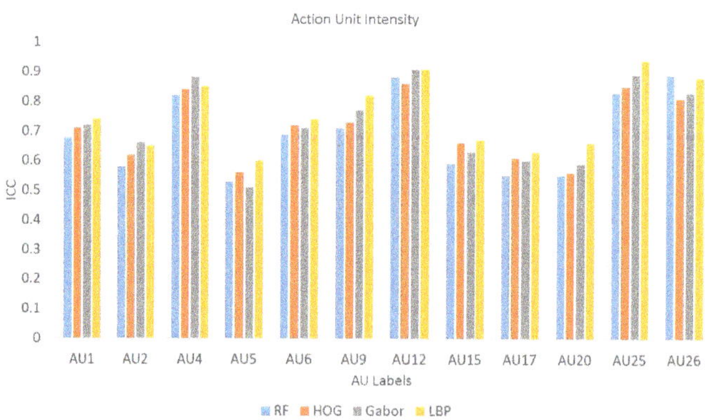

(**c**) Using RF, RF-HOG, RF-Gabor, and RF-LBP

Figure 6. Comparison of AU intensity Labels on Database.

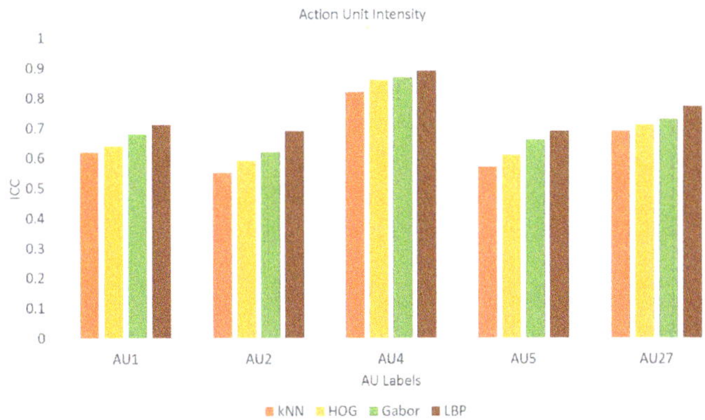

(**a**) Using kNN, kNN-HOG, kNN-Gabor, and kNN-LBP

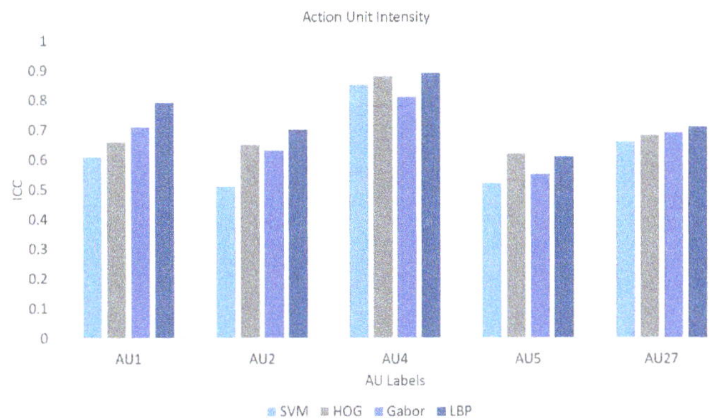

(**b**) Using SVM, SVM-HOG, SVM-Gabor, and SVM-LBP

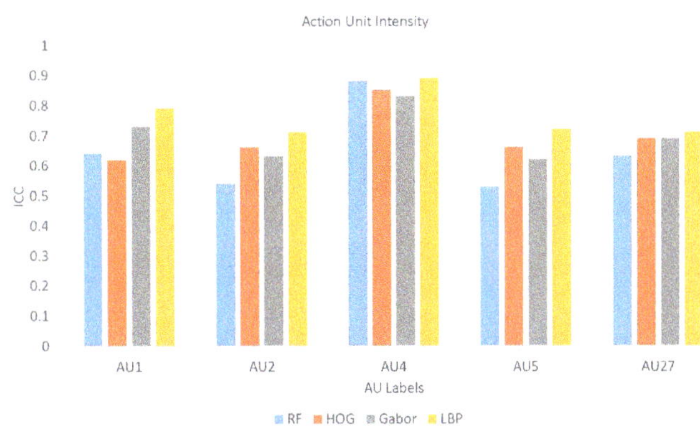

(**c**) Using RF, RF-HOG, RF-Gabor, and RF-LBP

Figure 7. Comparison of AU intensity Labels on Database.

(**a**) Using kNN, kNN-HOG, kNN-Gabor, and kNN-LBP

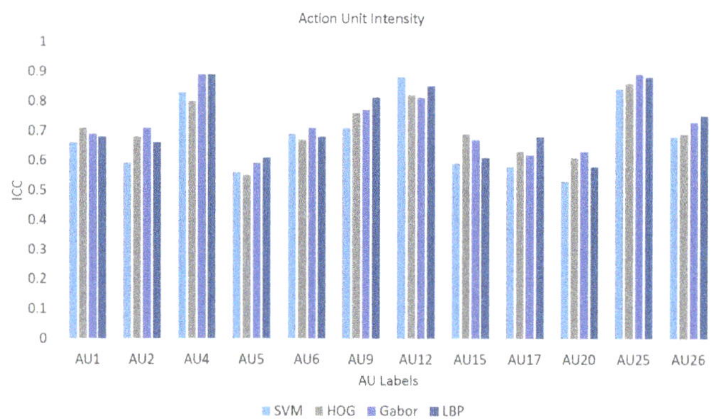

(**b**) Using SVM, SVM-HOG, SVM-Gabor, and SVM-LBP

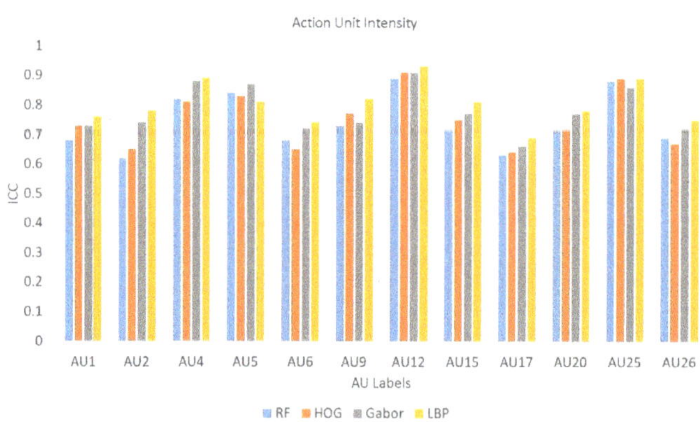

(**c**) Using RF, RF-HOG, RF-Gabor, and RF-LBP

Figure 8. Comparison of AU intensity Labels on Database.

In conclusion, it is evident from Table 3, LBP-kNN detects almost all AUs with high accuracy (>94%) while other techniques show this level of accuracy only for few AUs. Therefore, we LBP-SVM will recognize almost all emotions at all intensity levels better than the other studied techniques. It should also be noted that the average accuracy of detection of intensity of emotion decreases for all emotions with an increase in intensity; however, LBP+SVM still performs better than Gabor-SVM and HOG-SVM on average.

Table 8. Performance of proposed method for different Databases.

#	Databases	HOG	Gabor	LBP
1.	200 images	96.17%	97.08%	99.11%
2.	JAFFE	97.23%	98.07%	99.10%
3.	CK	98.69%	98.71%	98.14%
4.	B-DFE	97.31%	98.64%	98.02%

Nevertheless, in real-world applications, accuracy is also dependent on various other factors such as the image quality, environment it was captured in (controlled, uncontrolled), angle of the face, age of the person (fine lines on the face can make huge difference in accuracies), and lighting conditions. Also, accuracy differs from men to women, since women tend to express emotions more vividly than men. All these factors will come into consideration for emotion intensity as well as emotion detection for real-world face recognition systems and might significantly affect the performance of any technique.

5. Conclusions and Future Work

AUs are popularly used for measuring the facial emotion intensities from facial expressions. Use of an adequate amount of data is required for training and testing classifiers for its best performance in terms of accuracy. Majority of the databases consist of posed facial expressions or the emotion labels. Hence our research was focused on using the publicly available databases which have AUs that are annotated on a 6-point intensity scale. The experimentation was performed for both spontaneous and posed facial emotion intensity recognition, where we conclude that AU intensity is not always reliable and accurate in the case of spontaneous facial expressions. This happens due to the ambiguity and dynamic nature of the facial emotions when spontaneous expressions are taken into consideration. For measuring the intensity of emotions, it is not only required to improve the accuracy of feature extraction algorithms but also exploiting the facial actions. It is these spatiotemporal facial action interactions with synchronized and coherent actions that provide a full facial display. In our work, we presented a probabilistic model to calculate the ICC values and accuracies among the dynamic and semantic AU intensity levels. Also, AU intensity recognition is accomplished by integrating the images systematically with the proposed model. The accuracies for various algorithms (LBP, HOG, and Gabor) indicate that LBP achieves the highest accuracy in most cases. As a future work, in neural networks several hidden layers could be added to specifically handle each challenge in the spontaneous intensity of emotion recognition such as the head tilt and angle.

Author Contributions: Conceptualization, M.F.H.S. and A.Y.J.; Investigation, D.M.; Methodology, D.M.; Project administration, A.Y.J.; Resources, A.Y.J.; Supervision, A.Y.J.; Validation, M.F.H.S.; Writing—original draft, D.M.; Writing—review & editing, M.F.H.S. and A.Y.J.

Funding: Any funding agency did not support this work.

Acknowledgments: The authors are thankful to Paul A. Hotmer Family CSTAR (Cyber Security and Teaming Research) lab and the Electrical Engineering and Computer Science Department at the University of Toledo.

Conflicts of Interest: The authors declare no conflict of interest.

References

1. Wang, L. *Behavioral Biometrics for Human Identification: Intelligent Applications: Intelligent Applications*; IGI Global: Hershey, PA, USA, 2009.
2. Gamboa, H.; Fred, A. A behavioral biometric system based on human-computer interaction. *Proc. SPIE* **2004**, *5404*, 381–393.

3. Hess, U.; Banse, R.; Kappas, A. The intensity of facial expression is determined by underlying affective state and social situation. *J. Personal. Soc. Psychol.* **1995**, *69*, 280. [CrossRef]
4. Bronstein, A.M.; Bronstein, M.M.; Kimmel, R. Expression-invariant 3D face recognition. In *International Conference on Audio-and Video-Based Biometric Person Authentication*; Springer: Berlin/Heidelberg, Germany, 2003; pp. 62–70.
5. Hesher, C.; Srivastava, A.; Erlebacher, G. A novel technique for face recognition using range imaging. In Proceedings of the Seventh International Symposium on Signal Processing and Its Applications, Paris, France, 4 July 2003; Volume 2, pp. 201–204.
6. Lee, Y.; Yi, T. 3D face recognition using multiple features for local depth information. In Proceedings of the 4th EURASIP Conference focused on Video/Image Processing and Multimedia Communications, Zagreb, Croatia, 2–5 July 2003, Volume 1, pp. 429–434.
7. Moreno, A.B.; Sánchez, A.; Vélez, J.F.; Díaz, F.J. Face recognition using 3D surface-extracted descriptors. In Proceedings of the Irish Machine Vision and Image Processing Conference, Portrush, Northern Ireland, 3–5 September 2003.
8. Ekman, P.; Friesen, W.V. The repertoire of nonverbal behavior: Categories, origins, usage, and coding. *Semiotica* **1969**, *1*, 49–98. [CrossRef]
9. Mehta, D.; Siddiqui, M.F.H.; Javaid, A.Y. Facial Emotion Recognition: A Survey and Real-World User Experiences in Mixed Reality. *Sensors* **2018**, *18*, 416. [CrossRef] [PubMed]
10. Draper, B.A.; Baek, K.; Bartlett, M.S.; Beveridge, J.R. Recognizing faces with PCA and ICA. *Comput. Vis. Image Underst.* **2003**, *91*, 115–137. [CrossRef]
11. Liu, C.; Wechsler, H. Comparative assessment of independent component analysis (ICA) for face recognition. In Proceedings of the International Conference on Audio and Video Based Biometric Person Authentication, Washington, DC, USA, 22–23 March 1999.
12. Yan, W.Q. Biometrics for surveillance. In *Introduction to Intelligent Surveillance*; Springer: New York, NY, USA, 2017; pp. 107–130.
13. Bartlett, M.S.; Movellan, J.R.; Sejnowski, T.J. Face recognition by independent component analysis. *IEEE Trans. Neural Netw.* **2002**, *13*, 1450–1464. [CrossRef]
14. Mir, A.; Rubab, S.; Jhat, Z. Biometrics verification: A literature survey. *Int. J. Comput. ICT Res.* **2011**, *5*, 67–80.
15. Delac, K.; Grgic, M.; Grgic, S. Statistics in face recognition: Analyzing probability distributions of PCA, ICA and LDA performance results. In Proceedings of the 4th International Symposium on Image and Signal Processing and Analysis, ISPA 2005, Zagreb, Croatia, 15–17 September 2005; pp. 289–294.
16. Delac, K.; Grgic, M.; Grgic, S. Independent comparative study of PCA, ICA, and LDA on the FERET data set. *Int. J. Imaging Syst. Technol.* **2005**, *15*, 252–260. [CrossRef]
17. Friesen, E.; Ekman, P. Facial Action Coding System: A Technique for the Measurement of Facial Movement. Volume 1, 1978. Available online: https://scinapse.io/papers/69567567 (accessed on 20 April 2019).
18. Scherer K.R.; Ekman, P. Methods for measuring facial action. In *Handbook of Methods in Nonverbal Behavior Research*; Cambridge University Press: Cambridge, UK, 1982; pp. 45–90. [CrossRef]
19. Ekman, P.; Friesen, W.V.; O'sullivan, M.; Chan, A.; Diacoyanni-Tarlatzis, I.; Heider, K.; Krause, R.; LeCompte, W.A.; Pitcairn, T.; Ricci-Bitti, P.E.; et al. Universals and cultural differences in the judgments of facial expressions of emotion. *J. Personal. Soc. Psychol.* **1987**, *53*, 712. [CrossRef]
20. Zuckerman, M.; Lipets, M.S.; Koivumaki, J.H.; Rosenthal, R. Encoding and decoding nonverbal cues of emotion. *J. Personal. Soc. Psychol.* **1975**, *32*, 1068. [CrossRef]
21. Hall, J.A. Gender effects in decoding nonverbal cues. *Psychol. Bull.* **1978**, *85*, 845. [CrossRef]
22. Rosenthal, R.; DePaulo, B.M. Sex differences in eavesdropping on nonverbal cues. *J. Personal. Soc. Psychol.* **1979**, *37*, 273. [CrossRef]
23. Biele, C.; Grabowska, A. Sex differences in perception of emotion intensity in dynamic and static facial expressions. *Exp. Brain Res.* **2006**, *171*, 1–6. [CrossRef]
24. Kohler, C.G.; Turner, T.H.; Bilker, W.B.; Brensinger, C.M.; Siegel, S.J.; Kanes, S.J.; Gur, R.E.; Gur, R.C. Facial emotion recognition in schizophrenia: Intensity effects and error pattern. *Am. J. Psychiatry* **2003**, *160*, 1768–1774. [CrossRef]
25. Unoka, Z.; Fogd, D.; Füzy, M.; Csukly, G. Misreading the facial signs: specific impairments and error patterns in recognition of facial emotions with negative valence in borderline personality disorder. *Psychiatry Res.* **2011**, *189*, 419–425. [CrossRef] [PubMed]

26. Castelli, F. Understanding emotions from standardized facial expressions in autism and normal development. *Autism* **2005**, *9*, 428–449. [CrossRef]
27. Garman, H.D.; Spaulding, C.J.; Webb, S.J.; Mikami, A.Y.; Morris, J.P.; Lerner, M.D. Wanting it too much: An inverse relation between social motivation and facial emotion recognition in autism spectrum disorder. *Child Psychiatry Hum. Dev.* **2016**, *47*, 890–902. [CrossRef]
28. Lewis, M.B.; Dunn, E. Instructions to mimic improve facial emotion recognition in people with sub-clinical autism traits. *Q. J. Exp. Psychol.* **2017**, *70*, 2357–2370. [CrossRef]
29. Wingenbach, T.S.; Ashwin, C.; Brosnan, M. Diminished sensitivity and specificity at recognising facial emotional expressions of varying intensity underlie emotion-specific recognition deficits in autism spectrum disorders. *Res. Autism Spectr. Disord.* **2017**, *34*, 52–61. [CrossRef]
30. Lee, D.; Oh, K.J.; Choi, H.J. The chatbot feels you-a counseling service using emotional response generation. In Proceedings of the 2017 IEEE International Conference on Big Data and Smart Computing (BigComp), Jeju, Korea, 13–16 February 2017; pp. 437–440.
31. Oh, K.J.; Lee, D.; Ko, B.; Choi, H.J. A chatbot for psychiatric counseling in mental healthcare service based on emotional dialogue analysis and sentence generation. In Proceedings of the 2017 18th IEEE International Conference on Mobile Data Management (MDM), Daejeon, Korea, 29 May–1 June 2017; pp. 371–375.
32. Chiu, I.; Piguet, O.; Diehl-Schmid, J.; Riedl, L.; Beck, J.; Leyhe, T.; Holsboer-Trachsler, E.; Kressig, R.W.; Berres, M.; Monsch, A.U.; et al. Facial Emotion Recognition Performance Differentiates Between Behavioral Variant Frontotemporal Dementia and Major Depressive Disorder. *J. Clin. Psychiatry* **2018**, *79*. [CrossRef]
33. Huang, H.; Xie, Q.; Pan, J.; He, Y.; Wen, Z.; Yu, R.; Li, Y. An EEG-Based Brain Computer Interface for Emotion Recognition and Its Application in Patients with Disorder of Consciousness. *IEEE Trans. Affect. Comput.* **2019**. [CrossRef]
34. Lim, J.; Chong, H.J.; Kim, A.J. A comparison of emotion identification and its intensity between adults with schizophrenia and healthy adults: Using film music excerpts with emotional content. *Nord. J. Music. Ther.* **2018**, *27*, 126–141. [CrossRef]
35. Yang, S.; Zhou, P.; Duan, K.; Hossain, M.S.; Alhamid, M.F. emHealth: Towards emotion health through depression prediction and intelligent health recommender system. *Mob. Netw. Appl.* **2017**, *23*, 216–226. [CrossRef]
36. Devika, R.; Subramaniyaswamy, V. A Novel Model for Hospital Recommender System Using Hybrid Filtering and Big Data Techniques. In Proceedings of the 2018 2nd International Conference on I-SMAC (IoT in Social, Mobile, Analytics and Cloud) (I-SMAC) I-SMAC (IoT in Social, Mobile, Analytics and Cloud)(I-SMAC), Palladam, India, 30–31 August 2018; pp. 267–271.
37. Stai, E.; Kafetzoglou, S.; Tsiropoulou, E.E.; Papavassiliou, S. A holistic approach for personalization, relevance feedback & recommendation in enriched multimedia content. *Multimed. Tools Appl.* **2018**, *77*, 283–326.
38. Bobadilla, J.; Ortega, F.; Hernando, A.; Gutiérrez, A. Recommender systems survey. *Knowl.-Based Syst.* **2013**, *46*, 109–132. [CrossRef]
39. Carrer-Neto, W.; Hernández-Alcaraz, M.L.; Valencia-García, R.; García-Sánchez, F. Social knowledge-based recommender system. Application to the movies domain. *Expert Syst. Appl.* **2012**, *39*, 10990–11000. [CrossRef]
40. Winoto, P.; Tang, T.Y. The role of user mood in movie recommendations. *Expert Syst. Appl.* **2010**, *37*, 6086–6092. [CrossRef]
41. McNally, K.; O'Mahony, M.P.; Coyle, M.; Briggs, P.; Smyth, B. A case study of collaboration and reputation in social web search. *ACM Trans. Intell. Syst. Technol.* **2011**, *3*, 4. [CrossRef]
42. Huang, M.; Wang, Z.; Ying, Z. Facial expression recognition using stochastic neighbor embedding and SVMs. In Proceedings of the 2011 International Conference on System Science and Engineering (ICSSE), Macao, China, 8–10 June 2011; pp. 671–674.
43. Castro-Schez, J.J.; Miguel, R.; Vallejo, D.; López-López, L.M. A highly adaptive recommender system based on fuzzy logic for B2C e-commerce portals. *Expert Syst. Appl.* **2011**, *38*, 2441–2454. [CrossRef]
44. Consoli, D. A new concept of marketing: The emotional marketing. *Broad Res. Account. Negot. Distrib.* **2010**, *1*, 52–59.
45. Kratzwald, B.; Ilić, S.; Kraus, M.; Feuerriegel, S.; Prendinger, H. Deep learning for affective computing: Text-based emotion recognition in decision support. *Decis. Support Syst.* **2018**, *115*, 24–35. [CrossRef]

46. Byron, K.; Terranova, S.; Nowicki, S., Jr. Nonverbal emotion recognition and salespersons: Linking ability to perceived and actual success. *J. Appl. Soc. Psychol.* **2007**, *37*, 2600–2619. [CrossRef]
47. De Carolis, B.; de Gemmis, M.; Lops, P.; Palestra, G. Recognizing users feedback from non-verbal communicative acts in conversational recommender systems. *Pattern Recognit. Lett.* **2017**, *99*, 87–95. [CrossRef]
48. Patwardhan, A.S.; Knapp, G.M. Multimodal Affect Analysis for Product Feedback Assessment. *arXiv* **2017**, arXiv:1705.02694.
49. Bahreini, K.; Nadolski, R.; Westera, W. Towards multimodal emotion recognition in e-learning environments. *Interact. Learn. Environ.* **2016**, *24*, 590–605. [CrossRef]
50. Salmeron-Majadas, S.; Arevalillo-Herráez, M.; Santos, O.C.; Saneiro, M.; Cabestrero, R.; Quirós, P.; Arnau, D.; Boticario, J.G. Filtering of spontaneous and low intensity emotions in educational contexts. In *International Conference on Artificial Intelligence in Education*; Springer: Cham, Switzerland, 2015; pp. 429–438.
51. Hossain, M.S.; Muhammad, G.; Al-Qurishi, M.; Masud, M.; Almogren, A.; Abdul, W.; Alamri, A. Cloud-oriented emotion feedback-based Exergames framework. *Multimed. Tools Appl.* **2018**, *77*, 21861–21877. [CrossRef]
52. Müller, L.; Bernin, A.; Kamenz, A.; Ghose, S.; von Luck, K.; Grecos, C.; Wang, Q.; Vogt, F. Emotional journey for an emotion provoking cycling exergame. In Proceedings of the 2017 IEEE 4th International Conference on Soft Computing & Machine Intelligence (ISCMI), Port Louis, Mauritius, 23–24 November 2017; pp. 104–108.
53. Hossain, M.S.; Muhammad, G.; Song, B.; Hassan, M.M.; Alelaiwi, A.; Alamri, A. Audio–visual emotion-aware cloud gaming framework. *IEEE Trans. Circuits Syst. Video Technol.* **2015**, *25*, 2105–2118. [CrossRef]
54. Alhargan, A.; Cooke, N.; Binjammaz, T. Affect recognition in an interactive gaming environment using eye tracking. In Proceedings of the 2017 Seventh International Conference on Affective Computing and Intelligent Interaction (ACII), San Antonio, TX, USA, 23–26 October 2017; pp. 285–291.
55. Quintero, L.A.M.; Muñoz-Delgado, J.; Sánchez-Ferrer, J.C.; Fresán, A.; Brüne, M.; Arango de Montis, I. Facial emotion recognition and empathy in employees at a juvenile detention center. *Int. J. Offender Ther. Comp. Criminol.* **2018**, *62*, 2430–2446. [CrossRef]
56. Wu, Y.L.; Tsai, H.Y.; Huang, Y.C.; Chen, B.H. Accurate Emotion Recognition for Driving Risk Prevention in Driver Monitoring System. In Proceedings of the 2018 IEEE 7th Global Conference on Consumer Electronics (GCCE), Nara, Japan, 9–12 October 2018; pp. 796–797.
57. Alamri, A. Monitoring system for patients using multimedia for smart healthcare. *IEEE Access* **2018**, *6*, 23271–23276. [CrossRef]
58. Damacharla, P.; Mehta, D.; Javaid, A.Y.; Devabhaktuni, V. Study on State-of-the-art Cloud Systems Integration Capabilities with Autonomous Ground Vehicles. In Proceedings of the 2018 IEEE 88th Vehicular Technology Conference, Chicago, IL, USA, 27–30 August 2018; pp. 1–5.
59. Vural, E.; Çetin, M.; Erçil, A.; Littlewort, G.; Bartlett, M.; Movellan, J. Automated drowsiness detection for improved driving safety. In Proceedings of the ICAT 2008: International Conference on Automotive Technologies, Istanbul, Turkey, 13–14 November 2008.
60. Mavadati, S.M.; Mahoor, M.H.; Bartlett, K.; Trinh, P.; Cohn, J.F. DISFA: A spontaneous facial action intensity database. *IEEE Trans. Affect. Comput.* **2013**, *4*, 151–160. [CrossRef]
61. Yan, W.Q. Biometrics for surveillance. In *Introduction to Intelligent Surveillance*; Springer: New York, NY, USA, 2019; pp. 127–153.
62. Kanade, T.; Cohn, J.F.; Tian, Y. Comprehensive database for facial expression analysis. In Proceedings of the Fourth IEEE International Conference on Automatic Face and Gesture Recognition (Cat. No. PR00580), Grenoble, France, 28–30 March 2000; pp. 46–53.
63. Lucey, P.; Cohn, J.F.; Kanade, T.; Saragih, J.; Ambadar, Z.; Matthews, I. The extended cohn-kanade dataset (ck+): A complete dataset for action unit and emotion-specified expression. In Proceedings of the 2010 IEEE Computer Society Conference on Computer Vision and Pattern Recognition-Workshops, San Francisco, CA, USA, 13–18 June 2010; pp. 94–101.
64. Lyons, M.; Akamatsu, S.; Kamachi, M.; Gyoba, J. Coding facial expressions with gabor wavelets. In Proceedings of the Third IEEE International Conference on Automatic Face and Gesture Recognition, Nara, Japan, 14–16 April 1998; pp. 200–205.

65. Yin, L.; Wei, X.; Sun, Y.; Wang, J.; Rosato, M.J. A 3D facial expression database for facial behavior research. In Proceedings of the 7th International Conference on Automatic Face and Gesture Recognition (FGR06), Southampton, UK, 10–12 April 2006; pp. 211–216.
66. DISFA: Denver Intensity of Spontaneous Facial Actions. Available online: https://computervisiononline.com/dataset/1105138646 (accessed on 24 March 2019).
67. Mavadati, S.M.; Mahoor, M.H.; Bartlett, K.; Trinh, P. Automatic detection of non-posed facial action units. In Proceedings of the 2012 19th IEEE International Conference on Image Processing, Orlando, FL, USA, 30 September–3 October 2012; pp. 1817–1820.
68. Jens, S.A. Can You See It? Facial Expression Leakage in Response to Emotional Intensity. Undergraduate' Thesis. Paper 1124. 2017. Available online: https://scholarworks.wm.edu/honorstheses/1124 (accessed on 24 March 2019).
69. Chang, K.J.; Bowyer, K.W.; Flynn, P.J. Effects on facial expression in 3D face recognition. In Proceedings of the Biometric Technology for Human Identification II, Orlando, FL, USA, 28 March 2005; Volume 5779, pp. 132–144.
70. Kakadiaris, I.A.; Passalis, G.; Toderici, G.; Murtuza, M.N.; Lu, Y.; Karampatziakis, N.; Theoharis, T. Three-dimensional face recognition in the presence of facial expressions: An annotated deformable model approach. *IEEE Trans. Pattern Anal. Mach. Intell.* **2007**, *29*, 640–649. [CrossRef] [PubMed]
71. Pal, R.; Satsangi, C. Facial Expression Recognition Based on Basic Expressions and Intensities Using K-Means Clustering. *Int. J. Sci. Res.* **2016**, *5*, 1949–1952.
72. Song, K.T.; Chen, Y.W. A design for integrated face and facial expression recognition. In Proceedings of the IECon 2011-37th Annual Conference on IEEE Industrial Electronics Society, Melbourne, Australia, 7–10 November 2011; pp. 4306–4311.
73. Tang, H.; Huang, T.S. 3D facial expression recognition based on automatically selected features. In Proceedings of the IEEE Computer Society Conference on Computer Vision and Pattern Recognition Workshops, CVPRW'08, Anchorage, AK, USA, 23–28 June 2008, pp. 1–8.
74. Soyel, H.; Demirel, H. Facial expression recognition using 3D facial feature distances. In *International Conference Image Analysis and Recognition*; Springer: Berlin/Heidelberg, Germany, 2007; pp. 831–838.
75. Pan, G.; Han, S.; Wu, Z.; Wang, Y. 3D face recognition using mapped depth images. In Proceedings of the IEEE Computer Society Conference on Computer Vision and Pattern Recognition-Workshops, San Diego, CA, USA, 21–23 September 2005; pp. 175.
76. Zhang, H.; Luo, S.; Yoshie, O. Facial expression recognition by analyzing features of conceptual regions. In Proceedings of the 2013 IEEE/ACIS 12th International Conference on Computer and Information Science (ICIS), Niigata, Japan, 16–20 June 2013; pp. 529–534.
77. Ahmed, F.; Hossain, E. Automated facial expression recognition using gradient-based ternary texture patterns. *Chin. J. Eng.* **2013**, *2013*, 831747. [CrossRef]
78. Savran, A.; Sankur, B.; Bilge, M.T. Regression-based intensity estimation of facial action units. *Image Vis. Comput.* **2012**, *30*, 774–784. [CrossRef]
79. Pantic, M.; Rothkrantz, L.J. An expert system for recognition of facial actions and their intensity. In Proceedings of the 17th National Conference on Artificial Intelligence and 12th Conference on on Innovative Applications of Artificial Intelligence, Austin, TX, USA, 30 July–3 August 2000; pp. 1026–1033.
80. Mahoor, M.H.; Cadavid, S.; Messinger, D.S.; Cohn, J.F. A framework for automated measurement of the intensity of non-posed facial action units. In Proceedings of the IEEE Computer Society Conference on Computer Vision and Pattern Recognition Workshops, Miami, FL, USA, 20–25 June 2009; pp. 74–80.
81. Li, Y.; Mavadati, S.M.; Mahoor, M.H.; Zhao, Y.; Ji, Q. Measuring the intensity of spontaneous facial action units with dynamic Bayesian network. *Pattern Recognit.* **2015**, *48*, 3417–3427. [CrossRef]
82. Walecki, R.; Rudovic, O.; Pantic, M.; Pavlovic, V.; Cohn, J.F. A Framework for Joint Estimation and Guided Annotation of Facial Action Unit Intensity. In Proceedings of the IEEE Conference on Computer Vision and Pattern Recognition Workshops, Las Vegas, NV, USA, 26 June–1 July 2016; pp. 9–17.
83. Hess, U.; Blairy, S.; Kleck, R.E. The intensity of emotional facial expressions and decoding accuracy. *J. Nonverbal Behav.* **1997**, *21*, 241–257. [CrossRef]
84. N'diaye, K.; Sander, D.; Vuilleumier, P. Self-relevance processing in the human amygdala: Gaze direction, facial expression, and emotion intensity. *Emotion* **2009**, *9*, 798. [CrossRef] [PubMed]

85. Scherer, K.R.; Ellgring, H. Multimodal expression of emotion: Affect programs or componential appraisal patterns? *Emotion* **2007**, *7*, 158. [CrossRef]
86. Ekman, P. Universals and cultural differences in facial expressions of emotion. In *Nebraska Symposium on Motivation*; University of Nebraska Press: Lincoln, NE, USA, 1971.
87. Ekman, P. Darwin, deception, and facial expression. *Ann. N. Y. Acad. Sci.* **2003**, *1000*, 205–221. [CrossRef]
88. Delannoy, J.R.; McDonald, J. Automatic estimation of the dynamics of facial expression using a three-level model of intensity. In Proceedings of the FG'08. 8th IEEE International Conference on Automatic Face & Gesture Recognition, Amsterdam, The Netherlands, 17–19 September 2008; pp. 1–6.
89. Kim, M.; Pavlovic, V. Structured output ordinal regression for dynamic facial emotion intensity prediction. In *European Conference on Computer Vision*; Springer: Berlin/Heidelberg, Germany, 2010; pp. 649–662.
90. Valstar, M.F.; Pantic, M. Fully automatic recognition of the temporal phases of facial actions. *IEEE Trans. Syst. Man, Cybern. Part Cybern.* **2012**, *42*, 28–43. [CrossRef] [PubMed]
91. Dhall, A.; Goecke, R. Group expression intensity estimation in videos via gaussian processes. In Proceedings of the 2012 21st International Conference on Pattern Recognition (ICPR), Tsukuba, Japan, 11–15 November 2012; pp. 3525–3528.
92. Song, K.T.; Chien, S.C. Facial expression recognition based on mixture of basic expressions and intensities. In Proceedings of the 2012 IEEE International Conference on Systems, Man, and Cybernetics (SMC), Seoul, Korea, 14–17 October 2012; pp. 3123–3128.
93. Littlewort, G.; Bartlett, M.S.; Fasel, I.; Susskind, J.; Movellan, J. Dynamics of facial expression extracted automatically from video. *Image Vis. Comput.* **2006**, *24*, 615–625. [CrossRef]
94. Chang, W.Y.; Chen, C.S.; Hung, Y.P. Analyzing facial expression by fusing manifolds. In *Asian Conference on Computer Vision*; Springer: Berlin/Heidelberg, Germany, 2007; pp. 621–630.
95. Rudovic, O.; Pavlovic, V.; Pantic, M. Multi-output laplacian dynamic ordinal regression for facial expression recognition and intensity estimation. In Proceedings of the 2012 IEEE Conference on Computer Vision and Pattern Recognition (CVPR), Providence, RI, USA, 16–21 June 2012; pp. 2634–2641.

sensors

MDPI

Article

A Deep-Learning Model for Subject-Independent Human Emotion Recognition Using Electrodermal Activity Sensors

Fadi Al Machot [1,*], Ali Elmachot [2], Mouhannad Ali [3], Elyan Al Machot [4] and Kyandoghere Kyamakya [3]

1 Research Center Borstel—Leibniz Lung Center, 23845 Borstel, Germany
2 Faculty of Mechanical and Electrical Engineering, University of Damascus, Damascus, Syria; ali.elmachot@gmail.com
3 Institute for Smart Systems Technologies, Alpen-Adira University, 9020 Klagenfurt, Austria; mouhannad.ali@aau.at (M.A.); kyandoghere.kyamakya@aau.at (K.K.)
4 Carl Gustav Carus Faculty of Medicine, Dresden University of Technology, 01069 Dresden, Germany; Elyan.Al-Machot@uniklinikum-dresden.de
* Correspondence: fadi.almachot@aau.at; Tel.: +49-4537-188-2330

Received: 14 March 2019; Accepted: 3 April 2019; Published: 7 April 2019

Abstract: One of the main objectives of Active and Assisted Living (AAL) environments is to ensure that elderly and/or disabled people perform/live well in their immediate environments; this can be monitored by among others the recognition of emotions based on non-highly intrusive sensors such as Electrodermal Activity (EDA) sensors. However, designing a learning system or building a machine-learning model to recognize human emotions while training the system on a specific group of persons and testing the system on a totally a new group of persons is still a serious challenge in the field, as it is possible that the second testing group of persons may have different emotion patterns. Accordingly, the purpose of this paper is to contribute to the field of human emotion recognition by proposing a Convolutional Neural Network (CNN) architecture which ensures promising robustness-related results for both subject-dependent and subject-independent human emotion recognition. The CNN model has been trained using a grid search technique which is a model hyperparameter optimization technique to fine-tune the parameters of the proposed CNN architecture. The overall concept's performance is validated and stress-tested by using MAHNOB and DEAP datasets. The results demonstrate a promising robustness improvement regarding various evaluation metrics. We could increase the accuracy for subject-independent classification to 78% and 82% for MAHNOB and DEAP respectively and to 81% and 85% subject-dependent classification for MAHNOB and DEAP respectively (4 classes/labels). The work shows clearly that while using solely the non-intrusive EDA sensors a robust classification of human emotion is possible even without involving additional/other physiological signals.

Keywords: subject-dependent emotion recognition; subject-independent emotion recognition; electrodermal activity (EDA); deep learning; convolutional neural networks

1. Introduction

Emotion recognition plays an important role in various areas of life, especially in the field of Active and Assisted Living (AAL) [1] and Driver Assistance Systems (DAS) [2]. Recognizing emotions automatically is one of technical enablers of AAL, as it is considered to be a significant help for monitoring and observing the mental state of either old people or disabled persons.

Furthermore, it can be observed that according to the most recent related publications, the classification performance of emotion recognition approaches has been significantly improving and the opportunities for automatic emotion recognition systems are also getting higher.

Emotions can be recognized in various ways. The most well-known models for emotion recognition are the "discrete emotion model" proposed by Ekman [3] and the "emotion dimensional model" proposed by Lang [4]. The discrete emotion model categorizes emotions into six basic emotion states: surprise, anger, disgust, happiness sadness and fear [3]. These emotions are universal, biologically experienced by all humans and widely accepted as such in the research community. In contrast to the discrete emotional model, the dimensional model assumes that the emotions are a combination of several psychological dimensions. The most well-known dimensional model is the "valance-arousal dimensional model". The valance represents a form of pleasure level and ranges from negative to positive. However, the arousal indicates the physiological and/or psychological level of being awake and ranges from low to high [5].

Overall, researchers in the field have used two major approaches to recognize emotions. The first one consists of features engineering-based approaches [6] and the second one involves Deep Learning (DL) [7]. In the features engineering approach, human emotion recognition involves several steps ranging from collecting raw sensor data up to the final conclusion about the current emotional status. The steps thereby involved are the following ones [8]: (1) preprocessing of the raw data from sensor streams for handling incompleteness, eliminating noise and redundancy, and performing data aggregation and normalization; (2) feature extraction which means extracting the main characteristics of/from the raw signals (e.g., temporal and spatial information); (3) dimensionality reduction to decrease the number of features to increase their quality and reduce the computational effort needed for the classification task; and (4) classification based on machine-learning and reasoning techniques to recognize the effective emotion class.

On the other hand, DL does not require necessarily the feature engineering/extraction step, due to the fact that DL models do extract features internally and/or implicitly (within the training phase) [9]. Therefore, they have shown promising results while involving a combination of different physiological signals for human emotion recognition [10,11].

Additionally, DL showed promising results in other research fields for different applications, e.g., identification of gas mixture [12], classification of tea specimens [13] and cardiac arrhythmia detection [14,15].

Generally, subject-independent emotion recognition is a challenging field due to the facts that (a) physiological expressions of emotion depend on age, gender, culture and other social factors [16], and (b) it also depends on the environment in which a subject lives, (c) the subject-independent nature of human emotion recognition which means that the system has been trained on a group of subjects and tested on another different group, and (d) the lab-setting independent nature of emotion recognition is related to the fact that the classifier can/will be trained locally once using sensors of a given lab-setting and after that tested considering different datasets that are collected based on different lab settings. The motivation for developing a generalized model is that collecting training data each time for each subject is not a realistic task and is far from the practical reality.

Based on the previous facts, a concept to improve the performance of the subject-dependent and subject-independent human emotion recognition systems is required; in this paper we use solely EDA (electrodermal activity) biosignals based on a deep-learning model using convolutional neural networks (CNNs) that extracts the required features internally and performs well when this model is applied on new subjects. Although researchers have used CNN to classify human emotions using EDA, they did not propose the architecture that did perform better than the proposed model in this paper.

The contribution of this paper does significantly increase the performance of human emotion recognition approaches using only EDA sensors compared to the state-of-the-art approaches involving the same EDA signals. Furthermore, the results obtained suggest/underscore a novel fact and interesting situation that other (mostly "highly intrusive") physiological sensors might be replaced by the "only slightly intrusive" EDA-based sensors in this research field. The structure of the paper is as follows: Section 2 presents an overview of the state-of-the-art approaches. Section 3 introduces the datasets. Section 4 portrays the overall architecture of the proposed classification model. Sections 5 and 6 present the overall results and the related discussions respectively. The paper ends with a conclusion in Section 7.

2. Related Works

Regarding human emotion recognition based on EDA sensors which can be embedded in smart wearable devices, few works have been published so far. However, in [17], they proposed a system to recognize the driver's emotional state after transforming the EDA signals using a short-time Fourier transform. They considered three classes: neutral-stress, neutral-anger, and stress-anger.

Furthermore, in [18], they applied a convex optimization-based electrodermal activity (cvxEDA) framework and clustering algorithms to automatically classify the arousal and valence levels induced by affective sound stimuli.

In the literature, it has been proven that the stimuli nature plays an important role to increase the EDA response which helps to make the emotion recognition process less complex [19]. Furthermore, other works showed promising results when EDA responses are modulated by musical emotional [20,21]. Consequently, this result encouraged researchers to work on classifying arousal and valence levels induced by auditory stimuli.

In [22], authors used the AVEC 2016 dataset [23,24], they proposed a deep-learning model that consists of a CNN followed by a recurrent neural network and then fully connected layers. They showed that an end-to-end deep-learning approach directly depending on raw signals can replace feature engineering for emotion recognition purposes.

Moreover, the use of different physiological signals has been previously involved [25,26]. However, mounting different types of sensors on the human body is not preferred and nor well-accepted. In [26], authors fused different types of sensors, ECG (Electrocardiogram), EDA and ST (Skin Temperature) through a hybrid neural model which combines cellular neural networks and echo state neural networks to recognize four classes of valence and arousal, mainly, high valence high arousal, high valence low arousal, low valence high arousal, and low valence, low arousal. In [25], authors combined facial electromyograms, electrocardiogram, respiration, and EDA dataset which were collected during racing conditions. The emotional classes identified are high stress, low stress, disappointment, and euphoria. Support vector machines (SVMs) and adaptive neuro-fuzzy inference system (ANFIS) have been used for the classification.

In [27], the researchers reported results using only EDA to recognize four different states, joy, anger, sadness, pleasure using 193 features and a music and based on genetic algorithm and the K-neighbor methods.

Table 1 shows a summary of the state-of-the-art for human emotion recognition using physiological signals. More details regarding state-of-the-art experiments and obtained results can be found in Section 6.

The major limitations in the state-of-the-art can be summarized in three major points. First, the limitation regarding proposing generalized models to recognize human emotions based on EDA signals (i.e., published works do not comprehensively consider the lab-setting independence property of emotion classifiers for EDA signals). Second, the limitation concerning subject-independent human emotion recognition (i.e., published works do not comprehensively address the subject-independence property of emotion classifiers for EDA signals). Third, most published related works do focus mostly on classifying only 2 (active/passive) emotional states.

In this work, we focus on the second and the third limitation, due to the fact that classifying human emotion with respect to different lab settings is a research question which may need to adjust the raw data in a feature engineering level which is not the focus of this work where CNN does extract the desired features internally as it is a deep-learning model.

Table 1. Summary of the stare-of-the-art works for human emotion recognition using physiological signals.

Paper	Classifier	Features	Signals
[25]	SVM	Statistical Features	Facial electromyograms, electrocardiogram, respiration, and electrodermal activity
[27]	Genetic algorithm and K-NN	Statistical features	EDA
[25]	Neuro-fuzzy inference	Statistical Features	Facial electromyograms, electrocardiogram, respiration, and electrodermal activity
[18]	K-NN	Statistical features	EDA
[28]	SVM	Wrapper feature selection (WFS)	EDA
[29]	CNN	Raw data	Patient's movements XYZ + EDA
[22]	Deep learning (CNN+RNN)	Raw data	AVEC 2016
[26]	ESN-CNN	Statistical features	ECG (Electrocardiogram), EDA (Electrodermal activity) and ST (Skin Temperature)
[30]	Dynamic calibration + K-NN	Statistical features	EDA

SVM: Support Vector Machine, K-NN: K-Nearest Neighbor, CNN: Convolutional Neural Network, RNN: Recurrent Neural Network, ESN-CNN: Echo State Network - Cellular Neural Network.

3. Datasets

This study uses public benchmark datasets (MAHNOB and DEAP) of physiological signals to test our proposal for a robust emotion recognition system. However, for both solely the EDA related data will be used in the experiments for this paper.

3.1. MAHNOB

The dataset used is called MAHNOB and was collected by Soleymani Mohammad et al. [31]. The data is related to different physiological signals.

The data was collected from 30 young healthy adults who participated in the study. 17 of the participants were female and 13 of them were males. Their age varied between 19 to 40. The participants were shown 20 emotional video clips which were evaluated in terms of both valence and arousal by using the Self-Assessment Manikins (SAM) questionnaire [32]. SAM is a prominent tool that visualizes the degree of valence and arousal by manikins. The participants distinguished a scale from 1 to 9, see Figure 1.

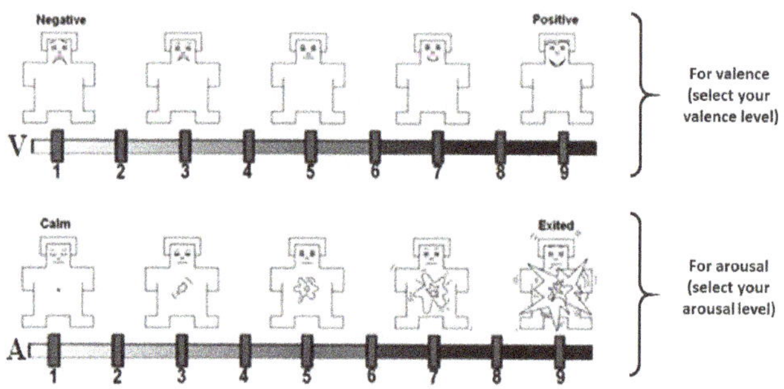

Figure 1. Self-assessment manikins scales for valence (above) and arousal (below) [32].

In the experiments for MAHNOB, electroencephalogram (EEG), blood volume pressure (BVP), respiration pattern, skin temperature, electromyogram (EMG), electrooculogram (EOG), electrocardiogram (ECG), and EDA of 30 participants were collected.

3.2. DEAP

DEAP [33] is a multimodal dataset used to analyze human emotional states.

The stimuli used in the experiments were chosen in different steps. First, they selected 120 initial stimuli that were selected both semi-automatically and manually. Second, a one-minute highlight part was specified for each stimulus. Third, through a web-based subjective assessment experiment, 40 final stimuli were chosen.

During the physiological experiment, 32 participants evaluated 40 videos via a web interface used for subjective emotion assessment in terms of the levels of arousal, valence, like/dislike, dominance, and familiarity. The age of participants varied between 19 to 37. Concerning the classes/labels for DEAP, we considered the same classes as same as in Section 4.1.

In the experiment, electroencephalogram (EEG), BVP, respiration pattern, ST, electromyogram (EMG), electrooculogram (EOG), electrocardiogram (ECG), and EDA of 32 participants were collected.

4. Classification Using a Convolution Neural Network—CNN

In this section, we present, the labelling of EDA signals, the design details of the proposed CNN for emotion classification and then, the evaluation metrics and evaluation.

4.1. Preprocessing and Labelling

First, raw data of EDA were scaled such that the distribution is centered around 0, with a standard deviation of 1. Additionally, after data normalization, two states [34] valence and arousal are addressed for emotion classification. In this regard, the scales (1–9) were mapped into 2 levels for each valence and arousal state according to the SAM ratings.

The valence scale of 1–5 was mapped to "negative" and 6–9 to "positive", respectively. The arousal scale of 1–5 was mapped to "passive" and 6–9 to "active", respectively.

- **High Valence/High Arousal (HVHA)**. This class includes positive emotions such as happy and excited.
- **High Valence/Low Arousal (HVLA)**. This class includes emotions such as relaxed, calm and pleased.
- **Low Valence/High Arousal (LVHA)**. This class includes emotions such as anger, fear and distressed.
- **Low Valence/Low Arousal (LVLA)**. This class includes negative emotions such as sad and depressed.

4.2. Classifiers

To perform the emotions classification task, we propose a deep-learning approach. A CNN is a kind of feedforward network structure that consists of multiple layers of convolutional filters followed by subsampling filters and ends with a fully connected classification layer. The classical LeNet-5CNN first proposed by LeCun et al. in [35] is the basic model for various CNN applications for object detection, localization, and prediction.

First, the EDA signals are converted into matrices whereby the goal is to make the application of CNN model possible (see Section 5).

As illustrated in Figure 2, the proposed CNN architecture has three convolutional layers (C1, C2, and C3), three subsampling layers in between (i.e., P1, P2, and P3), and an output layer F.

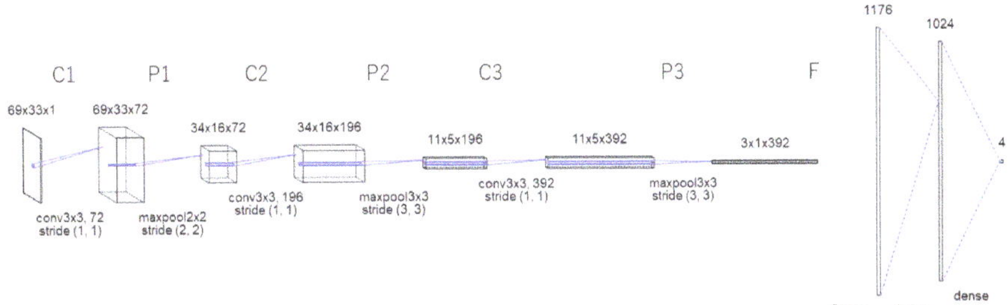

Figure 2. The proposed CNN model.

The convolutional layers generate feature maps using 72 (3 × 3) filters followed by a Scaled Exponential Linear Units (SELU) as an activation function, 196 (3 × 3) filters followed by a Rectified linear unit (ReLU) as an activation function and 392 (3 × 3) filters followed by a ReLU as an activation function.

Additionally, in the subsampling layers, the generated feature maps are spatially down-sampled. In our proposed model, the feature maps in layers C1, C2 and C3 are sub-sampled to a corresponding feature map of size 2 × 2, 3 × 3 and 3 × 3 in the subsequent layers P1, P2, and P3 respectively.

The output layer F is a fully connected neural model that performs the classification process, it consists of three layers. The first layer has 1176 nodes, each activated by a ReLU activation function. The second layer has 1024 nodes, each activated by a SELU activation function. The final layer is the SoftMax output layer C1.

The result of the mentioned layers is a 2D representation of extracted features from input feature map(s) based on the input EDA signals.

Since the dropout is a regularization technique to avoid over-fitting in neural networks based on preventing complex co-adaptations on training data [36], therefore, our dropout for each layer was 0.25 which is related to a fraction of the input units to drop. Table 2 shows parameters used for all the layers of the proposed CNN model.

Table 2. Parameters used for all the layers of the proposed CNN model.

Layer	Kernel, Units	Other Layers Parameters
C1	(3 × 3), 2	Activation = Selu, Strides = 1
P1	(2×2)	Strides = 2
C2	(3×3), 196	Activation = Selu, Strides = 1
P2	(3×3)	Strides=3
C3	(3×3)3, 92	Activation = Selu, Strides = 1
P3	(3×3)	Strides = 3

C is the convolution layer, P is the max-pooling layer and SELU is the Scaled Exponential Linear Unit activation function.

A grid search technique has been used to fine-tune the CNN model hyperparameters and to find out the optimal number of filters and layers needed to perform the emotion classification task. We have used the GridSearchCV class in Scikit-learn [37]. We have provided a dictionary of hyperparameters that should be checked during the performance evaluation. By default, the grid search uses one thread, but it can be configured to use all available cores to increase the calculation time. Then, the Scikit-learn class has been combined with Keras to find out what are the best hyperparameters values. Additionally, cross a validation is used to evaluate each individual model and the default of 10-fold cross-validation has been used.

All provided results have been obtained while using the following computer platform: Intel Corei7-7820HK processor Quad-Core 2.90 GHz, 16 GB DDR4 SDRAM, NVIDIA GeForce GTX 1080 with 8 GB dedicated storage.

Additionally, we examine several classifiers to compare the performance of the existing models with that of the here proposed one. In particular, Support Vector Machine (SVM) [38], K-Nearest Neighbor (KNN) [39], Naive Bayes [40] and Random Forest [41] are considered for benchmarking.

Based on Figures 3 and 4, selecting the previous classifiers has different advantages for comparison purposes. For example, the objective of random forests is that they consider a set of high-variance, low-bias decision trees and convert them into a model that has both low variance and low bias. On the other hand, KNNs is an algorithm which stores all the available cases and classifies new cases based on a similarity measure (e.g., distance functions). Therefore, KNN has been applied in statistical estimation and pattern recognition from the beginning of the 1970s on as a non-parametric technique [39]. Support Vector Machines are well-known in handling non-linearly separable data based on their non-linear kernel, e.g., the SVM with a polynomial kernel (SVM (poly)), and the SVM with a radial basis kernel (SVM (rbf)). Therefore, we classify the EDA data using three types of SVMs, namely the following ones: SVM (linear) (i.e., standard linear SVM), SVM (poly) and SVM (rbf). Finally, we used a simple probabilistic model which is the Naive Bayes. The purpose of using such a probabilistic model is to show how it behaves on EDA data. Table 3 shows the values of parameters of proposed CNN and other classifiers.

Table 3. Values of parameters of proposed CNN and other classifiers.

Model	Parameters
SVM (poly)	Degree of the polynomial kernel function = 3, $\gamma = \frac{1}{numberoffeatures}$
SVM (rbf)	$\gamma = \frac{1}{numberoffeatures}$
Random Forest	Number of estimators estimators = 10 trees, criterion = Gini impurity, The minimum number of samples required to split an internal node = 2
Naive Bayes	Prior = probabilities of the classes
KNN	Distance metric = 'minkowski', Power parameter for the Minkowski metric = 2, Number of neighbors = 3
Proposed (CNN)	Loss = categorical_crossentropy, optimizer = Adam, batch_size = 50, epochs = 1000

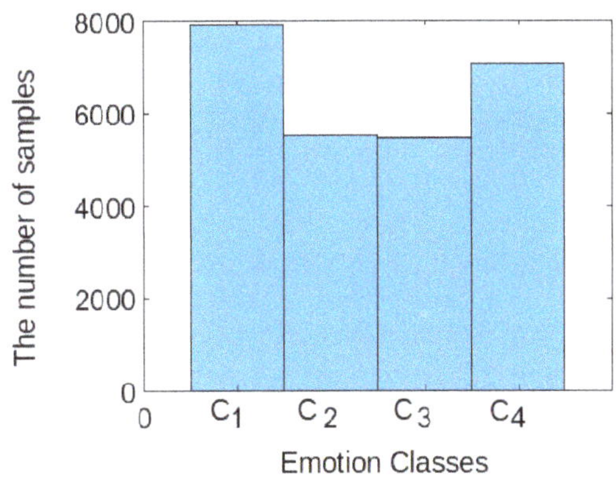

Figure 3. Overall emotion distribution for one Subject, where C1: High Valence/High Arousal (HVHA), C2: High Valence/Low Arousal (HVLA), C3: Low Valence/Low Arousal (LVLA) and C4: Low Valence/High Arousal (LVHA) based on a subject's data in MAHNOB.

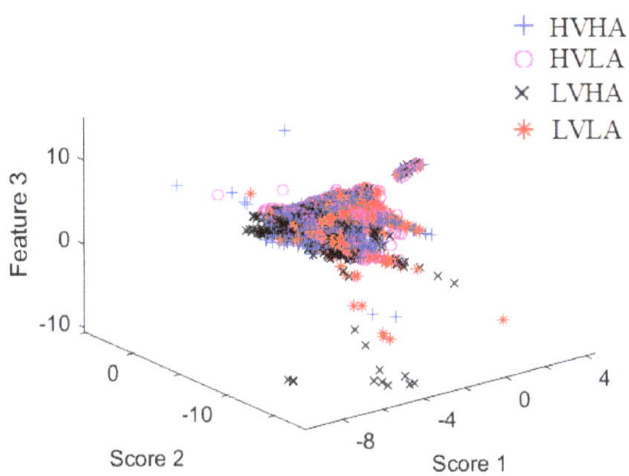

Figure 4. Scatter plot of the first three Fisher scores based on a subject's data in MAHNOB.

4.3. Evaluation Metrics and Validation Concept

To evaluate the overall performance of the classifiers, we consider several performance metrics. In particular, we use precision, recall, f-measure, and accuracy, as in [42].

The Equations (1)–(4) show mathematical expressions of the metrics precision, recall, accuracy, and f-measure respectively, where TP, TN, FP, and FN refer respectively to "True Positives", "True Negatives", "False Positives" and "False Negatives" respectively.

$$Precision = \frac{\mathrm{TP}}{\mathrm{TP} + \mathrm{FP}} \tag{1}$$

$$Recall = \frac{\mathrm{TP}}{\mathrm{TP} + \mathrm{FN}} \tag{2}$$

$$Accuracy = \frac{\mathrm{TP} + \mathrm{TN}}{\mathrm{TP} + \mathrm{TN} + \mathrm{FN} + \mathrm{FP}} \tag{3}$$

$$F1 = \frac{2 \cdot precision \cdot recall}{precision + recall} \tag{4}$$

Regarding the evaluation scenarios, we consider two cases. The subject-dependent and subject-independent cases. Subject-dependent means training and testing have been performed on the same subject. Subject-independent means the training has been performed on a group of subjects and testing has been performed on a totally new group of subjects.

5. Results

To have a deeper understanding of the performance of the proposed CNN model, MAHNOB, and DEAP datasets were used for testing the overall classification performance.

Moreover, the data distribution should be taken into consideration to choose a suitable classifier for comparison purposes. In this regard, a Fisher mapping [43] was used to define the three major scores in the samples that are investigated. Based on the output of Figures 3 and 4, it is concluded that the data is highly overlapped, and there is a kind of class imbalance problem.

In this assessment, 10 subjects were selected from the MAHNOB and DEAP datasets. Each dataset for each subject consists of four classes (see Sections 3.1 and 3.2). The average training time for each subject was approximately 21 min.

The length of the considered EDA signals is 2574 that are converted to matrices of size (39 $\times$ 66). All results are presented for ten-fold cross-validation.

Tables 4 and 5 present the average values for the precision, the recall, and the f-measure using DEAP and MAHNOB datasets respectively. The tables show the performance metrics values when training and testing are performed on the same subject. The tables show the average value of precision, recall, and f-measure with respect to each subject. The performance metrics values for each subject have been summed and divided by the total number of subjects. The major target of this experiment is to check out the overall performance for subject-dependent EDA-based emotion classification.

Tables 6 and 7 present the precision the recall, and the f-measure using DEAP and MAHNOB datasets respectively. The results are obtained when training and testing are performed on different subjects. The major target of this experiment is to check out the overall performance for subject-independent EDA-based emotion classification.

In all tables, the proposed CNN model shows the highest performance compared to K-NN and random forest which are hereby the best next two classifiers. When K-NN and random forest classifiers perform well, it indicates that the dataset is not easily separable, and the nonlinearity is high. This can be observed in Figure 4. Accordingly, the decision planes generated using other classifiers (see Tables 4–7) do not categorize some points in space to an inappropriate region as good as K-NN and random forest classifiers.

The performance metrics and the implementation are written in Python using Numpy (http://www.numpy.org/), Scikit-learn (https://scikit-learn.org/) and Keras (https://keras.io/). All performance metrics are calculated for each class and weighted taking the class imbalance into account. Accordingly, the evaluation metrics for each label have been calculated and their average has been weighted by the support measurement which is the number of true instances for each label.

Tables 8 and 9 show the confusion matrix for both MAHNOB and DEAP (the average performance results for training and testing on same subjects) and the confusion matrix for both MAHNOB and DEAP (the average performance results for training and testing on different subjects), respectively.

Table 4. Performance metrics for DEAP (the average performance results for training and testing on same subject).

Model	Accuracy	Precision	Recall	F-Measure
SVM (Linear)	0.46	0.41	0.46	0.42
SVM (poly)	0.41	0.53	0.43	0.33
SVM (rbf)	0.59	0.60	0.60	0.58
Random Forest	0.74	0.76	0.75	0.75
Naive Bayes	0.44	0.48	0.44	0.42
K-NN	0.80	0.80	0.80	0.80
Proposed CNN	**0.85**	**0.85**	**0.85**	**0.85**

Table 5. Performance metrics for MAHNOB (the average performance results for training and testing on same subject).

Model	Accuracy	Precision	Recall	F-Measure
SVM (Linear)	0.49	0.48	0.50	0.43
SVM (poly)	0.47	0.49	0.48	0.36
SVM (rbf)	0.55	0.53	0.56	0.51
Random Forest	0.68	0.70	0.70	0.70
Naive Bayes	0.37	0.43	0.39	0.35
K-NN	0.74	0.76	0.75	0.75
Proposed CNN	**0.81**	**0.81**	**0.81**	**0.81**

Table 6. Performance metrics for MAHNOB (the average performance results for training and testing on different subjects).

Model	Accuracy	Precision	Recall	F-Measure
SVM (Linear)	0.34	0.47	0.34	0.37
SVM (poly)	0.36	0.70	0.37	0.42
SVM (rbf)	0.41	0.53	0.42	0.45
Random Forest	0.64	0.65	0.65	0.65
Naive Bayes	0.27	0.43	0.27	0.33
K-NN	0.72	0.73	0.73	0.72
Proposed CNN	**0.78**	**0.78**	**0.78**	**0.78**

Table 7. Performance metrics for DEAP (the average performance results for training and testing on different subjects).

Model	Accuracy	Precision	Recall	F-Measure
SVM (Linear)	0.40	0.41	0.40	0.31
SVM (poly)	0.39	0.41	0.39	0.28
SVM (rbf)	0.44	0.50	0.44	0.40
Random Forest	0.69	0.70	0.69	0.69
Naive Bayes	0.36	0.31	0.36	0.28
K-NN	0.75	0.76	0.75	0.76
Proposed CNN	**0.82**	**0.83**	**0.82**	**0.83**

Table 8. Confusion matrix for both MAHNOB and DEAP (the average performance results for training and testing on same subjects).

Class	C1	C2	C3	C4
C1	0.861	0.057	0.071	0.046
C2	0.062	0.808	0.059	0.034
C3	0.039	0.050	0.878	0.017
C4	0.045	0.063	0.042	0.866

C1: High Valence/High Arousal (HVHA), C2: High Valence/Low Arousal (HVLA), C3: Low Valence/Low Arousal (LVLA) and C4: Low Valence/High Arousal (LVHA).

Table 9. Confusion matrix for both MAHNOB and DEAP (the average performance results for training and testing on different subjects).

Class	C1	C2	C3	C4
C1	0.762	0.177	0	0.146
C2	0.049	0.685	0	0.077
C3	0.004	0	0.705	0.017
C4	0.108	0.126	0.058	0.857

C1: High Valence/High Arousal (HVHA), C2: High Valence/Low Arousal (HVLA), C3: Low Valence/Low Arousal (LVLA) and C4: Low Valence/High Arousal (LVHA).

6. Discussion

Aiming at highlighting the contribution of this work, other works should be considered and analyzed. However, it is not easy to make such a comparison due to the fact that (a) other works may combine other types of physiological signals and they do not use only EDA, and (b) the reaction and the response of EDA does highly depend on the stimuli type, which showed better results when the stimuli is an acoustic one [18].

To our knowledge, this study shows for the first time that developing a subject-independent human emotion recognition using only EDA signals with a promising recognition rate is possible. It is also worthwhile noting that we were able to,

- increase the f-measure for subject-independent classification to 78% and 81% for MAHNOB and DEAP respectively (4 classes/labels).
- increase the f-measure for subject-dependent classification have been increased to 83% and 85% for MAHNOB and DEAP respectively (4 classes/labels).

In the state-of-the-art, researchers in [22] tested a deep-learning model which consists of RNN and CNN which showed a Concordance Correlation Coefficient (CCC) [44] of 0.10 on the arousal dimension and 0.33 on the valence dimension based on EDA only. They used AVEC 2016 dataset [23,24].

In addition, in [27], they reported an emotion recognition analysis using only the EDA signal for subject-dependent with an accuracy of 56.5% for the arousal dimension and 50.5% for the valence dimension based on four songs stimuli. In [18], authors suggested a system which can achieve a recognition accuracy of 77.33% on the arousal dimension, and 84% on the valence dimension based on three emotional states induced by affective sounds taken from IADS collection [45].

Furthermore, it should be mentioned that the binary classification (passive/active cases) of EDA signals showed high results as in [28] with an accuracy of 95% using SVM and an accuracy of 80% using CNNs in [29].

However, getting such a high performance for two classes is expected where other studies showed clearly that EDA signals for active and passive states form clear patterns compared to the 4 classes of arousal and valence for emotion recognition [46]. Table 10 shows a summary of the state-of-the-art for EDA-based emotion detection regarding, experiment, number of classes, used classifiers, and the reported accuracy.

Table 10. A summary of the state-of-the-art results using only EDA.

Paper	Experiment	Number of Classes	Classifier Used	Arousal	Valence	Accuracy (Both)
[27]	Subject-dependent	4	Genetic algorithm and K-NN	0.56	0.50	–
[18]	Subject-independent	3	K-NN	0.77	0.84	–
[28]	Subject-independent	2	SVM	–	–	0.95
[29]	Subject-dependent	2	CNN	–	–	0.80
[22]	Subject-independent	2	CNN	0.10	0.33	–
Proposed CNN	Subject-independent (DEAP)	4	CNN	–	–	**0.82**
Proposed CNN	Subject-independent (MAHNOB)	4	CNN	–	–	**0.78**
Proposed CNN	Subject-dependent (DEAP)	4	CNN	–	–	**0.85**
Proposed CNN	Subject-dependent (MAHNOB)	4	CNN	–	–	**0.81**

SVM: Support Vector Machine, K-NN: K-Nearest Neighbor, CNN: Convolutional Neural Network.

Additionally, analyzing the results of the state-of-art, clearly, feature engineering for subject-independent and subject-dependent human emotion detection based on EDA does not lead to high performance. In particular, when the number of classes is higher than two. This is because extracting the sympathetic response patterns which are part of each emotion is difficult. Furthermore, when trying to overcome this fact by analyzing more basic features such as level, response amplitude, rate, rise time, and recovery time, they discard flexible elicited behavior which might improve emotion recognition. Therefore, it has been proven in this work that DL can overcome this drawback quite well.

Regarding the point of testing the proposed model using different datasets from different labs, it is because human emotions do not form similar patterns. Consequently, the research community should develop generalized models to recognize human emotions, where subjects, elicitation materials, and physiological sensors brands are different from the ones involved in the initial training. Dealing with such research question has an important impact for human support in the frame of smart environments in different applications.

Concerning, human emotion recognition with respect to different lab–settings, in [30], authors showed that adjusting and manipulating the feature space to bring both datasets to a homogeneous feature space as a pre–processing step may increase the overall performance even when datasets come from different labs.

Moreover, in [47], they checked the ability of 504 school children aged between 8 and 11 years old to recognize the emotions of facial expressions based on pictures. The overall performance was approximately 86% to recognize anger, fear, sadness, happiness, disgust, and neutral facial expressions. It is impressive to see that the proposed automated EDA-based emotion recognition system is close to the performance of human capability to interpret the facial expressions.

7. Conclusions

This study can be considered to be a basic contribution in terms of overcoming the generalization problem for human emotion recognition. The aim was to show the feasibility and the possibility of building such generalized models for relevant application contexts. Furthermore, this study examined the less intrusive sensors based on statistical analyses in real-life datasets and reviewed various state-of-the-art approaches to human emotion recognition in smart home environments.

Additionally, emotion recognition is a cornerstone of advanced intelligent systems for monitoring a subject's comfort. Thus, information on a subject's emotion and stress level is a key component for the future of smart AAL environments.

In our future work, we will focus on human emotion recognition using EDA with respect to different lab–settings, which means, we will try to build a generalized approach which should be trained using lab–settings X and tested using lab–settings Y. Additionally, we plan to combine Stacked Sparse Auto Encoders with CNN. Moreover, CNN essentially learns local (spatial) features. On the other side, RNN does in essence rather learn temporal features. Consequently, combining both neural network concepts will result in a neuro-processor which can learn both contextual dependencies (i.e., spatial and temporal) from inputted local features. As a result, such a combination does potentially improve the overall performance.

Author Contributions: F.A.M. and A.E. conceived and designed the approach; E.A.M. and M.A. performed the formal analysis; F.A.M., K.K. wrote the paper.

Funding: This research received no external funding.

Conflicts of Interest: The authors declare no conflict of interest. The authors ensure that there are no personal circumstances, interest or sponsors that may be perceived as inappropriately influencing the representation or interpretation of reported research results.

References

1. Suryadevara, N.K.; Quazi, M.; Mukhopadhyay, S.C. Intelligent sensing systems for measuring wellness indices of the daily activities for the elderly. In Proceedings of the 2012 8th IEEE International Conference onIntelligent Environments (IE), Guanajuato, Mexico, 26–29 June 2012; pp. 347–350.
2. Al Machot, F.; Mosa, A.H.; Dabbour, K.; Fasih, A.; Schwarzlmuller, C.; Ali, M.; Kyamakya, K. A novel real-time emotion detection system from audio streams based on bayesian quadratic discriminate classifier for adas. In Proceedings of the 2011 Joint 3rd Int'l Workshop on IEEE Nonlinear Dynamics and Synchronization (INDS) & 16th Int'l Symposium on Theoretical Electrical Engineering (ISTET), Klagenfurt, Austria, 25–27 July 2011; pp. 1–5.
3. Krause, R. Universals and Cultural Differences in the Judgments of Facial Expressions of Emotion. *J. Personal. Soc. Psychol.* **1987**, *5*, 4–712.
4. Lang, P.J. The emotion probe: Studies of motivation and attention. *Am. Psychol.* **1995**, *50*, 372. [CrossRef] [PubMed]
5. Kim, J.; André, E. Emotion recognition based on physiological changes in music listening. *IEEE Trans. Pattern Anal. Mach. Intell.* **2008**, *30*, 2067–2083. [CrossRef]
6. Ali, M.; Mosa, A.H.; Al Machot, F.; Kyamakya, K. EEG-based emotion recognition approach for e-healthcare applications. In Proceedings of the 2016 IEEE Eighth International Conference on Ubiquitous and Future Networks (ICUFN), Vienna, Austria, 5–8 July 2016, pp. 946–950.
7. Kim, Y.; Lee, H.; Provost, E.M. Deep learning for robust feature generation in audiovisual emotion recognition. In Proceedings of the 2013 IEEE International Conference on Acoustics, Speech and Signal Processing (ICASSP), Vancouver, BC, Canada, 26–31 May 2013; pp. 3687–3691.
8. Zeng, Z.; Pantic, M.; Roisman, G.I.; Huang, T.S. A survey of affect recognition methods: Audio, visual, and spontaneous expressions. *IEEE Trans. Pattern Anal. Mach. Intell.* **2009**, *31*, 39–58. [CrossRef] [PubMed]
9. LeCun, Y.; Bengio, Y.; Hinton, G. Deep learning. *Nature* **2015**, *521*, 436. [CrossRef] [PubMed]
10. Zheng, W.L.; Lu, B.L. Investigating critical frequency bands and channels for EEG-based emotion recognition with deep neural networks. *IEEE Trans. Auton. Mental Dev.* **2015**, *7*, 162–175. [CrossRef]
11. Ranganathan, H.; Chakraborty, S.; Panchanathan, S. Multimodal emotion recognition using deep learning architectures. In Proceedings of the 2016 IEEE Winter Conference on Applications of Computer Vision (WACV), Lake Placid, NY, USA, 7–9 March 2016; pp. 1–9.
12. Pławiak, P.; Rzecki, K. Approximation of phenol concentration using computational intelligence methods based on signals from the metal-oxide sensor array. *IEEE Sens. J.* **2015**, *15*, 1770–1783.
13. Pławiak, P.; Maziarz, W. Classification of tea specimens using novel hybrid artificial intelligence methods. *Sens. Actuators B Chem.* **2014**, *192*, 117–125. [CrossRef]
14. Yıldırım, Ö.; Pławiak, P.; Tan, R.S.; Acharya, U.R. Arrhythmia detection using deep convolutional neural network with long duration ECG signals. *Comput. Biol. Med.* **2018**, *102*, 411–420. [CrossRef]
15. Pławiak, P.; Acharya, U.R. Novel Deep Genetic Ensemble of Classifiers for Arrhythmia Detection Using ECG Signals. Available online: https://www.researchgate.net/profile/Pawel_Plawiak/publication/329782366_Novel_Deep_Genetic_Ensemble_of_Classifiers_for_Arrhythmia_Detection_Using_ECG_Signals/links/5c1bad6792851c22a338cd02/Novel-Deep-Genetic-Ensemble-of-Classifiers-for-Arrhythmia-Detection-Using-ECG-Signals.pdf (accessed on 5 April 2019).
16. Soto, J.A.; Levenson, R.W. Emotion recognition across cultures: The influence of ethnicity on empathic accuracy and physiological linkage. *Emotion* **2009**, *9*, 874. [CrossRef]

17. Ooi, J.S.K.; Ahmad, S.A.; Chong, Y.Z.; Ali, S.H.M.; Ai, G.; Wagatsuma, H. Driver emotion recognition framework based on electrodermal activity measurements during simulated driving conditions. In Proceedings of the 2016 IEEE EMBS Conference on Biomedical Engineering and Sciences (IECBES), Kuala Lumpur, Malaysia, 4–7 December 2016; pp. 365–369.
18. Greco, A.; Valenza, G.; Citi, L.; Scilingo, E.P. Arousal and valence recognition of affective sounds based on electrodermal activity. *IEEE Sens. J.* **2017**, *17*, 716–725. [CrossRef]
19. Bradley, M.M.; Lang, P.J. Affective reactions to acoustic stimuli. *Psychophysiology* **2000**, *37*, 204–215. [CrossRef]
20. van der Zwaag, M.D.; Janssen, J.H.; Westerink, J.H. Directing physiology and mood through music: Validation of an affective music player. *IEEE Trans. Affect. Comput.* **2013**, *4*, 57–68. [CrossRef]
21. Ćosić, K.; Popović, S.; Kukolja, D.; Dropuljić, B.; Ivanec, D.; Tonković, M. Multimodal analysis of startle type responses. *Comput. Methods Programs Biomed.* **2016**, *129*, 186–202. [CrossRef] [PubMed]
22. Keren, G.; Kirschstein, T.; Marchi, E.; Ringeval, F.; Schuller, B. END-TO-END Learning for Dimensional Emotion Recognition from Physiological Signals. Available online: https://ieeexplore.ieee.org/document/8019533 (accessed on 5 April 2019).
23. Weber, R.; Barrielle, V.; Soladié, C.; Séguier, R. High-level geometry-based features of video modality for emotion prediction. In Proceedings of the 6th ACM International Workshop on Audio/Visual Emotion Challenge, Amsterdam, The Netherlands, 15–19 October 2016; pp. 51–58.
24. Povolny, F.; Matejka, P.; Hradis, M.; Popková, A.; Otrusina, L.; Smrz, P.; Wood, I.; Robin, C.; Lamel, L. Multimodal emotion recognition for AVEC 2016 challenge. In Proceedings of the 6th ACM International Workshop on Audio/Visual Emotion Challenge, Amsterdam, The Netherlands, 15–19 October 2016; pp. 75–82.
25. Katsis, C.D.; Katertsidis, N.; Ganiatsas, G.; Fotiadis, D.I. Toward emotion recognition in car-racing drivers: A biosignal processing approach. *IEEE Trans. Syst. Man Cybern. Part A Syst. Hum.* **2008**, *38*, 502–512. [CrossRef]
26. Ali, M.; Al Machot, F.; Mosa, A.H.; Kyamakya, K. CNN Based Subject-Independent Driver Emotion Recognition System Involving Physiological Signals for ADAS. In *Advanced Microsystems for Automotive Applications 2016*; Springer: New York, NY, USA, 2016; pp. 125–138.
27. Niu, X.; Chen, L.; Xie, H.; Chen, Q.; Li, H. Emotion pattern recognition using physiological signals. *Sens. Trans.* **2014**, *172*, 147.
28. Xia, V.; Jaques, N.; Taylor, S.; Fedor, S.; Picard, R. Active learning for electrodermal activity classification. In Proceedings of the 2015 IEEE Signal Processing in Medicine and Biology Symposium (SPMB), Philadelphia, PA, USA, 12 December 2015; pp. 1–6.
29. Paragliola, G.; Coronato, A. A Deep Learning-Based Approach for the Recognition of Sleep Disorders in Patients with Cognitive Diseases: A Case Study. FedCSIS Position Papers, 2017; pp. 43–48. Available online: https://annals-csis.org/Volume_12/drp/pdf/532.pdf (accessed on 5 April 2019).
30. Al Machot, F.; Ali, M.; Ranasinghe, S.; Mosa, A.H.; Kyandoghere, K. Improving Subject-independent Human Emotion Recognition Using Electrodermal Activity Sensors for Active and Assisted Living. In Proceedings of the 11th ACM PErvasive Technologies Related to Assistive Environments Conference, Corfu, Greece, 26–29 June 2018; pp. 222–228.
31. Soleymani, M.; Lichtenauer, J.; Pun, T.; Pantic, M. A multimodal database for affect recognition and implicit tagging. *IEEE Trans. Affect. Comput.* **2012**, *3*, 42–55. [CrossRef]
32. Bradley, M.M.; Lang, P.J. Measuring emotion: The self-assessment manikin and the semantic differential. *J. Behav. Ther. Exp. Psychiatry* **1994**, *25*, 49–59. [CrossRef]
33. Koelstra, S.; Muhl, C.; Soleymani, M.; Lee, J.S.; Yazdani, A.; Ebrahimi, T.; Pun, T.; Nijholt, A.; Patras, I. Deap: A database for emotion analysis; using physiological signals. *IEEE Trans. Affect. Comput.* **2012**, *3*, 18–31. [CrossRef]
34. Frijda, N.H. *The Emotions*; Cambridge University Press: Cambridge, UK, 1986; ISBN 0521301556.
35. LeCun, Y.; Bengio, Y. Convolutional networks for images, speech, and time series. *Handb. Brain Theory Neural Netw.* **1995**, *3361*, 1995.
36. Hinton, G.E.; Srivastava, N.; Krizhevsky, A.; Sutskever, I.; Salakhutdinov, R.R. Improving neural networks by preventing co-adaptation of feature detectors. *arXiv* **2012**, arXiv:1207.0580.

37. Pedregosa, F.; Varoquaux, G.; Gramfort, A.; Michel, V.; Thirion, B.; Grisel, O.; Blondel, M.; Prettenhofer, P.; Weiss, R.; Dubourg, V.; et al. Scikit-learn: Machine Learning in Python. *J. Mach. Learn. Res.* **2011**, *12*, 2825–2830.
38. Cortes, C.; Vapnik, V. Support-vector networks. *Mach. Learn.* **1995**, *20*, 273–297. [CrossRef]
39. Altman, N.S. An introduction to kernel and nearest-neighbor nonparametric regression. *Am. Stat.* **1992**, *46*, 175–185.
40. Webb, G.I. Naïve Bayes. In *Encyclopedia of Machine Learning and Data Mining*; Springer: New York, NY, USA, 2017; pp. 895–896.
41. Breiman, L. Random forests. *Mach. Learn.* **2001**, *45*, 5–32. [CrossRef]
42. Powers, D.M. Evaluation: From Precision, Recall and F-Measure to ROC, Informedness, Markedness and Correlation. 2011. Available online: https://dspace2.flinders.edu.au/xmlui/handle/2328/27165 (accessed on 5 April 2019).
43. Fukunaga, K. *Introduction to Statistical Pattern Recognition*; Academic Press: New York, NY, USA, 2013.
44. Lawrence, I.; Lin, K. A concordance correlation coefficient to evaluate reproducibility. *Biometrics* **1989**, 255–268. [CrossRef]
45. Bradley, M.M.; Lang, P.J. *The International Affective Digitized Sounds (IADS-2): Affective Ratings of Sounds and Instruction Manual*; University of Florida: Gainesville, FL, USA, 2007.
46. Gendolla, G.H.; Krüsken, J. The joint impact of mood state and task difficulty on cardiovascular and electrodermal reactivity in active coping. *Psychophysiology* **2001**, *38*, 548–556. [CrossRef] [PubMed]
47. Mancini, G.; Agnoli, S.; Baldaro, B.; Bitti, P.E.R.; Surcinelli, P. Facial Expressions of Emotions: Recognition Accuracy and Affective Reactions During Late Childhood. *J. Psychol.* **2013**, *147*, 599–617. [CrossRef] [PubMed]

sensors

MDPI

Article

Enhancing Mouth-Based Emotion Recognition Using Transfer Learning

Valentina Franzoni [1,*], Giulio Biondi [2], Damiano Perri [2] and Osvaldo Gervasi [1,*]

1 Department of Mathematics and Computer Science, University of Perugia, 06123 Perugia, Italy
2 Department of Mathematics and Computer Science, University of Florence, 50121 Firenze, Italy; giulio.biondi@unifi.it (G.B.); damiano.perri@unifi.it (D.P.)
* Correspondence: valentina.franzoni@dmi.unipg.it (V.F.); osvaldo.gervasi@unipg.it (O.G.)

Received: 16 July 2020; Accepted: 10 September 2020; Published: 13 September 2020

Abstract: This work concludes the first study on mouth-based emotion recognition while adopting a transfer learning approach. Transfer learning results are paramount for mouth-based emotion emotion recognition, because few datasets are available, and most of them include emotional expressions simulated by actors, instead of adopting real-world categorisation. Using transfer learning, we can use fewer training data than training a whole network from scratch, and thus more efficiently fine-tune the network with emotional data and improve the convolutional neural network's performance accuracy in the desired domain. The proposed approach aims at improving emotion recognition dynamically, taking into account not only new scenarios but also modified situations to the initial training phase, because the image of the mouth can be available even when the whole face is visible only in an unfavourable perspective. Typical applications include automated supervision of bedridden critical patients in a healthcare management environment, and portable applications supporting disabled users having difficulties in seeing or recognising facial emotions. This achievement takes advantage of previous preliminary works on mouth-based emotion recognition using deep-learning, and has the further benefit of having been tested and compared to a set of other networks using an extensive dataset for face-based emotion recognition, well known in the literature. The accuracy of mouth-based emotion recognition was also compared to the corresponding full-face emotion recognition; we found that the loss in accuracy is mostly compensated by consistent performance in the visual emotion recognition domain. We can, therefore, state that our method proves the importance of mouth detection in the complex process of emotion recognition.

Keywords: transfer learning; convolutional neural networks; emotion recognition

1. Introduction

Visual emotion recognition (ER) has been widely studied as one of the first affective computing techniques, based on visual features of the face, to combine features about the eyes, mouth and various facial elements at the same time. Several different approaches to visual recognition obtained different grades of classifications for different visual recognition techniques [1–3]. Recently, studies using only the mouth for facial emotion recognition obtained promising results, while still not gaining the proper recognition among the state-of-the-art. Such works used convolutional neural networks (CNNs) to detect basic emotions from innovative and ubiquitous devices, e.g., smartphone or computer cameras, to produce textual, audio or visual feedback for humans, or digital outputs to support other services, mainly for healthcare systems [4]. A neural network can obtain an excellent result with a relatively small dataset of images when trained on a single individual, e.g., to detect particular states needing immediate medical intervention, or changes over time indicating an underlying degenerative health condition.

Our analysis focused on emotion classification with the mouth only, to study how much the mouth is involved in emotional expression and can provide accuracy compared to full-face recognition. We analysed the position and curve of lips through CNNs. Recently, some of our preliminary works [1,4,5] obtained favourable results regarding emotion analysis using the mouth, and this is our first attempt to recap and complete our full study on mouth-based emotion recognition with extensive datasets. All our previous works used convolutional neural networks to reach their goals on the topic, using a self-collected dataset. In-depth work has been primarily done on three emotions (i.e., joy, disgust and neutral) [5], among which disgust is the less studied in the literature.

After proving that the mouth can be itself a promising element for emotion recognition, in this work, we focus on the mouth as a unique element for facial-expression-based emotion recognition using advanced deep learning techniques. The goal of this approach was to enhance the mouth-based approach to emotion recognition in order to generalise the previous experimentation on a multiple-user dataset. To that end, advanced deep learning techniques have been implemented, i.e., knowledge transfer with elements of continuous learning. Such techniques are particularly suitable to being applied in the healthcare environment.

For instance, connecting this architecture to appropriate services can help users to convey emotions in an automated way effectively, e.g., providing augmented emotional stimuli to users affected by autism or other conditions involving social relationship abilities wherein a user experiences difficulties in recognising emotions expressed by other people. Another example may be a system able to recognise severe conditions and call a human assistant for intervention, e.g., for hospitalised patients feeling intense pain or needing psychological support. Such applications may provide feedback from healthcare personnel to exploit continuous learning.

We tested the most promising neural networks [1] for face recognition and mouth recognition, e.g., lip reading [6], with previous training on widely used large datasets of images [7]. Knowledge transfer allowed our neural networks to be pre-trained on low-level features, e.g., edges, corners and colour distribution.

In the last layers of the CNN, we carried out ad-hoc training dedicated to human face recognition of emotional classes. This work concluded the experiments by testing and comparing several CNNs on a widely-used dataset [8] of human faces labelled with emotions from the Ekman model. As in the preliminary work, for the final dataset we provide a filtered version, wherein photos showing simulated emotions, i.e., non-spontaneous expressions, have been removed.

1.1. Data Availability Issues for Facial ER

The main problem in facial emotion recognition is the lack of proper datasets of images for training. Most of the available datasets take into consideration only the Ekman model [9] or its subsets [5], discarding more complex and complete models, such as Plutchik [10], or models based on emotional affordance [11]. Moreover, image datasets often contain non-genuine expressions (e.g., simulated by professional actors) rather than spontaneous and natural forms of facial expression [12]. Since deep learning can extract features not even recognisable by humans, items related to non-real emotions make it hard to train a neural network effectively for emotion recognition. For these reasons, our work focused on a subset of the Ekman model, augmented with the neutral expression as a state of control for recognition results on emotions [4], by investing a fair amount of effort into selecting proper images from the available datasets, both from internet sources and from self-produced material. Collecting emotional images is feasible for emotions which are easily triggered, e.g., joy and disgust, but it is quite challenging for other emotions wherein ethical issues are involved in the stimulus, e.g., anger and fear. In order to improve results with a relatively small set of images for each emotion, we used transfer learning.

1.2. Previous Works

Previous works focused on the acquisition of data from a single user, i.e., where researchers train the network on a specific user's face. This approach lets a user train the network with precision on his/her face, reaching advanced recognition performance [1,4]. Our work presents a more general approach using a multi-user dataset, containing images of users who were different regarding age, cultural background, gender and appearance. To that end, we used advanced methods of deep learning, i.e., transfer learning and continuous learning.

Multidisciplinary studies at the basis of our work in artificial intelligence (AI) stressed the importance of computer science for automated real-life tasks for assistive technologies [13,14]. Among them, labial detection and lip-reading [6] constitute our main background starting point [15].

One of the most promising advances of recent years for AI-assisted health care is the opportunity to develop mobile applications on widely-spread devices [4], e.g., smartphones, tablets and smartwatches, to support disabled users.

2. Problem Description and Proposed Solution

Our study exploits the high precision of CNNs by processing mouth images to recognise emotional states using the most recent advances in affective computing. Affective computing is a novel topic in artificial intelligence, defined for the first time in 2003 by Rosalind Picard [16], soon becoming one of the most trending multidisciplinary studies. Affective computing, involving the collaboration of several disciplines, e.g., psychology, physiology, neurology, liberal studies, computer science and robotics, recently stressed the importance of the extraction and recognition of affective mental states, e.g., emotions, moods, sentiments and personality traits. Notwithstanding the interest in such new research, most of the available research still focuses on trivial sentiment analysis or pure theoretical models (e.g., in psychology and social sciences) and product-based applications [17], mainly for marketing purposes. We prefer to focus on self-aid [18], health management and communication for understanding and supporting humans in real-life problems for any demanding task (e.g., due to disabilities [19], psychological states in emergencies or particular environments) with automated detectors and artificial assistants with machine emotional intelligence capabilities [20].

In our socially interconnected world, individuals already use and produce daily an overwhelming amount of heterogeneous data in a manageable and personalised subset of classified items. Data can be directly used for emotion recognition (ER), e.g., photos and text shared on social networks; or can contain elements indirectly inferable for emotional intelligence, such as physiological data collected by wearables—including sleep patterns, continuous heart-rate and movement tracking, emotions expressed by art [21] and data from users experiencing exciting games [22]. If sentiment analysis relates only to recognising the positiveness, negativeness or neutrality of sentiments, moods and emotions, the process of emotion recognition is still little studied, implying the recognition of specific emotions of an emotional model. Since scientists do not agree on all the available models, research on ER in any applicative domain starts from the widely recognised model of Ekman, which we chose for our study. Recent research underlines that Ekman's primary emotional states, including happiness, sadness, anger, disgust and neutral [23] can be recognised based on text [24] and physiological clues such as heart rate, skin conductance, gestures, facial expression and sound, which can be managed with a multidimensional approach [11] and compared.

Among all ER approaches, facial recognition is still predominant. Under that point of view, we decided to focus on the mouth as a particular facial element, almost always visible in any facial expression, even considering that some cultures underestimate the mouth's expressiveness more than others. Some tribes in South America, for instance, as Ekman himself underlined in his first experiments on facial emotions labelling [9], rely more on the upper part of the face. The new issue is that, in general, when comparing such original studies with our results in AI, the neutral emotion seems the most misunderstood both by humans and artificial agents; it tends to be labelled as anger or sadness. The fact that the same emotion is also the most mistaken by automatic mouth-based recognisers in the same classes of errors confirms that the automatic agent can recognise images correctly, based on human labels.

In this work, we tested the technique on multi-user datasets, to find solutions to the following research questions:

- With how much precision it is possible to recognise facial emotions solely from the mouth?
- Is the proposed technique capable of recognising emotions if trained on a generalised set of facial images?

In a user-centred implementation, the software supports personalised emotional feedback for each particular user: personal traits, such as scars or flaws, and individual variations in emotional feeling and expression, help the training to precise recognition. The system can recognise different users because it was trained on a comprehensive dataset, including images varying in ethnicity, age and gender.

In order to obtain optimised results, the ambient light setting should not require a particular setup. A consistent implementation should meet the following requirements:

- Robustnes. The algorithm must be able to operate even in the presence of low-quality data (e.g., low resolution or bad light conditions);
- Scalability. The user's position should not be necessarily fixed in front of the camera, in order to avoid constraining the person. Therefore, the software should be able to recognise the user despite the shot's point of view;
- Luminosity. Luminosity is an important issue because the variation of light hardly influences the recognition capabilities of a CNN-based automated system. Available datasets usually provide photos shot under a precise lighting setup. On the contrary, a sufficient number of training samples of each considered lighting (e.g., natural light, artificial bulbs and low-light conditions) should be provided for an efficient categorisation.

Our implementation of mouth-based ER exploits transfer learning (i.e., knowledge transfer), which uses general-purpose neural networks pre-trained on extensive datasets including different shapes, later fine-tuned on the classification domain (i.e., mouth-based emotion recognition). Our application can be easily adapted for continuous learning, given a domain where said method is useful, and appropriate data are available as feedback to track the weights of the neural network and prosecute the training. Such a method should allow enhancing the precision of emotion recognition in real-time situations, where the final weights of a previous training can be used in a subsequent time. The collateral effect of such a technique is the requirement of more exceptional computational capabilities and a semi-supervised system in order to stop and roll back in cases of glaring errors or overfitting in a specific environment, e.g., a particular camera resolution or light situation in a specific place or with certain timing. Continuous learning can continuously adapt and enhance the network's accuracy, leading on the go to a more reliable system, when used for a prolonged time.

3. The Framework Implementation

The proposed implementation has been developed using the Python programming language and the Keras framework [25]. The tests were carried out using the Google Colab platform, in which the Python code was developed and executed, exploiting the powerfulness of the Keras libraries. Data analysis and processing have been developed with a chain of operations. The implemented procedure can be described with the following steps:

1. Raw dataset import. In the first phase, the dataset (at the beginning, a raw dataset) is composed of RGB images and is imported into the system.
2. Raw dataset cleaning. All the faces in the photos of the dataset are manually scanned, and the images representing fake emotions (e.g., simulated by professional actors) are discarded.

3. Data set generation. From each image of the raw dataset, the portion representing the mouth of the subject is automatically extracted.
4. Training with data augmentation. At every step of the training, the system reads a partition of images from the dataset. At each reading, the system performs data augmentation [26] using random values of rotation degree within a given range. The model used for image analysis is exploited using transfer learning.
5. Model generation. At the end of the training, the structure and weights of the neurons that achieved the best performance in training are saved.
6. Results Evaluation. To evaluate the quality of the model, we analyse the accuracy of ER on the validation set.

The mouth extraction from the images of the raw dataset was carried out using as a pre-trained neural network the `shape_predictor_68_face_landmarks.dat` CNN [2,8,27], which produced in output 68 landmarks, detected for each image. The shape predictor was pre-trained on the ibug 300-W dataset. The landmarks are expressed as a series of coordinates identifying specific elements of a face, e.g., the positions of the mouth, eyes and cheekbones. Once we obtained the landmarks of a face, we used those identifying the area of the mouth and cropped the image (i.e., cutting out the original image to obtain only the part related to the mouth). All images of the mouth were transformed in size to 299 × 299 pixels.

In Figure 1, an example of mouth detection, cropping and resizing is shown. On the left side, an image taken from the dataset is visible; on the right, the corresponding extracted mouth. At the end of the process, the mouth images were stored in a directory structure associated with the labelled emotion; a final inspection of the mouth images dataset shows that the mouth is always correctly cut, whether it is open or not. We used data augmentation techniques [26] to increase the robustness of the neural network, thereby randomly transforming the images, i.e., with a flip and a random rotation within the [−4, +4] degree range.

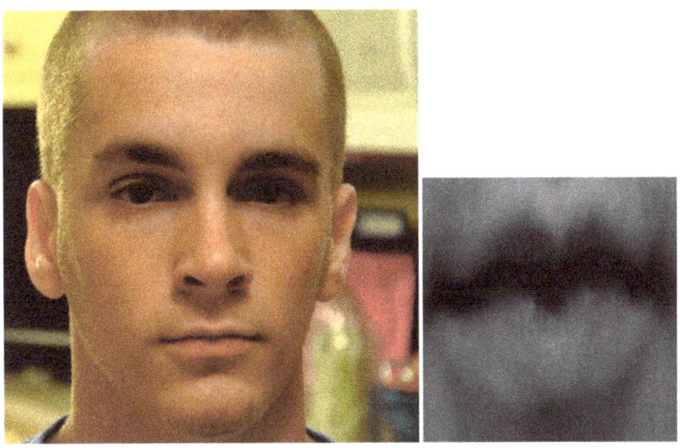

Figure 1. Mouth detection, cropping and resizing (source image from AffectNet database).

3.1. *Convolutional Neural Networks*

Convolutional neural networks constitute a class of deep neural networks which prove particularly efficient for different tasks on data organised in a grid topology, e.g., time series and visual inputs, and are among the most used for deep learning in image-based classification. Typical applications include image recognition, segmentation, detection and retrieval [28–30], on which CNNs achieved state-of-the-art performances. This significant breakthrough can be attributed to three critical factors boosted by CNNs, i.e., sparse interactions, parameter sharing and equivariant representation [31]; massive amounts of training samples can be processed with improved efficiency and significantly reduce training times, using deeper networks with millions of parameters to learn more complex and characteristic image features. An additional advantage is the possibility of having a variable input size, in contrast with traditional, fully-connected networks which require fixed-size input. In convolutional neural networks, some of the traditional fully-connected layers are replaced with convolution layers, which scan through the ordered, grid-like structured data to process the data in subsets, and multiply each subset by a kernel matrix (i.e., filter) to produce a feature map in output. The process resembles how individual neurons respond to visual inputs: each neuron is responsible for a small portion of the input, called its receptive field, and ignores additional information. Training a fully-connected neural network with the same feature recognition, and consequently, classification capabilities, would require much greater effort—e.g., for an image of size 1000 × 1000, training 1,000,000 weights for each neuron of a layer. In contrast, CNNs would have some trainable parameters dependent on the size of the applied kernel, but still in a much lower amount.

Typically, CNNs are interlaced sequences of three different types of layers: the previously described convolution layers, conventional fully-connected layers and pooling layers, used to aggregate multiple features into a single one, i.e., down-sampling the feature maps according to different strategies. The CNNs used in literature for reference tasks such as ImageNet classification, although varying in the number of layers, filter size and quantity, are mostly composed by the building blocks cited above.

3.2. CNN Settings

The experiments have been performed on four convolutional neural networks: VGG16 [32], InceptionResNetV2 [33], InceptionV3 [34] and Xception [35]. Among CNNs, these networks outperform AlexNet [28] on the widely used ImageNet [7] dataset, which is one of the largest image datasets used for the transfer learning pre-training phase.

All neural networks, whose general behaviour is described in Figure 2, have been tested using transfer learning (see Figure 3 showing the process for a general CNN).

Adopting transfer learning, we used the pre-trained neural networks, whose weights have been computed on the ImageNet [7] dataset. The first layers of the networks have been frozen for the training phase; i.e., the weights of the convolutional layers have been fixed and have not been altered during the fine-tuning phase on the mouth emotion dataset. This was done due to the high capability of the CNNs to recognise low-level features, e.g., points, lines, edges, corners and colour distribution. We replaced the final layers to fine-tune the network on mouth emotion recognition, using two dense layers with 64 neurons each and a SoftMax layer. This final layer ranks the likelihood of the most appropriate class, thereby returning the emotion classification. Only the final layers of the CNN networks, i.e., the fully-connected and the final SoftMax level, therefore, change and can be re-trained in the fine-tuning phase. The model was set up to save the weights only when they improve the emotion classification accuracy concerning the previous epoch, resulting in a final best neural network training configuration.

We tested two optimisers: Adam and SGD. Adam performed better with InceptionV3 and VGG16 with a learning rate equal to 0.001; and with Xception it had a learning rate of 0.01. SGD performed better with InceptionResNetV2 with a learning rate equal to 0.001, `momentum` = 0.9 and `nesterov` equal to true. The remaining parameters used were the following. Batch size equal to 25 and the maximum number of epochs equal to 100; however, we used the early stopping technique: if results do not improve, the training is stopped.

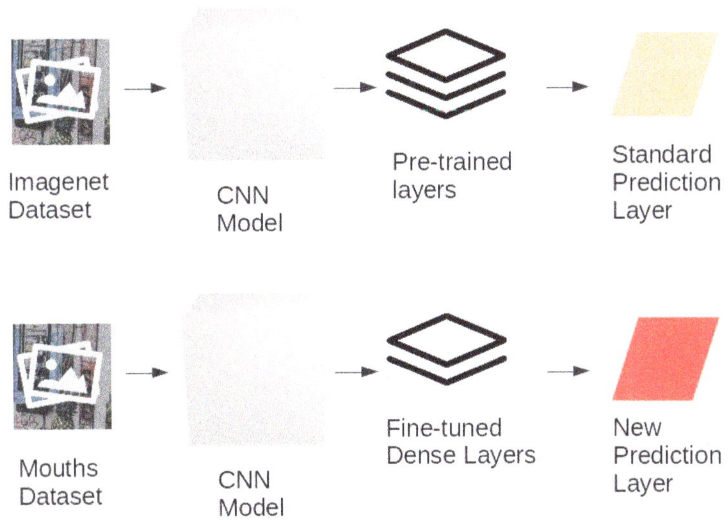

Figure 2. General scheme of the adapted transfer learning techniques with the considered CNNs.

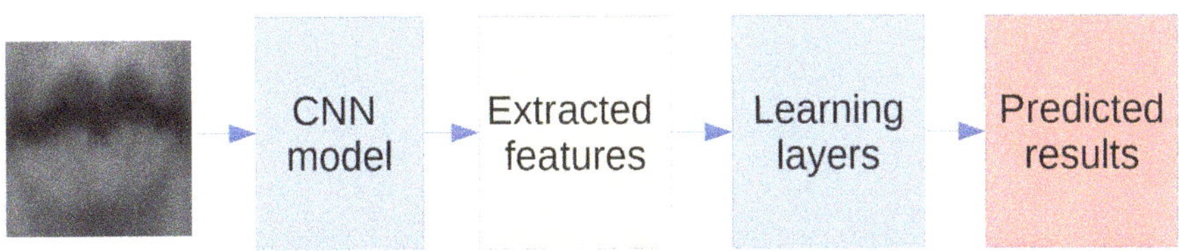

Figure 3. General scheme of our CNN.

4. Data Set and Filtering Phase

As introduced in Section 1.1, in the actual state-of-the-art, one of the most relevant issues for our goal is the total lack of a sufficient number of images of real emotions. Most of the available datasets, in fact, include a mixed set of real and fake emotions, performed by actors or pretended, which cannot be considered usable for a technique such as deep learning, which is able to base the classification on very detailed features, which are different between real and fake emotions even at a micro-expression level, and may vary in different cultures. A person who is striking a pose for a photograph (e.g., smiling voluntarily in front of the camera) often assumes an artificial, distorted expression that is not generated autonomously by the human brain, even if it can be appropriately identified and recognised by the human brain as a smile, thanks to mirror neurons. We must report that most of the available datasets provide images depicting actors; thus, a cleaning phase of the chosen dataset was mandatory. It should be considered that there are emotions such as anger and fear that are not easily replicated in a laboratory, whereas emotions such as joy or the neutral state, on the other hand, can be easily triggered.

Another relevant limitation is that many datasets include only images of a particular light state, perspective and image resolution, which leads to the system overfitting that particular environment and failing in others. Until a proper extensive dataset is ready for the test, all results have to be considered preliminary.

For the emotional training (i.e., fine-tuning of the CNNs) and classification test of our work, the dataset AffectNet provided by the University of Denver [36], USA, has been used. The raw dataset is composed of the images of a large number of faces, and a spreadsheet file used to assign to each image an emotion label. We analysed a subset of the Ekman emotions, i.e., neutral, happy, surprise and anger. A sample image for each class of emotion is shown in Figure 4.

Figure 4. Sample images from the filter dataset.

The dataset has been cleaned by analysing images and removing duplicates. The automated mouth recognition removed also all the items where the mouth was not visible. Though grainy or unsatisfactory resolution photographs can be removed, it is also essential to maintain significant differences in resolution, point of view and light conditions. This variety helps the network to train on general images and not on only those with particular light conditions, resolution and so on.

The primary filtering improvement of the dataset has exploited the contents of images, while avoiding all the photographs showing fake emotions, i.e., facial expressions apparently simulated.

In Table 1, the number of images per type of emotion in the cleaned dataset is shown.

Table 1. Number of images per type of emotion considered in our study.

	Neutral	Happy	Surprise	Anger
# *of images*	1239	562	463	478

5. Results and Discussion

The experimental work has been divided into two phases. A preliminary phase included experiments involving freezing the pre-trained convolutional layers and fine-tuning the fully-connected layers, as explained in Section 3. The results suggest discarding the CNN networks with the lowest accuracy percentages, i.e., VGG16, Inception V3 and Xception, while keeping InceptionResnetV2, which achieved the best performance. In the second phase, the freezing of InceptionResnetV2 was removed to further fine-tune all the layers of the network. For each network, the best model obtained in the training phase was saved, and we rolled back the weights to keep the highest accuracy. Iterating this process, we obtained a final accuracy of 79.5% on the validation set using the best network, as shown in Table 2. As shown, the network that performed worst was VGG16, while the best network was InceptionResnetV2; in Table 2, we included the final results for all the networks. Note that only InceptionResnetV2 has been fine-tuned in all its layers.

Table 2. Final results related to the considered CNNs.

Network	Accuracy
Vgg-16	71.8%
InceptionResNetV2	79.5%
Inception V3	77.0%
Xception	75.5%

Figures 5 and 6 show the loss and accuracy values of the model as a function of the training epochs. The number of epochs was different for each CNN, due to the use of the early stopping criterion described in Section 3.1. Early stopping is essential to eliminating or mitigating the overfitting effect that, as the epochs grow, would lead the accuracy on the training and validation sets, represented by the two curves (blue and orange), to diverge due to a loss of generalisation capability of the trained model. In agreement with the accuracy, InceptionResNetV2 shows the best trend for the loss function.

In Figure 6 we can observe for the InceptionResNetV2 a higher accuracy and a more regular shape of the function, compared to the other CNNs.

The confusion matrix of the training set for the InceptionResNetV2 network is reported in Figure 7.

The cells on the diagonal of the matrix shown in Figure 7, representing correctly classified images, show good performances, close to the perfect classification. Errors appear in classifying happiness, sometimes interpreted as anger; the same issues occur for surprise interpreted as neutral and anger. Finally, anger is sometimes misinterpreted as neutral. As introduced in Section 2, the misclassification of the neutral emotion presents a different issue, thereby requiring a separate discussion. Studies in cultural anthropology, including the first work of Paul Ekman on facial emotion recognition [9], show that humans sometimes tend to misclassify neutral as anger or sadness, with different results in different cultures. Such bias is clearly reflected in the AffectNet dataset because it is present in our automated recognition too. We can realistically suppose that our CNN correctly classified the mouths based on biased labels. In both the classification directions, neutral is sometimes misclassified, and the other classes are misclassified as neutral. In Table 3 the absolute values of misclassified images are reported for each class in the neutral evaluation. We can say that our evidence suggests that our networks behave similarly to the human brain, possibly learning the features associated with the human bias present in the considered dataset.

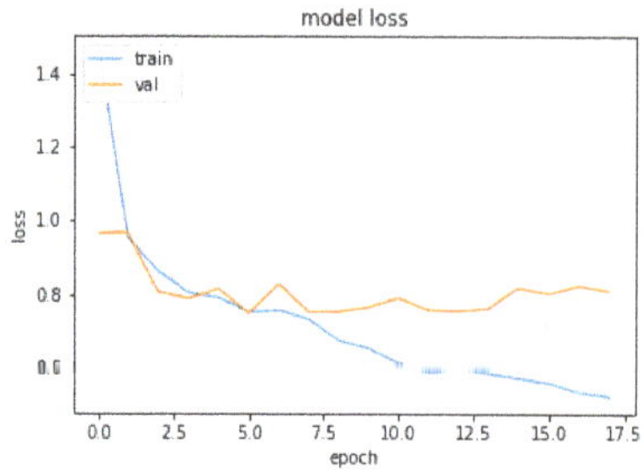

(**a**) VGG16

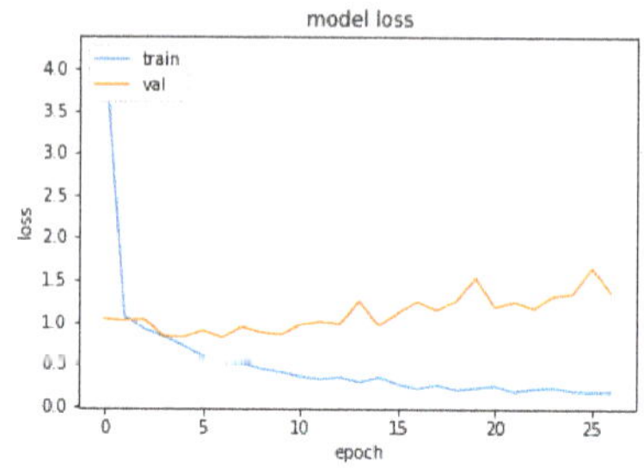

(**b**) Xception

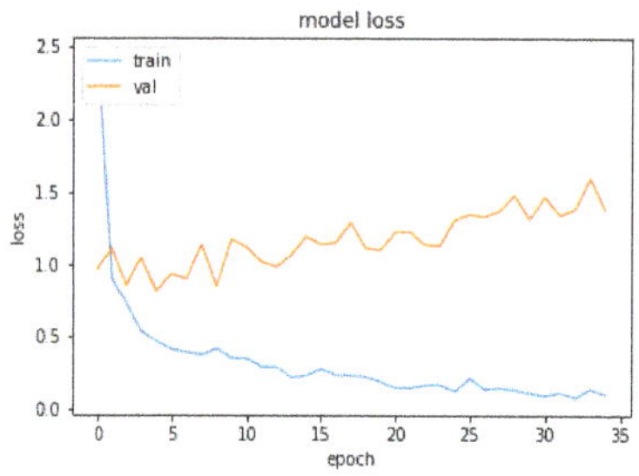

(**c**) Inception V3

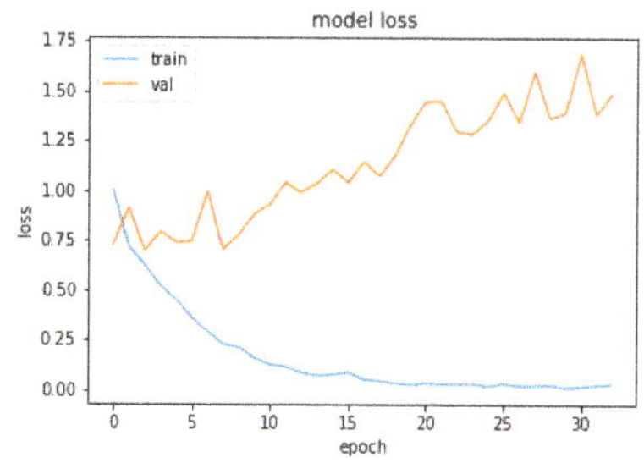

(**d**) InceptionResNetV2

Figure 5. Loss function evolution for each network as a function of the epochs.

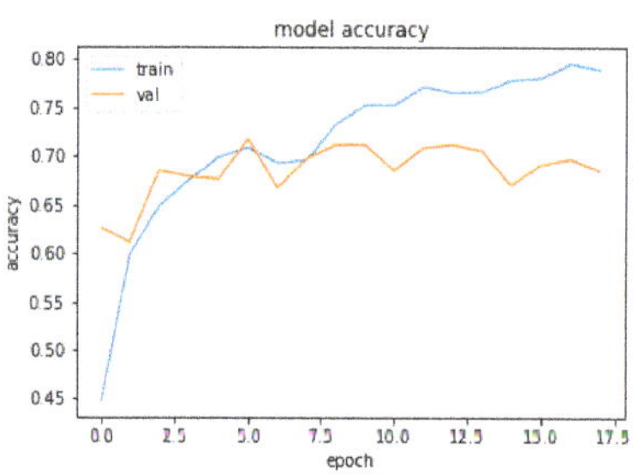

(**a**) VGG16

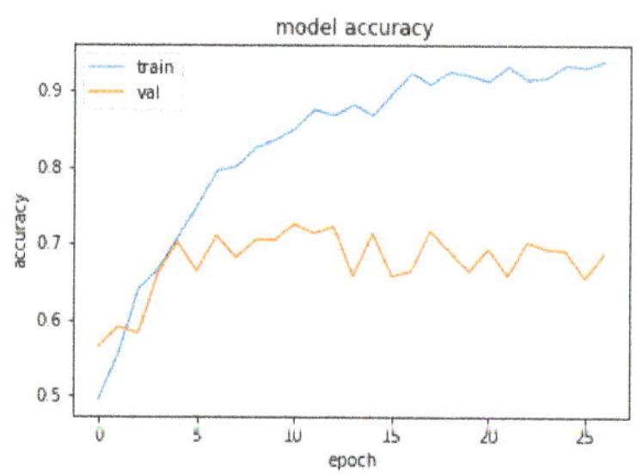

(**b**) Xception

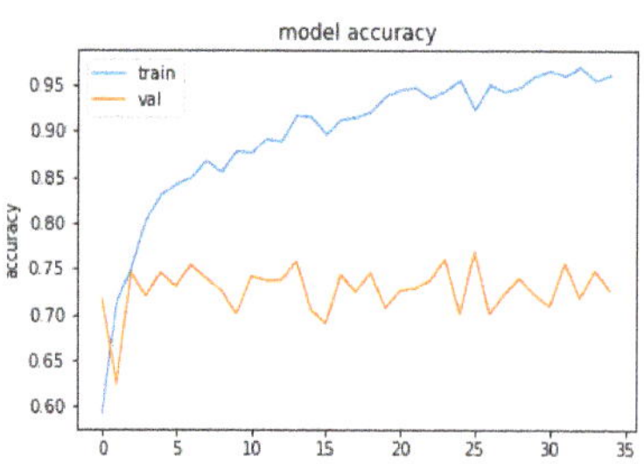

(**c**) Inception V3

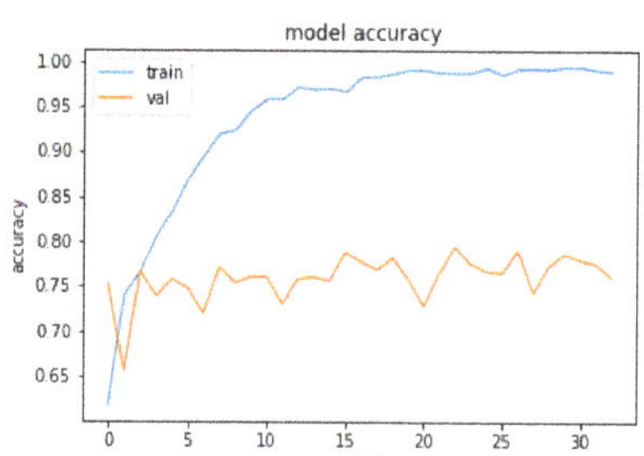

(**d**) InceptionResNetV2

Figure 6. Accuracy function evolution for each network as a function of the epochs.

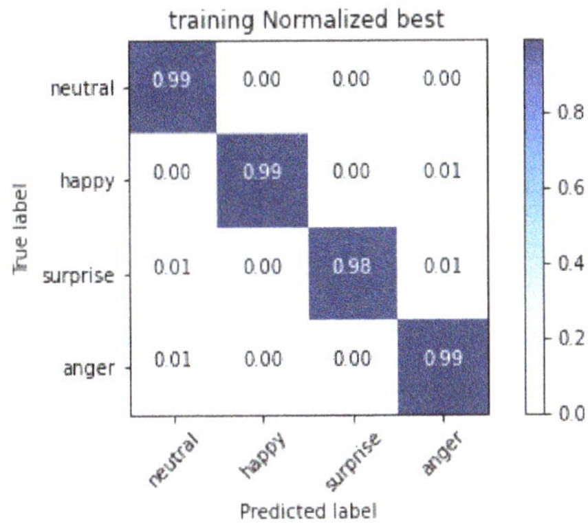

Figure 7. Confusion matrix of the training set for the network InceptionResNetV2.

The confusion matrix of the validation set relative to each tested network is shown in Figure 8.

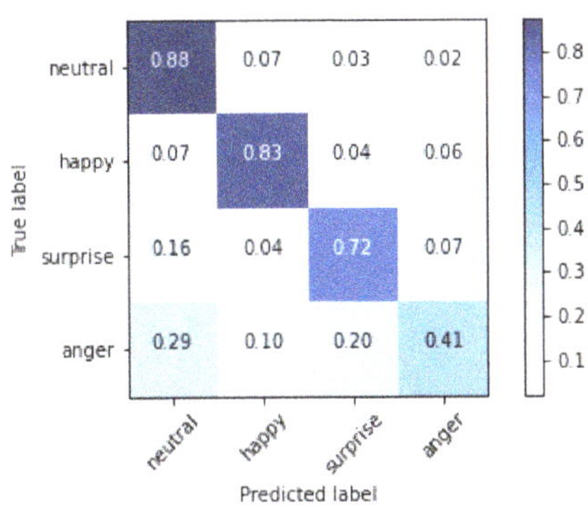

(**a**) VGG16

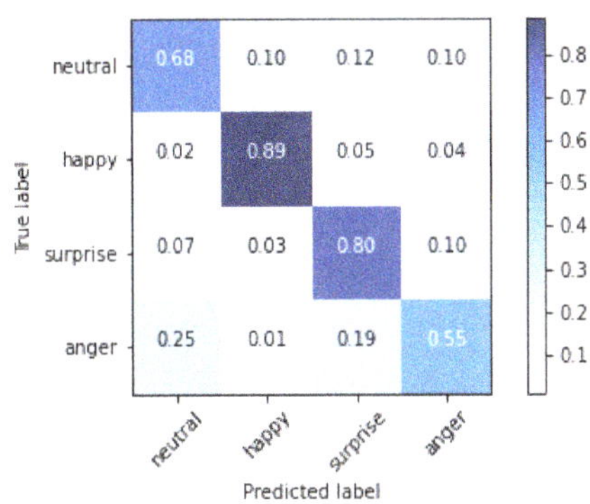

(**b**) Xception

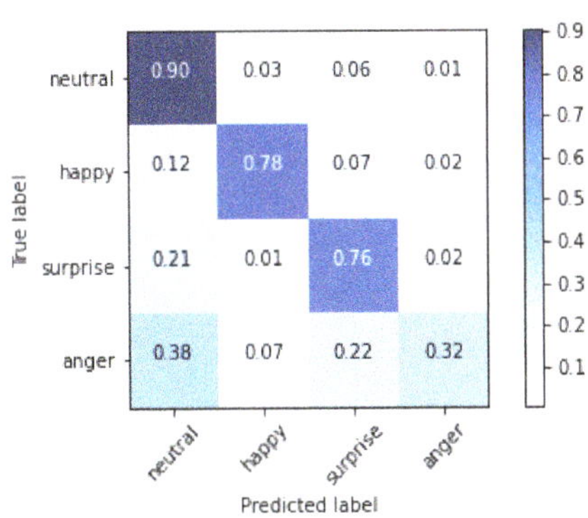

(**c**) Inception V3

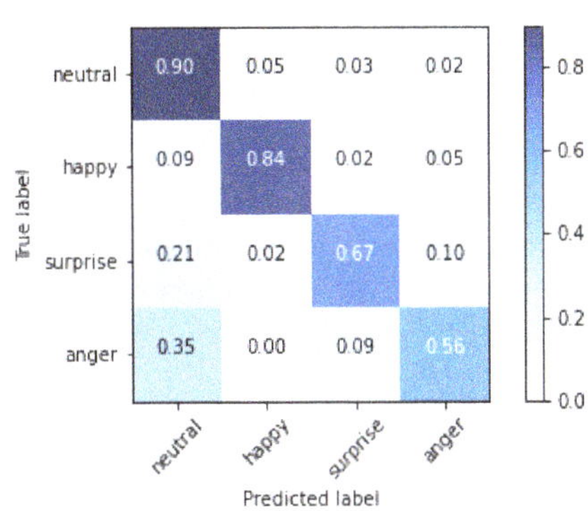

(**d**) InceptionResNetV2

Figure 8. Confusion matrices of the trained networks, with normalised values.

Table 3. Number of misclassified images of neutral faces in the three wrong categories over a total of 221 images.

	Happy	Surprise	Anger
# of images	11	7	4

The experiments have been replicated on images of the whole face, to study the contribution of the mouth; the resulting confusion matrix is shown in Figure 9.

Results can be summarised in the following points.

- The network that provided the best performance was the same as the one for mouth-based ER, i.e., InceptionResNetV2.
- Anger was again the most difficult emotion to classify.
- The performance improvement was only 5%, with an accuracy rate of 84.4%, thereby proving the consistent contribution of the mouth for ER.

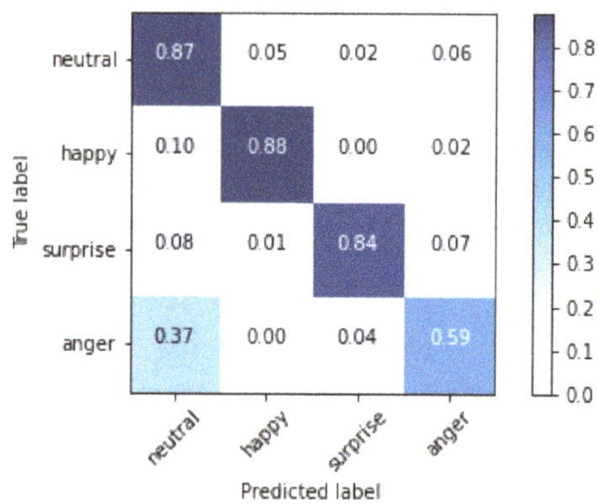

Figure 9. Confusion matrix of the validation set obtained while running InceptionResNetV2 to analyse the whole face images instead of the mouth portions of the images.

6. Conclusions and Future Developments

In this work, we presented the first consolidated results for the approach of mouth-based emotion recognition, gotten through the comparative analysis of four different CNNs using transfer learning.

We confirmed the significance of the mouth to classifying the user's emotions and provided inspiring and useful information to understand its contribution compared to full-face recognition.

As described in Section 5 we obtained with the InceptionResNetV2 neural network an accuracy of 79.5% on the validation set. With respect to the full face, results show a loss of accuracy of only 5%, which in our opinion is offset by the advantages of our method. The mouth is, in fact, a critical element of human face recognition, almost symmetric and usually visible from any perspective, and thus the ideal element to focus on in all those cases wherein the user can be shot from any point of view. Moreover, focusing on a smaller area of the image requires lesser computational capabilities.

Straightforward applications of our approach are emotion/pain recognition for healthcare management, and automated supervision of critical patients, e.g., those bedridden in hospitals. The system can support patients by using an ER system when direct human assistance is not available, e.g., during the night or in wards where assistance is not allowed, for advanced detection of an initial pained or discomfort state, and raising a signal and letting the sanitary staff be informed to react promptly, thereby avoiding the patient's suffering. Supportive systems can be planned using emotion recognition, e.g., to assist patients after car accidents, still in the emergency phase or post-surgery, in order to understand their pain levels. If the system recognises a critical situation, it can report the case to nurses or send an intervention/checkout request for the patient's room. Our approach could be easily extended to the scenario previously described, upon the availability of proper datasets of pain images. An additional application could be the early recognition of depressive states or the support of people with difficulties seeing or interpreting emotions, e.g., blind users, or people with autism spectrum disorders. Our emotion recognition system should in future give feedback to the user (e.g., text, emoticon, sound), or we could set up a channel for information transfer to software.

Author Contributions: The authors contributed equally to this work. All authors have read and agreed to the published version of the manuscript.

Funding: "This research was partially funded by the Italian Ministry of Research under PRIN Project "PHRAME" Grant number 20178XXKFY" and "The APC was funded by The Department of Mathematics and Computer Science, University of Florence, Italy and the Department of Mathematics and Computer Science, University of Perugia, Italy".

Acknowledgments: We acknowledge Google Inc. for the usage of the Google Colab infrastructure for carrying out our calculations. We acknowledge the University of Denver, CO, USA, for having provided access to the database of images we used for the present study.

Conflicts of Interest: The authors declare no conflict of interest.

References

1. Gervasi, O.; Franzoni, V.; Riganelli, M.; Tasso, S. Automating facial emotion recognition. *Web Intell.* **2019**, *17*, 17–27. [CrossRef]
2. Sagonas, C.; Tzimiropoulos, G.; Zafeiriou, S.; Pantic, M. A Semi-automatic Methodology for Facial Landmark Annotation. In Proceedings of the 2013 IEEE Conference on Computer Vision and Pattern Recognition Workshops, Portland, OR, USA, 23–28 June 2013; pp. 896–903.
3. Kazemi, V.; Sullivan, J. One millisecond face alignment with an ensemble of regression trees. In Proceedings of the 2014 IEEE Conference on Computer Vision and Pattern Recognition, Columbus, OH, USA, 23–28 June 2014; pp. 1867–1874.
4. Riganelli, M.; Franzoni, V.; Gervasi, O.; Tasso, S. EmEx, a Tool for Automated Emotive Face Recognition Using Convolutional Neural Networks. In *Lecture Notes in Computer Science, Proceedings of the International Conference on Computational Science and Its Applications, Trieste, Italy, 3–6 July 2017*; Springer International Publishing: Cham, Switzerland, 2017; Volume 10406, pp. 692–704.
5. Biondi, G.; Franzoni, V.; Gervasi, O.; Perri, D. An Approach for Improving Automatic Mouth Emotion Recognition. In *Lecture Notes in Computer Science, Proceedings of the Computational Science and Its Applications—ICCSA 2019, Saint Petersburg, Russia, 1–4 July 2019*; Misra, S., Gervasi, O., Murgante, B., Stankova, E., Korkhov, V., Torre, C., Rocha, A.M.A.C., Taniar, D., Apduhan, B.O., Tarantino, E., Eds.; Springer International Publishing: Cham, Switzerland, 2019; Volume 11619, pp. 649–664.
6. Gervasi, O.; Magni, R.; Ferri, M. A Method for Predicting Words by Interpreting Labial Movements. In *Lecture Notes in Computer Science, Proceedings of the Computational Science and Its Applications—ICCSA 2016, Beijing, China, 4–7 July 2016*; Gervasi, O., Murgante, B., Misra, S., Rocha, A.M.A., Torre, C.M., Taniar, D., Apduhan, B.O., Stankova, E., Wang, S., Eds.; Springer International Publishing: Cham, Switzerland, 2016; Volume 9787, pp. 450–464.
7. Deng, J.; Dong, W.; Socher, R.; Li, L.; Li, K.; Li, F.-F. ImageNet: A large-scale hierarchical image database. In Proceedings of the 2009 IEEE Conference on Computer Vision and Pattern Recognition, Miami, FL, USA, 20–25 June 2009; pp. 248–255.
8. Sagonas, C.; Antonakos, E.; Tzimiropoulos, G.; Zafeiriou, S.; Pantic, M. 300 Faces In-The-Wild Challenge: Database and results. *Image Vis. Comput.* **2016**, *47*, 3–18. [CrossRef]
9. Ekman, P. An Argument for Basic Emotions. *Cogn. Emot.* **1992**. [CrossRef]
10. Plutchik, R. A psychoevolutionary theory of emotions. *Soc. Sci. Inf.* **1982**, *21*, 529–553. [CrossRef]
11. Franzoni, V.; Milani, A.; Vallverdú, J. Emotional Affordances in Human-Machine Interactive Planning and Negotiation. In Proceedings of the International Conference on Web Intelligence, Leipzig, Germany, 23–26 August 2017; pp. 924–930. [CrossRef]
12. Franzoni, V.; Milani, A.; Vallverdú, J. Errors, Biases, and Overconfidence in Artificial Emotional Modeling. In Proceedings of the International Conference on Web Intelligence, Thessaloniki, Greece, 14–17 October 2019.

13. Franzoni, V.; Milani, A.; Nardi, D.; Vallverdú, J. Emotional machines: The next revolution. *Web Intell.* **2019**, *17*, 1–7. [CrossRef]
14. Gervasi, O.; Magni, R.; Macellari, S. A Brain Computer Interface for Enhancing the Communication of People with Severe Impairment. In *Lecture Notes in Computer Science, Proceedings of the Computational Science and Its Applications—ICCSA 2014, Guimarães, Portugal, 30 June–3 July 2014*; Murgante, B., Misra, S., Rocha, A.M.A.C., Torre, C., Rocha, J.G., Falcão, M.I., Taniar, D., Apduhan, B.O., Gervasi, O., Eds.; Springer International Publishing: Cham, Switzerland, 2014; Volume 8584, pp. 709–721.
15. Bastianelli, E.; Nardi, D.; Aiello, L.C.; Giacomelli, F.; Manes, N. Speaky for robots: The development of vocal interfaces for robotic applications. *Appl. Intell.* **2016**, *44*, 43–66. [CrossRef]
16. Picard, R.W. Affective Computing: Challenges. *Int. J. Hum. Comput. Stud.* **2003**, *59*, 55–64. [CrossRef]
17. Cieliebak, M.; Dürr, O.; Uzdilli, F. Potential and limitations of commercial sentiment detection tools. In Proceedings of the CEUR Workshop Proceedings, Valencia, Spain, 17–18 June 2013; pp. 47–58.
18. Franzoni, V.; Milani, A. Emotion Recognition for Self-aid in Addiction Treatment, Psychotherapy, and Nonviolent Communication. In *Lecture Notes in Computer Science, Proceedings of the Computational Science and Its Applications—ICCSA 2019, Saint Petersburg, Russia, 1–4 July 2019*; Misra, S., Gervasi, O., Murgante, B., Stankova, E., Korkhov, V., Torre, C., Rocha, A.M.A.C., Taniar, D., Apduhan, B.O., Tarantino, E., Eds.; Springer International Publishing: Cham, Switzerland, 2019; Volume 11620, pp. 391–404.
19. Hayes, G.R.; Hirano, S.; Marcu, G.; Monibi, M.; Nguyen, D.H.; Yeganyan, M. Interactive visual supports for children with autism. *Pers. Ubiquitous Comput.* **2010**, *14*, 663–680. [CrossRef]
20. Picard, R.W.; Vyzas, E.; Healey, J. Toward machine emotional intelligence: Analysis of affective physiological state. *IEEE Trans. Pattern Anal. Mach. Intell.* **2001**, *23*, 1175–1191. [CrossRef]
21. Bertola, F.; Patti, V. Emotional responses to artworks in online collections. In Proceedings of the CEUR Workshop Proceedings, Valencia, Spain, 17–18 June 2013; Volume 997.
22. Canossa, A.; Badler, J.B.; El-Nasr, M.S.; Anderson, E. Eliciting Emotions in Design of Games—A Theory Driven Approach. In Proceedings of the 4th Workshop on Emotions and Personality in Personalized Systems (EMPIRE), Boston, MA, USA, 16 September 2016.
23. Angelov, P.; Gu, X.; Iglesias, J.A.; Ledezma, A.; Sanchis, A.; Sipele, O.; Ramezani, R. Cybernetics of the Mind: Learning Individual's Perceptions Autonomously. *IEEE Syst. Man Cybern. Mag.* **2017**, *3*, 6–17. [CrossRef]
24. Biondi, G.; Franzoni, V.; Li, Y.; Milani, A. Web-Based Similarity for Emotion Recognition in Web Objects. In Proceedings of the 9th International Conference on Utility and Cloud Computing, Shanghai, China, 6–9 December 2016; pp. 327–332. [CrossRef]
25. Chollet, F. Keras. 2015. Available online: https://github.com/fchollet/keras (accessed on 14 July 2020).
26. Antoniou, A.; Storkey, A.; Edwards, H. Data Augmentation Generative Adversarial Networks. *arXiv* **2017**, arXiv:1711.04340.
27. Sagonas, C.; Tzimiropoulos, G.; Zafeiriou, S.; Pantic, M. 300 Faces in-the-Wild Challenge: The First Facial Landmark Localization Challenge. In Proceedings of the 2013 IEEE International Conference on Computer Vision Workshops, Sydney, Australia, 2–8 December 2013; pp. 397–403.
28. Krizhevsky, A.; Sutskever, I.; Hinton, G.E. ImageNet Classification with Deep Convolutional Neural Networks. *Commun. ACM* **2017**, *60*, 84–90. [CrossRef]
29. Perri, D.; Sylos Labini, P.; Gervasi, O.; Tasso, S.; Vella, F. Towards a Learning-Based Performance Modeling for Accelerating Deep Neural Networks. In *Lecture Notes in Computer Science Book, Proceedings of the Computational Science and Its Applications—ICCSA 2019, Saint Petersburg, Russia, 1–4 July 2019*; Misra, S., Gervasi, O., Murgante, B., Stankova, E., Korkhov, V., Torre, C., Rocha, A.M.A., Taniar, D., Apduhan, B.O., Tarantino, E., Eds.; Springer International Publishing: Cham, Switzerland, 2019; Volume 11619, pp. 665–676.
30. Farabet, C.; Couprie, C.; Najman, L.; LeCun, Y. Learning Hierarchical Features for Scene Labeling. *IEEE Trans. Pattern Anal. Mach. Intell.* **2013**, *35*, 1915–1929. [CrossRef] [PubMed]
31. Goodfellow, I.; Bengio, Y.; Courville, A. *Deep Learning*; MIT Press: Cambridge, MA, USA, 2016.
32. Simonyan, K.; Zisserman, A. Very Deep Convolutional Networks for Large-Scale Image Recognition. *arXiv* **2014**, arXiv:1409.1556.
33. Szegedy, C.; Ioffe, S.; Vanhoucke, V.; Alemi, A. Inception-v4, Inception-ResNet and the Impact of Residual Connections on Learning. *arXiv* **2016**, arXiv:1602.07261.
34. Szegedy, C.; Vanhoucke, V.; Ioffe, S.; Shlens, J.; Wojna, Z. Rethinking the Inception Architecture for Computer Vision. *arXiv* **2015**, arXiv:1512.00567.

35. Chollet, F. Xception: Deep Learning with Depthwise Separable Convolutions. In Proceedings of the 2017 IEEE Conference on Computer Vision and Pattern Recognition (CVPR), Honolulu, HI, USA, 21–26 July 2017; pp. 1800–1807.
36. Mollahosseini, A.; Hasani, B.; Mahoor, M.H. AffectNet: A Database for Facial Expression, Valence, and Arousal Computing in the Wild. *IEEE Trans. Affect. Comput.* **2019**, *10*, 18–31. [CrossRef]

sensors

MDPI

Article

Arousal Detection in Elderly People from Electrodermal Activity Using Musical Stimuli

Almudena Bartolomé-Tomás [1,2], Roberto Sánchez-Reolid [1,3], Alicia Fernández-Sotos [4], José Miguel Latorre [5] and Antonio Fernández-Caballero [1,3,6,*]

1. Instituto de Investigación en Informática de Albacete, Universidad de Castilla-La Mancha, 02071 Albacete, Spain; almuxbt@gmail.com (A.B.-T.); roberto.sanchez@uclm.es (R.S.-R.)
2. Conservatorio de Música de Cieza "Maestro Gómez Villa", Calle Cadenas, 6, 30530 Cieza, Spain
3. Departamento de Sistemas Informáticos, Universidad de Castilla-La Mancha, 02071 Albacete, Spain
4. Conservatorio de Música de Murcia, Calle Cartagena, 74, 30002 Murcia, Spain; alicia.fernandez5@murciaeduca.es
5. Departamento de Psicología, Universidad de Castilla-La Mancha, 02071 Albacete, Spain; Jose.Latorre@uclm.es
6. CIBERSAM (Biomedical Research Networking Centre in Mental Health), 28029 Madrid, Spain

* Correspondence: antonio.fdez@uclm.es; Tel.: +34-967-599-200

Received: 30 June 2020; Accepted: 22 August 2020; Published: 25 August 2020

Abstract: The detection of emotions is fundamental in many areas related to health and well-being. This paper presents the identification of the level of arousal in older people by monitoring their electrodermal activity (EDA) through a commercial device. The objective was to recognize arousal changes to create future therapies that help them to improve their mood, contributing to reduce possible situations of depression and anxiety. To this end, some elderly people in the region of Murcia were exposed to listening to various musical genres (flamenco, Spanish folklore, Cuban genre and rock/jazz) that they heard in their youth. Using methods based on the process of deconvolution of the EDA signal, two different studies were carried out. The first, of a purely statistical nature, was based on the search for statistically significant differences for a series of temporal, morphological, statistical and frequency features of the processed signals. It was found that Flamenco and Spanish Folklore presented the highest number of statistically significant parameters. In the second study, a wide range of classifiers was used to analyze the possible correlations between the detection of the EDA-based arousal level compared to the participants' responses to the level of arousal subjectively felt. In this case, it was obtained that the best classifiers are support vector machines, with 87% accuracy for flamenco and 83.1% for Spanish Folklore, followed by K-nearest neighbors with 81.4% and 81.5% for Flamenco and Spanish Folklore again. These results reinforce the notion of familiarity with a musical genre on emotional induction.

Keywords: arousal; aging adults; musical genres; electrodermal activity

---

1. Introduction

Understanding and recognizing human emotions has been identified as a main interest area in smart systems [1–5]. Such systems are being applied in many fields like well-being and healthcare [6–11], safe driving [12], smart cities [13] and smart environments [14,15], among others. Pleasure, arousal and dominance are three independent emotional dimensions to describe people's state of feeling [16,17]. Arousal was conceived as a mental activity describing the state of feeling along a single dimension ranging from sleep to frantic excitement and linked to adjectives such as stimulated–relaxed, excited–calm and wide awake–sleepy to define arousal [18].

The arousal level changes constantly, and it has a profound influence on performance during everyday activities [19]. Fluctuations in arousal are regulated by the autonomic nervous system, which is mainly controlled by the balanced activity of the parasympathetic and sympathetic systems [20]. Electrodermal activity (EDA; or skin conductance) has also frequently been used as a measure of arousal. The advantage of EDA is that it is unambiguous, given that it is innervated entirely by the sympathetic nervous system (SNS) [21]. Within the domain of music emotion research, physiological measures such as EDA, heart rate, respiration, and body temperature have been frequently used as correlates of emotional arousal. Among these, EDA is generally a preferred measure as it is highly sensitive and under strict control of the sympathetic nervous system and is therefore largely involuntary. Furthermore, a relationship between EDA as indicator of emotional arousal and experienced pleasure in response to music has previously been demonstrated [22]. At the same time, in their studies with volunteers the participants' feelings have been obtained by questionnaires in the form of Likert scales, self-assessment manikins (SAM) and free text [23–26].

This paper introduces arousal detection from EDA signals using musical stimuli. Several studies have reported that using music to elicit emotions is one of the most effective methods of emotion induction [27–30]. Music plays a key role in most people's lives, frequently being used to explore and regulate emotions. The proposal is linked to our current research elicitation of emotions in elderly people to trigger processes of emotional self-regulation [31–33]. Those processes should help elderly people to improve their mood and mental state. The importance of emotional self-regulation is related to the fact that older people, especially when living alone, are at high risk of suffering from diseases such as depression and anxiety [34,35].

Specifically, people over 60 years old from the region of Murcia, Spain, were recruited as participants to listen to a series of musical pieces similar to those played in their younger years in order to study the level of arousal produced by each musical genre. Although many protocols have investigated physiological responses to music, the present work explores the physiological responses to pieces of music composed specifically for this experiment. The use of original pieces of music, which had not been heard by the listener before, is a novel research technique that has yielded interesting results so far [27–30]. The use of this type of music fragments provides a high level of experimental control and allows knowledge of the influence of the independent variables on the dependent ones. Experimental control is especially important when analyzing physiological responses like EDA [36]. The signals collected during the experiment were used in conjunction with a SAM questionnaire to undergo a couple of studies oriented towards discriminating the arousal. One study analyzed some EDA features only, and the second, based on classifiers, examined possible correlations between the objective detection of the arousal level from processed physiological EDA signals and the level of arousal subjectively perceived by participants when answering the SAM questionnaire.

The remainder of the paper is as follows. Section 2 shows the materials and methods needed to perform the experiment successfully, as well as the investigation methods and metrics used. In Section 3, the results obtained are shown and a discussion about the results obtained in the context of the experiment is provided. Finally, in Section 4 the results obtained in this study are presented.

2. Materials and Methods

This section describes the methodology and materials required to carry out the proposed experiment. First, an introduction is made about the electrodermal activity as a biomarker of activation detection. Then, a description of the material used, and the processes required to detect the level of activation is made. Next, the methods of data collection (SAM questionnaires) and how they are used within the experiment are explained. Afterwards, a detailed explanation of the experiment is given. Finally, the process of data segmentation and feature extraction for further analysis is explained.

2.1. Electrodermal Activity

Electrodermal activity (EDA) reflects the output of the attentional and affective and motivational processes integrated within the central nervous system that act on the body [37]. When emotional arousal increases, the accompanying activation of the SNS results in increased sweat gland activity and skin conductance. The validity of EDA as a measure of emotional arousal has been established in studies showing that EDA varies linearly with self-reported arousal when viewing emotional pictures [38]. Therefore, EDA is outstanding in behavioral medicine as a biomarker of individual characteristics of emotional response. EDA monitoring has been used for multiple applications, including assessment of anxiety and stress, detection of orientation response, providing neurofeedback for epilepsy, recognition of emotional state, and many others. In addition, EDA can be very effective in discriminating patients with depression from healthy controls [39]. Specific patterns of electrodermal hypoactivity may be a reliable marker of a depressive state at population level, but they should be carefully combined with other physiological and non-physiological indicators when used for preventive and diagnostic purposes.

EDA covers the electrical variations that occur on the surface of the skin due to changes in sweat secretion. EDA signals are obtained by measuring the potential when a small constant current is applied between two metal electrodes (for example, chrome-silver electrodes). The skin usually responds to stress by producing an increase in sweat. Consequently, the skin's conductivity increases. On the other hand, sweat production stops and skin conductivity is reduced when a person is subjected to a calm or neutral induction. In this study, EDA is measured at the wrist, bearing in mind that wrist biosensors are being widely adopted in conventional and commercial devices. The bracelets provide excellent surfaces for attaching the electrodes to the skin. Ideally, the proposed system should be further miniaturized to record EDA in the areas of the palm where the activity of the skin conduction response (SCR) is most pronounced, without being intrusive or interfering with daily activities.

2.2. Data Acquisition and Empatica E4 Device

The commercial Empatica E4 wristband has been used to carry out our experiment. The Empatica E4 bracelet is a device that allows the collection and measurement of physiological signals such as EDA, blood volume pressure, temperature and acceleration. This device has been used with good results in some previous works [14,40,41]. In this work, we have used only the EDA signals to study the possibility of determining whether significant differences occur when a participant is subjected to different musical stimuli.

An essential component of our proposal is to acquire, process and obtain a set of data that will be used for identification of the listener's arousal. The Empatica E4 device must be firmly attached to the wrist so that the electrodes touch the skin correctly. Otherwise, if the device is not properly connected, the captured data are not valid due to manifold artefacts.

2.3. Participants

40 participants, all from the region of Murcia, Spain, were recruited for the experiment. These volunteers were 23 women and 17 men with an average age of 65 (SD = 6.3) and 68 (SD = 5.1), respectively. The volunteers were all in good health and cognitive conditions to perform the experiment. They were given two screening tests, the PROMIS (Patient-Reported Outcomes Measurement Information System) diagnostic test and the TYM (Test Your Memory) test for cognitive impairment. Those who scored above the cutoff point in depression and below in cognitive functioning did not participate in the study. No compensation was paid for the conduct of the study. In addition, participants were required to sign a consent form explaining the procedure and the risks that could arise from conducting the test.

The experiment had been previously validated by the Ethics Committee of the Universidad de Castilla-La Mancha in accordance with the Helsinki Declaration.

2.4. Self-Assessment Manikins

One way of quantifying and subsequently relating the signals obtained from EDA to each of the different musical stimuli is by using a self-assessment manikin (SAM) questionnaire [23,24]. This questionnaire is widely used in psychology to measure the subjectively felt intensity of emotions to compare with the emotional connotation of the different physiological signals captured by electrophysiological devices [42–44]. The questionnaire consists of a series of manikins representing different values of valence, activation and dominance [45]. In this experiment only the manikin for activation was used.

2.5. Music Stimuli

As mentioned above, in this experiment the key in provoking emotions is music. For this reason, eight music pieces have been specifically composed by a professional musician for this experiment. These compositions reflect some musical styles that older people listened to when they were young (more than 30 years ago). Thus, it was the first time the participants heard each of these original pieces. All eight pieces are characterized by a same main melody and eight variations according to eight musical styles. The duration of each variation was 60 s. Table 1 shows the eight selected variations of four musical genres, with each genre including two musical styles. They are "rock/jazz" (*twist* and *swing*), "Cuban" (*bolero* and *habanera*), "Spanish folklore" (*pasodoble* and *Murcian jota*) and "flamenco" (*fandango* and *petenera*), respectively.

Table 1. Musical genres and styles used in the experiment.

Musical Genre	Style
Rock/Jazz	*Twist* and *Swing*
Cuban	*Bolero* and *Habanera*
Spanish Folklore	*Pasodoble* and *Murcian jota*
Flamenco	*Fandango* and *Petenera*

The musical genres used in this experiment and their repercussion in the region of Murcia are briefly described below. First, flamenco, which has been widely disseminated on the radio and orally through simple songs, is a deeply rooted genre in Spain. The most cheerful, folkloric and festive flamenco styles were adopted, such as the *Fandango* and the *Petenera*, relegating everything related to the "jondo" singing to a secondary position [46]. Secondly, Spanish Folklore, mainly linked to moments of celebration, is characterized by its joyful and jovial character. Also profoundly anchored in the popular, its simplicity and the repetition of melodic-rhythmic elements give off energy and vitality. It is closely linked to dancing as a couple, allowing one to enjoy the social atmosphere and to relate the music to the parties and the cortege.

On its side, Cuban music evokes silent listening without movement or slow dancing in couples with direct physical contact. This musical genre has also been adopted by classical music and has been expanded mainly by the cinema and the radio due to its sentimental character. Finally, jazz and rock'n'roll imply a new way of listening and relating to music. The orchestration of this music that adds instruments and sounds unknown in their culture was novel to the participants. This music relies on simple and repetitive structures, as well as on melodic improvisation through instrumental or vocal solos. The dancing of this music is also new, in pairs but without physical contact, and with very rhythmic movements that sometimes are perceived as transgression.

2.6. Experimental Design

An appropriate experimental design is fundamental to achieving relevant results. The E-Prime software has been chosen to create the basic design of the experiment. This software is the most widely used in the field of psychology for setting up experimental trials. In fact, E-Prime is a very robust software tool for our proper study, since it allows us to randomize and synchronize the musical pieces that are played to the participants. Furthermore, it makes it possible to add the SAM questionnaire and to control/record different parameters that will be used the exploit the EDA signals acquired during music performance.

The design of the experiment has been carried out following the scheme shown in Figure 1. As it can be seen, the experiment has well-differentiated phases. In the first phase, the measuring instruments are placed on the participant. The EDA signals start to be collected when the participant is prepared, which means that he/she is in a neutral emotional state. To achieve this state, the participant remains silent looking at a black screen before the first piece of music is played. In the second phase, the participant listens to each of the musical pieces and, when the reproduction of each one of them is concluded, the person completes the SAM questionnaire. This process is carried out 8 times until all the musical pieces have been played.

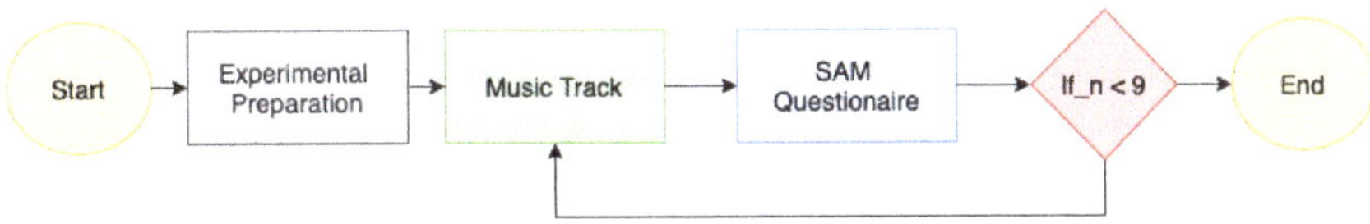

Figure 1. Flowchart of the experimental design

At the same time as the experiment was being conducted, the EDA signals were continuously collected, making possible further segmentation, preprocessing and analysis of the signal.

2.7. Electrodermal Activity Preprocessing

As discussed above, EDA has been measured by a non-invasive device. Concretely, the E4 Empatica bracelet measures the skin conductance (SC) in the form of EDA signals. These measurements are composed of two signals: a first signal that varies slowly, called the tonic driver or skin conductance level (SCL), and the second that varies rapidly, called the phase driver or skin conductance response (SCR). The SCL signal establishes the base level of the signal, while the SCR is directly associated with the activity of the sweat motor system which, in turn, is directly associated with the parasympathetic nervous system.

Within the process of processing the EDA signals, different phases are crossed during which the signals are transformed. These phases are usually preprocessing, filtering, artefact removal and discrete deconvolution. The preprocessing process is in charge of establishing the segments acquired in each of the phases of the experiment. Then, it is necessary to filter the SC signals to eliminate the artefacts and interference recorded during the acquisition phase. In our case, two different filters have been used: first, a low-pass filter with a 4 Hz cutoff frequency, and second, a Gaussian filter to smooth the signal and attenuate artefacts and noise.

The next step is the deconvolution process to separate the SCR from the SCL signals. This method makes it possible to minimize the effects that race, sex and age contribute to the SC signal. Figure 2 shows an outline of how this process has been performed. As can be seen, it is the SCR driver that can be used to detect the arousal level of the participant. For this sake, the MATLAB library called Ledalab 3.4.9 has been successfully used [47]. Mathematically, the sudomotor nerve activity can be considered a *Driver* containing a train of impulses that develop over time. This response is integrated in SC and, consequently also in SCR and SCL. The result is represented by a convolution (*) of the driver with the impulse-response function (IRF), which describes the flow of the impulse response over time, as shown in Equation (1).

$$SC = SC_{Driver} * IRF \tag{1}$$

The SC signal is composed of signals SCL and SCR, as shown in Equation (2).

$$SC = SCL + SCR \tag{2}$$

$$SC = (SCL_{Driver} + SCR_{Driver}) * IRF \tag{3}$$

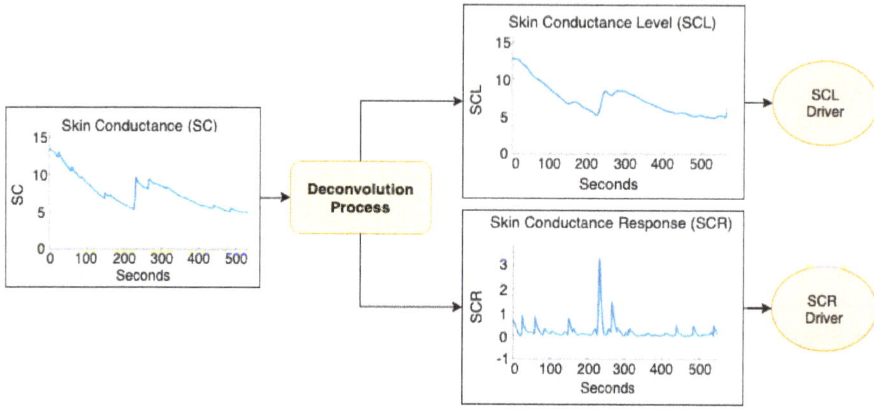

Figure 2. Flowchart of the deconvolution process

Thus, by deconvolution of Equation (3), the tonic signal driver is obtained as:

$$SCR_{Driver} = \frac{SC}{IRF} - SCL_{Driver} \tag{4}$$

At this point the resulting signals can be used in the following process, which is feature extraction and analysis.

2.8. Feature Extraction and Analysis

As commented above, to establish if there are differences between the EDA signals produced during the listening to the different music tracks, the SCR_{Driver} has been used. Figure 3 shows the feature extraction and analysis process, which aim is to assess those features (metrics) that characterize the signals. The SCR driver, obtained through the deconvolution process described above, is decomposed into a series of temporal, morphological, statistical and frequency features. These features are stored on a feature sheet for later analysis to investigate if there are differences in the arousal on the basis of each feature for each of the musical genres.

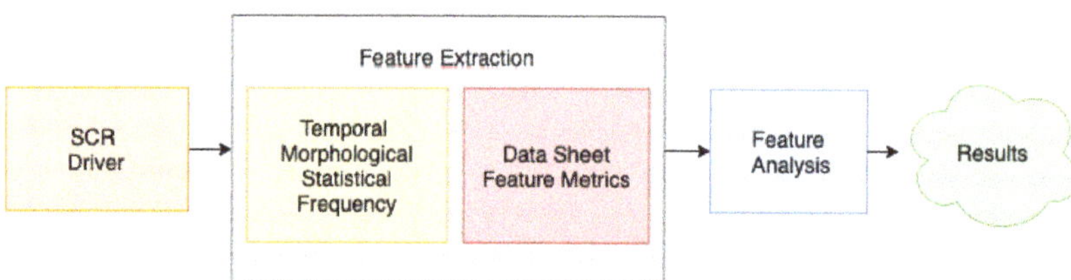

Figure 3. Flowchart of the feature extraction process

Notice that the human reaction against a specific stimulus is usually expressed as a peak or a burst of peaks in SCR_{Driver} as per the level of alertness involved. From a physiological perspective, the reactions against the stimuli are plotted on the signals as peaks proportional to the intensity, length and number of emotional events. The greater the disturbance caused, the greater the peak height produced in the SCR data. The number of peaks in SCR_{Driver} increase when the stimulus is maintained over time, which produces a series of sequential peaks.

Table 2 details the several features selected to characterize the different segments of the SCR_{Driver}. These features, which have been applied successfully in previous works [40,41,48], allow us to quantify each signal.

Table 2. Features obtained from skin conductance response (SCR)

Analysis	Features
Temporal	M, SD, MA, MI, DR, D1, D2, D1M, D2M, D1SD, D2SD
Morphological	AL, IN, AP, RMS, IL, EL
Statistical	SK, KU, MO
Frequency	F1, F2, F3

The temporal parameters are the mean value (M), standard deviation (SD), maximum and minimum peak value (MA and MI), and dynamic range (DR) establishing the difference between maximum and minimum. These parameters can provide globally significant feedback about the average and variability of the data series. They provide specific information about a higher or lower reaction obtained through the data, which may differ by the nature of the stimulus. Other temporal parameters used are the first and second derivative (D1, D2), their means (D1M, D2M) and their standard deviations (D1SD and D2SD). The use of these parameters is due to the fact that if the stimulus is intense it produces a greater slope than when it is less intense. It is, therefore, necessary to establish a criterion of speed and acceleration in the response. If the slope has reached its maximum, the time needed in the recovery produces a smoother and opposite sign gradient.

Within the morphological features there is arc length (AL), integral area (IN), normalized mean power (AP), root mean square (RMS), perimeter and area ratio (IL), and energy and perimeter ratio (EL). These parameters obey the need to understand the morphological differences in the shape of the SCR_{Driver}. There are not only peaks to be studied, but changes in the general morphology of the signals are of interest. Statistical features employed are skewness (SK), kurtosis (KU) and momentum (MO). These supply information about the distribution and variability of the data series. Finally, for the frequency domain the fast Fourier transform (FFT) for bandwidths F1 (0.1, 0.2), F2 (0.2, 0.3) and F3 (0.3, 0.4) has been chosen. Using these parameters enables discovering any variation in the frequency domain for each of the stimuli.

3. Results and Discussion

This section presents the results obtained in the experiment, broken down into two different studies. In the first study, a series of statistical tests were carried out to determine whether any significant statistical differences exist for each the temporal, morphological, statistical and frequency features described above in the EDA signals processed for each of the music genres. The objective was to identify the variations in arousal depending on the music genre, as well as to specify which features can confirm a significant statistical difference.

The second study consisted of analyzing whether there is a clear correspondence between the responses given by the participants in the SAM activation questionnaire and the physiological EDA signals acquired during listening to the music fragments. To this end, objective information on each of the EDA signal segments associated with each music genre was linked to the subjective response to the SAM questionnaire. Several classifiers were used to quantify whether there are differences between low and high excitation states. Our purpose was to check whether these classifiers can classify the states with good accuracy.

For the statistical analysis of both studies IBM SPSS Statistics version 23 was used. Please note that in all cases only a p-value < 0.05 was considered to be statistically significant.

3.1. *Direct Arousal Detection from Electrodermal Activity*

As mentioned before, first a statistical study was carried out to determine if there are any significant statistical differences for each of the features selected. This started by verifying whether the features obtained from the SCL driver signals satisfied the hypothesis of normality. This check defines whether a parametric or non-parametric test can be used. In our case, all the features were found to meet this criterion with a p-value < 0.05. Therefore, we chose to use the T-Student distribution to determine whether significant statistical differences existed. For each of the musical genres, the comparison was made with the values obtained at the beginning of the experiment, corresponding to each participant's neutral state (no music played). Table 3 shows the mean and the standard deviation of each of the features associated with the different musical genres. Hence, the p-value of each feature is provided for every musical genre in Table 4.

Table 3. Mean and standard deviation for the different features.

Type	Feature	Neutral	Flamenco	Cuban	Spanish Folklore	Rock/Jazz
Temp.	M	5.53 (4.00)	10.01 (6.62)	8.34 (6.60)	13.01 (8.62)	7.34 (1.80)
	SD	4.52 (4.76)	6.34 (2.79)	6.10 (1.70)	8.69 (5.50)	6.42 (3.40)
	MA	28.43 (26.67)	33.32 (6.23)	34.32 (5.20)	37.10 (1.82)	35.51 (4.24)
	MI	0.51 (0.13)	0.81 (0.60)	0.64 (0.51)	0.66 (0.38)	0.66 (0.38)
	DR	28.43 (6.67)	29.79 (6.21)	34.83 (17.21)	25.83 (2.69)	29.82 (6.43)
	D1	0.98 (0.14)	1.13 (0.45)	1.07 (0.23)	1.09 (0.28)	1.03 (0.07)
	D2	0.56 (0.20)	0.71 (0.38)	0.67 (0.51)	0.86 (0.54)	0.71 (0.54)
	D1M	0.86 (0.13)	0.90 (0.12)	0.74 (0.44)	0.93 (0.59)	0.74 (0.44)
	D2M	0.45 (0.17)	0.426 (0.02)	0.48 (0.17)	0.52 (0.26)	0.55 (0.31)
	D1SD	0.99 (0.96)	1.23 (0.36)	1.43 (0.29)	1.40 (0.21)	1.40 (0.21)
	D2SD	0.29 (0.01)	0.34 (0.12)	0.56 (0.39)	0.56 (0.39)	0.36 (0.12)
Morph.	AL	14,049.0 (99.8)	13,950.4 (388.4)	13,850.4 (606.3)	13,950.4 (348.6)	14,450.4 (890.0)
	IN	193.98 (148.38)	186.98 (64.76)	245.77 (86.67)	246.77 (115.67)	230.98 (75.34)
	AP	4.56 (9.33)	8.17 (1.97)	4.56 (9.33)	8.17 (3.12)	2.17 (2.12)
	RMS	7.14 (6.23)	8.96 (4.23)	10.96 (7.34)	9.80 (6.32)	8.25 (6.34)
	IL	5.50 (4.14)	6.32 (4.80)	5.17 (2.80)	7.43 (2.87)	6.96 (4.23)
	EL	0.065 (0.0013)	0.074 (0.049)	0.085 (0.054)	0.079 (0.039)	0.045 (0.099)
Stat.	SK	1.18 (0.98)	1.45 (0.89)	1.82(1.72)	1.69 (0.96)	1.82 (1.79)
	KU	1.65 (1.09)	2.67 (2.45)	1.87 (1.02)	1.89 (1.04)	1.40 (1.34)
	MO	2.10 (4.06)	4.21 (3.87)	3.44 (0.24)	4.01 (3.87)	3.8 (1.76)
Freq.	F1	2.90 (0.29)	3.60 (1.84)	3.28 (1.99)	2.78 (0.92)	3.04 (1.35)
	F2	0.15 (0.32)	0.20 (0.04)	0.29 (0.12)	0.24 (0.13)	0.19 (0.02)
	F3	0.92 (0.37)	0.79 (0.54)	0.98 (0.26)	0.98 (0.26)	1.15 (0.64)

Table 4. p-value for the different features.

Type	Features	Flamenco	Cuban	Spanish Folklore	Rock/Jazz
Temporal	M	**0.004**	**0.025**	**0.000**	**0.010**
	SD	**0.032**	**0.040**	**0.004**	**0.030**
	MA	**0.021**	**0.016**	**0.041**	0.116
	MI	**0.036**	0.120	**0.022**	**0.021**
	DR	0.340	**0.014**	**0.034**	0.320
	D1	**0.048**	**0.032**	0.260	**0.050**
	D2	**0.032**	0.211	**0.001**	0.116
	D1M	0.160	0.116	0.442	0.116
	D2M	0.320	0.424	0.120	0.070
	D1SD	**0.039**	0.074	**0.010**	0.098
	D2SD	**0.032**	**0.042**	**0.004**	**0.023**
Morphological	AL	**0.034**	**0.010**	**0.040**	**0.010**
	IN	1.100	0.065	0.080	0.766
	AP	**0.026**	**0.000**	**0.021**	0.135
	RMS	0.150	0.075	0.098	0.447
	IL	0.420	0.687	**0.012**	0.121
	EL	0.230	**0.021**	**0.034**	0.210
Statistical	SK	**0.023**	**0.045**	**0.019**	**0.051**
	KU	**0.018**	0.349	0.333	0.600
	MO	**0.013**	**0.042**	**0.034**	**0.010**
Frequential	F1	0.120	0.233	0.435	0.516
	F2	0.320	**0.011**	0.110	0.432
	F3	0.210	0.434	0.434	0.053

Moreover, Figure 4 visually displays the statistically significant features for each of the musical genres. From the previous figures and table, it can be observed that the musical genres with more statistically significant differences, according to the features employed, are Flamenco and Spanish Folklore. In contrast, there are far fewer statistically significant differences in Cuban and Rock/Jazz genres.

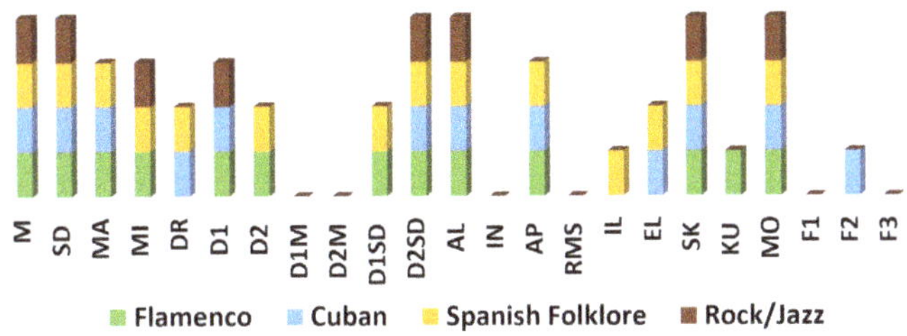

Figure 4. Statistically significant features for each of the musical genres according to their p-value.

In relation to the temporal features, M, SD and D2SD show significant differences for all four musical genres. Most other features also obtain statistically significant differences in two or three musical genres. Only for D1M and D2M there is no statistical evidence of a difference. For the group of morphological features there are only meaningful differences for all four musical genres in AL. AP presents meaningful differences in flamenco, Cuban and Spanish folklore. AP presents significant differences in flamenco, Cuban and Spanish folklore, followed by EL which has only Cuban and Spanish folk. For RM and IL no remarkable differences are found. Regarding statistical features, there are significant differences for all musical genres in SK and MO. On the contrary, for KU there are only differences in flamenco. Finally, in the category of frequency parameters, only F2 presents significant differences.

A plausible interpretation to the fact that more statistically significant differences are found in Flamenco and Spanish Folklore in contrast to Cuban and Rock/Jazz genres is provided next. Especially in the south of Spain, including the region of Murcia, flamenco is a genre that was strongly interpreted in the 60s and 70s, both in social life and in learning moments. We can say that there are many orally transmitted songs with a flamenco influence in the Spanish culture that over decades, have been sung and clapped in groups. Moreover, flamenco became a sign of identity of the purely Spanish [49]. On the other hand, through Spanish folklore, the choirs and dances, understood not as isolated elements of each Spanish region, but through musical bases common to the whole Spanish territory, were used for decades to strengthen the idea of unity of the homeland [50]. Moreover, the *Pasodoble* style and especially the *Murcian jota*, as its name indicates, are profoundly established in the region of Murcia.

On the other hand, in the 60s and 70s, and even earlier, foreign music, especially American music, was identified as the antithesis of Spanish music and as contrary to Spanish values and morality [51]. This led to the discrediting of these musical genres by the radio and the press. This was the case, although not to a high degree, of the Cuban genre. Finally, despite the media pressure of aversion towards foreign music, and mainly in foreign languages, there was an increase in fans of musical genres imported from the United States in the two great Spanish cities, Madrid and Barcelona. In small cities more rooted in traditional culture, such as the region of Murcia, these cultural manifestations had to wait a longer time [51].

3.2. Comparison of Arousal Detection and SAM Questionnaire Responses

The second study introduced the use of classifiers to verify that the differences between the two states (low and high arousal) mentioned above do exist. The classifiers were required to analyze possible correlations between the objective detection of the arousal level from processed physiological

EDA signals and the level of arousal subjectively perceived by participants when answering the SAM questionnaire.

It was decided to use different well-known classifiers, which were grouped into trees, ensemble, regression, discriminant, naïve Bayes, k-nearest neighbors (KNN) and support vector machines (SVM). In addition, several standard configurations were chosen [52–56]. More concretely, we used logistic regression and linear discriminant classifier. We tried with both Gaussian and Bayes distributions in the case of naïve Bayes. Three were the configurations used for trees, namely fine tree (Gini criterion and 4 splits), medium tree (Gini criterion and 20 splits) and coarse tree (Gini criterion and 100 splits). The kinds of ensemble trees were boosted, bagged, RUS boosted and subspace KNN. The KNN configurations used were fine (Euclidean distance and 2 neighbors), medium (Euclidean distance and 10 neighbors), coarse (Euclidean distance and 100 neighbors), cosine (angular distance and 10 neighbors) and weighted (Manhattan distance and 10 neighbors). Lastly for SVM the following configurations were studied: linear (polynomial kernel, grade 1), quadratic (polynomial kernel, grade 2), cubic (polynomial kernel, grade 3) and linear (radial basis function kernel), all of them with 10^5 iterations and MSE criterion.

As input parameters we used the different established features. As output we used the answers to the SAM excitation questionnaires completed during the experiment. Thirty iterations were performed for each of the classifiers, obtaining the precision (and its standard deviation) shown in Table 5. The dataset was randomly separated into 70% for training, 15% for testing and 15% for validation.

Table 5. Accuracy (%) of arousal assessment through different classifiers.

Classifier	Type	Flamenco	Cuban	Spanish Folklore	Rock/Jazz
Regression	Logistic	**67.0 (0.26)**	64.0 (0.19)	61.0 (1.20)	60.0 (1.01)
Discriminant	Linear	**57.0 (0.09)**	40.3 (0.03)	46.3 (0.73)	42.5 (1.47)
Naïve Bayes	Gaussian	**70.6 (0.01)**	**71.1 (0.50)**	**70.6 (0.01)**	**69.2 (0.11)**
	Standard	67.6 (0.45)	70.0 (0.00)	68.1 (0.82)	**69.2 (0.11)**
Tree	Fine	56.0 (0.12)	69.1 (0.03)	52.0 (0.02)	40.1 (0.10)
	Medium	**75.0 (0.20)**	67.1 (0.00)	**78.0 (0.06)**	45.1 (0.27)
	Coarse	70.1 (0.00)	**70.1 (0.00)**	62.1 (0.00)	52.1 (0.45)
Ensemble Tree	Boosted	72.3 (0.04)	69.7 (0.14)	76.85 (0.23)	67.3 (0.37)
	Bagged	71.0 (0.01)	67.9 (0.11)	72.0 (0.00)	68.1 (0.76)
	RUS boosted	73.0 (0.40)	70.1 (0.50)	70.9 (0.03)	**68.6 (1.20)**
	Subspace KNN	**74.5 (0.00)**	**71.43 (0.32)**	**72.1 (0.00)**	68.1 (0.20)
KNN	Fine	76.0 (0.09)	73.9 (0.10)	76.0 (0.09)	70.0 (0.00)
	Medium	82.3 (0.05)	**80.2 (0.04)**	**81.5 (0.00)**	**76.09 (1.20)**
	Coarse	80.4 (0.02)	79.1 (0.40)	77.1 (0.18)	71.09 (1.60)
	Cosine	**81.4 (0.13)**	77.1 (1.10)	77.1 (1.80)	68.18 (1.70)
	Weighted	80.9 (0.00)	79.2 (0.06)	80.9 (0.00)	75.0 (0.00)
SVM	Linear	78.0 (0.01)	73.3 (0.63)	79.1 (0.03)	**67.4 (0.60)**
	Quadratic	72.4 (0.13)	72.4 (0.13)	72.4 (0.13)	62.0 (0.13)
	Cubic	76.4 (0.60)	78.3 (0.54)	80.4 (0.53)	65.4 (0.30)
	Radial (RBF)	**87.4 (0.00)**	**81.4 (0.00)**	**83.1 (0.01)**	67.3 (0.20)

As a result, it can be seen that in the tree classifiers, for the Flamenco and Spanish Folklore genres, the tree that best classifies is the medium one with 75 and 78% respectively. On the contrary, for the rock/jazz genre, none of the trees exceed 50%, so we cannot consider that it is classified well enough. In the logistic regression classifier the results are between 60 and 67% for all music genres. One could argue that this is not a good classifier for this data set. For the linear discriminant, it was found that the best result obtained was for flamenco with 57%, which was not enough to accept it as a good classifier. Thus, this method of classification can be discarded. This is because this type of classifier works better with time series, as opposed to our proposal which is for the chosen features [57].

Naïve Naive Bayes only works well for the Gaussian configuration with an accuracy of 70.6%, 71.1% and 70.6% for flamenco, Cuban and Spanish folklore, respectively, and slightly worse for rock/jazz with 69.2%. The results of the above classifiers are in line with other studies carried out in recent years [58,59]. As for the ensemble trees, the configuration that performs the best classification is the subspace KNN. It classifies quite well the high versus low arousal states for the flamenco, Cuban genre and Spanish folklore with 74.5%, 71.43% and 72.1%, respectively. For rock/jazz the one that works better is the RUS boosted with 68.6% accuracy. The results are similar to those found in recent previous studies with EDA [60,61].

Among the KNN methods, the best classifier for flamenco is cosine KNN with an accuracy of 81.4%. For the remaining musical genres, the best is the medium configuration with an accuracy of 80.2, 81.5 and 76.09% for the Cuban, Spanish folklore and rock/jazz genres, respectively. Finally, for SVM the best classifier is the radial basis function kernel with 87.4, 81.4 and 83.1% accuracy for flamenco, Cuban genre and Spanish folklore, respectively. On the other hand, in the rock/jazz genre, the accuracy of the classifier increases to 67.4%, but it is not enough to conclude that it classifies well between the two states (low and high arousal) [62–64].

As is known from previous preliminary studies [40], kernel-based classifiers (SVM) perform better than the others because they can handle a larger number of features. Afterwards, distance-based classifiers of the k-NN type are the best for classifying this type of signals as may be seen from the results (see Table 5).

4. Conclusions

In this paper, we have presented a solution for the detection of the level of arousal from electrodermal signals (EDA) in people through their exposure to musical stimuli. For this purpose, participants over 60 years old from the region of Murcia, Spain, were recruited to listen to a series of musical pieces similar to those performed in their youth. During the playback of the music, the EDA of the participants was continuously monitored. The EDA signals acquired during the experiment were then used, along with a SAM questionnaire filled out by the participants, to conduct a couple of studies. A first study looked at the features of EDA and their ability to check for statistically significant differences for each feature extracted. A second study used well-known classifiers to analyze the potential correlation between the objective detection of the level of excitation of processed physiological EDA signals and the level of arousal subjectively perceived by participants when answering the SAM questionnaire.

The first study was based on the analysis of the existence of some kind of statistically significant difference in the selected features. The study found a greater number of statistically significant differences in the musical genres of Flamenco and Spanish Folklore, and much less in the genre of rock/jazz, which seems reasonable in the Spanish region under consideration. One of the most important factors determining musical preferences is familiarity. In accordance with our study, becoming familiar with a particular piece of music has demonstrated to increase a subject's level of enjoyment [65–68]. This is true for emotional and autobiographical memory experiences provoked by musical stimuli [68,69]. The use of new musical stimuli allows us to control familiarity, since they are stimuli that have not been heard before. In this work the use of the same neutral melodic base in all the musical fragments (own design of the study) on different musical styles was considered. The only variation in the experiment are the musical genres, so any differences we may find must be due to the styles and not to familiarity with the musical stimulus. Considering that EDA is very sensitive to familiarity and prior exposure, the use of the procedure used in this proposal provides an important advance in music psychology research.

The second study, based on classifiers, provided information on the ability to distinguish between low and high arousal levels using both the processed EDA signals and the responses to the SAM questionnaire completed by the participants. This second study concluded that SVM, KNN and ensemble trees are classifiers that work very well in this case. Other classifiers such as linear discriminant and logistic regression did not work well for any music genre. In relation to the second study, this work has some limitations both in terms of the number of participants and the selection of the EDA signal features. In first place, a larger number of participants would be necessary to reinforce the results obtained in this study. Second, despite the large number of features used during the machine learning process, overfitting was not detected in the experiment presented. Nonetheless, a more in-depth investigation on the reduction of the features would be of interest.

This study has another limitation that has to do with the evaluation of the participants' musical experience. Although stylistic variations of a new piece have been used, the previous exposure to the styles may not have been the same for the participants. Therefore, a system should be developed to evaluate the baseline of each participant in future studies.

The main contribution of this article has been the study of different music genres in older people to achieve a positive influence on their emotions and thus mitigate negative effects such as anxiety and depression. That contribution is based on the possibility of raising the arousal produced by memories evoked from their youth through the music heard at that time. The results of this work open the door to further studies on the fluctuations of EDA in older people with depression and/or cognitive impairment. We believe that these discoveries are expandable by developing new automated systems to help older people in their daily lives. Based on the results achieved in this experiment, we will be able to develop ambient intelligence systems to improve the quality of life and well-being of the elderly.

Author Contributions: Conceptualisation, A.B.-T. and A.F.-S.; methodology, R.S.-R.; software, R.S.-R.; validation, A.B.-T., A.F.-S., R.S.-R. and A.F.-C.; writing–original draft preparation, R.S.-R., A.B.-T. and A.F.-S.; writing–review and editing, A.F.-C. and J.M.L.; funding acquisition, A.F.-C. and J.M.L. All authors have read and agreed to the published version of the manuscript.

Funding: This work has been supported by Spanish Ministerio de Ciencia, Innovación y Universidades, Agencia Estatal de Investigación (AEI) / European Regional Development Fund (FEDER, UE) under DPI2016-80894-R grant, and by CIBERSAM of the Instituto de Salud Carlos III. Roberto Sánchez-Reolid holds BES-2017-081958 scholarship from Spanish Ministerio de Educación y Formación Profesional.

Conflicts of Interest: The authors declare no conflict of interest.

Abbreviations

The following abbreviations are used in this manuscript:

EDA	Electrodermal Activity
IRF	Impulse Response Function
KNN	K-Nearest Neighbor
SAM	Self-Assessment Mannequin
SC	Skin Conductance
SCL	Skin Conductance level
SCR	Skin Conductance response
SVM	Support Vector Machine

References

1. Sánchez-Reolid, R.; Martínez-Rodrigo, A.; López, M.T.; Fernández-Caballero, A. Deep support vector machines for the identification of stress condition from electrodermal activity. *Int. J. Neural Syst.* **2020**, *30*, 2050031. [CrossRef] [PubMed]
2. Picard, R.W.; Fedor, S.; Ayzenberg, Y. Multiple arousal theory and daily-life electrodermal activity asymmetry. *Emot. Rev.* **2016**, *8*, 62–75. [CrossRef]
3. Castillo, J.C.; Fernández-Caballero, A.; Castro-González, Á.; Salichs, M.A.; López, M.T. *A Framework for Recognizing and Regulating Emotions in the Elderly. Ambient Assisted Living and Daily Activities*; Pecchia, L., Chen, L.L., Nugent, C., Bravo, J., Eds.; Springer: Heidelberg, Germany, 2014; pp. 320–327.
4. Picard, R.W. *Affective Computing*; The MIT Press: Cambridge, MA, USA, 2000.
5. Górriz, J.M.; Ramírez, J.; Ortíz, A.; Martínez-Murcia, F.J.; Segovia, F.; Suckling, J.; Leming, M.; Zhang, Y.-D.; Álvarez-Sánchez, J.R.; Bologna, G. Artificial intelligence within the interplay between natural and artificial computation: Advances in data science, trends and applications. *Neurocomputing* **2020**, *410*, 237–270.
6. Jamil, F.; Ahmad, S.; Iqbal, N.; Kim, D.H. Towards a Remote Monitoring of Patient Vital Signs Based on IoT-Based Blockchain Integrity Management Platforms in Smart Hospitals. *Sensors* **2020**, *20*, 2195. [CrossRef]

7. Pala, D.; Caldarone, A.A.; Franzini, M.; Malovini, A.; Larizza, C.; Casella, V.; Bellazzi, R. Deep Learning to Unveil Correlations between Urban Landscape and Population Health. *Sensors* **2020**, *20*, 2105. [CrossRef]
8. Rathore, H.; Mohamed, A.; Guizani, M. A Survey of Blockchain Enabled Cyber-Physical Systems. *Sensors* **2020**, *20*, 282. [CrossRef]
9. Hazer-Rau, D.; Meudt, S.; Daucher, A.; Spohrs, J.; Hoffmann, H.; Schwenker, F.; Traue, H.C. The uulmMAC Database—A Multimodal Affective Corpus for Affective Computing in Human-Computer Interaction. *Sensors* **2020**, *20*, 2308. [CrossRef]
10. Pham, S.; Yeap, D.; Escalera, G.; Basu, R.; Wu, X.; Kenyon, N.J.; Hertz-Picciotto, I.; Ko, M.J.; Davis, C.E. Wearable sensor system to monitor physical activity and the physiological effects of heat exposure. *Sensors* **2020**, *20*, 855. [CrossRef]
11. Ying Wah, T.; Gopal Raj, R.; Lakhan, A. A Novel Cost-Efficient Framework for Critical Heartbeat Task Scheduling Using the Internet of Medical Things in a Fog Cloud System. *Sensors* **2020**, *20*, 441.
12. Steinberger, F.; Schroeter, R.; Watling, C.N. From road distraction to safe driving: Evaluating the effects of boredom and gamification on driving behaviour, physiological arousal, and subjective experience. *Comput. Hum. Behav.* **2017**, *75*, 714–726. [CrossRef]
13. Azami, P.; Jan, T.; Iranmanesh, S.; Ameri Sianaki, O.; Hajiebrahimi, S. Determining the Optimal Restricted Driving Zone Using Genetic Algorithm in a Smart City. *Sensors* **2020**, *20*, 2276.
14. Fernández-Caballero, A.; Martínez-Rodrigo, A.; Pastor, J.M.; Castillo, J.C.; Lozano-Monasor, E.; López, M.T.; Zangróniz, R.; Latorre, J.M.; Fernández-Sotos, A. Smart environment architecture for emotion recognition and regulation. *J. Biomed. Informatics* **2016**, *64*, 55–73. [CrossRef] [PubMed]
15. Lay-Ekuakille, A.; Mukhopadhyay, S.C. *Wearable and Autonomous Biomedical Devices and Systems for Smart Environment*; Springer: Heidelberg, Germany, 2010.
16. Mehrabian, A.; Russell, J.A. *An Approach to Environmental Psychology*; The MIT Press: Cambridge, MA, USA, 1974.
17. Russell, J.A. A circumplex model of affect. *J. Personal. Soc. Psychol.* **1980**, *39*, 1161. [CrossRef]
18. Bakker, I.; van der Voordt, T.; Vink, P.; de Boon, J. Pleasure, arousal, dominance: Mehrabian and Russell revisited. *Curr. Psychol.* **2014**, *33*, 405–421. [CrossRef]
19. Martínez-Rodrigo, A.; Zangróniz, R.; Pastor, J.M.; Fernández-Caballero, A. *Arousal Level Classification in the Ageing Adult by Measuring Electrodermal Skin Conductivity. Ambient Intelligence for Health*; Bravo, J., Hervás, R., Villarreal, V., Eds.; Springer: Heidelberg, Germany, 2015; pp. 213–223.
20. Wang, C.A.; Baird, T.; Huang, J.; Coutinho, J.D.; Brien, D.C.; Munoz, D.P. Arousal effects on pupil size, heart rate, and skin conductance in an emotional face task. *Front. Neurol.* **2018**, *9*, 1029. [CrossRef] [PubMed]
21. Dawson, M.E.; Schell, A.M.; Filion, D.L. The electrodermal system. In *Handbook of Psychophysiology*; Cambridge University Press: Cambridge, UK, 2007; Volume 1, pp. 159–181.
22. Salimpoor, V.N.; Benovoy, M.; Longo, G.; Cooperstock, J.R.; Zatorre, R.J. The rewarding aspects of music listening are related to degree of emotional arousal. *PLOS ONE* **2009**, *4*, 1–14. [CrossRef]
23. Bradley, M.M.; Lang, P.J. Measuring emotion: The self-assessment manikin and the semantic differential. *J. Behav. Ther. Exp. Psychiatry* **1994**, *25*, 49–59. [CrossRef]
24. Morris, J.D. Observations: SAM: The Self-Assessment Manikin; an efficient cross-cultural measurement of emotional response. *J. Advert. Res.* **1995**, *35*, 63–68.
25. Agrawal, A.; An, A. Unsupervised emotion detection from text using semantic and syntactic relations. In Proceedings of the 2012 IEEE/WIC/ACM International Conferences on Web Intelligence and Intelligent Agent Technology, Macau, China, 4–7 December 2012; pp. 346–353.
26. Canales, L.; Martínez-Barco, P. Emotion detection from text: A survey. In Proceedings of the Workshop on Natural Language Processing in the 5th Information Systems Research Working Days, Quito, Ecuador, 20–24 October 2014; pp. 37–43.
27. Fernández-Sotos, A.; Martínez-Rodrigo, A.; Moncho-Bogani, J.; Latorre, J.M.; Fernández-Caballero, A. Neural Correlates of Phrase Quadrature Perception in Harmonic Rhythm: An EEG Study Using a Brain–Computer Interface. *Int. J. Neural Syst.* **2018**, *28*, 1750054. [CrossRef]
28. Martínez-Rodrigo, A.; Fernández-Sotos, A.; Latorre, J.M.; Moncho-Bogani, J.; Fernández-Caballero, A. Neural Correlates of Phrase Rhythm: An EEG Study of Bipartite vs. Rondo Sonata Form. *Front. Neuroinformatics* **2017**, *11*, 29. [CrossRef]

29. Fernández-Sotos, A.; Fernández-Caballero, A.; Latorre, J.M. Influence of Tempo and Rhythmic Unit in Musical Emotion Regulation. *Front. Comput. Neurosci.* **2016**, *10*, 80. [CrossRef] [PubMed]
30. Fernández-Sotos, A.; Fernández-Caballero, A.; Latorre, J.M. Elicitation of Emotions through Music: The Influence of Note Value. In *Artificial Computation in Biology and Medicine*; Ferrández Vicente, J.M.; Álvarez-Sánchez, J.R., de la Paz López, F., Toledo-Moreo, F.J., Adeli, H., Eds.; Heidelberg, Germany, 2015; pp. 488–497.
31. Ratcliff, R. A theory of memory retrieval. *Psychol. Rev.* **1978**, *85*, 59. [CrossRef]
32. Serrano, J.P.; Latorre, J.M.; Gatz, M.; Montanes, J. Life review therapy using autobiographical retrieval practice for older adults with depressive symptomatology. *Psychol. Aging* **2004**, *19*, 272. [CrossRef]
33. Latorre, J.M.; Ricarte, J.J.; Serrano, J.P.; Ros, L.; Navarro, B.; Aguilar, M.J. Performance in autobiographical memory of older adults with depression symptoms. *Appl. Cogn. Psychol.* **2013**, *27*, 167–172. [CrossRef]
34. Charles, S.T.; Luong, G. Emotional experience across adulthood: The theoretical model of strength and vulnerability integration. *Curr. Dir. Psychol. Sci.* **2013**, *22*, 443–448. [CrossRef]
35. Fernández-Caballero, A.; González, P.; Navarro, E. Gerontechnologies–Current achievements and future trends. *Expert Syst.* **2017**, *34*, e12203. [CrossRef]
36. Siedlecka, E.; Denson, T.F. Experimental methods for inducing basic emotions: A qualitative review. *Emot. Rev.* **2019**, *11*, 87–97. [CrossRef]
37. Critchley, H.; Nagai, Y.; Electrodermal Activity (EDA). *Encyclopedia of Behavioral Medicine*; Springer: New York, NY, USA, 2013; pp. 666–669.
38. Lang, P.J.; Greenwald, M.K.; Bradley, M.M.; Hamm, A.O. Looking at pictures: Affective, facial, visceral, and behavioral reactions. *Psychophysiology* **1993**, *30*, 261–273. [CrossRef]
39. Sarchiapone, M.; Gramaglia, C.; Iosue, M.; Carli, V.; Mandelli, L.; Serretti, A.; Marangon, D.; Zeppegno, P. The association between electrodermal activity (EDA), depression and suicidal behaviour: A systematic review and narrative synthesis. *BMC Psychiatry* **2018**, *18*, 22. [CrossRef]
40. Sánchez-Reolid, R.; Martínez-Rodrigo, A.; Fernández-Caballero, A. Stress Identification from Electrodermal Activity by Support Vector Machines. In *Understanding the Brain Function and Emotions*; Springer: Heidelberg, Germany, 2019; pp. 202–211.
41. Zangróniz, R.; Martínez-Rodrigo, A.; Pastor, J.M.; López, M.T.; Fernández-Caballero, A. Electrodermal activity sensor for classification of calm/distress condition. *Sensors* **2017**, *17*, 2324. [CrossRef]
42. Posada-Quintero, H.F.; Chon, K.H. Innovations in Electrodermal Activity Data Collection and Signal Processing: A Systematic Review. *Sensors* **2020**, *20*, 479. [CrossRef] [PubMed]
43. Mohino-Herranz, I.; Gil-Pita, R.; Rosa-Zurera, M.; Seoane, F. Activity Recognition Using Wearable Physiological Measurements: Selection of Features from a Comprehensive Literature Study. *Sensors* **2019**, *19*, 5524. [CrossRef] [PubMed]
44. Silva Moreira, P.; Chaves, P.; Dias, R.; Dias, N.; Almeida, P.R. Validation of Wireless Sensors for Psychophysiological Studies. *Sensors* **2019**, *19*, 4824. [CrossRef] [PubMed]
45. Warriner, A.B.; Kuperman, V.; Brysbaert, M. Norms of valence, arousal, and dominance for 13,915 English lemmas. *Behav. Res. Methods* **2013**, *45*, 1191–1207. [CrossRef]
46. De Santiago Ortega, P.P. Flamenco: De la marginalidad social a la referencia cultural pasando por la apropiación política. *Revista de Investigación sobre Flamenco La Madrugá* **2018**, *15*, 91–115.
47. Benedek, M.; Kaernbach, C. A continuous measure of phasic electrodermal activity. *J. Neurosci. Methods* **2010**, *190*, 80–91. [CrossRef]
48. Martínez-Rodrigo, A.; Fernández-Aguilar, L.; Zangróniz, R.; Latorre, J.M.; Pastor, J.M.; Fernández-Caballero, A. Film mood induction and emotion classification using physiological signals for health and wellness promotion in older adults living alone. *Expert Syst.* **2020**, *37*, e12425. [CrossRef]
49. Piñeiro Blanca, J. Instrumentalización política de la música desde el franquismo hasta la consolidación de la democracia en España. *Revista del Centro de Estudios Históricos de Granada y su Reino* **2013**, *25*, 237–262.
50. Muñiz Velázquez, J.A. La música en el sistema propagandístio franquista. *Hist. Comun. Soc.* **1998**, *3*, 343–363.
51. Iglesias, I. (Re)construyendo la identidad musical española: el jazz y el discurso cultural del franquismo durante la Segunda Guerra Mundial. *Hist. Actual Online* **2010**, *23*, 119–135.
52. Al Machot, F.; Ali, M.; Ranasinghe, S.; Mosa, A.H.; Kyandoghere, K. Improving subject-independent human emotion recognition using electrodermal activity sensors for active and assisted living. In Proceedings of the

11th Pervasive Technologies Related to Assistive Environments Conference, Corfu, Greece, 25-29 June 2018; pp. 222–228.
53. Amalan, S.; Shyam, A.; Anusha, A.; Preejith, S.; Tony, A.; Jayaraj, J.; Mohanasankar, S. Electrodermal activity based classification of induced stress in a controlled setting. In Proceedings of the 2018 IEEE International Symposium on Medical Measurements and Applications, Rome, Italy, 11–13 June 2018; pp. 1–6.
54. Greco, A.; Valenza, G.; Citi, L.; Scilingo, E.P. Arousal and valence recognition of affective sounds based on electrodermal activity. *IEEE Sens. J.* **2016**, *17*, 716–725. [CrossRef]
55. Silveira, F.; Eriksson, B.; Sheth, A.; Sheppard, A. Predicting audience responses to movie content from electro-dermal activity signals. In Proceedings of the 2013 ACM International Joint Conference on Pervasive and Ubiquitous Computing, Zurich, Switzerland, 8–12 September 2013; pp. 707–716.
56. Wang, J.S.; Lin, C.W.; Yang, Y.T.C. A k-nearest-neighbor classifier with heart rate variability feature-based transformation algorithm for driving stress recognition. *Neurocomputing* **2013**, *116*, 136–143. [CrossRef]
57. Muñoz Expósito, J.; Galán, S.; Reyes, N.; Candeas, P.; Peña, F. Speech/music discrimination based on a new warped LPC-based feature and linear discriminant analysis. In Proceedings of the 7th International Conference on Digital Audio Effects, Naples, Italy, 5–8 October 2004.
58. Bandara, D.; Song, S.; Hirshfield, L.; Velipasalar, S. A more complete picture of emotion using electrocardiogram and electrodermal activity to complement cognitive data. In Proceedings of the 10th International Conference on Augmented Cognition, Toronto, ON, Canada, 17–22 July 2016; pp. 287–298.
59. Jang, E.H.; Park, B.J.; Kim, S.H.; Chung, M.A.; Park, M.S.; Sohn, J.H. Emotion classification based on bio-signals emotion recognition using machine learning algorithms. In Proceedings of the 2014 International Conference on Information Science, Electronics and Electrical Engineering, Sapporo City, Hokkaido, Japan, 26–28 April 2014; Volume 3, pp. 1373–1376.
60. Kim, A.Y.; Jang, E.H.; Kim, S.; Choi, K.W.; Jeon, H.J.; Yu, H.Y.; Byun, S. Automatic detection of major depressive disorder using electrodermal activity. *Sci. Rep.* **2018**, *8*, 1–9. [CrossRef] [PubMed]
61. Liu, Y.; Du, S. Psychological stress level detection based on electrodermal activity. *Behav. Brain Res.* **2018**, *341*, 50–53. [CrossRef]
62. Cavallo, F.; Semeraro, F.; Mancioppi, G.; Betti, S.; Fiorini, L. Mood classification through physiological parameters. *J. Ambient Intell. Humanized Comput.* **2019**, *106*, 1–14. [CrossRef]
63. Xin, S.Q.; Yahya, N.; Izhar, L.I. Classification of Neurological States from Biosensor Signals Based on Statistical Features. In Proceedings of the 2019 IEEE Student Conference on Research and Development, Perak, Malaysia, 15–17 October 2019; pp. 231–236.
64. Taylor, S.; Jaques, N.; Chen, W.; Fedor, S.; Sano, A.; Picard, R. Automatic identification of artifacts in electrodermal activity data. In Proceedings of the 2015 37th Annual International Conference of the IEEE Engineering in Medicine and Biology Society. IEEE, Milan, Italy, 25–29 August 2015; pp. 1934–1937.
65. Gregory, A.H.; Varney, N. Cross-cultural comparisons in the affective response to music. *Psychol. Music.* **1996**, *24*, 47–52. [CrossRef]
66. Demorest, S.M.; Morrison, S.J.; Jungbluth, D.; Beken, M.N. Lost in translation: An enculturation effect in music memory performance. *Music. Percept.* **2008**, *25*, 213–223. [CrossRef]
67. Pereira, C.S.; Teixeira, J.; Figueiredo, P.; Xavier, J.; Castro, S.L.; Brattico, E. Music and emotions in the brain: Familiarity matters. *PlOS ONE* **2011**, *6*, e27241. [CrossRef]
68. Platz, F.; Kopiez, R.; Hasselhorn, J.; Wolf, A. The impact of song-specific age and affective qualities of popular songs on music-evoked autobiographical memories (MEAMs). *Music. Sci.* **2015**, *19*, 327–349. [CrossRef]
69. Maksimainen, J.; Wikgren, J.; Eerola, T.; Saarikallio, S. The effect of memory in inducing pleasant emotions with musical and pictorial stimuli. *Sci. Rep.* **2018**, *8*, 1–12. [CrossRef]

sensors

MDPI

Article

Emotion Assessment Using Feature Fusion and Decision Fusion Classification Based on Physiological Data: Are We There Yet?

Patrícia Bota [1,*], Chen Wang [2], Ana Fred [1] and Hugo Silva [1]

1 Instituto Superior Técnico (IST), Department of Bioengineering (DBE) and Instituto de Telecomunicações (IT), Av. Rovisco Pais n. 1, Torre Norte-Piso 10, 1049-001 Lisbon, Portugal; afred@lx.it.pt (A.F.); hsilva@lx.it.pt (H.S.)

2 State Key Laboratory of Media Convergence Production Technology and Systems, Xinhua News Agency & Future Media Convergence Institute (FMCI), Xinhua Net, Jinxuan Building, No. 129 Xuanwumen West Street, Beijing 100031, China; wangchen@news.cn

* Correspondence: patricia.bota@tecnico.ulisboa.pt

Received: 28 July 2020; Accepted: 19 August 2020; Published: 21 August 2020

Abstract: Emotion recognition based on physiological data classification has been a topic of increasingly growing interest for more than a decade. However, there is a lack of systematic analysis in literature regarding the selection of classifiers to use, sensor modalities, features and range of expected accuracy, just to name a few limitations. In this work, we evaluate emotion in terms of low/high arousal and valence classification through Supervised Learning (SL), Decision Fusion (DF) and Feature Fusion (FF) techniques using multimodal physiological data, namely, Electrocardiography (ECG), Electrodermal Activity (EDA), Respiration (RESP), or Blood Volume Pulse (BVP). The main contribution of our work is a systematic study across five public datasets commonly used in the Emotion Recognition (ER) state-of-the-art, namely: (1) Classification performance analysis of ER benchmarking datasets in the arousal/valence space; (2) Summarising the ranges of the classification accuracy reported across the existing literature; (3) Characterising the results for diverse classifiers, sensor modalities and feature set combinations for ER using accuracy and F1-score; (4) Exploration of an extended feature set for each modality; (5) Systematic analysis of multimodal classification in DF and FF approaches. The experimental results showed that FF is the most competitive technique in terms of classification accuracy and computational complexity. We obtain superior or comparable results to those reported in the state-of-the-art for the selected datasets.

Keywords: emotion recognition; physiological signals; machine learning; signal processing

1. Introduction

Emotion is an integral part of human behaviour, exerting a powerful influence in mechanisms such as perception, attention, decision making and learning. Indeed, what humans tend to notice and memorise are usually not monotonous, commonplace events but the ones that evoke feelings of joy, sorrow, pleasure, or pain [1]. Therefore, understanding emotional states is crucial to understand human behaviour, cognition and decision making. The computer science field dedicated to the study of emotions is denoted as Affective Computing, whose modern potential applications include, among many others: (1) automated driver assistance—e.g., through an alert system monitoring and warning the user for sleepiness, unconscious or unhealthy states potentially hindering driving; (2) healthcare—e.g., through wellness monitoring applications identifying causes of stress, anxiety, depression or chronic diseases; (3) adaptive learning—e.g., through a teaching application able to adjust the content delivery rate and number of iterations according to the user enthusiasm and frustration level; (4) recommendation systems—e.g., assisting and asserting personalised content according to the user preferences as perceived by their response.

Emotions are communicated via external (facial or body expressions such as a smile, tense shoulders, and others) and internal body expressions (alterations in heart rate (HR), respiration rate, perspiration, and others). Such manifestations generally occur naturally and subconsciously, and their sentic modulation can be used to infer the subjects' current emotional state. Acquired in a systematic daily setting, it could be possible to infer the probability of a subjects' mood for the following day and their health condition.

External physical manifestations (e.g., facial expressions) are easily collected through a camera; however, they present low reliability since they depend highly on the user environment (if he is alone or in a group setting), or cultural background (if the subject grew up in a society promoting the externalisation or internalisation of emotion), and can be easily faked or manipulated according to the subject goals, compromising the assessment of the true emotional state [2]. On the other hand, for internal physical manifestations, these constraints are less prominent, since the subject has little control over his bodily states. Alterations in the physiological signals are not easily controlled by the subject, thus, these entitle a more authentic insight into the subject emotional experience.

Given these considerations, our work aims to perform a comprehensive study on automatic emotion recognition using physiological data, namely from Electrocardiography (ECG), Electrodermal Activity (EDA), Respiration (RESP), Blood Volume Pulse (BVP) sensors. This choice of modalities is due to three factors: (1) Data can be easily extracted from pervasive, discrete wearable technology, rather than more intrusive sensors (e.g., Electroencephalography (EEG), or Functional near-infrared spectroscopy (fNIRS)); (2) Widely reported in the recent state-of-the-art; (3) Publicly available multimodal datasets validated in literature. We use five public state-of-the-art datasets to evaluate two major techniques: Feature Fusion (FF) versus Decision Fusion (DF) on a feature-based representation, exploring also an extensive set of features comparatively to previous work. Furthermore, instead of the discrete model, the users' emotional response is assessed on the two-dimensional space: Valence (measuring how unpleasant or pleasant is the emotion), and Arousal (measuring the emotion intensity level).

The remaining of this paper is organised as follows: Section 2 presents a brief literature review on ER, with special emphasis on articles that describe the datasets used in our work. Section 3 describes the overall machine learning pipeline of the proposed methods. Section 4 evaluates our methodology in five public datasets. Lastly, in Section 5, the main conclusions of this work are presented along with future work directions.

2. State of the Art

In literature, human emotion processing is generally described using two models: One decomposing emotion in discrete categories divided into basic/primary (arriving from innate, fast and in response to "flight-or-fight" behaviour) and complex/secondary emotions (deriving from cognitive processes) [3,4]. On the other hand, the second model quantifies emotions into continuous dimensions. A popular model, proposed by Lang [5], suggested a Valence (unpleasant–pleasant level) versus Arousal (activation level) two-dimensional model [6], which we adopt in this work. Concerning affect elicitation, it is generally performed through films snippets [6], virtual reality [7], music [8], recall [9], or stressful environments [6], with no commonly established norm on which is the optimal methodology for ER elicitation.

Regarding the automated recognition of emotional states, it is usually performed based on two methodologies [2,10,11]: (1) Traditional Machine Learning (ML) techniques [12–14]; (2) Deep learning approaches [15–17]. Due to the limited size of existing datasets, most of the work focuses on traditional ML algorithms, in particular Supervised Learning (SL), such as Support Vector Machines (SVM) [18–20], k-Nearest Neighbour (kNN) [21–23], Decision Trees (DT) [24,25], and others [26,27], with the SVM method being the most commonly applied algorithm, showing overall good results and low computational complexity.

Many physiological modalities and features have been evaluated for ER, namely Electroencephalography (EEG) [28–30], Electrocardiography (ECG) [31–33], Electrodermal Activity (EDA) [34–36], Respiration (RESP) [26], Blood Volume Pulse (BVP) [26,35] and Temperature (TEMP) [26]. Multi-modal approaches have prevailed; however, there is still no clear evidence of which feature combinations and physiological signals are the most relevant. The literature has shown that the classification performance improves with the simultaneous exploitation of different signal modalities [2,8,10,37], and that modality fusion can be performed at two main levels: FF [24,38,39] and DF [8,26,37,40,41]. In the former, features are extracted from each modality and latter concatenated to form a single feature vector space, to be used as input for the ML model. On the other hand, in DF, from each modality, a feature vector is extracted to form a classifier prediction through a voting system. Hence, with k modalities, k classifiers will be created leading to k predictions that can be combined to yield a final result. Both methodologies are found in the state-of-the art [42], but it is unclear which is the best to use in the area of ER using multimodal physiological data obtained from non-intrusive wearable technology.

Detailed information on the current state-of-the-art in a more generalized perspective, we refer the reader to the surveys [2,11,43–47] and references therein, where a comprehensive review of the latest work on ER using ML and physiological signals can be found, highlighting the main achievements, challenges, take-home messages, and possible future opportunities.

The present work extends the state-of-the-art of ER through: (1) Classification performance analysis, in the arousal/valence space, of ER for five publicly available datasets that cover multiple elicitation methods; (2) Summarising the ranges of the classification accuracy reported across the existing literature for the evaluated datasets; (3) Characterising the results for diverse classifiers, sensor modalities and feature set combinations for ER using accuracy and F1-score as evaluation metrics (the later not being commonly reported albeit important to evaluate classification bias); (4) Exploration of an extended feature set for each modality, analyzing also their relevance through feature selection; (5) Systematic analysis of multimodal classification in DF and FF approaches, with superior or comparable results to those reported in the state-of-the-art for the selected datasets.

3. Methods

To evaluate the classification accuracy in ER from physiological signals, we adopted the two dimensional Valence/Arousal space. As previously mentioned, the ECG, RESP, EDA, and BVP signals are used, and we compare FF and DF techniques in a feature space based framework. In the forthcoming sub-sections, a more detailed description of each approach is presented.

3.1. Feature Fusion

As previously mentioned, when working with multi-modal approaches the exploitation of the different signal modalities can be performed resorting to different techniques. We start by testing the FF technique. In FF, the features are independently extracted from each sensor modality (in our case ECG, BVP, EDA, and RESP), and are concatenated afterwards to form a single, global, feature vector (570 features for EDA, 373 for ECG, 322 for BVP, and 487 for RESP, implemented and detailed in the BioSPPy software library https://github.com/PIA-Group/BioSPPy). Additionally, we applied sequential forward feature selection (SFFS) in order to preserve only the most informative features, and save time and computational power of the machine learning algorithm to be applied in the next step. All the presented methods were implemented in Python and made available as open source software https://github.com/PIA-Group/BioSPPy.

3.2. Decision Fusion

In contrast to FF, in DF, from each sensor signal, a feature vector is extracted and used independently to train and learn a classifier, so that each modality returns a set of predicted labels. Hence, with k modalities, k classifiers will be created returning k predictions per sample. The returned predictions are then combined to yield a final result, in our case, via a weighted majority voting system. In this voting system, the ensemble decides on the class that receives the highest number of votes taking into account all sensor modalities, and a weight (W) parameter per modality to give the more competent classifiers a greater power for the final decision. The weights were chosen for each modality according to the classifier accuracy on the validation set. In case of a draw in the class prediction, the selection is random.

3.3. Classifier

To perform the classification seven SL classifiers were tested: K-Nearest Neighbour (k-NN); Decision Tree (DT); Random Forest (RF); Support Vector Machines (SVM); AdaBoost (AB); Gaussian Naive Bayes (GNB); and Quadratic Discriminant Analysis (QDA). For more detail regarding these classifiers, the author refers the reader to [48] and references therein.

A comprehensive study of these classifiers performance and parameter tunning was performed using 4-fold Cross Validation (CV) to ensure a meaningful validation and avoiding overfitting. The value of 4 was selected to optimise the number of iterations and the homogeneity in number of the classes in the training and test set, since some of the datasets used were highly imbalanced. The best performing classifier was chosen using Leave-One-Subject-Out (LOSO) to be incorporated into the FF and DF frameworks.

To obtain a measurable evaluation of the model performance, the following metrics are computed: Accuracy—$\frac{TP+TN}{TP+TN+FP+FN}$; Precision—$\frac{TP}{TP+FP}$; Recall—$\frac{TP}{TP+FN}$; F1-score—the harmonic mean of precision and recall [49]. Nomenclature: TP—True Positive; FP—False Positive; FP—False Positive; FN—False negative.

4. Experimental Results

In this section, we start by introducing the datasets used in this paper, followed by an analysis and classification performance comparison of the FF and DF approaches.

4.1. Datasets

In the scope of our work we used five publicly available datasets for ER, commonly used in previous work for benchmarking:

1. **IT Multimodal Dataset for Emotion Recognition (ITMDER)** [7]: contains the physiological signals of interest to our work (EDA, RESP, ECG, and BVP) of 18 individuals using two devices based on the BITalino system [50,51] (one placed on the arm and the other on the chest of the participants), collected while the subjects watched seven VR videos to elicit the emotions: Boredom, Joyfulness, Panic/Fear, Interest, Anger, Sadness and Relaxation. The ground-truth annotations were obtained by the subjects self-report per video using the Self-Assessment Manikin (SAM), in the Valence-Arousal space. For more information regarding the dataset, the authors refer the reader to [7].
2. **Multimodal Dataset for Wearable Stress and Affect Detection (WESAD)** [6]: contains EDA, ECG, BVP, and RESP sensors data collected from 15 participants using a chest- and a wrist-worn device: a RespiBAN Professional (biosignalsplux.com/index.php/respiban-professional) and an Empatica E4 (empatica.com/en-eu/research/e4) under 4 main conditions: Baseline (reading neutral magazines); Amusement (funny video clips); Stress (Trier Social Stress Test (TSST) consisting of public speaking and a mental arithmetic task); and lastly, meditation. The annotations were obtained using 4 self-reports: PANAS; SAM in Valence-Arousal space;

State-Trait Anxiety Inventory (STAI); and Short Stress State Questionnaire (SSSQ). For more information regarding the dataset, the authors refer the reader to [6].

3. **A dataset for Emotion Analysis using Physiological Signals (DEAP) [8]:** contains EEG and peripheral (EDA, BVP, and RESP) physiological data from 32 participants, recorded as each watched 40 one-minute-long excerpts of music videos. The participants rated each video in terms of the levels of Arousal, Valence, like/dislike, dominance and familiarity. For more information regarding the dataset, the authors refer the reader to [8].
4. **Multimodal dataset for Affect Recognition and Implicit Tagging (MAHNOB-HCI) [52]:** contains face videos, audio signals, eye gaze data, and peripheral physiological data (EDA, ECG, RESP) of 27 participants watching 20 emotional videos, self-reported in Arousal, Valence, dominance, predictability, and additional emotional keywords. For more information regarding the dataset, the authors refer the reader to [52].
5. **Eight-Emotion Sentics Data (EESD) [9]:** contains physiological data (EMG, BVP, EDA, and RESP) from an actress during deliberate emotional expressions of Neutral, Anger, Hate, Grief, Platonic Love, Romantic Love, Joy, and Reverence. For more information regarding the dataset, the authors refer the reader to [9].

Table 1 shows a summary of the datasets used in this paper, highlighting their main characteristics. One should notice that the datasets are heavily imbalanced.

Table 1. Summary of the datasets information on: classes; ratio of number (N°) of samples per class label shown between parenthesis—N° samples per class label/total number of samples, for the classes 0 and 1, shown between parenthesis; demographic Information (DI)—number of participants; ages (years old) ± standard deviation, and Female (F)-Male (M) subject distribution; device used for this paper; and sampling rate. Dataset nomenclature: ITMDER—IT Multimodal Dataset for Emotion Recognition; WESAD—Multimodal Dataset for Wearable Stress and Affect Detection; DEAP—A dataset for Emotion Analysis using Physiological Signals; MAHNOB-HCI—Multimodal dataset for Affect Recognition and Implicit Tagging; EESD—Eight-Emotion Sentics Data.

Dataset	Classes	N° of Samples per Class	DI	Device	Sampling Rate
ITMDER	Low-high Arousal/Valence	Arousal: 0.54 (0); 0.46 (1) Valence: 0.12 (0); 0.88 (1)	18 23 ± 3.7 10 (F) – 13 (M)	Chest strap and armband based on BITalino [a]	1000
WESAD	Neutral, Stress, Amusement + 4 Questionnaires	Arousal: 0.86 (0); 0.14 (1) Valence: 0.07 (0); 0.93 (1)	15 27.5 ± 2.4 3 (F) – 12 (M)	RespiBAN Professional [b], Empatica E4	ECG and RESP: 700; EDA: 4; BVP: 64
DEAP	Arousal, Valence, Like/dislike, Dominance and Familiarity	Arousal: 0.41 (0); 0.59 (1) Valence: 0.43 (0); 0.57 (1)	32 16 (F) – 16 (M) 19 – 37	Biosemi Active II system [c]	128
MAHNOB-HCI	Arousal, Valence, Dominance	Arousal: 0.48 (0); 0.52 (1) Valence: 0.47 (0); 0.53 (1)	27 26.06 ± 4.39 17(F) – 13(M)	Biosemi Active II system	256
EESD	Neutral, Anger, Hate, Grief, Platonic love, Romantic Love, Joy, and Reverence	Arousal: 0.5 (0); 0.5 (1) Valence: 0.5 (0); 0.5 (1)	1 1 (F)	Thought Technologies ProComp prototype [d]	256

[a] https://bitalino.com/en/; [b] https://biosignalsplux.com; [c] https://www.biosemi.com; [d] http://thoughttechnology.com/index.php/procomp-infiniti-343.html.

4.2. Signal Pre-Processing

The raw data recorded from the sensors usually shows a low signal-to-noise ratio, thus, it is generally necessary to pre-process the data, namely filtering to remove motion artefacts, outliers, and further noise. Additionally, since different modalities were acquired, different filtering specifications are required according to each sensor modality. Considering what is typically found in the state-of-the-art [11], the filtering for which each modality was performed as follows:

- **Electrocardiography (ECG):** Finite impulse response (FIR) band-pass filter of order 300 and 3–45 Hz cut-off frequency.
- **Electrodermal Activity (EDA):** Butterworth low-pass pass filter of order 4 and 1 Hz cut-off frequency.
- **Respiration (RESP):** Butterworth band-pass filter of order 2 and 0.1–0.35 Hz cut-off frequency.
- **Blood Volume Pulse (BVP):** Butterworth band-pass filter of order 4 and 1–8 Hz cut-off frequency.

After noise removal, the data was segmented into 40 s sliding windows with 75% overlap. Lastly, the data was normalised per user, by subtracting the mean and dividing by the standard deviation, to values between 0–1 to remove subjective bias.

4.3. Supervised Learning Using Single Modality Classifiers

The ER classification is performed with a classifier tuned for Arousal and another for Valence. Table 2 presents the experimental results for the SL techniques.

As it can be seen, for the ITMDER dataset, the state-of-the-art results [7] were available for each sensor modality, which we display and, overall our methodology was able to achieve superior results. Additionally, altogether, we observe higher accuracy values in the Valence dimension compared to the Arousal scale. Thirdly, for the WESAD dataset, the F1-score drops significantly to 0.0, compared to the Accuracy score value. The F1-score low value derives from the fact, that the class labels were largely unbalanced, with some of the test sets having none of one of the labels. To conclude, overall, all the sensors modalities display competitive results with no individual sensor modality standing out as the optimal for ER.

We present the classifiers used per sensor modality and class dimension in Table 3. Additionally, the features obtained using the forward feature selection algorithm are displayed in Tables 4 and 5, for the Arousal and Valence dimensions, respectively. As shown, they explore similar correlated aspects in each modality.

Both the presented classifiers and features were selected via a 4-fold CV, to be used for the SL evaluation and for the DF algorithm, which is detailed in the next section. Hence, no classifier was generally able to emerge as the optimal for ER on the aforementioned axis. Lastly, concerning the features for each modality, we used 570, 373, 322, and 487 features respectively for the EDA, ECG, BVP, and RESP sensor data. However, such high dimension feature vector can be highly redundant and has many zero column features, therefore, we were able to reduce the feature vector without significant degradation of the classification performance.

Figure A1 in Appendix A displays two histograms merging the features used in the SL methodologies in all the datasets for the Arousal and Valence axis, respectively. The figure shows that most features are selected via the SFFS methodology, specifically for each dataset (a value of 1 means that the features were selected in just one dataset). The features EDA onsets spectrum mean value, and BVP signal mean are selected in 2 datasets for the Arousal axis; while, the features EDA onsets spectrum mean value (in 4), RESP signal mean (in 2), BVP (in 2) signal mean, and ECG NNI (NN intervals) minimum peaks value, are repeated for the Valence axis.

Table 2. Experimental results in terms of the classifier's Accuracy (1st row) and F1-score (2nd row) in %. All listed values are obtained using Leave-One-Subject-Out (LOSO). Nomenclature: SOA—State-of-the-art results; EDA H, EDA F—EDA obtained on a device placed on the hand and finger, respectively. The SOA column contains the results found in the literature [7]. The best results are shown in bold.

	ITMDER				WESAD		DEAP		MAHNOB-HCI		EESD	
	Arousal	**SOA**	**Valence**	**SOA**	**Arousal**	**Valence**	**Arousal**	**Valence**	**Arousal**	**Valence**	**Arousal**	**Valence**
EDA H	59.65 ± 13.46	0.572	89.26 ± 17.3	0.721	85.78 ± 16.55	92.86 ± 11.96	58.91 ± 15.21	56.56 ± 9.07	50.61 ± 21.84	56.43 ± 34.84	59.38 ± 16.24	**68.75 ± 18.75**
	40.74 ± 26.0		93.2 ± 12.37		0.0 ± 0.0	95.86 ± 6.99	72.91 ± 12.92	71.83 ± 7.42	47.53 ± 31.47	**64.63 ± 34.57**	56.82 ± 20.8	**66.71 ± 23.1**
EDA F	56.03 ± 11.0	0.572	90.91 ± 11.29	0.721								
	45.67 ± 20.01		91.24 ± 18.75									
ECG	**68.33 ± 5.58**	0.656	89.26 ± 17.3	0.7	85.75 ± 16.61	92.86 ± 11.96			49.36 ± 37.5	**59.15 ± 24.5**		
	58.79 ± 21.54		93.2 ± 12.37		0.0 ± 0.0	95.86 ± 6.99			53.0 ± 39.62	56.58 ± 32.61		
BVP	58.44 ± 12.69	0.660	89.35 ± 17.23	0.695	85.78 ± 16.55	**94.39 ± 9.98**	58.88 ± 15.19	56.56 ± 9.07			67.5 ± 13.35	66.25 ± 16.35
	45.91 ± 25.24		93.25 ± 12.34		0.0 ± 0.0	**96.68 ± 6.01**	72.9 ± 12.91	71.83 ± 7.42			66.98 ± 15.95	64.49 ± 22.07
RESP	62.37 ± 16.83	0.585	89.26 ± 17.3	0.629	85.78 ± 16.55	92.86 ± 11.96	58.83 ± 14.78	56.56 ± 9.07	50.62 ± 21.25	46.57 ± 20.67	**72.5 ± 12.87**	67.5 ± 10.0
	51.79 ± 23.16		93.2 ± 12.37		0.0 ± 0.0	95.86 ± 6.99	72.6 ± 12.74	71.83 ± 7.42	44.28 ± 31.66	48.27 ± 28.44	**70.12 ± 15.72**	57.92 ± 15.12

Table 3. Classifier used per dataset and sensor modality for the Arousal and Valence dimensions respectively used in the SL and DF methodologies, obtained using 4-fold CV. Nomenclature: K-Nearest Neighbour (k-NN); Decision Tree (DT); Random Forest (RF); Support Vector Machines (SVM); Gaussian Naive Bayes (GNB); and Quadratic Discriminant Analysis (QDA).

	ITMDER		WESAD		DEAP		MAHNOB-HCI		EESD	
	Arousal	**Valence**	**Arousal**	**Valence**	**Arousal**	**Valence**	**Arousal**	**Valence**	**Arousal**	**Valence**
EDA Hand	DT	RF	RF	RF	SVM	SVM	AdaBoost	SVM	AdaBoost	AdaBoost
EDA Finger	AdaBoost	QDA								
ECG	AdaBoost	RF	QDA	RF			RF	AdaBoost		
BVP	QDA	RF	AdaBoost	RF	RF	RF			AdaBoost	AdaBoost
Resp	AdaBoost	RF	RF	RF	AdaBoost	RF	QDA	AdaBoost	AdaBoost	QDA

Table 4. Features used per dataset and sensor modality for the Arousal dimension in the SL and DF methodologies, obtained using 4-fold CV.

	ITMDER	WESAD	DEAP	MAHNOB-HCI	EESD
EDA Hand	peaksOnVol_minpeaks EDRVolRatio_iqr onsets_temp_dev	EDA_onsets_spectrum_mean	onsets_spectrum_mean	half_rec_minAmp half_rec_rms amplitude_dist onsets_spectrum_statistic_hist43 rise_ts_temp_curve_distance phasic_rate_maxpeaks onsets_spectrum_meddiff EDRVolRatio_zero_cross	phasic_rate_abs_dev onsetspeaksVol_minpeaks
EDA Finger	onsets_spectrum_statistic_hist81 peaksOnVol_iqr six_rise_autocorr				
ECG	statistic_hist73, statistic_hist115 hr_sadiff statistic_hist7 statistic_hist137	mean rpeaks_medadev hr_meandiff		hr_mindiff	
BVP	hr_max hr_meandiff	mean	mean		spectral_skewness temp_curve_distance statistic_hist18 statistic_hist13 statistic_hist15 meddiff
RESP	exhale_counter inhExhRatio_iqr	statistic_hist0	mean	hr_total_energy meandiff statistic_hist95 inhale_dur_temp_curve_distance statistic_hist27 hr_meandiff	exhale_meanadiff max, zeros_mean

Table 5. Features used per dataset and sensor modality for the Valence dimension in the SL and DF methodology, obtained using 4-fold CV.

	ITMDER	WESAD	DEAP	MAHNOB-HCI	EESD
EDA Hand	onsets_spectrum_mean rise_ts_temp_curve_distance rise_ts_medadev	onsets_spectrum_mean	onsets_spectrum_mean	onsets_spectrum_mean	amplitude_mean onsets_spectrum_meanadev half_rise_medadev onsets_spectrum_statistic_hist9 EDRVolRatio_medadiff half_rec_minpeaks
EDA Finger	onset_peaks_Vol_max half_rise_mean, peaks_max onsets_spectrum_statistic_hist120 half_rec_meandiff onsets_spectrum_statistic_hist91 half_rise_var peaks_Onset_Vol_skewness				
ECG	nni_minpeaks	nni_minpeaks statistic_hist95		rpeaks_meandiff max mindiff	
BVP	statistic_hist44 meanadiff hr_meanadiff onsets_mean hr_meandiff	median minAmp	mean		mean statistic_hist16 statistic_hist5 statistic_hist31 meddiff
Resp	mean exhale_median statistic_hist196	mean	mean	hr_maxpeaks statistic_hist55 zeros_skewness statistic_hist36	iqr

4.4. Decision Fusion vs. Feature Fusion

In the current sub-section we present the experimental results for the DF and FF methodologies. Table 6 shows the experimental results in terms of Accuracy and F1-score for the Arousal and Valence dimensions in the 5 studied datasets, along with some state-of-the-art results. As it can be seen, once gain, both of our techniques outperform the results obtained for ITMDER [7], with more expression in the Valence dimension. Similarly for the DEAP dataset [8], where only for the Valence axis in terms of Accuracy we did not succeed, attaining, however, competitive results, and surpassing in terms of F1-score.

On the other hand, with the MAHNOB-HCI dataset [53], our proposal does not attain the literature results. For the EESD and the WESAD datasets, no state-of-the-art results are presented since it is yet, to the best of our knowledge, to be applied to ER. Thus, we denote as an un-explored annotation dimension which we evaluate in the present paper. Secondly, when comparing DF with FF, the former surpasses the latter for the EESD dataset in both the Arousal and Valence scale. For the remaining datasets, very competitive results are reached on both techniques. Regarding the computational time, FF is more competitive than DF, with an average execution time two orders of magnitude lower comparatively to DF (Language: Python 3.7.4; Memory: 16 GB 2133 MHz LPDDR3; Processor: 2.9 GHz Intel Core i7 quadruple core).

Table 7 presents the classifiers used per dataset and sensor modality for the Arousal and Valence dimension in the FF methodology.

The experimental results show that the selection was: 2 QDA; 1 SVM; 1 GNB; 1 DT (for the Arousal scale); and 2 RF; 1 SVM; 1 GNB; and 1 QDA (for the Valence scale). These results exhibit once again that, as for the SL techniques, no particular type of classifier was globally selected for all the datasets. Additionally, Table 8 displays the features used per dataset and sensor modality for the Arousal and Valence dimension in the FF methodology.

Results also showed that, similarly to the SL methodology, most features are specific per to a given dataset, with zero features being selected through the SFFS in common in all the datasets feature selection step.

In summary, this paper explored the datasets in new emotion dimensions and evaluation metrics yet to be reported in the literature, and attained similar or competitive results comparatively to the available state-of-the-art. The experimental results showed that between FF and DF using SL, very similar results are attained, and the best performing methodology is highly dependent on the dataset. These results are possibly due to the features being different for each dataset and sensor modality. In the SL classifier results, the best performing sensor modality is uncertain. While the DF methodology displayed the higher computation and time complexity. Therefore, considering these points, we select the FF methodology as the best modality fusion option since, with a single classifier, and pre-selected features, high results are reached with low processing time and computational complexity.

Table 6. Experimental results for the FF and DF methodologies in terms of Accuracy (A) and F1-score (F1), and time (T) in seconds, per dataset for the Arousal dimension in the FF methodology. Results obtained using LOSO. The SOA column contains the results found in the literature (ITMDER [7], DEAP [8], MAHNOB-HCI [53]). The best results are shown in bold.

	ITMDER			WESAD			DEAP				MAHNOB-HCI				EESD	
	Arousal	**SOA**	**Valence**	**SOA**	**Arousal**	**Valence**	**Arousal**	**SOA**	**Valence**	**SOA**	**Arousal**	**SOA**	**Valence**	**SOA**	**Arousal**	**Valence**
DF																
A	66.7 ± 9.0	58.1	89.3 ± 17.3	57.12	85.8 ± 16.5	92.9 ± 12.0	58.9 ± 15.2		56.6 ± 9.1		54.7 ± 13.3		58.1 ± 6.1		**75.0 ± 14.8**	**75.6 ± 17.9**
F1	**50.9 ± 23.5**		93.2 ± 12.4		0.0 ± 0.0	95.9 ± 7.0	72.9 ± 12.9		71.8 ± 7.4		63.8 ± 15.8		68.1 ± 8.9		**73.4 ± 16.4**	**72.4 ± 22.5**
T	1.5 ± 0.0		1.35 ± 0.0		2.04 ± 0.0	2.0 ± 0.0	1.58 ± 0.0		1.73 ± 0.0		1.1 ± 0.0		1.35 ± 0.0		0.6 ± 0.0	0.7 ± 0.0
FF																
A	**87.6 ± 16.7**		89.26 ± 17.3		**87.6 ± 16.7**	92.9 ± 12.0	60.0 ± 13.9	57.0	56.9 ± 8.2	62.7	55.2 ± 15.4	64.2 57 55 ± 3.9	56.0 ± 10.2	68.7 62.7 ± 3.9 57.5	60.0 ± 18.4	68.7 ± 22.2
F1	19.4 ± 34.4		93.2 ± 12.4		19.4 ± 34.4	95.9 ± 7.0	67.3 ± 23.8	53.3	70.7 ± 7.6	60.8	67.5 ± 16.6		59.0 ± 15.1		56.7 ± 22.5	67.7 ± 24.7
T	0.02 ± 0.0		0.02 ± 0.0		0.02 ± 0.0	0.07 ± 0.01	0.02 ± 0.01		0.02 ± 0.0		0.01 ± 0.0		0.01 ± 0.0		0.0 ± 0.0	0.01 ± 0.0

Table 7. Classifier used per dataset and sensor modality for the Arousal and Valence dimension in the FF methodology. Results obtained using 4-fold CV.

	ITMDER		**WESAD**		**DEAP**		**MAHNOB-HCI**		**EESD**	
	Arousal	Valence	Arousal	Valence	Arousal	Valence	Arousal	Valence	Arousal	Valence
Classifier	SVM	RF	QDA	SVM	QDA	GNB	GNB	QDA	DT	RF

Table 8. Features used per dataset and sensor modality for the Arousal and Valence dimension in the FF methodology. Results obtained using 4-fold CV.

ITMDER	WESAD	DEAP	MAHNOB-HCI	EESD
Arousal				
EDA_H_onsets_spectrum_mean	BVP_median ECG_min Resp_statistic_hist64	Resp_zeros_sadiff BVP_statistic_hist29 EDA_phasic_rate_total_energy EDA_rise_ts_mindiff Resp_statistic_hist25	Resp_inhExhRatio_maxpeaks EDA_phasic_rate_iqr Resp_inhExhRatio_zero_cross Resp_inhExhRatio_skewness ECG_rpeaks_meanadiff ECG_minpeaks Resp_meandiff EDA_onsets_spectrum_minAmp EDA_onsets_spectrum_statistic_hist22 ECG_hr_dist EDA_onsets_spectrum_statistic_hist62	Resp_exhale_max EDA_amplitude_kurtosis
Valence				
EDA_H_peaksOnVol_minAmp BVP_mean EDA_F_EDRVolRatio_total_energy EDA_H_onsets_spectrum_statistic_hist112	BVP_median ECG_dist ECG_zero_cross ECG_statistic_hist143	Resp_statistic_hist60 EDA_half_rise_dist BVP_statistic_hist10 BVP_statistic_hist39 EDA_half_rise_temp_curve_distance BVP_hr_maxAmp	ECG_meanadiff EDA_rise_ts_meandiff Resp_inhale_dur_dist EDA_onsets_spectrum_statistic_hist5	EDA_amplitude_mean BVP_statistic_hist35 Resp_rms Resp_zeros_meandiff EDA_onsets_spectrum_statistic_hist22

5. Conclusions and Future Work

Over the past decade, the field of affective computing has grown, with many datasets being created [6–9,52], however, a consolidation is lacking concerning: (1) What are the ranges of the expected classification performance; (2) The definition of the best sensor modality, SL classifier and features per modality for ER; (3) Which is the best technique to deal with multimodality and their limitations (FF or DF); (4) Selection of the classification model. Therefore, in this work, we studied the recognition of low/high emotional response in two dimensions: Arousal and Valence, for five publicly available datasets commonly found in literature. For this, we focus on physiological data sources easily measured from pervasive wearable technology, namely ECG, EDA, RESP and BVP data. Then, to deal with the multimodality, we analyse two techniques: FF and DF.

We extend the state-of-the-art with: (1) Benchmarking the ER classification performance for SL, FF and DF in a systematic way; (2) Summarising the accuracy and F1-score (important due to the imbalanced nature of the datasets); (3) Comprehensive study of SL classifiers and extended feature set for each modality; (4) Systematic analysis of multimodal classification in DF and FF approaches. We were able to obtain superior or comparable results to those found in literature for the selected datasets. Experimental results showed that FF is the most competitive technique.

For future work, we identified the following research lines: (1) Acquisition of additional data for the development of a subject-dependent model, since emotions are highly subject-dependent resulting, according to literature [11], in a higher classification performance; (2) Grouping the users by clusters of response might provide a look into sub-groups of personalities, a further parameter that must be taken into consideration when characterising emotion; (3) As stated in Section 4.3 we used a SFFS methodology to select the best feature set to use in all our tested techniques, however, it is not optimal, so the classification results using additional feature selection techniques should be tested; (4) Lastly, our work is highly conditioned on the extracted features, while lately, higher focus has been made to Deep Learning techniques, but in an approach where the feature extraction step is embedded in the neural network - ongoing work concerns the exploration and comparison of feature engineering and data representation learning approaches, with emphasis on performance and explainability aspects.

Author Contributions: Conceptualization, A.F.; Conceptualization, C.W.; Funding acquisition, C.W.; Methodology, A.F.; Project administration, A.F.; Project Administration, C.W.; Software, P.B.; Supervision, H.S.; Validation, P.B.; Writing—original draft, P.B.; Writing—review & editing, H.S. All authors have read and agreed to the published version of the manuscript.

Funding: This work has been partially funded by the Xinhua Net Future Media Convergence Institute under project S-0003-LX-18, by the Ministry of Economy and Competitiveness of the Spanish Government co-founded by the ERDF (PhysComp project) under Grant TIN2017-85409-P, and by FCT/MCTES through national funds and when applicable co funded EU funds under the project UIDB/EEA/50008/2020.

Conflicts of Interest: The authors declare no conflict of interest. The funders had no role in the design of the study; in the collection, analyses, or interpretation of data; in the writing of the manuscript, or in the decision to publish the results.

Appendix A

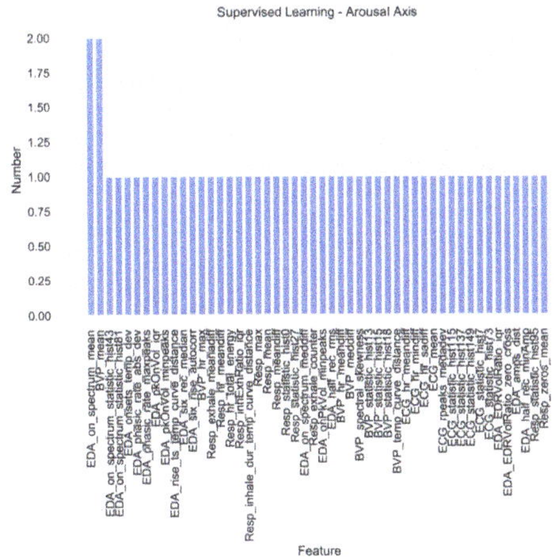

(a)

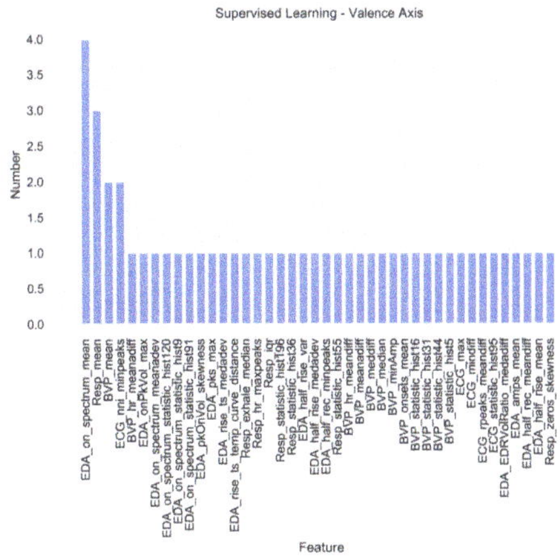

(b)

Figure A1. Histogram combining the features used in the SL (Supervised Learning) methodologies in all the datasets for the Arousal and Valence axis in (**a**,**b**), respectively. For information regarding the features the authors refer the reader to (https://github.com/PIA-Group/BioSPPy).

References

1. Greenberg, L.S.; Safran, J. Emotion, Cognition, and Action. In *Theoretical Foundations of Behavior Therapy*; Springer: Boston, MA, USA, 1987; pp. 295–311. [CrossRef]
2. Shu, L.; Xie, J.; Yang, M.; Li, Z.; Li, Z.; Liao, D.; Xu, X.; Yang, X. A Review of Emotion Recognition Using Physiological Signals. *Sensors* **2018**, *18*, 2074. [PubMed]
3. Paul, E. An argument for basic emotions. *Cogn. Emot.* **1992**, *6*, 169–200. [CrossRef]
4. Damasio, A.R. *Descartes' Error: Emotion, Reason, and the Human Brain*; G.P. Putnam: New York, NY, USA, 1994.
5. Lang, P.J. The emotion probe: Studies of motivation and attention. *Am. Psychol.* **1995**, *50*, 372–385. [CrossRef]
6. Schmidt, P.; Reiss, A.; Duerichen, R.; Marberger, C.; Van Laerhoven, K. Introducing WESAD, a Multimodal Dataset for Wearable Stress and Affect Detection. In Proceedings of the International Conference on Multimodal Interaction, Boulder, CO, USA, 16–20 October 2018; pp. 400–408. [CrossRef]
7. Pinto, J. Exploring Physiological Multimodality for Emotional Assessment. Master's Thesis, Instituto Superior Técnico, Rovisco Pais, Lisboa, Portugal, 2019.
8. Koelstra, S.; Muhl, C.; Soleymani, M.; Lee, J.; Yazdani, A.; Ebrahimi, T.; Pun, T.; Nijholt, A.; Patras, I. DEAP: A Database for Emotion Analysis using Physiological Signals. *IEEE Trans. Affect. Comput.* **2012**, *3*, 18–31. [CrossRef]
9. Picard, R.W.; Vyzas, E.; Healey, J. Toward machine emotional intelligence: Analysis of affective physiological state. *IEEE Trans. Pattern Anal. Mach. Intell.* **2001**, *23*, 1175–1191. [CrossRef]
10. Schmidt, P.; Reiss, A.; Duerichen, R.; Laerhoven, K.V. Wearable affect and stress recognition: A review. *arXiv* **2018**, arXiv:1811.08854.
11. Bota, P.J.; Wang, C.; Fred, A.L.N.; Plácido da Silva, H. A Review, Current Challenges, and Future Possibilities on Emotion Recognition Using Machine Learning and Physiological Signals. *IEEE Access* **2019**, *7*, 140990–141020. [CrossRef]
12. Liu, C.; Rani, P.; Sarkar, N. An empirical study of machine learning techniques for affect recognition in human-robot interaction. In Proceedings of the International Conference on Intelligent Robots and Systems, Edmonton, AB, Canada, 2–6 August 2005; pp. 2662–2667. [CrossRef]
13. Kim, S.M.; Valitutti, A.; Calvo, R.A. Evaluation of Unsupervised Emotion Models to Textual Affect Recognition. In Proceedings of the NAAL HLT Workshop on Computational Approaches to Analysis and Generation of Emotion in Text, Los Angeles, CA, USA, 5 June 2010; pp. 62–70.

14. Zhang, Z.; Han, J.; Deng, J.; Xu, X.; Ringeval, F.; Schuller, B. Leveraging Unlabeled Data for Emotion Recognition with Enhanced Collaborative Semi-Supervised Learning. *IEEE Access* **2018**, *6*, 22196–22209. [CrossRef]
15. Alhagry, S.; Fahmy, A.A.; El-Khoribi, R.A. Emotion Recognition based on EEG using LSTM Recurrent Neural Network. *Int. J. Adv. Comput. Sci. Appl.* **2017**, *8*. [CrossRef]
16. Zhang, J.; Chen, M.; Hu, S.; Cao, Y.; Kozma, R. PNN for EEG-based Emotion Recognition. In Proceedings of the International Conference on Systems, Man, and Cybernetics, Budapest, Hungary, 9–12 October 2016; pp. 2319–2323. [CrossRef]
17. Salari, S.; Ansarian, A.; Atrianfar, H. Robust emotion classification using neural network models. In Proceedings of the Iranian Joint Congress on Fuzzy and Intelligent Systems, Kerman, Iran, 28 February–2 March 2018; pp. 190–194. [CrossRef]
18. Vanny, M.; Park, S.M.; Ko, K.E.; Sim, K.B. Analysis of Physiological Signals for Emotion Recognition Based on Support Vector Machine. In *Robot Intelligence Technology and Applications 2012*; Kim, J.H., Matson, E.T., Myung, H., Xu, P., Eds.; Springer: Berlin/Heidelberg, Germany, 2013; pp. 115–125. [CrossRef]
19. Cheng, B. *Emotion Recognition from Physiological Signals Using Support Vector Machine*; Springer: Berlin/Heidelberg, Germany, 2012; Volume 114, pp. 49–52. [CrossRef]
20. He, C.; Yao, Y.J.; Ye, X.S. *An Emotion Recognition System Based on Physiological Signals Obtained by Wearable Sensors*; Springer: Singapore, 2017; pp. 15–25. [CrossRef]
21. Meftah, I.T.; Le Thanh, N.; Ben Amar, C. Emotion Recognition Using KNN Classification for User Modeling and Sharing of Affect States. In Proceedings of the Neural Information Processing, Doha, Qatar, 12–15 November 2012; Huang, T., Zeng, Z., Li, C., Leung, C.S., Eds.; Springer: Berlin/Heidelberg, Germany, 2012; pp. 234–242.
22. Li, M.; Xu, H.; Liu, X.; Lu, S. Emotion recognition from multichannel EEG signals using K-nearest neighbor classification. *Technol. Health Care* **2018**, *26*, 509–519. [CrossRef]
23. Kolodyazhniy, V.; Kreibig, S.D.; Gross, J.J.; Roth, W.T.; Wilhelm, F.H. An affective computing approach to physiological emotion specificity: Toward subject-independent and stimulus-independent classification of film-induced emotions. *Psychophysiology* **2011**, *48*, 908–922. [CrossRef] [PubMed]
24. Zhang, X.; Xu, C.; Xue, W.; Hu, J.; He, Y.; Gao, M. Emotion Recognition Based on Multichannel Physiological Signals with Comprehensive Nonlinear Processing. *Sensors* **2018**, *18*, 3886. [CrossRef] [PubMed]
25. Gong, P.; Ma, H.T.; Wang, Y. Emotion recognition based on the multiple physiological signals. In Proceedings of the International Conference on Real-time Computing and Robotics, Angkor Wat, Cambodia, 6–9 June 2016; pp. 140–143.
26. Ayata, D.; Yaslan, Y.; Kamasak, M.E. Emotion Recognition from Multimodal Physiological Signals for Emotion Aware Healthcare Systems. *J. Med. Biol. Eng.* **2020**, *40*, 149–157. [CrossRef]
27. Chen, J.; Hu, B.; Wang, Y.; Moore, P.; Dai, Y.; Feng, L.; Ding, Z. Subject-independent emotion recognition based on physiological signals: A three-stage decision method. *BMC Med. Informatics Decis. Mak.* **2017**, *17*, 167. [CrossRef] [PubMed]
28. Damaševičius, R.; Zhuang, N.; Zeng, Y.; Tong, L.; Zhang, C.; Zhang, H.; Yan, B. Emotion Recognition from EEG Signals Using Multidimensional Information in EMD Domain. *BioMed Res. Int.* **2017**, *2017*, 8317357. [CrossRef]
29. Lahane, P.; Sangaiah, A.K. An Approach to EEG Based Emotion Recognition and Classification Using Kernel Density Estimation. *Procedia Comput. Sci.* **2015**, *48*, 574–581. [CrossRef]
30. Qing, C.; Qiao, R.; Xu, X.; Cheng, Y. Interpretable Emotion Recognition Using EEG Signals. *IEEE Access* **2019**, *7*, 94160–94170. [CrossRef]
31. Xianhai, G. Study of Emotion Recognition Based on Electrocardiogram and RBF neural network. *Procedia Eng.* **2011**, *15*, 2408–2412. [CrossRef]
32. Xiefeng, C.; Wang, Y.; Dai, S.; Zhao, P.; Liu, Q. Heart sound signals can be used for emotion recognition. *Sci. Rep.* **2019**, *9*, 6486. [CrossRef]
33. Dissanayake, T.; Rajapaksha, Y.; Ragel, R.; Nawinne, I. An Ensemble Learning Approach for Electrocardiogram Sensor Based Human Emotion Recognition. *Sensors* **2019**, *19*, 4495. [CrossRef]
34. Shukla, J.; Barreda-Angeles, M.; Oliver, J.; Nandi, G.C.; Puig, D. Feature Extraction and Selection for Emotion Recognition from Electrodermal Activity. *IEEE Trans. Affect. Comput.* **2019**. [CrossRef]

35. Udovičić, G.; Đerek, J.; Russo, M.; Sikora, M. Wearable Emotion Recognition System Based on GSR and PPG Signals. In Proceedings of the 2nd International Workshop on Multimedia for Personal Health and Health Care, Mountain View, CA, USA, 23–27 October 2017; pp. 53–59. [CrossRef]
36. Liu, M.; Fan, D.; Zhang, X.; Gong, X. Human Emotion Recognition Based on Galvanic Skin Response Signal Feature Selection and SVM. In Proceedings of the 2016 International Conference on Smart City and Systems Engineering, Hunan, China, 25–26 November 2016; pp. 157–160. [CrossRef]
37. Wei, W.; Jia, Q.; Yongli, F.; Chen, G. Emotion Recognition Based on Weighted Fusion Strategy of Multichannel Physiological Signals. *Comput. Intell. Neurosci.* **2018**, *2018*, 1–9. [CrossRef] [PubMed]
38. Chen, J.; Hu, B.; Xu, L.; Moore, P.; Su, Y. Feature-level fusion of multimodal physiological signals for emotion recognition. In Proceedings of the International Conference on Bioinformatics and Biomedicine (BIBM), Washington, DC, USA, 9–12 November 2015; pp. 395–399. [CrossRef]
39. Canento, F.; Fred, A.; Silva, H.; Gamboa, H.; Lourenço, A. Multimodal biosignal sensor data handling for emotion recognition. In Proceedings of the 2011 IEEE Sensors Conference, Limerick, Ireland, 28–31 October 2011; pp. 647–650. [CrossRef]
40. Xie, J.; Xu, X.; Shu, L. WT Feature Based Emotion Recognition from Multi-channel Physiological Signals with Decision Fusion. In Proceedings of the Asian Conference on Affective Computing and Intelligent Interaction, Beijing, China, 20–22 May 2018; pp. 1–6.
41. Subramanian, R.; Wache, J.; Abadi, M.K.; Vieriu, R.L.; Winkler, S.; Sebe, N. ASCERTAIN: Emotion and Personality Recognition Using Commercial Sensors. *IEEE Trans. Affect. Comput.* **2018**, *9*, 147–160. [CrossRef]
42. Aguileta, A.A.; Brena, R.F.; Mayora, O.; Molino-Minero-Re, E.; Trejo, L.A. Multi-Sensor Fusion for Activity Recognition—A Survey. *Sensors* **2019**, *19*, 3808. [CrossRef] [PubMed]
43. Egger, M.; Ley, M.; Hanke, S. Emotion Recognition from Physiological Signal Analysis: A Review. *Electron. Notes Theor. Comput. Sci.* **2019**, *343*, 35–55. [CrossRef]
44. Doma, V.; Pirouz, M. A comparative analysis of machine learning methods for emotion recognition using EEG and peripheral physiological signals. *J. Big Data* **2020**, *7*, 18. [CrossRef]
45. Dzedzickis, A.; Kaklauskas, A.; Bucinskas, V. Human Emotion Recognition: Review of Sensors and Methods. *Sensors* **2020**, *20*, 592. [CrossRef]
46. Marechal, C.; Mikołajewski, D.; Tyburek, K.; Prokopowicz, P.; Bougueroua, L.; Ancourt, C.; Węgrzyn-Wolska, K. *High-Performance Modelling and Simulation for Big Data Applications: Selected Results of the COST Action IC1406 cHiPSet*; Springer International Publishing: Cham, Switzerland, 2019; pp. 307–324. [CrossRef]
47. Zhang, J.; Yin, Z.; Chen, P.; Nichele, S. Emotion recognition using multi-modal data and machine learning techniques: A tutorial and review. *Inf. Fusion* **2020**, *59*, 103–126. [CrossRef]
48. Duda, R.O.; Hart, P.E.; Stork, D.G. *Pattern Classification*, 2nd ed.; Wiley-Interscience: New York, NY, USA, 2000.
49. Pedregosa, F.; Varoquaux, G.; Gramfort, A.; Michel, V.; Thirion, B.; Grisel, O.; Blondel, M.; Prettenhofer, P.; Weiss, R.; Dubourg, V.; et al. Scikit-learn: Machine Learning in Python. *J. Mach. Learn. Res.* **2011**, *12*, 2825–2830.
50. Da Silva, H.P.; Fred, A.; Martins, R. Biosignals for Everyone. *IEEE Pervasive Comput.* **2014**, *13*, 64–71. [CrossRef]
51. Alves, A.P.; Plácido da Silva, H.; Lourenco, A.; Fred, A. BITalino: A Biosignal Acquisition System based on Arduino. In Proceedings of the International Conference on Biomedical Electronics and Devices (BIODEVICES), Barcelona, Spain, 11–14 February 2013.
52. Soleymani, M.; Lichtenauer, J.; Pun, T.; Pantic, M. A Multimodal Database for Affect Recognition and Implicit Tagging. *IEEE Trans. Affect. Comput.* **2012**, *3*, 42–55. [CrossRef]
53. Wiem, M.; Lachiri, Z. Emotion Classification in Arousal Valence Model using MAHNOB-HCI Database. *Int. J. Adv. Comput. Sci. Appl.* **2017**, *8*. [CrossRef]

sensors

MDPI

Review

Emotion Recognition in Immersive Virtual Reality: From Statistics to Affective Computing

Javier Marín-Morales *, Carmen Llinares, Jaime Guixeres and Mariano Alcañiz

Instituto de Investigación e Innovación en Bioingeniería, Universitat Politècnica de València, 46022 València, Spain; cllinare@omp.upv.es (C.L.); jaiguipr@i3b.upv.es (J.G.); malcaniz@i3b.upv.es (M.A.)

* Correspondence: jamarmo@i3b.upv.es

Received: 23 July 2020; Accepted: 8 September 2020; Published: 10 September 2020

Abstract: Emotions play a critical role in our daily lives, so the understanding and recognition of emotional responses is crucial for human research. Affective computing research has mostly used non-immersive two-dimensional (2D) images or videos to elicit emotional states. However, immersive virtual reality, which allows researchers to simulate environments in controlled laboratory conditions with high levels of sense of presence and interactivity, is becoming more popular in emotion research. Moreover, its synergy with implicit measurements and machine-learning techniques has the potential to impact transversely in many research areas, opening new opportunities for the scientific community. This paper presents a systematic review of the emotion recognition research undertaken with physiological and behavioural measures using head-mounted displays as elicitation devices. The results highlight the evolution of the field, give a clear perspective using aggregated analysis, reveal the current open issues and provide guidelines for future research.

Keywords: affective computing; emotion recognition; emotion elicitation; virtual reality; head-mounted display; machine learning

---

1. Introduction

Emotions play an essential role in rational decision-making, perception, learning and a variety of other functions that affect both human physiological and psychological status [1]. Therefore, understanding and recognising emotions are very important aspects of human behaviour research. To study human emotions, affective states need to be evoked in laboratory environments, using elicitation methods such as images, audio, videos and, recently, virtual reality (VR). VR has experienced an increase in popularity in recent years in scientific and commercial contexts [2]. Its general applications include gaming, training, education, health and marketing. This increase is based on the development of a new generation of low-cost headsets which has democratised global purchases of head-mounted displays (HMDs) [3]. Nonetheless, VR has been used in research since the 1990s [4]. The scientific interest in VR is due to the fact that it provides simulated experiences that create the sensation of being in the real world [5]. In particular, environmental simulations are representations of physical environments that allow researchers to analyse reactions to common concepts [6]. They are especially important when what they depict cannot be physically represented. VR makes it possible to study these scenarios under controlled laboratory conditions [7]. Moreover, VR allows the time- and cost-effective isolation and modification of variables, unfeasible in real space [8].

1.1. Virtual Reality Set-Ups

The set-ups that display VR simulations have been progressively integrated into studies as the relevant technologies have evolved. These consist of a combination of three objective features, formats, display devices and user interfaces.

The format describes the structure of the information displayed. The most common are two-dimensional (2D) multimedia and three-dimensional (3D) environments, and the main difference between them is their levels of interactivity [9]. 2D multimedia, including 360° panoramic images and videos, provide non-interactive visual representations. The validity of this format has been extensively explored [10]. Moreover, the latest advances in computer-generated images simulate light, texture and atmospheric conditions to such a degree of photorealism that it is possible to produce a virtual image that is indistinguishable, to the naked eye, from a photograph of a real-world scene [11]. This format allows scientists to test static computer-generated environments, with many variations, cheaply and quickly in a laboratory. On the other hand, 3D environments generate interactive representations which allow changes in the user's point of view, navigation and even interaction with objects and people [12]. Developing realistic 3D environments is more time consuming than developing 360° computer-generated photographs, and their level of realism is limited by the power of the hardware. However, the processing potency of GPUs (graphics processing units) is increasing every year, which will enhance the performance of 3D environments. Moreover, the interaction capacity of 3D environments, which facilitates the simulation of real-world tasks, is a key aspect in the application of virtual reality [2].

The display devices are the technological equipment used to visualise the formats. They are classified according to the level of immersion they provide, that is, the sensorimotor contingencies that they support. These are related to the actions that experimental subjects carry out in the perception process, for example, when they bend down and shift the position of their heads, and their gaze direction, to see underneath an object. Therefore, the sensorimotor contingencies supported by a system define a set of valid actions (e.g., turning the head, bending forward) that carry meaning in terms of perception within the virtual environment [13]. Since immersion is objective, one system is more immersive than another if it is superior in at least one characteristic while others remain equal. There are three categories of immersion system, non-immersive, semi-immersive and immersive [2]. Non-immersive systems are simpler devices which use a single screen, such as a desktop PC, to display environments [14]. Semi-immersive systems, such as the cave automatic virtual environment (CAVE), or the powerwall screen, use large projections to display environments on walls, enveloping the viewer [15,16]. These displays typically provide a stereo image of an environment, using a perspective projection linked to the position of the observer's head. Immersive devices, such as HMDs, are fully-immersive systems that isolate the user from external world stimuli [17]. These provide a complete simulated experience, including a stereoscopic view, which responds to the user's head movements. During the last two decades, VR has usually been displayed through desktop PCs or semi-immersive systems, such as CAVEs and powerwalls [18]. However, improvements in the performance and availability of the new generation of HMDs is boosting their use in research [19].

The user interfaces, which are exclusive to 3D environments which allow this level of interaction, are the functional connections between the user and the VR environment which allow him or her to interact with objects and navigate [20]. Regarding interaction with objects, manipulation tasks include: selection, that is, acquiring or identifying an object or subset of objects, positioning, that is, changing an object's 3D position, and rotation, that is, changing an object's 3D orientation. In terms of the navigation metaphors in 3D environments, virtual locomotion has been thoroughly analysed [21], and can be classified as physical or artificial. Regarding the physical, there are room-scale-based metaphors, such as real-walking, which allow the user to walk freely inside a limited physical space. These are normally used with HMDs, and position and orientation are determined by the position of the user's head. They are the most naturalistic of the metaphors, but are highly limited by the physical tracked area [22]. In addition, there are motion-based metaphors, such as walking-in-place or redirected walking. Walking-in-place is a pseudo-naturalistic metaphor where the user performs a virtual locomotion to navigate, for example, by moving his/her hands as if (s)he was walking, or by performing footstep-like movements, while remaining stationary [23]. Redirected walking is a technique where the user perceives (s)he is walking freely but, in fact, is being unknowingly manipulated by the virtual display: this allows navigation in an environment larger than the actual tracked area [24]. Regarding the artificial, controller-based metaphors allow users to control their movements directly through joysticks or similar devices, such as keyboards and trackballs [25]. In addition, teleportation-based metaphors allow the user to point where (s)he wants to go and teleport him or her there with an instantaneous "jump" [26]. Moreover, recent advancements in the latest generation HMD devices have increased the performance of navigation metaphors. Point-and-click teleport metaphors have become mainstream technologies implemented in all low-cost devices. However, other techniques have also increased in performance: walking-in-place metaphors have become more user-friendly and robust, room-scale-based metaphors now have increased coverage areas, provided by low-cost tracking methods, and controller-based locomotion now addresses virtual sickness through effective, dynamic field-of-view adjustments [27].

1.2. Sense of Presence

In addition to the objective features of the set-up, the experience of users in virtual environments can be measured by the concept of presence, understood as the subjective feeling of "being-there" [28]. A high degree of presence creates in the user the sensation of physical presence and the illusion of interacting and reacting as if (s)he was in the real world [29]. In the 2000s, the strong illusion of being in a place, in spite of the sure knowledge that one is not actually there, was characterised as "place illusion" (PI), to avoid any confusion that might be caused by the multiple meanings of the word "presence". Moreover, just as PI relates to how the world is perceived, and the correlation of movements and concomitant changes in the images that form perceptions, "plausibility illusion" (PsI) relates to what is perceived, in a correlation of external events not directly caused by the participant [13]. PsI is determined by the extent to which a system produces events that directly relate to the participant, and the overall credibility of the scenario being depicted in comparison with viewer expectations, for example, when an experimental participant is provoked into giving a quick, natural and automatic reply to a question posed by an avatar.

Although presence plays a critical role in VR experiences, there is limited understanding of what factors affect presence in virtual environments. However, there is consensus that exteroception and interoception factors affect presence. It has been shown that exteroception factors, such as higher levels of interactivity and immersion, which are directly related to the experimental set-up, provoke increased presence, especially in virtual environments not designed to induce particular emotions [30–32]. As to the interoception factors, which are defined by the content displayed, participants will perceive higher presence if they feel emotionally affected; for example, previous studies have found a strong correlation between arousal and presence [33]. Recent research has also analysed presence in specific contexts and suggested that, for example, in social environments, it is enhanced when the VR elicits genuine cognitive, emotional and behavioural responses, and when participants create their own narratives about events [34]. On the other hand, presence decreases when users experience physical problems, such as cybersickness [35].

1.3. Virtual Reality in Human Behaviour Research

VR is, thus, proposed as a powerful tool to simulate complex, real situations and environments, offering researchers unprecedented opportunities to investigate human behaviour in closely controlled designs in controlled laboratory conditions [33]. There are now many researchers in the field, who have published many studies, so a strong, interdisciplinary community exists [2].

Education and training is one field where VR has been much applied. Freina and Ott [36] showed that VR can offer great educational advantages. It can solve time-travel problems, for example, students can experience different historical periods. It can address physical inaccessibility, for example, students can explore the solar system in the first person. It can circumnavigate ethical problems, for example, students can "perform" serious surgery. Surgical training is now one of the most analysed research topics. Interventional surgery lacked satisfactory training methods before the advent of VR, except learning on real patients [37]. Bhagat, Liou and Chang [38] analysed improvements in military training. These authors suggested that cost-effective 3D VR significantly improved subjects learning motivation and outcomes and provided a positive impact on their live-firing achievement scores. In addition, besides enhancements in cost-effectivity, VR offers a safe training environment, as evidenced by the extensive research into driving and flight simulators [39,40]. Moreover, de-Juan-Ripoll et al. [41] proposed that VR is an invaluable tool for assessing risk-taking profiles and to train in related skills, due to its transferability to real-world situations.

Several researchers have also demonstrated the effectiveness of VR in therapeutic applications. It offers some distinct advantages over standard therapies, including precise control over the degree of exposure to the therapeutic scenario, the possibility of tailoring scenarios to individual patients' needs and even the capacity to provide therapies that might otherwise be impossible [42]. Taking some examples, studies using VR have analysed the improvement in the training in social skills for persons with mental and behavioural disorders, such as phobias [43], schizophrenia [44] and autism [45]. Lloréns, Noé, Colomer and Alcañiz [46] showed that VR-based telerehabilitation interventions promoted the reacquisition of locomotor skills associated with balance, in the same way as in-clinic interventions (both complemented with conventional therapy programmes). Moreover, it has been proposed as a key tool for the diagnosis of neurodevelopmental disorders [47].

In addition, VR has been applied transversally to many fields, such as architecture and marketing. In architecture, VR has been used as a framework within which to test the overall validity of proposed plans and architectural designs, generate alternatives and conceptualise learning, instruction and the design process itself [48]. In marketing, it has been applied in the analysis of consumer behaviour in laboratory-controlled conditions [49] and as a tool to develop emotionally engaging consumer experiences [50].

One of the most important topics in human behaviour research is human emotions, due to the central role that they play in many background processes, such as perception, decision-making, creativity, memory and social interaction [51]. Given the presence that VR provokes in users, it has been suggested as a powerful means of evoking emotions in laboratory environments [8]. In one of the first confirmatory studies into the efficacy of immersive VR as an affective medium, Baños et al. [30] showed that emotion has an impact on presence. Subsequently, many other similar studies showed that VR can evoke emotions, such as anxiety and relaxation [52], positive valence in obese children taking exercise [53], arousal in natural environments, such as parks [54], and different moods in social environments featuring avatars [55].

1.4. The Validity of Virtual Reality

Finally, it is crucial to point out that the usefulness of simulation in human behaviour research has been analysed through the validity concept, that is, the capacity to evoke a response from the user in a simulated environment similar to one that might be evoked by a physical environment [56]. Thus, there is a need to perform direct comparisons between virtual and real environments. Some comparisons have studied the validity of virtual environments by assessing psychological responses [57] and cognitive performance [58]. However, there have been fewer analyses of physiological and behavioural responses [59,60]. Heydarian et al. analysed user performance in office-related activities, for example, reading texts and identifying objects, and found that the participants performed similarly in an immersive virtual environment setting and in a benchmarked physical environment for all of the measured tasks [61]. Chamilothori, Wienold, and Andersen compared subjective perceptions of daylit spaces, and identified no significant differences between the real and virtual environments studied [62]. Kimura et al. analysed orienteering-task performance, where participants in a VR room showed less facility, suggesting that caution must be applied when interpreting the nuances of spatial cue use in virtual environments [63]. Higuera-Trujillo, López-Tarruella, and Llinares analysed psycho-physiological responses, through electrodermal activity (EDA), evoked by real-world and VR scenarios with different immersion levels, and demonstrated correlations in the physiological dynamics between real-world and 3D environments [64]. Marín-Morales et al. analysed the emotional responses evoked in subjects in a real and a virtual museum, and found no self-assessment differences, but did find differences in brain dynamics [65]. Therefore, further research is needed to understand the validity of VR in terms of physiological responses and behavioural performance.

1.5. Implicit Measures and the Neuroscience Approach

Traditionally, most theories of human behaviour research have been based on a model of the human mind that assumes that humans can think about and accurately verbalise their attitudes, emotions and behaviours [66]. Therefore, classical psychological evaluations used self-assessment questionnaires and interviews to quantify subjects' responses. However, these explicit measures have been demonstrated to be subjective, as stereotype-based expectations can lead to systematically biased behaviour, given that most individuals are motivated to be, or appear to be, nonbiased [67]. The terms used in questionnaires can also be differentially interpreted by respondents, and the outcomes depend on the subjects possessing a wide knowledge of their dispositions, which is not always the case [68].

Recent advances in neuroscience show that most of the brain processes that regulate our emotions, attitudes and behaviours are not conscious. In contrast to explicit processes, humans cannot verbalise these implicit processes [69]. In recent years, growing interest has developed in "looking" inside the brain to seek solutions to problems that have not traditionally been addressed by neuroscience. Thus, neuroscience offers techniques that can recognise implicit measurements not controlled by conscious processes [70]. These developments have provoked the emergence in the last decades of a new field called neuroeconomics, which blends psychology, neuroscience and economics into models of decision-making, rewards, risks and uncertainties [71]. Neuroeconomics addresses human behaviour research, in particular the brain mechanisms involved in economic decision-making, from the point of view of cognitive neuroscience, using implicit measures.

Several implicit measuring techniques have been proposed in recent years. Some examples of their applications in human behaviour research are: heart rate variability (HRV) has been correlated with arousal changes in vehicle drivers when detecting critical points on a route [72], electrodermal activity (EDA) has been used to measure stress caused by cognitive load in the workplace [73], electroencephalogram (EEG) has been used to assess engagement in audio-visual content [74], functional magnetic resonance imaging (fMRI) has been used to record the brain activity of participants engaged in social vs. mechanical/analytic tasks [75], functional near-infrared spectroscopy (fNIRS) has been used as a direct measure of brain activity related to decision-making processes in approach-avoidance theories [76], eye-tracking (ET) has been used to measure subconscious brain processes that show correlations with information processing in risky decisions [77], facial expression analysis (FEA) has been applied to detect emotional responses in e-learning environments [78] and speech emotion recognition (SER) has been used to detect depressive disorders [79]. Table 1 gives an overview of the implicit measuring techniques that have been used in human behaviour research.

Table 1. Overview of the main implicit techniques used in human behaviour research.

Implicit Technique	Biometric Signal Measured	Sensor	Features	Psychological or Behavioural Construct Inferred
EDA (electro dermal activity)	Changes in skin conductance	Electrodes attached to fingers, palms or soles	Skin conductance response, tonic activity and phasic activity	Attention and arousal [80]
HRV (heart rate variability)	Variability in heart contraction intervals	Electrodes attached to chest or limbs or optical sensor attached to finger, toe or earlobe	Time domain, frequency domain, non-linear domain	Stress, anxiety, arousal and valence [81,82]
EEG (electroencephalogram)	Changes in electrical activity of the brain	Electrodes placed on scalp	Frequency band power, functional connectivity, event-related potentials	Attention, mental workload, drowsiness, fatigue, arousal and valence [83,84]
fMRI (functional magnetic resonance imaging)	Concentrations of oxygenated vs. deoxygenated haemoglobin in the blood vessels of the brain	Magnetic resonance signal	blood-oxygen-level dependent	Motor execution, attention, memory, pain, anxiety, hunger, fear, arousal and valence [85]
fNIRS (functional near-infrared spectroscopy)	Concentrations of oxygenated vs. deoxygenated haemoglobin in the blood	Near-infrared light placed on scalp	blood-oxygen-level dependent	Motor execution, cognitive task (mental arithmetic), decision-making and valence [86]
ET (eye-tracking)	Corneal reflection and pupil dilation	Infrared cameras point towards eyes	Eye movements (gaze, fixation, saccades), blinks, pupil dilation	Visual attention, engagement, drowsiness and fatigue [87]
FEA (facial expression analysis)	Activity of facial muscles	Camera points towards face	Position and orientation of head. Activation of action units	Basic emotions, engagement, arousal and valence [88]
SER (speech emotion recognition)	Voice	Microphone	Prosodic and spectral features	Stress, basic emotions, arousal and valence [89]

In addition, recent studies have highlighted the potential of virtual reality environments for enhancing ecological validity in the clinical, affective and social neurosciences. These studies have usually involved the use of simple, static stimuli which lack many of the potentially important aspects of real-world activities and interactions [90]. Therefore, VR could play an important role in the future of neuroeconomics by providing a more ecological framework within which to develop experimental studies with implicit measures.

1.6. *Affective Computing and Emotion Recognition Systems*

Affective computing, which analyses human responses using implicit measures, has developed into an important field of study in the last decades. Introduced by Rosalind Picard in 1997, it proposed the automatic quantification and recognition of human emotions as an interdisciplinary field based on psychophysiology, computer science, biomedical engineering and artificial intelligence [1]. The automatic recognition of human emotion statements using implicit measures can be transversally applied to all human behaviour topics and complement classic explicit measures. In particular, it can be applied to neuroeconomic research as they share the same neuroscientific approach of using implicit measures, and due to the important relationship that has been found between emotions and decision-making [71]. Emotion recognition models can be divided into three approaches: emotional modelling, emotion classification and emotion elicitation.

The emotional modelling approach can be divided into the discrete and the dimensional. Discrete models characterise the emotion system as a set of basic emotions, which includes anger, disgust, fear, joy, sadness and surprise, and the complex emotions that result from combining them [91]. On the other hand, dimensional models propose that emotional responses can be modelled in a multidimensional space where each dimension represents a fundamental property common to all emotions. The most commonly used theory is the circumplex model of affect (CMA), which proposes a three-dimensional space consisting of: valence, that is, the degree to which an emotion is perceived as positive or negative, arousal, that is, the intensity of the emotion in terms of activation, from low to high, and dominance, which ranges from feelings of total lack of control or influence on events and surroundings to the opposite extreme of feeling influential and in control [92].

Affective computing uses biometric signals and machine-learning algorithms to classify emotions automatically. Many signals have been used, such as voice, face, neuroimaging and physiological [93]. It is noteworthy that one of the main emotion classification topics uses variables associated with central nervous system (CNS) and autonomic nervous system (ANS) dynamics [93]. First, human emotional processing and perception involve cerebral cortex activity, which allows the automatic classification of emotions using the CNS. EEG is one of the techniques most used in this context [94]. Second, many emotion recognition studies have used the ANS to analyse the changes in cardiovascular dynamics provoked by mood changes, where HRV and EDA are the most used techniques [95]. The combination of physiological features and machine-learning algorithms, such as in support vector machines, linear discriminant analysis, K-nearest neighbour and neural networks, has achieved high levels of accuracy in inferring subjects' emotional states [96].

Finally, emotion elicitation is the ability to reliably and ethically elicit affective states. This elicitation is a critical factor in the development of systems that can detect, interpret and adapt to human affect [97]. The many methods that elicit emotions in laboratories can be mainly divided into two groups, active and passive. Active methods involve directly influencing subjects, including behavioural manipulation [98], social interaction [99] and dyadic interaction [100]. Passive methods usually present external stimuli, such as images, sound or video. As to the use of images, the International Affective Picture System (IAPS) is among the databases most used as an elicitation tool in emotion recognition methodologies [95]. This includes over a thousand depictions of people, objects and events, standardised on the basis of valence and arousal [97]. As to audio, the International Affective Digitalised Sound System (IADS) database is the most commonly applied in studies which use sound to elicit emotions [101]. However, some studies directly use music or narrative to elicit emotions [102]. With respect to audio-visual stimuli, many studies have used film to induce arousal and valence [103]. These emotion elicitation methods have two important limitations. The set-ups used, mostly screens, are non-immersive devices, which provoke only a low level of presence in subjects [30]. Therefore, the stimuli do not evoke in the subjects a feeling of "being there", which is needed to analyse emotions in simulated real-world situations. In addition, the stimuli are non-interactive, so they do not allow the subjects to intervene in the scene, which would open the possibility to recognise emotional states during interactive tasks. These limitations can be overcome by using immersive VR as a new emotion elicitation method. Since the year 2000, VR has increasingly been used as affective stimulation, however the majority of the studies undertaken have applied classic statistical methods, such as hypotheses testing and correlation, to analyse subjects' physiological responses to different emotions [104]. However, in recent years, some research has started to apply affective computing paradigms with VR as the emotion elicitation method, combining implicit measures with machine-learning methods to develop automatic emotion recognition models [105].

This paper provides a systematic review of the literature on the use of head-mounted displays in implicit measure-based emotion recognition research, and examines the evolution of the research field, the emotions analysed, the implicit techniques, the data analysis, the set-ups and the validations performed.

2. Materials and Methods

Data Collection

We followed an adapted version of the Preferred Reporting Items for Systematic Reviews and Meta-Analyses (PRISMA) study selection guidelines [106]. This includes steps to identify literature, to screen the identified literature, to check the eligibility of the screened literature and, finally, to synthesise the literature. The screening and eligibility steps were performed simultaneously. The literature search was carried out on 25 March 2020. The Scopus database was queried using the following search string: TITLE-ABS-KEY ("virtual reality" OR "head-mounted display") AND TITLE-ABS-KEY ("emotion*" OR "affective*") AND DOCTYPE (ar OR re). The keywords virtual reality OR head-mounted display include all the studies on VR and, in particular, all that used HMDs. In addition, the keywords emotion* OR affective* include all the papers related to emotion. The combination of both requirements revealed the research that included virtual reality and emotions. The search was limited to articles in journals and reviews (for snowballing). A total of 1424 records were identified. Some 14 additional records were identified from other sources.

The screening and eligibility checks were undertaken as follows: (1) first, by investigating titles and abstracts, 13 duplicates were identified. (2) The manuscripts were superficially screened for a thematic match with virtual reality as emotion elicitation. A total of 1157 records were excluded for not matching with the topic, and 3 records because they were inaccessible. (3) We investigated 265 records to exclude those that did not fit, using a specific rejection order: that is, if they used HMDs, we moved on to the next filter criterion, implicit measures, if they used implicit measures, we moved on to the last criterion, the analysis of an emotion. Some 132 records were rejected for not using HMDs, 68 for not using implicit measures and 23 for not analysing an emotional dimension. Finally, 42 studies were included in the analysis which used virtual reality displayed in an HMD, in combination with any implicit measure to analyse or recognise emotional states. The summary of the procedure is depicted in Figure 1.

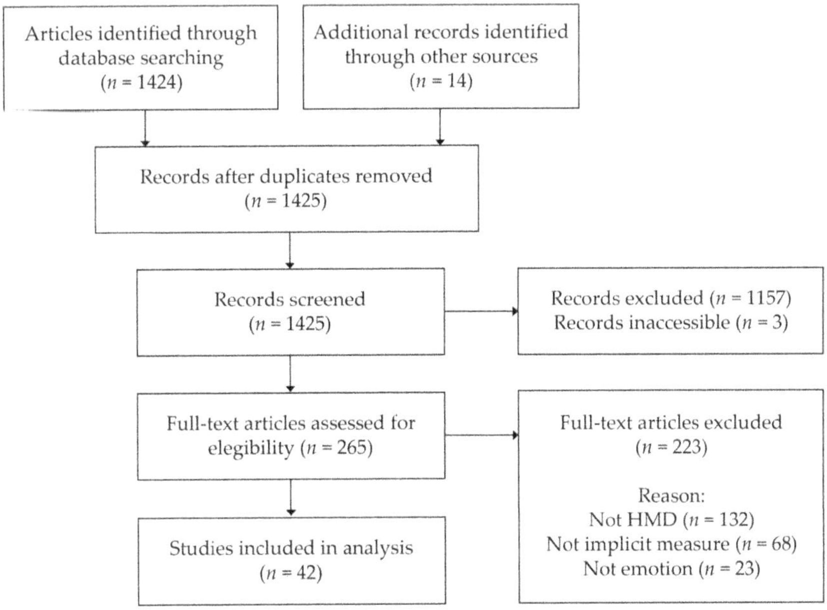

Figure 1. Scheme of the PRISMA procedure followed in the review.

3. Results

3.1. Summary of Previous Research

In recent years, studies have applied implicit measures to analyse emotions using immersive VR with HMDs. Table 2 provides a summary of the studies included in the analysis.

Table 2. Summary of previous research.

No	Author	Emotion	Signals	Features	Data Analysis	Subjects	HMD	VR Stimuli	Stimuli Comparison	Dataset Availability
1	Jang et al. (2002) [104]	Arousal	HRV, EDA	HR, HRV frequency domain, SCL, ST	*t*-test	11	VFX3D	3D flying and driving simulator	No	No
2	Meehan et al. (2005) [107]	Arousal	HRV, EDA	HR, SC, ST	*t*-test	67	Not reported	3D training room vs. pit room	No	No
3	Wilhelm et al. (2005) [108]	Anxiety	HRV, EDA	HR, SC	ANOVA, correlations	86	Not reported	3D height exposure	Partially (with a different real dataset)	No
4	Gorini et al. (2010) [109]	Anxiety	HRV, EDA	HR, SC	ANOVA	30 (20 with food disorders)	Not reported	3D photo and real food catering	VR vs. photo vs. real	No
5	Philipp et al. (2012) [110]	Valence	EMG	EMG	ANOVA	49	Virtual Research V8	3D room with IAPS pictures projected	No	No
6	Parsons et al. (2013) [111]	Arousal	HRV, EDA	HR, SC	ANOVA	50	eMagin Z800	3D high-mobility wheeled vehicle with Stroop task	No	No
7	Pallavicini et al. (2013) [112]	Stress	HRV, EMG, RSP	HR, SC, RR	ANOVA	39	Vuzix VR Bundle	3D classroom	No	No
8	Peperkorn et al. (2014) [43]	Fear	HRV, EDA	HR, SC	ANOVA	96 (48 spider-phobic)	eMagin Z800	3D virtual lab with time-varying threat (spiders and snakes)	No	No
9	Felnhofer et al. (2014) [113]	Anxiety	HRV	HR	ANOVA	75 (30 high anxiety)	eMagin Z800	3D lecture hall	No	No
10	Hartanto et al. (2014) [114]	Stress	HRV	HR	MANOVA	24 healthy subjects	eMagin Z800	3D stressful social environment	No	No
11	McCall et al. (2015) [115]	Arousal	HRV, EDA	HR, SC	Cross-correlations	306	NVIS nVisor SX60	3D room with time-varying threat (explosions, spiders, gunshots, etc.)	No	No
12	Felnhofer et al. (2015) [54]	Arousal	EDA	SCL	ANOVA	120	Sony HMZ-T1 3D	3D park with 5 variations (joy, sadness, boredom, anger and anxiety)	No	No
13	Notzon et al. (2015) [116]	Anxiety	HRV, EDA	HR, SC	ANOVA	83 (42 spider-phobic)	eMagin Z800	3D virtual lab with spiders	No	No
14	Hildebrandt et al. (2016) [117]	Arousal	HRV, EDA	RMSSD, SC	Regression	300	NVIS nVisor SX60	3D room with time-varying threats (explosions, spiders, gunshots, etc.)	No	No

Table 2. *Cont.*

No	Author	Emotion	Signals	Features	Data Analysis	Subjects	HMD	VR Stimuli	Stimuli Comparison	Dataset Availability
15	Higuera-Trujillo et al. (2016) [118]	Stress	EDA	SCR	Kruskall–Wallis Test and correlations	12	Oculus Rift DK2	3D rooms (neutral, stress and calm)	No	No
16	Bian et al. (2016) [119]	Arousal	HRV, EMG, RSP	HR, LF, HF, LF/HF, RR, RS	Regression	36	Oculus Rift DK2	3D Flight simulator	No	No
17	Shiban et al. (2016) [120]	Stress	HRV, EDA	HR, SC	ANOVA	45	NVIS nVisor SX60	3D Trier Social Stress Test	No	No
18	Chirico et al. (2017) [121]	Awe	HRV, EDA, EMG	HF, VLF, SC	ANOVA	42	Samsung Gear VR	360° neutral and awe videos	Immersive vs. non-immersive	No
19	Zou et al. (2017) [122]	Arousal	HRV, EDA	HRV time domain (AVNN, SDNN...) and frequency domain (LF, HF...), SC, SCL, SCR	*t*-test	40	Oculus Rift DK2	3D fire evacuation	No	No
20	Breuninger et al. (2017) [123]	Arousal	HRV, EDA	HR, HF, SC	*t*-test	51 (23 agoraphobics)	TriVisio VR Vision	3D car accident	No	No
21	van't Wout et al. (2017) [124]	Stress	EDA	SCR	MANOVA	44 veterans (19 with PTSD)	eMagin Z800	3D combat-related and classroom-related	No	No
22	Banaei et al. (2017) [125]	Arousal, Valence	EEG	PSD, ERSPs	MANOVA	17	Samsung Gear VR	3D rooms	No	No
23	Anderson et al. (2017) [126]	Stress	HRV, EDA	LF, HF, LF/HF, SC	MANOVA	18	Oculus Rift DK2	360° indoor vs. natural panoramas	No	No
24	Chittaro et al. (2017) [127]	Arousal	HRV	HR, LF, HF, LF/HF	ANOVA	108	Sony HMZ-T1 3D	3D cemetery and park	No	No
25	Higuera-Trujillo et al. (2017) [64]	Pleasantness	HRV, EDA	HF, SCR	Mann–Whitney U tests and correlations	100	Samsung Gear VR	3D, 360° and real retail store	real vs. 3D VR vs. 360° VR	No
26	Biedermann et al. (2017) [128]	Anxiety	HRV, EDA, RSP	HR, SC, RR	ANOVA	100	HTC Vive	Mixed reality (3D VR with real-world elements)	No	Yes
27	Tsai et al. (2018) [129]	Anxiety	HRV	HRV time domain (HR, RMSSD...) and frequency domain (HF, LF...)	ANOVA	30	eMagin Z800	3D VR claustrophobic environments	Augmented reality vs. VR	Upon request

Table 2. *Cont.*

No	Author	Emotion	Signals	Features	Data Analysis	Subjects	HMD	VR Stimuli	Stimuli Comparison	Dataset Availability
28	Marín-Morales et al. (2018) [105]	Arousal, Valence	EEG, HRV	PSD and functional connectivity, HRV Time (HR, RMSSD...), frequency (HF, LF...) and non-linear (SD1, SD2, Entropy...) domain	SVM	60	Samsung Gear VR	360° virtual rooms	No	Upon request
29	Kisker et al. (2019) [130]	Arousal	HRV	HR	*t*-test, correlations and regressions	30	HTC Vive	3D exposure to a high height	No	No
30	Gromer et al. (2019) [131]	Fear	HRV, EDA	HR, SC	ANOVA	49 (height-fearful)	HTC Vive	3D forest	No	Yes
31	Zimmer et al. (2019) [132]	Stress	HRV, salivary	HR, salivary cortisol responses, salivary alpha amylase	ANOVA	50	Oculus Rift DK2	3D Trier Social Stress Test	Replication of a real study	No
32	Lin et al. (2019) [133]	Stress	EDA, Navigation	SC, travel distance, travel time	Mann–Whitney U	60	HTC Vive	3D, building on fire	No	No
33	Schweizer et al. (2019) [134]	Stress	HRV, EDA	HR, SC	*t*-test and correlations	80	TriVisio VR Vision	3D neutral and trauma-related scene	No	No
34	Kim et al. (2019) [135]	Calm, sadness and joy	Gait Patterns	Step count, gait speed, foot planar pressure	ANOVA	12	HTC Vive	360° emotion-related videos	No	No
35	Uhm et at. (2019) [136]	Arousal	EEG	PSD	MANOVA	28	Samsung Gear VR	360° sport videos	No	No
36	Takac et al. (2019) [137]	Anxiety	HRV	HR	ANOVA	19	Oculus Rift	3D rooms with public audience	No	No
37	Marín-Morales et al. (2019) [65]	Arousal, Valence	HRV, EEG	PSD and functional connectivity, HRV Time (HR, RMSSD...), frequency (HF, LF...) and non-linear (SD1, SD2, Entropy...) domain	SVM	60	HTC Vive	3D art museum	Real museum vs. 3D museum	Upon request
38	Stolz et al. (2019) [138]	Fear	EEG	ERPs	ANOVA	29	Oculus Rift	3D room with angry avatars	No	No

Table 2. *Cont.*

No	Author	Emotion	Signals	Features	Data Analysis	Subjects	HMD	VR Stimuli	Stimuli Comparison	Dataset Availability
39	Granato et al. (2020) [139]	Arousal, Valence	HRV, EDA, EMG, RSP	HR, SC, SCL, SCR, EMG, RR	SVM, RF, Gradient Boosting, Gaussian Process Regression	33	Oculus Rift DK2	3D video games	No	Yes
40	Bălan et al. (2020) [140]	Fear	HRV, EDA, EEG	HR, SC, PSD	kNN, SVM, RF, LDA, NN	8	HTC Vive	3D acrophobia game	No	No
41	Reichenberger et al. (2020) [141]	Fear	Eye-tracking	Fixation counts, TTFF	ANOVA, *t*-test	53 (26 socially anxious)	HTC Vive	3D room with angry avatars	No	Upon request
42	Huang et al. (2020) [142]	Stress	EDA	SCL	MANOVA	89	Oculus Rift DK2	360° built vs. natural environments	No	Yes

Signals: electroencephalograph (EEG), heart rate variability (HRV), electrodermalactivity (EDA), respiration (RSP) and electromyography (EMG). Features: heart rate (HR), high frequency (HF), low frequency (LF), LF/HF (low/high frequency ratio), very low frequency (VLF), total skin conductance (SC), skin conductance tonic level (SCL), fast varying phasic activity (SCR), skin temperature (ST), respiratory rate (RR), respiratory depth (RS), power spectral density (PSD), event-related spectral perturbations (ERSPs), event-related potencials (ERPs) and time to first fixation (TTFF). Data analysis: support vector machines (SVM), k-nearest neighbors algorithm (kNN), random forest (RF), linear discriminant analysis (LDA) and neural networks (NN).

3.2. Evolution of the Research

Figure 2 shows the number of papers published each year which included the topics virtual reality and emotion analysis. This number of studies was calculated based on all the papers screened. In the 1990s, the average number of papers published annually was 6.4, the first being published in 1995. In the 2000s, the average number of papers published increased to 26.3. However, from 2010 to 2014, the average multiplied by three to 77.4. In the last five years, the curve has grown exponentially to 203 in 2019, and a predicted 278 in 2020.

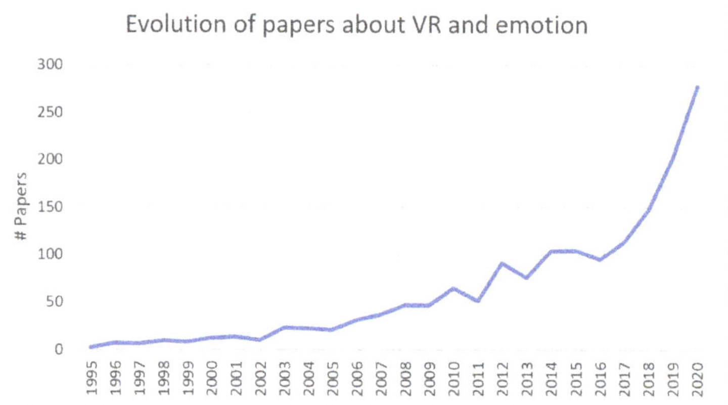

Figure 2. Evolution of the number of papers published each year on the topic of virtual reality and emotions. The total number of papers to be published in 2020 has been extrapolated using data up to 25 March 2020.

3.3. Emotions Analysed

Figure 3 depicts the evolution in the number of papers analysed in the review based on the emotion under analysis. Until 2015, the majority of the papers analysed arousal-related emotions, mostly arousal, anxiety and stress. From that year, some experiments started to analyse valence-related emotions, such as valence, joy, pleasantness and sadness, but the analysis of arousal-related emotions still predominated. Some 50% of the studies used CMA (arousal 38.1% [54] and valence 11.9% [125]), and the other 50% used basic or complex emotions (stress 23.8% [112], anxiety 16.7% [109], fear 11.9% [43], awe 2.4% [121], calmness 2.4% [135], joy 2.4% [135], pleasantness 2.4% [64] and sadness 2.4% [135]).

3.4. Implicit Technique, Features used and Participants

Figure 4 shows the evolution of the number of papers analysed in terms of the implicit measures used. The majority used HRV (73.8%) and EDA (59.5%). Therefore, the majority of the studies used ANS to analyse emotions. However, most of the studies that used HRV used very few features from the time domain, such as HR [115,120]. Very few studies used features from the frequency domain, such as HF, LF or HF/LF [119,126] and 2 used non-linear features, such as entropy and Poincare [65,105]. Of the studies that used EDA, the majority used total skin conductance (SC) [116], but some used tonic (SCL) [54] or phasic activity (SCR) [124]. In recent years, EEG use has increased, with 6 papers being published (14.3%), and the CNS has started to be used, in combination with HMDs, to recognise emotions. The analyses that have been used are ERP [138], power spectral density [140] and functional connectivity [65]. EMG (11.9%) and RSP (9.5) were also used, mostly in combination with HRV. Other implicit measures used were eye-tracking, gait patterns, navigation and salivary cortisol responses.

The average number of participants used in the various studies depended on the signal, that is, 75.34 ($\sigma = 73.57$) for EDA, 68.58 ($\sigma = 68.35$) for HRV and 33.67 ($\sigma = 21.80$) for EEG.

3.5. Data Analysis

Figure 5 shows the evolution of the number of papers published in terms of the data analysis performed. The vast majority analysed the implicit responses of the subjects in different emotional states using hypothesis testing (83.33%), correlations (14.29) or linear regression (4.76%). However, in recent years, we have seen the introduction of applied supervised machine-learning algorithms (11.90%), such as SVM [105], Random Forest [139] and kNN [140] to perform automatic emotion recognition models. They have been used in combination with EEG [65], HRV [105] and EDA [140].

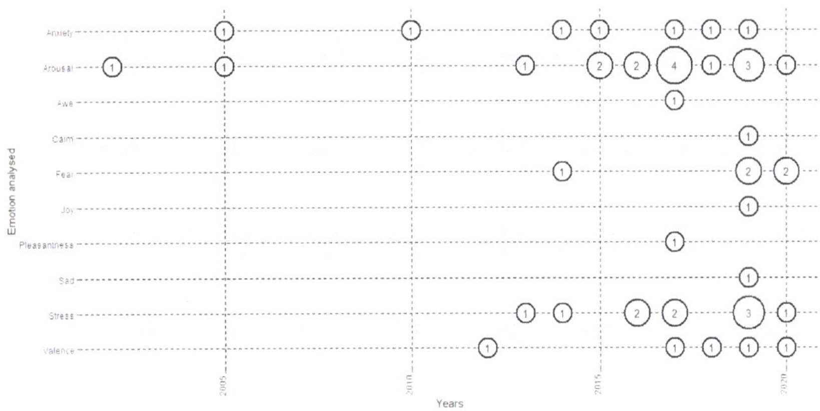

Figure 3. Evolution of the number of papers published each year based on emotion analysed.

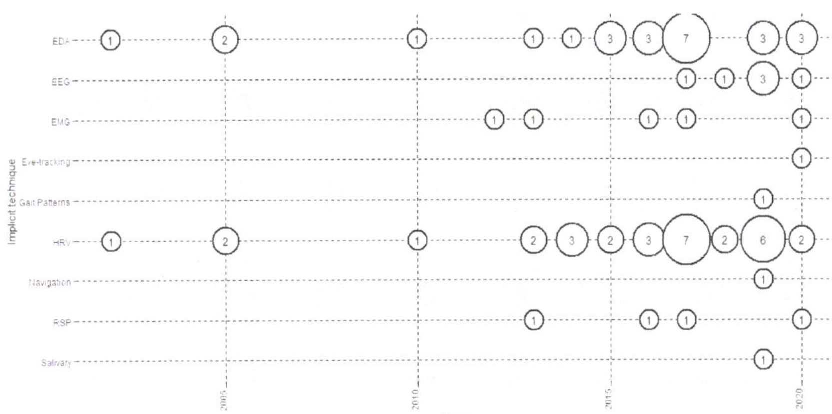

Figure 4. Evolution of the number of papers published each year based on the implicit measure used.

3.6. VR Set-Ups Used: HMDs and Formats

Figure 6 shows the evolution of the number of papers published based on HMD used. In the first years of the 2010s, eMagin was the most used. In more recent years, advances in HMD technologies have positioned HTC Vive as the most used (19.05%). In terms of formats, 3D environments are the most used [138] (85.71%), with 360° panoramas following far behind [142] (16.67%). One research used both formats [64].

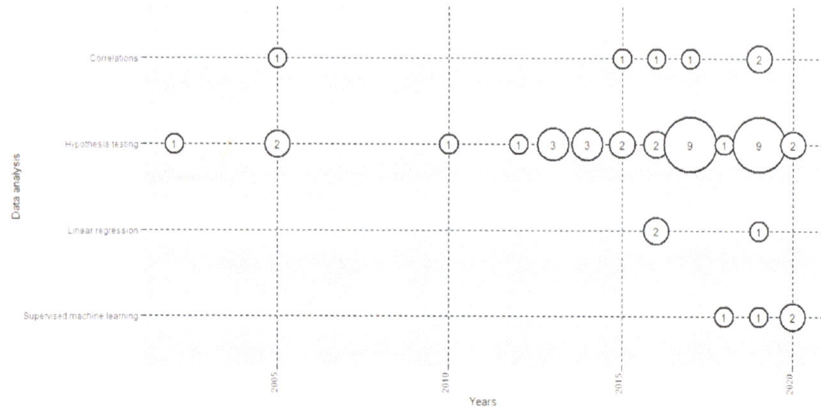

Figure 5. Evolution of the number of papers published each year by data analysis method used.

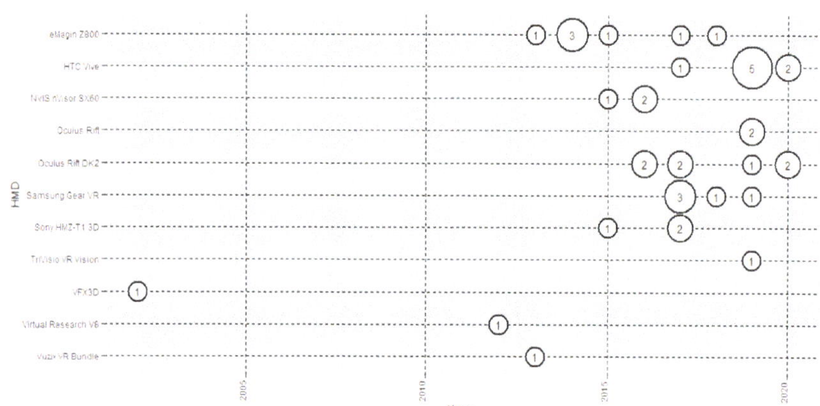

Figure 6. Evolution of the number of papers published each year based on head-mounted display (HMD) used.

3.7. Validation of VR

Table 3 shows the percentage of the papers that presented analyses of the validation of VR in an emotional research. Some 83.33% of the papers did not present any type of validation. Three papers included direct comparisons of results between VR environments and the physical world [64,65,109], and 3 compared, in terms of the formats used, the emotional reactions evoked in 3D VRs, photos [109], 360° panoramas [64] and augmented reality [129]. Finally, another compared the influence of immersion [121], the similarity of VR results with previous datasets [108] and one compared its results with a previous version of the study performed in the real world [132].

Table 3. Previous research that included analyses of the validation of virtual reality (VR).

Type of Validation	% of Papers	Number of Papers
No validation	83.33%	35
Real	7.14%	3
Format	7.14%	3
Immersivity	2.38%	1
Previous datasets	2.38%	1
Replication	2.38%	1

4. Discussion

This work highlights the evolution of the use of immersive VR, in particular using head-mounted displays, in emotion recognition research in combination with implicit measures. It provides a clear perspective based on a systematic review and aggregated analysis, focusing on the role that VR might play as an emotion elicitation tool in the coming years.

The evolution of scientific interest in VR and emotions has grown exponentially, to more than 200 papers per year (Figure 2). In particular, the performance improvements in the last few years in the latest generation of HMDs, in terms of resolution, field of view, immersion levels and the fall in their price, has boosted their use in emotion-related research. This accords with VR's increased application in recent years in other areas, such as rehabilitation, neurosurgery and therapy [2]. Therefore, the results suggest that the 2010s was the decade of the rapid growth of VR in emotion research using implicit measures, and the 2020s might be the decade when the field matures. Environmental simulations might, in the future, normally go beyond the paradigm of non-immersive/video-based 2D images to immersive VR scenarios, where subjects feel a very strong sense of presence and can interact with the stimuli presented.

In regard to HMDs and implicit measures in emotion analysis, there is no consensus about the use of CMA [92] or the Ekman theory of basic emotions [91], since both approaches are used in 50% of the research (Figure 3). The differences in the frameworks used causes some difficulties in comparing the results of different studies. The majority of the studies (90.5%) included analyses of arousal [54], or high-arousal-related discrete emotions, such as stress [112], anxiety [109] and fear [43]. On the other hand, only 23.9% of the studies analysed valence, or discrete emotions closely related to valence, such as awe [121], calm [135], joy [135], pleasantness [64] and sadness [135]. Therefore, although the whole sub-field of affective computing using HMDs is still in its first growth phase, valence recognition and its physiological dynamics, in particular, are under-researched. Recent research since 2017 has started to address this [65,139]. Dominance, a dimension of the CMA still not addressed in general affective computing research using pictures or videos [143], has also not been analysed in HMD set-up research. However, fear, a basic emotion closely related to the dominance dimension, was analysed in 11.9% of the studies examined in the review. In contrast to the fear that is felt when someone watches a horror film, which is based on the empathy of the viewer with the protagonist, the level of presence that immersive VR offers allows the analysis of fear directly felt by subjects based on scenarios they are viewing. Therefore, VR can boost the analysis of the dominance dimension in affective computing in the future. In addition, VR allows researchers to analyse emotional reactions to social stimuli, such as avatars [138], which might be the next stage in the application of classic 2D affective computing paradigms to simulated real-world situations, which can provide new insights with a social dimension.

In terms of the implicit techniques used to recognise emotions evoked through HMDs, ANS measurements are most used: specifically, HRV (73.8%) and EDA (59.5%), many times used in combination. However, until 2016, the majority of the papers featured only HR and SC (Table 2), sometimes in combination with EMG and RSP. From 2016, the research started to include HRV frequency domain and non-linear domain analyses [105,119], and EDA analyses, such as CDA, dividing the signals into tonic and phasic components [64]. In terms of the CNS, EEG research has been undertaken since 2016, including ERP [138], power spectral density [140] and functional connectivity analysis [65]. Other non-physiological implicit measures have been used since 2019, such as eye-tracking [141], gait patterns [135], navigation [133] and salivary cortisol responses [132]. The use of behavioural measures, such as eye-tracking, gait patterns and navigation, might be a very powerful approach where VR can contribute to affective computing research, as they provide high levels of interactivity with the simulated stimuli. This might open a new sub-field where emotional states can be assessed through behavioural measures in interactive, real situations.

However, the current weakest point of HMD-based emotion recognition systems is that only 11.90% of the studies, that is, four, used machine-learning algorithms to classify the emotions analysed. Since the early 2000s, when physiological signals, in combination with HMDs, were first applied to analyse emotions, until 2018, all studies used hypothesis testing and/or correlations to provide insights into the ANS oscillations produced during different affective states, except Reference [125], which used EEG. Although the classic statistical techniques obtained important and useful insights, they have some limitations: (i) hypothesis testing analyses differences between two populations based on means and deviations, but does not provide emotion recognition, (ii) it is difficult to analyse the effect of the combination of several features in datasets with large sets of variables and (iii) they do not take into account non-linear relationships. These limitations are being overcome with the use of machine-learning algorithms, as they can recognise emotions through the development of algorithms in classification problems, automatic feature selection procedures to recognise complex patterns inside data and offer non-linear kernels [143]. Marín-Morales et al. [105] presented the first emotion recognition system using SVM in combination with a large set of HRV features (time, frequency and non-linear domains) and EEG (PSD and mean phase coherence) in 360° emotional rooms, achieving a recognition rate of 75% in arousal and 71.21% in valence. Marín-Morales et al. [65] developed an emotion recognition system in a realistic 3D virtual museum, using SVM in combination with HRV and EEG, with rates of 75% and 71.08% of recognition in arousal and valence, respectively. Granato et al. [139] presented an arousal-valence emotion recognition model with subjects playing a VR racing game. This procedure collected physiological responses, that is, EDA, HRV, EMG and RSP. Bălan et al. [140] analysed the performance of a set of machine-learning and deep-learning techniques (kNN, SVM, RF, LDA, NN), which adapted their stimuli based on the level of fear recognised, in fear recognition in a 3D acrophobia game. The results showed recognition levels ranging from 42.5% to 89.5%. Therefore, the development of emotion recognition models in immersive VR is an open, fast-growing sub-field, which is moving from the classic statistical testing paradigm to supervised machine-learning.

As to the set-ups employed, Figure 6 shows the evolution of the HMDs used in implicit measure-based emotion research. Among the first-generation VR HMDs of the 2000s was VFX3D, which offers a resolution of 380 × 337 per eye. In the 2010s, the eMaginZ800 improved on the resolution of previous HMDs, offering 800 × 600 and 40° of field of view, followed by Oculus Rift DK2, which increased the resolution to 1080 × 960 and, in particular, the FOV to 90°. Finally, in the late 2010s, the HTC Vive offered an increase in resolution to 1600 × 1400 per eye, and democratised VR with its competitive price. Those increments in HMD performance are aligned with the exponential growth of the number of papers that have used HMD in emotion recognition research (Figure 2), and future HMDs, that might achieve 4K of resolution per eye, could boost the use of VR as a tool to recreate real situations in controlled laboratory environments.

The format most used overall was the 3D environment (85.71%)—360° panoramas were used in 16.67% of cases. This is probably due to the fact that 3D environments present a high level of interactivity, as 360° panoramas do not allow changes in point of view. However, both formats can be useful, depending on the aim of the experiment. The 360° panorama set-ups can be very effective for updating classic, closely controlled affective computing methodologies, in particular, when presenting users with a series of non-interactive stimuli, such as IAPS [95] and IADS [144], but increasing degrees of presence based on immersion level [30]. However, there is still a need to develop large datasets of validated immersive stimuli that cover a wide range of emotions, which could be used as general benchmarks to analyse physiological and behavioural dynamics in immersive VR. The 360° approach offers a good solution to this, as the interaction, for example, navigation, provokes uncontrolled variations during the emotional experience. The first dataset of stimuli published was by Marín-Morales et al. [105], which included 4 scenarios that recreated all quadrants of the CMA. On the other hand, the level of interactivity that 3D scenarios offer can be very useful in applied research, since they display more naturalistic and interactive environments, facilitating decision-making research and the analysis of daily situations. Taking some examples, Takac et al. [137] analysed the anxiety felt by speakers when faced by large audiences, Lin et al. [133] analysed the stress felt by individuals when in a building on fire scenario and Kisker et al. [130] analysed arousal in an exposure to a high height.

Immersive VR can be a very powerful tool to analyse human behaviour in controlled laboratory conditions, but we do not yet know the level of VR validity needed to allow the extrapolation to the real world of the insights gained in terms of physiological and behavioural responses. Indeed, 83.33% of the papers did not present any validation, and only 3 provided a direct comparison between the VR scene and the physical environment simulated. Gorini et al. [109] analysed anxiety through HRV and EDA with virtual and real food, Higuera-Trujillo et al. [64] analysed pleasantness through EDA responses in a 3D, 360° and real retail store, and Marín-Morales et al. [65] analysed arousal and valence oscillations with HRV and EEG in a virtual and physical museum. Other research analysed the influence of immersion [121] and other VR features. Thus, VR validation is still an open topic that needs to be more actively addressed. Understanding and isolating the intrinsic dynamics of VR will be key in future years for the validation of the insights obtained using HMDs.

Finally, the results suggest that VR will play a central role in the affective computing field. The research performed has increased its complexity and maturity during the last two decades, and this tendency is likely to continue during the next years. First, future research should extend the analysis of the physiological dynamics using VR as emotion elicitation in VR, to achieve a level of understanding at least as high as we have today using 2D pictures as stimulation. Subsequently, VR might open up many research opportunities that would be very difficult to assess with non-immersive stimuli. In particular, the inclusion of the dominance dimension, which is very closely related to the users' control of the environment, and impacts on very important features, such as sense of security. Moreover, the social dimension is a crucial factor in the understanding of the emotional dynamics of human beings. The future inclusion of responsive, realistic avatars will help increase the understanding of emotions evoked during social interactions, and the associated physiological responses, in controlled conditions.

5. Conclusions

This work analysed the current state-of-the-art in implicit measure-based emotion recognition elicited by HMDs, and gave a perspective using a systematic and aggregated analysis that can guide future research. After two decades of little research analysing emotions using HMDs in combination with implicit measures, mostly undertaken through the physiological arousal responses of the ANS, in recent years, an inflexion point has been reached. The number of papers published is increasing exponentially, and more emotions are being analysed, including valence-related states, more complex biomedical signal processing procedures are increasingly being performed, including EEG analyses and other behavioural measures, and machine-learning algorithms are being newly applied to develop automatic emotion recognition systems. The results suggest that VR might revolutionise emotion elicitation methods in laboratory environments in the next decade, and impact on affective computing research, transversely in many areas, opening new opportunities for the scientific community. However, more research is needed to increase the understanding of emotion dynamics in immersive VR and, in particular, its validity in performing direct comparisons between simulated and real environments.

Author Contributions: Conceptualisation, J.M.-M.; methodology, J.M.-M.; formal analysis, J.M.-M.; investigation, J.M.-M.; writing—original draft preparation, J.M.-M.; writing—review and editing, J.M.-M., C.L., J.G. and M.A.; visualisation, J.M.-M.; supervision, C.L., J.G. and M.A.; project administration, J.M.-M.; funding acquisition, J.G. and M.A. All authors have read and agreed to the published version of the manuscript.

Funding: This research was funded by European Commission, grant number H2020-825585 HELIOS.

Conflicts of Interest: The authors declare no conflict of interest.

References

1. Picard, R.W. *Affective Computing*; MIT Press: Cambridge, MA, USA, 1997.
2. Cipresso, P.; Chicchi, I.A.; Alcañiz, M.; Riva, G. The Past, Present, and Future of Virtual and Augmented Reality Research: A network and cluster analysis of the literature. *Front. Psychol.* **2018**, *9*, 2086. [CrossRef]
3. Castelvecchi, D. Low-cost headsets boost virtual reality's lab appeal. *Nature* **2016**, *533*, 153–154. [CrossRef] [PubMed]
4. Slater, M.; Usoh, M. Body centred interaction in immersive virtual environments. *Artif. Life Virtual Real.* **1994**, *1*, 125–148.
5. Giglioli, I.A.C.; Pravettoni, G.; Martín, D.L.S.; Parra, E.; Alcañiz, M. A novel integrating virtual reality approach for the assessment of the attachment behavioral system. *Front. Psychol.* **2017**, *8*, 1–7. [CrossRef] [PubMed]
6. Kwartler, M. Visualization in support of public participation. In *Visualization in Landscape and Environmental Planning: Technology and Applications*; Bishop, I., Lange, E., Eds.; Taylor & Francis: London, UK, 2005; pp. 251–260.
7. Vince, J. *Introduction to Virtual Reality*; Media, Springer: Berlin/Heidelberg, Germany, 2004.
8. Alcañiz, M.; Baños, R.; Botella, C.; Rey, B. The EMMA Project: Emotions as a Determinant of Presence. *PsychNology J.* **2003**, *1*, 141–150.
9. Mengoni, M.; Germani, M.; Peruzzini, M. Benchmarking of virtual reality performance in mechanics education. *Int. J. Interact. Des. Manuf.* **2011**, *5*, 103–117. [CrossRef]
10. Stamps, A.E., III. Use of photographs to simulate environments: A meta-analysis. *Percept. Mot. Ski.* **1990**, *71*, 907–913. [CrossRef]
11. Morinaga, A.; Hara, K.; Inoue, K.; Urahama, K. Classification between natural and graphics images based on generalized Gaussian distributions. *Inf. Process. Lett.* **2018**, *138*, 31–34. [CrossRef]
12. Siriaraya, P.; Ang, C.S. The Social Interaction Experiences of Older People in a 3D Virtual Environment. In *Perspectives on Human-Computer Interaction Research with Older People*; Sayago, S., Ed.; Springer: Cham, Switzerland, 2019; pp. 101–117. ISBN 978-3-030-06076-3.
13. Slater, M. Place illusion and plausibility can lead to realistic behaviour in immersive virtual environments. *Philos. Trans. R. Soc. B Biol. Sci.* **2009**, *364*, 3549–3557. [CrossRef]
14. Kober, S.E.; Kurzmann, J.; Neuper, C. Cortical correlate of spatial presence in 2D and 3D interactive virtual reality: An EEG study. *Int. J. Psychophysiol.* **2012**, *83*, 365–374. [CrossRef]
15. Borrego, A.; Latorre, J.; Llorens, R.; Alcañiz, M.; Noé, E. Feasibility of a walking virtual reality system for rehabilitation: Objective and subjective parameters. *J. Neuroeng. Rehabil.* **2016**, *13*, 68. [CrossRef] [PubMed]
16. Clemente, M.; Rodriguez, A.; Rey, B.; Alcañiz, M. Assessment of the influence of navigation control and screen size on the sense of presence in virtual reality using EEG. *Expert Syst. Appl.* **2014**, *41*, 1584–1592. [CrossRef]
17. Borrego, A.; Latorre, J.; Alcañiz, M.; Llorens, R. Comparison of Oculus Rift and HTC Vive: Feasibility for Virtual Reality-Based Exploration, Navigation, Exergaming, and Rehabilitation. *Games Health J.* **2018**, *7*. [CrossRef] [PubMed]
18. Vecchiato, G.; Jelic, A.; Tieri, G.; Maglione, A.G.; De Matteis, F.; Babiloni, F. Neurophysiological correlates of embodiment and motivational factors during the perception of virtual architectural environments. *Cogn. Process.* **2015**, *16*, 425–429. [CrossRef]
19. Jensen, L.; Konradsen, F. A review of the use of virtual reality head-mounted displays in education and training. *Educ. Inf. Technol.* **2017**, *11*, 1–15. [CrossRef]
20. Riecke, B.E.; LaViola, J.J., Jr.; Kruijff, E. 3D user interfaces for virtual reality and games: 3D selection, manipulation, and spatial navigation. In Proceedings of the ACM SIGGRAPH 2018 Courses, Vancouver, BC, Canada, 12–16 August 2018; p. 13.
21. Templeman, J.N.; Denbrook, P.S.; Sibert, L.E. Virtual locomotion: Walking in place through virtual environments. *Presence* **1999**, *8*, 598–617. [CrossRef]
22. Bozgeyikli, E.; Bozgeyikli, L.; Raij, A.; Katkoori, S.; Alqasemi, R.; Dubey, R. Virtual reality interaction techniques for individuals with autism spectrum disorder: Design considerations and preliminary results.

In Proceedings of the International Conference on Human-Computer Interaction, Florence, Italy, 11–15 July 2016; pp. 127–137.
23. Tregillus, S.; Folmer, E. Vr-step: Walking-in-place using inertial sensing for hands free navigation in mobile vr environments. In Proceedings of the 2016 CHI Conference on Human Factors in Computing Systems; ACM: New York, NY, USA; pp. 1250–1255.
24. Nescher, T.; Huang, Y.-Y.; Kunz, A. Planning redirection techniques for optimal free walking experience using model predictive control. In Proceedings of the 2014 IEEE Symposium on 3D User Interfaces (3DUI), Minneapolis, MN, USA, 29–30 March 2014; pp. 111–118.
25. Nabiyouni, M.; Saktheeswaran, A.; Bowman, D.A.; Karanth, A. Comparing the performance of natural, semi-natural, and non-natural locomotion techniques in virtual reality. In Proceedings of the 2015 IEEE Symposium on 3D User Interfaces (3DUI), Arles, France, 23–24 March 2015; pp. 3–10.
26. Bozgeyikli, E.; Raij, A.; Katkoori, S.; Dubey, R. Locomotion in virtual reality for individuals with autism spectrum disorder. In Proceedings of the 2016 Symposium on Spatial User Interaction, Tokyo, Japan, 15–16 October 2016; pp. 33–42.
27. Boletsis, C. The New Era of Virtual Reality Locomotion: A Systematic Literature Review of Techniques and a Proposed Typology. *Multimodal Technol. Interact.* **2017**, *1*, 24. [CrossRef]
28. Slater, M.; Wilbur, S. A Framework for Immersive Virtual Environments (FIVE): Speculations on the Role of Presence in Virtual Environments. *Presence Teleoperators Virtual Environ.* **1997**, *6*, 603–616. [CrossRef]
29. Heeter, C. Being There: The Subjective Experience of Presence. *Presence Teleoperators Virtual Environ.* **1992**, *1*, 262–271. [CrossRef]
30. Baños, R.M.; Botella, C.; Alcañiz, M.; Liaño, V.; Guerrero, B.; Rey, B. Immersion and Emotion: Their Impact on the Sense of Presence. *CyberPsychol. Behav.* **2004**, *7*, 734–741. [CrossRef]
31. Slater, M.; Usoh, M.; Steed, A. Depth of Presence in virtual environments. *Presence Teleoperators Virtual Environ.* **1994**, *3*, 130–144. [CrossRef]
32. Usoh, M.; Arthur, K.; Whitton, M.C.; Bastos, R.; Steed, A.; Slater, M.; Brooks, F.P. Walking > walking-in-place > flying, in virtual environments. In *Proceedings of the 26th Annual Conference on Computer Graphics and Interactive Techniques-SIGGRAPH '99*; Waggenspack, W., Ed.; ACM Press/Addison-Wesley Publishing: New York, NY, USA, 1999; pp. 359–364.
33. Diemer, J.; Alpers, G.W.; Peperkorn, H.M.; Shiban, Y.; Mühlberger, A. The impact of perception and presence on emotional reactions: A review of research in virtual reality. *Front. Psychol.* **2015**, *6*, 1–9. [CrossRef] [PubMed]
34. Riches, S.; Elghany, S.; Garety, P.; Rus-Calafell, M.; Valmaggia, L. Factors Affecting Sense of Presence in a Virtual Reality Social Environment: A Qualitative Study. *Cyberpsychol. Behav. Soc. Netw.* **2019**, *22*, 288–292. [CrossRef] [PubMed]
35. Kiryu, T.; So, R.H.Y. Sensation of presence and cybersickness in applications of virtual reality for advanced rehabilitation. *J. NeuroEng. Rehabil.* **2007**, *4*, 34. [CrossRef]
36. Freina, L.; Ott, M. A literature review on immersive virtual reality in education: State of the art and perspectives. In Proceedings of the International Scientific Conference eLearning and Software for Education, Bucharest, Italy, 23–24 April 2015; Volume 1, p. 133.
37. Alaraj, A.; Lemole, M.G.; Finkle, J.H.; Yudkowsky, R.; Wallace, A.; Luciano, C.; Banerjee, P.P.; Rizzi, S.H.; Charbel, F.T. Virtual reality training in neurosurgery: Review of current status and future applications. *Surg. Neurol. Int.* **2011**, *2*, 52. [CrossRef]
38. Bhagat, K.K.; Liou, W.-K.; Chang, C.-Y. A cost-effective interactive 3D virtual reality system applied to military live firing training. *Virtual Real.* **2016**, *20*, 127–140. [CrossRef]
39. Yavrucuk, I.; Kubali, E.; Tarimci, O. A low cost flight simulator using virtual reality tools. *IEEE Aerosp. Electron. Syst. Mag.* **2011**, *26*, 10–14. [CrossRef]
40. Dols, J.F.; Molina, J.; Camacho, F.J.; Marín-Morales, J.; Pérez-Zuriaga, A.M.; Garcia, A. Design and development of driving simulator scenarios for road validation studies. *Transp. Res. Procedia* **2016**, *18*, 289–296. [CrossRef]
41. de-Juan-Ripoll, C.; Soler-Domínguez, J.L.; Guixeres, J.; Contero, M.; Gutiérrez, N.Á.; Alcañiz, M. Virtual reality as a new approach for risk taking assessment. *Front. Psychol.* **2018**, *9*, 1–8. [CrossRef]
42. Bohil, C.J.; Alicea, B.; Biocca, F.A. Virtual reality in neuroscience research and therapy. *Nat. Rev. Neurosci.* **2011**, *12*, 752–762. [CrossRef]

43. Peperkorn, H.M.; Alpers, G.W.; Mühlberger, A. Triggers of fear: Perceptual cues versus conceptual information in spider phobia. *J. Clin. Psychol.* **2014**, *70*, 704–714. [CrossRef] [PubMed]
44. Park, K.-M.; Ku, J.; Choi, S.-H.; Jang, H.-J.; Park, J.-Y.; Kim, S.I.; Kim, J.-J. A virtual reality application in role-plays of social skills training for schizophrenia: A randomized, controlled trial. *Psychiatry Res.* **2011**, *189*, 166–172. [CrossRef] [PubMed]
45. Didehbani, N.; Allen, T.; Kandalaft, M.; Krawczyk, D.; Chapman, S. Virtual reality social cognition training for children with high functioning autism. *Comput. Hum. Behav.* **2016**, *62*, 703–711. [CrossRef]
46. Lloréns, R.; Noé, E.; Colomer, C.; Alcañiz, M. Effectiveness, usability, and cost-benefit of a virtual reality—Based telerehabilitation program for balance recovery after stroke: A randomized controlled trial. *Arch. Phys. Med. Rehabil.* **2015**, *96*, 418–425. [CrossRef] [PubMed]
47. Alcañiz, M.L.; Olmos-Raya, E.; Abad, L. Use of virtual reality for neurodevelopmental disorders. A review of the state of the art and future agenda. *Medicina* **2019**, *79*, 77–81.
48. Portman, M.E.; Natapov, A.; Fisher-Gewirtzman, D. To go where no man has gone before: Virtual reality in architecture, landscape architecture and environmental planning. *Comput. Environ. Urban Syst.* **2015**, *54*, 376–384. [CrossRef]
49. Bigné, E.; Llinares, C.; Torrecilla, C. Elapsed time on first buying triggers brand choices within a category: A virtual reality-based study. *J. Bus. Res.* **2015**. [CrossRef]
50. Alcañiz, M.; Bigné, E.; Guixeres, J. Virtual Reality in Marketing: A Framework, Review, and Research Agenda. *Front. Psychol.* **2019**, *10*, 1–15. [CrossRef]
51. Picard, R.W. Affective Computing: Challenges. *Int. J. Hum. Comput. Stud.* **2003**, *59*, 55–64. [CrossRef]
52. Riva, G.; Mantovani, F.; Capideville, C.S.; Preziosa, A.; Morganti, F.; Villani, D.; Gaggioli, A.; Botella, C.; Alcañiz, M. Affective Interactions Using Virtual Reality: The Link between Presence and Emotions. *CyberPsychol. Behav.* **2007**, *10*, 45–56. [CrossRef]
53. Guixeres, J.; Saiz, J.; Alcañiz, M.; Cebolla, A.; Escobar, P.; Baños, R.; Botella, C.; Lison, J.F.; Alvarez, J.; Cantero, L.; et al. Effects of virtual reality during exercise in children. *J. Univers. Comput. Sci.* **2013**, *19*, 1199–1218.
54. Felnhofer, A.; Kothgassner, O.D.; Schmidt, M.; Heinzle, A.K.; Beutl, L.; Hlavacs, H.; Kryspin-Exner, I. Is virtual reality emotionally arousing? Investigating five emotion inducing virtual park scenarios. *Int. J. Hum. Comput. Stud.* **2015**, *82*, 48–56. [CrossRef]
55. Lorenzo, G.; Lledó, A.; Pomares, J.; Roig, R. Design and application of an immersive virtual reality system to enhance emotional skills for children with autism spectrum disorders. *Comput. Educ.* **2016**, *98*, 192–205. [CrossRef]
56. Rohrmann, B.; Bishop, I.D. Subjective responses to computer simulations of urban environments. *J. Environ. Psychol.* **2002**, *22*, 319–331. [CrossRef]
57. Bishop, I.D.; Rohrmann, B. Subjective responses to simulated and real environments: A comparison. *Landsc. Urban Plan.* **2003**, *65*, 261–277. [CrossRef]
58. de Kort, Y.A.W.; Ijsselsteijn, W.A.; Kooijman, J.; Schuurmans, Y. Virtual laboratories: Comparability of real and virtual environments for environmental psychology. *Presence Teleoperators Virtual Environ.* **2003**, *12*, 360–373. [CrossRef]
59. Yeom, D.; Choi, J.-H.; Zhu, Y. Investigation of the Physiological Differences between Immersive Virtual Environment and Indoor Enviorment in a Building. *Indoor Built Environ.* **2017**, 1–17. [CrossRef]
60. van der Ham, I.J.; Faber, A.M.; Venselaar, M.; van Kreveld, M.J.; Löffler, M. Ecological validity of virtual environments to assess human navigation ability. *Front. Psychol.* **2015**, *6*, 637. [CrossRef]
61. Heydarian, A.; Carneiro, J.P.; Gerber, D.; Becerik-Gerber, B.; Hayes, T.; Wood, W. Immersive virtual environments versus physical built environments: A benchmarking study for building design and user-built environment explorations. *Autom. Constr.* **2015**, *54*, 116–126. [CrossRef]
62. Chamilothori, K.; Wienold, J.; Andersen, M. Adequacy of Immersive Virtual Reality for the Perception of Daylit Spaces: Comparison of Real and Virtual Environments. *LEUKOS J. Illum. Eng. Soc. N. Am.* **2018**, 1–24. [CrossRef]
63. Kimura, K.; Reichert, J.F.; Olson, A.; Pouya, O.R.; Wang, X.; Moussavi, Z.; Kelly, D.M. Orientation in Virtual Reality Does Not Fully Measure Up to the Real-World. *Sci. Rep.* **2017**, *7*, 6–13. [CrossRef] [PubMed]

64. Higuera-Trujillo, J.L.; López-Tarruella, J.; Llinares, M.C. Psychological and physiological human responses to simulated and real environments: A comparison between Photographs, 360° Panoramas, and Virtual Reality. *Appl. Ergon.* **2017**, *65*, 398–409. [CrossRef] [PubMed]
65. Marín-Morales, J.; Higuera-Trujillo, J.L.; Greco, A.; Guixeres, J.; Llinares, C.; Gentili, C.; Scilingo, E.P.; Alcañiz, M.; Valenza, G. Real vs. immersive-virtual emotional experience: Analysis of psycho-physiological patterns in a free exploration of an art museum. *PLoS ONE* **2019**, *14*, e0223881. [CrossRef] [PubMed]
66. Brief, A.P. *Attitudes in and Around Organizations*; Sage: Thousand Oaks, CA, USA, 1998; Volume 9.
67. Payne, B.K. Prejudice and perception: The role of automatic and controlled processes in misperceiving a weapon. *J. Pers. Soc. Psychol.* **2001**, *81*, 181. [CrossRef] [PubMed]
68. Schmitt, N. Method bias: The importance of theory and measurement. *J. Organ. Behav.* **1994**, *15*, 393–398. [CrossRef]
69. Barsade, S.G.; Ramarajan, L.; Westen, D. Implicit affect in organizations. *Res. Organ. Behav.* **2009**, *29*, 135–162. [CrossRef]
70. Lieberman, M.D. Social cognitive neuroscience: A review of core processes. *Annu. Rev. Psychol.* **2007**, *58*, 259–289. [CrossRef]
71. Camerer, C.; Loewenstein, G.; Prelec, D. Neuroeconomics: How neuroscience can inform economics. *J. Econ. Lit.* **2005**, *43*, 9–64. [CrossRef]
72. Riener, A.; Ferscha, A.; Aly, M. Heart on the road: HRV analysis for monitoring a driver's affective state. In Proceedings of the 1st International Conference on Automotive User Interfaces and Interactive Vehicular Applications, Essen, Germany, 21–22 September 2009; pp. 99–106.
73. Setz, C.; Arnrich, B.; Schumm, J.; La Marca, R.; Tröster, G.; Ehlert, U. Discriminating stress from cognitive load using a wearable EDA device. *IEEE Trans. Inf. Technol. Biomed.* **2009**, *14*, 410–417. [CrossRef] [PubMed]
74. Berka, C.; Levendowski, D.J.; Lumicao, M.N.; Yau, A.; Davis, G.; Zivkovic, V.T.; Olmstead, R.E.; Tremoulet, P.D.; Craven, P.L. EEG correlates of task engagement and mental workload in vigilance, learning, and memory tasks. *Aviat. Space Environ. Med.* **2007**, *78*, B231–B244. [PubMed]
75. Jack, A.I.; Dawson, A.J.; Begany, K.L.; Leckie, R.L.; Barry, K.P.; Ciccia, A.H.; Snyder, A.Z. fMRI reveals reciprocal inhibition between social and physical cognitive domains. *Neuroimage* **2013**, *66*, 385–401. [CrossRef] [PubMed]
76. Ernst, L.H.; Plichta, M.M.; Lutz, E.; Zesewitz, A.K.; Tupak, S.V.; Dresler, T.; Ehlis, A.-C.; Fallgatter, A.J. Prefrontal activation patterns of automatic and regulated approach–avoidance reactions–a functional near-infrared spectroscopy (fNIRS) study. *Cortex* **2013**, *49*, 131–142. [CrossRef]
77. Glöckner, A.; Herbold, A.-K. An eye-tracking study on information processing in risky decisions: Evidence for compensatory strategies based on automatic processes. *J. Behav. Decis. Mak.* **2011**, *24*, 71–98. [CrossRef]
78. Bahreini, K.; Nadolski, R.; Westera, W. Towards multimodal emotion recognition in e-learning environments. *Interact. Learn. Environ.* **2016**, *24*, 590–605. [CrossRef]
79. Huang, K.-Y.; Wu, C.-H.; Su, M.-H.; Kuo, Y.-T. Detecting unipolar and bipolar depressive disorders from elicited speech responses using latent affective structure model. *IEEE Trans. Affect. Comput.* **2018**. [CrossRef]
80. Prokasy, W. *Electrodermal Activity in Psychological Research*; Elsevier: Amsterdam, The Netherland, 2012.
81. Kim, H.-G.; Cheon, E.-J.; Bai, D.-S.; Lee, Y.H.; Koo, B.-H. Stress and heart rate variability: A meta-analysis and review of the literature. *Psychiatry Investig.* **2018**, *15*, 235. [CrossRef] [PubMed]
82. Kreibig, S.D. Autonomic nervous system activity in emotion: A review. *Biol. Psychol.* **2010**, *84*, 394–421. [CrossRef]
83. Lotte, F.; Bougrain, L.; Cichocki, A.; Clerc, M.; Congedo, M.; Rakotomamonjy, A.; Yger, F. A review of classification algorithms for EEG-based brain–computer interfaces: A 10 year update. *J. Neural Eng.* **2018**, *15*, 31005. [CrossRef]
84. Gruzelier, J.H. EEG-neurofeedback for optimising performance. I: A review of cognitive and affective outcome in healthy participants. *Neurosci. Biobehav. Rev.* **2014**, *44*, 124–141. [CrossRef]
85. Thibault, R.T.; MacPherson, A.; Lifshitz, M.; Roth, R.R.; Raz, A. Neurofeedback with fMRI: A critical systematic review. *Neuroimage* **2018**, *172*, 786–807. [CrossRef]
86. Naseer, N.; Hong, K.-S. fNIRS-based brain-computer interfaces: A review. *Front. Hum. Neurosci.* **2015**, *9*, 3. [CrossRef] [PubMed]
87. Meißner, M.; Oll, J. The promise of eye-tracking methodology in organizational research: A taxonomy, review, and future avenues. *Organ. Res. Methods* **2019**, *22*, 590–617. [CrossRef]

88. Calvo, M.G.; Nummenmaa, L. Perceptual and affective mechanisms in facial expression recognition: An integrative review. *Cogn. Emot.* **2016**, *30*, 1081–1106. [CrossRef] [PubMed]
89. Schuller, B.W. Speech emotion recognition: Two decades in a nutshell, benchmarks, and ongoing trends. *Commun. ACM* **2018**, *61*, 90–99. [CrossRef]
90. Parsons, T.D. Virtual Reality for enhanced ecological validity and experimental control in the clinical, affective and social neurosciences. *Front. Hum. Neurosci.* **2015**, *9*, 660. [CrossRef]
91. Ekman, P. Basic Emotions. *Handb. Cogn. Emot.* **1999**, 45–60. [CrossRef]
92. Russell, J.A.; Mehrabian, A. Evidence for a three-factor theory of emotions. *J. Res. Pers.* **1977**, *11*, 273–294. [CrossRef]
93. Calvo, R.A.; D'Mello, S. Affect detection: An interdisciplinary review of models, methods, and their applications. *IEEE Trans. Affect. Comput.* **2010**, *1*, 18–37. [CrossRef]
94. Valenza, G.; Greco, A.; Gentili, C.; Lanata, A.; Sebastiani, L.; Menicucci, D.; Gemignani, A.; Scilingo, E.P. Combining electroencephalographic activity and instantaneous heart rate for assessing brain–heart dynamics during visual emotional elicitation in healthy subjects. *Philos. Trans. R. Soc. A Math. Phys. Eng. Sci.* **2016**, *374*, 20150176. [CrossRef]
95. Valenza, G.; Lanata, A.; Scilingo, E.P. The role of nonlinear dynamics in affective valence and arousal recognition. *IEEE Trans. Affect. Comput.* **2012**, *3*, 237–249. [CrossRef]
96. Zangeneh Soroush, M.; Maghooli, K.; Setarehdan, S.K.; Motie Nasrabadi, A. A Review on EEG Signals Based Emotion Recognition. *Int. Clin. Neurosci. J.* **2018**, *4*, 118–129. [CrossRef]
97. Kory Jacqueline, D. Sidney Affect Elicitation for affective Computing. In *The Oxford Handbook of Affective Computing*; Oxford University Press: New York, NY, USA, 2014; pp. 371–383.
98. Ekman, P. The directed facial action task. In *Handbook of Emotion Elicitation and Assessment*; Oxford University Press: New York, NY, USA, 2007; pp. 47–53.
99. Harmon-Jones, E.; Amodio, D.M.; Zinner, L.R. Social psychological methods of emotion elicitation. *Handb. Emot. Elicitation Assess.* **2007**, *25*, 91–105.
100. Roberts, N.A.; Tsai, J.L.; Coan, J.A. Emotion elicitation using dyadic interaction task. *Handb. Emot. Elicitation Assess.* **2007**, *01*, 106–123.
101. Nardelli, M.; Valenza, G.; Greco, A.; Lanata, A.; Scilingo, E.P. Recognizing emotions induced by affective sounds through heart rate variability. *IEEE Trans. Affect. Comput.* **2015**, *6*, 385–394. [CrossRef]
102. Kim, J. Emotion Recognition Using Speech and Physiological Changes. *Robust Speech Recognit. Underst.* **2007**, *29*, 265–280.
103. Soleymani, M.; Pantic, M.; Pun, T. Multimodal emotion recognition in response to videos (Extended abstract). In Proceedings of the ACII 2015: International Conference on Affective Computing and Intelligent Interaction, Xi'an, China, 21–24 September 2015; Volume 3, pp. 491–497.
104. Jang, D.P.; Kim, I.Y.; Nam, S.W.; Wiederhold, B.K.; Wiederhold, M.D.; Kim, S.I. Analysis of physiological response to two virtual environments: Driving and flying simulation. *CyberPsychol. Behav.* **2002**, *5*, 11–18. [CrossRef] [PubMed]
105. Marín-Morales, J.; Higuera-Trujillo, J.L.; Greco, A.; Guixeres, J.; Llinares, C.; Scilingo, E.P.; Alcañiz, M.; Valenza, G. Affective computing in virtual reality: Emotion recognition from brain and heartbeat dynamics using wearable sensors. *Sci. Rep.* **2018**, *8*, 13657. [CrossRef] [PubMed]
106. Moher, D.; Liberati, A.; Tetzlaff, J.; Altman, D.G.; The PRISMA Group. Preferred reporting items for systematic reviews and meta-analyses: The PRISMA statement. *PLoS Med.* **2009**, *6*, e1000097. [CrossRef]
107. Meehan, M.; Razzaque, S.; Insko, B.; Whitton, M.; Brooks, F.P. Review of four studies on the use of physiological reaction as a measure of presence in stressful virtual environments. *Appl. Psychophysiol. Biofeedback* **2005**, *30*, 239–258. [CrossRef]
108. Wilhelm, F.H.; Pfaltz, M.C.; Gross, J.J.; Mauss, I.B.; Kim, S.I.; Wiederhold, B.K. Mechanisms of virtual reality exposure therapy: The role of the behavioral activation and behavioral inhibition systems. *Appl. Psychophysiol. Biofeedback* **2005**, *30*, 271–284. [CrossRef]
109. Gorini, A.; Griez, E.; Petrova, A.; Riva, G. Assessment of the emotional responses produced by exposure to real food, virtual food and photographs of food in patients affected by eating disorders. *Ann. Gen. Psychiatry* **2010**, *9*, 30. [CrossRef] [PubMed]
110. Philipp, M.C.; Storrs, K.R.; Vanman, E.J. Sociality of facial expressions in immersive virtual environments: A facial EMG study. *Biol. Psychol.* **2012**, *91*, 17–21. [CrossRef] [PubMed]

111. Parsons, T.D.; Courtney, C.G.; Dawson, M.E. Virtual reality Stroop task for assessment of supervisory attentional processing. *J. Clin. Exp. Neuropsychol.* **2013**, *35*, 812–826. [CrossRef] [PubMed]
112. Pallavicini, F.; Cipresso, P.; Raspelli, S.; Grassi, A.; Serino, S.; Vigna, C.; Triberti, S.; Villamira, M.; Gaggioli, A.; Riva, G. Is virtual reality always an effective stressors for exposure treatments? Some insights from a controlled trial. *BMC Psychiatry* **2013**, *13*, 52. [CrossRef]
113. Felnhofer, A.; Kothgassner, O.D.; Hetterle, T.; Beutl, L.; Hlavacs, H.; Kryspin-Exner, I. Afraid to be there? Evaluating the relation between presence, self-reported anxiety, and heart rate in a virtual public speaking task. *Cyberpsychol. Behav. Soc. Netw.* **2014**, *17*, 310–316. [CrossRef]
114. Hartanto, D.; Kampmann, I.L.; Morina, N.; Emmelkamp, P.G.M.; Neerincx, M.A.; Brinkman, W.-P. Controlling social stress in virtual reality environments. *PLoS ONE* **2014**, *9*, e92804. [CrossRef]
115. McCall, C.; Hildebrandt, L.K.; Bornemann, B.; Singer, T. Physiophenomenology in retrospect: Memory reliably reflects physiological arousal during a prior threatening experience. *Conscious. Cogn.* **2015**, *38*, 60–70. [CrossRef]
116. Notzon, S.; Deppermann, S.; Fallgatter, A.; Diemer, J.; Kroczek, A.; Domschke, K.; Zwanzger, P.; Ehlis, A.C. Psychophysiological effects of an iTBS modulated virtual reality challenge including participants with spider phobia. *Biol. Psychol.* **2015**, *112*, 66–76. [CrossRef]
117. Hildebrandt, L.K.; Mccall, C.; Engen, H.G.; Singer, T. Cognitive flexibility, heart rate variability, and resilience predict fine-grained regulation of arousal during prolonged threat. *Psychophysiology* **2016**, *53*, 880–890. [CrossRef]
118. Higuera-Trujillo, J.L.; Marín-Morales, J.; Rojas, J.C.; López-Tarruella-Maldonado, J. Emotional maps: Neuro architecture and design applications. 6th International Forum Design as a Processes. *Syst. Des. Beyond Process. Think.* **2016**, 677–685. [CrossRef]
119. Bian, Y.; Yang, C.; Gao, F.; Li, H.; Zhou, S.; Li, H.; Sun, X.; Meng, X. A framework for physiological indicators of flow in VR games: Construction and preliminary evaluation. *Pers. Ubiquitous Comput.* **2016**, *20*, 821–832. [CrossRef]
120. Shiban, Y.; Diemer, J.; Brandl, S.; Zack, R.; Mühlberger, A.; Wüst, S. Trier Social Stress Test in vivo and in virtual reality: Dissociation of response domains. *Int. J. Psychophysiol.* **2016**, *110*, 47–55. [CrossRef] [PubMed]
121. Chirico, A.; Cipresso, P.; Yaden, D.B.; Biassoni, F.; Riva, G.; Gaggioli, A. Effectiveness of Immersive Videos in Inducing Awe: An Experimental Study. *Sci. Rep.* **2017**, *7*, 1–11. [CrossRef] [PubMed]
122. Zou, H.; Li, N.; Cao, L. Emotional response—Based approach for assessing the sense of presence of subjects in virtual building evacuation studies. *J. Comput. Civ. Eng.* **2017**, *31*, 4017028. [CrossRef]
123. Breuninger, C.; Sláma, D.M.; Krämer, M.; Schmitz, J.; Tuschen-Caffier, B. Psychophysiological reactivity, interoception and emotion regulation in patients with agoraphobia during virtual reality anxiety induction. *Cognit. Ther. Res.* **2017**, *41*, 193–205. [CrossRef]
124. van't Wout, M.; Spofford, C.M.; Unger, W.S.; Sevin, E.B.; Shea, M.T. Skin conductance reactivity to standardized virtual reality combat scenes in veterans with PTSD. *Appl. Psychophysiol. Biofeedback* **2017**, *42*, 209–221. [CrossRef]
125. Banaei, M.; Hatami, J.; Yazdanfar, A.; Gramann, K. Walking through Architectural Spaces: The Impact of Interior Forms on Human Brain Dynamics. *Front. Hum. Neurosci.* **2017**, *11*, 1–14. [CrossRef]
126. Anderson, A.P.; Mayer, M.D.; Fellows, A.M.; Cowan, D.R.; Hegel, M.T.; Buckey, J.C. Relaxation with immersive natural scenes presented using virtual reality. *Aerosp. Med. Hum. Perform.* **2017**, *88*, 520–526. [CrossRef]
127. Chittaro, L.; Sioni, R.; Crescentini, C.; Fabbro, F. Mortality salience in virtual reality experiences and its effects on users' attitudes towards risk. *Int. J. Hum. Comput. Stud.* **2017**, *101*, 10–22. [CrossRef]
128. Biedermann, S.V.; Biedermann, D.G.; Wenzlaff, F.; Kurjak, T.; Nouri, S.; Auer, M.K.; Wiedemann, K.; Briken, P.; Haaker, J.; Lonsdorf, T.B.; et al. An elevated plus-maze in mixed reality for studying human anxiety-related behavior. *BMC Biol.* **2017**, *15*, 125. [CrossRef]
129. Tsai, C.-F.; Yeh, S.-C.; Huang, Y.; Wu, Z.; Cui, J.; Zheng, L. The Effect of Augmented Reality and Virtual Reality on Inducing Anxiety for Exposure Therapy: A Comparison Using Heart Rate Variability. *J. Healthc. Eng.* **2018**, *2018*, 27–36. [CrossRef] [PubMed]
130. Kisker, J.; Gruber, T.; Schöne, B. Behavioral realism and lifelike psychophysiological responses in virtual reality by the example of a height exposure. *Psychol. Res.* **2019**, 1–14. [CrossRef] [PubMed]

131. Gromer, D.; Reinke, M.; Christner, I.; Pauli, P. Causal Interactive Links Between Presence and Fear in Virtual Reality Height Exposure. *Front. Psychol.* **2019**, *10*, 141. [CrossRef] [PubMed]
132. Zimmer, P.; Wu, C. Same same but different? Replicating the real surroundings in a virtual Trier Social Stress Test (TSST-VR) does not enhance presence or the psychophysiological stress response. *Physiol. Behav.* **2019**, *212*, 112690. [CrossRef]
133. Lin, J.; Cao, L.; Li, N. Assessing the influence of repeated exposures and mental stress on human wayfinding performance in indoor environments using virtual reality technology. *Adv. Eng. Inform.* **2019**, *39*, 53–61. [CrossRef]
134. Schweizer, T.; Renner, F.; Sun, D.; Becker-Asano, C.; Tuschen-Caffier, B. Cognitive processing and regulation modulates analogue trauma symptoms in a Virtual Reality paradigm. *Cognit. Ther. Res.* **2019**, *43*, 199–213. [CrossRef]
135. Kim, Y.; Moon, J.; Sung, N.-J.; Hong, M. Correlation between selected gait variables and emotion using virtual reality. *J. Ambient Intell. Humaniz. Comput.* **2019**, *8*, 1–8. [CrossRef]
136. Uhm, J.-P.; Lee, H.-W.; Han, J.-W. Creating sense of presence in a virtual reality experience: Impact on neurophysiological arousal and attitude towards a winter sport. *Sport Manag. Rev.* **2019**. [CrossRef]
137. Takac, M.; Collett, J.; Blom, K.J.; Conduit, R.; Rehm, I.; De Foe, A. Public speaking anxiety decreases within repeated virtual reality training sessions. *PLoS ONE* **2019**, *14*, e0216288. [CrossRef]
138. Stolz, C.; Endres, D.; Mueller, E.M. Threat-conditioned contexts modulate the late positive potential to faces—A mobile EEG/virtual reality study. *Psychophysiology* **2019**, *56*, e13308. [CrossRef]
139. Granato, M.; Gadia, D.; Maggiorini, D.; Ripamonti, L.A. An empirical study of players' emotions in VR racing games based on a dataset of physiological data. *Multimed. Tools Appl.* **2020**, 1–30. [CrossRef]
140. Bălan, O.; Moise, G.; Moldoveanu, A.; Leordeanu, M.; Moldoveanu, F. An investigation of various machine and deep learning techniques applied in automatic fear level detection and acrophobia virtual therapy. *Sensors* **2020**, *20*, 496. [CrossRef] [PubMed]
141. Reichenberger, J.; Pfaller, M.; Mühlberger, A. Gaze Behavior in Social Fear Conditioning: An Eye-Tracking Study in Virtual Reality. *Front. Psychol.* **2020**, *11*, 1–12. [CrossRef] [PubMed]
142. Huang, Q.; Yang, M.; Jane, H.-A.; Li, S.; Bauer, N. Trees, grass, or concrete? The effects of different types of environments on stress reduction. *Landsc. Urban Plan.* **2020**, *193*, 103654. [CrossRef]
143. Shu, L.; Xie, J.; Yang, M.; Li, Z.; Li, Z.; Liao, D.; Xu, X.; Yang, X. A review of emotion recognition using physiological signals. *Sensors* **2018**, *18*, 2074. [CrossRef]
144. Greco, A.; Valenza, G.; Citi, L.; Scilingo, E.P. Arousal and valence recognition of affective sounds based on electrodermal activity. *IEEE Sens. J.* **2016**, *17*, 716–725. [CrossRef]

sensors MDPI

Review

EEG-Based BCI Emotion Recognition: A Survey

Edgar P. Torres [1], Edgar A. Torres [2], Myriam Hernández-Álvarez [1,*] and Sang Guun Yoo [1]

1 Escuela Politécnica Nacional, Facultad de Ingeniería de Sistemas, Departamento de Informática y Ciencias de la Computación, Quito 170143, Ecuador; edgar.torres@epn.edu.ec (E.P.T.); sang.yoo@epn.edu.ec (S.G.Y.)
2 Pontificia Universidad Católica del Ecuador, Quito 170143, Ecuador; etorresh777@gmail.com
* Correspondence: myriam.hernandez@epn.edu.ec

Received: 24 June 2020; Accepted: 25 August 2020; Published: 7 September 2020

Abstract: Affecting computing is an artificial intelligence area of study that recognizes, interprets, processes, and simulates human affects. The user's emotional states can be sensed through electroencephalography (EEG)-based Brain Computer Interfaces (BCI) devices. Research in emotion recognition using these tools is a rapidly growing field with multiple inter-disciplinary applications. This article performs a survey of the pertinent scientific literature from 2015 to 2020. It presents trends and a comparative analysis of algorithm applications in new implementations from a computer science perspective. Our survey gives an overview of datasets, emotion elicitation methods, feature extraction and selection, classification algorithms, and performance evaluation. Lastly, we provide insights for future developments.

Keywords: emotion recognition; emotion elicitation; datasets; emotion representation; feature selection; feature extraction; classification; computer science; artificial intelligence; affective computing

1. Introduction

Affective computing is a branch of artificial intelligence. It is computing that relates to, arises from, or influences emotions [1]. Automatic emotion recognition is an area of study that forms part of affective computing. Research in this area is rapidly evolving thanks to the availability of affordable devices for capturing brain signals, which serve as inputs for systems that decode the relationship between emotions and electroencephalographic (EEG) variations. These devices are called EEG-based brain-computer interfaces (BCIs).

Affective states play an essential role in decision-making. Such states can facilitate or hinder problem-solving. Emotion recognition takes advantage of positive affective states, enhances emotional intelligence, and consequently improves professional and personal success [2]. Moreover, emotion self-awareness can help people manage their mental health and optimize their work performance. Automatic systems can increase our understanding of emotions, and therefore promote effective communication among individuals and human-to-machine information exchanges. Automatic EEG-based emotion recognition could also help enrich people's relationships with their environment. Besides, automatic emotion recognition will play an essential role in artificial intelligence entities designed for human interaction [3].

According to Gartner's 2019 Hype Cycle report on trending research topics, affective computing is at the innovation trigger stage, which is evidenced by the field's copious publications. However, there are still no defined standards for the different components of the systems that recognize emotions using EEG signals, and it is still challenging to detect and classify emotions reliably. Thus, a survey that updates the information in the emotion recognition field, with a focus on new computational developments, is worthwhile.

This work reviews emotion recognition advances using EEG signals and BCI to (1) identify trends in algorithm usage and technology, (2) detect potential errors that must be overcome for better results, and (3) identify possible knowledge gaps in the field. The aim is to distinguish what has already been done in systems implementations and catch a glimpse of what could lie ahead. For context, our study is a survey from 2015 to 2020.

The present article gives an overview of datasets, emotion elicitation methods, feature extraction and selection, classification algorithms, and in general terms, computer intelligence techniques used in this field. We present a brief review of the components of an EEG-based system to recognize emotions and highlight trends showing statistics of their use in the literature. We deliver a compilation of papers describing new implementations, analyzing their inputs, tools, and considered classes. This up-to-date information could be used to discover and suggest new research paths.

The present survey followed the guidelines of [4]. We used Semanticscholar.org for searches of sources because it links to the major databases that contain journals and conferences proceedings. The search criteria were the keywords linked to our review's objectives.

We extracted articles from journals and conferences that present new implementations of computational intelligence techniques. Concretely, the analyzed papers' primary objectives were computational systems that applied algorithms for the detection and classification of emotions using EEG-based BCI devices. Such studies also included performance measures that allowed a comparison of results while taking into account the classified number of emotions.

As a result, we obtained 136 journal articles, 63 conference papers, and 15 reviews. Each whole article was read to have complete information to guide the application of inclusion and exclusion filters. The inclusion criteria were: (1) The articles were published in the considered period in peer-reviewed journals and conferences, (2) they constitute emotion recognition systems that used EEG-based BCI devices with a focus on computational intelligence applications, and (3) they include experimental setups and performance evaluations. Lastly, we applied additional exclusion criteria and eliminated review articles and other studies that have a different perspective as medical studies for diagnosis or assessment.

With these considerations, we selected 36 journal studies and 24 conference papers. From this group, we extracted statistical data about computational techniques to detect trends and perform a comparative analysis. Finally, from these 60 papers, we chose a sample of 31 articles to show a summary of technical details, components, and algorithms. It should be noted that according to generally accepted practices, 31 observations are sufficient for statistically valid conclusions due to the central limit theorem. Then, from this subsample of articles, we obtained some additional data and tendencies.

This document is organized as follows: Section 1 presents an introduction of the topic, with an overview of BCI devices, emotion representations, and correlations among brain locations, frequency bands, and affective states. Section 2 shows the structure of EEG-based BCI systems for emotion recognition. Their principal components are revised: (1) Signal acquisition, (2) preprocessing, (3) eature extraction, (4) feature selection, (5) classification, and (6) performance evaluation. Section 3 analyzes the components of our chosen research pieces and discusses trends and challenges. Section 4 presents future work. Section 5 features the conclusions or this survey.

1.1. EEG-Based BCI in Emotion Recognition

Many studies suggest that emotional states are associated with electrical activity that is produced in the central nervous system. Brain activity can be detected through its electrical signals by sensing its variations, locations, and functional interactions [5] using EEG devices. EEG signals have excellent temporal resolution and are a direct measurement of neuronal activity. These signals cannot be manipulated or simulated to fake an emotional state, so they provide reliable information. The challenge is to decode this information and map it to specific emotions.

One affordable and convenient way to detect EEG signals is through EEG-based BCI devices that are non-invasive, low cost, and even wearable, such as helmets and headbands. The development of these tools has facilitated the emergence of abundant research in the emotion recognition field.

Some scientists predict that EEG-based BCI devices will soon improve their usability. Therefore, shortly, they could be used on an everyday basis for emotion detection with several purposes, such as emotion monitoring in health care facilities, gaming and entertainment, teaching-learning scenarios, and for optimizing performance in the workplace [6], among other applications.

1.2. Emotion Representations

Emotions can be represented using different general models [7]. The most used are the discrete model and the dimensional models. The discrete model identifies basic, innate, and universal emotions from which all other emotions can be derived. Some authors state that these primary emotions are happiness, sadness, anger, surprise, disgust, and fear [8]. Some researchers consider that this model has limitations to represent specific emotions in a broader range of affective states.

Alternatively, dimensional models can express complex emotions in a two-dimensional continuous space: Valence-arousal (VA), or in three dimensions: Valence, arousal, and dominance (VAD) [9]. The VA model has valence and arousal as axes. Valence is used to rate positive and negative emotions and ranges from happy to unhappy (or sad). Arousal measures emotions from calm to stimulated (or excited). Three-dimensional models add a dominance axis to evaluate from submissive (powerless) to empowered emotions. This representation distinguishes emotions that are jointly represented in the VA model. For instance, fear and anger have similar valence-arousal representations on the VA plane. Thus, three-dimensional models improve "emotional resolution" through the dominance dimension. In this example, fear is a submissive feeling, but anger requires power [10]. Hence, the dominance dimension improves the differentiation between these two emotions.

Figure 1 shows a VA plane with the representation of basic emotions. The horizontal axis corresponds to valence dimensions, from positive to negative emotions. Likewise, the vertical axis corresponds to arousal. These two variables can be thought of as emotional state components [5]. Figure 2 presents the VAD space with a representation of the same basic emotions.

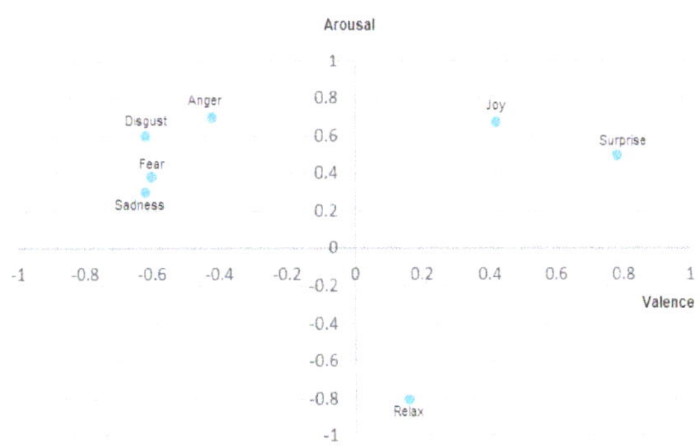

Figure 1. Emotional states in the Valence-Arousal space [11].

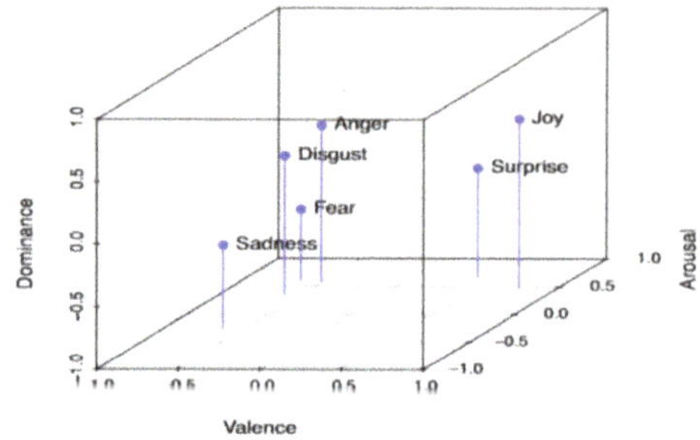

Figure 2. Emotional states in the Valence-Arousal-Dominance space [12].

Table 1 shows that some researchers studying EEG-based functional connectivity in the brain have reported a relationship between specific brain areas and emotional states. Studies that take at-single-electrode-level analysis into account have shown that asymmetric activity at the frontal site in the alpha band is associated with emotion. Ekman and Davidson found that enjoyment generated an activation of the brain's left frontal parts [13]. Another study found a left frontal activity reduction when volunteers adopted fear expressions [14]. Increased power in theta bands at the frontal midline is associated with pleasurable emotions, and the opposite has been observed with unpleasant feelings [15].

Table 1. Frequency bands associations [16,17].

Band	State Association	Potential Localization	Stimuli
Gamma rhythm (above 30 HZ)	Positive valence. These waves are correlated with positive spiritual feelings. Arousal increases with high-intensity visual stimuli.	Different sensory and non-sensory cortical networks.	These waves appear stimulated by the attention, multi-sensory information, memory, and consciousness.
Beta (13 to 30 Hz)	They are related to visual self-induced positive and negative emotions. These waves are associated with alertness and problem-solving.	Motor cortex.	They are stimulated by motor activity, motor imagination, or tactile stimulation. Beta power increases during the tension of scalp muscles, which are also involved in frowning and smiling.
Alpha (8 to 13 Hz)	They are linked to relaxed and wakeful states, feelings of conscious awareness, and learning.	Parietal and occipital regions. Asymmetries reported: rightward-lateralization of frontal alpha power during positive emotions, compared to negative or withdrawal-related emotions, originates from leftward-lateralization of prefrontal structures.	These waves are believed to appear during relaxation periods with eyes shut while remaining still awake. They represent the visual cortex in a repose state. These waves slow down when falling asleep and accelerate when opening the eyes, moving, or even when thinking about the intention to move.
Theta (4 to 7 Hz)	They appear in relaxation states, and in those cases, they allow better concentration. These waves also correlate with anxious feelings.	The front central head region is associated with the hippocampal theta waves.	Theta oscillations are involved in memory encoding and retrieval. Additionally, individuals that experience higher emotional arousal in a reward situation reveal an increase of theta waves in their EEG [17]. Theta coma waves appear in patients with brain damage.
Delta (0 to 4 Hz)	They are present in deep NREM 3 sleep stage. Since adolescence, their presence during sleep declines with advancing age.	Frontal, temporal, and occipital regions.	Deep sleep. These waves also have been found in continuous attention tasks [18].

Several studies confirm that frequency bands are related to affective responses. However, emotions are complex processes. The authors in [15] assert that the recognition of different emotional states may be more valid if EEG-based functional connectivity is examined, rather than a single analysis at the electrode level. Correlation, coherence, and phase synchronization indices between EEG electrode pairs are used to estimate functional connectivity between different brain locations. Likewise, differential entropy (DE), and its derivatives like differential asymmetry (DASM), rational asymmetry (RASM), and differential caudality (DCAU) measure functional dissimilarities. Such features are calculated through logarithmic power spectral density for a fixed-length EEG sequence, plus the differences and ratios between DE features of hemispheric asymmetry electrodes [19].

The growing consensus seems to be that a simple mapping between emotions and specific brain structures is inconsistent with observations of different emotions activating the same structure, or one emotion activating several structures [20]. Additionally, functional connectivity between brain regions or signal complexity measures may help to detect and describe emotional states [21].

2. EEG-Based BCI Systems for Emotion Recognition

Figure 3 presents the structure of an EEG-based BCI system for emotion recognition. The processes of signal acquisition, preprocessing, feature extraction, feature selection, classification, and performance evaluation can be distinguished and will be reviewed in the following subsections.

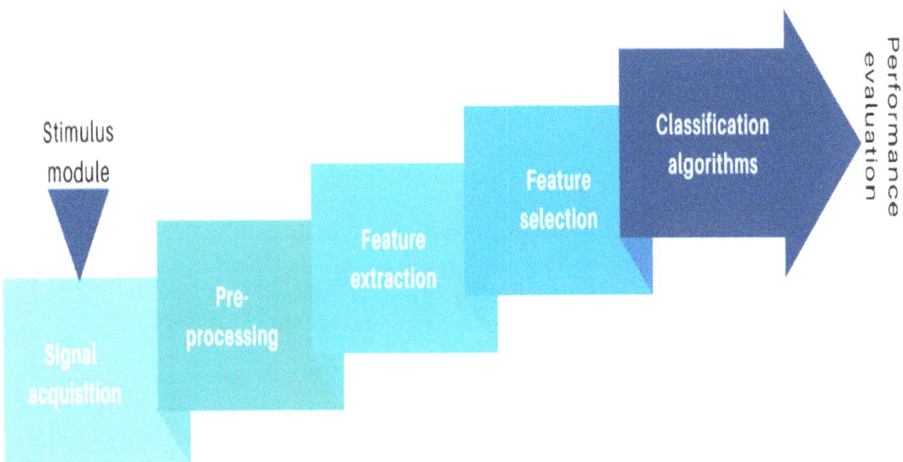

Figure 3. Components of an EEG-based BCI for emotion recognition.

2.1. *Signal Acquisition*

Inexpensive wearable EEG helmets and headsets that position noninvasive electrodes along the scalp can efficiently acquire EEG signals. The clinical definition of EEG is an electrical signal recording of brain activity over time. Thus, electrodes capture signals, amplify them, and send them to a computer (or mobile device) for storage and processing. Currently, there are various low-cost EEG-based BCI devices available on the market [22]. However, many current models of EEG-based BCI become incommodious after continued use. Therefore, it is still necessary to improve their usability.

2.1.1. Public Databases

Alternatively, there are also public databases with EEG data for affective information. Table 2 presents a list of available datasets related to emotion recognition. Such datasets are convenient for research, and several emotion recognition studies use them.

Table 2. Publicly available datasets.

Source	Dataset	Number of Channels	Emotion Elicitation	Number of Participants	Target Emotions
[19]	DEAP	32 EEG channels	Music videos	32	Valence, arousal, dominance, liking
[23]	eNTERFACE'06	54 EEG channels	Selected images from IAPS.	5	Calm, positive, exciting, negative exciting
[24]	headIT	-	Recall past emotions	31	Positive valence (joy, happiness) or of negative valence (sadness, anger)
[25]	SEED	62 channels	Film clips	15	Positive, negative, neutral
[26]	SEED-IV	62 channels	72 film clips	15	Happy, sad, neutral, fear
[27]	Mahnob-HCI-tagging	32 channels	Fragments of movies and pictures.	30	Valence and arousal rated with the self-assessment manikin
[28]	EEG Alpha Waves dataset	16 channels	Resting-state eyes open/closed experimental protocol	20	Relaxation
[29]	DREAMER	14 channels	Film clips	23	Rating 1 to 5 to valence, arousal, and dominance
[30]	RCLS	64 channels	Native Chinese Affective Video System	14	Happy, sad, and neutral

2.1.2. Emotion Elicitation

The International Affective Picture System (IAPS) [31] and the International Affective Digitized Sound System (IADS) [32] are the most popular resources for emotion elicitation. These datasets provide emotional stimuli in a standardized way. Hence, it is useful for experimental investigations.

IAPS consists of 1200 images divided into 20 sets of 60 photos. Valence and arousal values are tagged for each photograph. IADS' latest version provides 167 digitally recorded natural sounds familiar in daily life, with sounds labeled for valence, arousal, and dominance. Participants labeled the dataset using the Self-Assessment Manikin system [12]. IAPS and IADS stimuli are accessible with labeled information, which is convenient for the construction of a ground-truth for emotion assessment [33].

Other researchers used movie clips, which have also been shown capable of provoking emotions. In [34], the authors state that emotions using visual or auditory stimuli are similar. However, results obtained through affective labeling of multimedia may not be generalizable to more interactive situations or everyday circumstances. Thus, new studies using interactive emotional stimuli to ensure the generalizability of results for BCI would be welcomed.

Numerous experiments stimulated emotions in different settings, but they do not use EEG devices. However, they collected other physiological indicators as heartrate, skin galvanic changes, and respiration rate, among others. Conceptually, such paradigms could be useful if they are replicated for EEG signal acquisition. Possible experiments include stress during interviews for the detection of anger, anxiety, rejection, and depression. Exposure to odorants triggers emotions, such as anger, disgust, fear, happiness, sadness, and surprise. Harassment provokes fear. A threat of short-circuit, or a sudden backward-tilting chair elicits fear. A thread of shock provokes anxiety. Naturally, these EEG-based BCIs experiments should take into account ethical considerations.

To our knowledge, only a few studies have used more interactive conditions where participants played games or used flight simulators to induce emotions [35,36]. Alternatively, some authors have successfully used auto-induced emotions through memory recall [37].

2.1.3. Normalization

EEG signals vary widely in amplitude depending on age, sex, and other factors like changes in subjects' alertness during the day. Hence, it is necessary to normalize measured values to deal with this variability.

There are three possible approaches to normalization. The first is to record reference conditions without stimulus on the subject. The values obtained can be normalized by subtracting the reference value, then dividing by the reference value (or subtracting the reference value), and then dividing by that same value. The second approach also requires reference conditions. Those values are included in the feature vector, which will have twice the characteristics that make up the "baseline matrix". The third approach normalizes the data separately by obtaining a specific range, for example, between −1 and 1. This method applied to each feature independently ensures that all characteristics have the same value ranges [38,39].

The effect of normalization and its influence on the entire process of emotion recognition is not yet evident. However, some studies show that normalization allows the characteristics to be generalized so that they can be used in cross-subject emotion recognition. Tangentially, data normalization helps machine learning algorithms' efficiency due to faster convergence.

2.2. Preprocessing

EEG signals' preprocessing relates to signal cleaning and enhancement. EEG signals are weak and easily contaminated with noise from internal and external sources. Thus, these processes are essential to avoid noise contamination that could affect posterior classification. The body itself may produce electrical impulses through blinking, eye or muscular movement, or even heartbeats that blend with EEG signals. It should be carefully considered whether these artifacts should be removed because they may have relevant emotional state information and could improve emotion recognition algorithms' performance. If filters are used, it is necessary to use caution to apply them to avoid signal distortions.

The three commonly used filter types in EEG are (1) low-frequency filters, (2) high-frequency filters (commonly known by electrical engineers as low-pass and high-pass filters), and (3) notch filters. The first two filters are used to filter frequencies between 1 and 50–60 Hz.

For EEG signal processing, filters, such as Butterworth, Chebyshev, or inverse Chebyshev, are preferred [39]. Each of them has specific features that need to be analyzed. A Butterworth filter has a flat response in the passband and the stopband but also has a wide transition zone. The Chebyshev filter has a ripple on the passband, and a steeper transition, so it is monotonic on the stopband. The inverse Chevishev has a flat response in the passband, is narrow in the transition, and has a ripple in the stopband. A Butterworth phase zero filter should be used to prevent a phase shift because this filter goes forward and backward over the signal to avoid this problem.

Another preprocessing objective is to clean the noise that may correspond to low-frequency signals generated by an external source, such as power line interference [40]. Notch filters are used to stop the passage of a specific frequency rather than a frequency range. This filter is designed to eliminate frequencies originated by electrical networks, and it typically ranges from 50 to 60 Hz depending on the electrical signal's frequency in the specific country.

All of these filters are appropriate for artifact elimination in EEG signals. However, as previously noted, care must be taken when using filters. Generally, filters could distort the EEG signal's waveform and structure in the time domain. Hence, filtering should be kept to a minimum to avoid loss of EEG signal information.

Nevertheless, preprocessing helps to separate different signals and sources. Table 3 shows methods used for preprocessing EEG signals [41] and the percentage in which they are mentioned in the literature as used from 2015 to 2020. Independent Component Analysis (ICA) and Principal Component Analysis (PCA) are tools that apply blind source analysis to isolate the source signal from noise when using multi-channel recordings so they can be used for artifact removal and noise reduction. Common Average Reference (CAR) is right for noise reduction. SL is applied for spatial filtering to improve the signal's spatial resolution. The Common Spatial Patterns (CSP) algorithm finds spatial filters that could serve to distinguish signals corresponding to muscular movements.

Table 3. Frequently used pre-preprocessing methods of EEG signals.

Preprocessing Method	Main Characteristics	Advantages	Limitations	Literature's Usage Statistics % (2015–2020)
Independent component analysis (ICA) [42]	ICA separates artifacts from EEG signals into independent components based on the data's characteristics without relying on reference channels. It decomposes the multi-channel EEG data into temporal separate and spatial-fixed components. It has been applied for ocular artifact extraction.	ICA efficiently separates artifacts from noise components. ICA decomposes signals into temporal independent and spatially fixed components.	ICA is successful only under specific conditions where one of the signals is of greater magnitude than the others. The quality of the corrected signals depends strongly on the quality of the artifacts.	26.8
Common Average Reference (CAR) [43,44]	CAR is used to generate a reference for each channel. The algorithm obtains an average or all the recordings on every electrode and then uses it as a reference. The result is an improvement in the quality of Signal to Noise Ratio.	CAR outperforms standard types of electrical referencing, reducing noise by >30%.	The average calculation may present problems for finite sample density and incomplete head coverage.	5.0
Surface Laplacian (SL) [45–49]	SL is a way of viewing the EEG data with high spatial resolution. It is an estimate of current density entering or leaving the scalp through the skull, considering the volume conductor's outer shape and does not require details of volume conduction.	SL estimates are reference-free, meaning that any EEG recording reference scheme will render the same SL estimates. SL enhances the spatial resolution of the EEG signal. SL does not require any additional assumptions about functional neuroanatomy.	It is sensitive to artifacts and spline patterns.	0.4
Principal Component Analysis (PCA) [35,50–55]	PCA finds patterns in data. It can be pictured as a rotation of the coordinate axes so that they are not along with single time points. Still, along with linear combinations of sets of time points, collectively represents a pattern within the signal. PCA rotates the axes to maximize the variance within the data along the first axis, maintaining their orthogonality.	PCA helps in the reduction of feature dimensions. The ranking will be done and helps in the classification of data.	PCA does not eliminate noise, but it can reduce it. PCA compresses data compared to ICA and allows for data separation.	50.1
Common Spatial Patterns (CSP) [55–57]	CSP applies spatial filters that are used to discriminate different classes of EEG signals. For instance, those corresponding to different motor activity types. CSP also estimates covariance matrices.	CSP does not require a priori selection of sub-specific bands and knowledge of these bands.	CSP requires many electrodes. Changes in electrode location may affect classification accuracies.	17.7

Therefore, each of the most widely used preprocessing algorithms has its benefits. In Table 3, we can observe from the percentage of the usage column that the most utilized algorithms for preprocessing are PCA (50.1%), ICA (26.8%), and CSP (17.7%).

2.3. *Feature Extraction*

Once signals are noise free, the BCI needs to extract essential features, which will be fed to the classifier. Features can be computed in the domain of (1) time, (2) frequency, (3) time-frequency, or (4) space, as shown in Table 4 [31,38,39]. This table presents the most popular techniques used for feature extraction, their domain, advantages, and limitations.

Time-domain features include the event-related potential (ERP), Hjorth features, and higher-order crossing (HOC) [58–60], independent component analysis (ICA), principal component analysis (PCA), and Higuchi's fractal dimensions (FD) as a measure of signal complexity and self-similarity in this domain. There are also statistical measures, such as power, mean, standard deviation, variance, skewness, kurtosis, relative band energy, and entropy. The latter evaluates signal randomness [61].

Among frequency-domain methods, the most popular is the fast Fourier transform (FFT). Auto-regressive (AR) modeling is an alternative to Fourier-based methods for computing the frequency spectrum of a signal [62,63].

The time-frequency domain exploits variations in time and frequency, which are very descriptive of the neural activities. For this, wavelet transform (WT) and wavelet packet decomposition (WPD) are used [62].

The spatial information provided in the description of EEG signals' characteristics is also considered in a broader approach. For this dimension, signals are referenced to digitally linked ears (DLE) values, which are calculated in terms of the left and right earlobes as follows:

$$V_e{}^{DLE} = V_e - \frac{1}{2}(V_{A1} + V_{A2}), \quad (1)$$

where V_{A1} and V_{A2} are the reference voltages on the left and right earlobe. Thus, EEG data is broken down, considering each electrode. Consequently, each channel contains spatial information of the location pertinent to its source.

For spatial computation, the surface Laplacian (SL) algorithm reduces volume conduction effects dramatically. SL also improves EEG spatial resolution by reducing the distortion produced by volume conduction and reference electrodes [47].

Figure 4 shows EEG signals in the time domain, the frequency domain, and spatial information.

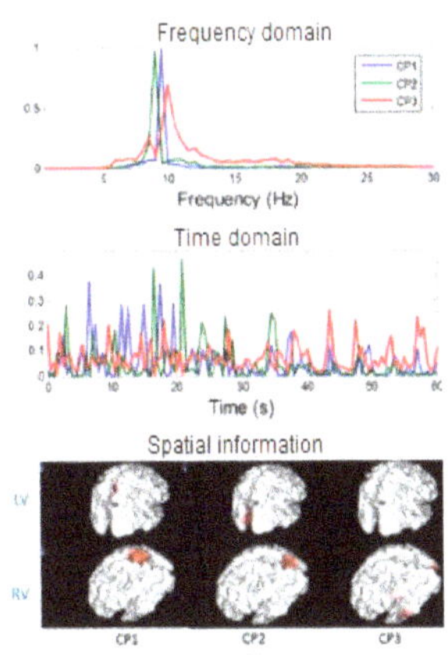

Figure 4. Frequency domain, time domain, and spatial information [63].

Table 4. Feature extraction algorithms.

Feature Extraction Method	Main Characteristics	Domain	Advantages	Limitations	Literature's usage statistics % (2015–2020)
ERP [18,40,64–69]	It is the brain response to a sensory, cognitive, or motor event. Two sub-classifications are (1) evoked potentials and (2) induced potentials.	Time	It has an excellent temporal resolution. ERPs provide a measure of the processing between a stimulus and a response.	ERP has a poor spatial resolution, so it is not useful for research questions related to the activity location.	2.9
Hjorth Features [52,59,60]	These are statistical indicators whose parameters are normalized slope descriptors. These indicators are activity (variance of a time function), mobility (mean frequency of the proportion of standard deviation of the power spectrum), and complexity (change in frequency compared to the signal's similarity to a pure sine wave).	Time	Low computational cost appropriate for real-time analysis.	Possible statistical bias in signal parameter calculations	17.0
Statistical Measures [39,40,42,52,61–70]	Signal statistics: power, mean, standard deviation, variance, kurtosis, relative band energy.	Time	Low computational cost.	-	8.6
DE [1,10,11,15,59,68,71–84]	Entropy evidences scattering in data. Differential Entropy can reflect spatial signal variations.	Time–spatial	Entropy and derivate indexes reflect the intra-cortical information flow.		4.9
HOC [1,2,42,63,85–88]	Oscillation in times series can be represented by counts of axis crossing and its differences. HOC displays a monotone property whose rate of increase discriminates between processes.	Time	HOC reveals the oscillatory pattern of the EEG signal providing a feature set that conveys enough emotion information to the classification space.	The training process is time-consuming due to the dependence of the HOC order on different channels and different channel combinations [60].	2.0
ICA [20,37,53,69,89–91]	ICA is a signal enhancing method and a feature extraction algorithm. ICA separates components that are independent of each other based on the statistical independence principle.	Time. There is also a FastICA in the frequency domain.	ICA efficiently separates artifacts from noise components. ICA decomposes signals into temporal independent and spatially fixed components.	ICA is only useful under specific conditions (one of the signals is of greater magnitude than the others). The quality of the corrected signals depends strongly on the quality of the isolated artifacts.	11.3
PCA [33,40,52,69,92–95]	The PCA algorithm is mostly used for feature extraction but could also be used for feature extraction. It reduces the dimensionality of the signals creating new uncorrelated variables.	Time	PCA reduces data dimensionality without information loss.	PCA assumes that the data is linear and continuous.	19.7
WT [48]	The WT method represents the original EEG signal with secured and straightforward building blocks known as wavelets, which can be discrete or continuous.	Time-frequency	WT describes the features of the signal within a specified frequency domain and localized time domain properties. It is used to analyze irregular data patterns. Uses variable windows, wide for low frequencies, and narrow for high frequencies.	High computational and memory requirements.	26.0

Table 4. *Cont.*

Feature Extraction Method	Main Characteristics	Domain	Advantages	Limitations	Literature's usage statistics % (2015–2020)
AR [48]	AR is used for feature extraction in the frequency domain. AR estimates the power spectrum density (PSD) of the EEG using a parametric approach. The estimation of PSD is achieved by calculating the coefficients or parameters of the linear system under consideration.	Frequency domain	AR is used for feature extraction in the frequency domain. AR limits the leakage problem in the spectral domain and improves frequency resolution.	The order of the model in the spectral estimation is challenging to select. It is susceptible to biases and variability.	1.6
WPD [95]	WPD generates a sub-band tree structuring since a full binary tree can characterize the decomposition process. WPD decomposes the original signals orthogonally and independently from each other and satisfies the law of conservation of energy. The energy distribution is extracted as the feature.	Time-frequency	WPD can analyze non-stationary signals such as EEG.	WPD uses a high computational time to analyze the signals.	1.6
FFT [48]	FFT is an analysis method in the frequency domain. EEG signal characteristics are reviewed and computed by power spectral density (PSD) estimation to represent the EEG samples signal selectively.	Frequency	FFT has a higher speed than all the available methods so that it can be used for real-time applications. It is a useful tool for stationary signal processing.	FFT has low-frequency resolution and high spectral loss of information, which makes it hard to find the actual frequency of the signal.	2.2
Functional EEG connectivity indices [15]	EEG-based functional connectivity is estimated in the frequency bands for all pairs of electrodes using correlation, coherence, and phase synchronization index. Repeated measures of variance for each frequency band were used to determine different connectivity indices among all pairs.	Frequency	Connectivity indices at each frequency band can be used as features to recognize emotional states.	Difficult to generalize and distinguish individual differences in functional brain activity.	1.3
Rhythm [14,56]	Detection of repeating patterns in the frequency band or "rhythm".	Frequency	Specific band rhythms contribute to emotion recognition.	-	0.1
Graph Regularized Sparse Linear Regularized GRSLR [30]	This method applies a graph regularization and a sparse regularization on the transform matrix of linear regression	Frequency	It can simultaneously cope with sparse transform matrix learning while preserving the intrinsic manifold of the data samples.	-	0.2
Granger causality [63,96]	This feature is a statistical concept of causation that is based on prediction.	Frequency	The authors can analyze the brain's underlying structural connectivity.	These features only give information about the linear characteristics of signals.	0.6

According to [97], emotions emerge as the synchronization of various subsystems. Several authors use synchronized activity indexes in different parts of the brain. The efficiency of these indexes has been demonstrated in [98], calculating the correlation dimension of a group of EEG signals. In [98], other methods were used to calculate the synchronization of different areas of the brain. Synchronized indexes are a promising method for emotion recognition that deserves further research.

Table 4 shows the most commonly used algorithms and their respective mention percentages in the literature: (1) WT (26%), (2) PCA (19.7%), (3) Hjorth (17%), (4) ICA (11.3%), and (5) statistical measures (8.6%).

2.4. Feature Selection

The feature selection process is vital because it obtains the signal's properties that best describe the EEG characteristics to be classified. In BCI systems, the feature vector generally has high dimensionality [99]. Feature selection reduces the number of input variables for the classifier (not to be confused with dimensionality reduction). While both processes decrease the data's attributes, dimensionality reduction combines features to reduce their quantity.

A feature selection method does not change characteristics but excludes some according to specific usefulness criteria. Feature selection methods aim to achieve the best results by processing the least amount of data. It serves to remove attributes that do not contribute to the classification because they are irrelevant (or redundant) for simpler classification models (which are faster and have better performance). Additionally, feature selection methods reduce the overfitting likelihood in regular datasets, flexible models, or when the dataset has too many features but not enough observations.

One classification of feature selection methods based on the number of variables divides them into two classes: (1) Univariate and (2) multivariate. Univariate methods consider the input features one by one. Multivariate methods consider whole groups of characteristics together.

Another classification distinguishes feature selection methods as filtering, wrapper, and built-in algorithms.

- Filter methods evaluate features using the data's intrinsic properties. Additionally, most of the filtering methods are univariate, so each feature is self-evaluated. These methods are appropriate for large data sets because they are less computationally expensive.
- Wrapping methods depend on classifier types when selecting new features based on their impact on characteristics already chosen. Only features that increase accuracy are selected.
- Built-in methods run internally in the classifier algorithms, such as deep learning. This type of process requires less computation than wrapper methods.

Examples of Feature Selection Algorithms

The following are some examples of algorithms for feature selection:

- Effect-size (ES)-based feature selection is a filter method. ES-based univariate: Cohen's is an appropriate effect size for comparisons between two means [100]. So, if two groups' means do not differ by 0.2 standard deviations or more, the difference is trivial, even if it is statistically significant. The effect size is calculated by taking the difference between the two groups and dividing it by the standard deviation of one of the groups. Univariate methods may discard features that could have provided useful information. ES-based multivariate helps remove several features with redundant information, therefore selecting fewer features, while retaining the most information [58]. It considers all the dependencies between characteristics when evaluating them. For example, calculating the Mahalanobis distance using the covariance structure of the noise. Min-redundancy max-relevance (mRMR) is a wrapper method [101]. This algorithm compares the mutual information between each feature with each class at the output. Mutual information between two random variables x and y is calculated as:

$$I(x;y) = \iint p(x,y) \log \frac{p(x,y)}{p(x)p(y)} dxdy, \quad (2)$$

where $p\ (x)$ and $p\ (y)$ are the marginal probability density functions of x and y, respectively, and $p\ (x, y)$ is their joint probability function. If $I\ (x, y)$ equals zero, the two random variables x and y are statistically independent [58]. mRMR maximizes $I\ (x_i, y)$ between each characteristic xi and the target vector y; and minimizes the average mutual information $I\ (x_i, y_i)$ between two characteristics.

- Genetic algorithms allow the dimensionality of the feature vector to be reduced using evolutionary methods, leaving only more informative feature [2,86,97].
- Stepwise discriminant analysis SDA [74]. SDA is the extension of the statistical tool for discriminant analysis that includes the stepwise technique.
- Fisher score is a feature selection technique to calculate interrelation between output classes and each feature using statistic measures [101].

Table 5 shows feature selection algorithms and their percentage of usage in the literature. Genetic algorithms are frequently used (32.3%), followed by SDA (17.7%), wrapper methods (15.6%), and mRMR (11.5%).

Table 5. Feature selection methods used in the literature (2015–2020) in percentages (%).

Feature Selection Method	Literature's Usage Statistics % (2015–2020)
min-Redundancy Max-Relevance mRMR	11.5%
Univariate	6.3%
Multivariate	6.3%
Genetic Algorithms	32.3%
Stepwise Discriminant Analysis SDA	17.7%
Fisher score	7.3%
Wrapper methods	15.6%
Built-in methods	3.1%

2.5. Classification Algorithms

Model frameworks can categorize classification algorithms [56,57]. The model's categories may be (1) generative-discriminative, (2) static-dynamic, (3) stable-unstable, and (4) regularized [102–104].

There are two different selection approaches for the classifier that works best under certain conditions in emotion recognition [56]. The first identifies the best classifier for a given BCI device. The second specifies the best classifier for a given set of features.

For synchronous BCIs, dynamic classifiers and ensemble combinations have shown better performances than SVMs. For asynchronous BCIs, the authors in this field have not determined an optimal classifier. However, it seems that dynamic classifiers perform better than static classifiers [56] because they handle better the identification of the onset of mental processes.

From the second approach, discriminative classifiers have been found to perform better than generative classifiers, principally in the presence of noise or outliers. Dynamic classifiers like SVM generally handle high dimensionality in the features better. If there is a small training set, simple techniques like LDA classifiers may yield satisfactory results [58].

2.5.1. Generative Discriminative

These classifier models generally have supervised learning problems that fit the data's probability. A generative model specifies the distribution of each class using the joint probability distribution p(x,y) and Bayes theorem. A discriminative model finds the decision boundary between the categories using the conditional probability distribution p(y|x). Such a model includes the following classifiers: Naïve Bayes, Bayesian networks, Markov random fields, and hidden Markov models (HMM).

2.5.2. Static-Dynamic Classification

Static-dynamic classification takes into account the training method's time variations. A static model trains the data once and then uses the trained model to classify a single feature vector. In a dynamic model, the system is updated continually. Thus, dynamic models can obtain a sequence of feature vectors and catch temporal dynamics.

Multilayer perceptron (MLP) can be considered a static classifier. Likewise, an example of a dynamic classifier is hidden Markov methods (HMM) because it can classify a sequence of feature vectors.

2.5.3. Stable Unstable

Stable classifiers usually have low complexity and do not affect their performance with small variations of the training set. For example, k Nearest Neighbors (kNN) is a common stable classifier. Unstable classifiers have high complexity and present considerable changes in performance with minor variations of the training set. Examples of unstable classifiers are linear support vector machine (SVM), multi-layer perceptron (MLP), and bilinear recurrent neural network (BLR-NN).

2.5.4. Regularized

Regularization consists of carefully controlling classifier complexity to prevent overtraining. These classifiers have excellent generalization performance. Regularized's Fisher LDA (RF-LDA), linear SVM, and radial basis function kernel for support vector machine (RBF-SVM) are examples of regularized classifiers.

2.5.5. General Taxonomy of Classification Algorithms

Another taxonomy divides classifiers using their properties to distinguish them into general types of algorithms as linear, neural networks, nonlinear Bayesian, nearest neighbor classifiers, and combinations of systems (ensemble). Most of the more specialized algorithms can be generated from these general types. Table 6 shows this taxonomy criterion with five different categories of general classifiers: (1) Linear, (2) neural networks, (3) nonlinear Bayesian, (4) nearest neighbor classifiers, and (5) combinations of classifiers or ensemble [44,56,58].

All general classifiers have characteristics of each of the previously mentioned framework models. For instance, SVM is discriminant, static, stable, and regularized; HMM is generative, dynamic, unstable, and not regularized; and kNN is discriminant, static, stable, and not regularized.

Consequently, the suggested guidelines for classifier selection are also applicable in this categorization. Table 6 presents the usage statistics of these classifiers in the 2015–2020 literature. The following are the most noteworthy classifiers: Neural networks CNN (46.16%), Linear classifiers SVM (30.3%), and LDA (5.5%), Nearest Neighbors kNN (4.5%), and Ensembled classifier AdaBoost (3.9%).

Table 6. Categories of general classifiers.

Category of Classifier	Description	Examples of Algorithms in the Category	Advantages	Limitations	Literature's Usage Statistics % (2015–2020)
Linear	Discriminant algorithms that use linear functions (hyperplanes) to separate classes.	Linear Discriminant Analysis LDA [65]. Bayesian Linear Discriminant Analysis. Support Vector Machine SVM [105,106]. Graph Regularized Sparse Linear Regularized GRSLR [30].	These algorithms have reasonable classification accuracy and generalization properties.	Linear algorithms tend to have poor outcomes in processing complex nonlinear EEG data.	5.50 1.40 30.30 0.02
Neural networks (NN)	NN are discriminant algorithms that recognize underlying relationships in a set of data resembling the human brain operation.	Multilayer Perceptron MLP [107]. Long Short-term Memory Recurrent Neural Network LSTM-RNN [66–69]. Domain Adversarial Neural Network DANN [108]. Convolutional Neural Network CNN [68,70–73,109–111]. Complex-Valued Convolutional Neural Network CVCNN [105]. Gated-Shape Convolutional Neural Network GSCNN [105]. Global Space Local Time Filter Convolutional Neural Network GSLTFCNN [105]. CapsNet-NN Genetic Extreme Learning Machine GELM–NN [82].	NN generally yields good classification accuracy	Sensitive to overfitting with noisy and non-stationary data as EEGs.	1.60 1.10 0.20 46.16 0.40 0.40 0.02 0.10 0.10
Nonlinear Bayesian classifier	Generative classifiers produce nonlinear decision boundaries.	Bayes quadratic [110]. Hidden Markov Model HMM [50,112].	Generative classifiers reject uncertain samples efficiently.	For Bayes quadratic, the covariance matrix cannot be estimated accurately if the dimensionality is vast, and there are not enough training sample patterns.	0.10 0.30
Nearest neighbor classifiers	Discriminative algorithms that classify cases based on its similarity to other samples	k-Nearest Neighbors kNN [113]. Mahalanobis Distance [114].	kNN has excellent performance with low-dimensional feature vectors. Mahalanobis Distance is a simple but efficient classifier, suitable even for asynchronous BCI.	kNN has reduced performance for classifying high dimension feature vectors or noise distorted features.	4.5 0.1
Combination of classifiers (ensemble-learning)	Combined classifiers using boosting, voting, or stacking. Boosting consists of several cascading classifiers. In voting, classifiers have scores, which yield a combined score per class, and a final class label. Stacking uses classifiers as meta-classifier inputs.	Ensemble-methods can combine almost any type of classifier [115]. Random Forest [10,116]. Bagging Tree [111,115]. XGBoost [117] AdaBoost [118]	Variance reduction that leads to increase of classification accuracy.	Quality measures are application dependent.	2.1 1.1 0.2 0.4 3.9

2.6. Performance Evaluation

Results must be reported consistently so that different research groups can understand and compare them. Hence, evaluation procedures need to be chosen and described accurately [119]. The evaluation of the classifier's execution involves addressing performance measures, error estimation, and statistical significance testing [120]. Performance measures and error estimation configure the fulfillment rate of the classifier's function. The most recommended performance evaluation measures are shown in Table 7. They are confusion matrix, accuracy, error rating, and other measures obtained from the confusion matrix, such as the recall, specificity, precision, Area Under the Curve (AUC), and F-measure. Other performance evaluation coefficients are Cohen's kappa (k) [121], information transfer rate (ITR) [65], and written symbol rate (WSR) [121].

Performance evaluation and error estimation may need to be complemented with a significance evaluation. This is because high accuracies can be of little impact if the sample size is too small, or classes are imbalanced (labeled EEG signals typically are). Therefore, significance classification is essential. There are general approaches that can handle arbitrary class distributions to verify accuracy values that lie significantly above certain levels. Used methods are the theoretical level of random classification and adjusted Wald confidence interval for classification accuracy.

The theoretical level of random classification test classification results for randomness is the sum of the products between the experimental results' classification probability and the probability calculated if all the categorization randomly occurs (p_0 = classification accuracy of a random classifier). This approach can only be used after the classification has been performed [122].

Adjusted Wald confidence interval gives the lower and upper confidence limits for the probability of the correct classification, which specifies the intervals for the classifier performance evaluation index [123].

Table 7. Conventional performance evaluation methods for BCI.

Performance Evaluation	Main characteristics	Advantages	Limitations
Confusion matrix	The confusion matrix presents the number of correct and erroneous classifications specifying the erroneously categorized class.	The confusion matrix gives insights into the classifier's error types (correct and incorrect predictions for each class). It is a good option for reporting results in M-class classification.	Results are difficult to compare and discuss. Instead, some authors use some parameters extracted from the confusion matrix.
Accuracy and error rate	The accuracy p is the probability of correct classification in a certain number of repeated measures. The error rate is e = 1 − p and corresponds to the probability that an incorrect classification has been made.	It works well if the classes are balanced, i.e., there are an equal number of samples belonging to each class.	Accuracy and error rate do not take into account whether the dataset is balanced or not. If one class occurs more than another, the evaluation may appear with a high value for accuracy even though the classification is not performing well. These parameters depend on the number of classes and the number of cases. In a 2-class problem the chance level is 50%, but with a confidence level depending on the number of cases.
Cohen's kappa (k)	k is agreement evaluation between nominal scales. This index measures the agreement between a true class compared to a classifier output. 1 is a perfect agreement, and 0 is pure chance agreement.	Cohen's kappa returns the theoretical chance level of a classifier. This index evaluates the classifier realistically. If k has a low value, the confusion matrix would not have a meaningful classification even with high accuracy values. This coefficient presents more information than simple percentages because it uses the entire confusion matrix.	This coefficient has to be interpreted appropriately. It is necessary to report the bias and prevalence of the k value and test the significance for a minimum acceptable level of agreement.
Sensitivity or Recall	Sensitivity, also called Recall, identifies the true positive rate for describing the accuracy of classification results. It evaluates the proportion of correctly identified true positives related to the sum of true positives plus false negatives.	Sensitivity measures how often a classifier correctly categorizes a positive result.	The Recall should not be used when the positive class is larger (imbalanced dataset), and correct detection of positives samples is less critical to the problem.
Specificity	Specificity is the ability to identify a true negative rate. It measures the proportion of correctly identified true negatives over the sum of the true negatives plus false positives. The False Positive Rate (FPR) is then equal to 1 − Specificity.	Specificity measures how often a classifier correctly categorizes a negative result.	Specificity focuses on one class only, and the majority class biases it.
Precision	Precision also referred to as Positive Predicted Value, is calculated as 1 − False Detection Rate (F). False detection rate is the ratio between false positives over the sum of true positives plus false positives.	Precision measures the fraction of correct classifications.	Precision should not be used when the positive class is larger (imbalanced dataset), and correct detection of positives samples is less critical to the problem.

Table 7. *Cont.*

Performance Evaluation	Main characteristics	Advantages	Limitations
ROC	The ROC curve is a Sensitivity plot as a function of the False Positive Rate. The area under the ROC curve is a measure of how well a parameter can distinguish between a true positive and a true negative.	ROC curve provides a measure of the classifier performance across different significance levels.	ROC is not recommended when the negative class is smaller but more important. The Precision and Recall will mostly reflect the ability to predict the positive class if it is larger in an imbalanced dataset.
F-Measure	F-Measure is the harmonic mean of Precision and Recall. It is useful because as the Precision increases, Recall decreases, and vice versa.	F-measure can handle imbalanced data. F-measure (like ROC and kappa) provides a measure of the classifier performance across different significance levels.	F-measure does not generally take into account true negatives. True negatives can change without affecting the F-measure.
Pearson correlation coefficient	Pearson's correlation coefficient (r), quantifies the degree of a ratio between the true and predicted values by a value ranking from −1 to +1.	Pearson's correlation is a valid way to measure the performance of a regression algorithm.	Pearson's correlation ignores any bias which might exist between the true and the predicted values.
Information transfer rate (ITR)	As BCI is a channel from the brain to a device, it is possible to estimate the bits transmitted from the brain. ITR is a standard metric for measuring the information sent within a given time in bits per second.	ITR is a metric that contributes to criteria to evaluate a BCI System.	ITR is often misreported due to inadequate understanding of many considerations as delays are necessary to process data, to present feedback, and clear the screen. TR is best suited for synchronous BCIs over user-paced BCI.

3. Literature Review of BCI Systems that Estimate Emotional States

In recent years, several research papers have been published in emotion recognition using BCI devices for data capture. Such publications use different models and strategies that produce a wide range of frameworks. Table 8 offers a summary of the research in this field from 2015 to 2020.

The following components characterize the systems presented in Table 8: (1) Stimulus type; (2) databases, generated by the paper's authors or publicly available; (3) the number of participants; (4) extraction and selection of characteristics; (5) features; (6) classification algorithms; (7) number and types of classes; and (8) performance evaluation.

The applied preprocessing methods are mostly similar in the reviewed studies. Their primary preprocessing methods are standard, so this information was omitted in Table 8.

3.1. Emotion Elicitation Methods

This article analyzes research papers that used different resources to provoke emotions in their subjects. These stimuli are music videos, film clips, music tracks, self-induced disgust (produced by remembering an unpleasant odor), and risky situations in a flight simulator as an example of active elicitation of emotions. EEG-based BCI systems frequently use the public DEAP and SEED databases that apply music videos and film clips as stimuli, respectively. Different stimuli provoke emotions that affect different areas of the brain and produce EEG signals that can be recognized concerning specific emotions. Figure 5 shows the frequency in which different emotion elicitation methods are applied to generate datasets used in the reviewed systems.

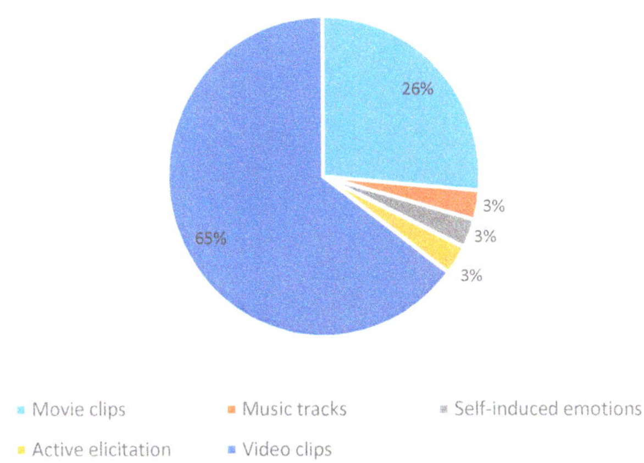

Figure 5. Emotion elicitation methods.

Few research papers resort to more elaborate platforms to provoke "real life" emotions. However, such methods have been applied to other physiological responses (other than EEG like skin conductance, respiration, electrocardiogram (ECG), facial expressions, among others) [124]. Some authors state that stimuli that provoke wide-ranging emotions could make it challenging to explore the brain's mechanisms activated for specific emotion generation. In this sense, focusing on a particular emotion could improve our understanding of such mechanisms. For our research sample, we highlighted research pieces that study emotions, such as dislike, and disgust separately [37,125].

Table 8. Summary of emotion recognition systems using BCI [1].

Reference/Year	Stimuli	EEG Data	Feature Extraction	Feature Selection	Features	Classification	Emotions	Accuracy
[126]/2016	-	DEAP	Computation in the time domain, Hjorth, Higuchi, FFT	mRMR	Statistical features, BP, Hjorth, FD	RBF NN SVM	3 class/Arousal 3 class/Valence	Arousal/60.7% Valence/62.33%
[85]/2015	15 movie clips	Own dataset/15 participants	DBN	-	DE, DASM, RASM, DCAU, from Delta, Theta, Alpha, Beta, and Gamma.	kNN LR SVM DBNs	Positive Neutral Negative.	SVM/83.99% DBN/86.08%
[37]/2015	Self-induced emotions	Own dataset/10 participants	WT	PCA	Eigenvalues vector	SVM	Disgust	Avg. 90.2%
[127]/2018	Video clips	Own dataset/10 participants	Higuchi	-	FD	RBF SVM	Happy Calm Angry	Avg. 60%
[128]/2017	Video clips	Own dataset/30 participants	SFTT, ERD, ERS	LDA	PSD	LIBSVM	Joy Amusement Tenderness Anger Disgust Fear Sadness Neutrality	Neutrality 81.26% 3 Positive emotions 86.43% 4 Negative emotions 65.09%
[125]/2020	-	DEAP	DFT, DWT	-	PSD, Logarithmic compression of Power Bands, LFCC, PSD, DW	NB CART kNN RBF SVM SMO	Dislike	Avg. SMO/81.1% NB/63.55% kNN/86.73% CAR/74.08%
[86]/2019	-	DEAP and SEED-IV	Computations in time domain, FFT, DWT	-	PSD, Energy, DE, Statistical features	SVM	HAHV HALV LALV LAHV	Avg DEAP/79% Avg.SEED/76.5%
[14]/2016	Music tracks	Own dataset/30 participants	SFTT, WT	-	PSD, BP Entropy, Energy, Statistical features, Wavelets	SVM MLP kNN	Happy Sad Love Anger	Avg. SVM/75.62% MLP/78.11% kNN/72.81%
[79]/2017	-	SEED	FFT, and electrode location	Max Pooling	DE, DASM, RASM, DCAU	SVM ELM Own NN method	Positive Negative Neutral	Avg. SVM/74.59% ELM/74.37% Own NN/86.71%
[48]/2019	Video clips	Own dataset/16 participants	SFTT, WT, Hjorth, AR	-	PSD, BP, Quadratic mean, AR Parameters, Hjorth	SVM	Happy Sad Fear Relaxed	Avg. 90.41%
[129]/2019	-	DEAP	WT	-	Wavelets	LSTM RNN	Valence Arousal	Avg. 59.03%
[130]/2018	-	SEED	LSTM to learn context information for each hemispheric data	-	DE	BiDANN	Positive Negative Neutral	Avg. 92.38%

Table 8. *Cont.*

Reference/Year	Stimuli	EEG Data	Feature Extraction	Feature Selection	Features	Classification	Emotions	Accuracy
[111]/2019	-	DEAP	Signal computation in the time domain, and FFT		Statistical characteristics. PSD	BT SVM LDA BLDA CNN	Valence Arousal	Avg. for combination features AUC BT/0.9254 BLDA/0.8093 SVM/0.7460 LDA/0.5147 CVCNN/0.9997 GSCNN/1 GSCNN/1
[118]/2017	-	DEAP	Computation in the time domain, and FFT	GA	Statistical characteristics, PSD, and nonlinear dynamic characteristics	AdaBoost	Joy Sadness	95.84%
[131]/2019	-	DEAP	SFTT, NMI	-	Inter-channel connection matrix based on NMI	SVM	HAHV HALV LALV LAHV	Arousal/73.64% Valence/74.41%
[74]/2018	-	SEED	FFT	SDA	Delta, Theta, Alpha, Beta, and, Gamma	LDA	Positive Negative Neutral	Avg. 93.21%
[112]/2019	-	SEED	FFT	-	Electrodes-frequency Distribution Maps (EFDMs)	CNN	Positive Negative Neutral	Avg. 82.16%
[80]/2019	-	SEED/ DEAP/ MAHNOB-HCI	Computation in the time domain, and FFT	Fisher-score, classifier-dependent structure (wrapper), mRMR, SFEW	EEG based network patterns (ENP) PSD, DE, ASM, DASM, RASM, DACU, ENP, PSD + ENP, DE + ENP	SVM GELM	Positive Negative Neutral	Best feature F1 SEED/DE+ENP gamma 0.88 DEAP/PSD+ENP gamma 0.62 MAHNOB/PSD+ENP Gamma 0.68
[96]/2019	-	DEAP	Tensorflow framework	Sparse group lasso	Granger causality feature	CapsNet Neural Network	Valence-arousal	Arousal/87.37% Valence/88.09%
[30]/2019	Video clips	Own dataset RCLS/14 participants. SEED	Computation in the time domain, WT	-	HOC, FD, Statistics, Hjorth, Wavelets	GRSLR	Happy Sad Neutral	81.13%
[132]/2019	-	DEAP	Computation in the time domain, FFT, WT	Correlation matrix, information gain, and sequential feature elimination	Statistical measures, Hjorth, Autoregressive parameters, frequency bands, the ratio between frequency bands, wavelet domain features	XGBoost	Valence, arousal, dominance, and liking	Valence/75.97% Arousal/74.20% Dominance/75.23% Liking 76.42%

Table 8. *Cont.*

Reference/Year	Stimuli	EEG Data	Feature Extraction	Feature Selection	Features	Classification	Emotions	Accuracy
[133]/2015	-	DEAP	Frequency phase information	Sequential feature elimination	Derived features of bispectrum	SVM	Low/high valence, low/high arousal	Low-high arousal/64.84% Low-high valence/61.17%
[134]/2016	-	DEAP	Higuchi, FFT	-	FD, PSD	SVM	Valence, arousal	Valence/86.91% Arousal/87.70%
[135]/2017	-	DEAP	DWT	-	Discrete wavelets	kNN	Valence, arousal	Valence/84.05% Arousal/86.75%
[136]/2015	-	DEAP	RBM	-	Raw signal-6 channels	Deep-Learning	Happy, calm, sad, scared	Avg. 75%
[137]/2017	-	DEAP	DWT	Best classification performance for channel selection	Discrete wavelets	MLP kNN	Positive, negative	MLP/77.14% kNN/72.92%
[138]/2017	-	DEAP	-	-	-	LSTM NN	Low/high valence, Low/high arousal, Low/high liking	Low-high valence/85.45% Low-high arousal/85.65% Low-high liking/87.99%
[139]/2018	-	DEAP	-	-	-	3D-CNN	Valence, arousal	Valence/87.44% Arousal/88.49%
[140]/2018	-	DEAP	FFT, phase computations, Pearson correlation	-	PSD, phase, phase synchronization, Pearson correlation	CNN	Valence	Valence/96.41%
[36]/2019	Flight simulator	Own dataset/8 participants	Computation in time domain, and WT	-	Statistical measures, DE, Wavelets	ANN	Happy, Sad, Angry, Surprise Scared	Avg. 53.18%

[1] Autoregressive Parameter (AR). Bagging Tree (BT). Band Power (BP). Bayesian linear discriminant analysis (BLDA). Bi-hemispheres Domain Adversarial Neural Network (BiDANN). Convolutional Neural Network (CNN). Complex-Valued Convolutional Neural Network (CVCNN). Gated-Shape Convolutional Neural Network (GSCNN). Global Space Local Time Filter Convolutional Neural Network (GSLTFCNN). Deep Belief Networks (DBNs). Differential entropy (DE). DE feature Differential Asymmetry (DASM). DE feature Rational Assimetry (RASM). DE feature Differential Caudality (DCAU). Electrooculography (EOG). Electromyogram (EMG). Event-Related Desynchronization (ERD) and Synchronization (ERS). Feature selection and weighting method (SFEW). Fractal dimensions (FD). Genetic Algorithm (GA). Graph regularized Extreme Learning Machine (GELM) NN. Graph Regularized Sparse Linear Regularized (GRSLR). High Order Crossing (HOC). Linear Discriminant Analysis (LDA). Logistic Regression (LR). Long short-term memory Recurrent Neural Network (LSTM RNN). Minimum-Redundancy-Maximum-Relevance (mRMR) Normalized Mutual Information (NMI). Principal Component Analysis (PCA). Radial Basis Function (RBF). Short-Time Fourier Transform (STFT). Stepwise Discriminant Analysis (SDA). Support Vector Machine (SVM). Wavelet Transform (WT).

3.2. Number of Participants to Generate the System Dataset

Figure 6 presents the number of participants in the experiments to obtain EEG datasets to train and test the emotion recognition systems. Most of the systems used a number of subjects in a range from 31–40 (53%), and 11–20 (31%). The targeted studies used EEG data from healthy individuals.

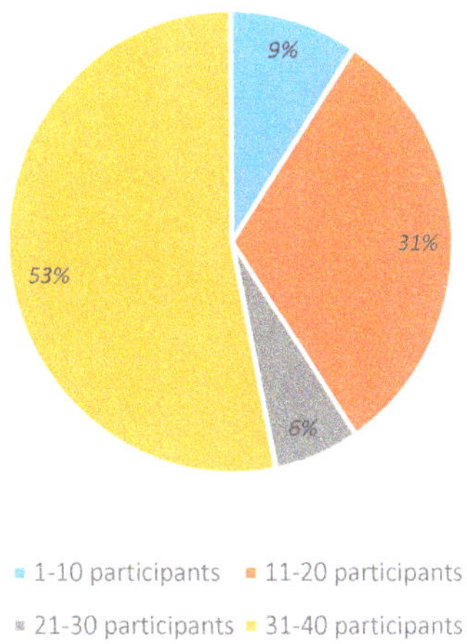

Figure 6. Number of participants in EEG datasets.

3.3. Datasets

Figure 7 presents the usage percentage of datasets used in emotion recognition. DEEP and SEED are publicly available databases, and are the most frequently used (49% and 23% of applications, respectively). Sometimes, other studies used self-generated datasets (23%), which are typically not freely accessible. The MAHNOB-HCI and RCLS public datasets appeared in our research sample, with a participation of 3% each.

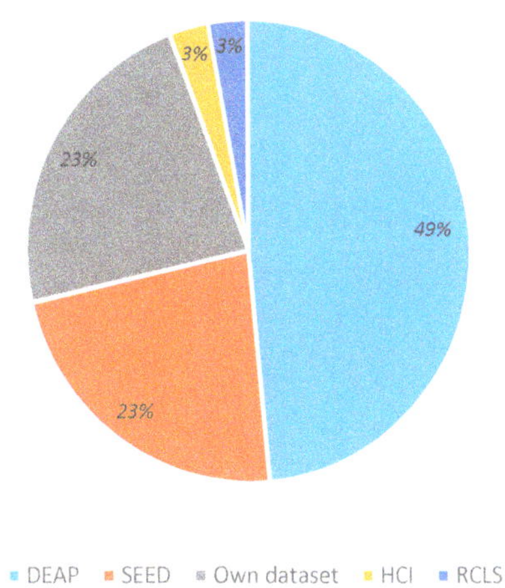

Figure 7. EEG datasets for emotion recognition.

Systems that use public databases offer some comparability, but contrast is limited even if the same characteristics are handled. Still, such public databases could eventually lead to findings if objective comparisons are performed.

3.4. Feature Extraction

Most systems use feature extraction methods in the time, frequency, time-frequency, or space domains. A small percentage of works evaluate the functional connectivity (or differences) in the observed activity between brain regions when emotions are provoked. Features with non-redundant information combined from different domains yield better classification results. However, it is still unclear if features work better alone or in combination with each other, or which type of features are more relevant for emotion recognition.

In our review, we found that researchers addressed these issues through the development of feature extraction algorithms that outperform the classic frequency bands and extract as much information as possible from brain signals. We believe that further developments should be connected to a comprehensive understanding of the brain's neurophysiology.

Figure 8 presents the domains of the used features. Frequency domain features are the most frequently used, and appear nearly twice as often as time domain or time-frequency domain features. Asymmetry characteristics between electrode pairs (by each hemisphere) are increasingly being used—likewise, electrodes' location data in different brain sections. Additionally, raw data (without features) is used as inputs for deep learning classifiers.

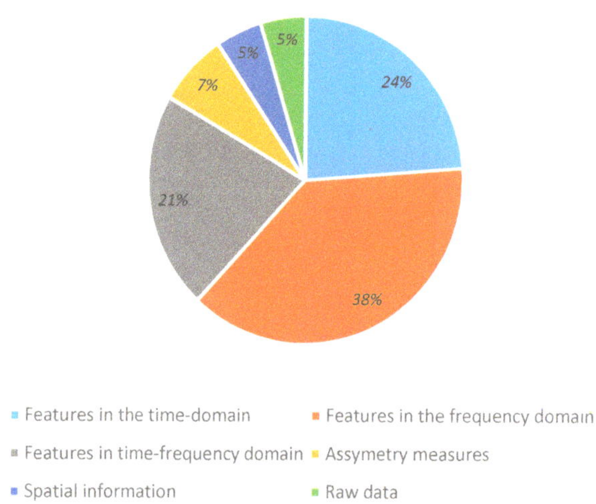

Figure 8. Domain of used features.

Figure 9 shows the usage percentage of various algorithms for feature extraction computed in the 31 papers shown in Table 8. We found that FFT, SFFT, and DFT are the most commonly used tools for characteristic extraction in the frequency domain (27.9%). AR is used less frequently to estimate the spectrum (4.7%). WT and DWT appear in 23.3% of the systems in our sample. These algorithms are applied to obtain features in the time-frequency domain. Likewise, data from channel or electrode specific locations are less frequent (4.7%). Researchers also use statistics and computed parameters in the time domain (9.3%), normalized mutual information NMI (2.3%), ERS (2.3%), and ERD (2.3%).

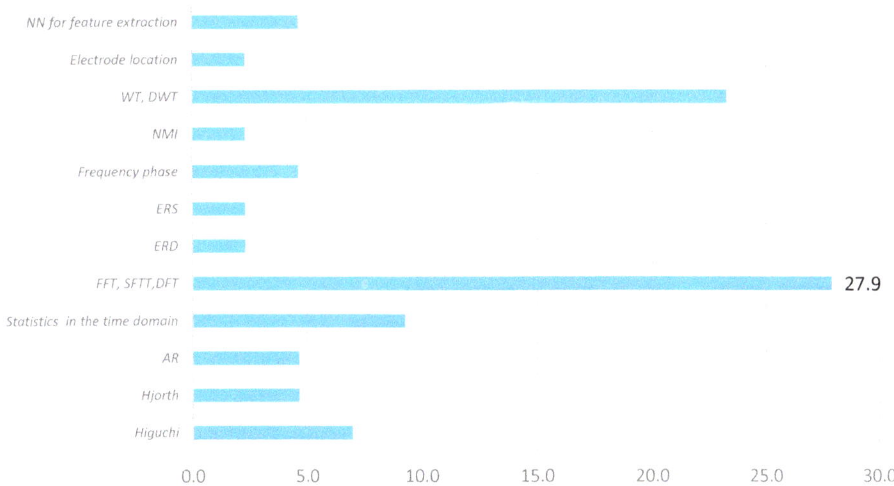

Figure 9. Percentage of the use of algorithms for feature extraction from Table 8.

We observed an increasing presence of algorithms embedded in neural networks like RBN, DBN, TensorFlow functions, and LSTM (4.7%) that are used to extract signal features automatically from raw data. This approach yields a good enough classifier performance, probably because it preserves information and avoids the risk of removing essential emotion-related signal features.

3.5. Feature Selection

It is worth noting that 61.3% of the systems presented in Table 8 do not use a feature selection method. Table 9 lists the systems that utilized feature selection algorithms. Interestingly, virtually every system uses a different algorithm except for the methods minimum redundancy maximum relevance (mRMR) and recursive feature elimination, which are utilized for two different schemes.

Table 9. Systems in Table 8 using feature selection algorithms.

Feature Selection Algorithm	Reference
mRMR	[80,126]
PCA	[38]
LDA	[128]
Max Pooling	[79]
Genetic Algorithm	[118]
SDA	[75]
Fisher-score	[80]
SFEW	[80]
Sparse group lasso	[96]
Correlation matrix	[132]
Information gain	[132]
Recursive feature elimination	[132,133]
Best classification performance for channel selection	[137]

3.6. Classifiers

Figure 10 shows that most classifiers were linear (48%) and neural networks (41%); a few papers used nearest neighbors (7%) and ensemble methods (5%). Consequently, it is worth mentioning that the following algorithms have become increasingly popular for EEG-based emotion recognition applications:

- Linear classifiers, such as naïve Bayes (NB), logistic regression (LR), support vector machine (SVM), linear discriminant analysis (LDA) (48% of use); and
- Neural networks like multi-layer perceptron (MLP), radial basis function RBF, convolutional neural network (CNN), deep belief networks (DBN), extreme learning method (ELM), graph regularized extreme learning machine (GELM), long short term memory (LSTM), domain adversarial neural network (DANN), Caps Net, and graph regularized sparse linear regularized (GRSLR) (41% of use).
- Ensemble classifiers like random forest, CART, bagging tree, Adaboost, and XGBoost are less used (5%). The same situation occurs with the kNN algorithm despite their consistently good performance results, probably because it works better with a simpler feature vector (7%).

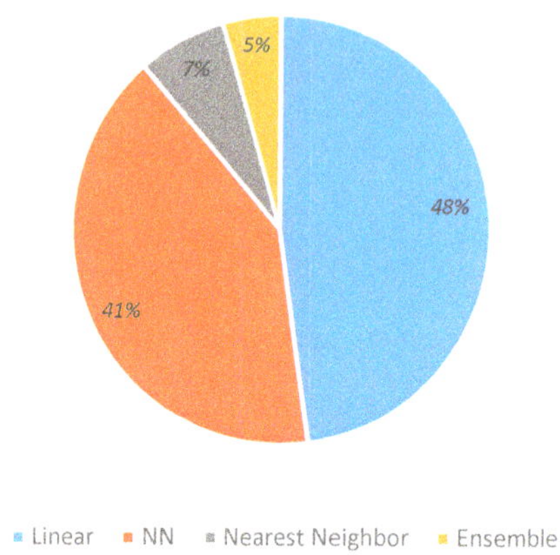

Figure 10. Classifiers' usage.

During our considered period, this review did not find studies that applied non-linear Bayesian classifiers as hidden Markov models (HMM).

3.7. Performance vs. the Number of Classes-Emotions

The performance of almost all systems was evaluated using accuracy, except for two systems in which one used area under the curve (AUC), and the other one presented an F1 measure. Unfortunately, EEG datasets are usually unbalanced, with one or two labeled emotions more numerous than the others, which is somewhat problematic for this approach. Thus, this situation could lead to biased classifications. Moreover, EEG datasets are typically unbalanced, and performance measures should be calculated to contextualize their outcomes. In our view, this is why such results are not entirely comparable among different studies.

In Figure 11, we present the relationship between systems and the number of classified emotions. Most systems use the VA or VAD spaces and classify each dimension as a bi-class (for instance, valence positive and negative; arousal high-value and low value) or tri-class problem (for example, valence positive, neutral, and negative; arousal and dominance high-value and low-value).

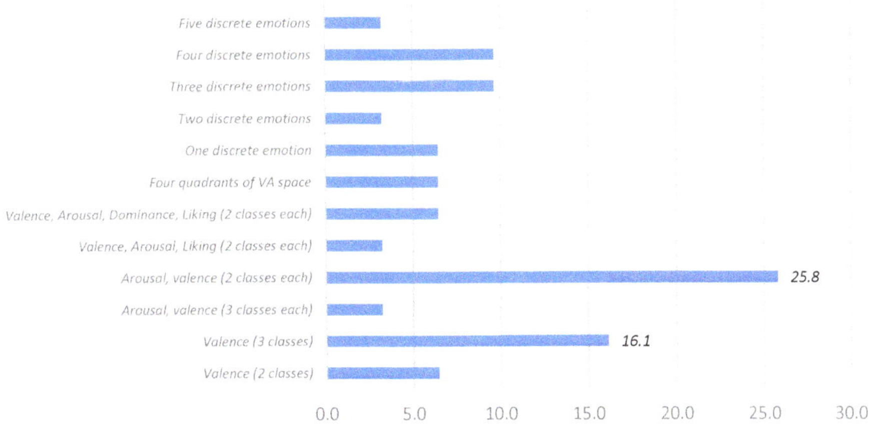

Figure 11. Percentage of systems with different numbers of classified emotions.

Arousal and valence have the highest usage percentages (25.8%). On the other hand, 16.1% categorized valence with three classes: Positive, neutral, and negative. Then, 9.7% classified three discrete emotions like sadness, love, and anger. Moreover, lastly, 6.5% ranked valence as two classes (positive and negative), four discrete emotions (happy, sad, fear, and relaxed), one discrete emotion (disgust), or emotions located in one of four quadrants of the VA space (high valence-high arousal, high valence–low arousal, low valence–high arousal, and low valence–low arousal).

Classifier performance should be evaluated, taking into account that accuracy would be inversely proportional to the number of detected emotions. In other words, classification accuracy should be higher than a random classification process (equal chance for each class). Thus, as classification classes increase, a random classification process would yield a lower accuracy. For instance, a two-class random classification process would be 50% accurate. Likewise, three classes would imply a 33% classification accuracy for a random classification process, and so on. Therefore, such accuracy metrics should provide the classification performance benchmark for our evaluations.

Although the results of the performance of the systems depend on many factors, it is possible to find some relationship between the number of classes, the type of emotions classified, and the accuracy obtained (Figure 12). The best results are obtained with two classes, either as discrete emotions or as positive or negative values in a dimensional space. The second-best value is found for the recognition of one negative discrete emotion like dislike or disgust. The result that the classification of one emotion does not obtain the best performance value could be explained by the fact that in our review, we observed that negative emotions are more challenging to classify and tend to yield smaller performance values.

Comparing approaches and results obtained through different BCI-based systems is complex. This is because each system uses diverse experimental methods for emotion elicitation, protocols to detect EEG signals, datasets, extraction and selection of features, classification algorithms, and generally speaking, each implementation has different settings. Ideally, systems should be tested under similar conditions, but that scenario is not yet available. However, we can perform a comparative analysis to extract trends, bearing in mind such limitations.

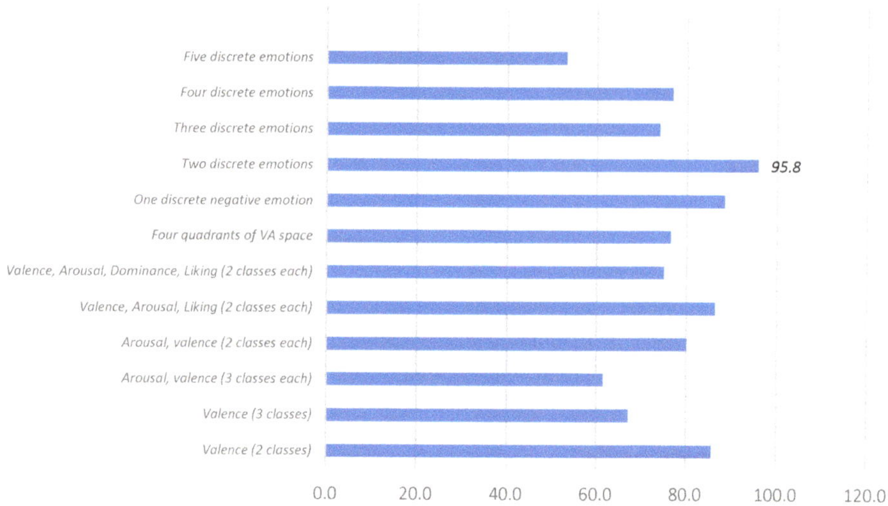

Figure 12. Accuracy vs. types and number of classified emotions.

4. Future Work

Datasets developed for specific applications use passive methods to provoke emotions, such as IAPS, IADS, music videos, and film clips. Public databases, such as DEAP and SEED, use emotion elicitation through music videos and film clips, respectively. Few studies implement active emotion methods for provoking emotions, such as video games and flight simulators.

Going forward, we expect the generation of datasets that use active elicitation methods because these techniques simulate "real life" events better, and are more efficient at emotion induction. However, the implementation of such types of studies requires a significantly more complex experimental setup.

Furthermore, the study of individual emotions has been recently trending. Some works include fear detection, an analysis that has applications in phobia investigation, and other psychiatric disorders. It is worth mentioning that our survey found that negative emotions are more challenging to detect than positive ones.

We did not find in the literature the EEG-based emotion recognition of mixed feelings that combine positive and negative affects sensed at the same moment, for instance, bittersweet feelings. These mixed emotions are interesting because they are related to the study of higher creative performance [141].

Feature extraction and selection are EEG-based BCI system components, which are continuously evolving. They should be designed based on a profound understanding of the brain's biology and physiology. The development of novel features is a topic that can contribute significantly to the improvement of results for emotion recognition systems. For instance, time-domain features are combined with frequency, time-frequency characteristics, channel location, and connectivity criteria. The development of novel feature extraction methods includes asymmetry discoveries in different functioning brain segments, new electrode locations that provide more information, connectivity models (between channels), and correlations needed for understanding functionality.

These evolving features contend that EEG signals and their frequency bands are related to multiple functional and connectivity considerations. The study of the relationship between EEG and biological or psycho-emotional elements should improve going forward. Improved features could better capture individual emotion dynamics and also correlate characteristics across individuals and sessions.

A particularly interesting trend in feature extraction is to use deep neural networks. These systems receive raw data to avoid loss of information and take advantage of the neural networks functioning to obtain relevant features automatically.

The overall reported system accuracy results range from 53% to 90% for the classification of one or more emotions. However, there likely is a gap between real-world applications performed in real time, which presents enormous challenges compared to experiments conducted in a laboratory. Some authors suggest that training datasets should be generated on a larger scale to overcome those challenges. Indeed, we believe it is reasonable that larger datasets could catalyze the research in this field. It is worth mentioning that a similar dynamic played out in the area of image recognition, which experienced a rapid expansion due to the generation of massive databases. Nevertheless, this effort for EEG datasets would likely require collaboration between various research groups to achieve emotions triggered by active elicitation methods.

Overall, we believe systems should be trained with larger sample sizes (and samples per subject), plus the use of real-time data. With such improved datasets, unsupervised techniques could be implemented to obtain comprehensive models. Moreover, these robust systems might allow for transfer learning, i.e., general models that can be applied successfully to particular individuals.

5. Conclusions

EEG signals are reliable information that cannot be simulated or faked. To decode EEG and relate these signals to specific emotion is a complex problem. Affective states do not have a simple mapping with specific brain structures because different emotions activate the same brain locations, or conversely, a single emotion can activate several structures.

In recent years, EEG-based BCI emotion recognition has been a field affecting computing that has generated much interest. Significant advances in the development of low-cost BCI devices with increasingly better usability have encouraged numerous research studies.

In this article, we reviewed the different algorithms and processes that can be part of EEG-based BCI emotion recognition systems: (1) Emotion elicitation, (2) signal acquisition, (3) feature extraction and selection, (4) classification techniques, and (5) performance evaluation. For our survey of this topic, we mined different databases and selected 60 studies carried out under a computer science perspective to gain insight into state of the art and suggest possible future research efforts.

As seen in this review, computational methods still do not have standards for various applications. Researchers continuing to look for new solutions in an ongoing effort. The study of the relationship between brain signals and emotions is a complex problem, and novel methods and new implementations are continuously presented. We expect that many of the existing challenges will soon be solved and will pave the way for a vast area of possible applications using EEG-based emotion recognition.

Author Contributions: Conceptualization and Investigation as part of his Ph.D. research, E.P.T.; E.A.T., M.H.-Á., and S.G.Y. contributed with overall supervision, review editing. All authors have read and agreed to the published version of the manuscript.

Funding: Escuela Politécnica Nacional funded the publication of this article.

Conflicts of Interest: The authors declare no conflict of interest.

References

1. Picard, R.W. Affective Computing for HCI. In Proceedings of the HCI International 1999-Proceedings of the 8th International Conference on Human-Computer Interaction, Munich, Germany, 22–26 August 1999.
2. Elfenbein, H.A.; Ambady, N. Predicting workplace outcomes from the ability to eavesdrop on feelings. *J. Appl. Psychol.* **2002**, *87*, 963–971. [CrossRef] [PubMed]
3. Goenaga, S.; Navarro, L.; Quintero, C.G.M.; Pardo, M. Imitating human emotions with a nao robot as interviewer playing the role of vocational tutor. *Electronics* **2020**, *9*, 971. [CrossRef]
4. Kitcheman, B. Procedures for performing systematic reviews. *Comput. Sci.* **2004**, 1–28. Available online: http://www.inf.ufsc.br/~{}aldo.vw/kitchenham.pdf (accessed on 26 May 2020).
5. Salzman, C.D.; Fusi, S. Emotion, cognition, and mental state representation in Amygdala and prefrontal Cortex. *Annu. Rev. Neurosci.* **2010**, *33*, 173–202. [CrossRef]

6. Konar, A.; Chakraborty, A. *Emotion Recognition: A Pattern Analysis Approach*; John Wiley & Sons: Hoboken, NJ, USA, 2015; ISBN 9781118910566.
7. Panoulas, K.J.; Hadjileontiadis, L.J.; Panas, S.M. *Brain-Computer Interface (BCI): Types, Processing Perspectives and Applications*; Springer: Berlin/Heidelberg, Germany, 2010; pp. 299–321.
8. Ekman, P. Ekman 1992.pdf. *Psychol. Rev.* **1992**, *99*, 550–553. [CrossRef] [PubMed]
9. Verma, G.K.; Tiwary, U.S. Affect representation and recognition in 3D continuous valence–arousal–dominance space. *Multimed. Tools Appl.* **2017**, *76*, 2159–2183. [CrossRef]
10. Bălan, O.; Moise, G.; Moldoveanu, A.; Leordeanu, M.; Moldoveanu, F. Fear level classification based on emotional dimensions and machine learning techniques. *Sensors* **2019**, *19*, 1738. [CrossRef] [PubMed]
11. Russell, J.A. A circumplex model of affect. *J. Pers. Soc. Psychol.* **1980**, *39*, 1161. [CrossRef]
12. Bradley, M.M.; Lang, P.J. Measuring emotion: The self-assessment manikin and the semantic differential. *J. Behav. Ther. Exp. Psychiatry* **1994**, *25*, 49–59. [CrossRef]
13. Ekman, P.; Davidson, R.J. Voluntary smiling changes regional brain activity. *Psychol. Sci.* **1993**, *4*, 342–345. [CrossRef]
14. Bhatti, A.M.; Majid, M.; Anwar, S.M.; Khan, B. Human emotion recognition and analysis in response to audio music using brain signals. *Comput. Human Behav.* **2016**, *65*, 267–275. [CrossRef]
15. Lee, Y.Y.; Hsieh, S. Classifying different emotional states by means of eegbased functional connectivity patterns. *PLoS ONE* **2014**, *9*, e95415. [CrossRef]
16. Zheng, W.L.; Guo, H.T.; Lu, B.L. Revealing critical channels and frequency bands for emotion recognition from EEG with deep belief network. *Int. IEEE/EMBS Conf. Neural Eng. NER* **2015**, *2015*, 154–157. [CrossRef]
17. Knyazev, G.G.; Slobodskoj-Plusnin, J.Y. Behavioural approach system as a moderator of emotional arousal elicited by reward and punishment cues. *Pers. Individ. Dif.* **2007**, *42*, 49–59. [CrossRef]
18. Kirmizi-Alsan, E.; Bayraktaroglu, Z.; Gurvit, H.; Keskin, Y.H.; Emre, M.; Demiralp, T. Comparative analysis of event-related potentials during Go/NoGo and CPT: Decomposition of electrophysiological markers of response inhibition and sustained attention. *Brain Res.* **2006**, *1104*, 114–128. [CrossRef]
19. Hyvarinen, A. New Approximations of differential entropy for independent component analysis and projection pursuit. In Proceedings of the Advances in Neural Information Processing Systems, Denver, CO, USA; 1998; pp. 273–279.
20. Hamann, S. Mapping discrete and dimensional emotions onto the brain: Controversies and consensus. *Trends Cogn. Sci.* **2012**, *16*, 458–466. [CrossRef]
21. Davidson, R.J.; Ekman, P.; Saron, C.D.; Senulis, J.A.; Friesen, W.V. Approach-withdrawl-and-cerebral-asymmetry-emotional-expres davidson 1990.pdf. *J. Pers. Soc. Psychol.* **1990**, *58*, 330–341. [CrossRef]
22. Peterson, V.; Galván, C.; Hernández, H.; Spies, R. A feasibility study of a complete low-cost consumer-grade brain-computer interface system. *Heliyon* **2020**, *6*. [CrossRef]
23. Savran, A.; Ciftci, K.; Chanel, G.; Mota, J.C.; Viet, L.H.; Sankur, B.; Akarun, L.; Caplier, A.; Rombaut, M. Emotion detection in the loop from brain signals and facial images. *eNTERFACE* **2006**, *6*, 69–80.
24. Onton, J.; Makeig, S. High-frequency broadband modulations of electroencephalographic spectra. *Front. Hum. Neurosci.* **2009**, *3*, 1–18. [CrossRef]
25. Yadava, M.; Kumar, P.; Saini, R.; Roy, P.P.; Prosad Dogra, D. Analysis of EEG signals and its application to neuromarketing. *Multimed. Tools Appl.* **2017**, *76*, 19087–19111. [CrossRef]
26. Zheng, W.L.; Liu, W.; Lu, Y.; Lu, B.L.; Cichocki, A. EmotionMeter: A Multimodal framework for recognizing human emotions. *IEEE Trans. Cybern.* **2019**, *49*, 1110–1122. [CrossRef] [PubMed]
27. Soleymani, M.; Lichtenauer, J.; Pun, T.; Pantic, M. A multimodal database for affect recognition and implicit tagging. *IEEE Trans. Affect. Comput.* **2012**, *3*, 42–55. [CrossRef]
28. Grégoire, C.; Rodrigues, P.L.C.; Congedo, M. *EEG Alpha Waves Dataset*; Centre pour la Communication Scientifique Directe: Grenoble, France, 2019.
29. Katsigiannis, S.; Ramzan, N. DREAMER: A database for emotion recognition through EEG and ECG signals from wireless low-cost off-the-shelf devices. *IEEE J. Biomed. Heal. Informatics* **2018**, *22*, 98–107. [CrossRef]
30. Li, Y.; Zheng, W.; Cui, Z.; Zong, Y.; Ge, S. EEG emotion recognition based on graph regularized sparse linear regression. *Neural Process. Lett.* **2019**, *49*, 555–571. [CrossRef]
31. Lang, P.J.; Bradley, M.M.; Cuthbert, B.N. International affective picture system (IAPS): Technical manual and affective ratings. *NIMH Cent. Study Emot. Atten.* **1997**, *1*, 39–58.

32. Yang, W.; Makita, K.; Nakao, T.; Kanayama, N.; Machizawa, M.G.; Sasaoka, T.; Sugata, A.; Kobayashi, R.; Hiramoto, R.; Yamawaki, S.; et al. Affective auditory stimulus database: An expanded version of the International Affective Digitized Sounds (IADS-E). *Behav. Res. Methods* **2018**, *50*, 1415–1429. [CrossRef]
33. Mühl, C.; Allison, B.; Nijholt, A.; Chanel, G. A survey of affective brain computer interfaces: Principles, state-of-the-art, and challenges. *Brain Comput. Interfaces* **2014**, *1*, 66–84. [CrossRef]
34. Zhou, F.; Qu, X.; Jiao, J.; Helander, M.G. Emotion prediction from physiological signals: A comparison study between visual and auditory elicitors. *Interact. Comput.* **2014**, *26*, 285–302. [CrossRef]
35. Pallavicini, F.; Ferrari, A.; Pepe, A.; Garcea, G. Effectiveness of virtual reality survival horror games for the emotional elicitation: Preliminary insights using Resident Evil 7: Biohazard. In *International Conference on Universal Access in Human-Computer Interaction*; Springer: Cham, Switzerland, 2018. [CrossRef]
36. Roza, V.C.C.; Postolache, O.A. Multimodal approach for emotion recognition based on simulated flight experiments. *Sensors* **2019**, *19*, 5516. [CrossRef] [PubMed]
37. Iacoviello, D.; Petracca, A.; Spezialetti, M.; Placidi, G. A real-time classification algorithm for EEG-based BCI driven by self-induced emotions. *Comput. Methods Programs Biomed.* **2015**, *122*, 293–303. [CrossRef] [PubMed]
38. Novak, D.; Mihelj, M.; Munih, M. A survey of methods for data fusion and system adaptation using autonomic nervous system responses in physiological computing. *Interact. Comput.* **2012**, *24*, 154–172. [CrossRef]
39. Bustamante, P.A.; Lopez Celani, N.M.; Perez, M.E.; Quintero Montoya, O.L. Recognition and regionalization of emotions in the arousal-valence plane. In Proceedings of the 2015 Milano, Italy 37th Annual International Conference of the IEEE Engineering in Medicine and Biology Society (EMBC), Milano, Italy, 25–29 August 2015; Volume 2015, pp. 6042–6045.
40. Sanei, S.; Chambers, J.A. *EEG Signal Processing*; John Wiley & Sons: Hoboken, NJ, USA, 2013; ISBN 9780470025819.
41. Abhang, P.A.; Suresh, C.; Mehrotra, B.W.G. *Introduction to EEG-and Speech-Based Emotion Recognition*; Elsevier: Amsterdam, The Netherlands, 2016; ISBN 9780128044902.
42. Jardim-Gonçalves, R.; Universidade Nova de Lisboa. Faculdade de Ciências e Tecnologia; Institute of Electrical and Electronics Engineers; IEEE Technology Engineering and Management Society. In Proceedings of the IEEE International Technology Management Conference, Madeira Islands, Portugal, 27–29 June 2017; International Conference on Engineering, Technology and Innovation (ICE/ITMC) : "Engineering, technology & innovation management beyond 2020: New challenges, new approaches" : Conference proceedings. ISBN 9781538607749.
43. Alhaddad, M.J.; Kamel, M.; Malibary, H.; Thabit, K.; Dahlwi, F.; Hadi, A. P300 speller efficiency with common average reference. *Lect. Notes Comput. Sci.* **2012**, *7326 LNAI*, 234–241. [CrossRef]
44. Alhaddad, M.J.; Kamel, M.; Malibary, H.; Thabit, K.; Dahlwi, F.; Hadi, A. P300 speller efficiency with common average reference. In Proceedings of the International Conference on Autonomous and Intelligent Systems, Aveiro, Portugal, 25–27 June 2012; pp. 234–241.
45. Murugappan, M.; Nagarajan, R.; Yaacob, S. Combining spatial filtering and wavelet transform for classifying human emotions using EEG Signals. *J. Med. Biol. Eng.* **2011**, *31*, 45–51. [CrossRef]
46. Murugappan, M.; Murugappan, S. Human emotion recognition through short time Electroencephalogram (EEG) signals using Fast Fourier Transform (FFT). In Proceedings of the Proceedings-2013 IEEE 9th International Colloquium on Signal Processing and its Applications, Kuala Lumpur, Malaysia, 8–10 March 2013; pp. 289–294.
47. Burle, B.; Spieser, L.; Roger, C.; Casini, L.; Hasbroucq, T.; Vidal, F. Spatial and temporal resolutions of EEG: Is it really black and white? A scalp current density view. *Int. J. Psychophysiol.* **2015**, *97*, 210–220. [CrossRef]
48. Mazumder, I. An analytical approach of EEG analysis for emotion recognition. In Proceedings of the 2019 Devices for Integrated Circuit (DevIC), Kalyani, India, 23 March 2019; pp. 256–260. [CrossRef]
49. Subasi, A.; Gursoy, M.I. EEG signal classification using PCA, ICA, LDA and support vector machines. *Expert Syst. Appl.* **2010**, *37*, 8659–8666. [CrossRef]
50. Lee, H.; Choi, S. Pca + hmm + svm for eeg pattern classification. In Proceedings of the Seventh International Symposium on Signal Processing and Its Applications, Paris, France, 4 July 2003; pp. 541–544.
51. Doma, V.; Pirouz, M. A comparative analysis of machine learning methods for emotion recognition using EEG and peripheral physiological signals. *J. Big Data* **2020**, *7*. [CrossRef]

52. Shaw, L.; Routray, A. Statistical features extraction for multivariate pattern analysis in meditation EEG using PCA. In Proceedings of the 2016 IEEE EMBS International Student Conference ISC, Ottawa, ON, Canada, 31 May 2016; pp. 1–4. [CrossRef]
53. Symposium, I.; Analysis, I.C.; Separation, B.S. 4th International Symposium on Independent Component Analysis and Blind Signal Separation (ICA2003), April 2003, Nara, Japan. *Analysis* **2003**, 975–980.
54. Liu, J.; Meng, H.; Li, M.; Zhang, F.; Qin, R.; Nandi, A.K. Emotion detection from EEG recordings based on supervised and unsupervised dimension reduction. *Concurr. Comput.* **2018**, *30*, 1–13. [CrossRef]
55. Yong, X.; Ward, R.K.; Birch, G.E. Robust common spatial patterns for EEG signal preprocessing. In Proceedings of the 30th Annual International Conference of the IEEE Engineering in Medicine and Biology Society, EMBS'08-"Personalized Healthcare through Technology", Boston, MA, USA, 30 August 2011; pp. 2087–2090.
56. Li, X.; Fan, H.; Wang, H.; Wang, L. Common spatial patterns combined with phase synchronization information for classification of EEG signals. *Biomed. Signal Process. Control* **2019**, *52*, 248–256. [CrossRef]
57. Interfaces, B. A Tutorial on EEG signal processing techniques for mental state recognition in brain-computer interfaces. *Guid. Brain Comput. Music Interfacing* **2014**. [CrossRef]
58. Jenke, R.; Peer, A.; Buss, M. Feature extraction and selection for emotion recognition from EEG. *IEEE Trans. Affect. Comput.* **2014**, *5*, 327–339. [CrossRef]
59. Al-Fahoum, A.S.; Al-Fraihat, A.A. Methods of EEG signal features extraction using linear analysis in frequency and time-frequency domains. *ISRN Neurosci.* **2014**, *2014*, 1–7. [CrossRef] [PubMed]
60. Petrantonakis, P.C.; Hadjileontiadis, L.J. Emotion recognition from brain signals using hybrid adaptive filtering and higher order crossings analysis. *IEEE Trans. Affect. Comput.* **2010**, *1*, 81–97. [CrossRef]
61. Torres, E.P.; Torres, E.A.; Hernandez-Alvarez, M.; Yoo, S.G. Machine learning analysis of EEG measurements of stock trading performance. In *Advances in Artificial Intelligence, Software and Systems Engineering*; Springer Nature: Lodon, UK, 2020.
62. Kubben, P.; Dumontier, M.; Dekker, A. Fundamentals of clinical data science. *Fundam. Clin. Data Sci.* **2018**, 1–219. [CrossRef]
63. Karahan, E.; Rojas-Lopez, P.A.; Bringas-Vega, M.L.; Valdes-Hernandez, P.A.; Valdes-Sosa, P.A. Tensor analysis and fusion of multimodal brain images. *Proc. IEEE* **2015**, *103*, 1531–1559. [CrossRef]
64. Winkler, I.; Debener, S.; Muller, K.R.; Tangermann, M. On the influence of high-pass filtering on ICA-based artifact reduction in EEG-ERP. In Proceedings of the 2015 37th Annual International Conference of the IEEE Engineering in Medicine and Biology Society (EMBC), Milan, Italy, 25–29 August 2015. [CrossRef]
65. Zhang, Y.; Zhou, G.; Zhao, Q.; Jin, J.; Wang, X.; Cichocki, A. Spatial-temporal discriminant analysis for ERP-based brain-computer interface. *IEEE Trans. Neural Syst. Rehabil. Eng.* **2013**, *21*, 233–243. [CrossRef]
66. Brouwer, A.M.; Zander, T.O.; Van Erp, J.B.F.; Korteling, J.E.; Bronkhorst, A.W. Using neurophysiological signals that reflect cognitive or affective state: Six recommendations to avoid common pitfalls. *Front. Neurosci.* **2015**, *9*, 1–11. [CrossRef]
67. Wu, Z.; Yao, D.; Tang, Y.; Huang, Y.; Su, S. Amplitude modulation of steady-state visual evoked potentials by event-related potentials in a working memory task. *J. Biol. Phys.* **2010**, *36*, 261–271. [CrossRef]
68. Abootalebi, V.; Moradi, M.H.; Khalilzadeh, M.A. A new approach for EEG feature extraction in P300-based lie detection. *Comput. Methods Programs Biomed.* **2009**, *94*, 48–57. [CrossRef] [PubMed]
69. Bhise, P.R.; Kulkarni, S.B.; Aldhaheri, T.A. Brain computer interface based EEG for emotion recognition system: A systematic review. In Proceedings of the 2020 2nd International Conference on Innovative Mechanisms for Industry Applications (ICIMIA), Bangalore, India, 5–7 March 2020; ISBN 9781728141671.
70. Li, X.; Song, D.; Zhang, P.; Zhang, Y.; Hou, Y.; Hu, B. Exploring EEG features in cross-subject emotion recognition. *Front. Neurosci.* **2018**, *12*. [CrossRef] [PubMed]
71. Liu, Y.; Sourina, O. EEG databases for emotion recognition. In Proceedings of the 2013 International Conference on Cyberworlds, Yokohama, Japan, 21–23 October 2013; pp. 302–309.
72. Hossain, M.Z.; Kabir, M.M.; Shahjahan, M. Feature selection of EEG data with neuro-statistical method. In Proceedings of the 2013 International Conference on Electrical Information and Communication Technology (EICT), Khulna, Bangladesh, 13–15 February 2014. [CrossRef]
73. Bavkar, S.; Iyer, B.; Deosarkar, S. Detection of alcoholism: An EEG hybrid features and ensemble subspace K-NN based approach. In Proceedings of the International Conference on Distributed Computing and Internet Technology, Bhubaneswar, India, 10–13 January 2019; pp. 161–168.

74. Pane, E.S.; Wibawa, A.D.; Pumomo, M.H. Channel Selection of EEG Emotion Recognition using Stepwise Discriminant Analysis. In Proceedings of the 2018 International Conference on Computer Engineering, Network and Intelligent Multimedia (CENIM), Surabaya, Indonesia, 26–27 November 2018; pp. 14–19. [CrossRef]
75. Musselman, M.; Djurdjanovic, D. Time-frequency distributions in the classification of epilepsy from EEG signals. *Expert Syst. Appl.* **2012**, *39*, 11413–11422. [CrossRef]
76. Xu, H.; Plataniotis, K.N. Affect recognition using EEG signal. In Proceedings of the 2012 IEEE 14th International Workshop on Multimedia Signal Processing (MMSP), Banff, AB, Canada, 17 September 2012; pp. 299–304. [CrossRef]
77. Wu, X.; Zheng, W.-L.; Lu, B.-L. Investigating EEG-Based Functional Connectivity Patterns for Multimodal Emotion Recognition. 2020. Available online: https://arxiv.org/abs/2004.01973 (accessed on 26 May 2020).
78. Zheng, W.-L.; Zhu, J.-Y.; Lu, B.-L. Identifying Stable Patterns over Time for Emotion Recognition from EEG. *IEEE Trans. Affect. Comput.* **2017**, *10*, 417–429. [CrossRef]
79. Yang, Y.; Wu, Q.M.J.; Zheng, W.L.; Lu, B.L. EEG-based emotion recognition using hierarchical network with subnetwork nodes. *IEEE Trans. Cogn. Dev. Syst.* **2018**, *10*, 408–419. [CrossRef]
80. Li, P.; Liu, H.; Si, Y.; Li, C.; Li, F.; Zhu, X.; Huang, X.; Zeng, Y.; Yao, D.; Zhang, Y.; et al. EEG based emotion recognition by combining functional connectivity network and local activations. *IEEE Trans. Biomed. Eng.* **2019**, *66*, 2869–2881. [CrossRef]
81. Li, Z.; Tian, X.; Shu, L.; Xu, X.; Hu, B. Emotion recognition from EEG using RASM and LSTM. In Proceedings of the International Conference on Internet Multimedia Computing and Service, Qingdao, China, 23–25 August 2017; pp. 310–318.
82. Mowla, M.R.; Cano, R.I.; Dhuyvetter, K.J.; Thompson, D.E. Affective brain-computer interfaces: A tutorial to choose performance measuring metric. *arXiv* **2020**, arXiv:2005.02619.
83. Lan, Z.; Sourina, O.; Wang, L.; Scherer, R.; Muller-Putz, G.R. Domain adaptation techniques for eeg-based emotion recognition: A comparative study on two public datasets. *IEEE Trans. Cogn. Dev. Syst.* **2019**, *11*, 85–94. [CrossRef]
84. Duan, R.N.; Zhu, J.Y.; Lu, B.L. Differential entropy feature for EEG-based emotion classification. In Proceedings of the International IEEE/EMBS Conference on Neural Engineering, San Diego, CA, USA, 6–8 November 2013; pp. 81–84.
85. Zheng, W.L.; Lu, B.L. Investigating critical frequency bands and channels for eeg-based emotion recognition with deep neural networks. *IEEE Trans. Auton. Ment. Dev.* **2015**, *7*, 162–175. [CrossRef]
86. Assistant Professor, T.S.; Ravi Kumar Principal, K.M.; Nataraj, A.; K Students, A.K. Analysis of EEG Based Emotion Detection of DEAP and SEED-IV Databases Using SVM. Available online: https://papers.ssrn.com/sol3/papers.cfm?abstract_id=3509130 (accessed on 26 May 2020).
87. Wang, X.H.; Zhang, T.; Xu, X.M.; Chen, L.; Xing, X.F.; Chen, C.L.P. EEG Emotion Recognition Using Dynamical Graph Convolutional Neural Networks and Broad Learning System. In Proceedings of the 2018 IEEE International Conference on Bioinformatics and Biomedicine (BIBM), Madrid, Spain, 3–6 December 2018; pp. 1240–1244. [CrossRef]
88. Li, J.; Qiu, S.; Du, C.; Wang, Y.; He, H. Domain adaptation for eeg emotion recognition based on latent representation similarity. *IEEE Trans. Cogn. Dev. Syst.* **2020**, *12*, 344–353. [CrossRef]
89. Petrantonakis, P.C.; Hadjileontiadis, L.J. Emotion recognition from EEG using higher order crossings. *IEEE Trans. Inf. Technol. Biomed.* **2010**, *14*, 186–197. [CrossRef] [PubMed]
90. Kim, M.K.; Kim, M.; Oh, E.; Kim, S.P. A review on the computational methods for emotional state estimation from the human EEG. *Comput. Math. Methods Med.* **2013**, *2013*. [CrossRef] [PubMed]
91. Yoon, H.J.; Chung, S.Y. EEG-based emotion estimation using Bayesian weighted-log-posterior function and perceptron convergence algorithm. *Comput. Biol. Med.* **2013**, *43*, 2230–2237. [CrossRef] [PubMed]
92. Hosni, S.M.; Gadallah, M.E.; Bahgat, S.F.; AbdelWahab, M.S. Classification of EEG signals using different feature extraction techniques for mental-task BCI. In Proceedings of the ICCES'07-2007 International Conference on Computer Engineering and Systems, Cairo, Egypt, 27–29 November 2007; pp. 220–226.
93. Xing, X.; Li, Z.; Xu, T.; Shu, L.; Hu, B.; Xu, X. SAE+LSTM: A new framework for emotion recognition from multi-channel EEG. *Front. Neurorobot.* **2019**, *13*, 1–14. [CrossRef] [PubMed]

94. Navarro, I.; Sepulveda, F.; Hubais, B. A comparison of time, frequency and ICA based features and five classifiers for wrist movement classification in EEG signals. *Conf. Proc. IEEE Eng. Med. Biol. Soc.* **2005**, *2005*, 2118–2121.
95. Ting, W.; Guo-zheng, Y.; Bang-hua, Y.; Hong, S. EEG feature extraction based on wavelet packet decomposition for brain computer interface. *Meas. J. Int. Meas. Confed.* **2008**, *41*, 618–625. [CrossRef]
96. Guo, J.; Fang, F.; Wang, W.; Ren, F. EEG emotion recognition based on granger causality and capsnet neural network. In Proceedings of the 2018 5th IEEE International Conference on Cloud Computing and Intelligence Systems (CCIS), Nanjing, China, 23–25 November 2018; pp. 47–52. [CrossRef]
97. Sander, D.; Grandjean, D.; Scherer, K.R. A systems approach to appraisal mechanisms in emotion. *Neural Netw.* **2005**, *18*, 317–352. [CrossRef]
98. Chanel, G.; Kierkels, J.J.M.; Soleymani, M.; Pun, T. Short-term emotion assessment in a recall paradigm. *Int. J. Hum. Comput. Stud.* **2009**, *67*, 607–627. [CrossRef]
99. Lotte, F.; Congedo, M.; Lécuyer, A.; Lamarche, F.; Arnaldi, B.; Anatole, L.; Lotte, F.; Congedo, M.; Anatole, L.; Abdulhay, E.; et al. A review of classification algorithms for EEG-based brain–Computer interfaces To cite this version: A review of classification algorithms for EEG-based brain-computer interfaces. *Hum. Brain Mapp.* **2018**, *38*, 270–278. [CrossRef]
100. Jenke, R.; Peer, A.; Buss, M. Effect-size-based Electrode and Feature Selection for Emotion Recognition from EEG. In Proceedings of the 2013 IEEE International Conference on Acoustics, Speech and Signal Processing, Vancouver, Canada, 26–31 May 2013; pp. 1217–1221.
101. Hassanien, A.E.; Azar, A.T. (Eds.) *Intelligent Systems Reference Library 74 Brain-Computer Interfaces Current Trends and Applications*; Springer: Berlin/Heidelberg, Germany, 2015.
102. Zhang, L.; Xiong, G.; Liu, H.; Zou, H.; Guo, W. Time-frequency representation based on time-varying autoregressive model with applications to non-stationary rotor vibration analysis. *Sadhana Acad. Proc. Eng. Sci.* **2010**, *35*, 215–232. [CrossRef]
103. Hill, N.J.; Wolpaw, J.R. Brain–Computer Interface. In *Reference Module in Biomedical Sciences*; Elsevier: Amsterdam, The Netherlands, 2016.
104. Rashid, M.; Sulaiman, N.P.P.; Abdul Majeed, A.; Musa, R.M.; Ab. Nasir, A.F.; Bari, B.S.; Khatun, S. Current Status, Challenges, and Possible Solutions of EEG-Based Brain-Computer Interface: A Comprehensive Review. *Front. Neurorobot.* **2020**, *14*. [CrossRef] [PubMed]
105. Vaid, S.; Singh, P.; Kaur, C. EEG signal analysis for BCI interface: A review. In Proceedings of the International Conference on Advanced Computing and Communication Technologies, Haryana, India, 21 February 2015; pp. 143–147.
106. Ackermann, P.; Kohlschein, C.; Bitsch, J.Á.; Wehrle, K.; Jeschke, S. EEG-based automatic emotion recognition: Feature extraction, selection and classification methods. In Proceedings of the 2016 IEEE 18th international conference on e-health networking, applications and services (Healthcom), Munich, Germany, 14–16 September 2016. [CrossRef]
107. Atangana, R.; Tchiotsop, D.; Kenne, G.; DjoufackNkengfac k, L.C. EEG signal classification using LDA and MLP classifier. *Heal. Inform. An Int. J.* **2020**, *9*, 14–32. [CrossRef]
108. Srivastava, S.; Gupta, M.R.; Frigyik, B.A. Bayesian quadratic discriminant analysis. *J. Mach. Learn. Res.* **2007**, *8*, 1277–1305.
109. Cimtay, Y.; Ekmekcioglu, E. Investigating the use of pretrained convolutional neural network on cross-subject and cross-dataset eeg emotion recognition. *Sensors* **2020**, *20*, 34. [CrossRef] [PubMed]
110. Tzirakis, P.; Trigeorgis, G.; Nicolaou, M.A.; Schuller, B.W.; Zafeiriou, S. End-to-end multimodal emotion recognition using deep neural networks. *IEEE J. Sel. Top. Signal Process.* **2017**, *11*, 1301–1309. [CrossRef]
111. Chen, J.X.; Zhang, P.W.; Mao, Z.J.; Huang, Y.F.; Jiang, D.M.; Zhang, Y.N. Accurate EEG-based emotion recognition on combined features using deep convolutional neural networks. *IEEE Access* **2019**, *7*, 44317–44328. [CrossRef]
112. Zhang, W.; Wang, F.; Jiang, Y.; Xu, Z.; Wu, S.; Zhang, Y. *Cross-Subject EEG-Based Emotion Recognition with Deep Domain Confusion*; Springer International Publishing: Berlin/Heidelberg, Germany, 2019; ISBN 9783030275259.
113. Lechner, U. *Scientific Workflow Scheduling for Cloud Computing Environments*; Springer International Publishing: Berlin/Heidelberg, Germany, 2019; ISBN 9783030053666.

114. Babiloni, F.; Bianchi, L.; Semeraro, F.; Millán, J.D.R.; Mouriño, J.; Cattini, A.; Salinari, S.; Marciani, M.G.; Cincotti, F. Mahalanobis distance-based classifiers are able to recognize EEG patterns by using few EEG electrodes. *Annu. Reports Res. React. Institute, Kyoto Univ.* **2001**, *1*, 651–654. [CrossRef]
115. Sun, S.; Zhang, C.; Zhang, D. An experimental evaluation of ensemble methods for EEG signal classification. *Pattern Recognit. Lett.* **2007**, *28*, 2157–2163. [CrossRef]
116. Fraiwan, L.; Lweesy, K.; Khasawneh, N.; Wenz, H.; Dickhaus, H. Automated sleep stage identification system based on time-frequency analysis of a single EEG channel and random forest classifier. *Comput. Methods Programs Biomed.* **2012**, *108*, 10–19. [CrossRef]
117. Kumar, P.; Valentina, M.; Balas, E.; Kumar Bhoi, A.; Chae, G.-S. *Advances in Intelligent Systems and Computing 1040 Cognitive Informatics and Soft Computing*; Springer: Berlin/Heidelberg, Germany, 2020.
118. Lv, T.; Yan, J.; Xu, H. An EEG emotion recognition method based on AdaBoost classifier. In Proceedings of the 2017 Chinese Automation Congress (CAC), Jinan, China, 20–22 October 2017; pp. 6050–6054.
119. Ilyas, M.Z.; Saad, P.; Ahmad, M.I. A survey of analysis and classification of EEG signals for brain-computer interfaces. In Proceedings of the 2015 2nd International Conference on Biomedical Engineering (ICoBE), Penang, Malaysia, 30–31 March 2015; pp. 30–31. [CrossRef]
120. Japkowicz, N.; Shah, M. *Evaluating Learning Algorithms*; Cambridge University Press: Cambridge, UK, 2011; ISBN 9780511921803.
121. Biological and medical physics, biomedical engineering. In *Towards Practical Brain-Computer Interfaces*; Allison, B.Z.; Dunne, S.; Leeb, R.; Del R. Millán, J.; Nijholt, A. (Eds.) Springer: Berlin/Heidelberg, Germany, 2013; ISBN 978-3-642-29745-8.
122. Combrisson, E.; Jerbi, K. Exceeding chance level by chance: The caveat of theoretical chance levels in brain signal classification and statistical assessment of decoding accuracy. *J. Neurosci. Methods* **2015**, *250*, 126–136. [CrossRef]
123. Bonett, D.G.; Price, R.M. Adjusted Wald Confidence Interval for a Difference of Binomial Proportions Based on Paired Data. *J. Educ. Behav. Stat.* **2012**, *37*, 479–488. [CrossRef]
124. Kreibig, S.D. Autonomic nervous system activity in emotion: A review. *Biol. Psychol.* **2010**, *84*, 394–421. [CrossRef]
125. Feradov, F.; Mporas, I.; Ganchev, T. Evaluation of features in detection of dislike responses to audio–visual stimuli from EEG signals. *Computers* **2020**, *9*, 33. [CrossRef]
126. Atkinson, J.; Campos, D. Improving BCI-based emotion recognition by combining EEG feature selection and kernel classifiers. *Expert Syst. Appl.* **2016**, *47*, 35–41. [CrossRef]
127. Kaur, B.; Singh, D.; Roy, P.P. EEG based emotion classification mechanism in BCI. *Procedia Comput. Sci.* **2018**, *132*, 752–758. [CrossRef]
128. Liu, Y.J.; Yu, M.; Zhao, G.; Song, J.; Ge, Y.; Shi, Y. Real-time movie-induced discrete emotion recognition from EEG signals. *IEEE Trans. Affect. Comput.* **2018**, *9*, 550–562. [CrossRef]
129. Yan, J.; Chen, S.; Deng, S. A EEG-based emotion recognition model with rhythm and time characteristics. *Brain Informatics* **2019**, *6*. [CrossRef]
130. Li, Y.; Zheng, W.; Cui, Z.; Zhang, T.; Zong, Y. A novel neural network model based on cerebral hemispheric asymmetry for EEG emotion recognition. *IJCAI Int. Jt. Conf. Artif. Intell.* **2018**, *2018*, 1561–1567. [CrossRef]
131. Wang, Z.M.; Hu, S.Y.; Song, H. Channel selection method for eeg emotion recognition using normalized mutual information. *IEEE Access* **2019**, *7*, 143303–143311. [CrossRef]
132. Parui, S.; Kumar, A.; Bajiya, R.; Samanta, D.; Chakravorty, N. Emotion recognition from EEG signal using XGBoost algorithm. In Proceedings of the 2019 IEEE 16th India Council International Conference (INDICON), Rajkot, Gujarat, 13–15 December 2019; pp. 1–4. [CrossRef]
133. Kumar, N.; Khaund, K.; Hazarika, S.M. Bispectral analysis of EEG for emotion recognition. *Procedia. Comput. Sci.* **2016**, *84*, 31–35. [CrossRef]
134. Liu, Y.; Sourina, O. EEG-based subject-dependent emotion recognition algorithm using fractal dimension. In Proceedings of the 2014 IEEE International Conference on Systems, Man, and Cybernetics (SMC), San Diego, CA, USA, 5–8 October 2014; pp. 3166–3171. [CrossRef]
135. Thammasan, N.; Moriyama, K.; Fukui, K.I.; Numao, M. Familiarity effects in EEG-based emotion recognition. *Brain Inform.* **2017**, *4*, 39–50. [CrossRef]
136. Technology, I.; Yongbin, G.; Hyo, J.L.; Raja, M.M. Deep Learn. *Eeg. Signals Emot. Recognit.* **2015**, *2*, 1–5. [CrossRef]

137. Özerdem, M.S.; Polat, H. Emotion recognition based on EEG features in movie clips with channel selection. *Brain Inform.* **2017**, *4*, 241–252. [CrossRef] [PubMed]
138. Alhagry, S.; Aly, A.A.R. Emotion recognition based on EEG using LSTM recurrent neural network. *Int. J. Adv. Comput. Sci. Appl.* **2017**, *8*, 8–11. [CrossRef]
139. Salama, E.S.; El-Khoribi, R.A.; Shoman, M.E.; Wahby Shalaby, M.A. EEG-based emotion recognition using 3D convolutional neural networks. *Int. J. Adv. Comput. Sci. Appl.* **2018**, *9*, 329–337. [CrossRef]
140. Moon, S.E.; Jang, S.; Lee, J.S. Convolutional neural network approach for eeg-based emotion recognition using brain connectivity and its spatial information. In Proceedings of the 2018 IEEE International Conference on Acoustics, Speech and Signal Processing (ICASSP), Calgary, ON, Canada, 15–20 April 2018; pp. 2556–2560.
141. Kung, F.Y.H.; Chao, M.M. The impact of mixed emotions on creativity in negotiation: An interpersonal perspective. *Front. Psychol.* **2019**, *9*, 1–15. [CrossRef]

sensors

MDPI

Letter

Comparison of Regression and Classification Models for User-Independent and Personal Stress Detection

Pekka Siirtola * and Juha Röning

Biomimetics and Intelligent Systems Group, University of Oulu, P.O. BOX 4500, FI-90014 Oulu, Finland; juha.roning@oulu.fi

* Correspondence: pekka.siirtola@oulu.fi

Received: 2 July 2020; Accepted: 5 August 2020; Published: 7 August 2020

Abstract: In this article, regression and classification models are compared for stress detection. Both personal and user-independent models are experimented. The article is based on publicly open dataset called AffectiveROAD, which contains data gathered using Empatica E4 sensor and unlike most of the other stress detection datasets, it contains continuous target variables. The used classification model is Random Forest and the regression model is Bagged tree based ensemble. Based on experiments, regression models outperform classification models, when classifying observations as stressed or not-stressed. The best user-independent results are obtained using a combination of blood volume pulse and skin temperature features, and using these the average balanced accuracy was 74.1% with classification model and 82.3% using regression model. In addition, regression models can be used to estimate the level of the stress. Moreover, the results based on models trained using personal data are not encouraging showing that biosignals have a lot of variation not only between the study subjects but also between the session gathered from the same person. On the other hand, it is shown that with subject-wise feature selection for user-independent model, it is possible to improve recognition models more than by using personal training data to build personal models. In fact, it is shown that with subject-wise feature selection, the average detection rate can be improved as much as 4%-units, and it is especially useful to reduce the variance in the recognition rates between the study subjects.

Keywords: stress detection; wearable sensors; regression; classification

---

1. Introduction and Related Work

Wearable sensors are commonly used to monitor human motion based on inertial sensors such as accelerometer, gyroscope, and magnetometer. However, wearables can also include sensors to measure biosignals. In fact, nowadays wrist-worn wearable devices can include a wide range of biosensors, including photoplethysmography to measure blood volume pulse, thermometer to measure body temperature, and electrodermal activity sensor to measure galvanic skin response. Based on these, it is possible not only to monitor human motion, but they also enable a possibility to monitor other aspects of human behaviour and things happening inside human body.

Recently, there has been a lot of attention to stress and affect recognition using wearable sensors. For instance, in [1] four affect states (normal, happy, sad, fear, and anger) were recognized based galvanic skin response and blood volume pulse signals, and in [2] eight affect states were detected based on acceleration, electrocardiogram, blood volume pulse, and body temperature signals.

One especially interesting affect state to recognize is stress. There is evidence that well-being at work and working efficiency are connected [3]. This is because employees that are feeling well at work have better engagement, are more motivated and take fewer sick days. One of the factors leading to reduced work well-being and working efficiency is stress. To increase productivity, causes of stress at work should be studied, and before this can be done, methods to measure stress needs to be studied. Moreover, chronic long-term stress can impact on brain structures involved in cognition and mental health [4] and cause many other problems as well for instance by effecting to the immune system [5].

Stress detection based on wearable sensors have been studied a lot, probably more than detection of any other affect state. For instance, in [6] a classifier to detect high and low stress, as well as, non-stressful situations at laboratory conditions based on wearable wrist-work sensors. The result of the paper was these two classes can be detected with the accuracy of 83%. Moreover, in [7] a binary classifier to stress and non-stressed state was trained, and it was noted that stress can be detected using sensors of commercial smartwatches. There are also several other studies showing that stress detection based on classification models can be performed with high accuracy using user-independent models (see for instance [8,9]). However, according to some studies, the accuracy can be further improved by using models based on personal data ([10,11]).

What is noticeable is that most of the stress and affect detection studies are based on classification methods commonly used in human activity recognition, for instance [6,12]. These methods include feature extraction using sliding window technique and usage of the same classifiers. The reason for this is probably that the devices used in both studies are similar, or at least partly similar, as they both are using wearable devices and the data obtained by the sensors of these devices. While the recognition accuracies shown in stress detection articles are good, the problem is that they are based on classification of discrete target variables. However, stress and other emotions are not discrete phenomenons as for instance the level of the stress of a person can be high or low or anything between these. In [6], it was suggested that discrete classification results could be transformed as continuous based on posterior values. Thus, still also in this case, the original classification to stress/non-stressed would be based on discrete target values. In fact, one problem is that in affect recognition studies it is already decided in the data gathering phase simplify the studied phenomenon by transforming a continuous phenomenon as discrete due to difficulty to gather continuous target values. Of course, transforming continuous phenomenon as discrete is far from optimal solution, and this type of problem simplification can cause problems in the modelling phase. In fact, this means that a problem that originally was regression problem is transformed as classification problem. Therefore, a better option would be to base stress and affect detection to continuous target values.

While the most of the stress detection studies are based on discrete target values, there are some attempts to gather and analyze continuous target values for stress detection. In [13], stress while driving a car was studied. In the study, continuous target variables for stress level were created based on video recorded to analyze driver's facial expression, body motion and road conditions. Another study, where continuous target values for stress detection were gathered is [14,15]. Furthermore, in [14] the study concentrates on analyzing stress while driving, and during the data gathering, driver constantly estimated his/her own stress level. However, while in these two datasets the data contains continuous target values, in both cases the final analysis is based on discrete targets created based on continuous targets. On the other hand, there is also studies where stress and other affective states have analyzed using regression model (see for instance [16,17]; however, these are rare compared to studied based on classification methods.

As stress is not discrete phenomenon, it should be analyzed based on continuous target values. This means that instead of classification, the data should be analyzed using regression methods. This article is the first one where regression and classification models for stress detection are properly compared to experiment regression methods really out perform classification method. In addition, the cons and pros of classification and regression models are discussed. Moreover, user-independent and personal recognition models are compared. The study is based on a publicly open dataset making the reproduction of the presented results easy.

The article is organized as follows: Section 2 introduces the used dataset and Section 3 introduces the methods used in the experiments. Experimental setup and results are explained in Section 4, and finally, the discussion and conclusions are in Section 5.

2. Experimental Dataset

This study is based on publicly open data set called AffectiveROAD [14]. It contains data from nine participants (from now on these participants are called NM, RY, BK, MT, EK, KGS, AD, GM and SJ) measured using Empatica E4 wrist-worn device [18]. However, three participants performed data gathering session more than one's, and therefore, the dataset contains data from 13 data gathering sessions. Participants wore this device on both wrists. E4 includes accelerometers (ACC), as well as, sensors to measure skin temperature (ST), electrodermal activity (EDA), blood volume pulse (BVP), heart rate (HR), and heart rate variability (HRV). In addition to E4 data, AffectiveROAD includes data from Zephyr Bioharness 3.0 chest belt. However, in this study, only Empatica E4 data are used as the focus of this article is in wrist-worn sensors. Moreover, to be able to better compare the results of the study to the results of previous studies, only data from right wrist-worn Empatica was used.

In the data gathering session, the task of the study subject was to drive car in a normal daily traffic. However, the session started with a rest period where study subject was sitting and resting in a car, eyes closed and engine running. This can be considered as baseline data. The actual driving consisted of driving at two types of roads: city driving and driving at highway. City driving is assumed to be stressful as it contains traffic lights, a lot of vehicles, pedestrians and bikes. On the other hand, a highway is a smooth road and driving there is assumed to be less stressful.

What makes this dataset special and unique is that it includes continuous subjective stress estimates which were collected by the experimenter sitting at rear seat while study subject was driving. Moreover, the driver validated these estimates after driving session. The scale for estimates was from 0 (=no stress) to 1 (=maximum stress). However, there are no stress estimates available from baseline. In this study, it is assumed that during the whole baseline session, the level of stress is 0. As this dataset contains continuous target variables, these data can be used for regression analysis as well as for classification after discretizing target values. More detailed information about the dataset can be found from [14].

For the model training signals from the data gathering sessions were divided into windows, and from these windows, features were extracted. Window size of 60 s was used in the experiment, which is the same as used in [7,19] for stress detection. The used slide was between two adjacent windows was 0.5 s. The features extracted from these windows were the same as the one's used in [19]. Therefore, four types of features were extracted: features from the ACC signal, EDA signal, BVP signal and ST signal. Altogether 119 features were extracted.

3. Methods

Classification was is in this study based on Random Forest, as it was found as the best classification algorithm for stress detection in our previous study [7]. However, to found the most appropriate regression model for the purpose, several regression models were compared. The comparison was made using Matlab's (version 2018b) Regression Learner application which enables fast experimenting with multiple regression algorithms. Regression Learner contains two options for dividing data into training and testing: cross-validation which randomly divides data into desired number of groups, and holdout which randomly divides data into two groups. As the aim of this study is to build user-independent models, leave-one-subject-out -method needs to be used in the final model training process, and therefore, neither of the approaches provided by Regression Learner are valid for the purpose. Moreover, both approaches lead to over-trained models as same persons data can be in training and testing sets. However, though Regression Learner cannot be used to calculate the final outcomes of this paper, it can be used to compare the performance of different regression algorithms when predicting the amount of stress, and help to select the right regression algorithm to calculate the final outcomes of this article.

In the end, 13 regression algorithms were compared by training regression models using 5-fold cross-validation, see Table 1. As a result of this comparison, it was decided to use Bagged tree based ensemble model in the experiments as the root mean square error (RMSE) is the lowest using it.

Table 1. Comparison of regression models using 5-fold cross-validation.

Method	RMSE
Linear regression	0.23
Robust linear regression	0.24
Fine tree	0.05
Medium tree	0.06
Coarse tree	0.08
SVM with linear kernel	0.24
SVM with quadratic kernel	0.14
SVM with cubic kernel	0.7
SVM with fine Gaussian kernel	0.14
SVM with medium Gaussian kernel	0.8
SVM with coarse Gaussian kernel	0.21
Boosted tree based ensemble	0.16
Bagged tree based ensemble	0.03

4. Experiments

Experiments were made using classification and regression models. Moreover, user-independent and personal models were compared. The accuracies presented in this section are balanced accuracies, which is calculated by separately calculating accuracy for both classes and the reported balances accuracy is mean from these. Balanced accuracy was selected as the performance metric as it is not as vulnerable to unbalanced data as accuracy.

4.1. User-Independent Stress Detection

To compare regression and classification models, binary Random Forest classifier was trained and results of it were compared to Bagged Tree based ensemble regression model. Both were trained using leave-one-subject-out -method, meaning that in turn one person's data were used for testing and others data for training. For classification model, target values were transformed as 1 and 0, so that data from baseline are labelled as 0 and data from driving as 1. However, regression model was trained using continuous target values. Moreover, as the outputs of the regression model are continuous values, they were transformed as binary by finding an optimal threshold for each person to divide outputs as stress and non-stress that maximizes the accuracy. This was done by analyzing the obtained continuous prediction values study subject-wise, and such stress level value was searched which divides the prediction values as stressed and non-stressed so that the obtained balanced accuracy rate is as high high as possible. This of course over-estimates the performance of regression model as the threshold is calculated based on personal data but still it shows the potential of regression model based stress detection.

The results of user-independent recognition rates using regression and classification algorithms and different sensor combinations are presented in Table 2. Using leave-one-subject-out cross-validation method, a own recognition model was trained for each study subject and the performance of the model was calculated using balanced accuracy, sensitivity and specificity. The values presented in Table 2 shows the mean balanced accuracies, sensitivities and specificities calculated over all nine study subjects, and standard deviation between the study subjects is presented in parenthesis.

Table 2. Average recognition results accuracies, sensitivities and specificities (standard deviation in parentheses) using regression and classification model and sensor combinations.

Regression			
Sensors	Balanced accuracy	Sensitivity	Specificity
ACC+EDA+ST+BVP	89.0 (13.3)	86.9 (22.7)	93.6 (5.8)
EDA+ST+BVP	75.0 (16.0)	71.6 (27.3)	90.9 (9.2)
EDA+BVP	75.1 (15.8)	73.5 (24.5)	88.3 (7.8)
BVP+ST	82.3 (17.0)	76.6 (23.8)	91.0 (6.1)
EDA+ST	72.3 (16.7)	70.9 (24.2)	84.0 (11.0)
EDA	73.5 (15.2)	72.0 (23.7)	87.5 (10.8)
BVP	78.5 (16.7)	72.1 (23.7)	89.0 (7.2)
ST	68.6 (11.8)	56.1 (19.3)	83.5 (9.0)
ACC	94.3 (5.5)	93.2 (7.0)	96.2 (3.4)
Classification			
Sensors	Balanced accuracy	Sensitivity	Specificity
ACC+EDA+ST+BVP	85.2 (14.2)	82.4 (22.9)	91.7 (9.4)
EDA+ST+BVP	65.4 (22.8)	61.6 (33.9)	76.7 (15.7)
EDA+BVP	69.3 (16.9)	57.9 (33.3)	78.7 (11.4)
BVP+ST	74.1 (16.7)	66.0 (33.5)	85.6 (10.9)
EDA+ST	64.6 (23.1)	54.1 (31.8)	73.2 (17.5)
EDA	64.5 (14.6)	55.8 (23.9)	78.1 (12.8)
BVP	70.4 (20.4)	64.0 (38.2)	82.9 (11.0)
ST	61.3 (14.9)	50.5 (27.6)	76.5 (12.0)
ACC	90.2 (10.1)	91.4 (9.2)	96.0 (6.4)

Interestingly, according to Table 2 in both cases (regression and classification), the best recognition rates are obtained using only accelerometer features. This is not in line with previous stress detection studies, where it has been noted that features extracted biosignals are more useful for detecting stress than accelerometer features (see, for instance [19]). However, the reason for this is that, in this case, accelerometer-based models do not recognize stress at all; instead, they recognize the mode of transportation as baseline signal and stress signal are performed under different activity (sitting in non-moving car vs. driving a car). In fact, it is shown that these two activities can be recognized based on accelerometer data [20]. Therefore, as the aim of the study is to recognize stress, only models based on biosignals (=EDA, BVP, and ST) are worth studying.

Table 2 shows the potential of regression models. It outperforms classification algorithm with each sensor combination, and the highest recognition rates can be obtained using BVP and ST features. With this setting, the average balanced accuracy was 74.1% with classification model and 82.3% using regression model. In addition, sensitivity and specificity values are the highest using this combination. Confusion matrix for regression model for regression model using these features is shown in Table 3, where it can be seen that both classes are detected with reasonable accuracy. Moreover, according to Table 4 where user-independent accuracies are studied subject-wise using BVP and ST features, it can be noted that regression model performs better than classification model with all nine study subjects. Partly, the great performance of regression model is due to optimizing the threshold to classify outputs as stressed/non-stressed based on personal data but this is not the only reason for good performance of regression model. Another reason for it is that regression models gets more information as an input than classification model. In fact, classification models gets binary class labels as input while regression model's inputs are continuous targets. This results show that stress detection benefits from continuous targets and is as a nature a regression problem instead of classification problem. This was expected, as the level of the stress of a person can be high or low or anything between these.

Table 3. Confusion matrix, when user-independent regression model based on BVP and ST features is used for stress detection.

True/Predicted	Non-Stressed	Stressed
Non-stressed	88.0%	12.0%
Stressed	26.6%	73.4%

Table 4. Subject-wise balanced accuracy using user-independent model and a combination of BVP and ST features.

Subject	Classification (%)	Regression (%)
NM	64.3	84.6
RY	90.8	93.7
BK	87.0	88.3
MT	49.0	50.5
EK	59.4	60.7
KGS	88.3	93.7
AD	96.3	99.7
GM	69.9	95.0
SJ	62.2	74.7
Mean	74.1 (STD 16.7)	82.3 (STD 17.0)

While regression models outperform classification models in stress detection, the real benefit of regression models is that, unlike classification, they can be used to estimate the level of stress. Figure 1 shows how well regression models estimate the level of the stress compared to user reported stress level values. The figure shows that in most cases the regression model has approximately managed to predict the level of the stress. However, this approximation is not very accurate, and it is inconsistent. This suggests that the the trained regression models is very sensitive to small changes in feature values, and due to this, the value of predicted stress level can rapidly change. Moreover, the approximation does not work at all for some study subjects (see study subjects MT, Figure 1d and SJ, Figure 1i). In the case of MT, the model has not managed to recognize the non-stressed stage, and in the case of SJ, the recognition of stressed stage has caused problems. These exceptions suggest that stress does not cause similar reactions to every person. Moreover, Figure 1 shows the personal threshold used to divide regression model outputs as stressed and non-stressed (black horizontal line). In most cases, the value of this threshold is between 0.3 and 0.5. However, also in this case MT and SJ are exceptions. In the case of MT, the value of threshold is close to 1, and in the case of SJ, the value is close to 0. This underlines how special these the cases are compared to other seven study subjects.

Table 5 shows root mean square error (RMSE) of predicted and user reported values for each study subject. As the stress level of the baseline was not measured during the data gathering, it was assumed that during the baseline measurement the level of the stress was constant and zero. This of course is just a best guess, and therefore, cannot be considered as fact. In fact, also the predictions shown in Figure 1 suggests that the level of the stress is not constant during the baseline and neither it is zero. Due to this, RMSE values of Table 5 are calculated from two parts of the signal: the second column shows the RMSE value from the whole signal and the third column shows the RMSE calculated only from driving part of the signal as only it contained target values defined by the driver. For instance, in the case of subject MT, there is a big difference between these two. Moreover, the R-Squared value was calculated from each subject, and they are shown in Table 5. These also underline the fact that prediction does not work for each subject. In fact, R-Squared value is zero for three study subjects indicating that for these persons user-independent model is not able to predict any variation on response. In overall, the results of the Table 5 show the potential of predicting the level of the stress using regression models. However, it also shows that at this point of the research, it only works for some study subjects and the prediction can only give a rough estimation of the stress value, and it needs to be further studied how the exactness of the prediction could be improved.

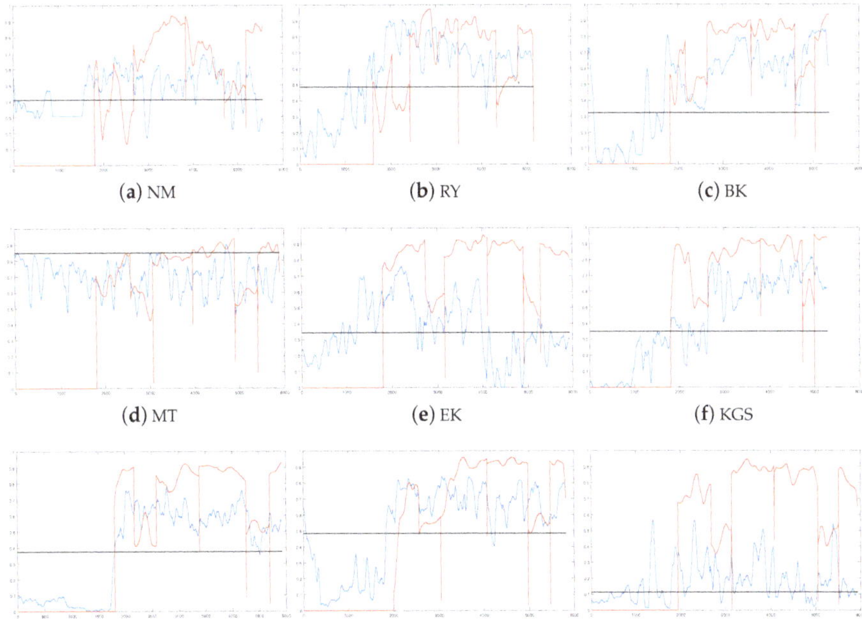

(a) NM **(b)** RY **(c)** BK **(d)** MT **(e)** EK **(f)** KGS **(g)** AD **(h)** GM **(i)** SJ

Figure 1. Predicted stress level (blue line) vs. subject estimated continuous target value for stress level (orange line). Personal threshold used to divide outputs as stressed and non-stressed is shown using black horizontal line.

Table 5. Subject-wise RMSE and R-Squared using user-independent model and a combination of BVP and ST features.

Subject	RMSE Total	RMSE Stress	R-Squared Total
NM	0.31	0.26	0.09
RY	0.27	0.26	0.49
BK	0.24	0.18	0.60
MT	0.43	0.18	0
EK	0.46	0.52	0
KGS	0.25	0.28	0.60
AD	0.20	0.25	0.75
GM	0.22	0.21	0.64
SJ	0.48	0.60	0
Mean	0.32	0.31	0.35

The combination of BVP and ST features produced on average the highest recognition rates using user-independent regression model. However, this combination was not the best for each subject. Table 6 shows subject-wisely which sensor combination produces the best result. It can be seen that for most subjects, BVP + ST produces the best results but there is also exceptions. For instance for subject EK, a different selection of sensors improves the recognition rate over 20%-units (59.4 vs. 83.2, respectively). However, it has even bigger effect on standard deviation which is much smaller when features are selected subject-wise (17.0 vs. 11.7, respectively). Therefore, possibility to select features subject-wise, and this way personalize recognition model, could potentially have a significant positive effect to the recognition rates.

Table 6. Best subject-wise recognition rates obtained using user-independent regression model, and which sensor combination was used to obtain this result.

Subject	Balanced Accuracy (%)	Sensor Combination
NM	84.6	BVP+ST
RY	93.7	BVP+ST
BK	91.1	EDA+BVP+ST
MT	64.3	EDA+BVP+ST
EK	83.2	BVP+EDA
KGS	93.7	BVP+ST
AD	99.7	BVP+ST
GM	95.0	BVP+ST
SJ	70.8	ST
Mean	86.3 (STD 11.7)	

4.2. Personal Stress Detection

The dataset contains more that one data gathering session from three study subjects (three sessions from subject NM (named NM1, NM2 and NM3), two from RY (named RY1 and RY2) and GM (named GM1 and GM2). Therefore, these three study subjects can be used to experiment if the stress recognition models trained using personal data are more accurate than user-independent models, as it is suggested in some of the papers. The data gathering protocol was the same in each of these session, and on each session, the study subjects drove the same route. However, what makes each session unique is that traffic was different in each case as the study subjects drove on public roads.

User-dependent recognition rates for NM, RY and GM are shown in Table 7 using different sensor combinations, and classification and regression models. In each case, models are trained with data from one session and tested with data from another session from the same user. NM1, RY1 and GM1 were used for testing, as these were also used in previous section to train and test user-independent models. According to Table 4 using user-independent model and a combination of BVP and ST features, which based on the Table 2 provide the highest average accuracy, Subject NM's stress can be detected with balanced accuracy of 64.3%/84.6% (classification vs. regression), RY's 90.8%/93.7%, and GM's 69.9%/95.0%. When these are compared to personal recognition rates based on the same features, it can be noted that personal training data do not have a big effect on the recognition rates. In fact, in the case of RY, the results based on personal model are much worse than the results based on user-independent model. Moreover, according to Table 7, the results using personal models are surprisingly bad no matter which sensor combination is used. There are some exceptions, though. RY's results are really good when using BVP features, and GM's stress can be recognized with high accuracy using a combination of BVP and ST features. However, the results from these users were already really good using user-independent model and usage of personal data do not have a big effect compared to those.

There can be several reasons for low personal recognition rates. For instance, in the case of data from subject GM, there seems to be problems with EDA signal, and thus, prediction using it always leads to really bad results. Another reason for low personal recognition rates can be that the size of the training data was not big enough, as in the case of personal model training data consist of measurement from one session and in user-independent case it consisted of measurements from eight sessions. In fact, when NM1 was trained using data from two personal data gathering sessions (NM2 and NM3), the results were much better than the one's reported in Table 7, and over 98% balanced accuracy was obtained using BVP and ST features, see Table 8. However, also in this case it can be noted that the recognition rate is highly dependent on which data are used for training and which for validation. It can be noted that when NM2 and NM3 are used for training and NM3 for testing, the recognition rates are really bad no matter which sensor combination is used. This shows that there are a lot of variation between data gathering sessions, even if the the data are collected from the same person.

Table 7. Recognition rates of users NM, RY, and GM using models trained using personal data.

Sensors	Valid/Train NM1/NM2 Classif./Regr.	Valid/Train NM1/NM3 Classif./Regr.	Valid/Train RY1/RY2 Classif./Regr.	Valid/Train GM1/GM2 Classif./Regr.
EDA+ST+BVP	69.3/70.1	49.2/68.0	50.0/50.0	50.0/ 71.4
EDA+BVP	69.9/79.5	50.0 /50.0	54.8/71.2	50.0/50.0
BVP+ST	69.3/88.0	50.8/67.9	50.0/65.0	83.7/91.6
EDA+ST	69.3/70.1	80.9/91.4	50.0/50.6	50.0/50.0
EDA	69.2/83.5	49.4/68.8	51.1/50.5	50.0/50.0
BVP	83.3/84.9	51.4/50.8	95.0/94.4	57.8/91.1
ST	69.3/70.1	69.3/69.4	50.0/84.7	50.0/78.7

Table 8. Cross-validation of three datasets collected from study subject NM.

Sensors	Valid/Train NM1/NM2 + NM3 Classif./Regr.	Valid/Train NM2/NM1 + NM3 Classif./Regr.	Valid/Train NM3/NM1 + NM2 Classif./Regr.
EDA+ST+BVP	74.4 /98.3	77.5/99.3	50.1/52.6
EDA+BVP	62.4/77.2	32.0 /50.0	51.0/52.4
BVP+ST	85.5/98.8	50.0/60.4	52.3/52.1
EDA+ST	69.3/85.6	80.6/99.4	48.2/52.6
EDA	63.8/80.1	32.9/68.3	49.6/51.0
BVP	74.5/78.7	64.8/87.4	51.9/50.9
ST	69.3/69.4	50.0/51.1	56.9/61.2

5. Discussion and Conclusions

In this article, regression and classification models were compared for stress detection. Both personal and user-independent models were experimented. The article was based on publicly open dataset [14], which unlike most of the other stress detection datasets, contained continuous target variables. The used classification model was Random Forest and the regression model was Bagged tree based ensemble.

The study shows that regression models are superior to classification models when it comes to user-independent stress detection, if continuous target values are available. The best results were obtained using a combination of BVP and ST features, and using these the average balanced accuracy was 74.1% with classification model and 82.3% using regression model. In fact, basically no matter which sensor combination was used, the results using regression models were better than the one's obtained using classification models. This is because during the training process, regression model gets more information than classification model: regression model gets continuous targets as input while classification models inputs are discrete. Moreover, the results show that stress is not binary problem, a person can be highly stress, not stressed at all or anything between these two. Therefore, the results of this article show that prediction model should be trained so that all the available can be given as an input to it.

The main advantage of using regression models is that they can not only be used to recognize whether the person is stressed or not, they can also be used to predict the level of the stress. However, based on our experiments this prediction is not very accurate, and for some study subjects it does not work at all. In fact, the main goal of the future work is to study how the quality of this prediction could be improved. Due to this, different regression models needs to be compared to find out which is the most capable to predict the level of the stress. This also includes comparison of different quantitative indicators to measures the goodness of the prediction. Unfortunately, continuous target variables are rarely available for the datasets which is the reason why regression models are not often used for stress detection, and why stress detection based on classification models is an important topic to study also in the future.

Stress detection based on personal training data was also studied using data from three persons. The results obtained using personal models were surprisingly bad. The situation was better when more personal data were used for training but still there was a lot of variation in the recognition rates depending on how data gathering sessions were divided for training and testing. This shows that biosignals have a lot of variation not only between the study subjects but also between the session gathered from the same person. Therefore, personal recognition models should be able to constantly adapt to changes in human body. Due to this, for instance incremental learning-based method should be experimented [21]. On the other hand, it was shown that subject-wise feature selection for user-independent model can be a better approach than usage of personal training data to personalize and improve recognition models. It was noted that with subject-wise feature selection, the average detection rate improved 4%-units, and it was especially useful to reduce the variance in the recognition rates between the study subjects.

Moreover, it can be noted that the recognition rates obtained in this article are not as high as the one's obtained in several other stress detection studies where the same features are used ([7,19]). The reason for this can be used data gathering protocol. In many other studies, the stress data are collected from contexts (giving public speech, performing difficult arithmetic tasks, etc.) that most likely are more stressful than driving a car. Therefore, the dataset used in this study does as big difference between relaxed and stressed state as other datasets making it more difficult to analyze. It is also possible that the type of the stress is different in these contexts. In fact, part of the future work is to make experiments using different datasets. In fact, to extend this study, these datasets should include data from new types of sensors (for instance electroencephalography (EEG)), to show that the proposed method is not dependent on the used sensor. The expected result is that the proposed method works similar to this study if appropriate features are extracted from the user sensor. In fact, extracted features needs to be selected sensor-wise, and though, the same features should not be extracted from each sensor. In addition, the experimented datasets should contain data not only from stress, but also from other affect states to show that the proposed method is not dependent on the studied affect state. The final aim is to detect multiple affect states from at same time, for example a person can be stressed and angry. This could be done for instance by training an own regression model for each affect state.

Lastly, it should be noted that when dealing data from stress and affect states which is labeled by the study subjects themselves, it should be noted that study subjects do not necessarily know what they feel [22]. Therefore, the target variables obtained from the study subjects can be unreliable. Due to this, while the preliminary results presented in this article are encouraging, more experiments with more datasets should be carried out to get a better understanding of how accurate the presented method actually is.

Author Contributions: Conceptualization, P.S.; Formal analysis, P.S.; Supervision, J.R.; Validation, P.S.; Writing–original draft, P.S.; Writing–review & editing, J.R. Both authors have read and agreed to the published version of the manuscript.

Funding: This research is supported by the Business Finland funding for Reboot IoT Factory-project (www.rebootiotfactory.fi).

Acknowledgments: Authors are thankful for Infotech Oulu. Authors would also like to thank Neska El Haouij for collecting a great dataset and helping us to connect the right observations for the right target values.

Conflicts of Interest: The authors declare no conflict of interest.

References

1. Rathod, P.; George, K.; Shinde, N. Bio-signal based emotion detection device. In Proceedings of the 2016 IEEE 13th International Conference on Wearable and Implantable Body Sensor Networks (BSN), San Francisco, CA, USA, 14–17 June 2016; pp. 105–108.
2. Zenonos, A.; Khan, A.; Kalogridis, G.; Vatsikas, S.; Lewis, T.; Sooriyabandara, M. HealthyOffice: Mood recognition at work using smartphones and wearable sensors. In Proceedings of the 2016 IEEE International Conference on Pervasive Computing and Communication Workshops (PerCom Workshops), Sydney, Australia, 14–18 March 2016; pp. 1–6.
3. Renee Baptiste, N. Tightening the link between employee wellbeing at work and performance: A new dimension for HRM. *Manag. Decis.* **2008**, *46*, 284–309. [CrossRef]
4. Lupien, S.J.; McEwen, B.S.; Gunnar, M.R.; Heim, C. Effects of stress throughout the lifespan on the brain, behaviour and cognition. *Nat. Rev. Neurosci.* **2009**, *10*, 434–445. [CrossRef] [PubMed]
5. Nagaraja, A.S.; Sadaoui, N.C.; Dorniak, P.L.; Lutgendorf, S.K.; Sood, A.K. SnapShot: Stress and disease. *Cell Metab.* **2016**, *23*, 388–388. [CrossRef] [PubMed]
6. Gjoreski, M.; Gjoreski, H.; Luštrek, M.; Gams, M. Continuous Stress Detection Using a Wrist Device: In Laboratory and Real Life. In Proceedings of the 2016 ACM International Joint Conference on Pervasive and Ubiquitous Computing: Adjunct, Heidelberg, Germany, 12–16 September 2016; pp. 1185–1193. [CrossRef]
7. Siirtola, P. Continuous stress detection using the sensors of commercial smartwatch. In Proceedings of the 2019 ACM International Joint Conference on Pervasive and Ubiquitous Computing and Proceedings of the 2019 ACM International Symposium on Wearable Computers, London, UK, 9–13 September 2019; pp. 1198–1201.
8. Schmidt, P.; Reiss, A.; Dürichen, R.; Laerhoven, K.V. Wearable-Based Affect Recognition—A Review. *Sensors* **2019**, *19*, 4079. [CrossRef] [PubMed]
9. Can, Y.S.; Arnrich, B.; Ersoy, C. Stress detection in daily life scenarios using smart phones and wearable sensors: A survey. *J. Biomed. Inform.* **2019**, *92*, 103139. [CrossRef] [PubMed]
10. Hernandez, J.; Morris, R.R.; Picard, R.W. Call center stress recognition with person-specific models. In Proceedings of the International Conference on Affective Computing and Intelligent Interaction, Memphis, TN, USA, 9–12 October 2011; pp. 125–134.
11. Stewart, C.L.; Folarin, A.; Dobson, R. Personalized acute stress classification from physiological signals with neural processes. *arXiv* **2020**, arXiv:2002.04176.
12. Mishra, V.; Hao, T.; Sun, S.; Walter, K.N.; Ball, M.J.; Chen, C.H.; Zhu, X. Investigating the role of context in perceived stress detection in the wild. In Proceedings of the 2018 ACM International Joint Conference and 2018 International Symposium on Pervasive and Ubiquitous Computing and Wearable Computers, Singapore, 8–12 October 2018; pp. 1708–1716.
13. Healey, J.; Picard, R. Detecting stress during real-world driving tasks using physiological sensors. *IEEE Trans. Intell. Transp. Syst.* **2005**, *6*, 156–166. [CrossRef]
14. Haouij, N.E.; Poggi, J.M.; Sevestre-Ghalila, S.; Ghozi, R.; Jaïdane, M. AffectiveROAD System and Database to Assess Driver's Attention. In Proceedings of the 33rd Annual ACM Symposium on Applied Computing, Pau, France, 9–13 April 2018; pp. 800–803. [CrossRef]
15. El Haouij, N. Biosignals for Driver's Stress Level Assessment: Functional Variable Selection and Fractal Characterization. Ph.D. Thesis, Paris Saclay, Saint-Aubin, France, 2018.
16. Novak, D.; Mihelj, M.; Munih, M. A survey of methods for data fusion and system adaptation using autonomic nervous system responses in physiological computing. *Interact. Comput.* **2012**, *24*, 154–172. [CrossRef]
17. Wei, J.; Chen, T.; Liu, G.; Yang, J. Higher-order multivariable polynomial regression to estimate human affective states. *Sci. Rep.* **2016**, *6*, 23384. [CrossRef] [PubMed]
18. Empatica E4. 2019. Available online: https://www.empatica.com/e4-wristband (accessed on 27 May 2019).
19. Schmidt, P.; Reiss, A.; Duerichen, R.; Marberger, C.; Van Laerhoven, K. Introducing WESAD, a Multimodal Dataset for Wearable Stress and Affect Detection. In Proceedings of the 2018 on International Conference on Multimodal Interaction, Boulder, CO, USA, 16–20 October 2018; pp. 400–408.
20. Siirtola, P.; Röning, J. Recognizing human activities user-independently on smartphones based on accelerometer data. *IJIMAI* **2012**, *1*, 38–45. [CrossRef]

21. Siirtola, P.; Röning, J. Incremental Learning to Personalize Human Activity Recognition Models: The Importance of Human AI Collaboration. *Sensors* **2019**, *19*, 5151. [CrossRef] [PubMed]
22. Schmidt, P.; Dürichen, R.; Reiss, A.; Van Laerhoven, K.; Plötz, T. Multi-target affect detection in the wild: An exploratory study. In Proceedings of the 23rd International Symposium on Wearable Computers, London, UK, 9–13 September 2019; pp. 211–219.

MDPI
St. Alban-Anlage 66
4052 Basel
Switzerland
Tel. +41 61 683 77 34
Fax +41 61 302 89 18
www.mdpi.com

Sensors Editorial Office
E-mail: sensors@mdpi.com
www.mdpi.com/journal/sensors

www.ingramcontent.com/pod-product-compliance
Lightning Source LLC
Chambersburg PA
CBHW041933260926
51850CB00018B/113
9783036511382